LES BROWN'S ENCYCLOPEDIA OF TELEVISION

Books By Les Brown

Televi$ion: The Business Behind the Box
Electric Media
*The New York Times Encyclopedia of
    Television*
Keeping Your Eye On Television

LES BROWN'S ENCYCLOPEDIA OF TELEVISION

Les Brown

New York Zoetrope

1982

A NOTE ON THE TYPE

Les Brown's Encyclopedia of Television was set in Garamond No. 3 (Garamond Old Style) by Expertel, Inc. in New York City. The display type is ITC Garamond Light.

The book was printed and bound by R.R. Donnelley & Sons Company at their Harrisonburg, Virginia plant.

Editing: Ann Watson
Production: Susan Cohn

Copyright © 1977, 1982 Les Brown.
All rights reserved.

Reproduction by any means, electronic or otherwise, is strictly forbidden under U.S. and international copyright laws without the express written consent of the publisher.

Library of Congress Cataloguing in Publication Data:

    Brown, Les, 1928-
      Les Brown's Encyclopedia of television.

      Expanded ed. of: New York Times Encyclopedia
of television. 1977.
      Bibliography: p.
      1. Television broadcasting—Dictionaries.
    I. Title.   II. Title: Encyclopedia of television.
    PN1992.18.B7 1982   791.45'03   82-7867
    ISBN 0-918432-28-6   AACR2
    ISBN 0-918432-29-4 (pbk.)

New York Zoetrope
80 East 11th Street
New York 10003
Printed in the United States of America.
First printing: May 1982.
5 4 3 2 1

For Becky, Josh, Jessica and Jean

Table of Contents

Preface to the New Edition

Five years have elapsed since the original edition of this encyclopedia was published, and as it happened they were the most turbulent years in the history of broadcasting. For forces of change were being unleashed everywhere in the electronic environment, ending the first age of television. In a mighty wave powered by technology and business, and encouraged by government, Television II emerged—the age of cable, pay-cable, Qube, satellites, STV, MDS, DBS, videotext, video recorders, video discs, superstations, low-power stations, and a rainstorm of cable-satellite networks. During these five years television, as a field, was advancing on fast-forward.

Cable spread to the urban centers and reached a national penetration of nearly 30 percent. The three major networks, which used to do battle with cable, jumped in as providers of cable services. Former network presidents signed on as cable-network presidents. Large newspaper companies took the plunge in cable, as did such non-media corporations as Getty Oil and American Express. Ted Turner, a maverick entrepreneur, became cable's first superstar.

Meanwhile, CBS-TV underwent several reorganizations and finally rebounded from its ratings slump to reclaim leadership among the networks. Fred Silverman came and went at NBC—noisily in each instance—leaving the company in disarray. ABC's aggressiveness in news sparked changes at the other networks, in the executive suites as well as on the screen. News people moved as never before from one network to another. Walter Cronkite stepped down as anchor at CBS to make way for youth; John Chancellor did the same at NBC. Public television went from an all-time high, having finally pulled itself together in the wake of the second Carnegie Commission report, to an all-time low: its survival threatened by harsh cutbacks in Federal support and competition from the new electronic media.

The Federal Communications Commission, after a changing of the guard, went full steam ahead towards deregulation. The White House Office of Telecommunications Policy was dismantled and replaced by the National Telecommunications and Information Administration. Rep. Lionel Van Deerlin produced two versions of his bill to rewrite the Communications Act, failed to sell the concept to his heavily-lobbied subcommittee, and lost his office in the 1980 elections. Ironically, some key provisions of his bill were then enacted in other legislation.

License periods for television were lengthened to five years, from three. The courts knocked out restrictive rules for cable, helping spur its growth. Television went on trial for murder in the Zamora case. The Supreme Court ruled that the states are not prohibited by the Constitution from allowing television cameras in courtrooms. The broadcast reform movement went into decline as organizations representing the Religious New Right came to the fore primarily to denounce sex and violence on television.

Change on such a scale and at such a pace is unkind to the compiler of an encyclopedia such as this. Each new development affects the updating of numerous entries, and some articles had to be revised several times in a period of weeks.

In all, there are close to four hundred new articles in this edition, and hundreds of updated and rewritten entries from the original book, which was entitled *The New York Times Encyclopedia of Television*. Several of the new pieces make up for previous oversights, but most earned inclusion as newly significant matters of information, standing for enterprises, institutions, rulings, programs, or persons who are considered to have attained reference value.

During the interval between publications, the author heard from people pointing out mistakes about themselves or calling attention to small factual errors and clerical lapses. Those have now been set right; happily, all were minor.

The author is enormously indebted to Ann Watson and Clare Lynch O'Brien who, at different stages, took on the arduous task of organizing the manuscript (and reorganizing much of it), entering the alterations, and contributing to the writing and research. He is also indebted to Bob Knight, veteran reporter for *Variety*; Philip E. Clapp, formerly of the National Cable Television Association and now a Congressional aide; Geraldine Newton at CBS; and Dawson (Tack) Nail, executive editor of *Television Digest*, the invaluable trade weekly, for their assistance in updating the entries and contributing new material. Christopher Lee and Mac McCorkle were research assistants for this new volume.

Still represented in this edition are contributions to the earlier version made by Richard C. Block, Stuart Sucherman, David Lachenbruch, Jack Pitman, Kathryn Moody, Bob Knight, Avra Fliegelman, John Abrams and Karen Farr.

Les Brown
April, 1982

INTRODUCTION

A few years ago, I received a call from a certain Washington bureaucrat who seemed to want to turn himself in for an old misdeed, one committed in 1953 while he was working in the press department of NBC. He confessed that it was he who cooked up the unfounded claim that more people watched Maurice Evans's single performance of *Hamlet* on the *Hallmark Hall of Fame* that year than saw it in all its stage performances since Elizabethan times. The claim was, he pleaded, an innocent conceit by a young publicity man writing an inspired press release; he never dreamed it would enter the scriptures of television and still be quoted as one of the fundamental truths in the 70s, on Emmy Awards' telecasts or network anniversary specials.

Of course, there is no way of computing how many performances of *Hamlet* had been staged in the 350 years prior to the telecast, but, if the publicity man's assertion had gone unchallenged all those years (except by him), it was because he made the kind of lucky guess that was logical enough to be accurate. Television's mass audience is of such incredible size that the viewers for a single episode of an ordinary, low-rated TV series could fill the largest Broadway theater to capacity every performance for 20 years. A smash like *Dallas* has drawn an audience in one evening large enough to keep the same Broadway theater filled for half a century. Since *Hamlet,* even in its most glorious stage mountings, rarely runs for more than a few weeks at a time, it's conceivable that the sum total of its theater audiences did not equal what the extraordinary medium of television could deliver in a one-shot.

The extraordinary power of American television, and the character of its programming, rise from a single remarkable fact: it doesn't have to generate its immense audience. Plays, movies, phonograph records, books and newspapers all have to find their markets and motivate attendance or purchases. But in commercial television the audience delivers itself, in fairly predictable numbers each hour of the day, no matter what is being televised, a gaggle of soap operas or a Presidential address. The audience actually precedes the program; it is the most dependable element in television. And how that existing audience is divided up among competing broadcasters— how skillfully programmers play to the viewers in each time period—is what the game of television is all about.

This was, to be sure, how matters evolved. In the earliest days of mass-audience television—the late 40s after World War II—building the stations and transmitting their signals were the primary concerns. In the first managerial wave were the electronics engineers, who implemented the technology. After that, the mission was to make people want television, and since most people do not buy technology for itself but rather for what it provides, leadership in the medium was passed on to program experts. Mainly they came from radio, and they not only drew off the stuff of mass-audience radio—the stars and program forms, even the programs themselves—but also radio's economic system of advertiser sponsorship.

Until the mid-60s, the advertisers controlled in a direct way most of the programs that aired on local and network television. It was no accident that serious dramas and news documentaries were basic fare during the early years of the medium; these played to the tastes of the wealthier households, the ones most able to afford television sets. Milton Berle's place in history is that he popularized the medium with his special brand of low comedy on NBC's Tuesday night *Texaco Star Theater.* And as television entered more and more of the poorer homes, the level of entertainment established by Berle became the standard.

Meanwhile, having siphoned off the contents of commercial radio, television proceeded to doom neighborhood movie houses, the popular slick and pulp magazines, comic books and local newspapers, by providing essentially what they survived on. By then, the reins had passed on to a new managerial breed—the salesman, conduit to the advertising dollars, the fuel commercial television runs on.

For a variety of reasons, one of them the rising costs of programs, advertising sponsorship ended in the mid-60s. For most advertisers, it became more prudent to buy participating spots scattered over a number of programs than to concentrate their budgets on specific shows that might, or might not, succeed. This development marked a significant change in the television process, for it threw the control of television program schedules entirely into the hands of the networks; they determined what would be selected for broadcast and what the alignment of shows would be.

Most advertisers now buy audience rather than programs. From computer projections, they buy viewers on a cost-per-thousand basis, and when the projections fall short the networks make up the difference with additional spots. Thus, the advertiser is able to get exactly what he bargained for, millions of women or men in the 18-to-49 age bracket, at an agreed upon rate which might range from $8 to $14 a thousand.

With the risk element removed from television advertising, and with advertisers of mass-marketed products almost compelled to be in television, the business became failure-proof. Even with a run of disappointing programs, a network could still make record profits. The managerial baton in the industry then passed on to accountants and wizards of finance charged with expanding and diversifying their businesses.

By the mid-70s, when broadcast advertising was lush and incremental profits swung on shares of audience, the showmen—or program experts—made a comeback at the top. This was signaled by the elevation of Frederick S. Pierce to president of ABC Television in 1974. A few years later, NBC recruited Fred Silverman as its president and chief executive officer.

The managerial cycles provide one way to trace the history of American commercial television. Another is to track the complex relationship that developed between the broadcast and motion picture industries. The two have become so intertwined that it is difficult to speak of them as separate industries, or even of movies and television as discrete art forms. Indeed, by the 70s, former television executives were heading such film companies as Paramount, Columbia and MGM.

Like newspapers and magazines, the picture companies at first abominated television because they were threatened by it. But, in the late 50s, Leonard H. Goldenson, ABC's top official, who had come from the motion picture exhibition business, coaxed first Walt Disney Studios and then Warner Brothers into producing for television. The rest of the Hollywood film industry then discovered what is now a truism: that institutions which adapt to television tend to survive while those which fight it do so at their own peril. By an ironic turn, just when most of the major film companies were establishing television production divisions, scores of writers, producers, directors and actors who developed their reputations in television were leaving the medium to pursue careers in motion pictures. A good many of them never returned to TV.

Television, which began its life as the enemy of motion pictures, became the medium that was to give eternal life to movies. Some films are likely to be aired forever. Television also brought into being such hybrids as the made-for-

TV movie and the miniseries, the latter a form of movie freed from the normal time restrictions required for theatrical exhibition.

In the course of their ambiguous relationship, television changed the movies and the movies television. When NBC introduced theatrical movies into its prime-time schedule in the mid-60s, it unleashed a form of programming that devastated all conventional television competition. By 1966, each network had scheduled one or two nights a week of movies, and with that they created their own crisis. For a time it seemed that nothing produced expressly for television could survive against the big feature films. New television programs had to be produced with what was called "motion picture quality," and the costs for TV software began to soar.

During the late 60s, when advertisers became obsessed with demographics and established that they would pay more to reach the youth market, the television industry went full tilt in pursuit of viewers in the age range of 18 to 35. As it happened, that was precisely the audience that was patronizing movie houses. Competition between the media heated up again.

To survive, the movies concentrated on whatever was taboo in the family medium of television—themes of sex and violence, mainly. But these films were eventually to find their way onto television, and, as the networks liberalized their standards to accept them in the early 70s, they opened the new era of permissiveness in TV programming.

Yet for all their influence on the content of television, movies have played only a minor role in shaping the television system that has developed in America. It is a system built upon stations licensed to serve their communities, with an overlay of three powerful national networks providing lucrative program services. But while that remains the framework, the system is nowhere the same as that which operated in the 50s or 60s, or even that of five years ago.

Always there are outside forces at work to alter the course and economic system of television—advances in technology, for example. The introduction of video tape in 1956 sent television in a new direction, as did the industrywide adoption of color television in 1965. ENG, the technology of portable video cameras tied to microwave relays, created a revolution in electronic newsgathering, just ahead of the revolution brought on by communications satellites. The superstation, scourge of the broadcast industry, is a creature of technology, as of course are all the steadily advancing new communications media.

Government has been a leading source of change, whether through policy-making, judicial decisions, regulatory actions or political manipulations. Public television modified itself in the Nixon era, along the lines prescribed by the Administration, after the White House vetoed a Federal appropriation; and commercial television, which aired Vice President Agnew's famous Des Moines speech on Nov. 13, 1969, was not the same the morning after. The quiz-show scandals, the Blue Book, the Red Lion Decision, the WLBT case, the WHDH case, the cigarette-advertising ban, the prime time-access rule, the cross-ownership rules and the ascertainment primer all brought changes to the television business.

Broadcast reform groups and special-interest pressure groups—the voices of the people, beyond the ratings—have also had an impact on broadcast practices and policies. At their best, these groups hold broadcasters to their responsibilities in the public interest; at their worst, they have been responsible for acts of censorship and repression.

Certainly the most dramatic episodes of citizen involvement have been in the area of children and television. Campaigns by such groups as Action for Children's Television for reforms in television's service to juveniles succeeded in causing the industry to revise its code and the networks to alter their advertising and programming policies. These campaigns were aided by public television's *Sesame Street* which, at the critical moment, demonstrated that it is possible for a children's program to be at once entertaining and edifying.

The civil rights movement in the 60s brought blacks into prominent roles on television for the first time, without catastrophe to the industry. The crowning irony was that the television adaptation of Alex Haley's *Roots,* a miniseries dealing with the wretched history of blacks in America, turned out to be the most popular entertainment program ever. The women's movement, the gay rights' activists and groups representing Hispanics and Asian-Americans have played what must be considered positive roles in making television a more representative medium than it had been and in increasing its responsiveness to the needs and interests of its varied publics.

On the other hand, in the negative manifestation of the public's direct involvement in the television process, certain political and moral watchdog groups have attempted to suppress points of view that varied with their own. Their chief weapon has been the actual or threatened boycott of advertisers. Television's darkest hour may well have been that period in the early 50s when it reacted to *Red Channels* and the crusades of Aware Inc. by creating blacklists that destroyed the reputations and careers of people suspected of being sympathetic to left-wing causes.

But of all the agents of change in the medium, the most volatile is that called *fashion*. When most households had but a single TV set, the industry programmed each hour according to its notions of which member of the family typically had control of the dial. But when it became the fashion in the mid-60s to have more than one TV set, television was redefined. No longer an all-family medium, it played at once on its various channels to several different audiences residing in the same house. Usually those audiences were differentiated by generation and by sex.

On a more subtle level, fashion dictates television's program trends, its prime cycles. When the "golden age" of drama went out of fashion, the quizz show reigned. In time, it gave way to the western, which later yielded to the private eye. Next, came the lawyer and doctor melodramas, then spy fiction and movies, then social sitcoms, then police-actions shows and then comedies and dramas about working-class people.

For television's first quarter century, the unshakeable principle at the networks was that stars drew the audience. CBS ruled the air for 20 years on star power, while ABC was mired in last place largely because the big stars refused to do business there, wishing to avoid the taint of a loser. Without warning, and for reasons that can only be guessed at in retrospect, the star system died in the 70s. The most popular shows weren't those built around the great names in show business; they were shows that began with ensembles of nobodies out of which new stars emerged overnight: Carroll O'Connor, Farrah Fawcett, Robin Williams, Suzanne Somers, John Travolta, and John Belushi. The end of the old star system figured importantly in ABC's rise to power.

Perhaps the most momentous change in television has been the willingness of people to pay for what they wish to see. The first evidence that people would actually buy television came with the voluntary "membership" donations to public television stations. But in 1975, when Home Box Office went national by satellite transmission, people everywhere began to demand the television service they would have to pay for. The success of HBO spurred the expansion of cable into the major cities and stimulated the development of STV, MDS, video discs, videotext, and direct-broadcast satellites. With each day now, television becomes a larger bonfire.

I have been wandering through television history here with a purpose, with two purposes actually: first to provide for the reader a context for the thousands of moveable parts of television detailed in this encyclopedia; second to point up what has kept me following the medium, as a journalist, since the early 50s—its ever-changing nature.

Something is always new with television, in, on, about or behind the box. Inevitably, the new renders something else old, or even obsolete. A program captures America—*The Flip Wilson Show* say—then meets a competitor that becomes overpowering—*The Waltons*—and a national love affair ends even as a new one begins. Transiency is in the essence of television.

Always, because of the pace of change, the insistent question is, where is it all going? Yet, because the clues to the future tend to lie in the past, it is well to examine where it has all been.

A

AA RATING (Average Audience)—a measure of the size of a program's audience; it is expressed as the percentage of all TV households in the survey area tuned to a telecast in the average minute. Thus, an AA rating of 24.2 means that an average of 24.2% of all TV households in the survey area were viewing at any point in the program.

Because the universe is constant—all TV households in the survey area, whether it be the national market or a local one—an AA rating can be translated directly into the number of households that were tuned in. Advertisers usually refer to the AA rating to determine how many people watched their commercials.

The TA (Total Audience) rating differs from the AA in that it represents the total number of *different* homes that watched all or part of a telecast for six minutes or longer. Unlike an AA rating, the TA is not an average but rather is a measure of the cumulative audience—that is, all those who watched some of the program, even if they tuned out.

A comparison of the TA and AA ratings for any given telecast usually indicates how well it was liked. When there is no great disparity between the two rating indices (the TA will always be somewhat larger), it may be concluded that viewers who tuned in enjoyed the program because they remained with it. A wide difference, however, suggests that the program did not satisfy viewer expectations.

See also Cume, Pocketpiece, Rating.

AAA RATE—a station's designation for its most expensive commercial positions, usually those in the highest-rated time periods. The conventional price classifications are passing from use, however, because of the rapidly fluctuating prices for commercial time. The practice at most stations now is to quote rates according to specific rating and demographic information.

(AAAA) AMERICAN ASSN. OF ADVERTISING AGENCIES—national organization for ad agencies founded in 1917 to improve agency business and advance the cause of advertising as a whole. Today it has a membership of 385 agencies, all of which must adhere to the 4-A codes for agency service and standards and practices.

AARON, CHLOE—senior v.p. of programming for PBS (1976–80) until she left to start an independent production company, the Television Corporation of America in partnership with Nancy Dickerson. She was succeeded at PBS by Suzanne Weil. The five previous years she was director of the public media program of the National Endowment for the Arts. Earlier, she had been a freelance writer specializing in film and TV.

AARON, JOHN A. (d. 1972)—co-producer with Jesse Zoussmer of Edward R. Murrow's *Person To Person* (1953–60). Earlier he had been a journalist with CBS.

ABC—youngest of the networks and for most of its years the weak sister of the three, although its fortunes began to change in 1976 with a new management team that brought in a rash of hit programs in prime time. By the end of the 1976–77 season, ABC had become the No. 1 network in the peak viewing hours with the largest rating advantage ever amassed by a network over its competitors. The swing from third place to first was estimated to have been worth around $60 million in profits.

The network grew even stronger in the subsequent years, with a roster of youth-appeal series—*Happy Days, Laverne & Shirley, Three's Company, Mork & Mindy, Eight Is Enough, Charlie's Angels* and *Love Boat*—that crushed the competition and dominated the Nielsen Top 10 consistently. In 1979, ABC was making sharp gains in all program areas, including news, and for the first time became a serious challenger to NBC for second place in news circulation. Most importantly, ABC improved its overall distribution strength by raiding CBS and NBC for major stations, particularly in markets with only one or two VHF stations, where ABC had been left to depend on a UHF outlet.

But for most of its history, ABC perennially ran last and was forced continually to find new ways to compete. These efforts on several occasions created dramatic changes in television: It was ABC that brought the major Hollywood film studios into TV production, causing all television to jettison live programs for film; and it was ABC that opened prime time to participating advertising in the 60s, signaling the end of program sponsorships.

Even while enjoying extraordinary success, ABC broke new ground in the medium in February 1977 with the unusual scheduling tactic for *Roots,* the 12-hour serialization of the new novel by Alex Haley which attracted the largest single-night audience, as well as cumulative audience, for any entertainment program in the history of television. By airing the full adaptation over eight consecutive nights, ABC demonstrated that the practice of playing a series in weekly installments was only a tradition and not a commandment of television etched in stone.

ABC is the offspring of two government-ordered divestitures, the breakup of NBC's two-network system of radio in 1943 and the divorcement of Paramount Pictures from its theaters division, creating United Paramount Theaters. The newly liberated theater chain and the young network merged in 1953 and proceeded to build a TV network, scrambling with the equally weak DuMont network for affiliates from the relatively few stations not linked with CBS and NBC. DuMont eventually folded.

The history of ABC has been the history of a struggle for economic survival against the two powerful and well-entrenched networks that simply carried over their strongest affiliations from radio to TV.

For the first two decades of its existence, ABC was the lone captive of network television's poverty cycle. Having fewer stations than its rivals and fewer powerhouse shows to bring over from radio, it drew smaller audiences and was perpetually on the lean end of what was then called "a 212-network economy." The best-known movie, radio and recording stars who decided to go into TV invariably chose to go with CBS or NBC. Packagers and advertising agencies that developed shows which promised to be hits rarely risked taking them to the third-place network. Thus, in the years when star power was what counted in prime time, ABC had no access to it to climb out of the depths.

The network survived on expedient moves and the exploitation of fads. Unable to sign stars, it turned instead to the major film studios for products and made a significant impact in 1955 with shows from Walt Disney and Warner Bros. It also developed an unwanted tradition of bringing in shows that had great appeal one season and very little a season or two later (*Batman,* for example), the result of aiming them at teenagers, whose tastes are notoriously fickle.

Burdened always with a large roster of low-rated shows, ABC was saved finally by the prime time–access rule, which confined the networks to three hours of programming in the peak hours, forcing them to cut back their schedules by 30 minutes a night in 1971. This enabled ABC to slough off seven failures and shore up its finances. The network then was able to build its evenings around its most dependable programs, principally movies, *Movie of the Week,* and Monday night football.

History—Edward J. Noble, the maker of Life Savers Candy, created ABC in 1943 when he purchased the Blue Network, which NBC was forced to shed. Ten years later, the network entered into a $25 million merger with United Paramount Theaters, whose president, Leonard Goldenson, became president of the new company, AB-PT. (In the late 60s, the corporate name was changed to American Broadcasting Companies, Inc., of which the American Broadcasting Co. was a division.) Noble became chairman of the finance committee and Robert E. Kintner remained president of the broadcast division.

At the time, the company owned two TV stations in major markets and shared ownership in a third. Problems abounded on every front—programs, sales and affiliates, although the last was eased somewhat when the DuMont Network dropped out in 1955 and as more and more new stations went on the air (unfortunately for ABC, many of them on UHF). Goldenson's deal with Disney—his motion picture background paying off—gave the network its first big lift. The first show, *Disneyland* (which later became *The Wonderful World of Disney* when Kintner joined NBC and lured it away), was a substantial hit, and the second, *The Mickey Mouse Club,* created a sensation with children in the early evening and gained secondary affiliations for ABC in most of the two-station markets committed to CBS and NBC.

With the Warner Bros. array—*Cheyenne, 77 Sunset Strip, Hawaiian Eye, Surfside 6, Maverick*—the ABC network began making a profit for the first time, but soon enough it was beaten at the film game by the other networks. NBC came up strong with programs from Universal, and CBS with shows from an assortment of Hollywood studios.

ABC's attempt to keep up with its rivals in news, meanwhile, proved costly. From 1963 through 1971 the network failed to make money in any single year, and in fact lost more than $100 million before taxes in that span, although its five owned stations were extremely prosperous as a separate division. The network's financial problems were exacerbated by the need to equip for color in 1965, in order to keep step with the other networks. As long as NBC had been the only all-color network, it was acceptable for ABC to remain black & white; but when CBS joined NBC, the third network was forced to go along.

The enormously expensive conversion to color prompted ABC to sell off some of its properties and its investment in Disneyland, and the hard-pressed company in 1965 agreed to a merger with the giant international conglomerate ITT. The FCC approved the merger by a 4–3 vote in December 1966, but the Justice Department called for a reopening of the case and asked the commission to hold hearings. The FCC ordered a rehearing in March 1967, and in the meantime ITT extended ABC a loan of $25 million to help it through its critical cash shortage. The FCC again voted to approve the merger but ordered a 30-day delay for Justice to decide whether to appeal. When the JD decided it would appeal—essentially objecting to a multinational company controlling a powerful news medium—ITT decided, in January 1968, that it had had enough of government obstruction and canceled the merger.

This made ABC ripe for attempted takeovers. Reclusive billionaire Howard Hughes offered $50 million in 1968 for working control of ABC and was rebuffed. Goldenson was also able to thwart a move for control by industrialist Norton Simon. ABC reorganized the management of its broadcast operations, naming Elton H. Rule president of the network, and raised financing through a convertible bond issue.

Throughout the difficult period, ABC continued to improve its competitive standing in a single program area, sports. The aggressive and imaginative leadership of Roone Arledge, president of ABC Sports, was largely the reason; another was the logistical genius of Julius Barnathan, head of engineering and operations who designed and arranged most of the coverage. The ABC program envied most by the other networks, and the one they were unable to copy adequately, was the popular fringe-time omnibus *Wide World of Sports*. In 1970, Arledge ran a large gamble by signing a long-term contract for *Monday Night Football*, dealing ABC into the prestigious NFL events. Football had been a sport for weekend afternoons, and no one could be sure that the games would attract the necessary audience on Monday night. They did, and *Monday Night Football* proved a tremen-

dous coup for ABC, which never before had been able to mount a successful Monday night schedule.

Although it was NBC that innovated the made-for-TV movies, with Universal TV producing them, ABC for once beat another network at its own game. With Barry Diller as the executive in charge (he later went on to become chairman of Paramount Pictures), ABC initiated *Movie of the Week* as a 90-minute entry, commissioning films from a range of major studios and independents at half what NBC paid Universal for the 2-hour shows. It was so successful that the network added a second *Movie of the Week*.

But ABC's progress in these areas came to naught when the 1974 season opened, with ABC's schedule a colossal disappointment. In November, the network's cagiest programmer and strategist, Frederick Pierce, was elevated to president of ABC Television, and with a number of deft maneuvers he immediately reversed the network's downward trend. The following year, Pierce hired away from CBS its program chief, Fred Silverman—notable for the fact that top executives almost never moved "backward" to ABC. Together, Pierce and Silverman improved ABC's ratings in the fall of 1975 and then made midseason changes which shot the network ahead of its rivals. Wall Street became bullish about ABC's prospects.

Early in 1978, Silverman left to become president of NBC, and Anthony D. Thomopoulos was named to succeed him. It was clear during the next year that ABC's power was not diminished by the change.

Structure—ABC Inc. owns the network; five TV stations which comprise the most profitable o&o station group; ABC Radio with four network services, 7 AM and 7 FM stations; a chain of 266 motion picture theaters; a recording company and a chain of retail shops; a scenic and wildlife division operating tourist attractions; a motion picture production division; and a publishing division specializing in farm magazines, *Prairie Farmer, Wallace's Farmer* and *Wisconsin Agriculture*. Through acquisitions, the company also publishes *Modern Photography* and *High Fidelity* magazines.

Corporate management consists of the chairman, Goldenson; the president, Rule; an executive v.p., Pierce; and a senior v.p., Everett Erlick. On the next tier are the various divisions, whose presidents report to Rule.

The president of ABC Television supervises the network, ABC News, the owned TV stations, ABC Sports and ABC Entertainment, and ABC Operations, each of which has its own president.

"ABC CLOSE-UP"—a series of monthly ABC News documentaries begun in 1973 under the supervision of Av Westin and passing in 1976 to Marlene Sanders. Titles included *Crashes: The Illusion of Safety; Danger in Sports: Paying the Price; Food: The Crisis of Price; Fire!; The CIA;* and *Prime Time: The Decision Makers.*

In 1978, Pamela Hill was given charge of the documentary unit, and the programs grew bolder, more venturesome, and more controversial. Notable examples were *The Politics of Torture,* examining U.S. support of repressive regimes in Third World countries in the light of Carter Administration human rights rhetoric; *Terror in the Promised Land,* a documentary on Palestinian terrorists and their mystique of martyrdom; *Arson: Fire for Hire!,* a Brit Hume opus on the fastest-growing crime in the U.S.; *Youth Terror: The View from Behind the Gun,* a cinema verite film, with no written narration, on youth crime, produced and directed by Helen Whitney; and *The Shooting of Big Man: Anatomy of a Criminal Case,* a two-hour examination of an entire criminal case, much of it shot in the courtroom, produced and directed by Eric Saltzman, a trial attorney associated with Harvard University, with the cooperation of Charles Nesson, a professor and assistant dean of Harvard Law School. See also Hill, Pamela; Sanders, Marlene; Westin, Av.

ABC OWNED TELEVISION STATIONS—a group founded by the American Broadcasting Co. in 1948 consisting of stations in five major cities, all with an identical channel number, 7. Those channels were available because, in the early years, many broadcasters believed that "high band" channels (those beginning with Channel 7) were inferior to those on the low band and undesirable. In fact, the ABC o&os were highly profitable and supported the ABC network through the two decades during which it operated in the red. They became even more profitable in the 70s, when their convivial approach to newscasting—a style known as "eyewitness news" or, sometimes derisively referred to as "happy talk news"—elevated them to leading positions in their markets.

The stations are WABC-TV New York, WLS-TV Chicago, WXYZ-TV Detroit, KABC-TV Los Angeles, and KGO-TV San Francisco. The first three went on the air in 1948, the latter two in 1949. Richard A. O'Leary has been president of the division since 1970.

"ABC STAGE 67"—hour-long weekly anthology series created in the fall of 1966 for experimentation with new forms of programming and new talent. Little came of the venture, however, since its offerings received only moderate critical approval and generally unimpressive ratings. The series opened with an original film drama featuring Alan Arkin, *The Love Song of Barney Kempinski,* and over its course presented programs with such titles as *The Kennedy Wit, The Bob Dylan Show, The Anthony Newley Show, Rogers & Hart Today, Noon Wine, The Many Worlds of Mike Nichols, The Legend of Marilyn Monroe* and *An Essay on the American Jew.* Hubbell Robinson, the former head of programs for CBS who had

been responsible for *Playhouse 90,* was executive producer of the series.

ABC VIDEO ENTERPRISES— Division formed by ABC in 1979 to concentrate on ventures in the new fields of television spawned by technology. After creating a unit to market programming on video cassettes and video disks, ABC entered the field of cable programming in 1981 with a nightly cultural service known as ARTS (Alpha Repertory Television Service) carried on the Warner Amex Nickelodeon channel. Next, in partnership with Hearst Corp., it created a four-hour weekday magazine called *Daytime* drawing on the resources of such Hearst magazines as *Good Housekeeping, Cosmopolitan, Town & Country, House Beautiful*, and *Harper's Bazaar*. Transmitted, like ARTS, on the Satcom I satellite, it made its debut in March 1982. Then, with Westinghouse Broadcasting Co. as its partner, it developed two Satellite News Channels, both advertiser-supported and operating 24 hours a day. News Channel One, which started in the spring of 1982, is devoted to continuously updated newscasts in 18-minute cycles. News Channel Two, scheduled for a 1983 debut, is to be less rigidly formatted and will specialize in news analysis and extensive coverage of stories. In a joint venture with Getty Oil, ABC Video entered pay-cable in 1982 with a subscription service featuring a major sports event each month. Herbert Granath is president of ABC Video.

Elie Abel

ABEL, ELIE—NBC diplomatic correspondent during the 60s, who resigned in 1970 to become Dean of the Columbia University Graduate School of Journalism. He gave up the post in 1979 to become Norman Chandler Professor of Communications at Stanford University.

During the period Abel served as head of NBC's London bureau (1965–67), his book, *The Missile Crisis,* was published. Later he collaborated on separate books with Marvin Kalb and Averell Harriman. Abel had a long career as a print journalist before joining NBC, working variously as a national and foreign correspondent for the *Montreal Gazette, The New York Times* and the *Detroit News.*

ABOVE THE LINE/BELOW THE LINE

ABOVE THE LINE/BELOW THE LINE—terms for the general types of television production costs. On the budgeting sheet, a line separates the constants and the variables. Technical expenses are listed below the line and generally are determined by union requirements, so that they are virtually fixed costs; creative expenses are written above the line and may range widely according to terms made with writers, directors and producers.

For most network television programs of a specified length, the below-the-line costs are standard. They involve the basic technical staff, camera unit, electrical unit, music, wardrobe, makeup, props, set design, special effects, editing, laboratory processing, transportation, and the cost of film stock.

The venturesome area is above the line. A script property may involve the payment of literary rights or may be a modest original purchased at Writers Guild scale. The director may be famous and highly expensive or a journeyman who works for the going rate. The program may have a small cast with few salaries or a large cast with many. The difference between a superstar and a relatively unknown actor can amount to $20,000 or more per program, or even a percentage of the profits. The producer may be extravagant or resourceful with regard to his concept of the production and his administrative style. Location and night shooting are extra, as are elaborate special effects.

For a typical half-hour television program of modest pretensions, the above-the-line costs will exceed the below-the-line by around 25 per cent.

The costs for promotion, program testing and the rental of facilities are listed separately as a third general category.

ABRAMS, FLOYD

ABRAMS, FLOYD— probably the most prominent lawyer specializing in First Amendment issues. A partner in the New York law firm of Cahill, Gordon & Reindel, he was co-counsel in the Pentagon Papers case, represented Myron Farber for *The New York Times* in connection with a famous New Jersey murder case, and argued such cases as *Landmark Communications v. Virginia, Nixon v. Warner Communications* and *Herbert v. Lando* (the latter for CBS News) in the Supreme Court.

ABRAHAMS, MORT

ABRAHAMS, MORT—TV producer in the 50s and 60s. He was executive producer of *Producers Showcase* on NBC (1956) and of *Suspicion* (1958). His producing credits included *Target, The Corrupters* (1961) and *Route 66.*

ACADEMY AWARDS TELECAST ("OSCAR SHOW")

ACADEMY AWARDS TELECAST ("OSCAR SHOW") —biggest of the TV awards shows and annually one of the medium's most potent special events, nearly always ranking among the top-rated shows of the year, despite the late hour of the broadcast in the East, usually 10 or 10:30 P.M.

The first national telecast of the Oscar ceremonies was provided by NBC on March 19, 1953, although the awards had been covered on local Los Angeles stations in years before, as well as on network radio. NBC continued to televise the event through 1960, when ABC acquired the rights. After 10 years on ABC, the Oscars went back to NBC for five (1971–75) and then returned to ABC.

The Oscars have had the appeal of suspense, glamour, the unpredictability of live TV and a procession of stars whose dress or deportment might be the stuff of gossip. Although the TV Emmy Awards would seem to have comparable attraction, in fact those ceremonies have been less an event in audience terms and on the whole less successful as programs. Largely this is because the quality and variety of TV programs make for an excess of categories, which tends to dilute the suspense, even when not causing confusion.

While the Oscar presentations have undergone changes year by year, the performer most closely associated with the ceremonies has been Bob Hope, who has been emcee more than 15 times, including the years before TV coverage, when the awards were broadcast on radio. The comedian had served as host for five consecutive telecasts (1958–62) and sporadically thereafter.

A rating history of the Oscarcasts, at five-year intervals but including 1976–1982 follows:

Date	Network	Total Audience Ratings	Share	Total Audience Homes*
3/19/53	NBC	49.7	76	10.9
3/26/58	NBC	53.4	76	22.7
4/8/63	ABC	51.5	71	25.6
4/10/68	ABC	47.1	67	26.4
3/27/73	NBC	50.1	68	32.5
3/26/76	ABC	52.0	64	36.1
3/28/77	ABC	50.0	63	35.6
4/3/78	ABC	36.3	68	26.5
4/9/79	ABC	34.6	63	25.8
4/14/80	ABC	33.7	55	25.7
3/31/81	ABC	31.0	58	24.1
3/29/82	ABC	33.6	53	27.3

ACADEMY OF TELEVISION ARTS & SCIENCES—
See Emmy Awards, NATAS.

ACE, GOODMAN—one of the radio era's most successful comedy writers. He made the transition to television in 1952, working first for Milton Berle and later for the Perry Como and Sid Caesar shows. That he was adaptable to the vastly differing styles of Berle and Como—one brash and slapstick, the other relaxed and low-key—was the essence of Ace's success. He did little TV writing after the 1950s.

Ace gained his fame in radio chiefly as the straight-man opposite his wife Jane in their popular programs *The Easy Aces* and *Mr. Ace and Jane,* both of which he wrote. The programs were noted for the malapropisms freely dropped by Jane Ace, such as "a ragged individualist" and "The Ten Amendments."

ACE AWARDS—annual awards conferred by the National Cable Television Assn. in recognition of outstanding achievement in original programs for cable television. Initiated in 1979, the awards reflected a recognition in the cable industry that the medium's future depended on the variety of new, specialized programming made possible by cable technology, rather than on the traditional retransmission of broadcast signals. The awards were made both to national cable-TV program suppliers and to local cable systems for innovative programming, including that on public access channels.

ACKERMAN, HARRY—head of production for Screen Gems from 1958 through the 60s and previously a program executive for CBS. At Screen Gems he was executive producer of *Bewitched, Hazel, Temperature's Rising,'The Paul Lynde Show* and numerous others.

ACTION FOR CHILDREN'S TELEVISION —a national citizens group formed in 1968 by four women in Newton, Mass., which from the beginning has been extraordinarily effective in pressuring the television industry for reforms in children's programming and advertising. Largely through ACT's early efforts in calling attention to the violence and commercialism in Saturday morning children's programming, the networks shifted to more wholesome shows and cut back commercial announcements from 16 minutes an hour eventually to 9.5 minutes.

ACT produced literature, games and consumer kits designed to discourage heavy television viewing of inferior shows and(to alert parents and children to healthier kinds of food than that advertised on television.

In 1974, having failed to persuade the FCC to adopt a set of hard and fast rules for children's television, ACT formally petitioned the commission to turn its policy statement for children's television into a set of regulations that would affect license renewals.

In reply, the commission indicated that its guidelines were adequate. After four years, during which the issue lay dormant, ACT again filed a petition with the FCC, alleging that self-regulation had proved to be an unsatisfactory solution. In response to ACT's 1978 petition, the FCC voted to reopen the inquiry to ascertain whether licensees are in compliance with the 1974 guidelines, to evaluate the effectiveness of self-regulations, and to assess the viability of alternatives to previously sanctioned policies. See also Children's Television, Broadcast Reform Movement.

"ACTION NEWS"—a style of local newscasting in the 70s that placed a heavy emphasis on newsfilm and on the "realism" of the newsroom: newsmen in shirtsleeves and reporters typing in the background. Some also feature an ombudsman or problem-solving service for viewers. "Action News" formats usually involved joshing among the reporters and informal exchanges that strived for humor. Although it was not TV's noblest contribution to journalism, the style was popular and won audiences for stations. See Eyewitness News.

George Grizzard & Kathryn Walker as John & Abigail Adams

"ADAMS CHRONICLES, THE"—ambitious historical-drama series on PBS produced by WNET New York for the Bicentennial; it not only won critical praise but drew the largest audiences ever for a PBS series. In addition, the series inspired a college-credit extension course designed by Coast Community College District in Costa Mesa, Calif., and offered by around 400 colleges around the country.

A series of 13 one-hour programs on four generations of the family descended from John and Abigail Adams, *Chronicles* covered American history, from the viewpoint of the Adamses, from 1750 to 1900. It premiered on PBS Jan. 20, 1976 and was repeated that fall. The series was produced with meticulous attention to historical and cultural detail, using as its source The Adams Papers, letters, diaries and journals written by members of the Adams family and preserved by the Massachusetts Historical Society. Various scholars quarreled with the historical accuracy of the series, but the producers maintained that, while it was never intended as objective history, it was faithful to The Adams Papers.

The program was conceived and produced by Virginia Kassel through WNET New York, which aspired to create a dramatic series equal in scale and quality to those imported from the BBC. The project raised $5.2 million in grants from the National Endowment for the Humanities, the Andrew Mellon Foundation and the Atlantic Richfield Co.

"ADAM 12"—half-hour, low-key police series on NBC (1968–75) about the day-to-day activities of two Los Angeles Police Department officers who ride a patrol car identified as Adam-12. It featured Martin Milner and Kent McCord and was produced by Jack Webb's Mark VII Productions and Universal TV.

ADAMS, DAVID C.—leading corporate figure at NBC, member of its board of directors since 1958, chairman from 1972–74, and vice-chairman thereafter, until his retirement in the spring of 1979. Disdaining public prominence, Adams preferred to serve in the background as high councillor and grey eminence, with titles such as vice-chairman or executive v.p. His judgment was brought to bear on a wide range of NBC management matters and general industry problems, as well as in formulating NBC's position when it was called to testify before congressional committees or the FCC.

A lawyer, he came to NBC in 1947 from the staff of the FCC. Working initially in the legal area, he shortly moved into the corporate sphere and in 1956 became executive v.p. for corporate relations. Except for a year's leave of absence in 1968, he had been involved in NBC policy decisions for 23 years.

"ADDAMS FAMILY, THE"—live-action situation comedy based on the macabre cartoon characters of Charles Addams. It had a moderately successful run on ABC (1964–66). Produced by Filmways, it featured Carolyn Jones, John Astin and Jackie Coogan.

AD HOC NETWORKS—temporary hook-ups of stations, nationally or regionally, for a single specific purpose, usually the distribution of a sports event or entertainment program in behalf of a sponsor. Such occasional networks enlist affiliates of ABC, CBS and NBC as well as independent stations, the lineup differing for each event. Network affiliates are usually won by the inherent appeal of the event and by the fact that ad hoc networks either barter the shows or pay a higher rate of compensation than the commercial networks, usually 40% or more of ratecard for the time period. Among the leading ad hoc networks are Operation Prime Time, the Mobil Network and SFM's Family Network. Others are set up periodically by the Robert Wold Company, Syndicast, TVS, Mizlou and the Hughes Television Network. See also Operation Prime Time.

ADI (AREA OF DOMINANT INFLUENCE)—a means by which a television station's market is defined for ratings and sales purposes. Each ADI market consists of all counties in which the home market stations receive a preponderance of their viewing. Each county in the U.S. belongs only to one ADI, so that the total of all ADIs represents the total number of TV households in the country.

The system of ADI allocations was introduced in 1966 by Arbitron, one of the rating services, and the concept has become standard. ADI rankings and population data (households, TV households, persons, penetration of cable-TV, multi-set households, etc.) are revised every year to reflect population shifts. The A.C. Nielsen Co.'s equivalent of the ADI is the DMA (Designated Market Area).

In the 1974–75 report, the Kansas City ADI consisted of 30 counties in both Missouri and Kansas; it ranked 23d among the 207 designated ADIs in the U.S., and its potential audience represented .91% of all viewing households in the country. In the 1976–77 report, there were 208 designated ADIs, and Kansas City dropped to 26th in rank.

New York City, the No. 1 ADI, with 29 counties in New York, New Jersey, Connecticut and Pennsylvania, represents 9.01% of the viewing households in the U.

The largest ADIs in the 1976–77 report are, in order, New York, Los Angeles, Chicago, Philadelphia, San Francisco, Boston, Detroit, Washington, D.C., Cleveland and Dallas-Ft. Worth. Together, the ten cover slightly more than one-third of the potential viewing population in the U.S. See also DMA.

ADMINISTRATIVE LAW JUDGE (previously known as hearing examiner)—a member of the FCC staff responsible for conducting hearings for mutually exclusive license applications involving a problem at renewal time. After hearing testimony from parties of interest, usually with the aid of counsel, and on considering the evidence submitted, the ALJ issues an Initial Decision, which is reviewed directly by the commission. The ALJ decisions are not binding on the commission and frequently are overturned in the final decision.

"ADVENTURES IN PARADISE"—an hour-long ABC adventure series (1959–62) about the skipper of a small schooner in the South Pacific and based loosely on the fiction of James Michener. It starred Gardner McKay and was produced by 20th Century-Fox TV. McKay later became a writer.

"ADVENTURES OF JIM BOWIE"—ABC western series (1956–58) on the famed pioneer and frontiersman who invented the Bowie knife. Scott Forbes played the title role, and Louis Edelman produced for Jim Bowie Enterprises.

The Nelson Family: Harriet, David, Ricky & Ozzie

"ADVENTURES OF OZZIE AND HARRIET, THE"—long-running family situation comedy on ABC (1952–66) whose principals were actually a family, that of Ozzie Nelson, a former bandleader, and Harriet Hilliard, a former band singer. Their two sons, Ricky and David, grew up on the TV series and eventually were joined by their wives. When the series ended, the sons were launched on show business careers of their own.

"ADVENTURES OF RIN TIN TIN"—ABC series (1954–57) based on the movies of the canine hero, produced by Screen Gems and featuring Rin Tin Tin, Lee Aaker, Jim Brown and Joe Sawyer. The series was re-run on ABC (1959–61) and then on CBS (1962–64). Tinted in sepia with

new wraparounds, it was revised for syndication in 1976 and bartered by SFM Media Services.

"ADVENTURES OF ROBIN HOOD"—one of the first British series to play successfully on American TV. Starring Richard Greene, it was produced by Sapphire Films Ltd. in England and was carried by CBS (1955–58).

"ADVOCATES, THE"—essentially a debate series, produced for PBS on an alternating basis by KCET Los Angeles and WGBH Boston, from 1969–73. The series sprung amid criticism from the Nixon Administration that TV, in general, was coloring the issues with a liberal bias and that public TV was not giving adequate voice to conservative views. A creature of those pressures, *The Advocates* allowed both sides of a public issue to be argued in courtroom fashion, live, by expert representatives. The series was produced on grants from the Ford Foundation. It was revived by WGBH in 1978 for another season.

"AFRICA"—four-hour documentary by ABC News examining Africa in the modern world, to which the network devoted a full evening in 1967 as the opening night event of the fall season. The longest single program produced for U.S. TV at the time, it had been more than a year in production with six film crews assigned to the project. Although lavishly promoted, and despite the popular appeal of Gregory Peck, who narrated, it drew modest ratings.

James Fleming was executive producer and Blaine Littell, who had covered Africa as a correspondent, was the project producer. Others who figured prominently in the production were Eliot Elisofon, Richard Siemanowski, Leon Gluckmas, Edward Magruder Jones and William Peters. Alex North composed the original score. The program was later syndicated by Worldvision as four one-hour specials.

"AFTERSCHOOL SPECIALS"—occasional series on ABC of quality productions for 8- to 14-year-old children. Begun in 1972 with eight telecasts, seven new dramas and seven repeats were broadcast each subsequent season. The stories are built around the problems of youngsters in which the resolution is brought about by the actions of children rather than intervening adults. Themes include divorce, first love, parent-child relationships, physical handicaps and children's rights.

AGNEW'S ATTACK ON TV NEWS—a televised speech by Vice-President Spiro T. Agnew (Nov. 13, 1969) before a Republican party conference in Des Moines which became the opening shot in a continuing assault on the credibility and integrity of network news by the Nixon Administration. Agnew, who up to that point was scarcely known to the public, became thereafter the leading White House critic of the media.

The networks actually were responsible for the wide circulation the Des Moines speech received. In the belief that they would be ventilating an important issue that concerned them, the three networks—in carrying the speech live—in effect force-fed it to the viewers. And the fact that Agnew was addressing a partisan group which applauded everything he said undoubtedly heightened his effectiveness with the TV audience.

"The purpose of my remarks tonight is to focus your attention on this little group of men who not only enjoy a right of instant rebuttal to every presidential address, but, more importantly, wield a free hand in selecting, presenting and interpreting the great issues of our nation," Agnew said.

"Is it not fair and relevant to question concentration (of power) in the hands of a tiny, enclosed fraternity of privileged men elected by no one and enjoying a monopoly sanctioned and licensed by government?

"The views of the majority of this fraternity do not—and I repeat, not—represent the views of America."

The address, written by one of President Nixon's speechwriters, Patrick J. Buchanan, conveyed strong hints that the networks were controlled by an Eastern establishment with a decidedly liberal bias, a theme which was to become a motif in Agnew's later speeches, as well as those of Buchanan and Clay T. Whitehead, director of the Office of Telecommunications Policy.

The Des Moines speech also denounced the networks' practice of following a President's speech with instant analysis by newsmen, without taking into account the fact that newspaper pundits also write their analyses immediately after such a speech, although those do not appear until the following morning. The networks did not follow Agnew's speech with an analysis, and not long afterwards CBS chairman William S. Paley banned instant analysis on his network, although he restored the practice later. See also Instant Analysis; Whitehead, Clay T.

AGRONSKY, MARTIN—veteran TV newsman working between commercial and public television. After a network career with NBC, ABC and CBS, spanning the period 1952–69, he became an anchorman and commentator for the Post-Newsweek Stations, based at WTOP-TV in Washington, and simultaneously for PTV's Eastern Educational Network. His principal programs were *Agronsky and Company, Evening Edition,* and *Agronsky At Large.*

Agronsky joined NBC in 1952 as a foreign correspondent and later became a Washington correspondent for ABC, performing his most distinguished work in covering the

activities of Sen. Joseph McCarthy. He rejoined NBC in 1956 and covered the Eichmann trial in Jerusalem (he had once been a reporter for the Palestine *Post* in that city) in 1961, among other assignments. In 1965, he went to CBS and produced several noted documentaries (including *Justice Black and the Constitution*) before becoming Paris bureau chief (1968–69).

AILES, ROGER—independent newsfilm producer who became Richard Nixon's TV advisor for the 1968 campaign and in 1974 news editor of TVN, the short-lived syndicated news service financed by Joseph Coors. Before joining the Nixon campaign, Ailes had been producer of *The Mike Douglas Show* in Philadelphia.

AIM (ACCURACY IN MEDIA)—a Washington-based citizens organization founded in 1969 that watches over print and broadcast media for instances of what it judges to be biased, slanted or unbalanced reporting. The organization, which could be characterized as politically conservative, reports its specific complaints to the FCC, the National News Council and the broadcaster (or publisher) involved, and to the public as well, through its newsletter, the *AIM Report.* Its charges of one-sidedness and journalistic unfairness in NBC's documentary, *Pensions: The Broken Promise* (aired on Sept. 12, 1972), were upheld by the FCC. In 1974, however, the Court of Appeals for the District of Columbia reversed the FCC's decision to avoid inhibiting investigative journalism. See also *Pensions: The Broken Promise.*

AIR TIME INC.—syndication and program distribution company which set up an ad hoc network for the live presentation of the Cannes Film Festival Awards in the spring of 1979 and also financed and syndicated *The Unknown War,* a series of 20 documentaries on World War II from footage shot by the Russians, assembled by Ike Kleinerman and narrated by Burt Lancaster. The company also syndicates such diversified fare as a two-hour production of *Giselle* by the American Ballet Theater, the World Championship Tennis matches and *The Race for the White House,* a series of 10 one-hour programs recreating elections of the twentieth century.

The coverage of the Cannes Festival Awards, the first ever for U.S. television, went awry when the French engineers union staged a wildcat strike and cut off the satellite transmission. The two-hour telecast, carried by 83 stations in America, wound up 43 minutes shy, losing the presentation of the key awards. Most of the stations carried a rebroadcast on tape a day or two later.

ALABAMA ETV LICENSES—case in which the FCC denied the renewal of Alabama's eight public TV licenses but then permitted the licensee to reapply for them.

The licenses of the Alabama Educational Television Commission (AETC), an agency of the state of Alabama, were denied renewal in 1975 when the FCC determined that the stations followed a racially discriminatory policy in their overall programming practices. Citing "pervasive neglect" of Alabama's black population, the commission maintained that the stations failed to meet adequately the needs of the public they were licensed to serve.

The FCC's opinion was based on AETC's conduct during the 1967–70 license term, during which the licensee rejected most of the black-oriented programming available to it and failed to give blacks adequate representation on the air or in the production or planning operations at the stations. The commission said that while it recognized the vital service of educational TV, it would not condone AETC's dereliction and deficiency simply because it was engaged in public broadcasting.

However, the commission noted that improvements had occurred since 1970 and that there was a pressing need for public television in Alabama. It ruled, therefore, that the public interest would be served by granting AETC interim authority to continue operating the eight stations.

The commission also held that in light of the fact that the licensee was a state agency, and that it had demonstrated a greater responsiveness to the special needs of Alabama's black citizens since 1970, AETC should not be ruled ineligible to file applications for the stations. In this connection, the commission waived its rules governing the filing of repetitious applications, and AETC resubmitted its eight license applications.

In 1976, the FCC granted five of the AETC license applications. The other three were designated for hearing along with applications filed by a nonprofit group, made up predominantly of blacks. The hearing is still pending before the FCC.

ALBERG, MILDRED FREED—producer of prestigious dramatic series in the late 60s, including *Playhouse 90* and the *Our American Heritage* series. She was also executive producer of *The Story of Jacob and Joseph* and *The Story of David* in the mid-70s.

ALDA, ALAN—star of the successful CBS situation comedy *M*A*S*H* who also wrote and directed some episodes of that series. His wry, detached, intellectual style and crisp delivery of wisecracks became hallmarks. Alda developed into one of television's distinctive actors.

The popularity of *M*A*S*H* made him one of the richest actors in television. His earnings from the syndication of the series reportedly came to $30 million.

He was director of the TV production of the play *6 Rms Riv Vu* (1974) and creator and co-executive producer of *We'll Get By,* a situation comedy on CBS (1974) which had a brief run.

See also, *M*A*S*H.*

ALEXANDER, DAVID—director whose credits range from *U.S. Steel Hour, Climax* and *Studio One* in the 50s to episodes of *Marcus Welby, M.D., Emergency, F. Troop, My Favorite Martian, Gunsmoke, Please Don't Eat the Daisies* and other comedy and action series in the 60s and 70s.

ALEXANDERSON, ERNST F.W. (Dr.) (d. 1975)—engineer of the General Electric Co. (U.S.) whose invention of the alternator made possible long-distance radio communications and who presented the first home and theater demonstrations of television. The first home reception of television took place in 1927 in his home in Schenectady, N.Y., and the theater demonstration was held the following year in the same city. Both demonstrations used perforated scanning discs and high-frequency neon lamps to originate and reproduce the picture. In the theater demonstration the picture was flashed on a seven-foot screen.

Alfred Hitchcock & friends

"ALFRED HITCHCOCK PRESENTS"—popular series of mystery and suspense stories hosted by Alfred Hitchcock, famed producer of classic suspense movies. Hitchcock's wry delivery and his sardonic closing speech were important assets to the series, which ran from 1955 to 1962 in half-hour form and three additional seasons in a one-hour version, *The*

Alfred Hitchcock Hour. Produced by Shamley Productions, the half-hour series began on CBS (1955–59) and switched to NBC (1960–62). Similarly, the hour version began on CBS in the fall of 1962 and moved to NBC in 1964.

"ALIAS SMITH AND JONES"—ABC Western series (1971–73) featuring Ben Murphy and Pete Duel as two young outlaws trying to earn amnesty. When Duel died in December 1971, he was replaced by Roger Davis. The hour-long Universal TV series was produced by Glen Larson, with Roy Huggins as executive producer.

Linda Lavin as Alice

"ALICE"—half-hour videotaped situation comedy on CBS (1976–   ) loosely based on the film, *Alice Doesn't Live Here Anymore,* concerning a widow waiting tables in a diner to support her young son. Linda Lavin is featured in the title role, with Vic Tayback, Beth Howland and Celia Weston in the supporting cast. The series is by Warner Bros., with Bob Carroll Jr. and Madelyn Davis as executive producers. It became a hit in a Sunday night parlay with *All in the Family* and regularly made the top 10 as a mainstay of the strong Sunday lineup at CBS.

ALL-CHANNEL LAW—passed by Congress in 1962, legislation designed to help faltering UHF broadcasting by giving the FCC the power to require that all television sets shipped in interstate commerce be "capable of adequately receiving" all 82 channels—the 70 UHF channels as well as the 12 VHF channels.

The FCC implemented the law by setting minimum standards for UHF tuners and requiring that they be included in all receivers manufactured after April 30, 1964. Since that time, the FCC has added new rules designed to

make UHF channels as easy to tune as VHF. As the result of the law, more than 90% of all television-equipped homes now have sets with UHF tuners. See also UHF.

Fred Allen

ALLEN, FRED (d. 1956)—one of radio's leading comedians who made the switch to TV in 1950, although he disapproved of the new medium and continually railed against it. A satirist and social commentator who had become a darling of the intellectuals, Allen made his first TV appearances in the *Colgate Comedy Theater* as one of the stars in the rotation. He became emcee of *Judge For Yourself* (1953–54), starred in *Fred Allen's Sketchbook* in 1954 and became a member of the *What's My Line?* panel in 1955, serving until his death on March 7, 1956. He was never to recapture the glory of his radio days, and his own show, *Sketchbook,* was knocked off the air by competition with one of the big giveaway shows of the time. He made guest appearances on numerous shows, and his last special effort for NBC was to narrate "The Jazz Age" for the *Project 20* series. It aired nine months after his death.

ALLEN, IRWIN—a leading producer of special effects movies (*Poseidon Adventure, The Towering Inferno*) whose TV series lend themselves to the disaster situations that are his specialty. His TV offerings included *Voyage To the Bottom of the Sea, Swiss Family Robinson* and *Code Red.*

ALLEN, MEL—sportscaster prominent from the late 30s to the early 70s, whose authoritative style and polished delivery won him network assignments for major baseball and football events. The regular commentator for the N.Y. Yankees games for three decades, he was also announcer for most of the annual All-Star games during that period, as well as for many of the football bowl games. He was also commentator for the Fox Movietone Newsreel (1946–64) and host of *Jackpot Bowling* (1959). In the 70s, he narrated the syndicated *This Week in Major League Baseball.*

Steve Allen

ALLEN, STEVE—many-talented TV personality who hosted several successful shows, including the first *Tonight Show* on NBC, and a Sunday evening variety series that was competitive with *The Ed Sullivan Show* for a few seasons. He was also a pianist, composer, recording artist, fiction writer and political activist.

The Steve Allen Show began as a late-night program in New York on July 27, 1953 and went on the NBC network in September 1954 as *The Tonight Show.* Aired on weeknights for 90 minutes, it was a potpourri of music, comedy, interviews, and inventive sketches with the resident cast, many of whom went on to become stars—Andy Williams, Eydie Gorme, Steve Lawrence, Don Knotts and Bill Dana, among them.

Simultaneously, Allen served as a panelist on *What's My Line* on CBS from 1953 to 1955 and starred in spectaculars for Max Liebman. He left *Tonight* in January 1957 after having begun the Sunday evening *Steve Allen Show* (1956–60), a jazz-accented variety hour pitted against Sullivan on CBS. It featured his wife, Jayne Meadows.

He later did several syndicated programs, including a popular talk show for Westinghouse Broadcasting, and a series for PBS in 1977, *Meeting of Minds.*

A new prime time venture on NBC in 1980, *The Steve Allen Comedy Hours,* was short-lived.

"ALL IN THE FAMILY"—a landmark series that changed the nature of situation comedies, opening them to realistic characters, mature themes and frank dialogue. Written by Norman Lear and produced by Yorkin-Lear Productions, the series was introduced on CBS in January 1971 as a second season entry and did poorly in the ratings at the outset. Rare among TV programs, it developed its great popularity during the summer reruns.

The Bunkers: Jean Stapleton & Carroll O'Connor

Based on an immensely popular and controversial British series, *Till Death Us Do Part,* created by Johnny Speight for a limited run on BBC-TV, *Family* is built upon the clashes of a working-class bigot, Archie Bunker, with his neighbors and his liberal son-in-law. (His British counterpart was named Alf Garnett.) Bunker has become so well established as representative of an American type that his name bids fair to enter the language.

The half-hour series was developed by ABC, but when that network rejected two versions of the pilot, Lear took it to CBS and landed a berth. Fortuitously, it was in a time when CBS was actively searching for programs relevant to contemporary life and to the liberal-conservative rift in American attitudes.

The series made stars of its principals—Carroll O'Connor as Archie, Jean Stapleton as his wife Edith, Sally Struthers as their daughter and Rob Reiner as their live-in son-in-law—none of whom was well known before *Family* began. It also spun off two other successful series, *Maude* and *The Jeffersons.*

Reiner and Miss Struthers quit the series after the 1977–78 season to pursue other ventures, but it maintained its popularity without them. When Miss Stapleton also withdrew as a regular a year later, the concept of the series was altered and the basic setting changed to a bar. It opened the 1979–80 season with a new title, *Archie Bunker's Place* and continued strong in the popular *CBS* Sunday lineup. See also Wood, Robert D; *Archie Bunker's Place.*

"ALL MY CHILDREN"—ABC soap opera created by Agnes Nixon which premiered Jan. 5, 1970 and concerns two families, the Martins and the Tylers, and their romantic entanglements and marital stresses. The cast is headed by Mary Fickett, Ruth Warrick, Kay Campbell, Hugh Franklin, Ray McConnell, Frances Heflin, Eileen Herlie, Susan Luccie, James Mitchell, Harriet Hall, Taylor Miller, and Richard Van Fleet.

"ALL'S FAIR"—half-hour videotaped CBS situation comedy from the Norman Lear organization built upon the romantic relationship between political opposites, an arch-conservative male and an ultra-liberal female. The leads were portrayed by Richard Crenna and Bernadette Peters, with J.A. Preston, Jack Dodson and Judy Kahan featured. Created by Rod Parker, Bob Schiller and Bob Weiskopf, it was developed and produced by Lear's T.A.T. Productions, with Parker as executive producer and script supervisor and Bob Claver as director. It lasted one season, 1976–77.

"ALMOST ANYTHING GOES"—prime-time action game show on ABC (1976), derived from a European show, *It's a Knockout,* in which teams of citizens representing different countries compete with each other in bizarre contests. In the U.S. version, which had a successful summer tryout in 1975, the competing teams represented small towns with populations of less than 20,000. Regis Philbin hosted, and the action was covered in the manner of a sportscast by Charlie Jones and Lynn Shackelford. A Saturday night hour entry, it was co-produced by Bob Banner Associates and the Robert Stigwood Organization. When the nighttime program failed, a children's version was mounted, *Junior Anything Goes.*

"AMAHL AND THE NIGHT VISITORS"—Gian Carlo Menotti opera on a Christmas theme, the first opera to be written expressly for TV; it premiered in a two-hour production by the NBC Opera Company on Christmas Eve, 1951, and has been a seasonal offering nearly every year since. The 1953 production, presented in the *Hallmark Hall of Fame* series, was the first sponsored show to be televised in color. Samuel Chotzinoff produced the early telecasts of the opera.

The role of Amahl, the crippled shepherd boy, was performed the first year by Chet Allen and in many of the subsequent productions by Bill McIver. Rosemary Kuhlmann performed the role of the boy's mother in the original and the later productions.

In 1978, NBC offered a new filmed production of the opera with Teresa Stratas as the mother and Robert Sapolsky as Amahl, and with Giorgio Tozzi, Nico Castel and Willard White as the Three Kings.

AMATEAU, ROD—producer-director, very active in the 50s, best known for creative supervision of *The Burns and Allen Show.* More recently, he was line producer for NBC's ambitious 1979 series, *Supertrain. He segued from that fiasco to supervising producer of Dukes of Hazzard,* a ratings hit on CBS.

Amahl and the Night Visitors

"AMERICA" (ALISTAIR COOKE'S "AMERICA")—13-hour series produced by the BBC as a sort of personal essay by journalist Alistair Cooke on the history of the United States. NBC televised it simultaneously with the BBC in 1972; then it went to PBS in 1974 in a different form—26 half-hour episodes underwritten by Xerox.

Cooke, an Englishman who became a U.S. citizen, called the series a "personal interpretation" of American history and was its writer as well as narrator. The shows, which won numerous honors, traced the country's growth from Indian times to the present. Michael Gill was producer, and Time-Life Films co-financed the production with the BBC.

"AMERICA AFTER DARK"—an attempt by NBC at a new concept in late-night television, that of surveying live the key population centers for coverage of the social and celebrity functions. The nightly program, which began in January 1957 in the 11:30 P.M. time-slot, was poorly received and lasted only seven months.

Columnists Hy Gardner, Earl Wilson and Bob Considine covered the New York segments, while Paul Coates went on for the West Coast and Irv Kupcinet for Chicago. Jack Lescoulie served as moderator. The program gave way to the vastly more successful *Tonight* show, with Jack Paar.

"AMERICA ALIVE!"—NBC daytime magazine program (1978) which switched around the country for interviews and features to convey a sense of covering life-styles in America on a daily basis. Jack Linkletter was host and Bruce Jenner, Janet Langhart and Pat Mitchell traveling co-hosts. The series began in July 1978 and was dropped by the end of the year because of lean ratings. Woody Fraser, who had left ABC's *Good Morning, America* to take on the project, was executive producer.

"AMERICA 2NIGHT"—syndicated series via the Norman Lear shop satirizing the crassness of television talk shows. It ran one season (1977–78), growing out of a summer entry, *Fernwood 2Night,* which in turn was a spin-off of *Mary Hartman, Mary Hartman.*

When *Mary Hartman,* as a comic soap opera, could not sustain the grind of five shows a week, Lear's T.A.T. Productions came up with the idea of *Fernwood 2Night*—a fictional local TV talk show using some of the cast and the mythical locale of *Mary Hartman*—as a summer replacement. The concept was broadened to a national talk show with *America 2Night.* While it had a loyal and appreciative following, the audience was by TV standards a small one. Martin Mull played the talk-show host and Fred Willard his sidekick.

"AMERICAN BANDSTAND"—a TV disk jockey show that originated on a local station, WFIL-TV, in Philadelphia in the mid 50s and made a national figure of host Dick Clark when it went on the ABC network in 1957. The program was televised daily in the afternoons and soon added a Saturday evening version, helping to introduce such rock stars of the era as Chubby Checker, Paul Anka and Frankie Avalon. In 1963, it was cut back to the Saturday edition, which is still running. Meanwhile, Clark himself branched into program-packaging and other TV ventures.

AMERICAN FEDERATION OF MUSICIANS —large and powerful union (AFL-CIO) with 725 chapters in the U.S. and Canada. Its TV contracts involve few individual stations and are mainly with networks and group owners. During the radio era, AFM had fought a famous battle to prohibit stations from using prerecorded music but lost when a Federal court, in 1940, ruled that a musician's rights to recorded music ended with the sale of the record.

AMERICAN FEDERATION OF TELEVISION AND RADIO ARTISTS (AFTRA)—broadcast performers union (AFL-CIO) founded in August 1937 as AFRA (radio artists), adding the "T" in 1952 after a merger with the Television Authority. The Television Authority had been formed by the American Guild of Variety Artists, Actors' Equity, Chorus Equity and the American Guild of Music

Artists to represent their membership in the new electronic medium.

With AFTRA representing performers in live TV and Screen Actors Guild those in filmed TV programs, a jurisdictional dispute developed with the advent of video tape. SAG contended it was a new form of film, and AFTRA insisted the intent of tape was to preserve a live performance. AFTRA prevailed, but the two unions have been discussing the possibility of a merger ever since.

AFTRA staged a 13-day strike against the networks in March 1967, chiefly over a contract for new employees at the network-owned stations, and in the fall of that year the union briefly honored the picket lines of a technical union, NABET. In 1980, a strike by AFTRA and SAG, running from mid-July to October, halted much television production and caused delays in the new season premieres. The issue was the actors' stake in the new electronic media. See also Wolff, Sanford .

"AMERICAN NEWSSTAND"—an attempt by ABC to create a daily five-minute newscast for the young audience in the late afternoon. It featured news of particular interest to teenagers but lasted only a year (1961–62).

"AMERICAN SHORT STORY"—public television anthology of short-story adaptations, which made its debut in 1977 with nine films and then received funding for eight more productions airing in the spring of 1980. Among its distinctions, the series was the first major work for American public TV to be purchased by the BBC. (It was a source of some embarrassment to PBS that although it imported heavily from the BBC, the British public broadcaster chose to import programs from American commercial television and rejected even PBS's proudest offerings, such as *Sesame Street* and *The Adams Chronicles.*

The short-story series was produced by Robert Geller, a former high school English teacher, through his company, Learning In Focus Inc. The series was presented on PBS by the South Carolina ETV network and WGBH Boston.

In the first group of stories were Sherwood Anderson's "I'm a Fool," Stephen Crane's "The Blue Hotel," F. Scott Fitzgerald's "Bernice Bobs Her Hair," Ernest Hemingway's "Soldier's Home," Ambrose Bierce's "Parker Adderson, Philosopher," Henry James's "The Jolly Corner," Flannery O'Connor's "The Displaced Person," John Updike's "The Music School" and Richard Wright's "Almost a Man."

The second group, for which $2.6 million in grants were made by the National Endowment for the Humanities, the Corporation for Public Broadcasting and Xerox Corp., includes: Willa Cather's "Paul's Case," William Faulkner's "Barn Burning," Ernest Gaines's "The Sky Is Gray," Nathaniel Hawthorne's "Rappaccini's Daughter," Ring

Lardner's "The Golden Honeymoon," Katherine Anne Porter's "The Jilting of Granny Weatherall," James Thurber's "The Greatest Man in the World" and Mark Twain's "The Man That Corrupted Hadleyburg."

The critical success of the series won Geller an opportunity to produce for NBC *Too Far To Go,* a two-hour adaptation of a series of stories by John Updike. The program aired in March of 1979, and it too was a critical success.

AMERICAN WOMEN IN RADIO AND TELEVISION

(AWRT)—nonprofit professional organization of women working in broadcasting and allied fields. Its purpose has been to encourage cooperation between, and to enhance the role of, females in the industry. Its national and regional conferences are forums for the discussion of industry issues.

Established in 1951, AWRT now has more than 50 chapters in the U.S. and some 2,500 members. In addition, there are campus groups assisted by the chapters known as College Women in Broadcasting. An AWRT educational foundation formed in 1960 finances broadcast industry forums, career clinics, international study tours and closed-circuit programs for hospitalized children.

AMLEN, SEYMOUR—v.p. of ABC Entertainment concentrating on program planning and scheduling. He joined ABC in 1955 as a ratings analyst, rose to become v.p. in charge of research and then moved into programming as a strategist.

"AMOS 'N' ANDY"—TV version of the enormously popular radio series, produced and carried by CBS (1951–53). It ceased production not for lack of audience but because black organizations such as NAACP objected to it for depicting blacks in a demeaning and stereotyped manner. When Blatz Beer yielded to the organizations' campaigns and withdrew its sponsorship, CBS took the show off the network, but the reruns continued to be syndicated by CBS Films until 1966. During the civil rights movement, CBS responded to protests by removing the show from both domestic and overseas sale and making it unavailable for any purpose. The series featured Alvin Childress and Spencer Williams, with Tim Moore in the focal role of "Kingfish."

AMPTP (ASSOCIATION OF MOTION PICTURE AND TELEVISION PRODUCERS)—trade association

representing about 70 production companies. As the collective bargaining arm of the Motion Picture Assn. of America, it handles contract negotiation and labor relations in television for the member companies. It also conducts training, apprenticeship and technical research programs.

AMST (ASSOCIATION OF MAXIMUM SERVICE TELECASTERS)—an organization of local broadcasters

formed in 1956 to preserve the coverage, range and power of their stations against proposals by the FCC to increase the number of VHF stations by reducing the distances covered by existing stations. The AMST is essentially concerned with protecting the maximum effective radiated power permitted for stations by the FCC rules adopted July 1956. The organization has been instrumental in securing passage of the All-Channel Receiver legislation in 1962 and has mounted resistance to proposed reallocations of the television broadcast spectrum to nonbroadcast frequency use.

An American Family: The Louds

"AN AMERICAN FAMILY"—a PBS documentary series (1973) on the affluent William C. Loud family of Santa Barbara, Calif., which some critics considered a brilliant television venture and others a grotesque use of the medium. Producer Craig Gilbert and his cameras in effect moved in with the family from May 1971 to the following New Year's Eve to record the life-style, values and relationships of Bill and Pat Loud and their five children, Lance, Kevin, Grant, Delilah and Michele. The cinema verite portrait that resulted was provocative because it captured highly personal moments in their lives, including the break-up of the parents and the eldest son's flamboyant involvement in the New York homosexual scene.

When the series aired, the Louds went on talk shows and gave newspaper interviews contending that the 12-part documentary presented a distorted picture, emphasizing the bad moments in their lives rather than the good. Meanwhile, the press and public debated whether the film was valid as a revelation of American family life or was merely a form of peeping tomism with exhibitionists. The series originated as a project of NET and was completed when that organization was absorbed by WNET New York.

ANDREWS, CHARLES—producer of live television shows in the early years of TV. He had a long association with Arthur Godfrey (as his producer) at CBS.

ANDREWS, EAMONN—a leading British TV personality who since 1968 has been a dominant on-camera figure for Thames Television. He has been one of the co-hosts of the Thames *Today* show, which is televised in the evening, and also host of *This Is Your Life* (a program quite different in the U.K. from its U.S. counterpart). Before signing on with Thames, he had been a BBC performer, hosting the British edition of *What's My Line?*, which ran for 12 years, and doubling as a sportscaster on BBC Radio. An Irishman, he broke into radio in Dublin as a boxing commentator at the age of 16.

"ANDY GRIFFITH SHOW, THE"—a countrified situation comedy concerning a small town sheriff and his friends. A Monday night fixture on CBS for nearly a decade (1960–68), it featured Don Knotts, Ronny Howard and Frances Bavier. When Griffith left the series, it continued in its CBS time period as *Mayberry R.F.D.*, with Ken Berry in the principal role as a small town councilman. It fell, finally, not to poor ratings but to the CBS decision to undo what had become the network's rural image. The syndicated reruns of the Andy Griffith version carried the title *Andy of Mayberry*. The series was by Mayberry Productions.

"ANDY WILLIAMS SHOW, THE"—title of four different music-variety series starring one of the leading popular singers of the 60s. Longest-running of the four was Williams's first NBC variety hour (1962–67), which featured The New Christy Minstrels and a singing group of young boys, The Osmond Brothers. It was by Barnaby Productions. Two years later NBC mounted a new series for Williams with a different supporting cast that lasted two seasons, until 1971.

A 1959 *Andy Williams Show* was a summertime venture for CBS. In 1976 Williams began a new half-hour variety series for prime-access syndication, produced by Pierre Cossette.

"ANGIE"—one of several ABC sitcoms that clicked in an especially good year for the network, the 1978–79 season, where it was a midseason entry. The show may well have made it on its own merits, but it was undeniably helped—frequently into the Nielsen Top 10—by its placement following the network's runaway hit, *Mork & Mindy*. The episodes turn on the marriage of a waitress from a blue-collar family to the scion of a wealthy Philadelphia family. Donna Pescow and Robert Hays were the principals, with Sharon Spelman, Doris Roberts, Debralee Scott and Diane Robin featured. It was cancelled in the 1980–'81 season.

The series, via Paramount Television, was created by Dale McRaven and Garry Marshall. McRaven and Bob Ellison were executive producers. Thomas L. Miller and Edward K. Milkis received billing as executive consultants.

ANGLIA TELEVISION—commercial licensee for the East of England region, based in Norwich and founded in 1961. Survival Anglia Ltd., its production subsidiary, made its mark globally with the *Survival* series of nature and animal shows. See also Trident Anglia.

ANIK—domestic satellite, launched by Canada in November 1972 and used for transcontinental Canadian television links and for TV and telephone links with remote Arctic settlements. Anik 2 was launched in April 1973 and Anik 3 in May 1975. All were manufactured by Hughes Aircraft Co. and operated by Telesat Canada. The satellites' name is the Eskimo word for "brother." See also Satellite, Communications.

ANN, DORIS—executive producer of NBC's Television Religious Programs unit since 1950, in charge of its regular series and special productions. She joined the network in 1944 working in its personnel department before becoming a producer of religious programs.

ANNAN COMMISSION—See Britain's Broadcasting Commissions.

ANNENBERG, WALTER H.—head of a media empire largely concentrated in Philadelphia and eastern Pennsylvania. It was dismantled after his appointment as President Nixon's Envoy to the Court of St. James's. Annenberg's broadcast group, Triangle Stations, was broken up in 1970, with a total of nine stations sold to Capital Cities and the remainder to former employees under the banner of Gateway Communications. Earlier he had divested himself of his two newspapers, the *Philadelphia Inquirer* and the *Philadelphia Daily News,* but retained the magazines *TV Guide* and *Seventeen* and the racing publications the *Morning Telegraph* and the *Daily Racing Form.*

The Triangle TV group had consisted of WFIL-TV Philadelphia; WFBG-TV Altoona, Pa.; WLYH-TV Lancaster-Lebanon, Pa.; WNBF-TV Binghamton, N.Y.; WNHC-TV New Haven, Conn.; and KFRE-TV Fresno, Calif.

Annenberg was also founder and president of the Annenberg Schools of Communications at the University of Pennsylvania and the University of Southern California and contributed to the communications program at Temple University. In 1977, he proposed to establish a multimillion dollar communications facility and school at the Metropolitan Museum of Art in New York but withdrew the offer when it proved controversial with the Museum's board. Two years later, he made a multi-million dollar gift to public TV for education.

"ANN SOTHERN SHOW, THE"—moderately successful CBS situation comedy (1958–61) which was part of a parade of shows in the 50s that absorbed movie actors into TV. Actually, Miss Sothern had made the transition earlier with *Private Secretary* (1952–54), whose reruns went into syndication under the title *Susie*. In both series she played opposite Don Porter. The first series cast her as a secretary in a talent agency, the second as an assistant manager at a New York hotel. In 1965 Miss Sothern performed the voice of the car in the short-lived fantasy comedy *My Mother, The Car.*

"ANOTHER WORLD"—NBC soap opera which premiered in 30-minute format on May 4, 1964, was expanded to an hour a day in 1975 and to 90 minutes in 1979. Meanwhile, in 1970, it had spun off an unsuccessful soap opera, *Another World: Somerset,* which later took the title *Somerset.* The stars include Hugh Marlowe, David Canary, Rose Gregorio, Anne Meacham, Paul Stevens, Constance Ford, Irene Dailey, Douglass Watson, Beverly Penberthy and Victoria Wyndham.

ANTENNA FARM—a location set aside for all or most of the television transmitting antennae in a community or area. The use of antenna farms is considered preferable to locating antennae in various different areas because it reduces air traffic hazards and usually makes possible better home television reception by permitting all receiving antennae to be oriented in the same direction.

ANTHOLOGY—nonepisodic program series constituting an omnibus of different programs that are related only by genre. The studio drama series of the 50s—*U.S. Steel Hour, Philco Playhouse,* etc.—are notable examples, as in another sphere is ABC's *Wide World of Sports.* Often a host is used to provide a sense of weekly continuity, as with *Alfred Hitchcock Presents, Boris Karloff's "Thriller," Dick Powell's Zane Grey Theater* and Rod Serling's *Twilight Zone.* The longest-running anthology in prime time is NBC's *Wonderful World of Disney,*

although it is exceeded for longevity by the Sunday morning religious shows.

The so-called Golden Age of TV drama ended and anthologies in general went into a decline in the early 60s when network programmers determined that TV's heavy viewers—the habit viewers—were more likely to embrace episodic shows with familiar elements than series whose casts changed every week. Through most of the 60s the networks usually avoided the anthology series as a high-risk item except in one form, theatrical movies.

Audience criteria changed around 1970, however, sparking the return of the anthology. The advertising industry's preoccupation with demographics sent the networks in quest of viewers in the 18–49 age range; this ruled out the habit viewer, who was perceived as either very young or over 50. The 18–49 group was the movie-going group, and the networks catered to it with made-for-TV movies which, of course, were merely telefilm anthologies going by the names of *World Premiere* and *Movie of the Week.* Other anthologies such as *Police Story* and the news-magazine *60 Minutes* soon after established themselves in prime time.

ANTI-BLACKOUT LAW—legislation enacted in 1972 prohibiting local TV blackouts of major professional sports events if they are sold out 72 hours ahead of game time. The law, as initially passed, was contingent on a three-year trial basis, the result of public disaffection at being denied telecasts of sold-out home games. The blackouts occurred, according to team spokesmen, because owners feared that the availability of home games on TV would discourage ticket sales. Annual studies by the FCC, however, have proved otherwise, prompting Congress to consider legislation to make the original law permanent.

ANTIOPE— Videotext system developed by the French communications industry under government leadership which now has a lead over competing British (Prestel) and Canadian (Telidon) systems. Antiope uses a mosaic pattern to produce graphics rather than a geometric pattern, like Telidon, and thus is less expensive and quicker. Currently being tested by First Bank, Minneapolis, and other companies.

ANTONOWSKY, MARVIN—NBC program chief for one year (1975–76), after which he became a v.p. for Universal TV. In June 1979, he joined Columbia Pictures as senior v.p. and assistant to the president, Frank Price. A former research executive for ABC and earlier for several major advertising agencies, he approached programming with a heavy reliance on program testing and other statistical data. His tendency to make swift alterations in the schedule during his single disappointing season at NBC earned him the nickname "the mad programmer."

Antonowsky had been recruited from ABC in July 1973 to become NBC's v.p. of program development, and he moved rapidly into other posts until he was named to succeed Larry White as v.p. in charge of programming. He had joined ABC in 1969 from the J. Walter Thompson ad agency.

APPELL, DON—director-writer whose credits include *Apple's Way, Love Story, Love, American Style, Arnie* and *Run For Your Life,* among scores of other shows.

"APPLE'S WAY"—an hour-long dramatic series about a contemporary idealistic family, which was an attempt by CBS to duplicate its success with *The Waltons*. Like *Waltons,* it was created by Earl Hamner and produced by Lee Rich's Lorimar Productions. The series premiered as a second season replacement in February 1974 and fared well enough to be renewed for the following fall. However, it faltered, or cloyed, and became an early casualty of that season. Featured were Ronny Cox, Lee McCain, Malcolm Atterbury and Vincent Van Patten.

"APPOINTMENT WITH DESTINY"—series of specials by the Wolper Organization using documentary techniques in dramatic reenactments of historical events. Six were televised by CBS between 1971 and 1973 and a seventh, *They've Killed President Lincoln,* by NBC.

The CBS group included *The Crucifixion of Jesus,* narrated by John Huston; *Surrender at Appomattox,* narrated by Hal Holbrook; *Showdown at O.K. Corral,* narrated by Lorne Greene; *The Plot to Murder Hitler,* narrated by James Mason; *The Last Days of John Dillinger,* narrated by Rod Serling; and *Peary's Race to the North Pole,* narrated by Greene. The Lincoln episode was narrated by Richard Basehart. Nicholas Webster and Robert Guenette separately produced and directed various episodes.

ARBITRON—the rating service of the American Research Bureau, which is the prime competitor to the A.C. Nielsen Co. in gathering audience data at the local market level. A subsidiary of Control Data Corp., it is the preferred service of some advertising agencies and stations for measuring local audiences.

Before it incorporated electronic metering of viewing, when the data was obtained entirely through diaries, the ratings were known as *ARB's*. The Arbitron Television Market Report provides individual surveys of TV viewing, with demographic breakdowns, in the 208 marketing areas defined by its table of Areas of Dominant Influence. The

Arbitrons also offer estimates of the numbers of households subscribing to cable-TV, owning color receivers, owning more than one TV set, capable of receiving UHF and other such information.

Arbitron's ADI geographic specifications coincide for the most part with the DMA (Designated Market Area) data of the Nielsen Co., but stations are more likely to refer to their ADI in discussing their coverage areas than to their DMA, possibly because the ADI was developed first.

ARCHER, NICHOLAS—ABC News executive who became v.p. of Television News Services in 1975. He was responsible for the planning and implementation of the network's coverage throughout the world and was supervisor of the assignment desk. He joined ABC as assignment manager in 1963, after working for the N.Y. *Journal-American*.

"ARCHIE BUNKER'S PLACE"— Carroll O'Connor vehicle on CBS which represents the remains of *All In the Family* after the three other principals, Jean Stapleton, Sally Struthers, and Rob Reiner, left the cast. Started in 1979, the new series has Bunker (O'Connor), now a widower, tending a neighborhood saloon with his business partner (initially Martin Balsam), and raising his young niece. Bunker without his family proved nearly as popular as before and helped maintain the CBS dominance of Sunday night. Via Tandem Productions and Ugo Productions, with Joseph Gannon as producer, the series features Danielle Brisbois, Allan Melvin, Denise Miller, Steven Hendrickson, Jason Wingreen, Barbara Meek, Bill Quinn, Anne Meara, and Abraham Alvarez. See also *All In the Family*.

ARLEDGE, ROONE—president of ABC News and Sports since the spring of 1977 and previously president of ABC Sports. He has been a leading figure in TV sports since the 60s when, as v.p. for ABC Sports, he moved aggressively to acquire rights to top events, developed omnibus programs such as *Wide World of Sports* and *The American Sportsman* and stimulated innovations in live production. He was named president of ABC Sports in 1968 and shortly afterwards successfully negotiated for the Monday night *NFL Football* series, giving an important lift to the network's prime-time schedule.

His promotion to head of News, at a time when ABC was prospering on the entertainment front, was alarming to many in the industry since Arledge had no background in journalism. It was feared that in his zeal to improve the network's news circulation, Arledge would let high news standards fall to showmanship and electronic razzle-dazzle. To a certain degree he did, at first, as with the pilot for the news magazine *20/20* and with his multiple anchor format to enliven the evening newscast. But despite some early lapses, ABC News maintained respectability and gained in stature in Arledge's administration, and by the second year the ratings for the evening newscast, *World News Tonight,* had for the first time caught up with those for NBC's *Nightly News* and on several occasions surpassed them. By 1981, ABC News had become a full-fledged competitor to the other networks, and equally respected.

Arledge outbid his rivals for Olympic events in 1968, 1972 and 1976 and won honors for ABC with distinguished coverage. By using the Atlantic satellite he arranged live coverage of numerous international sporting events, including heavyweight title fights, the U.S.-Russia Track Meet from Kiev, the Le Mans endurance race in 1965, the Irish Sweepstakes and the World Figure Skating Championships.

His skill in producing elaborate live sports events brought him into the show business realm in the mid-70s as producer of live specials with Frank Sinatra and Barbra Streisand, and later as executive producer of the variety series *Saturday Night Live With Howard Cosell.* None, however, scored impressively in the ratings.

Arledge joined ABC in 1960 in a production capacity after six years at NBC, during which time he produced news and special events broadcasts and the series *Hi Mom.* He created *Wide World of Sports* in 1961 and, with that, started ABC Sports on its period of rapid growth.

Roone Arledge

"ARMY-McCARTHY HEARINGS"—an historic congressional proceeding which began April 22, 1954; it marked the end of Sen. Joseph McCarthy's career as a ruthless hunter of communists and subversives in government, largely because it was televised. The hearings spanned two months and occupied a total of 35 broadcast days. ABC, which had a meager daytime schedule at the time, carried the hearings in full, as did the fading DuMont Network. CBS and NBC carried excerpts.

As he had on Edward R. Murrow's *See It Now* programs weeks before the hearings began, Sen. McCarthy projected the image of a callous and opportunistic villain to millions in the viewing audience. Largely as a result of the hearings, the Senate voted 67 to 22 to condemn McCarthy for his tactics and irresponsible charges, and conventional wisdom insists that it was this television exposure that turned public opinion against the Wisconsin senator and was his real undoing.

ARNAZ, DESI—co-star of the *I Love Lucy* series and subsequently producer and occasional director of *Lucy* and *The Lucille Ball Show* while he was married to the comedienne. He also headed their Desilu Studios prior to its sale to Paramount TV. Before embarking on a TV career, Arnaz, a Cuban, had been leader of a Latin band.

ARNESS, JAMES—star of *Gunsmoke* (1955–75) on CBS, through which he became one of TV's best-known and highest paid actors, while the character he portrayed, Marshal Matt Dillon, became a national folk hero. A relatively obscure screen actor, he won the role because he bore a resemblance to John Wayne, who had been envisioned for the part. Arness eventually became part-owner of the series. He later starred in a two-hour series, *How the West Was Won* (1976–79), which was aired by ABC on Monday nights in place of the NFL games when the football season ended.

Arness returned to a new NBC Series in 1981–82, McClain's Law, with Eril Bercovili as executive producer. But it was one of that season's many casulaties.

"ARNIE"—a domestic situation comedy on CBS (1970–72) about a Greek-American, blue-collar worker who becomes an executive. Produced by 20th Century-Fox TV, it starred Herschel Bernardi and featured Sue Ane Langdon, Roger Bowen, Stephanie Steele and Del Russel.

ARNOLD, DANNY—comedy writer-producer whose series credits include *Barney Miller* (which he created with Theodore Flicker and of which he is executive producer) and *My World and Welcome to It.*

He began as a film editor, worked briefly as a standup comedian and actor in films with Dean Martin and Jerry Lewis and, in 1956, became a writer for the *Tennessee Ernie Ford Show* and later the *Rosemary Clooney Show.* He was producer and story editor for *The Real McCoys* (1962), and producer of *Bewitched* (1963–67) and of *That Girl* (1967–69). He formed his own company, Four D Productions, in 1974.

The company is named for his wife, his two sons, and himself, all of whose names begin with D.

"ARREST AND TRIAL"—90-minute series introduced by ABC (1963–64), which was two shows in one: a 45-minute police show involving the capture of a suspect, followed by a 45-minute courtroom drama on the prosecution of the case. The joining of two popular modes of melodrama did not catch the public fancy, however, and the series lasted only one season. Ben Gazzara starred in the "Arrest" segment as a police detective and Chuck Connors played the criminal lawyer in the "Trial" portion. The series was by Universal TV.

"ARTHUR MURRAY DANCE PARTY, THE"—prime-time ballroom dance show which ran through the 50s variously on four networks. It featured the head of the famous dance-instruction chain but actually starred his wife, Kathryn Murray, as hostess. The series, which involved skits, demonstrations of dance steps, and dance contests among celebrities began on the DuMont Network in 1950 and the following year switched to ABC. Later it went to NBC, CBS and back to NBC.

ARTS (Alpha Repertory Television Service)— see Cable Networks.

ASCAP (AMERICAN SOCIETY OF COMPOSERS, AUTHORS AND PUBLISHERS)—founded in 1914, the oldest of the U.S. music licensing organizations. As a non-profit membership society, it collects copyright fees from users of music and distributes the royalty payments to writers and publishers. Radio and TV stations, and networks, are permitted to use ASCAP music under a blanket annual license fee; ASCAP then surveys the air plays to determine what each song has earned in royalties.

ASCAP's ability to control absolutely the copyright fees for music during the 1930s so concerned the broadcast industry that it created the National Assn. of Broadcasters to fight the anticipated increases in licensing fees. In 1939, the broadcast industry created a new music licensing organization as a counter-force, BMI (Broadcast Music, Inc.), which was controlled by broadcasters. Lacking the big-name established song writers, BMI endeavored to sign up young new composers; many were drawn to BMI because its formulas for royalty payments were more generous to writers of current hits than were those of ASCAP, oriented as it was then to the established writers.

In 1941, when broadcasters would not agree to ASCAP's demands for higher royalty payments, all ASCAP music was barred from the air—the current popular hits, as well as the standards by Berlin, Kern, Porter, Gershwin and others. Radio stations were left to play public-domain and BMI music until the matter was settled, but in holding out ASCAP actually helped BMI to grow stronger. ASCAP's new rival developed initially in the peripheral country music, jazz and rhythm & blues fields, and BMI came to full flower with the emergence of rock in the mid-50s. See also BMI, SESAC.

"ASCENT OF MAN, THE"—outstanding 13-hour series on the history of man's ideas from prehistoric times to the present, prepared and presented by the late Dr. Jacob Bronowski. A co-production of the BBC and Time-Life Films, it had its U.S. premiere on PBS Jan. 7, 1975 on underwriting from Mobil Oil.

Dr. Bronowski, a noted scholar and leader in the movement of Scientific Humanism, proved an engaging TV performer, warm and infectiously enthusiastic; these appealing personal qualities made the difference between a remarkable mustering of knowledge and a television triumph. More than informative, the series was inspirational because of Bronowski's concern not with the historical discoveries themselves—"the great moments of human invention"—but with what they revealed of man.

Bronowski died of a heart attack in California, where he was a fellow at the Salk Institute, shortly before the series aired on PBS. But he had already assisted in the design of the college-credit courses that were to be based on the programs, courses which marked a breakthrough in the use of television for off-campus education.

ASCERTAINMENT PRIMER—a spelling out by the FCC of requirements and procedures to assist stations in their ascertainment activities, issued on Dec. 19, 1969 as *Primer On Ascertainment of Community Needs*. It was amended and clarified by a Report and Order of Feb. 23, 1971. The primer, whose leading proponents on the commission were Robert Taylor Bartley and Kenneth Cox, made ascertainment of the needs and interests of a licensee's market an ongoing procedure rather than one performed only at license-renewal time.

With the primer, the FCC hoped to create a formal mechanism by which to put all broadcasters in continual touch with all significant elements of the communities they were licensed to serve. Following the guidelines prescribed, each station was required to show what it ascertained of community needs and how it responded to those needs on the air. The primer called for key station personnel, including management, to meet on a regular basis with community leaders and to conduct surveys of the general population.

The commission's ascertainment procedures came in stages, beginning with its 1960 Programming Policy Statement, which expressed the agency's interest in licensees making a concerted effort to seek out the community issues that would lead to programming. In 1968, the FCC Public Notice *Ascertainment of Community Needs by Broadcast Applicants* asserted that long residency in an area was not necessarily an indication of familiarity with the programming needs of the community, and that a survey of needs was mandatory. The primer that resulted was hatched in part by the Federal Communications Bar Assn. (FCBA). In 1981, the primer was discarded by Mark Fowler, the FCC chairman, as a step towards deregulation.

ASHER, WILLIAM—producer-director whose credits include the *Fibber McGee & Molly* TV series and *The Shirley Temple Show* in the late 60s, as well as *Bewitched, Temperature's Rising* and the *Paul Lynde Show.*

ASHLEY, TED—noted talent agent who became chairman of Warner Bros. in 1969 when Kinney Services Inc. acquired the motion picture company; in 1975, he gave up the day-to-day operational responsibilities to Frank Wells and became active on a part-time basis as co-chairman.

A force in show business since the early 50s, Ashley was an agent for William Morris for six years before starting his own agency at about the time commercial TV was beginning to grow. His agency expanded through acquisitions, becoming known as the Ashley Famous Agency, and was instrumental in packaging and selling to the networks around 100 series, including *The Carol Burnett Show, The Doris Day Show* and *Mission: Impossible.* Ashley represented such artists as Tennessee Williams, Arthur Miller, Perry Como, Burt Lancaster, Rex Harrison and Ingrid Bergman.

The agency was acquired by Kinney in 1967 and Ashley became a director and member of the executive committee. Two years later, when Kinney acquired Warner Bros., it sold off the talent agency to avoid the conflict of interests in both representing talent and producing but it retained Ashley and named him head of the motion picture subsidiary of what became Warner Communications Inc.

ASI (AUDIENCE STUDIES, INC.)—program testing system in Hollywood used by NBC, ABC and the production studios to determine whether a program will be popular and whether the storylines and cast members have appeal. CBS rarely uses ASI because it operates its own testing system, "Little Annie." ASI testing is conducted in a plush Sunset Strip theater, Preview House, with a seating capacity of 400, to which the randomly accumulated participants are invited free. Each fills out forms to provide breakdown of the

group by age, sex, education and income and is asked to operate a dial at his/her seat to indicate when something in the program is pleasing or displeasing. The dial has degrees for either reaction. To set a norm for each night's audience, a Mister Magoo cartoon is run before the program. Oscilloscopes in the control room produce graphs of the audience reactions as expressed in dial-turning, and this information is collated with data from questionnaires and personal interviews. Although ASI had been owned by Screen Gems in the 60s, other studios and the networks use it and trust its results. See also Stanton-Lazarsfeld Program Analyzer.

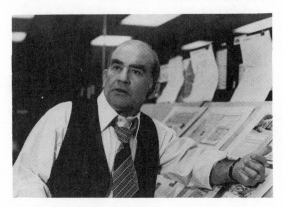

Ed Asner as Lou Grant

ASNER, EDWARD—television star and president of Screen Actors Guild (1981), who became controversial as a liberal political activist during the Reagan Administration. An issue developed over whether the causes he championed—the Equal Rights Amendment, opposition to American involvement in El Salvador, opposition to nuclear arms—were perceived by the public as the point of view of his television persona, the sympathetic crusading newspaper editor, Lou Grant. Thus it became a journalistic issue, not unlike the issue of patent medicine endorsements by actors portraying doctors on television. Asner argued that actors on the right were able to endorse political causes with impunity, while liberals taking a position were considered out of line.

Asner became a star, and winner of numerous Emmy awards, through his role as the hard-boiled but sentimental boss of the newsroom in *The Mary Tyler Moore Show* (1970–77). When that sitcom folded, Asner's character, Lou Grant, was kept alive in a new CBS series by that title under the same production auspices, MTM Productions. There were two switches, however: In the new series, Grant became head of a newspaper's newsroom instead of a television station's and in the one-hour format he became a dramatic rather than a comic character. For all Asner's popularity, *Lou Grant* was a slow starter and didn't establish itself firmly as a hit until its third season on the air. It became one of the most respected shows in television and was acknowledged as being generally accurate in its portrayal of newspaper life.

As a television character, Lou Grant grew into a classic—not a leading man, but an everyman—a beefy, blustering, blundering, middle-aged, all-around fellow working in rolled-up shirtsleeves. The good, decent boss: gruff but, underneath it all, humane.

During the actor's strike of 1980, Asner, for all his success and wealth, became a leading activist and spokesman for the workers' cause. A year later he was elected president of SAG, the post that started Ronald Reagan on his political career. Asner was at political odds with Reagan, and he was outspoken in opposing administration policies that did not concern actors or the union directly. These activities made him a controversial figure in 1982.

Asner had been a serious actor on the stage before going into television. He made his mark first with Chicago's Playwrights Theatre Company and later appeared off-Broadway as Mr. Peachum in *The Three Penny Opera*. On television, in addition to his Lou Grant roles, he played Capt. Davies in *Roots* in 1977 and in the same year starred in the drama, *The Gathering*. He also was featured in the 1980 motion picture *Fort Apache, the Bronx* and starred in the 1981 teleplay, *A Small Killing*.

He was born and raised in Kansas City.

ASPEN INSTITUTE PROGRAM ON COMMUNICATIONS AND SOCIETY—project of the Aspen Institute for Humanistic Studies to propose policies and action on major issues relating to the communications media. The institute's initial areas of priority have been government and the media, public broadcasting, television and social behavior and the humanistic uses of cable and the new technologies. Begun in 1971 with Douglass Cater as director, the program has consisted of seminars and conferences and has produced a number of books, policy papers and special reports. Michael Rice became director of the communication project in 1978.

It was at the instigation of the Aspen Institute that the FCC late in 1975 reinterpreted the Equal Time Law, classifying debates between political candidates and their press conferences as exempt from equal time considerations if they are covered by TV and radio as external, on-the-spot news events. Thus the institute helped to make possible the 1976 presidential and vice-presidential debates. See also Carter-Ford Debates.

"ASPHALT JUNGLE, THE"—crime series on ABC (1961) based loosely on W.R. Burnett's best seller and on the John Huston 1950 movie of that title. It lasted only 13 weeks. Produced by MGM-TV, it featured Jack Warden, Arch Johnson and Bill Smith.

ASTAIRE, FRED—indestructible dancing star who achieved stardom in movies during the 30s and contributed three dazzling TV specials almost a quarter-century later. His first, *An Evening With Fred Astaire*, was an artful blockbuster in 1958, winning a passel of Emmy Awards. Barrie Chase was his dancing partner, Hermes Pan the choreographer and Bud Yorkin the producer, with the Jonah Jones Quartet also featured. Astaire encored the following year with *Another Evening With Fred Astaire* and in 1960 offered *Astaire Time*.

Aside from the specials, he hosted *Alcoa Premiere* in 1962, a drama anthology that went into syndication as *Fred Astaire Presents*.

"AS THE WORLD TURNS"—CBS daytime serial which premiered April 2, 1956 and which for almost a decade was the No. 1 daytime program. In fact, its ratings often equaled those of certain nighttime shows. The stories concern the romantic trials of residents of mythical Oakdale, principally the families of doctors and lawyers. In December 1975 its format was expanded from 30 minutes to an hour.

Members of the cast have included Eileen Fulton, Henderson Forsythe, Don Hastings, Anthony Herrera, Lisa Loring, Elaine Prince, Don MacLaughlin, Larry Bryggman, Kathryn Hays and Colleen Zenk.

ATLASS, H. LESLIE (d. 1960)—a pioneer in radio and TV who founded the WBBM stations that became the powerful and lucrative CBS base in Chicago. Atlass had been owner of WBBM Radio until it was purchased by CBS in 1933. He was then named v.p. in charge of the network's central division and in the early 50s acquired a TV station for CBS, which became WBBM-TV. He then became v.p. and general manager of the Chicago TV, AM and FM stations and was a forceful figure in the company until he reached the mandatory retirement age in 1959. At WBBM Atlass groomed numerous executives who rose to high positions at CBS.

ATS-6 (APPLICATIONS TECHNOLOGY SATELLITE 6)—an experimental NASA satellite launched in May 1974 with the largest antenna yet devised for space and powerful enough to transmit directional signals to small and inexpensive ground receivers, making it a prototype for the futuristic satellite-to-home form of television. ATS-6, in its first year, was used for communications experiments, among them health and education transmissions to Appalachia,

Alaska and the Rocky Mountain region. In 1975 the satellite was repositioned for use by the Government of India for an instructional television experiment. In late 1976, it was moved back to the Western hemisphere, where it has resumed its original experimental services. See also Satellites, Communications.

ATV—British independent licensed to the Midlands region, with headquarters in Birmingham and other production studios in Borehamwood, Herts. Named for its parent company, Associated Television Corp., headed by Lord Lew Grade, ATV was a major producer for the commercial network in Britain and one of the U.K.'s largest exporters of TV shows. The station received a mandate from the Independent Broadcasting Authority in 1981 to operate as a "dual" station with separate facilities in Birmingham and Nottingham. It changed its name to Central Independent Television in 1982. Its distribution arm in the U.S. is known as Independent Television Corp. (ITC).

AUBREY, JAMES T.—president of CBS-TV from December 1959 to February 1965. His administration was probably the most competitively successful in TV history but was marked by hints of scandal. A cum laude graduate of Princeton, Aubrey was a champion of unsophisticated entertainment programs of strong rural appeal (*Mister Ed, The Andy Griffith Show, The Beverly Hillbillies, Petticoat Junction*) and a leading exponent of the "habit" theory of television, which holds that most people want the same series in the same time periods every week, without interruption by specials. In the years before demographic ratings, when reaching the greatest numbers of viewers was the objective, his approach proved sound and his programming prowess made him the most powerful albeit most arrogant of network presidents.

His business style—which included an ability to tell a star of long standing, simply and coldly, "You're through"—earned him the nicknames of "Jungle Jim" and "The Smiling Cobra." Still he was held in awe for his ability to win the ratings race decisively, and for his assertions of power. Rumors began to circulate about his allegedly bizarre personal life; about his acceptance of an apartment in Manhattan and private limousine from Filmways, a studio supplying programs to CBS; and mainly about his strange business association with Keefe Brasselle, an actor of modest achievement who professed a connection with the underworld.

Brasselle's company, Richelieu Productions, with no record of achievement in television, suddenly landed three major shows on CBS in the 1964 season, all without benefit of a pilot *The Reporter, The Baileys of Balboa* and *The Cara Williams Show*, the latter expected to be the successor to the Lucille Ball show. The arrangement with Brasselle raised suspicions within CBS as well as at the FCC, and both conducted investigations. All three Richelieu programs were failures, and the CBS rating lead shrunk for the first time in Aubrey's presidency. Over a weekend, in late February 1965, Aubrey was fired and was replaced at the network by John A. Schneider, who had been v.p. and general manager of WCBS-TV, the flagship station. Aubrey went into independent production, then became president of MGM for a time and became an independent again. In recent years, he returned to TV as an independent producer, mostly of made–for–TV movies, and had a series sale as well—*Shannon* on CBS in the 1981–82 season.

Aubrey started with CBS in 1948 as a salesman for KNX-Radio and KNXT Los Angeles, later becoming general manager of the television station and of what was then the CBS Pacific Network. In 1956 he became manager of network programs for CBS-TV in Hollywood, then skipped to ABC in New York as a v.p. in charge of programs and talent. There he introduced a number of hit shows, *77 Sunset Strip, The Rifleman, The Real McCoys, Hawaiian Eye*, demonstrating that the wave of the future in TV was the Hollywood film series and not the New York game- or variety-show. CBS rehired him in 1958 as v.p. of creative services, having apparently decided then that he was to be the next president. Within a year he had the job.

AUDIENCE FLOW—the movement of the audience between programs and stations during successive time periods. By scheduling programs consecutively which appeal to similar audiences, the networks and local stations attempt to achieve a flow of audience to minimize the tuning away to other channels. Program schedules are devised with a mind to audience flow, because programmers have found that when two shows are contiguous that appeal to different kinds of viewers neither will do well.

AUDILOG—Nielsen's television viewing diary used for both national and local market (NSI) estimates of audience size and demographics. Households agreeing to keep a diary for Nielsen fill out how long the set was on, to which program or channel the viewing was given, and which household members were watching. For national ratings, the diaries are kept during designated weeks on a continuing basis. For NSI, the local market viewing data is gathered from three to eight times annually (depending on the size of the market). NSI households keep a diary only for one week, and they are selected at random from telephone listings.

Diaries are used by the ratings services to find out *who* in the household was viewing. These data are collated with those from metered households, because the meter can only record how long each TV set was on and to which channel it was tuned.

AUDIMETER—a device patented by the A. C. Nielsen Co. which is attached to television sets in sample households to record minute-by-minute viewing. The device was acquired by Nielsen in the late 30s and was first used to measure national radio audiences in 1942. Within the standard Audimeter, now passing from use, a slowly revolving cartridge of film records the set-on, set-off information and the channel tuned to. The results, from approximately 1,200 Audimeter households for the national sample, were tabulated and projected for the Nielsen weekly and bi-weekly television reports.

Starting with the 1973 season, Nielsen put into national operation the Storage Instantaneous Audimeter (SIA), an electronic information storage and retrieval system capable of reading up to four TV sets in a household, including battery-powered portables. Three markets have been metered with SIA for overnight ratings—New York, Los Angeles and Chicago. In addition to those, there are approximately 1,170 other SIA households nationally, each connected to a special telephone line. At least twice a day a central office computer dials the household units and retrieves the stored viewing information. Unlike the standard Audimeter, which required the removal and mailing of the cartridge by members of the sample household, the automatic process requires no such work.

The SIA makes possible daily national ratings (available two days after broadcast) for each day of the year, and the daily overnight ratings for the three largest markets. See also Nielsen, A.C., Co.; Pocketpiece; Ratings.

AUER, JOHN H. (d. 1975)—producer-director whose credits in the early years of TV included *Whirlybirds, U.S. Marshal* and *Harrigan & Sons*.

AURTHUR, ROBERT ALAN (d. 1978)—a prolific writer during the "golden age" of live TV drama, best known for his *A Man Is Ten Feet Tall*. He later doubled as an executive producer of drama specials, such as *What Makes Sammy Run?* for NBC's *Sunday Showcase* in 1959. During the 60s and 70s he worked mainly in motion pictures.

AUSTRALIA, TELEVISION IN—consists of a national public channel (no advertising) operated by the Australian Broadcasting Commission, plus three commercial networks operating in the key cities, and more than 30 so-called "country" stations in the provinces which are the equivalent of unaffiliated independents in the U.S.

Australian TV began with the first public station in 1956. The first commercial station came in the early 60s. The medium went to 625-line PAL color in March 1975.

ABC, the public broadcaster, is patterned very much after Britain's BBC, both in programming terms and as a kind of quasi-optical patron of culture. It sponsors, for instance, several symphonic orchestras. Governed by a nine-member commission appointed by the governor-general, the ABC is funded by annual government appropriation. It was created in 1932, the successor to the old Australian Broadcasting Co., which has been operating public radio in the country since 1923.

The three commercial chains are the 7 Network, the National 9 Network, and the 0–10 Network, each consisting of independently owned affiliates in Sydney, Melbourne, Adelaide and Brisbane.

The "country" stations, though unaffiliated with any of the three chains, buy much of their programming from them.

Commercial stations are licensed by the Australian Broadcasting Control Board, which comes under a ministry of the Federal Government and has ultimate control over everything aired by the independents.

AUSTRIA, TELEVISION IN—two channels operated by Osterreichischer Rundfunk, supported by a mix of spot advertising and license fees. The color standard is 625-line PAL, incepted in 1969. Total set count is about 2,250,000. Television households pay the equivalent of $126 a year in license fees for color TV or black and white.

The service attempts to defray the cost of expensive special productions by offering them to underwriters on a single sponsorship basis, as in public TV in the U.S. It provides for billboards at the start and close of the show only, but assures the sponsor's identification with the program.

"AUTOBIOGRAPHY OF MISS JANE PITTMAN, THE"—two-hour dramatic film on CBS (premiere: Jan. 31, 1974) tracing the life of a black woman in the south from slavery to the civil rights movement of the 60s. A poignant made-for-TV movie, it won nine Emmy Awards and at once seemed destined for annual repeats (which so far have occurred). The fictional film proved a triumphant vehicle for actress Cicely Tyson, affording a role in which she advanced from young womanhood to old age. It climaxes with a powerful scene in which the ancient woman makes a protest by drinking from a water fountain marked "Whites Only."

The film was written by Tracy Keenan Wynn, based on a novel by Ernest J. Gaines. Robert W. Christiansen and Rick Rosenberg were the producers for Tomorrow Entertainment, and John Korty was the director. Featured were Odetta,

Josephine Premice, Ted Airhart, Richard Dysart, Roy Poole and Valerie O'Dell.

Cicely Tyson as Jane Pittman

AUTRY, GENE—singing cowboy of the movies and early western star of TV who became a station tycoon as majority owner and chairman-president of Golden West Broadcasters, a West Coast group whose flagship is KTLA-TV Los Angeles. Through his Flying A Productions, he produced *The Gene Autry Show* (1947–54), with Pat Buttram as his sidekick, which enjoyed particular popularity with children. In 1966, at KTLA, he produced the syndicated series *Gene Autry's Melody Ranch*. But the era of the singing cowboy had already ended with the arrival of the more sophisticated westerns, such as *Cheyenne, Gunsmoke* and *Rawhide*.

AVCO BROADCASTING CORP.—once powerful station group whose properties were sold off in the early 70s when the parent company, Avco Corporation, experienced financial difficulties. Avco, during the early 60s, had acquired the venerable Cincinnati-based group that had been known as Crosley Broadcasting, founded by Powell Crosley in 1922. Until they were divested, the Avco TV stations had been WLWT Cincinnati, WLWD Dayton, WLWC Columbus, Ohio, WLWI Indianapolis, and KMOL San Antonio. John T. Murphy was president.

"AVENGERS, THE"—secret agent series produced in England by Associated British Pathe, which played on ABC in interrupted runs from March 1966 to September 1969. Of the 83 films made, 57 were in color. Patrick Macnee starred as Steed, the stylish independent sleuth. His original female partner, with a talent for karate, was Honor Blackman; she was succeeded by Diana Rigg and then by Linda Thorson.

In marketing terms, at least, *The Avengers* was probably the most successful British-flag series, sold in around 120 world markets. After a hiatus of several years, producers Brian Clemens and Albert Fennell reactivated the show, with Macnee again in the lead, and with financing by a French syndicator.

The Avengers: Patrick Macnee & Diana Rigg

AVERBACK, HY—producer-director of comedy series, whose credits include *Needles and Pins* and *Big Eddie*. Before TV he was a prominent radio announcer associated with the Bob Hope show, among others.

AYLESWORTH, MERLIN HALL ('DEAC') (d. 1952)—first president of NBC, joining on Nov. 15, 1926, after having been managing director of the National Electric Light Assn. Initially, he concentrated on promoting the sale of radio sets for NBC's parent, RCA, but toward the end of his nine-year term he was a dominant figure in the entertainment and advertising fields and had played a key role in the development of the Radio City complex and the Radio City Music Hall. He left the network on Dec. 31, 1935 to become chairman of RKO-Radio Pictures, which RCA had acquired several years earlier, and in 1938 he became publisher of the *New York World-Telegram*.

AZCARRAGA, EMILIO (d. 1973)—cofounder (with Romulo O'Farrill) of Telesistema Mexicana and head of that television monopoly in Mexico for two decades, until his death. Azcarraga, who earlier had dabbled in radio after working as a representative for RCA Records, built his fortune in TV and expanded his business interests with cable-television, automobile sales, hotels and the ownership of a soccer team.

Azcarraga's system controlled four channels in Mexico City, one of which served as a national commercial network, and other stations in Monterrey, Guadalajara and Tijuana. He also established Televisa, a production company for the stations, which sells its programs throughout Latin America and to the Spanish International Network in the U.S. See also Mexico, Television In; Televisa.

B

"BAA BAA BLACK SHEEP"—hour-long NBC action-adventure series (1976) based on the World War II exploits of Marine Corps Major Gregory (Pappy) Boyington and his crew of nonconformist fighter pilots. After its second week on the air, the series drew a formal complaint from CBS for allegedly violating the "family viewing time" code in glorifying characters who engage in brawling, drinking, wenching and violence. NBC explained that the series had not initially been conceived for "family time" (before 9 P.M.) and that subsequent episodes had been cleaned up to meet the code standard. CBS withdrew the complaint but the publicity it generated helped to build an audience for the NBC show, although the ratings later flagged.

Produced by Universal TV, with Stephen J. Cannell as executive producer, the series featured Robert Conrad as Boyington. It was canceled after one season, but then was given a second chance as a midseason replacement in 1977–78 with the new title, *Black Sheep Squadron,* and in a 9 P.M. time period. Again it failed, against tough competition from ABC's *Charlie's Angels.*

BABBIN, JACQUELINE—New York-based producer-writer specializing in drama. Her first major credits were as an adapter of plays and novels for the David Susskind productions in the "golden age" on TV drama. In 1962 she produced the *DuPont Show of the Week* and in the 70s a series of dramas for ABC's *Wide World of Entertainment.* In 1975 she became producer of the weekly CBS series *Beacon Hill,* and in 1976 she produced the three-hour *Sybil* for NBC's *The Big Event.*

After working in network programming on the west coast for a number of years, she gave up her job as vice president at ABC in early 1982 to return to New York as producer of Agnes Nixon's *All My Children* daytime serial.

"BACKSTAIRS AT THE WHITE HOUSE"—well-received nine-hour NBC miniseries in 1979 presenting a docu-drama view into the private lives of America's First Families from the Tafts to the Eisenhowers. The story of the domestic lives of the Presidents was told from the perspective of two generations of black maids who between them worked at the White House for more than 50 years. In the background were such major events as World Wars I and II and the Great Depression.

The production was based on material from the 1961 best-selling memoir, *My Thirty Years at the White House,* by Lillian Rogers Parks and Frances Spatz Leighton. The husband and wife team of Gwen Bagni and Paul Dubov adapted the story for television. Ed Friendly produced, and Michael O'Herlihy directed the drama. The cast included Leslie Uggams, Olivia Cole, Cloris Leachman, George Kennedy, Robert Vaughn, Harry Morgan, Paul Winfield, Robert Hooks, Louis Gossatt, Jr. and Ed Flanders.

"BACHELOR FATHER"—successful situation comedy about an eligible bachelor raising his teenage niece. It premiered on CBS in September 1957, then moved to NBC (1958–60) and finally to ABC (1961–62). By Bachelor Productions, it starred John Forsythe and featured Noreen Corcoran and Sammee Tong.

BACKE, JOHN D.—president of CBS Inc. since October 1976, receiving the appointment upon chairman William S. Paley's sudden dismissal of Arthur R. Taylor, who had been Paley's heir-apparent since 1972. He received the additional title of chief executive officer in May 1977. But he was ousted suddenly by Paley in the Spring of 1980 and was succeeded by Thomas Wyman. The following year he joined Tomorrow Entertainment, an independent production company, as president.

Backe's background had been in publications. He joined CBS in 1973 as corporate vice-president of the CBS Publishing Group (Holt, Rinehart and Winston, W.B. Saunders, Popular Library and magazines such as *Field and Stream, Road and Track* and *Popular Tennis*). Just before his promotion to corporate president, he helped to engineer, with Taylor, the acquisition of Fawcett Publications.

Earlier, Backe had been president and chief executive of General Learning Corp., a joint venture of General Electric and Time Inc., rising to that position after having headed its textbook publishing subsidiary Silver Burdett Co. See also CBS.

BAILEY, JACK (d. 1979)—veteran game-show emcee, best known for *Truth or Consequences* and *Queen For a Day.* Earlier he had been an announcer for network radio programs.

"BAILEYS OF BALBOA, THE"—CBS situation comedy (1964) whose flop contributed to ending James Aubrey's reign as president of the network. Produced by Richelieu Productions, it featured Paul Ford, John Dehner, Sterling Holloway and Judy Carne. See also Aubrey, James. T.

BAIRD, JOHN LOGIE (d. 1946)—Scottish inventor who developed the first television system in full-scale use. He demonstrated a mechanically scanned television system which showed objects in outline in 1924, recognizable faces in 1925. In 1928 he transmitted a television signal from England which was received in the U.S. The BBC adopted Baird's mechanical system in 1936 for regular transmissions but it was replaced the next year by the Marconi Company's all-electronic system using a cathode-ray tube.

In 1939 Baird demonstrated color television using a picture tube and shortly before his death he had completed research on a stereoscopic television system. In the 20s Baird demonstrated the first videodisc (using a standard wax phonograph record) and the first projection television system.

BAKER, WILLIAM F. (Dr.)— President of Westinghouse Broadcasting Co.'s Television Group since March 1979 and simultaneously, chairman of Group W Satellite Communications since 1981, he thus is in overall charge of Group W's television stations, the company's production and syndication subsidiary, and, in addition, heads the unit developing new program services for cable to be distributed by satellite. Baker, who earned a Ph.D. degree from Case Western Reserve University, advanced to the presidency through the Group W organization.

Lucille Ball

BALL, LUCILLE—probably TV's biggest star, familiar to viewers throughout the world through her classic situation comedy *I Love Lucy,* and the subsequent *Lucy* series that ran almost continuously on CBS from 1951 until her retirement from weekly television (ratings still high) in the fall of 1974. She created such a vast library of episodes that it was typical, in the 70s, for the *Lucy* reruns to have four or five different airings over the course of a day in large markets like New York. On a network laden with stars in the 60s, Miss Ball remained the heart of the CBS schedule, her show the one "sure thing" in the lineup.

Her TV run had a brief hiatus in 1960 after her divorce from her husband and partner Desi Arnaz. She emerged from the divorce president of their production company Desilu, which had grown into one of the large independents, supplying a number of shows to the networks. She doubled as an executive and star when she resumed weekly television with *The Lucy Show* in 1962, until Paramount purchased the company in 1967.

Although she had previous stage and film credits, Miss Ball was scarcely known to most of the TV audience when *I Love Lucy* premiered on Oct. 15, 1951. Arnaz, a Cuban bandleader whom she had married in 1940, had only a slight reputation in U.S. show business. But the zany series was an instant hit and rose quickly to become the most popular show on TV. Miss Ball portrayed a beautiful, well-meaning schemer who seemingly never lived a day on earth without becoming involved in an outlandish predicament. The situations gave full range to her talents for sight comedy, but remarkably—through all the absurd disguises and broad physical antics—she was able to preserve a lady-like persona.

There were strong contributions from William Frawley and Vivian Vance, who portrayed Fred and Ethel Mertz, neighbors and close friends of Lucy and Ricky Ricardo, the leads.

When Miss Ball became pregnant in 1952, producer Jess Oppenheim decided that television's Lucy would become pregnant too. The dual event, Miss Ball's real-life delivery of Desi Arnaz Jr., and the arrival of the baby on the program, Ricky Jr., was nationally awaited, and the birth episode achieved an enormous rating.

I Love Lucy ended its run June 24, 1957 when the stars decided to concentrate on specials. In November 1957, they began a one-hour monthly telecast, *The Lucille Ball-Desi Arnaz Show*, a comedy hour with guest stars. The series ended with their divorce, and Miss Ball left TV to star in the Broadway musical *Wildcat*, which was both a hit and a personal triumph for her.

She returned to TV with a weekly series, *The Lucy Show* (1962–68), featuring Vivian Vance and Gale Gordon. The following year the series was changed to include her children Desi Jr. and Lucie Arnaz, and given the title *Here's Lucy* (1968–74). Gordon remained as a principal, and Desi left the series after three seasons.

All three *Lucy* sitcoms played Mondays at 9 P.M., giving CBS dominance of the night through virtually every season Miss Ball was with the network.

In 1979, she was recruited by NBC president Fred Silverman—with whom she had worked at CBS—to develop comedy series and talent for the network through her own new production company, and also to star in comedy specials. This resulted in little more than publicity, however, although two unsuccessful pilots were made, neither starring Miss Ball.

BALMUTH PETITION—a futile attempt by Hollywood craft unions in the early 70s to persuade the FCC to adopt rules limiting the volume of network reruns in prime time to 25% a year from more than 40%. Such a restriction, the petition argued, would stimulate creativity, boost employment, give the viewer a wider choice of fare and curtail the networks' ability to use prime time as they please for their greatest economic gain.

The original petition was filed in 1972 by Bernard A. Balmuth, a film editor, and was staunchly supported by the Screen Actors Guild, the Writers Guild of America, West,

and the Hollywood Film Council, representing 28 labor organizations, in addition to others. In July 1976, while acknowledging that many people felt strongly about the saturation of reruns, the FCC rejected the plea in a 7–0 decision. It reasoned that the Government had no right to control such matters and that the issue was better resolved by market forces.

The commission also refused to require stations and networks to identify repeat programs as reruns on the ground that to do so would be to increase the "clutter of announcements."

Balmuth noted that in the 50s the networks played 39 firstruns and 13 reruns a year for practically every series and that the pattern eroded gradually until, in the 70s, the firstruns were reduced to 22 a year in most cases. The networks argued that reruns give the viewers additional opportunities to see programs they missed during the first airing.

An amicus brief filed by the unions pointed out that the rerun rate in prime time is more than 80% in June and July, more than 75% in May and August and about 67% in April, concluding that for five months of the year there are more reruns than original material.

In denying the petition, the FCC reasoned also that the limiting of reruns would increase network production costs, reduce their profits and probably result in cheaper programming overall. The commission added that to engage in regulatory action for the purpose of improving employment was not within its province. It said, finally, that it lacked the authority to regulate the type and quality of programming unless a substantial public benefit was certain.

"BANACEK"—NBC series (1972) about an investigator dealing in the recovery of lost or stolen property, which rotated as part of the *Wednesday Mystery Movie*. George Peppard starred in the 16 episodes produced by Universal TV.

BANDWIDTH—a section of the frequency spectrum needed to transmit visually, aurally or both. The bandwidth of the average television channel is 6 million cycles per second (6 MHZ).

"BANKS AND THE POOR"—controversial public TV documentary in 1970 critical of moneylending institutions and which held the banking industry to blame for perpetuating slum conditions. The program ended with a crawl that listed 98 members of Congress who were either shareholders or directors of banks, while *The Battle Hymn of the Republic* ironically played on the soundtrack.

Such theatrical flourishes made the film vulnerable to journalistic criticism. But the documentary caused nervousness among the PBS stations because the congressmen named on the crawl (including Sen. John O. Pastore, chairman of the Senate Communications subcommittee) would be voting on the next federal appropriation for public TV. Many of the stations, besides, had board members who were bankers. Several stations postponed the showing and a few declined to show it at all. PBS acted against its own policy by inviting representatives of the banking industry for pre-screenings.

The hour was produced by NET for the *Realities* series, which was canceled after that season.

BANNER, BOB—independent producer-director, whose credits include *The Dinah Shore Chevy Show* in the mid-50s, *The Garry Moore Show* in the late 50s and *Candid Camera* (as executive producer) in the mid-60s. Early in his career he was director for NBC of *Garroway At Large* and *Omnibus*. His independent company, Bob Banner Associates, has concentrated primarily on music-variety specials, such as *Carnegie Hall Salutes Jack Benny, Julie and Carol at Carnegie Hall, Here's Peggy Fleming, The Kraft Music Hall* and the John Davidson shows, with occasional ventures into made-for-TV movies. More recently, his company produced Perry Como's seasonal special for ABC, and *Solid Gold* for Operation Prime Time syndication.

BANZHAF CASE—[John F. Banzhaf III *v.* FCC/405 F2d 1082 (1968)] the Fairness Doctrine applied to the broadcast of cigarette commercials. In December 1966, John Banzhaf, a New York lawyer acting as a private citizen, requested free time from WCBS-TV New York to answer its cigarette commercials under the Fairness Doctrine. When the station rejected the request, Banzhaf filed a complaint with the FCC. The commission ruled, six months later, that a station presenting such advertisements had the duty of informing the public of the hazards of smoking because the promotion of smoking was proved not to be in the public interest. Banzhaf's bid for counter-commercial time approximately equal to that for cigarette ads was rejected, however; instead, the FCC proposed a 3 to 1 ratio.

A number of appeals were instituted. The NAB filed for review in the Court of Appeals in Richmond, Va. Banzhaf filed for review in the District Court of Appeals claiming that while the commission awarded significant time he had not been afforded equal time. The tobacco and broadcasting industries' argument was that by passing the Cigarette Labeling Act of 1965, the Congress meant to forbid any additional regulation addressed to the problem of danger to health.

The two cases were joined in the District Court of Appeals in Washington, and in 1968 the court upheld the commission's ruling. The court rejected the intent of Con-

gress's argument and also concluded that cigarette advertisements were not constitutionally protected speech. The court did not accept Banzhaf's equal time claim nor the cigarette manufacturers' claim for rebuttal time to answer antismoking messages.

The cigarette companies appealed, but certiorari was denied by the Supreme Court. Antismoking commercials became familiar on TV until cigarette advertising was barred from broadcast media by the Public Health Cigarette Smoking Act of 1969, which went into effect in January 1971. The constitutionality of the law was upheld in 1971 by a special three-judge court in Capitol Broadcasting v. Mitchell [333 F Supp. 582 (1971)] with a strong dissent from Judge J. Skelly Wright. See also Counter Commercials.

BAR (BROADCAST ADVERTISERS REPORTS)—company engaged in monitoring commercials televised on networks and stations; its syndicated reports are used by advertisers and agencies as proof-of-performance for their commercials and by stations to keep abreast of advertising purchased at competing stations.

BAR monitors 262 stations in 75 TV markets by means of off-air tape recordings. The markets covered represent around 80% of the ADI households and 85% of spot TV dollar expenditures. New York and Los Angeles are monitored on a full-time basis while the remaining markets are monitored one sample week each month.

In monitoring the networks, BAR is able to keep a running score on the estimated revenues for each.

"BARBARA McNAIR SHOW, THE"—half-hour music-variety series produced in Canada for U.S. syndication in the prime time–access periods (1969–71). Winters-Rosen Productions taped 52 programs, but the series was not widely accepted by stations.

"BARBARA STANWYCK THEATRE, THE"—30-minute anthology series on NBC (1960–61) with Barbara Stanwyck as hostess and frequent star. It was by E.S.W. Enterprises.

BARBER, RED (WALTER L.)—one of the most popular and respected sportscasters in the medium whose career spanned the period 1934–66. Chiefly associated with baseball, he was dismissed by the Yankees management after a dozen years as the team's play-by-play announcer for reporting that home attendance was low and then directing the TV cameras to scan the empty seats. The incident raised the issue of whether TV sportscasters in the employ of teams should be identified as hired boosters rather than being allowed to pose as objective journalists. Barber did the play-by-play for the Cincinnati Reds (1934–39), then became the "voice" of the Brooklyn Dodgers (1939–54) and then of the Yankees (1954–66). His refined Mississippi accent contributed to his distinctive style, and he popularized such Southern expressions as "catbird seat" (sitting high up in an advantageous position) and "rhubarb" (an argument on the playing field). After his dismissal he wrote, freelanced occasionally in broadcasting and went into retirement around 1970.

Red Barber

"BAREFOOT IN THE PARK"—NBC situation comedy (1970) drawn from the Neil Simon play of that title but with an all-black cast. The series fared poorly and ran less than half a season. Featured were Scoey Mitchlll, Tracey Reed, Thelma Carpenter and Nipsey Russell. It was by Paramount TV.

"BARETTA"—ABC series about an unorthodox undercover police detective portrayed by Robert Blake. The series had an unusual history. It began in 1973 as *Toma*, a series with Tony Musante based on the adventures of a real-life cop who relied on his wits and imaginative disguises, but ended after one season when Musante became disenchanted with the role. Because the ratings were promising, it was revived in January 1975 with Blake, and although the title was changed to *Baretta* it was given the Friday night slot originally held by *Toma*. For several weeks the series fared poorly, but when ABC switched it to Wednesdays the ratings began to soar. *Baretta* went on to demolish *Cannon* on CBS and to become a mainstay of the ABC lineup. It was cancelled in Spring 1978. Among the regulars in the series were Tom Ewell and Dana Eclar. Bernard L. Kowalski was executive

producer and Jo Swerling Jr. producer for The Public Arts, Roy Huggins Productions and Universal TV.

BARISH, SHERLEE—founder and president of Broadcast Personnel Agency, which specializes in finding newscasters and news directors for local stations around the country. Miss Barish started her business in 1961 after traveling extensively among stations as a saleswoman for Officials Films. In that capacity she found herself recommending talent she was impressed with in one city to station operators elsewhere who were looking for new anchormen or sportscasters. Her agency receives a commission from the station of 20% of the employee's annual salary above $20,000 or 15% below $20,000.

BARKER, BOB—TV emcee who hosted Ralph Edwards' *Truth or Consequences*, first on NBC and then in syndication, for 18 seasons (1956–74). In 1972, he also took on *The New Price Is Right* on CBS and then the syndicated nighttime version of that show. Since 1966, he has been the emcee for both the Miss Universe and the Miss USA beauty pageants; since 1967 he has hosted the Indianapolis 500 Parade; and since 1969 he has narrated the Rose Bowl Parade on CBS. Through his own production company, he has also produced the *Pillsbury Bake-Off* specials.

BARKER, CECIL—producer chiefly associated with the *Red Skelton Show* (1955–61). His producing credits as program executive with ABC and then CBS included *Space Patrol*, *You Asked For It*, the *Burns and Allen Show*, *That's My Boy*, *Lineup* and *Shower of Stars*. He also produced specials for Skelton and Jack Benny.

"BARNABY JONES"—hour-long private-eye series which held its own nicely in a variety of time period on CBS from 1973–80. Starring Buddy Ebsen, who overcame the typecasting of his previous hit series, *The Beverly Hillbillies*, it featured Lee Meriwether and Mark Shera and was by Quinn Martin Productions.

BARNATHAN, JULIUS—for two decades a key ABC executive, whose quick intelligence and adeptness at solving problems gained him the most diverse appointments of any official at that network. At various points he has been head of research, head of affiliate relations, head of the owned TV stations, general manager of the network and head of engineering and broadcast operations.

In the latter capacity, he has been responsible for the planning, designing and acquiring of broadcast facilities and equipment for all areas of the company, radio as well as TV, and for directing the technical operations for major special events coverage, from political conventions to the Olympic Games.

Barnathan was named v.p. for engineering in 1965 to supervise the equipping of the network and owned stations for color TV. For the three previous years he had been v.p. and general manager of ABC-TV, and for a brief period earlier was president of the ABC o&os. From 1959–62 he was v.p. for affiliated TV stations. He began with ABC in 1954 as supervisor of ratings and swiftly won promotions in the research department until, in 1959, he became v.p. in charge. In 1976 he received the title of president of engineering and operations.

"BARNEY MILLER"—ABC half-hour comedy centering on an ethnic-rich squad room of New York precinct detectives. After an unsteady start in January 1975, the series developed into a hit. Hal Linden starred in the title role, with Barbara Barrie as his wife. Featured as other detectives in the squad were Abe Vigoda, Max Gail, Gregory Sierra, Jack Soo and Ron Glass. The series was created by Danny Arnold and Theodore Flicker, with Arnold as executive producer for Four D. Productions. Producers were Chris Hayward and Arne Sultan, and Norm Pitlik directed most of the episodes. In 1977 Vigoda's character was spun off into a new ABC situation comedy, *Fish,* but the show was unsuccessful.

television and chaired the film division in the School of Arts. During much of that period he was also active in the broadcast industry, chiefly as a writer and at one time headed the Writers Guild of America (1957–59). He also produced and wrote several films for National Educational Television.

His books include *Mass Communication* (1956), the trilogy comprising the *History of Broadcasting: I. A Tower In Babel* (1966), *II. The Golden Web* (1968), and *III. The Image Empire* (1970); as well as *Tube of Plenty* (1975), and *The Sponsor* (1978).

"BARON, THE"—British action-adventure series slotted by ABC as a midseason replacement in January 1966 with unimpressive results. Based on mystery stories by John Creasey, it starred Steve Forrest and was by ATV.

BARRETT, RONA—Hollywood gossip columnist carrying on, via TV, the tradition of Hedda Hopper and Louella Parsons. In the late 60s she began doing two-minute reports for ABC o&os to use in their newscasts, and in 1969 she started a daily syndicated TV "column" distributed by Metromedia. In 1976 she became a regular contributor to ABC's *Good Morning, America* and later that year hosted the primetime ABC special *Rona Barrett Looks At the Oscars.*

NBC hired her away in 1981 to team her Tom Snyder on the *Tomorrow* show, but the two quarreled over turf, and the match was never consummated. Barrett wound up ith a short-lived prime time series of her own, *Television Inside and Out,* in the winter of 1981.

Erik Barnouw

Chuck Barris

BARNOUW, ERIK— the leading broadcasting historian, author of the three-volume *History of Broadcasting In the United States,* and chief of the Motion Picture, Broadcasting, and Recorded Sound Division of the Library of Congress (1978–81). From 1946 to 1973, he taught courses in film and

BARRIS, CHUCK—game-show packager whose long list of entries began with *The Dating Game, The Newlywed Game* and *Operation Entertainment* for ABC, the network with which he had previously been associated as a program executive. His Chuck Barris Productions was also responsible for *The Game Game* and others in that perishable daytime genre, and it provided the financial backing for the Blye-Bearde production of the syndicated *Bobby Vinton Show* (1975). Barris cast himself as host of the daytime NBC series *The Gong Show* in 1976.

During 1979, Barris had five syndicated programs in production: *The Gong Show, Three's a Crowd, The $1.98 Beauty Show, The Newlywed Game* and *The Dating Game*. A year or so later most of them were gone.

BARRON, ARTHUR—freelance producer, director and writer generally associated with documentaries, although in latter years his work extended to movies and some TV drama. Between 1975 and 1977, Barron produced two episodes of the *Six American Families* series for Group W, adaptations of short stories by Henry James and Ambrose Bierce for the PBS *American Short Story* series, and a theatrical movie for Warner Bros., *Brothers*.

While on the staff of CBS News in the 60s Barron was responsible for such documentaries as *Sixteen in Webster Groves, Webster Groves Revisited* and *The Berkeley Rebels*. He also had worked for NBC News and Metromedia. As a freelancer, he produced, wrote and directed *Birth and Death, Factory* and *An Essay on Loneliness* for PBS. His credits in the sphere of TV drama include *The Child Is Father of the Man* for CBS and *It Must Be Love 'Cause I Feel So Dumb*, an ABC *Afterschool Special*.

BARRY, JACK—prominent game-show producer and host in the 50s who was forced to drop out of television for more than a decade after being implicated in the quiz-show scandals of 1958. Barry and his partner Jack Enright had been the producers of *21*, one of the programs that congressional witnesses had said was rigged. Later, their application for radio station licenses was challenged on the ground of character, but the FCC ruled those objections invalid. Having served his term of penance, Barry was admitted back into television as a game show producer in 1970 and made his comeback with such successful shows as *The Joker's Wild* and *Break the Bank*. Among the earlier programs he had produced and appeared on were *Winky Dink, Tic Tac Dough* and *Juvenile Jury*. See also Quiz Show Scandals.

BARRY, PHILIP JR.—executive producer chiefly of made-for-TV movies. Son of the famous playwright, he began his television career in the early days of the medium working on such prestige drama series as *Goodyear Television Playhouse*. In the early 70s he served for CBS as executive producer of its made-for-TV movies and then moved on to Tomorrow Entertainment in a similar capacity.

BARTELME, JOE—veteran NBC News executive who in May 1979 became executive producer of *Today*. He had been v.p. of NBC News, responsible for all regularly scheduled news programs, which included *Today* as well as the *Nightly News*. Previously (1974–77), he had been v.p. of news for the NBC o&os and before that the network's West Coast news director. He came to NBC in 1971 after having been news director for WCCO-TV Minneapolis.

BARTER—a form of advertising sale in which the advertiser gives a program or program matter to a station or network in exchange for commercial spots. The number of spots and the time periods in which they are played are subject to negotiation. While barter was common in the early years of television, such as when the companies that manufacture bowling equipment produced bowling shows with built-in plugs and offered them to stations gratis, the practice fell into disrepute for many years.

It was revived on a large scale in 1971 when the prime time-access rule went into effect and many stations hesitated to invest in programing. Here advertisers seeking bargains in choice viewing hours provided stations with reasonably attractive programs containing two or three minutes of their commercials. The stations' profits came from the sale of up to three additional minutes in those half-hour programs.

Colgate-Palmolive bartered *Police Surgeon* in this manner for several seasons, and Chevrolet did the same with such series as *Stand Up and Cheer, The Golddiggers, The Jonathan Winters Show* and *The Henry Mancini Show*.

Some advertising agencies bartered programs for more than a single client and thus became involved in the production of TV programs again. Other advertisers bartered program matter, such as syndicated news or feature pieces, in exchange for commercial time.

Some preferred to barter on a time-bank principle, giving programs to stations in exchange for credit so that the advertiser was owed commercial time to dispose of as and whenever he chose. See also Syndication.

BARTLESVILLE TEST—an early pay-TV experiment begun in the fall of 1957 in which a Bartlesville, Okla. theater owner sent movies by wire to subscribing homes for a monthly fee. It ended the following spring, deemed a failure.

BARTLEY, ROBERT T.—FCC commissioner who served three terms (1952–72). He was concerned particularly with media monopolies and the concentration of control, and he had cast a dissenting vote in the proposed ABC-ITT merger essentially on the ground that a vast international conglomerate with its impersonal approach to business could have little sense of local community needs. He was the FCC's leading advocate of ascertainment rules for license renewals. A Democrat from Texas and generally considered a liberal, he had been administrative assistant to Speaker of the House Sam Rayburn. He was also an executive with the old Yankee Network and for five years a staff member of NAB.

BARUCH, RALPH M.—major figure in both the cable-TV and syndication industries as president and chief executive officer of Viacom International Inc., a company that in 1979 began expanding also into broadcast station ownership and network program production.

Baruch was one of the CBS executives who was spun out of the company when CBS spun off Viacom in June 1971, after the FCC ordered the networks out of syndication and cable ownership. He joined CBS in sales in 1954, after five years with the DuMont Television Network, and in 1959 moved into syndication as v.p. of international sales for what then was called CBS Enterprises. He became president of the division in 1970, just before divestiture was ordered. Viacom has flourished under his stewardship and expanded into cable networking with Showtime, station ownership and network TV production.

BASEBALL ON TV—a huge ratings-getter in October at World Series time, and earlier for the divisional playoffs and the All-Star game, but otherwise not the national draw during the course of the season that professional football has been. Indeed it was television that reduced baseball from national pastime to probably the number two sport in America, amplifying as it does the relative slowness of the game and the length of the season.

Locally and regionally, however, baseball has been a powerful TV attraction, especially for teams that are both colorful and pennant contenders. But the number of games televised locally varies at the discretion of the individual teams. At one extreme, the entire home and away schedule of the Chicago Cubs is televised by WGN-TV in Chicago; at the other, coverage is limited by some clubs to only a handful of games each season out of concern for ticket sales at the stadium.

Network rights to baseball have in recent years been divided into packages. Under the 1979 contracts with NBC and ABC, for example, the networks alternate coverage year by year of the All-Star, league playoff and World Series games, while ABC carries a schedule of Monday night games during the season and NBC Saturday or Sunday games, with the right to increase coverage for critical games as the pennant races come down to the wire. The four-year network contracts reportedly total around $200 million.

From 1960 to 1970, the cost to the networks of major-league baseball rights rose from $3.2 million a year to $16.6 million. From 1975–79, total rights between NBC and ABC came to $23 million annually. This contrasted with the $52 million paid for NFL football rights in 1975 alone.

In 1979, baseball found an additional source of revenue from cable-TV. UA-Columbia Cablevision, which handles the Madison Square Garden cable network, arranged to televise a total of 40 Thursday night games a season via satellite. It was the first national cable agreement by a major professional sports organization. However, the distribution was restricted to areas where the televised games would not compete with local baseball games on commercial TV. The games are offered free to cable subscribers, but cable operators are charged one cent a subscriber per game and are given nine minutes of advertising time to sell locally. The network takes 11 advertising minutes per game.

BAST, WILLIAM—screenwriter and TV scripter, working for periods of time in Britain, whose recent credits include *The Man In the Iron Mask* for NBC and *The Legend of Lizzie Borden* for ABC, in addition to episodes of various weekly series. A close friend of the late actor James Dean, whom he had met at college, Bast wrote the TV special, *James Dean—Portrait of a Friend,* which aired on NBC in 1976. In the U.K., he wrote and adapted several plays for the BBC and a number for Granada TV, the best known of which was probably *The Myth Makers.*

Jim Bouton in *Ball Four*

"BATMAN"—popular ABC prime-time series (1966–68) that, in spoof style, was a live-action representation of the famous comic book creations of Bob Kane. Batman was played by Adam West and Robin by Burt Ward. When ratings began to slip, Yvonne Craig was introduced in the third season as Batgirl. Recurring villains were portrayed by such guest stars as Burgess Meredith, Vincent Price, Caesar

Romero, Frank Gorshin and Zsa Zsa Gabor. Among notable gimmicks of style was the superimposition over the fight scenes of comic book words like "Pow" and "Bam." Via Greenway Productions and 20th Century-Fox TV, the half-hour series continues in syndication.

"BAT MASTERSON"—western series on NBC (1958–60) about a fashionplate marshal who wears a derby and carries a cane. It starred Gene Barry and was produced by United Artists TV.

"BATTLESTAR GALACTICA"—high-budgeted ABC science-fiction series (1978–79) that attempted, without success, to trade on the popularity of the movie *Star Wars*. The special effects and costumes of the Universal hour-long series resembled those of the 20th Century-Fox movie, and John Dykstra, who managed the motion picture's special effects, was employed to do the same for *Galactica* and to be producer.

The series, a space-age mutation of the traditional Western, concerning a group of homeless pioneers making their way in a caravan of space vessels to find a new frontier, made a soaring start in the ratings but soon went into a decline and eventually became earthbound when CBS moved *All in the Family* against it.

Lorne Greene starred as commander of the fleet, with Richard Hatch, Dirk Benedict, Herb Jefferson Jr., Maren Jensen, Noah Hathaway and Terry Carter featured. Glen Larson, whose company produced the series in association with Universal, was executive producer. Leslie Stevens had the title of supervising producer.

"BAXTERS, THE"—unusual 1979 syndicated series combining the entertainment elements of situation comedy with local public-affairs discussion. This odd mix was achieved by a format that provided for a 12-minute prepackaged dramatic scene followed by a locally produced segment in which viewers commented on the issues raised by the fictional material. The program was developed as a local series by Hubert Jessup at WCVB-TV Boston and went into syndication when it caught the fancy of Norman Lear, the noted Hollywood producer of social comedy.

Lear's T.A.T. Communications Co. created the provocative opening sitcom sequences, leaving it to the stations themselves to produce, each in its own way, the group discussion segments. The fictional scenes focused on a middle-class family named Baxter and contained standard sitcom ingredients developed around such themes as the effects of inflation, the problems of the elderly, responsibility for birth control and a variety of marital and family stresses. Featured in the regular cast were Larry Keith, Anita Gillette,

Terri Lynn Wood, Derin Altay and Chris Petersen. Lear was executive producer and Fern Field the producer. It was via T.A.T. in association with Boston Broadcasters Inc.

BAZELON, DAVID L. (JUDGE)—chief judge of the U.S. Court of Appeals for the District of Columbia Circuit since 1962, in which capacity he has been a force in striking down numerous regulations and decisions of the FCC. Heading the court most active in the judicial review of FCC actions, Judge Bazelon, a liberal with a strong sense of the public interest in broadcast matters, served as a counterbalance to the commission's tendency to accommodate the industry it is supposed to regulate. He figured prominently in negating the FCC's renewal of the WLBT license and in denying the commission's policy statement on license renewals; in speeches he was openly critical of the quality of broadcasting and of broadcast regulation in the U.S. He was appointed to the court in 1949, having been nominated by President Truman, and became chief judge through seniority 13 years later. He retired as chief judge in 1979.

Downstairs on *Beacon Hill*

"BEACON HILL"—ambitious CBS dramatic serial (1975) about the intertwined lives of a wealthy Irish-American family and their staff of servants in Boston just after World War I. Aspiring to the critical success of *Upstairs, Downstairs*, the British series from which it borrowed its form and concept, *Beacon Hill* failed to win either acclaim or a large enough audience to last more than three months. Produced in New York by the Robert Stigwood Organization, with Beryl Vertue as executive producer and Jacqueline Babbin as producer, it featured Nancy Marchand, Beatrice Straight, George Rose, Stephen Elliott, David Dukes, Paul Rudd, Kathryn Walker and Kitty Winn. Fielder Cook was director.

BEATTY, MORGAN (d. 1975)—NBC News correspondent and commentator from 1941 to his retirement in 1967. He had worked mainly in radio.

BEAUDINE, WILLIAM, JR.—producer of *Lassie* who also produced and directed episodes of *Wonderful World of Disney* and other programs.

BECKER, ARNOLD—v.p. of national television research for CBS-TV since 1977 and for many years a key advisor to the network's programmers. He joined CBS in 1959, after holding research posts with Lennen & Newell Advertising and ABC-TV, and steadily advanced in the research department ranks. When CBS moved its program unit to the West Coast in 1977, Becker was transferred with it. His father, the late I.S. (Zac) Becker, had been v.p. of business affairs for the CBS Radio network during the 50s.

BELGIUM, TELEVISION IN—probably the most TV-intensive nation in Europe, given that its two state-controlled channels are augmented by an extensive cable system that regularly pulls in programing from Britain, France, Germany, Holland and Luxembourg. That gives Belgians a theoretical choice comparable to that enjoyed by viewers in the New York City and Los Angeles metropolitan areas. The Belgian anomaly is abetted by set manufacturers in that nearly all color receivers are equipped to pick up both PAL and SECAM.

The two domestic services reflect the country's bilingual society, with one operating in Flemish, the other in French, both for about five hours per day. Both are operated by state-appointed boards. No advertising is carried; license fees are the only source of income to cover all costs. The fees are $46.17 a year for black & white sets and the equivalent of $71.82 for color sets.

BELL & HOWELL "CLOSE-UP"—distinguished series of documentary specials produced for ABC by Robert Drew Associates and sponsored by Bell & Howell. It premiered in 1960 and extended through four seasons. Among the notable entries were *Walk in My Shoes* and *Yanqui, No*. All featured cinema verite techniques.

BELL, WILLIAM J.—one of the leading writers of daytime TV soap operas for more than two decades. Working from Chicago where his wife Lee Phillip has been a leading personality on WBBM-TV, Bell in the mid-50s became a protege of the late Irna Phillips (who was known as "queen of the soaps") and began writing for *The Guiding Light*

and then *As the World Turns*. In the 60s he created *Another World* for NBC and then became story editor for *Days of Our Lives,* another serial on that network. He is responsible for the long-term story-lines which are then embellished and scripted by a team of writers.

BELLISARIO, DONALD— creator and executive producer of *Magnum, P.I.,* one of the few hits of the 1981–82 season. Previously he was writer and producer of the *Black Sheep Squadron* series.

BELSON, JERRY—comedy writer whose long string of credits includes *The Dick Van Dyke Show.* With Garry Marshall he adapted Neil Simon's play *The Odd Couple* as a TV series and served with Marshall as co-executive producer of the early episodes.

BEM CASE [Business Executives' Move for Vietnam Peace *v.* FCC/Post-Newsweek Stations *v.* Business Executives' Move for Vietnam Peace/412 U.S. 94 (1973)]— legal test on rights of broadcasters to deny the sale of time for the discussion of controversial issues. The Supreme Court in 1973 determined that broadcasters had such a right.

In June 1969 BEM, an ad hoc organization of 2,700 business executives opposed to the U.S. involvement in the Vietnam War, was thwarted by station policy in its attempt to buy a series of one-minute antiwar spots on WTOP-TV Washington. Like many other stations and the three TV networks, WTOP refused to sell air time for editorial advertising. BEM then filed a complaint with the FCC claiming that its First Amendment rights were violated by the licensee's policy.

The FCC upheld the broadcaster, deeming it unnecessary for a station to sell editorial advertising if a broadcaster was fulfilling his duty under the Fairness Doctrine by adequately covering all sides of the Vietnam debate.

But the D.C. Court of Appeals, joining the case with that of the Democratic National Committee which sought to buy time on CBS to reply to President Nixon's policies on the war, ruled in 1971 that a flat ban on paid public issue announcements was in violation of the First Amendment. The case was remanded to the FCC to develop reasonable procedures and regulations determining how to implement editorial advertisements on the air. Essentially, the court found that broadcasters could not retain total editorial control.

The case was appealed to the Supreme Court, which in five complex and multifaceted opinions, reversed the Court of Appeals.

"BEN CASEY"—very popular, hour-long medical series in the early 60s whose appeal centered on the refreshingly different antihero personality of the title character, a surly and haunted but gifted surgeon. It played on ABC from 1961–66. Vince Edwards, who was propelled to stardom by the series, oddly was unable to carry his popularity from that show to others. Featured were Sam Jaffe, Bettye Ackerman, Jeanne Bates, Nick Dennis and, for a time, Franchot Tone.

BENDICK, ROBERT—producer of *Wide, Wide World* and the Dave Garroway *Today* show in the late 50s and earlier a program and special events executive with CBS. Trained as a documentary cameraman, he became one of the producers of *This is Cinerama* and a director of *Cinerama Holiday*.

BENDIX, WILLIAM (d. 1964)—character actor best known in TV for his portrayal of Chester Riley, the bumbling father of the popular situation comedy *The Life of Riley* (1953–58). He performed a wide range of dramatic roles, however, including one in the western series *Overland Trail* (1960).

BENJAMIN, BURTON—veteran CBS News producer who became v.p. and director of news in 1978, succeeding Bill Small who was named Washington v.p. for CBS Inc. For the preceding three years, Benjamin had been executive producer of the *CBS Evening News with Walter Cronkite.* Earlier, he worked chiefly in documentaries.

From a background with the Cleveland *News,* UPI, NEA and RKO-Pathe, he joined CBS in 1957 as executive producer of *The Twentieth Century,* an occasional documentary series that ran nine years. Later, as senior executive producer for CBS News, he produced numerous episodes for *CBS Reports* and such specials as *Mr. Justice Douglas* (1972), *The Rockefellers* (1973) and *Solzhenitsyn* (1974).

"BENJAMIN FRANKLIN"—four-part series of 90-minute dramatic specials highlighting the life of Franklin. Offered by CBS (1974–76) for the Bicentennial, each episode had a different writer and a different actor playing Franklin. The role was taken variously by Eddie Albert, Lloyd Bridges, Richard Widmark and Melvyn Douglas. The executive producer was Lewis Freedman, the producers Glenn Jordan and George Lefferts and the director Jordan.

BENJAMIN, ROBERT S. (d. 1979)—lawyer and film company executive who was a charter member of the Corporation for Public Broadcasting and became its chairman in 1975. In 1977, after serving nine years as a director, he resigned and was elected chairman emeritus of CPB. He was also chairman of United Artists Pictures Corp. and a partner in the law firm of Phillips, Nizer, Benjamin and Krim.

BENNETT, HARVE—series producer associated with Universal TV who was executive producer of *The Six Million Dollar Man, The Invisible Man,* and the *Rich Man, Poor Man* miniseries. Bennett had been an ABC program v.p. in Hollywood during the 60s, leaving in 1968 to become co-producer of *Mod Squad* for Thomas-Spelling Productions. During his childhood, he was one of the prodigies on radio's *Quiz Kids* and was known then as Harve Fischman.

Jack Benny & Eddie Anderson

BENNY, JACK (d. 1974)—one of the great radio comedians who made the transition to television in the 50s and had a weekly series on CBS-TV from 1950 to 1965. With a kind of continuing sketch comedy, in which he represented himself as an aging and somewhat pompous bachelor who was an outstanding tightwad, Benny held forth in radio and TV for more than 40 years. Through most of it he carried the same troupe of supporting players, which included his wife, Mary Livingstone, announcer Don Wilson, singer Dennis Day, band leader Phil Harris and character comedians Eddie (Rochester) Anderson and Mel Blanc, all of whom served him as foils.

Benny was not a joke-telling or slapstick comedian. A peer described him aptly as "not one who said funny *things* but one who said things *funny."* The character he created was always the source of the humor, the butt of insults, made funnier by Benny's catalog of mannerisms and responses—a martyr-like stare, with hand against chin; facial expressions of disbelief or frustration; and the utterances, "Well!," "Hmm" and "Now cut that out!" These became comedy motifs savored in their weekly repetition by huge audiences.

Fellow comedians admired Benny's expert timing and his ability to mine laughter from glowering silence.

Familiar trappings of the Benny shows were the antiquated car, a Maxwell, and the violin on which he regularly essayed an inept version of "Love In Bloom." He made famous the birthday on which he annually turned 39, and he maintained a mock feud with the dry-witted comic Fred Allen, through which they cross-plugged each other's programs.

Benny's radio series began in 1932 on NBC, and from 1934 through 1936 it led the popularity polls; after that it was seldom out of the top 10. CBS hired Benny away in 1948, and he continued on radio until 1955. In the meantime, Benny and his cast made sporadic television appearances and eventually were able to transfer the basic elements of their radio success to the new medium. When his radio series ended, Benny increased his television work from occasional programs to a bi-weekly series and finally to a weekly.

After 1965, his television performances were limited to a few specials each year. His last *Jack Benny's Second Farewell*, was on Jan. 24, 1974. He died of cancer of the pancreas, at 80, the following December.

BENSLEY, RUSS—director of special events for CBS News since 1974, prior to which he was executive producer of certain special broadcasts. Earlier he had been producer of *The Evening News with Walter Cronkite* (1964–71) and then executive producer (1971–72).

BERCOVICI, ERIC— veteran writer, often in collaboration with Jerry Ludwig, specializing in writing TV series pilots and made-for-TV movies. He came into prominence as producer and adapter of James Clavell's *Shogun*, the hit mini-series. Later, he created two series for NBC, *McClain's Law* and *Chicago Story*, and was executive producer of both.

BERG, DICK—executive producer of the *Bob Hope Chrysler Theatre* in the 60s and otherwise active as a producer of dramatic programs from that decade to the present.

BERG, GERTRUDE (d. 1966)—character actress identified with her role as Molly Goldberg, the matriarch of a Jewish family in the Bronx, in the situation comedy *The Goldbergs* (1949, revived in 1956). Miss Berg created the series and also wrote for it.

BERGEN, EDGAR (d. 1978)—ventriloquist whose most famous creation was his dummy Charlie McCarthy. Their popular radio series led to movies and frequent appearances on TV variety shows. In the 50s they hosted the daytime quiz show *Do You Trust Your Wife?*, which later became *Who Do You Trust?* (with Johnny Carson as host).

BERGMAN, JULES—since 1961 ABC News science editor who covered all U.S. manned spaceflights and many of the key developments in fields ranging from medicine to aeronautics. His documentary credits include the six-part series *What About Tomorrow* (1973), *Closeup on Fire* (1973), *Closeup on Crashes: The Illusion of Safety* (1974), *Earthquake* (1972) and *SST: Super Sound and Fury.* He joined ABC News as a newswriter in 1952 after brief stints with *Time* magazine and CBS.

BERGMANN, TED—managing director of the DuMont Television Network (1953–56), after which he entered the advertising industry. He worked previously for NBC in the international division and joined DuMont in 1947 as a sales executive.

Mr. Television, Milton Berle

BERLE, MILTON—a practitioner of broad and noisy comedy who, through his popularity from 1948 to 1956, earned the sobriquet of "Mr. Television." His program, beginning in a time when TV was a luxury enjoyed chiefly by the wealthier families, helped to spur the sale of television sets to working-class homes. Always presented on Tuesday nights on NBC, his *Texaco Star Theater,* as it was originally called, underwent several changes of title and sponsorship over its eight-year run.

Although a champion of buffoonery and lavish production, Berle was later to surprise viewers with his adeptness at serious drama when he began accepting occasional roles in TV plays. At the height of his popularity, he became known affectionately as "Uncle Miltie" and was chided by other comedians as "The Thief of Badgags." Such was his popularity that NBC offered him a contract through 1981, under which he has been paid since his series ended.

In the 60s, there was an unsuccessful attempt to revive his comedy series. Under a modification of the contract, he was able to host a prime-time bowling series for ABC, which was also unsuccessful.

Leonard Bernstein

BERNSTEIN, LEONARD—conductor and composer who introduced millions of children to classical music through the *Young People's Concert* series on CBS. His extraordinary ability to explain the complexities of symphonic music to the uninitiated, and his highly theatrical style in conducting, served to make the series one of the most popular cultural offerings on TV. Bernstein became music director of the New York Philharmonic in 1958 with the reputation of "wunderkind." The Philharmonic had been doing "Young People's Concerts" since 1922, but it was Bernstein who put them on TV. He left the orchestra in 1969 but continued to conduct the educational concerts on CBS until 1972, after which the TV baton passed to Michael Tilson Thomas.

"BEST SELLERS"—umbrella title for a weekly NBC series (1976–77) of dramatization of best-selling novels, each book serialized over several weeks. The program was inspired by the success of ABC's miniseries *Rich Man, Poor Man,* a TV adaptation of a novel by Irwin Shaw. First of the *Best Sellers* was Taylor Caldwell's *Captains and the Kings,* whose cast included Richard Jordan, Barbara Parkins, Charles Durning, David Huffman and Jenny Sullivan. Others were Anton Myrer's *Once An Eagle,* with Sam Elliott, Cliff Potts, Darleen Carr, Amy Irving and Glenn Ford; Norman Bognar's *Seventh*

Avenue, with Steven Keats, Kristoffer Tabori, Jane Seymour, Dori Brenner, Alan King, Eli Wallach and Jack Gilford; and Robert Ludlum's *The Rhinemann Exchange,* with Stephen Collins and Lauren Hutton.

The series was produced by Universal TV, with Charles Engel as executive in charge. Executive producers of the serials included Roy Huggins, William Sackheim and David Victor.

BET (Black Entertainment Network)— see Cable Networks.

The Clampett Clan: Buddy Ebsen, Donna Douglas, Irene Ryan & Max Baer

"BEVERLY HILLBILLIES, THE"—situation comedy about kindly country bumpkins, The Clampetts, who move to posh Beverly Hills after striking oil. A smash hit when it premiered on CBS in 1962, it ran through 1971 despite abuses heaped on it by the critics. Created and produced by Paul Henning for Filmways, it starred Buddy Ebsen, Irene Ryan, Donna Douglas and Max Baer and featured Raymond Bailey and Nancy Kulp. Reruns were stripped daytime on CBS, 1968–72.

"BEWITCHED"—highly successful ABC fantasy situation comedy (1964–72) about an ordinary fellow married to a beautiful witch. It starred Elizabeth Montgomery and Dick York (who left after the fifth season to be replaced by Dick Sargent). Other regulars were Agnes Moorehead, David White and Alice Pearce. It was by Screen Gems.

"BICENTENNIAL MINUTES"—series of one-minute programs broadcast every evening in prime time on CBS from July 4, 1974 through the end of 1976 as a Bicentennial salute. Each minute offered a vignette of an occurrence 200 years earlier on that date, related to the birth of the U.S.; each also featured a different well-known personality. Among those who appeared on the *Minutes* were Charlton Heston, Deborah Kerr, Rise Stevens, Beverly Sills, Kirk Douglas, Alfred Hitchcock, Zsa Zsa Gabor and Walter Cronkite, in addition to senators, generals and scientists. Lewis Freedman was succeeded by Bob Markell as executive producer. Shell Oil sponsored.

"BIG BLUE MARBLE, THE"—Emmy- and Peabody-award-winning children's magazine series devoted to the life-styles of children around the world. Each half-hour includes seven- to ten-minute portraits of children, a regular "Dear Pen Pal" feature that arranges correspondence between American children and English-speaking youngsters in other countries, and five-part serialized dramas related to the general theme of international understanding.

Designed to run without commercial interruption, the series was underwritten by ITT as a public service. It premiered in September of 1974 and is carried on 180 commercial and public stations in the U.S. and in 60 countries and areas abroad. It was created by Alphaventure and produced by that company from 1974 to 1978. Since 1978 the program has been a Blue Marble Co. Production.

"BIG EVENT, THE"—NBC's attempt in the fall of 1976 to carve out a weekly two- or three-hour block on Sunday nights for varied blockbuster special programs as the keystone of its commitment to "event television" (i.e., unique and momentous specials). The effort had uneven results but was continued the following season and then dropped.

Alvin Cooperman had been brought back to the network to concentrate exclusively on securing properties for the time period, but he was dismissed before the season ended. The opening program, *The Big Party,* was conceived as a salute to the start of various show business and sports seasons through the device of switching live among three parties arranged by NBC in New York. The program fizzled as entertainment, a fact which was reflected in the ratings. *The Big Event* rallied from that embarrassment with the showing of the movie, *Gone With the Wind;* the presentation of *The Moneychangers,* an adaptation of a best-selling novel; and a star-laden four-hour extravaganza celebrating NBC's 50th anniversary in broadcasting. But overall, the timeslot failed to deliver on its grandiose billing.

"BIG MARCH, THE"—CBS News special report on highlights of the civil rights rally known as the March On Washington of Aug. 28, 1963, at the Lincoln Memorial.

The hour special that night included Dr. Martin Luther King delivering his famous "I Have a Dream" speech.

"BIG TOP, THE"—CBS circus show which debuted in 1950, with Jack Sterling as ringmaster. Ed McMahon, who later became Johnny Carson's announcer and sidekick, played one of the clowns.

"BIG TOWN"—NBC series (1950–56) based on a successful radio series about the crusading editor of *The Illustrated Press.* Mark Stevens played the lead the final two seasons and Patrick McVey the first four. The role of Lorelei, the society reporter, was played by a number of actresses, among them Julie Stevens, Jane Nigh and Beverly Tyler. The series was produced by Gross-Krasne Inc. and in syndication went by such titles as *Heart of the City, City Assignment, Headline Story* and *Byline—Steve Wilson.*

"BIG VALLEY, THE"—an hour-long saga of a family living in the heartland of California in the 1870s that starred Barbara Stanwyck as a strong-willed widow and leader of her powerful family of three sons and a daughter. Co-starring Richard Long, Peter Breck, Lee Majors and Linda Evans, the series premiered on ABC Sept. 16, 1965, and ran until 1969 when it was put into syndication by its producer, Four Star Productions.

BILBY, KENNETH W.—v.p. of public relations for NBC (1954–60) who moved up to the parent company RCA, first as v.p. of public affairs and then as an executive v.p. He retired from RCA in 1979, but returned in 1981 to help the new chairman, Thornton Bradshaw, through a transition period.

Early in his career he had been a foreign correspondent for the *New York Herald-Tribune.*

"BILL COSBY SHOW, THE"—situation comedy on NBC (1969–71) in which Cosby played a gym teacher in a big city high school. It was produced by Jemin Co.

BINDER, STEVE—director associated with comedy shows, including *The Danny Kaye Show* and *Steve Allen Comedy Hour* for CBS, Allen's syndicated show for Westinghouse and network specials with Jack Paar, Liza Minelli, Bob Newhart, Petula Clark and Lucille Ball. In recent years, he doubled as a producer with his own production company.

"BING CROSBY SHOW, THE"—domestic situation comedy on ABC (1964) starring Crosby and featuring Beverly Garland, Frank McHugh and Dianne Sherry. It was produced by Bing Crosby Productions.

"BIOGRAPHY OF A BOOKIE JOINT"—extraordinary CBS documentary by investigative reporter Jay McMullen on the operations of an illegal Boston betting parlor which posed as a key-maker's shop; it aired Nov. 30, 1961 and brought about the resignation of Boston's police commissioner. Working with Palmer Williams, McMullen spent several months filming the entrance to the key shop from a room across the street, catching numerous visits by police officers. The documentary also contained footage of the bookie joint in operation, photographed by McMullen with a concealed 8 mm camera.

"BIONIC WOMAN"—successful spinoff of ABC's *Six Million Dollar Man* introduced in January 1976 with Lindsay Wagner in the role of a reconstituted woman with superhuman powers. Featuring Martha Scott, Richard Anderson and Ford Rainey, it was by Harve Bennett Productions and Universal TV, with Bennett as executive producer.

Although it ranked in the Nielsen Top 10, the series was dropped by ABC for the 1977 fall schedule, but it was immediately picked up by NBC. Its run ended in 1978.

"BIRD'S EYE VIEW"—NBC situation comedy (1970) about a British airline stewardess and a zany crew assigned to international flights; it was an early and unsuccessful attempt at co-production between an American and a British company (Sheldon Leonard Productions and ATV) and had only a brief run. Featured were Millicent Martin and Patte Finley.

"BJ AND THE BEAR"—one-hour NBC adventure series centering on an independent trucker who travels with a pet chimpanzee (The Bear), borrowing its premise from the movie *Every Which Way But Loose*. A midseason replacement on NBC in the 1978–79 season, it had modest success in a time when hit shows were scarce on that network, and it earned a renewal for the fall schedule. Greg Evigan played the lead. Glen Larson was the producer. It was cancelled in the 1980–81 season.

BLACKLIST—See Faulk, John Henry; *Goldbergs, The;* Red Channels.

"BLACK JOURNAL"—long-running PBS magazine series produced by WNET New York (1968–77) dealing with black issues, history and culture. A 60-minute weekly series, it has been produced and hosted by Tony Brown, with Billy Taylor as music director. For the 1975 season it attempted the *Tonight* show format but then returned to the magazine presentation. In 1977 the show went into commercial syndication as *Tony Brown's Journal* under Pepsi-Cola sponsorship. See also *Tony Brown's Journal.*

Adam Wade & Tony Brown with guest Percy Sutton

"BLACK PERSPECTIVE ON THE NEWS"—weekly public affairs series on PBS that began in 1973 and covers current events from a black point of view or national issues of specific interest to blacks. Originating at WHYY, the Wilmington-Philadelphia PTV station, the series is produced by Acel Moore of the Philadelphia *Inquirer* and Reginald Bryant, who is also host-moderator.

BLACK WEEK—one of four weeks during the year (a fifth black week occurs every fifth year) when the definitive Nielsen rating report, known as the pocketpiece, takes a rest. The black weeks usually occur in December (during the week of Christmas), April, June and August. Until the Nielsen overnight and fast weekly ratings began on a 52-week basis in the early 70s, the black weeks were periods when no audience data was gathered, and a programming tradition grew up around them.

As unrated weeks, and therefore noncompetitive for the networks, they became the most suitable time for documentaries, cultural programs and public affairs—offerings that stood to receive low ratings under normal conditions. The networks also made a practice of scheduling reruns during black weeks, since there was no point in wasting firstrun episodes of series in a time when Nielsen was not counting the audience.

For practical purposes, the black weeks vanished when the overnight and weekly services began, although the pocketpiece is still limited to 48-week coverage. And although the audience is now being counted during the remaining four weeks by the *fast* reports, the programming tradition of black week remains. There is still a tendency at the networks to concentrate programs of low rating potential in those weeks.

Black weeks were adopted by Nielsen to enable the company to review its internal tabulations and to create vacation time for its field staff. See also Sweeps.

BLAIR, FRANK—newscaster for NBC's *Today* show who was with the program from its premiere in January 1952 until his retirement in 1975. For the first two years Blair was *Today's* Washington correspondent; then he moved to New York to read the four daily news summaries. Blair retired to Charleston, S.C., where he had begun his broadcast career in 1935 at radio station WCSC.

BLEIER, EDWARD—executive v.p. of Warner Bros. TV since 1976 and head of its New York office. He joined the company in 1969 after 14 years as an ABC executive variously in sales, programming, planning and public relations. In the 50s he worked in sales and programming for several New York stations and for the DuMont Network.

"BLIND AMBITION"—eight-hour dramatic serial based on the personal accounts of John W. Dean, III, and his wife Maureen of their years in the Nixon Administration and the effects of Watergate upon their lives. It was presented on CBS on four consecutive nights from 9:00 to 11:00 P.M., May 20 to May 23, 1979, and drew excellent ratings.

In the Time-Life Television production, Martin Sheen portrayed Dean, Theresa Russell his wife and Rip Torn President Nixon. Others in the cast were Michael Callan as Charles Colson, Lonny Chapman as L. Patrick Gray, William Daniels as G. Gordon Liddy, Fred Grandy as Donald Segretti, Graham Jarvis as John Ehrlichman, Lawrence Pressman as H.R. Haldeman, John Randolph as John Mitchell, William Schallert as Herbert Kalmbach, James Sloyan as Ron Ziegler and William Windom as Richard Kleindienst.

Written by Stanley R. Greenberg from material in the books *Blind Ambition* by John Dean and *Mo* by Maureen Dean, the program was produced and directed by George Schaefer, with Renee Valente as coproducer and David Susskind as executive producer.

BLINN, WILLIAM—writer and producer. His writing credits include *Brian's Song* and the pilot script for *S.W.A.T.* He was also co-producer of *The Rookies* for two years and of *The New Land*. In the 1982–82 season, he was executive producer of NBC's *Fame* series.

BLOCK, RICHARD C.—pioneer in UHF broadcasting as head of the Kaiser station group from 1958 to 1974. After leaving Kaiser, he became a consultant to Columbia Pictures Television, Lorimar Productions and PBS. He was also chairman of the Council for UHF Broadcasting and has been an instructor in communication at Stanford University since 1964.

BLOCK PROGRAMMING—the bunching of shows similar in type into a number of consecutive time periods for the purpose of serving a single audience over the span. A children's block may run as long as six hours, a sports block may consume an afternoon or an evening. Protracted periods may also be blocked out for news, game shows, soap operas or public affairs.

Program blocks differ from *ghettos* in that they are created affirmatively and represent a technique for maintaining audience, while ghettos connote a dumping ground for programs not likely to attain large audiences.

"BLONDIE"—situation comedy on CBS (1968–69) that was an attempt by the network to resurrect the successful radio and movie series that had starred Penny Singleton and Arthur Lake. The TV version starred Pat Harty, Will Hutchins, Jim Backus and Henny Backus. It was not successful. Based on the popular comic strip by Chic Young, it was produced by King Features and Kayro Productions.

BLOOM, HAROLD JACK—a writer of drama series during the "golden age" who has since divided his time between films and creating series pilots. Through his Thoroughbred Productions he created and sometimes produced the *Hec Ramsey* series for NBC.

BLUE BOOK—a controversial FCC report issued in 1946 with the formal title of *Public Service Responsibility of Broadcast Licensees,* which set forth program criteria for license renewals. Although its supporters on the commission considered it nothing more than the enunciation of minimum standards for broadcasting in the public interest, the document was immediately attacked by broadcasters and some members of Congress as an instrument of censorship representing an attempt by the FCC to control programming.

The combination of these attacks and public indifference to the issue rendered the Blue Book ineffective. But it was never rescinded and presumably could be made to apply to TV licensees today if the commission saw fit.

The Blue Book came by its nickname from the blue paper cover on the mimeographed report. In essence it maintained that though the licensee had the primary responsibility to determine its own programming, the FCC had a duty to consider a station's program service in determining whether it was operating in the public interest. The basic criteria to be applied in license renewal decisions concerned (1) a station's commitment to sustaining (noncommercial) programs in the interest of a balanced program structure; (2) its use of local talent; (3) the presentation of programs dealing with important public issues; and discretion in the amount of advertising carried.

Heightening the controversy over the report was the fact that its author, Dr. Charles Siepmann, was an Englishman who had been with the BBC. He came to the U.S. at the behest of FCC chairman Paul A. Porter specifically to direct a study and draw up proposed criteria by which the FCC might evaluate program service. Porter was acting in response to congressional criticism of the FCC for its low requirements from broadcasters that made license renewals almost automatic. Not long after the Blue Book was issued, Siepmann published a hard-cover book, *Radio's Second Chance,* which echoed the FCC report and articulated the rationale for standards. Dr. Siepmann's book gave broadcasters the opportunity to charge him with opportunism.

Ironically, the much respected chairman Porter had left the FCC before the Blue Book was released. It fell to Charles R. Denny Jr., the young new chairman, to defend the report. But its real champion was Commissioner Clifford J. Durr, a liberal from Alabama who was vilified by the industry for his vigorous support of the document.

Although the Blue Book was of little practical use, its criticisms of broadcast practices prompted the industry to tighten its own codes somewhat along the lines prescribed.

"BLUE KNIGHT, THE"—initially a four-hour miniseries on NBC (1973) and then a weekly series on CBS (1976–77). At both networks it was based on the best-selling novel by Joseph Wambaugh. The limited series starred William Holden and Lee Remick and was a straight adaptation of the book, airing on four consecutive nights (Nov. 13–16, 1973); it was rerun in the spring of 1975 on consecutive nights as two two-hour programs. The CBS series was built around Wambaugh's character Bumper Morgan, portrayed by George Kennedy. Both series were by Lorimar, with Lee Rich as executive producer.

BLYE, ALLAN & BEARDE, CHRIS—successful team of comedy writer-producers whose credits include the *Andy Williams Show, Ray Stevens Show, Sonny & Cher Comedy Hour, Sonny Comedy Revue* and the first season of *That's My Mama.* The partnership dissolved in 1975, and Blye teamed up with Bob Einstein to form Blye-Einstein Productions.

BMI (BROADCAST MUSIC INC.)—music licensing organization created by the broadcast industry in 1939 as a weapon against ASCAP, the licensing society which enjoyed a monopoly and which was, at the time, threatening to raise the blanket license fees paid by stations for the use of copyrighted music. Because ASCAP's formula for the payment of royalties tended to favor established songwriters and penalize the newer composers, BMI was able to build rapidly a rival group of new publishers and writers. By the time the negotiations with ASCAP were resolved in 1941, BMI was firmly established. It grew to be the largest of the music licensing organizations, with nearly 45,000 writer and publisher affiliates and 850,000 licensed works as of the mid-70s.

Giving impetus to BMI's catalog was the denial to broadcasters of ASCAP music for several months when they resisted ASCAP's demands for an increase in fees (stations had been paying 2⅛% of their gross revenues from time sales for their music licenses up to that time). During that period, only public-domain and BMI music was heard on the air.

The emergence of BMI changed the nature and the flavor of popular music in the U.S. since it provided for royalty payments to be made to every kind of songwriter, even those composing specialized music restricted to a locale or ethnic group. Under the ASCAP system before 1941, broadcast royalties were not paid for recorded music, and the monies were distributed solely on the basis of live performance during evening hours on the country's four radio networks. Since payments to writers of country music and rhythm and blues was therefore scant, those fields had remained in the background of pops for lack of economic encouragement. BMI devised a system that would cover recorded as well as live music, and all air performances, whether national, regional or local.

This change in broadcast royalty procedures stimulated activity in regional and ethnic music, and much of it found its way into the mainstream. These new influences were synthesized into rock 'n' roll in the 50s.

As do ASCAP and SESAC, the other two licensing societies, BMI issues a license to stations for the use of its music and collects fees based on station income. Stations are required to keep logs of the music they play; to determine how the money is to be distributed to songwriters, BMI projects the number of air performances for each song from a representative sample of the logs. In television, producers maintain cue sheets listing all music performed in a program for the licensing societies.

BMI operates as a nonprofit society, and its board of directors is made up exclusively of broadcasters. See also ASCAP, SESAC.

BMS—symbol used in Nielsen rating reports for programs with too small an audience to be rated. The letters stand for "below minimum standards" and often, but not necessarily, refer to a rating below 0.5. The actual BMS level varies by market and time period and is determined by Nielsen according to the sample size in the survey area.

"BOB & CAROL & TED & ALICE"—short-lived ABC situation comedy (1973) derived from a movie of greater sophistication. Produced by Screen Gems it featured Bob Urich, Anne Archer, David Spielberg and Anita Gillette.

"BOBBY GOLDSBORO SHOW, THE"—syndicated country music-variety series in the mid-70s notably successful in areas where country music has greatest acceptance. Bill Hobin, Jane Dowden and Reginald Dunlap produced for Show Biz Inc., with Bill Graham as executive producer.

"BOB HOPE PRESENTS THE CHRYSLER THEATER" —hour series of 114 filmed dramas hosted by Bob Hope. Produced by Hope Pictures and Universal, the series played on NBC (1963–67). See also Hope, Bob.

"BOB NEWHART SHOW, THE"—domestic situation comedy involving a psychologist that premiered in 1972 and became part of the winning CBS Saturday night lineup. Produced by MTM Enterprises, it featured Suzanne Pleshette, Peter Bonerz, Bill Daily and Marcia Wallace.

As a new comedian, in the early 60s, Newhart had had a show of his own that chiefly involved comedy sketches. Following that unsuccessful effort, he became one of the rotating stars of the NBC series *The Entertainers*.

BOCHCO, STEVEN—writer who created *Sarge* and collaborated on the pilot for *The Six Million Dollar Man*. He also produced the series, *Griff*, and the *Delvecchio* series, and was executive producer of James Earl Jones' *Paris* series, before hitting it be as co-creator and co-executive producer of *Hill Street Blues*.

BOCK, LOTHAR—impresario from West Germany who served as go-between with the Soviet negotiating team in gaining for NBC the exclusive U.S. rights for the 1980 Moscow Olympics. Bock, who had had a number of previous theatrical and TV dealings with the Soviets, initially had been engaged by CBS to help land the Olympics plum. When CBS decided to drop out of the bidding early in 1977, Bock, who had already worked out most of the arrangements for an $85 million deal, immediately offered his services to NBC and promptly wrapped up the prize. For his efforts, he was paid a commission of $1 million and also received commitments from NBC for TV specials over a period of years.

BOGART, PAUL—noted director for such prestige drama anthologies as *Armstrong Circle Theatre* during the early years of TV. During the 60s he divided his time between films and occasional TV drama specials. In the 70s he directed a number of productions in public TV's *Hollywood Television Theatre* series and such commercial specials as *The House Without a Christmas Tree*. He became director of *All In the Family* for the 1975–76 season.

"BOLD JOURNEY"—half-hour series on the travel adventures of scientists and explorers who showed and discussed their own 16 mm films. The series, by Advenco productions and Julian Lesser, played on ABC (1956–58) and then in syndication. John Stevenson was host initially and then Jack Douglas.

"BOLD ONES, THE"—umbrella title for several rotating adventure series on NBC (1969–72) produced by Universal TV. From season to season, some rotating elements were dropped and others added. Included were *The Doctors*, with E.G. Marshall, David Hartman and John Saxon; *The Lawyers*, with Joseph Campanella, Burl Ives and James Farentino; *The Senator*, with Hal Holbrook; *Protectors*, with Leslie Nielsen and Hari Rhodes; and *Sarge*, with George Kennedy.

BOLEN, LIN—v.p. of daytime programs for NBC (1972–76), the highest position held by a woman at any TV network up to that time. She left to establish her own independent production company with certain exclusive ties to NBC. Her first series, *W.E.B.*, an episodic melodrama set at a mythical TV network, was a flop in 1978.

As daytime chief Miss Bolen added dazzle and larger cash prizes to the network's game shows and broke the 30-minute tradition of soap operas in expanding such serials as *Another World* and *Days of Our Lives* to an hour's length. In early 1982, she joined Fred Silverman's independent InterMedia Entertainment as head of creative affairs.

BOOSTER—an unattended, low-powered TV rebroadcasting station that picks up the signal of a conventional station and amplifies and rebroadcasts it; the booster is generally used to fill in gaps within a station's assigned coverage area.

The Cartwrights: Dan Blocker, Lorne Greene, Pernell Roberts &
Michael Landon

"BONANZA"—a great Sunday night hit on NBC for most of 14 years (1959–73) that many critics considered to be a western soap opera, dealing as it did with the concerns and adventures of a widower and his sons on the prosperous Ponderosa ranch. Dozens of competing shows were overwhelmed by its popularity, as the fictional Cartwrights became part of Americana. Lorne Greene portrayed the patriarch and Michael Landon, Dan Blocker and Pernell Roberts his sons. Roberts left the cast after six seasons, and Blocker died in the 13th year. Other regulars were David Canary, Mitch Vogel, Tim Matheson and Victor Sen Yung. Produced by David Dortort for NBC Productions, it began as an advertising vehicle for Chevrolet.

"BOOK BEAT"—half-hour PBS series of interviews with authors hosted by Robert Cromie, book editor of the *Chicago Tribune*. Emanating from Chicago public station WTTW, the weekly series began in 1965 and was considered by many publishers to be the most valuable TV exposure on the book-promotion circuit.

BORDER TELEVISION—commercial television licensee in U.K. for the region covering Southern Scotland, Cumbria, the Isle of Man and North Northumberland,

including Berwick-upon-Tweed. Its main studios are in Carlisle.

BORDER WAR—TV conflict between the U.S. and Canada growing out of Canada's attempts to curtail the penetration of American TV so that its native TV systems might have better opportunity to develop and thereby serve to strengthen a Canadian national identity.

The popularity of U.S. TV was reflected in the fact that more than 40% of Canadian households subscribed to cable-TV chiefly to receive the American networks. Meanwhile, U.S. border stations were beaming directly into Canada, fragmenting the audiences for its stations and drawing off an estimated $20 million a year in advertising revenues. The stations of Buffalo, N.Y., blanketing the Toronto market, were estimated to be earning $9 million a year from Canadian audiences, while WVOS-TV in Bellingham, Wash., beaming into Vancouver, alone earned close to $8 million a year from serving Canada.

Under the chairmanship of Pierre Juneau in the early 70s, the CRTC took steps to inhibit the U.S. spillover, the major one being its order to cable systems to delete commercials from the U.S. programs they carried. U.S. stations protested, calling the interception of their signals without the commercials tantamount to piracy, and asked the U.S. State Department to intervene.

Also upsetting to U.S. broadcasters is the amendment to the Income Tax Act (known as Bill C-58) that bars deductions for Canadian companies for advertising purchased out of the country, specifically advertising purchased in U.S. media directed at Canadian audiences. The purpose was to keep Canadian ad dollars in Canada for the support of the country's own media.

In retaliation the Buffalo stations threatened to jam their own signals to keep them from entering Canada. Senators and the Secretary of State, as well as the chairman of the FCC, have met with Canadian officials at various points in efforts to resolve the problem. See also Canada, Television In.

"BORN FREE"—NBC hour-long series (1974) based on the best-selling book and motion picture of that title about naturalists in Africa. Although the movie had several successful runs on television, the series failed and yielded only 13 episodes. Featuring Gary Collins and Diana Muldaur, it was by CPT in association with David Gerber Productions.

"BORN INNOCENT"—made-for-TV movie which, when it aired on NBC in 1974, triggered the congressional concern that brought on the industry's adoption of "family viewing time" in 1975. The two-hour film, which featured

Linda Blair as a 14-year-old in a juvenile detention home, contained a violent sexual scene in which the girl is raped with a broomhandle by other female inmates. Since the film was scheduled at 8 P.M. it had a large audience of juveniles, and the network received protests from its own affiliated stations as well as from the public.

Making matters worse, soon thereafter a young child in California was raped by other children in a manner resembling that in the film; the child's parents sued the network and lost the case. Three congressional committees subsequently demanded that the FCC take action to protect children from the excesses of sex and violence on television. Prohibited by the Communications Act from engaging in any form of censorship, the FCC held meetings with the heads of the networks to suggest ways in which the industry might, on its own, take corrective steps. CBS later proposed to create a "family viewing" hour at 8 P.M., and the other networks followed.

Linda Blair in *Born Innocent*

Despite the graphic controversial scene, *Born Innocent* won commendations from critics. It was produced by Tomorrow Entertainment, with Rick Rosenberg and Robert Christiansen as executive producers.

The film was rerun by NBC the following season but at a later hour, with some editing of the rape scene and with advisories for parental guidance. See also Family Viewing Time.

"BOSTON BLACKIE"—syndicated private-eye series drawn from the movie series. It was produced by United Artists TV from 1951 to 1953 and played long afterward. It featured Kent Taylor, Lois Collier and Frank Orth.

"BOURBON STREET BEAT"—hour-long ABC series (1959) about a team of private detectives involved in offbeat cases in New Orleans. By Warner Bros. TV, it featured Richard Long, Andrew Duggan, Van Williams and Arlene Howell.

BOURGHOLTZER, FRANK—NBC News correspondent since 1946. He became head of the Los Angeles bureau in 1969 after having been the network's bureau chief in Paris (1953–55), Bonn (1955–56), Vienna (1957–58) and Moscow (1961–63), with returns to Paris and Moscow in the 60s. At intervals, he was also a Washington correspondent for NBC.

BOWIE, NOLAN A.— public interest lawyer who headed Citizen's Communications Center from 1977 through 1980. Later he became consultant in communications law and policy to the Georgetown University Law Center's Institute for Public Representation. Before joining CCC, Bowie was with the Watergate Special Prosecution Force of the Department of Justice.

BOYAJIAN, ARAM—producer-director for ABC News whose credits include *The Reasoner Report* (1973–74) and such documentaries as *The American Indian: This Land Was His Land, Fellow Citizen, A. Lincoln* and *Heart Attack!* He was also a producer-director of *The Great American Dream Machine* (1971–72) for PBS.

BOYD, WILLIAM (d. 1972)—western actor better known by the name of the character he created, Hopalong Cassidy, a TV rage in the early 50s. See also *Hopalong Cassidy.*

BOYLE, HARRY J.—a leading broadcast theorist (and onetime practitioner) in Canada who became chairman of the Canadian Radio-Television Commission in January 1976 after serving as vice-chairman since 1968. In a career that began in 1931, Boyle had been variously a radio commentator, writer and executive for the CBC, then a TV producer (*Across Canada* and *The Observer*), a weekly columnist for the Toronto *Telegram* (1957–68), lecturer and author of numerous novels, essays and books on broadcasting.

BOXING ON TV—a prime-time staple of the 50s, with regular telecasts on Wednesdays and Fridays, which all but disappeared from the medium after 1960. A chief reason was that the matches promoted for television were arranged for exigencies of the medium and therefore were not as consequential as those arranged for the normal progress of a fighter; too often, also, they were of poor quality.

Among the leading boxing announcers of the period were Jack Drees, Jimmy Powers, Russ Hodges and Don Dunphy. Dr. Joyce Brothers, a winner of *The $64,000 Question* for her knowledge of the sport, occasionally contributed her observations.

In 1976, Barry Frank, then the new v.p. of CBS Sports, made arrangements to revive the sport on TV, on an occasional basis, Saturday afternoons or Friday evenings. He maintained that CBS would carry only boxing events that were independently promoted and not matches conceived for TV exhibition.

ABC then invested $1.5 million for a weekly elimination tournament, the United States Boxing Championship, put together for television by Don King Productions and designed to develop American champions capable of challenging for world titles. The tournament began in January 1977 but was suspended the following April amid charges of kickbacks, falsified ring ratings for the fighters, phony won-lost records and other irregularities. ABC Sports met the scandal by appointing Michael Armstrong, former counsel to the Knapp Commission which had helped to expose police corruption in New York City in 1972, to direct its investigation of the tournament.

Meanwhile, both ABC and CBS had begun to sign the most promising boxers to exclusive contracts for TV coverage of their bouts.

Jerry Cooney: Boxing on TV

"BOZO"—clown character around whom numerous successful children's shows have been built by local stations since 1959. Larry Harmon, who developed and marketed the property, created a library of cartoon films featuring Bozo and also franchised the character as a live local program host. Those who portrayed the local Bozo had to be trained by Harmon, and the programs were required to feature the Bozo film shorts. In 1966 Bozo was on more than 70 stations, half of them carrying a program furnished by Harmon, the other half building their own shows around the live clown. At WGN-TV Chicago, the locally produced *Bozo's Circus* was a daily one-hour extravaganza of circus acts and was consistently popular with children.

"BRACKEN'S WORLD"—hour-long dramatic series on NBC (1969-70), fictively portraying the glamour and intrigues of the Hollywood movie colony. Produced by 20th-Fox TV, it featured Leslie Nielsen, Peter Haskell, Linda Harrison and Laraine Stephens. Eleanor Parker was in the first 16 episodes.

Ed Bradley

BRADLEY, ED— CBS News correspondent and, since March 1981, one of the four principals on *60 Minutes*. He succeeded Dan Rather in the high-rated series when Rather became anchor of the *Evening News*. Previously, Bradley was principal correspondent for *CBS Reports* (1978-81) and anchor of the Sunday night news.

Bradley was named a CBS correspondent in 1973 while working as a reporter in the Saigon bureau. After being assigned to the Washington bureau in 1975, he volunteered to return to Indochina to cover what became the fall of Cambodia and Vietnam. He later became White House correspondent (1976-78). With CBS Reports he worked on such notable documentaries as *The Boat People* (1979), *Blacks In America: With All Deliberate Speed?* (1979), and *The Defense of the U.S.* (1981).

Before joining CBS News, Bradley was a reporter for WCBS Radio in New York (1967-71), having come from WDAS Radio in Philadelphia.

BRADLEY, TRUMAN (d. 1974)—TV-radio announcer and commercial spokesman. During the late 50s he hosted *Science Fiction Theater* on TV.

"BRADY BUNCH, THE"—comedy series about a young widow and widower merging their families in a second marriage. It enjoyed a good run on ABC (1969–74), with Robert Reed and Florence Henderson as the parents and Ann B. Davis as housekeeper. The Brady kids were Maureen McCormick, Eve Plumb, Susan Olsen, Barry Williams, Christopher Knight and Mike Lookinland. It was via Paramount TV.

Late in 1976 the cast was reassembled for a variety show special on ABC. It was so well received that ABC installed *The Brady Bunch Hour* in a regular Monday night slot in March 1977 for a number of weeks.

"BRANDED"—briefly popular Western series on NBC (1964–66) about an Army officer discharged as a coward but dedicated to proving that he is not. It starred Chuck Connors and was by Goodson-Todman Productions.

BRASSELLE, KEEFE (d. 1981)—one-time movie actor who made a stir in 1964 as an independent producer of series for CBS through his production company Richelieu. All three series failed. Brasselle, who had been a friend of then CBS-TV president James T. Aubrey, wrote a bitter and poorly disguised novel about the experience, *The CanniBalS*. See Aubrey, James T.

BRAVERMAN, CHUCK—independent TV producer and director perhaps best known for delivering a montage history of the U.S. in three minutes in a film called *American Time Capsule*. He made a similar film on the history of TV. His credits include specials or special sequences for programs on ABC and CBS. He has also entered the field of pay-cable production, creating a series of specials entitled *What's Up, America?* for the Showtime pay-cable network.

BRAVO —see Cable Networks.

BRC (BROADCAST RESEARCH COUNCIL)—organization of professionals in broadcast audience measurement and research. It represents all parts of the industry concerned with the examination and improvement of techniques in the field. The organization was formed after the congression(al investigation of TV ratings in the 60s. The BRC "approved" symbol is carried on all data reports passed by the group.

"BREAKING POINT"—an ABC dramatic series dealing with psychiatric patients and their doctors. A spin-off of *Ben Casey*, it ran one season in 1963. It starred Paul Richards and Eduard Franz and was produced by Bing Crosby Productions.

BRESNAN, WILLIAM J.—chairman and chief executive officer of Teleprompter Cable, the second largest cable concern, since December 1981. Prior to that he was president and chief operating officer since July 1972 of Teleprompter Cable, largest of the cable-TV concerns. During that tenure he was instrumental in reorganizing and rehabilitating the company after the dual trauma of the bribery conviction of Irving Kahn and the financial bust that resulted from the company's having overextended itself.

Bresnan's promotion to chairman came after the company's acquisition by Westinghouse Braodcasting Co. He was also elected to the board of the new parent company.

Bresnan entered the cable field as an engineer in 1958 and worked for a number of small systems until he was hired in 1965 as chief engineer of American Cablevision Co. in Los Angeles. Three years later the company merged with H&B American Corp. and Bresnan was named president of the cable-TV subsidiary, H&B Cablevision Co. When H&B was acquired by Teleprompter in 1970, Bresnan became western v.p. and rose eventually to president. Throughout his career, he was active in numerous capacities with NCTA, including a stint as board chairman (1972–73).

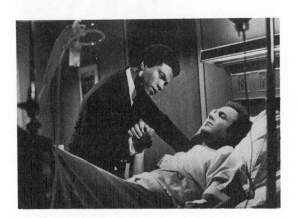

Billy Dee Williams & James Caan as
Gayle Sayers & Brian Piccolo

"BRIAN'S SONG"—poignant and popular ABC TV movie (premiere: Nov. 11, 1971) based on the real-life relationship between Brian Piccolo and Gale Sayers, professional football players and roommates with the Chicago Bears. The drama centers on the illness that eventually takes Piccolo's life and is heightened by the fact that one roommate is white, the other black. Written by William Blinn and produced by Paul Junger Witt for Screen Gems, it featured James Caan, Billy Dee Williams, Jack Warden, Shelley Fabares and Judy Pace, with some football stars playing themselves.

Brideshead Revisited with Anthony Andrews & Jeremy Irons

"BRIDESHEAD REVISITED"— opulent 11 episode adaptation of Evelyn Waugh's novel of that title, which drew critical praise and enjoyed an intense following on American public television during its initial run early in 1982. Produced by Granada Television International of England, on the largest production budget ever for a British commerical series—$13 million—it was a smash in its native country in two runs prior to its American debut. The 11 episodes are of uneven lengths and consist of 12½ hours of film.

John Mortimer wrote the adaptation of the Waugh novel, concerning an Oxford student who gets drawn into the extravagant, careless way of life of the young nobility in the years between the wars. The serial was filmed entirely on location in England, Venice, Malta, and the island of Gozo. Brideshead Castle was represented by Castle Howard in Yorkshire.

Heading the cast were Jeremy Irons, Anthony Andrews, and Diana Quick. Laurence Olivier, Claire Bloom, and John Gielgud received guest star credit, along with Stephane Audran, Mona Washbourne, and John Le Mesurier.

Derek Granger was producer and Charles Sturridge director. Michael Lindsay-Hogg, sharing the directing credit, had been the original director but withdrew for other commitments when a strike halted the production for several months in its early stages. Original music was composed by Geoffrey Burgon. WNET New York and NDR West Germany produced in association with Granada Television. The

series aired in the U.S. as part of Exxon's *Great Performances* series.

"BRIDGET LOVES BERNIE"—situation comedy about Irish-Jewish newlyweds. Adapting the *Abie's Irish Rose* premise to modern times, it ran one season on CBS (1972), and featured Meredith Baxter, David Birney, Harold J. Stone, Bibi Osterwald and Audra Lindley. Screen Gems produced, in association with Douglas S. Cramer Co. and Thornhill Productions.

BRIDGING—a programming maneuver to cripple a competing show by starting 30 minutes earlier, thus gaining the advantage of being in progress when the other begins. An hour-long program scheduled at 8:30 thus would *bridge* the opening of a 9 o'clock show on another channel, presumably keeping an audience that might have been inclined to watch the competing show.

David Brinkley

BRINKLEY, DAVID—leading NBC News figure for more than two decades who left for ABC News in 1981 after being shunted out of the main newscasts. At ABC, where he represented a prize catch, he was given a Sunday morning news program, *This Week*.

Brinkley shot into prominence in 1956 when he was teamed with Chet Huntley for the political conventions and later for the evening news. Ever the consummate journalist, however, Brinkley adhered to professional standards despite the commercial opportunities afforded by his public acclaim and recognition. Wit and a dry, sardonic delivery were the hallmarks of his style, and his baritone intonations and distinctive phrasing were imitated by dozens of reporters at NBC.

While the *Huntley-Brinkley Report* was at its height in the 60s, Brinkley ventured into the documentary field with a distinguished series, *David Brinkley's Journal*. When Huntley retired in 1970, Brinkley was teamed with other network correspondents for a time. In 1971 he gave up the anchor position to become a commentator for what then was called the *NBC Nightly News* and to work on news specials. During 1976 he co-anchored the political conventions with John Chancellor and proved as appealing to viewers as he had been 20 years before. Following the conventions, he became co-anchorman with Chancellor of the *Nightly News*.

BRITAIN'S BROADCASTING COMMISSIONS—The history of broadcasting in Great Britain has been punctuated and shaped by a series of ad hoc committees appointed by the government of the day. There have been seven such bodies in all, and their brief has been to formulate national policy with respect to the structure and performance of broadcasting in the U.K.

Some of these committees, such as the one headed by Lord Crawford in 1925, have been more significant than others. The Crawford committee, convened in the days when radio in Britain was still a private enterprise, laid the foundations for the royally chartered British Broadcasting Corp. in 1927.

The first of these blue ribbon committees was created in 1923 and chaired by Sir Frederick Sykes. It produced some rudimentary ideas about public service radio that were to be more fully developed two years later by the Crawford committee.

Subsequent committees were headed by Lord Selsdon (1934), Lord Ullswater (1935), Lord Beveridge (1949) and Sir Harry Pilkington (1960). The most recent committee, invoked in 1974, was headed by Lord Annan.

Spadework by the Selsdon panel led to Britain incepting, in 1936, the world's first high-definition television service on a regular basis. The daily service was interrupted by the outbreak of war in 1939, resuming in mid-1946.

The advent of a commercial television medium in competition with the BBC was "seeded" by proposals of the Beveridge committee. These were subsequently endorsed by two government white papers, and in 1954 an act of Parliament created the Independent Television Authority (ITA) as the instrument that would assign and administer commercial station licenses. The first of these independent stations began operating in 1955.

One of the more famous committees was the one headed by Sir Harry Pilkington. It proposed more air time and more channels, described British broadcasting as a potent national asset and gave the broadcast establishment a vote of approval. None of its specific recommendations was acted on, however, although the Pilkington Report was said to have prompted the shake-up in British Commercial TV that created new licensees and struck down some of the regional broadcasters.

Potentially one of the more significant committees was the one most recently constituted under Lord Annan. Its brief was of the most fundamental kind—to study and make proposals for the structure of broadcasting in Britain in the 1980s, taking into account the new technology ranging from cassettes to cable and satellites.

On its recommendations might hang the future not only of radio and television services in the country, but also the structure of the BBC itself, with the possibility that it might be split into two separate entities, one for radio and the other for television. Another possibility: that all broadcast services would be consolidated under a single administrative authority.

BRITISH BROADCASTING CORPORATION (BBC) —since 1927 the public broadcasting entity operating in the United Kingdom of England, Scotland, Wales, Northern Ireland, the Channel islands and the Isle of Man in the Irish Sea. BBC was the first to operate a regularly scheduled television service, starting in 1936, though operations were suspended in September 1939 for the duration of World War II.

The BBC is not state-owned, as is widely believed, but operates independent of government control under a Royal charter (first granted in 1927) and a license from the secretary of state for home affairs as stipulated by the terms of the wireless telegraphy acts of the U.K.

Financial support for the corporation derives mainly from an annual tax called a license fee, payable by all who maintain a radio or TV receiver—the equivalent of $13.68 for black & white sets and $30.78 for color TV. These fees are collected for the company by the post office. The company, whose radio and television services have never carried paid advertising (unlike some broadcast systems that are state-owned), also realized some revenue from the production and distribution of recordings (from bird calls to music), and from various publications including weekly magazines such as *Radio Times*, which carries radio and TV log listings and feature articles, and *The Listener*, which offers opinion, book reviews and depth pieces keyed to current BBC programming. The company also publishes books derived from, or related to, its broadcast output, e.g., *Alistair Cooke's America*, Dr. Jacob Bronowski's *The Ascent of Man*, Nigel Calder's *Violent Universe*, etc.

In 1932, the BBC added an external services division, subsidized by the government, dominantly for the worldwide dissemination of news but also including programs of music, drama and discussion. This offshore arm currently beams daily in 39 languages, in addition to an English service that operates around the clock.

Domestically, the BBC operates two television channels (BBC-1 and BBC-2), as well as four national radio channels and local radio stations in London and 19 other key cities. British television (including the independent advertiser-supported stations—commercial television in the U.K. started in 1956) uses the German-developed 625-line PAL color system.

Organizationally, BBC is headed by a board of governors appointed by the Queen, on the advice of her ministers, for five-year terms. The permanent staff, which currently numbers more than 23,000, is headed by a director-general as the chief executive officer of the company.

The unique constitutional position of the BBC, broadly unaltered since the first Royal charter was granted, was determined largely by the policy and operational practices adopted by its predecessor, the British Broadcasting Company, which inaugurated regular radio service in Britain in 1922. That original company was formed, at the invitation of the postmaster general, by the principal manufacturers of radio receivers, and their only brief was to provide a service "to the reasonable satisfaction" of the postmaster general.

To run the service for them, the manufacturers recruited the late J.C.W. Reith (later to become a peer of the realm), a strict Scottish Calvinist who believed in *noblesse oblige*. He was to become the first director-general of the successor broadcast corporation in 1927, and the single most influential figure in the development of public broadcasting in Britain.

In the first years of British radio, starting in 1922, the postmaster general was the ultimate arbiter of what was suitable for the nation to hear. The fledgling company had no charter from the crown, and no statutory sanction from Parliament. Reith, in command of the service, believed the new medium had a great capacity for public service. He saw it providing not only entertainment, but also information and enlightenment.

In 1925, evidently pleased with the way broadcasting had evolved under Reith's stewardship, the government appointed a blue ribbon committee under Lord Crawford whose assignment was to frame the guidelines for the future structure of broadcasting in Britain. In time, the committee came up with the recommendation that the nation's broadcasting service should henceforth be operated by an independent public corporation "acting as trustee for the national interest." And so the present-day BBC was born.

Over the years the BBC was to become a national institution, an instrument for uniting the British people in times of stress, and for transmitting the ideas and values that collectively make up the British way of life. By intent or otherwise, the BBC became a national arbiter of taste and standards. Admittedly by design, there was even an attempt over many years to impose standards of language that came to be known as "BBC English." The attempt was clearly doomed to failure, and the corporation today appears to have retreated from the position that all voices with access to its microphones should speak as one in oval tones. Contrarily, these days the richness and diversity of British dialects and accents is acknowledged and encouraged.

Besides its broadcasting functions, the BBC has a number of subsidiary roles in British life. It is an archivist, Britain's voice to the world, and a patron of the arts. The corporation maintains 11 orchestras and dancebands, chief of which is the 101-piece BBC Symphony, which regularly performs in public concert at home and abroad with some of the world's leading conductors and soloists. It is also a significant force in education, producing more than 3,000 hours of classroom programming for radio and television each year. More recently this has come to include a unique experiment in adult education, with credit courses and the awarding of recognized diplomas, known as Open University.

In the breadth and variety of its program output, both for radio and television, the BBC is probably unrivaled. Its drama output alone, both in anthology and serial form, totals more than 500 hours annually for TV, and many of these productions are seen around the world.

Because the company is such a prodigious consumer of self-produced film, much of which figures in its drama content, it decided several years ago on the need for its own film facility, and acquired the old and distinguished Ealing studios in west London where many a classic British theatrical movie was made. It is the busiest film facility in Britain, and probably one of the busiest anywhere.

BROADCASTERS PROMOTION ASSN. —national organization of station and network personnel engaged in promotion and public relations, whose annual seminars are a major forum for trends and practices in the field. Founded in 1955 by a group of Chicago TV and radio station promotion managers, with initial underwriting from *Advertising Age,* BPA is supported now by its membership.

BROADCAST INFORMATION BUREAU —a television library and research organization, founded in 1950 by Joseph Koehler and owned now by Film Service International. BIB publishes annual fact books on television programs and movies, principally to guide stations in their buying; conducts research surveys on the industry; and maintains computer banks of program information and history.

BROADCASTING—according to the FCC definition, radio communication designed for reception by the general public. Television is a form of radio, involving synchronous transmissions that are both visual and aural. The television transmitter may be regarded as two separate units, one sending out visual signals by AM (Amplitude Modulation), the other sound by FM (Frequency Modulation).

BROADCAST REFORM MOVEMENT—an effort by citizens groups and public interest organizations to counterbalance the economic drive of broadcasters by working actively to safeguard the public's rights in TV and radio. With broadcasting the common focus of the various groups, although they represent diverse special interests, a movement began to form in the late 1960s to keep watch on individual stations negligent of their responsibilities to the total community and also to influence regulatory policies to achieve an equitable and properly responsive broadcast system. Among the motivating concerns of the groups have been excessive violence on TV, insensitivity toward children in programming and advertising, lack of access to broadcasting for minorities, discriminatory hiring practices and concentration of control of the mass media.

Central to the movement have been national organizations working for change through both legal and nonlitigative procedures, notably:

The National Citizens Committee for Broadcasting (NCCB) in Washington, D.C., headed by Ralph Nader, aiming to coordinate all citizens groups efforts and to represent the views of the reform movement at the FCC and in congressional hearings;

Office of Communication, United Church of Christ, in New York, led by the Rev. Dr. Everett Parker, devoted to correcting injustices to minorities in broadcasting;

Action for Children's Television , based in Boston, and headed by Peggy Charren, working to upgrade children's programming;

Citizens Communications Center and Ralph Nader's Center for the Study of Responsive Law, both in Washington, providing legal assistance and materials to those in taking action against offending stations;

Council on Children, Media and Merchandising, headed by Robert Choate, active in areas relating to nutrition and TV ads;

National Assn. for Better Broadcasting, in Los Angeles, which publishes an annual critique of programs on the air and has negotiated with stations to eliminate violent children's programs.

Minority rights organizations—the National Organization for Women , Gay Media Task Force, National Black Media Coalition, among others—also participate in the movement.

The movement received its greatest impetus from a 1969 landmark case in which the U.S. Court of Appeals ordered the revocation of the license for WLBT-TV in Jackson, Miss., for discriminating against the interests of the black community, which constituted 45% of its potential viewing population. The case established the right of citizens' groups to participate in license-renewal proceedings. The Office of Communication of the United Church of Christ had joined with two black civil rights leaders from Jackson to file a *petition to deny* against WLBT. When the court upheld the validity of the petition, the grassroots participation movement had the legal tool that would foster its growth.

Around the country, stations entertained dialogues with dissatisfied groups in preference to facing possible petitions to deny.

When a local chapter of NABB filed a petition to deny the license renewal of Metromedia's KTTV in Los Angeles, the station negotiated with the group and in 1973 agreed to ban 42 specified programs declared as too violent for children—including *Batman* and *Superman*. KTTV also agreed to post a warning to parents about violent content with 81 other programs available to it through syndication. With the agreement, NABB withdrew its petition. But in 1975, the FCC disallowed the settlement on the ground that a broadcaster has a nondelegable responsibility for what he puts on the air. It was also considered to represent a form of censorship.

In Lansing, Michigan, the Citizens for Better Broadcasting, a local group, succeeded in persuading WJIM-TV to remove reruns of *Wild, Wild West* from its afterschool timeslot and in general to improve its service to children. A petition to deny was the wedge for negotiations and was withdrawn when the grievance was settled.

NOW used the license lever to pry agreements from KCST San Diego and WXYZ Detroit to hire quotas of women, provide serious women's programming and accept women's advisory councils.

ACT petitioned the FCC to adopt regulations for children's television. Hearings on the petition elicited around 100,000 letters from concerned parents around the country. In response to pressures, pharmaceutical companies ceased advertising vitamin pills in the manner of candy, the NAB amended its code to prohibit selling by those who host children's programs and the FCC issued a set of guidelines on children's television.

The networks, in response to efforts by broadcast reform groups, reduced violence-oriented programs on Saturday mornings, created children's programs with pro-social messages, cut back the number of commercials carried in children's programming, altered their hiring practices to include more members of minority groups, gave ethnic identities to protagonists in entertainment programs and accelerated the promotion of women and blacks to managerial posts.

The FCC responded by making community ascertainment a condition of license renewal. See also ACT, KTTV Agreement, WLBT Case.

"BROADWAY OPEN HOUSE"—first late-night series by a network, essayed by NBC in the summer of 1950, which featured variety acts and the broad, brash comedy of Jerry Lester, with Dagmar as his sexy, dumb sidekick. It ended its run as a weeknight entry in August 1951.

BRODKIN, HERBERT—producer of quality drama, one of the few based in New York who has remained active in television since the early years of the medium. In the 50s he produced for *Studio One, Motorola Hour, Playhouse 90* and other showcases, his credits including such plays as *Judgment at Nuremberg* and *Child of Our Times*. Later, he produced such series as *The Defenders, The Nurses, Shane* and *Coronet Blue*. A drama he produced for CBS in the 60s, *The People Next Door,* was subsequently made into a theatrical movie, which Brodkin also produced. His principal work in the 70s was for *ABC Theatre* and included the documentary-dramas, *Pueblo* and *The Missiles of October,* and the biographical dramas, *F. Scott Fitzgerald and the Last of the Belles* and *F. Scott Fitzgerald in Hollywood*. In 1978 he produced the highly successful miniseries, *Holocaust;* in 1981, *Skokie*.

BROKAW, TOM—co-anchor, with Roger Mudd, of the *NBC Nightly News* since April 1982. The two succeeded John Chancellor.

Tom Brokaw

Brokaw became a leading condidate for the anchor post after six distinguished years as host of NBC's *Today*. He joined the morning program as Barbara Walters' replacement in August 1976 when she was hired away by ABC News. Brokaw had been NBC's White House correspondent for three years and in addition was anchorman of the NBC Saturday night newscast. Brokaw continued to be given some top news assignments while hosting *Today,* and he was rated NBC's leading candidate for an anchor position on the *Nightly News*

when one became available. He became a network correspondent in 1973 after having anchored the 11 o'clock news for NBC's Los Angeles station, KNBC.

"BROKEN ARROW"—western series on ABC (1956–60), via 20th Century-Fox TV, which starred John Lupton and Michael Ansara.

BROKERING TIME—the practice of selling air time to an entrepreneur, who creates his own programs and keeps the revenues from the advertising he sells. More prevalent in the early years when television was not such an easy sale to advertisers, it remains a fairly widespread practice among foreign-language or ethnic-oriented stations.

BROOKHAVEN CABLE-TV V. KELLY [*aff'd* 573 F.2d 765 (2nd Cir. 1978) *cert denied* 47 U.S. L.W. 3005 (U.S. July 11, 1978) 77-1845)]—case in which the U.S. Court of Appeals for the Second Circuit upheld the FCC's authority to preempt state and local regulation of pay-cable television rates. At issue were rules issued by the New York State Commission on Cable Television requiring municipal authorities to regulate the rates charged subscribers for pay-cable channels despite a clearly stated FCC policy that such rates should remain unregulated.

Brookhaven Cable-TV brought the action against New York Cable Commission chairman Robert F. Kelly and the four other state commissioners after the commission ordered local authorities in 1976 to (1) include regulation of pay-cable rates in all new cable television franchises and require cable companies to submit notice of their current rates within two months. Brookhaven challenged the order under the supremacy clause of the Constitution, arguing that the FCC's policy that pay-cable rates should remain unregulated took precedence over state regulations. Although the complaint also sought a judgment under the First, Fourteenth and Fifteenth Amendments as well, a federal district court granted summary judgment only on the supremacy clause arguments, and the Court of Appeals upheld the ruling solely on that basis.

The court found that the FCC's preemption of pay-cable price regulation was "reasonably ancillary" to its responsibilities for regulation of television broadcasting under the Communications Act of 1934, in line with UNITED STATES V. SOUTHWESTERN CABLE CO. (1968). "A policy of permitting development free of price restraints at every level is reasonably ancillary to the objective of increasing program diversity" and is therefore within the FCC's authority, the court found.

BROOKS, JAMES L., AND BURNS, ALLAN—the resident geniuses of the MTM Enterprises situation comedy factory, as a producing-writing team. Creators and executive producers of *The Mary Tyler Moore Show,* they added Paul Sand in *Friends and Lovers* and *Rhoda.* Brooks, by himself, was producer and writer of *Thursday's Game,* a made-for-TV movie. In 1977 he signed an exclusive contract with ABC, working through Paramount TV, a pact that led to the creation, with associates, of *Taxi* and *The Associates.*

BROOKS, MEL—comedy writer, producer and occasional performer, who first attracted attention as a member of the writing stable for *Your Show of Shows.* He later was co-creator (with Buck Henry) of the hit series, *Get Smart.* In 1975, while enjoying success as a movie producer, he became executive producer and co-creator of the ABC series *When Things Were Rotten.*

BROOKSHIER, TOM—sportscaster who joined CBS and its owned station in Philadelphia, WCAU-TV, following his retirement as a star defensive back for the Philadelphia Eagles. Since 1964 he has been an analyst for CBS in its pro football telecasts, also covers boxing and serves as host of *CBS Sports Spectacular.* For several years he hosted two syndicated series, *This Is the NFL* and *Sports Illustrated.*

BROTHERS, JOYCE—quiz-show contestant of the 50s who briefly became a national celebrity through winning the top prizes on both *The $64,000 Question* and *The $64,000 Challenge.* An attractive blonde, she was at the time a clinical psychologist and the newly wed wife of a doctor, but her field on the quiz shows was boxing. After her dual triumph in 1957, she was signed by NBC to co-host *Sports Showcase* and then did radio pieces for *Emphasis* and *Monitor.* Later she conducted syndicated TV programs as a psychologist, variously entitled, *Dr. Joyce Brothers, Consult Dr. Brothers, Tell Me, Dr. Brothers* and *Ask Dr. Brothers.*

BROWN, TYRONE—brilliant young FCC commissioner, appointed in November 1977 at the age of 35 to replace Benjamin L. Hooks in what probably has become the permanent black seat on the commission. Brown was the FCC's second black member, Hooks the first. A lawyer, Brown had worked previously as v.p. for legal affairs for the Post-Newsweek Stations, as staff director of the Senate Intergovernmental Relations Subcommittee and as law clerk to former Chief Justice Earl Warren. He was appointed to a Democratic seat on the FCC and became part of the liberal minority, with chairman Ferris and commissioner Fogarty.

He left the ICC to return to law practice in 1981 after the defeat of President Carter.

BROWNING, KIRK—veteran director of cultural programming. With NBC he directed the *Bell Telephone Hour* and productions of the NBC Opera. In public television he helped develop the techniques for live telecasts from concert halls that were unobtrusive to the paying customers. These techniques made possible *Live From Lincoln Center,* for which he directed the New York City Opera's telecast of *Don Carlo, Handel's Messiah,* and the Beverly Sills farewell gala. He also directed the drama, *Big Blonde* for PBS's *Great Performances.*

BUCHANAN, PATRICK J.—hard-line conservative and former speechwriter for President Nixon, credited with writing Vice-President Agnew's famous attack on network news, delivered in Des Moines in November 1969. Buchanan, who was also in charge of Mr. Nixon's morning news summary and was responsible for monitoring the media's coverage of the White House, personally kept up the public assaults on what he considered to be the liberal bias of TV journalism. A former newspaperman, he became a Media Watch columnist for *TV Guide* when the Nixon Administration ended and wrote a syndicated column for *The New York Times's* special features subsidiary.

BUNKER, ARCHIE—fictional TV character (*All In the Family*) whose name has entered the language as the label for a ridiculous blue-collar bigot. He was created by Norman Lear and portrayed by Carroll O'Connor.

BURCH, DEAN—activist FCC chairman (1969–74) who, after initially seeming intent on politicizing the office, won the respect of broadcasters and citizens' groups alike for his even-handed policies and personal fairness. The radical decisions made by the agency during his administration belied the fact that Burch was a political conservative—he had formerly been an aide to Sen. Barry Goldwater (R-Ariz.) and then his campaign manager for the 1964 Presidential race.

It was not Burch's way to push controversial issues onto the backburner or to postpone the voting on items that he knew would leave him in the minority. Thus, the prime time–access rule came into being and the WHDH license was denied to the Boston *Herald-Traveler,* two of the most momentous developments in the industry in years. Burch's chairmanship was marked also by periodic clashes with his ideological opposite number, the gadfly commissioner, Nicholas Johnson, although their disagreements were as often over style as over ideologies.

Burch had alarmed many during his first weeks on the commission when he openly praised Vice-President Agnew's attack on liberal bias of network journalism and followed by remarking to the industry, "Physician, heal thyself!" He also personally made calls to the networks to question certain news decisions. But he quickly ceased those activities, recognizing their impropriety and First Amendment implications, and began to apply himself vigorously to the work of the commission.

During the Watergate episode, memoranda surfaced from the early years of the Nixon Administration suggesting that Burch would exercise control over news programming. But Burch said he had never discussed such matters with the White House and that even if he had he would never abide the idea of government interference in news.

Under Burch, the FCC approved the full-scale release of cable-TV for expansion into the major cities, gave the long-delayed go-ahead to over-the-air pay-TV (STV), tentatively adopted the one-to-a-market rule for ownership of media in new situations, levied a license fee on broadcasters to help defray the commission's expenses and reaffirmed the Fairness Doctrine after a long inquiry and study.

Also notable was a decision that the congressional opponents to administration policies were entitled to free air time to respond to the President's prime-time speech, in the interest of keeping the executive branch from attaining excessive power through TV.

In February 1974 Burch left the commission to become counselor to the President, with cabinet rank. He served in that capacity ten months, staying on when Nixon resigned to assist President Ford through the transition period. At the end of the year, Burch left government service to join the Washington firm of Pierson, Ball and Dowd, which specializes in communications law.

BURDEN OF PROOF (Fairness Issue)—[Allen C. Phelps/21 FCC 2d 12 (1969)] FCC determination that the initial burden in the fairness area rests with the complainant. The case involved a charge that WTOP Washington presented only liberal positions on several issues, but the complainant otherwise made no showing on his general allegation. The FCC, in dismissing the complaint, listed among the elements necessary for a prima facie Fairness Doctrine complaint a reasonable basis for the complainant's conclusion that the licensee failed in its overall programming to present a reasonable opportunity for opposing views. While this decision was never directly appealed, it was cited with approval by the Supreme Court in the BEM Case. See also BEM Case.

BURDETTE, WINSTON—veteran CBS News correspondent who joined the network in 1941 while on freelance assignments in Europe during the outbreak of World War II. Although he later worked briefly for CBS in New York, Washington and London, most of his career has been spent on the European continent, with Rome as his base. From the Rome bureau, which he joined in 1956, Burdette has covered events in Europe, Africa and Asia.

"BURNS AND ALLEN SHOW"—See *George Burns and Gracie Allen Show.*

"BURKE'S LAW"—hour series for ABC (1963–65) about a wealthy and sophisticated detective who attracts beautiful women. It featured Gene Barry in the lead, and Gary Conway and Regis Toomey, and it was produced by Four Star-Barbety. For the 1965 season, the lead became a daring secret agent and the series was retitled, *Amos Burke, Secret Agent,* to no avail.

BURR, RAYMOND—actor who starred in two series that were mammoth hits, *Perry Mason* (1957–66) and *Ironside* (1967–75). Earlier, he was featured in numerous TV dramas. In March 1977 he essayed another series, *Kingston: Confidential,* on NBC, but it was short-lived.

BURRUD, BILL—prolific producer and host of syndicated outdoor, travel and animal series. His Hollywood-based Bill Burrud Productions has turned out such skeins as *Animal World, Safari to Adventure* and *World of the Sea,* along with dozens more over two decades.

"BUS STOP"—ABC series (1961–62) loosely based on the William Inge play of the title, one of whose episodes led to the downfall of a network president. The hour-long dramatic series dealt with strangers passing through a small western community, the focus of which was a bus stop luncheonette run by Marilyn Maxwell. In the fateful episode "Told by an Idiot," which was based on a novel by Tom Wicker, the pop singer Fabian portrayed a psychopath whose sadistic excesses in the show raised a storm of public protest that led to one of the several congressional investigations into television violence. Oliver Treyz, then president of ABC, was fired shortly after the congressional inquiry in what many believe was a corporate action to place the responsibility for the episode upon him. The series, by 20th Century-Fox TV, otherwise was of no distinction. It featured besides Miss Maxwell, Rhodes Reason, Richard Anderson, Joan Freeman and guest performers.

BUTTERFIELD THEATERS v. FCC—[237 F2d 552 (D.C. Circ. 1956)] case involving a license applicant that substantially modified its license proposals after winning the license in a comparative hearing.

Three applicants sought a license to construct and operate a TV station in Flint, Mich.—WJR, Butterfield and Trebit. After a comparative hearing, the FCC found WJR best qualified, basing its decision on a number of factors, including the proposed location of the WJR transmitter, the WJR programming proposals and its proposed studio construction. Butterfield and Trebit asked the FCC to rehear the argument but the commission declined. Ten days later WJR filed a petition to alter its application.

In its modification petition, WJR proposed to move its transmitter farther away from the city and to a lower altitude than it had originally proposed, affiliate with a network other than the one initially mentioned, offer different kinds of local live programming than it had promised at first and purchase a one-story building for studios at a cost of $125,000 instead of constructing a two-story building at a cost of $776,000. On the basis of this petition, Trebit and Butterfield again sought to reopen the hearing. Again the commission refused, and the losing parties appealed.

The D. C. Court of Appeals held that the commission had erred by not reopening the hearing. The court stated that the FCC unquestionably had the power to call a rehearing and that it was an abuse of discretion not to do so. The changes proposed by WJR were directed at the very reasons it had been given preference in the first place. An alteration of these proposals mandated a review of the commission to determine if WJR's application was still superior to those of the competing applicants.

On remand the FCC determined that the new transmitter site was as good as the initial one and that it was as good as those proposed by Trebit and Butterfield. The FCC additionally found that the changes for programming were not substantial, and that the less expensive studio building did not affect the preference given to WJR in that category. Finally, the commission decided that the proposed modification did not reflect adversely upon WJR's character and fitness as a licensee. Therefore, the commission affirmed its original decision to award the license to WJR.

BUTTONS, RED—a former burlesque comedian who had a brief fling at TV stardom with a half-hour comedy variety series on CBS (1952–54), *The Red Buttons Show,* and then became a dramatic actor in serious roles.

BUTTRAM, PAT—countrified comedian who became known in TV as the sidekick in *The Gene Autry Show* (1949–54) and later was featured in *Green Acres* (1965–71).

BUXTON, FRANK—producer-director and writer, whose credits include *The Odd Couple, Love, American Style, The Bob Newhart Show, Hot Dog* and *Children's Letters to God.*

BYINGTON, SPRING (d. 1971)—comedy actress in movies who became a TV star in 1954 with her situation comedy *December Bride.* It ran five seasons on CBS and the reruns remained popular in syndication.

C

CABLECASTING—televised programming originating on cable channels, including locally originated programs produced by the cable system or public access groups, programs on leased channels, and satellite-transmitted national programming made specially for cable-TV. The term distinguishes this programming from a cable system's retransmission of broadcast signals.

CABLE FRANCHISE—authorization to a cable-TV company by the issuing authority, usually municipalities or other local governments, defining conditions under which it may construct, operate and maintain a cable service. The agreed-upon contract amounts to a license to operate and ensures due process and public involvement. Because of the localized nature of cable-TV, the franchising authority has the principal responsibility for the regulation of the service. The license establishes construction timetables, technical standards, extension of services, rates, channel utilization not expressly forbidden by the FCC, consumer services, channel leasing and access channels.

CABLE HEALTH NETWORK—see Cable Networks.

CABLE NETWORKS—national program services transmitted by satellite to cable systems; nearly all of them specialize in a single kind of programming, for example, sports, news, movies, or children's fare. These networks began proliferating in the late 70s after HBO pioneered program distribution by satellite in 1975. By 1982, there were 43 operating networks and 18 others in planning stages. Most of the cable networks were on the Satcom I satellite, but the growing traffic caused a spillover to the Comstar and Westar satellite systems.

The spur to the development of satellite networks was the expansion of cable systems (with channel capacities ranging from 60 to more than 100) to the major population centers.

The networks are of several types; some are pay services, some advertiser-supported, some noncommercial. Some charge the cable operator fees ranging from one cent to 20 cents per subscriber each month; some are given free to any cable system that will accept them; a few pay the cable operator for carrying the service. The pay-cable networks share part of their fees from subscribers with the cable system.

The following are among the better-known networks:

ARTS (Alpha Repertory Television Service), cultural service created by ABC in 1981, operating 9 P.M. to midnight and intended to be advertising-supported. The service is carried on Warner Amex's Nickelodeon channel, which goes dark in the evenings.

BET (Black Entertainment Network), ad-supported service for black audiences which started in 1979 on a limited schedule Friday nights on the USA Network channel. Taft Broadcasting bought into the company in 1982 and expanded the schedule substantially on another transponder.

BRAVO, pay-cable cultural service founded in 1980 providing performing arts fare and international films. Initially it was offered as a package with Escapade, an "adult movie" service, but later was separated as an independent weeklong network. It is owned by a consortium of cable systems.

CBN (Christian Broadcasting Network), an ad-supported network specializing in religious programming formed in 1977 by Pat Robertson. The network is given free to cable systems.

CBS CABLE, ad-supported channel of performing arts and other cultural programming founded by CBS in 1981. The programs are presented in three-hour blocks that are repeated from 4:30 P.M. to 4:30 A.M. The fare is a mix of original production and programs acquired from foreign television systems.

CINEMAX, companion network to HBO; a movie channel with a different selection of films, offered as HBO's second-tier pay service.

CNN (Cable News Network), 24-hour news network founded in 1980 by Ted Turner and operated from Atlanta. It immediately became one of the basic services offered by all new cable systems. Although CNN lost millions of dollars during its first two years, it quickly established itself as a major news organization. When Group W and ABC announced that they would create two news networks in 1982 to compete with CNN—one a straight newscast service, the other covering news in greater depth—and would distribute them to cable systems without charge, Turner answered the threat by creating an all-newscast network of his own, CNN-2, and beating his rivals to the punch. Reese Schonfeld is president of CNN.

C-SPAN (Cable Satellite Public Affairs Network), a part-time noncommercial network organized by Brian Lamb in 1979 and partially funded by large cable companies. This service provides live coverage of the House of Representatives sessions and other special features.

THE ENTERTAINMENT CHANNEL, a pay service headed by Arthur Taylor, former president of CBS Inc., and financed by Rockefeller Center and RCA. This network which began broadcasting in 1982, offers 24-hour service devoted to BBC programming, movies, theater productions, cultural presentations, and children's fare. Unlike the cultural channels, it is designed for broad appeal.

ESCAPADE, pay channel specializing in R-rated movies, televised from 8 P.M. to 6 A.M. *Playboy* magazine became a partner in 1981, adding *Playboy* entertainment programs to the fare. Escapade was formed in 1980 as a stablemate of Bravo by Rainbow Programming Service, owned by a consortium of cable companies.

ESPN (Entertainment & Sports Programming Network), a 24-hour service devoted entirely to sports coverage and sports-related programming. It was founded in 1979 in Connecticut and attracted an $18 million investment by Getty Oil. Its president is Chet Simmons, former president of NBC Sports. ESPN proved one of the most successful cable networks at attracting advertising.

GALAVISION, a pay-cable network formed in 1979 by Spanish International Network featuring movies, variety, and sports programs for Spanish-speaking subscribers.

HBO (Home Box Office), father of the cable network revolution and by far the most profitable of the pay-cable services. Founded by Time Inc. in the early seventies as a regional pay-cable operation in the east, HBO hit the jackpot with the decision to go national, by satellite, in 1975. Its success triggered a new boom in cable. By the spring of 1982, HBO had close to 8 million subscribers.

HTN (Home Theater Network), a pay service begun in 1978 offering family entertainment, travel features, and G- and PG-rated movies. It is headquartered in Portland, Me.

THE MOVIE CHANNEL, Warner Amex's answer to HBO, differing in that it offers movies exclusively, and on a 24-hour basis. It was created as a pay service in 1979 and is part of the WASEC (Warner Amex Satellite Entertainment Corp.) group of networks headed by John A. Schneider.

MSN (Modern Satellite Network), advertising-supported service providing informational and public service programming during the daytime hours weekdays. A main feature is *The Home Shopping Show*.

MTV (Music Television), ad-supported service devoted entirely to pop music, which started in August 1981 and quickly became a solid hit with youthful viewers. It is another of the WASEC networks.

NICKELODEON, noncommerical network of high-grade children's programming founded by Warner Amex in 1979. It proved a highly desirable service for new cable systems to offer. Operating without commercials from 8 A.M. to 9 P.M., Nickelodeon makes its money from fees charged to cable systems.

PRIVATE SCREENINGS, service providing pornographic films to paying subscribers on Friday and Saturday nights, from midnight to 6 A.M. It started in December 1980 and is owned by Satori Productions Inc.

PTL (People That Love), a religious network originating in Charlotte, N.C. offering Christian entertainment and news. It started in 1979 and is given without charge to cable systems.

SHOWTIME, a pay-cable network directly competitive with HBO, created in 1978 by Viacom. A few years later Teleprompter, one of the largest cable companies, purchased a half interest in Showtime and began offering it instead of HBO on many of its larger systems. Operating 24 hours a day and offering entertainment specials as well as movies, Showtime had close to 3 million subscribers by 1982. Its president, Mike Weinblatt, is former head of programs for NBC.

SIN (Spanish International Network), Spanish-language network which carries some direct transmissions from Mexico and offers Latin American serials, sporting events, specials, variety shows, and news. Begun in the 60s on commerical UHF stations in cities with large Hispanic populations, it became also a cable satellite network in 1976. It carries advertising and is offered to cable systems free.

SPN (Satellite Program Network), a Tulsa-based network that provides paid access to a variety of programmers offering family entertainment, public affairs, and foreign fare. Its offerings include *Telefrance USA, Ireland's Eye, Vision of Asia* and an Italian program, *Studio I.*

TBN (Trinity Broadcasting Network), a network of Christian programming with headquarters in Santa Ana, Calif. It began in 1978 and is given to cable systems without charge.

USA NETWORK, one of the few major cable networks offering a wide spectrum of programming, from sports to fashions and children's shows. Begun in the late 70s as a partnership of Madison Square Garden and UA-Columbia Cablevision, it became the USA Network in 1980 with Kay Koplovitz as president and with headquarters in Glen Rock, N.J. Time Inc. purchased the UA Columbia interest in 1981. The channel carries Madison Square Garden sports, Thursday night baseball, Monday night hockey, *The English Channel* (programs from Britain's Granada Television), *Calliope* (children's programming), *Night Flight* (rock music), and *Alive and Well* (health and physical fitness). It is supported by advertising and small fees charged to cable systems for offering it as a basic service.

While not strictly networks, there are, in addition, the three so-called *superstations*—WTBS Atlanta, WGN Chicago, and WOR New York—which become national stations by nature of being distributed in cable systems by satellite. Cable operators pay fees to offer them as part of the basic service. Mainly they are popular because they carry sports and vintage movies. Ted Turner's WTBS, the most widely-carried superstation, went on the satellite in 1976 by way of Southern Satellite Systems. United Video put WGN on the satellite in 1978, and Eastern Microwave began transmitting WOR in 1979.

There are also several educational cable networks, among them American Educational Television Network (AETN) and Appalachian Community Service Network (ACSN). Several other satellite services carry news wires and other textual information. These include UPI Cablenews Wire, North American Newsline, Dow Jones Cable News, and CableText.

At deadline for this edition of the encyclopedia, several new networks were preparing to launch.

DISNEY CHANNEL, a pay-cable service for family viewing drawing most of its programming from the vast Walt Disney library. The network is a joint venture of Group W Satellite Communications and Walt Disney Productions.

Also on tap were the CABLE HEALTH NETWORK, headed by Jeffrey Reiss; CINEAMERICA, a channel of varied programming for people over 45; DAYTIME, based on materials from the Hearst magazines, aimed at women, a partnership of ABC Video and Hearst Corp.; DELTA, an all-

sports network being set up by ABC Video and Getty Oil; THE NASHVILLE NETWORK, featuring country music programming; PENTHOUSE ENTERTAINMENT TELE-VISION NETWORK, owned and inspired by *Penthouse* magazine; the satellite news channels of Group W and ABC; UTV (Ultimedia Television), a channel of consumer interest programming emphasizing subscriber participation; and THE WEATHER CHANNEL, continuous coverage of national, regional and local weather.

Under the partnership, Fox would contribute to the programming of CBS Cable while CBS Theatrical Films would gain an established outlet for home video. Also envisioned is the creation of a CBS Fox pay-cable service.

CABLE RELAY STATIONS (CARS)—microwave relay facilities requiring FCC approval that are constructed for the passing of television signals from one cable system to another. They are used both for interconnection and for the delivery of distant market signals to the cable system.

CABLE-TELEVISION—technology of transmitting program signals by coaxial cable, whose broadband capacity makes possible a vast number of channels in a variety of modes. Because cable-TV not only retransmits ordinary broadcast signals to improve their reception but also makes possible a multitude of specialized and localized services, it is the technology most threatening to conventional broadcasting with its spectrum scarcity and mass-appeal standards.

After several false starts in its quarter-century of development, cable caught fire in the mid-70s chiefly for the appeal to consumers of pay-cable channels that deliver movies to the home, uncut and without commercials. The spur to the expansion of cable in the cities was the conversion of Home Box Office (HBO) into an instant national network when it began distributing its programming by satellite in 1975. Its wide acceptance was a signal to entrepreneurs of a monumental cultural change: the willingness of people to pay for television.

A rainstorm of other cable networks quickly followed, some offering pay services, some advertiser-supported programing. But the importance of the development was that cable at last had something to sell in the major cities—something more than better television reception. By the 1980s, cable was in demand everywhere in the U.S., and virtually every large city was either being wired for cable or in the thick of the franchising process. Most cities were demanding large-capacity systems with two-way capability.

By 1982, there were more than 4,700 cable systems in operation with more than 23 million subscribers, or close to 30% of all television households. Cable thus became a medium taken seriously by advertisers, and the satellites became saturated with cable networks.

While over-the-air television is restricted by mileage separation standards to no more than 7 VHF stations in a market and by economics to a sprinkling of UHFs, cable immediately lends itself to 30, or even 100 channels. And when cable is supplanted by the lightwave technology of fiber optics, as is expected, the capacity easily increases to 1,000 channels. Moreover, sophisticated interactive systems are being constructed that permit two-way communications. In these systems, the TV viewer may send digital responses to the origination point of the transmission. This is useful for polling, adult education, playing games, the ordering of merchandise, and home security services.

Cable communications may be offered in the broadcast mode, reaching all subscribers at once, or point-to-point, as in the interconnection of a medical center to an outlying clinic. Its multiplicity of channels opens television to common carrier use and in theory provides long-sought access to the medium for minorities, independent producers and private parties who have felt shut out of conventional TV.

But cable-TV must be purchased, its subscribers charged a monthly fee ranging from $5 to $10 for the basic service, while over-the-air TV is delivered free. The reluctance toward paying for TV when free service sufficient for most people already enters the home, and the great cost of constructing cable systems, were among the reasons for the relatively slight penetration of cable in the U.S. its first quarter century. Aside from these factors, the growth of cable had been hampered by the lack of a broad and enlightened national policy. Cable-TV struggled against the complexities of regulation at three tiers of government—federal, state and municipal—and against the political pressures exerted by the vested interests that are most wary of cable, commercial broadcasting and the telephone companies.

The FCC, which finally assumed regulatory jurisdiction over cable-TV in 1966 at the urging of Congress, is also the agency regulating the broadcast and telephone industries. Its policies for cable have in the main been forged in attempts to be even-handed with all three industries.

Cable's beginnings were humble. In the late 40s, when it was known as CATV, it was nothing more formidable than a master antenna system for communities with little or no TV reception. Signals picked off the air by an antenna situated on high ground were carried to homes by cable systems concentrated in small-town America.

Not until the 1960s did the realization dawn that cable might ultimately serve as more than a retransmission system—that it might function as an independent medium in its own right. The original systems used unsophisticated cable which could carry only a few channels, mostly devoted to picking up nearby broadcast signals. But during the late 50s and early 60s, technical improvements made it possible for systems to carry as many as 12 channels of programming. As a result, cable operators began to seek new sources of programming to fill their new channels, turning to origination of their own local programs and, more extensively, to broadcast television signals from distant cities. Suddenly, cable television was transformed from a reception-improving system into a medium which could offer subscribers more program choices than were available off the air even in areas with a full complement of broadcast stations.

The transformation launched the industry into a new wave of growth between 1960 and 1965. Some of the original "Mom & Pop" cable systems were absorbed into larger companies, which became known as MSOs (multiple system operators). The greater revenues of these companies and their ability to attract heftier financing enabled them to actively seek new franchises in communities spread across the country and to build larger and more sophisticated systems. Cable began to infiltrate suburban areas around some major cities—areas of *good* local television reception.

Not only did cable-TV attract venture capital, it excited interest among futurists, city-planners, social scientists and arts promoters because it promised a cornucopia of educational, cultural and social services. Foreseeably, on bi-directional systems, those services were likely to include schooling, banking and shopping by TV, fire and burglar alarm protection (the TV set in effect watches the house), automated meter reading, facsimile print-outs of newspapers and even the delivery of mail. But more immediately, the channel capacity of the prevailing one-way systems made possible separate channels for use by local government, school systems and business institutions, stations for the elderly conveying health-care information, neighborhood stations, ethnic stations, free-speech public access stations, closed channels for data transmission and channels carrying pay-television.

In the practical world, however, cable-TV proliferated on the strength of the entertainment channels it provided. Canada, for example, has experienced greater cable penetration than any country in the world—by 1979, more than half of the country was hooked up—and the single reason for cable's rapid growth there was that it brought in the over-the-air signals of U.S. stations.

Cable systems are also able to import an unlimited number of distant radio signals. Some cable systems import as many as 15 or 20 radio stations.

Cable began to spread rapidly in the U.S. in the 60s, not for its promise of new communications services but for its ability to import distant TV signals and to equalize VHF and UHF (making them equally accessible on the tuner). Communities that were unable to receive ABC and had no local independent stations (desirable for their movies and sports programming) were prime markets for cable. During the 60s the industry burgeoned from the 70-odd systems of the previous decade to more than 800 cable systems, and it enlisted more than 1 million subscribers. The MSOs then moved upon the big cities, but just as that frontier was opening the FCC effectively froze the development of urban cable with stiff regulations on the importation of distant signals while it pondered rules for cable in the top 100 markets.

The rules were promulgated finally in the 1972 Report and Order, but developments in the meantime had dampened the enthusiasm of cable entrepreneurs and lending institutions for wiring the cities. The MSOs that had begun building big-city cable systems before the freeze found striking differences between urban and small-town cable, and several were in a financial struggle as a result. The two Manhattan cable systems, which had expected to produce a bonanza, were each running up operating losses of $3 million a year. Along with unforeseen construction costs and impediments created by the telephone company, they encountered landlords seeking payoffs for entry to their property, neighbors of subscribers stealing the service by tapping into the feeder line, thefts of the converters, acts of vandalism and innumerable parking tickets during service calls.

But even more significant was the indifference of most New Yorkers to the offer of cable service, even with the inducement of a channel presenting Madison Square Garden events, including professional hockey and basketball.

By 1974 it was clear that basic cable would not suffice for the cities and that a special mass-appeal service was needed. That role fell to pay-cable, which was developing nicely in California and in a three-state area on the Atlantic coast. It was reasoned that urban residents would subscribe to cable for the opportunity to receive major boxoffice events— chiefly movies and sports—despite the fact that they would be faced with two fees, one for cable service and another for the pay channel. A selling point was that the pay-cable movies were presented numerous times, so that they might be watched at the subscriber's convenience, and without editing or interruption for commercials.

At that point, the cable industry appealed to the FCC to ease its restrictions on pay-cable, and was joined in the effort by representatives of the motion picture industry who saw in cable a potentially lucrative new market for their products. Among other restrictions, the existing FCC rules had prohibited cable from using new movies after they were more than two years out of theatrical release, which caused many to be lost to them entirely.

The request unleashed the broadcast industry's heaviest propaganda barrage against cable. With a war chest of around $1 million, the NAB—backed by campaigns of the networks—lobbied before the public, warning that if the pay-cable rules were eased the new medium would siphon away commercial TV's most popular programs, and people would be made to pay for what they now received free. The industry bore down heavily on the theme that the poor, the children and the elderly would suffer most. Typical was a CBS ad with a cartoon of a little boy saying, "Daddy, can I have a dollar to buy 'Gunsmoke'?" The FCC's modified rules for pay-cable issued in 1975 satisfied neither the broadcast nor the cable industries, and both appealed the commission's action in court. In March 1977 the D.C. Court of Appeals held that the rules were unconstitutional and ordered them vacated; the Supreme Court upheld the ruling.

Meanwhile, an eastern pay-cable distributor, Home Box Office, which presented a daily channel of fare for a monthly fee, arranged to disseminate its programming to all parts of the country by domestic satellite. Agreements from several major companies, including Teleprompter, to put HBO on their local systems gave birth to the first national pay-cable network.

The introduction of satellite transmission and its growth between 1975 and 1979 marked a new stage of development for cable-TV. One of the key impediments to the growth of original cable programming throughout the 1970s had been the lack of a low-cost national distribution system. Since 1975, however, a variety of new program services have been developed for satellite transmission, ranging from the Cable-Satellite Public Affairs Network, which offers gavel-to-gavel coverage of the House of Representatives from Washington, to Nickelodeon, a full channel of original children's programming.

Although the estimated costs for wiring a city are in the $14–$20 million range, and the cost for wiring the entire nation has been estimated at $123 billion, few doubt the eventuality of wired cities and a wired nation. The question is not whether but when.

Cable has raised a swarm of problems. The wire leading into the home will tell what people are watching, opening a Pandora's box on the right of privacy. Copyright questions arise with the importation of signals. The common carrier virtues of the medium—giving people the uncensored right to broadcast for whomever may be watching—has the negative side of bringing offensive material into homes, which is sure to create enemies for cable. On the other hand, if cable is not permitted to be a common carrier, the owners of the systems stand to become more powerful than any media barons ever dreamed of. Finally, the threat to conventional television is not to be sneered at for the "blue sky" promises of cable that are likely to be cast aside later for lack of profitability.

Cable enthusiasts have called the medium a better mousetrap and the most significant development in communications since the telephone. Nicholas Johnson, the former FCC commissioner, said that broadcasting is unto cable what the garden hose is to Niagara Falls. The landmark Sloan Commission Report on Cable Communications (1971) hailed the medium as "the television of abundance" and concluded that the encouragement of cable-TV's growth would be in the public interest.

See also Pay-Cable.

CABLE-TELEVISION INFORMATION CENTER OF THE URBAN INSTITUTE (CTIC)—advisory group in Washington, D.C., representing the public interest in matters involving cable-television. Established in 1971 by grants from the Ford and Markle Foundations, the Center continues to provide technical assistance to states and municipalities on cable development, particularly in assisting officials involved in the process of franchising systems, evaluating ordinances and analyzing information to deal with the legal and technical complexities of cablecasting.

CABLE-TELEVISION RULES—with the Cable Television Report and Order, Feb. 12, 1972, the FCC provided the first comprehensive body of rules for television-by-wire, ending the four-year freeze on cable in the top 100 markets. In July, the rules were amended slightly by the FCC's Reconsideration of Report and Order, and there were other revisions in 1974 and 1976.

Prior to 1972, the commission had regulated cable for about ten years in an ad hoc fashion that tended to restrict the growth of the new telecommunications industry. The 1972 rules were designed to remove some of the restrictions on cable while still minimizing the impact on related industries, particularly broadcast television.

The commission divided the rules into two categories—those signals a cable system "must carry" and those it "may carry." The must-carry signals are those of all TV stations licensed within 35 miles of the cable system, noncommercial stations within whose Grade B contour the cable system's community is located, and other stations whose signals are "significantly viewed" in the cable system's community.

In the may-carry category, the FCC limited the number of distant signals—those not emanating from the immediate area. In the top 50 markets, cable systems were to bring in distant signals sufficient to fill the complement of three full network stations and three independent (nonaffiliated) stations. In TV markets 50–100, three networks and two independents. In smaller markets, three networks and one independent.

No limits were imposed on the importation of foreign language stations or educational (public) stations, provided that, in the latter case, there is no objection from the local noncommercial authority. The initial rules specified that independent stations imported to fill the complement had to be taken from the two closest of the 25 top markets. These so-called "leapfrogging" rules were deleted in 1976, and cable systems may now carry network-affiliated stations or independents from anywhere in the country.

For the protection of local broadcasters, the report in 1972 included a "program exclusivity" rule under which the local TV station could require the blackout of an imported program on cable that was already being carried locally by the station. Cable systems in the top 50 markets were not to carry syndicated programming on a distant signal for a period of one year from the date that such programs are sold for the first time in the U.S. Nor could they carry a syndicated program to which a commercial station in the community had exclusive broadcast rights. These rules no longer exist.

But the rules that protect a local station's sports blackout still stand. They prohibit cable systems from carrying a local sports event carried on a distant signal which could not be broadcast locally under a blackout policy.

All major market systems were required to be built with a capacity for at least 20 channels and a technical capability for two-way service (nonvoice return communications); systems are not required to activate the bi-directional capability but are to have the line space available for possible future use. These rules were also abandoned later.

For program origination, major-market systems were required under the rules to provide access channels for use by the public, local government and educational institutions, but these rules were struck down by the Supreme Court in 1979 as beyond the commission's authority. However, local governments still have the authority to require access channels and production facilities in offering new franchises, and cable systems operating under existing franchises which specify access requirements must continue to fulfill them.

The 1972 Report and Order was under scrutiny for two years in the FCC's *Inquiry into the Economic Relationship between Broadcasting and Cable Television,* in which the commission made its first serious attempt to determine the extent of cable-TV's potential impact on conventional television. As a result of that inquiry, the commission initiated a proceeding in April 1979, which would eliminate its signal carriage and syndicated exclusivity rules, allowing cable systems to import programming on distant-city television signals without restriction. In 1980, the FCC struck down these rules. See also Cable-TV Regulation, Inquiry into the Economic Relationship Between Broadcasting and Cable-Television, Pay-Cable.

CABLE-TV REGULATION—more complex than that for conventional television because it is accomplished on the municipal level as well as the federal, and in some cases even on the state level.

Municipalities are involved because they, logically, are best qualified to award franchises to serve the specific needs of their communities and because they govern the use of the streets and ways required for building cable systems. Local governments, through their power to grant franchises, thus are in a position to determine who may enter the cable business.

Federal regulation began effectively in 1966, under pressure from local broadcast stations and regulators concerned with issues of copyright, technical standards and the protection of over-the-air television. Although the commission has no express authority over cable-TV under the Communications Act of 1934, written before the medium's inception, the Supreme Court in 1968 ruled that the FCC could regulate the industry to the extent that it was "ancillary to broadcasting.

Between 1966 and 1972, the commission followed a zig-zag regulatory course in an attempt to discover some scheme which would protect broadcast TV from the assumed dangers of cable competition but at the same time allow the medium to grow. Finally, in 1972, the commission issued a comprehensive *Report and Order* which protected local broadcast stations by limiting the number of distant signals cable systems could carry and requiring the blackout of distant programming under contract to local stations; required that cable systems provide access channels and production facilities for use by the public, local government and educational institutions; and established uniform technical standards for systems to be constructed.

Since 1977, the commission has taken significant steps toward deregulating the industry. After two years of study, the FCC, in April 1979, concluded that cable-TV could compete freely with broadcasters without endangering continued over-the-air service and began proceedings to eliminate its distant signal limits and blackout requirements, called syndicated exclusivity rules. In addition, the commission in 1978 dropped its cable television certification procedure—somewhat similar to broadcast licensing—in favor of simple registration by new cable systems.

In 1979, the Supreme Court struck down the 1972 access rules as beyond the commission's statutory authority. However, cable systems operating under franchises which require them to provide access channels and maintain production facilities for public use must continue to honor the franchise provisions. In addition, neither states nor local governments are restricted from requiring access facilities and channels in new franchises or through legislation.

By 1980, the FCC's role in cable had been pared down to a handful of rules. It eliminated rules requiring systems to be built with two-way capability, a minimum of 20 channels, and provisions for public access,leaving those to the franchising process. But it kept the rules banning duplication of network programs and the "must carry" rules which require cable to carry all "significantly viewed" local stations.It also continued to impose a limit of 5% on franchise fees.

State commissions began to be formed after the rash of cases involving bribes, payoffs and trafficking in licenses. By 1976 twelve states had established commissions or agencies to regulate cable. New York, Massachusetts and Minnesota created cable commissions, while Alaska, Connecticut, Delaware, Hawaii, Idaho, Nevada, New Jersey, Rhode Island and Vermont undertook regulation through agencies of their public utilities commissions.

State commissions have set standards to prevent the unsophisticated deals made by gullible municipalities in the early years—such as granting franchises for 60 years or more—and to protect consumers by requiring direct citizen involvement in the franchising process. The rationale for state regulation is that it provides balance to the FCC's over-protection of conventional broadcasting and that it has the management and technical resources that are lacking in most municipalities. See Cable-TV Rules, Pay-Cable, Inquiry into the Economic Relationship between Broadcasting and Cable-Television, Midwest Video Case.

"CADE'S COUNTY"—hour-long western series on CBS (1971) starring Glenn Ford as a marshal and featuring Edgar Buchanan, Peter Ford and Sandra Ego. It was produced by 20th Century-Fox TV.

CAESAR, SID—sketch comedian, pantomimic and satirist who reigned as one of TV's stars from 1949 to 1954, chiefly on *Your Show of Shows,* a 90-minute Saturday night series on NBC. A comic of uncommon versatility and ingenuity, he had a special appeal to the literate viewer and may have been, like studio drama, a casualty of the proliferation of TV receivers into the lower income, lesser educated homes during the mid-50s.

If Caesar was celebrated for his comic range and inventiveness, he was faulted for overworking his funniest concepts until they palled. While he was at his height his female co-star was Imogene Coca. The popularity of both performers declined when their partnership ended. Miss Coca was succeeded on Caesar's shows by Nanette Fabray, Janet Blair and Gisele MacKenzie. Also regularly featured with Caesar were Carl Reiner, Howard Morris and Marguerite Piazza.

An attempt to reunite Caesar and Coca in 1958 was unsuccessful and Caesar's TV appearances since then have been chiefly guest shots.

"CAIN'S HUNDRED"—series about a young lawyer who heads a government law enforcement agency. It aired on NBC in 1961 and featured Mark Richman. Produced by MGM-TV with Paul Monash as executive producer, it yielded 30 episodes.

"CALL IT MACARONI"—quality children's series produced by Group W for syndication starting in January 1975. Aimed at the age group of 10–12, the series explored the diversity of America and the varying life-styles of children around the country. Episodes focused on a traveling tent circus, an Indian village, a climb up Mt. Hood and dogsledding in Alaska. George Moynihan was executive producer.

CALL SIGNS (Call Letters)—distinctive combinations of letters by which stations are identified, normally consisting of four letters (followed by -TV), although stations descended from the earliest radio operations may retain three. By international agreement, the broadcast stations of a nation were identified by the first letter, or the first two letters in some cases, of their assigned call signal. Under the agreement, the U.S. was apportioned call letters beginning with K or W. (Canadian stations begin their call signs with C and those of Mexico with X.

In the U.S. most stations east of the Mississippi River must begin their call signs with W, those west of the Mississippi with K. (KDKA Pittsburgh is one of several exceptions, its calls having been grandfathered because they were assigned before the system went into effect.) Some call signs contain the initials of people (WPLG—Philip L. Graham), some code initials for the ownership group (WKBD, WKBG, WKBF, all Kaiser Broadcasting stations) and some stand for a slogan (WGN, World's Greatest Newspaper—it was founded by the *Chicago Tribune*). Other call signs are selected because they spell out words for easy identification, e.g., KAKE, KISS, KEEN, WAVY. Call signs are authorized by the FCC and may be changed with FCC approval if they do not duplicate another or do not create confusion with a similar-sounding call in the service area.

The suffix "TV" is required with common AM radio and TV call letters; in other instances stations may opt to use the "TV" suffix, upon FCC approval.

"CALUCCI'S DEPARTMENT"—situation comedy centering on the lives of lowly employees in a large concern. Created by Renee Taylor and Joseph Bologna, and produced in New York by Sullivan Productions, it had only a 13-week run on CBS in 1973. Featured were James Coco, Candy Azzara, Jose Perez, Jack Fletcher, Peggy Pope and Bill Lazarus.

"CAMEL NEWS CARAVAN"—NBC's 15-minute early evening newscast presented by John Cameron Swayze that ran from 1948 to 1956 when it was replaced by *The Huntley-Brinkley Report*. Swayze, dapper and suave, became known for the lines used in every program: "A good good-evening to you," "hopscotching the globe" and "Glad we could be together." The program, sponsored by Camel cigarettes, competed with CBS's *Douglas Edwards with the News*.

The termination of the *Camel News Caravan* marked a turning point in TV journalism. After it came the network

policies dissociating advertisers from the news product and the practice of using working reporters as newscasters.

"CAMERA THREE"—half-hour CBS cultural series on Sunday mornings (11 A.M.) that has maintained a high standard of quality on a spartan budget since its beginning as a local show on WCBS-TV New York in 1953. It became a network series in 1956 and soon won the admiration of critics for its innovative and imaginative presentations. Devoted to all the arts—dance, poetry, opera, theater, cinema—the series over the years has been a showcase for numerous performers who were later to become celebrated, among them Beverly Sills, Twyla Tharp and scores of actors. Under its original producer-director Robert Herridge, who helped to shape the series when it had to be presented live, the show leaned heavily to drama and poetry readings, and it undertook such ambitious projects as a six-part adaptation of Dostoevsky's *Crime and Punishment*. In latter years, its executive producer has been John Musilli.

It was canceled by CBS early in 1979, along with the long-running religious series, *Lamp Unto My Feet* and *Look Up and Live*, to make way for the new 90-minute newscast, *CBS News Sunday Morning*. Ironically, it was the only CBS program to win a Peabody Award in 1978.

Camera Three became a prime-time PBS series underwritten by Arco, in the fall of 1979. Musilli and writer Stephan Chodorov formed their own company, Camera Three Productions, to produce new segments, while CBS made available tapes of the old programs without charge. But the program had only a brief run on PBS.

CAMPBELL, NORMAN—Canadian director and producer of specials for the CBC. Associated with musicals and light cultural productions, he directed for CBS *The Gershwin Years* and for NBC *The Golddiggers in London*. In the area of ballet, he directed productions of *Sleeping Beauty, Cinderella* and *Romeo and Juliet,* which were shown in a number of countries.

"CAMP RUNAMUCK"—unsuccessful NBC situation comedy (1965) set in a children's summer camp. It featured David Ketchum, Arch Johnson and Nina Wayne and was by Screen Gems.

CANADA, TELEVISION IN—a multinetwork system operating on fewer than 100 stations. It has scores of translators and satellite stations and greater cable-TV penetration than any country in the world. Although vast in area, Canada has only around 8 million TV homes, a majority of them in a fairly narrow geographical strip along the southern border.

The chief networks are the CBC (Canadian Broadcasting Corp.), the state-supported service which grew out of CBC Radio, and CTV, the commercial network owned by independent stations. In the French-speaking regions, the CBC maintains a separate French network as well as an English-language network, and there are also smaller networks such as Global TV and provincial educational networks.

Because Canada has no national newspapers and few magazines, and because culturally it is overshadowed by the U.S., radio and TV are regarded by the government as the means to a distinct Canadian identity and to the unification of the country. CBC operates on an annual budget of $200 million—most of it by government appropriation, and feeds out approximately 40 hours a week of its own programming. Although hockey and variety programs tend to draw the highest ratings, CBC has attempted to nurture Canadian drama by setting up a special fund to support Canadian TV playwrights. CBC and its stations derive part of their funds from the sale of commercials.

Both CBC and CTV have been heavy purchasers of U.S. programming, but the reliance on imported fare prompted the regulatory agency, CRTC, to impose a quota system. It requires "Canadian content" for 50% of a station's overall programming and 60% during the prime-time hours. In general, U.S. programs—which are sold to Canada on a pre-release basis—have been among the strongest attractions on the Canadian networks.

Indeed, the desire for American programs is the chief reason for the proliferation of cable in Canada, which in 1976 passed 60% of the country's homes. Competition with U.S. stations (some of which are received over the air in certain Canadian cities) has made the Canadian broadcast mandate difficult to carry out and has been a source of great frustration to the CRTC.

With much of their audience and advertising revenues lost to U.S. broadcasters, the Canadian networks and stations have been unable to amass budgets to produce programs on a scale to compete with those of Hollywood. Moreover, scores of performers and directors have quit Canadian TV at the first opportunity to work in U.S. TV. The drain has been particularly great among producers-directors of variety shows.

During the 70s, the CRTC took steps to reverse the tide, including ordering cable systems to delete commercials in U.S. programs and substitute Canadian commercials instead —an act that was called "piracy" by American broadcasters and government officials. The conflict with U.S. TV escalated into a "border war" when the Parliament, to encourage the purchase of Canadian time, considered and then passed a bill denying business deductions to native advertisers who bought time or space in U.S. media.

See also Border War, CRTC, and CTV Television Network.

Mr. Candid Camera, Allen Funt

"CANDID CAMERA"—unusual CBS series (1960–67) which sought to reveal human foibles by using a hidden camera to catch the reactions of ordinary people to unusual situations. The situations were devised by Allen Funt, producer and creator of the series who also participated in the humorous deceptions, his ordinary looks and con-artist gifts making him perfect for the role. It was revived briefly in the early 70s in a new format and then was produced in half-hour form for syndication as *The New Candid Camera*. A new round of production for syndication began in 1979.

The show originated on radio in 1947 as *Candid Microphone* and became so popular that it spawned a series of movie shorts using film cameras in place of concealed mikes. The TV version began in 1948 with the title, *Candid Mike*, changed the following year to *Candid Camera*. It played sporadically until 1960 when it became a regular series on CBS. For several years it ranked in the Nielsen Top 10.

During that run, which lasted until 1967, Funt had a succession of co-hosts, among them Arthur Godfrey, Durward Kirby and Bess Myerson. CBS carried the series as a daytime strip in 1967. It was by Bob Banner Associates and Allen A. Funt Productions.

CANNELL, STEPHEN J.—writer and producer who polished his reputation on *The Rockford Files*, then became executive producer of *Baa Baa Black Sheep* and its successor, *Black Sheep Squadron*. Later, with his own production company, he created and produced *Tenspeed and Brown Shoe* and *The Greatest American Hero* for ABC. Although neither succeeded in the ratings, they led to an exclusive agreement with ABC to produce one series a year.

"CANNON"—hour-long private-eye series on CBS (1971–76) whose hero was distinctively overweight and middle-aged. William Conrad, who portrayed Cannon, made a comeback as an actor with the series. In the radio era, he had originated the role of Marshal Dillon in *Gunsmoke* but was unable to continue when that series moved to TV because he was the wrong physical type. He became a producer-director in TV and films until Quinn Martin chose him for the role in *Cannon*.

CANTOR, EDDIE (d. 1964)—radio entertainer who made the transition to television in 1950 as one of the rotating stars on *The Colgate Comedy Hour* on NBC. He remained with the show through four seasons despite suffering a heart attack after a performance during the second year.

A former Ziegfeld star, he became nationally famous on radio and in movies as a singer and funnyman. Cantor made an early television appearance, in March 1944, on the *Philco Relay Program*, which was beamed locally in New York by NBC. After the Colgate series he appeared frequently on TV as a guest and as the subject of tributes.

CAPICE, PHILIP—independent Hollywood producer associated with Lorimar Television, and executive producer of such Lorimar series as *Eight Is Enough, Dallas* and *Married: First Year.* A former CBS-TV program executive, he served as president of Lorimar for a year (1978–79), giving up the post voluntarily to return to production with his own company, Raven's Claws Productions, which has exclusive ties to Lorimar.

CAPITAL CITIES COMMUNICATIONS INC.—one of the largest and most prosperous broadcast groups, publicly owned, with less than 9% of outstanding stock held by its officers and directors. Founded in 1954 by Frank M. Smith, Lowell Thomas, Ellen B. Elliott and Alger B. Chapman, it makes its headquarters in New York City and added to its potency in 1971 by acquiring several stations of Walter Annenberg's Triangle group. In 1980 it purchased RKO Cablecom for $139 million, making its entry into cables. The TV group consists of WPVI Philadelphia, an ABC affiliate; WKBW Buffalo, ABC; KTRK Houston, ABC; WTNH New Haven, ABC; WTVD Durham, N.C., CBS; and KFSN Fresno, Calif., CBS. Officers include Thomas S. Murphy, chairman; Daniel B. Furke, president and chief operating officer; and Joseph P. Dougherty, John B. Fairchild and John B. Sias, executive vice-presidents.

"CAPITOL"—half-hour CBS daytime serial, built around politics and dealing with the public lives and private ambi-

tions of two prominent Washington, D.C. families. It premiered in March 1982 with John Conboy, who helped launch *The Young and the Restless* on CBS as executive producer. The cast includes Carolyn Jones, Ed Nelson, Rory Calhoun, Constance Towers, Kimberly Beck-Hilton, Bill Beyers and Leslie Graves.

"CAPTAIN AND TENNILLE, THE"—hour-long ABC music-variety series (1976–77) featuring the pop singing duo of Daryl and Toni Dragon, a husband-wife team who billed themselves as the Captain and Tennille on their hit recordings ("Love Will Keep Us Together"). At the start the series drew higher ratings than the competing *Rhoda* and *Phyllis,* which previously had been among the highest rated situation comedies on CBS. By Bob Henry Productions and Moonlight and Magnolias Productions, it was produced by Henry with Alan Bernard as executive producer and Tony Charmoli as director.

Bob Keeshan, Captain Kangaroo

"CAPTAIN KANGAROO"—a pillar of pre-schooler programming since its debut on CBS, Oct. 3, 1955, running more than 24 years without substantial change on weekdays in the 8 A.M. hour. For most of those years it was the only daily program for children on the networks.

The program was cut back to 30 minutes in the fall of 1981 in favor of the news program, *Morning,* which preceded it in the schedule. A few months later, *Kangaroo* was shifted to 6:30 A.M. to allow *Morning* to expand to two hours, os that it might be fully competitive with *Today* and *Good Morning America.* The shift was costly to *Kangaroo,* since a number of key affiliates refused to clear the new time period.

Hosted from its inception by Bob Keeshan, a low-keyed performer with a straightforward manner, the programs presented a form of situation comedy for the very young, involving other live performers, animals and puppets. In that, it represented a departure from children's programming that attempted to seduce an audience through loudness, silliness and animated cartoons.

Captain Kangaroo, as portrayed by Keeshan, was an avuncular, heavyset and easily gulled master of a bachelor household, identified by his uniform and cap and ample grey mustache. Seeming formless, the programs made use of music, dance and conversation, and they dealt both with fantasy and aspects of daily living.

A long-time producer of the show, David Connell, left in 1968 to produce *Sesame Street* for public television and took with him knowledge gained from *Captain Kangaroo* in communicating with young children effectively.

"CAPTAIN VIDEO"—early space-adventure series featuring the far-out weaponry and costumes reminiscent of Flash Gordon. Captain Video was played by Al Hodge. The series premiered on the DuMont Network in 1949. It ended in 1955 with the demise of the network. See also, Hodge, Al.

CAPTIONING—superimposing subtitles on TV programs at the bottom of the screen for the benefit of an estimated 13.4 million persons in the U.S. who are deaf or have hearing impairments. Broadcasters generally have resisted using "open" captions—lettering visible to all—because they tend to annoy viewers with normal hearing. But in the early 70s PBS began experimenting with electronic systems developed for "closed" captions—those that come into view only on specially equipped sets—and in 1976 received approval from the FCC for full-scale use of the system.

ABC, NBC and PBS all agreed to begin offering closed-caption programs in 1980 on a limited basis, using the same technology. CBS declined to go along, saying it was holding out for the development of a more sophisticated device that could also be used for teletext services. NBC dropped out in 1982.

The PBS system involves the use of Line 21 on the TV screen, the first nonvisual line above the picture, for the transmission of the captions, which are dropped onto the screen by the special decoder. In four years of experimentation with prototype decoders placed at various sites accessible to the hearing-impaired, PBS captioned such programs as *Upstairs, Downstairs* and *The Adams Chronicles*. They were unnoticed by viewers without the special unit.

Meanwhile, appeals from associations for the hearing-impaired have resulted in some use of open captions or in the use of inserts on the screen carrying a sign-language interpreter. These found their widest use in local five-minute morning newscasts.

WGBH, Boston's public TV station, took a major step in captioning in the early 70s by securing the right from ABC to repeat the network's evening newscast at midnight with bottom-of-the-screen open captions. The agreement also permitted the commercials to be deleted.

The captioned newscast then was sent out over the Eastern Educational Network, but the PBS stations in those markets were free to carry it only if they received permission from the local ABC affiliate.

"CARA WILLIAMS SHOW, THE"—a 1965 CBS situation comedy which failed to establish Cara Williams as heiress to the comedy mantle of the network's mainstay, Lucille Ball. Miss Williams somewhat resembled Miss Ball but the series fell short of her standard. It featured Frank Aletter, Paul Reed and Jeanne Arnold and was produced by Richelieu Productions in association with CBS.

"CAR 54, WHERE ARE YOU?"—half-hour comedy series about a team of inept police officers who share a patrol car. Shot in New York, it featured Joe E. Ross and Fred Gwynne and was produced by Nat Hiken for Euopolis Productions. It aired on NBC (1961–63).

CARLIN "DIRTY WORDS" DECISION—See WBAI Indecency Case.

CARLIN, STEVE—game-show producer who, in the mid-50s, mounted *The $64,000 Question* and *The $64,000 Challenge*. He revived *Question* for syndication in the 1976 season.

CARLINER, MARK—producer of TV movies, including *Nightmare, The Strangers in 7A, A Death of Innocence* and *Revenge*.

CARLINO, LEWIS JOHN—writer who first attracted attention on drama anthologies such as the *CBS Television Workshop* in the 60s. Although he worked primarily in motion pictures, his occasional TV ventures included the pilot script for *Doc Elliot*.

CARLISLE, WILLIAM—v.p. of government relations for NAB since 1974 and employed by the association in various capacities since 1954. Earlier he had been a radio executive in New Hampshire.

CARNEGIE COMMISSION REPORT—the publication of a study by a 15-member commission whose recommendations led to the Public Broadcasting Act of 1967 and caused educational television to become public television. The report, *Public Television: A Program for Action,* was the product of a two-year investigation of ETV begun in 1965 by a distinguished panel headed by Dr. James R. Killian, Jr., chairman of the corporation of MIT. The commission was funded by the Carnegie Corp.

Recommended was a system that would be devoted in the broadest sense to public service and cultural enrichment and that would have a national sweep while being essentially local-oriented. Central to the recommendations was the proposed creation of a corporation to serve the system (the Corporation for Public Broadcasting) and increased appropriations by all levels of government to support it. The report estimated that $270 million a year would be needed to maintain a strong national system, and it suggested an excise tax on the manufacture of TV sets to help in the funding. The Carnegie Report remained the gospel for public TV. A second Carnegie Commission, formed in 1977, examined the eleven-year history of public television and recommended a revamping of the system.

Other members of the original commission were James B. Conant, former president of Harvard; Lee A. DuBridge, president of the California Institute of Technology; Ralph Ellison, author; John S. Hayes, Ambassador to Switzerland; David D. Henry, president of the University of Illinois; Ovetta Culp Hobby, chairman of the Houston Post Co.; J. C. Kellam, president of Texas Broadcasting Corp.; Edwin H. Land, president of Polaroid Corp.; Joseph H. McConnell, president of Reynolds Metal Co.; Franklin Patterson, president of Hampshire College; Terry Sanford, former Governor of North Carolina; Robert Saudek, TV producer; Rudolf Serkin, concert pianist; and Leonard Woodcock, v.p. of United Automobile Workers of America. See also Public Television.

CARNEGIE II—short for the Carnegie Commission on the Future of Public Broadcasting, whose report issued in January 1979 called the system created by the first Carnegie Commission a failure and proceeded to make recommendations for a legislated restructuring of the public broadcasting apparatus. To head off any such legislation, the Corporation for Public Broadcasting and PBS both took steps to reorganize themselves somewhat along the lines charted by Carnegie II.

Headed by Dr. William J. McGill, then president of Columbia University, the 17-member commission also called for a commitment by the Federal Government to a larger, better insulated and better-financed noncommercial system than the existing one. It put the proper level of total support for public TV and radio at $1.2 billion annually, of which the Federal Government's share should be $590 million by the mid-80s. It recommended that some of this money come from a spectrum-use fee imposed on commercial broadcasters and other users of the public airwaves.

The commission's report, the product of an 18-month examination and evaluation of public broadcasting, urged that the CPB be eliminated and replaced by the Public Telecommunications Trust, as an organization serving as fiduciary agent for the system with no involvement whatever in programming. It proposed the creation of the Program Services Endowment, which would be a semi-autonomous division of the trust, to control a budget of $190 million a year that would be used almost exclusively for the development and financing of national programming.

Carnegie II was established in June 1977 on a $1 million grant from the Carnegie Corporation of New York. Its report was published as a paperback by Bantam Books under the title *The Public Trust.* The commission's members, in addition to Dr. McGill, were Stephen K. Bailey, Red Burns, Henry Cauthen, Peggy Charren, Wilbur Davenport, Virginia Duncan, Eli N. Evans, John Gardner, Alex Haley, Walter Heller, Josie Johnson, Kenneth Mason, Bill Moyers, Kathleen Nolan, Leonard Reinsch and Tomas Rivera.

CARNEY, ART—versatile actor who enjoyed great popularity as Jackie Gleason's sidekick Ed Norton, the sewer-worker, in *The Honeymooners.* Carney was able to surmount a legendary identification with that role and to perform serious roles in the drama showcases of the 50s. He played the lead in a TV production of *Harvey* and was host in a production of *Peter and the Wolf.* He went on to star in movies and Broadway shows, yet for all his success he never starred in a TV series of his own, although he was a regular in NBC's *Snoop Sisters* which starred Helen Hayes.

Carol Burnett

"CAROL BURNETT SHOW, THE"—durable comedy-variety series on CBS in which Carol Burnett established her preeminence in the vaudeville style of comedy. Premiering in the fall of 1967, the program held up steadily in the ratings and enjoyed a harvest of Emmy Awards. Major support came from comedian Harvey Korman. Other regulars were Lyle Waggoner, Vicki Lawrence and the Ernest Flatt Dancers, with sketch comedian Tim Conway joining in the 1975 season. Executive producer was the star's husband Joe Hamilton, and the producer was Ed Simmons. It was via Punkin Productions.

Korman left the show in 1977, but his place in the regular cast was taken by Dick Van Dyke. The show was terminated in 1978 and the reruns successfully syndicated, edited down to a 30-minute form.

"CARRASCOLENDAS"—bilingual (Spanish-English) children's series for PBS produced for the system by KRLN-TV San Antonio-Austin, Texas, on funds from the U.S. Office of Education. It premiered in 1972 with Aida Barrera as executive producer and Raoul Gonzales as story editor.

CARR, MARTIN—producer of muckraking documentaries for CBS News (1963–69) and NBC News (1969–74). Earlier he produced cultural documentaries and children's public affairs shows for CBS. Among his notable works were *Migrant* for NBC (1970) and *Hunger In America* for CBS (1968). He also produced and directed *Five Faces of Tokyo, Search for Ulysses* and *Gaughin in Tahiti,* among other films.

CARROLL, SIDNEY—writer of dramatic specials, noted for adaptations of classic material. His credits include the teleplay for the *Beacon Hill* pilot.

CARRUTHERS, BILL—producer-director with credits in *The Newlywed Game, The Dating Game, Operation Entertainment* and others; he was executive producer of *The Johnny Cash Show* on ABC in 1970. In 1976 he became TV advisor to President Ford.

CARSEY, MARCIA—senior v.p. of prime-time series for ABC Entertainment from June 1979 to late 1981, when she went into independent production. She is credited with having participated in the incubation of such shows as *Barney Miller, Taxi, Three's Company* and *Mork & Mindy.* She joined ABC in 1974 as a general program executive, after having worked for Tomorrow Entertainment as executive story editor, William Esty Advertising as program supervisor and NBC's *Tonight.*

Johnny Carson

CARSON, JOHNNY—perennially boyish comedian whose major TV assignments were all outside prime time, a fact which didn't prevent him from becoming one of the medium's most popular personalities and consistently its highest paid talent. By the 70s NBC was paying Carson more than $2 million a year to continue as host of *The Tonight Show,* a job he began in 1962. With each contract period the network made greater concessions to him for vacation time.

Most of his earlier shows were daytime series in the 50s, principally standard quiz programs such as *Who Do You Trust?* and *Earn Your Vacation.* His qualifications for *Tonight* were demonstrated, however, in a daytime *Johnny Carson Show,* which began in 1955.

Witty, urbane and cool, Carson proved an even more effective host of the late-night show than his predecessor Jack Paar. A chief strength was his opening monologue; another, from a ratings standpoint, his penchant for slightly off-color material. Carson came to represent on TV the clean-cut Midwesterner with a bad-boy streak, just mild enough to make him endearing. He carefully avoided intellectual pretensions and usually drew his guests from show business.

In the spring of 1979, Carson elevated the crises NBC was facing on a number of program fronts by announcing that he intended to leave before the termination of his contract in 1981. The news made the front pages across the country and was covered in the network newscasts and on *60 Minutes.* The episode served to illuminate Carson's importance to NBC as a major source of income both to the network and its affiliated stations. It was estimated that his nightly program produced 17% of the network's total revenues. The concession in Carson's new contract in 1980 was that the program be cut down from 90 to 60 minutes a night. Early in 1982, Carson's ratings began to slip, and

NBC was faced with trying to prevent affiliates from delaying the program in favor of syndicated fare. See also *Tonight*.

"CARTER COUNTRY"—ABC sitcom (1977–79), set in a red-neck town, which was picked for the schedule—perhaps coincidentally, perhaps not—shortly after President Carter took office. The situation concerned a young New York City policeman, who is black, going to work for an old-fashioned police chief in a southern burg. Kene Holliday played the Yankee and Victor French the Southerner. Featured were Barbara Cason, Guich Koock, Richard Paul and Harvey Vernon. It was via T.O.Y. Productions, with Bud Yorkin, Saul Turteltaub and Bernie Orenstein as executive producers, and Douglas Arango and Phillip Doran as producers.

CARTER-FORD DEBATES—a series of televised confrontations in 1976 between the major-party Presidential candidates—the incumbent President Gerald R. Ford and his Democratic challenger, Gov. Jimmy Carter—modeled somewhat on the Kennedy-Nixon "Great Debates" of 1960 and staged under auspices of the League of Women Voters. While the Nixon-Kennedy debates required a special act of Congress suspending the Equal Time Law just for that campaign, the Carter-Ford debates were born under a new interpretation by the FCC of the equal time requirements.

Just before the election year opened, the FCC determined that political debates could be exempt from equal time if the confrontations were arranged by independent organizations and the debates held outside the television studios. If two candidates agreed to meet to debate the issues before a live audience, the rationale went, it would be appropriate for television to cover the occurrence as a legitimate news event. The FCC stipulated, however, that television would have to cover the event live and in its entirety—no delayed telecasts and no presentation of excerpts.

With Jim Karayn spearheading the effort, the League of Women Voters offered to become the organization inviting the debates. The networks, which were still seeking another suspension of the equal time law, had to give up trying when the candidates accepted the League's proposal. All three commercial networks and PBS carried the debates simultaneously and among them reached 90 percent of the households (per Nielsen TA ratings) with the four telecasts. Three of the four involved Ford and Carter; debate No. 3 was between the Vice-Presidential candidates, Robert Dole and Walter Mondale.

The first debate, airing on Thursday, Sept. 23, from Philadelphia, ran two hours rather than the scheduled 90 minutes because of a protracted audio outage caused by the failure of a simple electronic part. ABC provided the pool coverage. The telecast had a total audience of 51.6 million

households. According to Nielsen, the average duration of tuning was 88 minutes.

The second debate, on Wednesday, Oct. 6, from San Francisco, ran 90 minutes and had a total audience of 46.5 million households. The fourth, on Oct. 22 from Williamsburg, Va., also ran 90 minutes and drew 42.5 million households. The third, between the Vice-Presidential candidates, aired on Friday, Oct. 15 from Houston and was produced for the pool by PBS. It had a total audience of 35.6 million households and ran 75 minutes.

While the debates were generally considered to have been a draw, the eventual election of Carter was probably foretold in the ratings for the paid political programs each candidate bought on all three networks on election eve. Carter and Ford purchased alternate half hours from 8 to 11 P.M., playing the same program once on each of the networks at different hours. Carter had a gross rating of 34.2 for his three telecasts and Ford 25.3. In each hour, regardless what was competing on the other networks, Carter beat Ford.

CARTER-MONDALE DECISION—case which affirmed the right of "reasonable access" to television by candidates for federal offices and established that broadcasters may not decide for candidates when their campaign periods shall begin.

In the fall of 1979, the Carter-Mondale committee was turned down by all three networks when it sought to buy 30 minutes of prime time for a special program to kick-off the President's reelection campaign. The networks, which were in the midst of their own autumnal ratings campaigns, argued that the request was unreasonable, since political campaigns normally begin during the year in which the elections are held and not an entire year before.

The question of "reasonableness" was key, because Congress in 1971 had amended the Communications Act to prevent broadcasters from restricting access to candidates. The amendment gave the FCC authority to revoke the license of a broadcaster who would deny a "reasonable" request for access from a candidate for a federal office.

When the FCC ruled that the Carter-Mondale request was reasonable and that the candidates' needs were paramount to those of the broadcasters', the case went to the Court of Appeals for the District of Columbia, which concurred with the commission. Later the Supreme Court upheld the FCC ruling and the 1971 amendment—Section 312 (a) (7)—which states that the statutory right of access properly balances the First Amendment rights of federal candidates, the public, and broadcasters.

CARTER MOUNTAIN DECISION [Carter Mountain Transmission Corp./32 FCC 459 (1962)/Off'd 321 F2d (D.C. Cir. 1963), cert. den./375 U.S. 951 (1963)]—case in which the FCC refused to authorize a CATV system that would import distant signals to a rural area where the sole existing television station would, as a result, be driven out of business. The D.C. Court of Appeals affirmed the commission's decision.

The Carter Mountain Transmission Corp. applied to the FCC for a permit to construct a microwave radio communication system to transmit signals received from TV stations in distant cities to CATV systems in Riverton, Lander and Thermopolis, Wyo. The only television station in the area, KWRB-TV in Riverton, protested, saying that the importation of distant stations would cause its demise.

After examining the evidence the FCC concluded that KWRB's protest had substance and that the importation of distant signals would leave the residents of the three towns without a local TV station. The commission decided that it was better to have a single local TV station than several distant ones which would only be available to subscribers. When the FCC denied Carter's application Carter appealed.

The Court of Appeals held that the FCC had the power to deny the application and that the denial did not amount to censorship. The Supreme Court refused to hear the case.

"CASE AGAINST MILO RADULOVICH, THE"—one of the famous Edward R. Murrow documentaries in the *See It Now* series. It aired on CBS Oct. 20, 1953, preceding by almost six months Murrow's devastating half-hour *Report On McCarthy*. McCarthyism was actually the theme of this 30-minute program, which focused on the case of an Air Force Reserve officer forced as a security risk to resign his commission because of anonymous charges that his father and sister harbored pro-Communist sympathies. Murrow's editorial stand was bold and controversial, and his defense of Lt. Radulovich against guilt-by-association reasoning led the Air Force to reverse itself and reinstate him as an officer. The program, produced by Murrow and Fred W. Friendly, was a prelude to the expose on McCarthy. See also Murrow, Edward R.

CASTLE, WILLIAM—movie producer *(Rosemary's Baby)* who has drifted into TV from time to time as an executive producer, such as for NBC's *Ghost Story* series (later retitled *Circle of Fear*).

CATES, GILBERT—producer-director of TV, films and the stage. For TV he was producer in the 70s of such CBS dramas as *To All My Friends on Shore* and *I Never Sang For My Father* and Arthur Miller's *After the Fall* (NBC). Earlier he was producer-director of *International Showtime* (1962–65) on NBC and *All-Star Circus* for Timex.

CATES, JOSEPH—director specializing in music-variety whose credits include specials with Anne Bancroft, Gene Kelly, Victor Borge, Ethel Merman and Yves Montand. He was associated with his brother, Gil Cates, in *International Showtime,* which had a three-year run on NBC in the early 60s. On Broadway, he produced *Joe Egg, What Makes Sammy Run?* and *Spoon River.* In recent years he has been involved in country music TV specials originating in Nashville.

CATHODE RAY TUBE (CRT)—an electron tube designed for graphic display, of which the television picture tube is one type. Invented in 1897 by Karl Ferdinand Braun of the University of Strassburg, the CRT was in the 20s converted by Philo T. Farnsworth and Vladimir Zworykin, working independently, into a device to display television pictures.

A black-and-white picture tube has three principal components: an electron gun, a deflection system and a phosphor-coated screen. A stream of electrons flows from the gun to the screen and is deflected from left to right and from top to bottom. The beam activates one tiny spot of phosphor at a time, causing it to glow, but the persistence of the phosphor makes the entire screen appear to glow. The intensity of the electron beam determines the brightness with which the appropriate phosphor spot glows.

The color picture tube has three electron guns, or three "barrels" in a single gun—one for each of color television's three primary colors (red, blue and green). A "shadow mask" (also called aperture mask) is located between the electron gun and the screen. The shadow mask is perforated with tiny round holes and the screen is coated with triads of phosphor dots which glow red, blue and green activated by the electron beam. The positioning of the mask with respect to the electron guns and phosphor screen is such that the beam from the "red" gun lands only on red phosphor dots, the beam from the "green" gun only on green phosphor, the beam from the "blue" gun only on blue phosphor.

A variation of the color picture tube is called the "slot-mask" type, which has alternate red, blue and green phosphor stripes on the faceplate and a mask with vertical slits or slots. The principle of operation is the same as that of conventional color tubes. See also Television, Technology of.

"CATHOLIC HOUR, THE"—an NBC series presented by the Catholic Church that, through dramatic and documentary techniques, deals with social dilemmas such as mental illness, death and racism. See also Religious Television.

"CATHOLICS"—adaptation by Brian Moore of his own short novel for *CBS Playhouse 90* (Nov. 29, 1973) concerning the conflict between an aging abbot and a young American priest with new ideas. It starred Trevor Howard and Martin Sheen and was filmed in Ireland. Sidney Glazier was executive producer, Barry Levinson producer (Lewis Freedman producer for CBS) and Jack Gold director. The program was repeated on the network Aug. 1, 1975.

CATV—acronym for Community Antenna Television, the early name for cable-TV when it provided essentially an antenna service for households with poor TV reception or those in under-serviced markets needing additional stations. The term "CATV" is still in use, interchangeably with cable-TV, although the latter better suggests the range of sophisticated communications services afforded by the broadband technology, including pay-cable.

As a community antenna for retransmission of broadcast stations, CATV draws TV signals off the air from an advantageous location, amplifies them and distributes them by a wire-thin coaxial cable to TV sets. A single cable may carry up to 40 channels, although 20 is typical today. In rural areas the cable is strung over telephone poles; in cities the trunk cable is laid in underground telephone ducts.

When the incoming cable is attached directly to the antenna input of the TV set, which is the simplest method, the normal dial is used for channel selection. This, however, limits the subscriber to the 12 VHF channels on the dial. In order to receive the full channel-carrying capacity of the cable, tunable converters must be attached to television sets to replace the limited tuners.

CATV subscribers are charged a monthly fee for the basic service and a small rental fee for the converters. See also Cable-TV.

CAUCUS FOR PRODUCERS, WRITERS AND DIRECTORS—an organization of members of the Hollywood creative community formed in 1975 to give those who create TV programs and motion pictures a voice in industry issues that are beyond the scope and concern of the guilds. The essential aims of the Caucus are to protect the standards and integrity of the creative work of its members and to allow producers, writers and directors to assume a more direct responsibility to the viewing public in network programming and films.

CAVETT, DICK—urbane comedian and talk show host who, although praised by many critics as witty and intellectual, failed to establish a competitive program on ABC-TV in a number of tries from 1968 through 1974. He was more successful on PBS with his nightly interview series which ran from 1977 to 1982.

A onetime comedy writer for Jack Paar and others, Cavett made his debut as a program host on March 4, 1968 with a 90-minute daytime talk show on ABC entitled *This Morning,* which soon was changed to *The Dick Cavett Show.* It won an Emmy Award but was terminated for low ratings on Jan. 24, 1969. In the summer of that year Cavett conducted a prime-time series on ABC three nights a week and, on Dec. 29, was given the network's late-night program, previously hosted by Joey Bishop. The program ran five nights a week, opposite *Tonight* with Johnny Carson on NBC and *The Merv Griffin Show* on CBS. Cavett won another Emmy but ran third in the competition.

In December 1972 his nightly series was cut back to one week in four, where it alternated with a week of Jack Paar (in his return to the late-night arena) and two weeks of varied programming. Two years later when ABC adopted *Wide World of Entertainment* as its late-night entry, Cavett was again cut back, this time to two programs a month (alternate Thursdays). He won a third Emmy. But when it appeared that ABC might not renew his contract, late in 1974, he signed with CBS for specials and other projects, and also did occasional programs for PBS, including *V.D. Blues* and *Feelin' Good.*

No permanent program for Cavett developed at CBS, and in the fall of 1977 he began a nightly interview series for PBS, produced by WNET New York. The program became one of the bread-and-butter entries of public television until the stations voted it out of the lineup in 1982.

A native of Nebraska, Cavett worked initially in summer stock in the east and became a comedy writer when Paar accepted his material for his opening monologues. He later wrote for Merv Griffin, Jerry Lewis and Johnny Carson, before developing the nightclub act that led to his career in TV.

Dick Cavett

CBN—(Christian Broadcasting Network) see Cable Networks.

CBS—the leading network in the ratings through virtually all the years of three-network competition until ABC's ascendancy in 1976; also, the most glamorous of the three, partly an effect of its traditional star-consciousness and partly of a carefully cultivated public image. Though rarely as innovative as the other networks, CBS has been the perfectionist among them, maintaining high production standards and providing nearly always for the inclusion of quality and cultural programming somewhere in an otherwise highly commercial schedule.

CBS was for many years the runaway leader in such lucrative areas as daytime and Saturday mornings, as well as in prime time. The only two areas it has never dominated have been the early morning and the late evening, which NBC has controlled since the mid-50s with *Today* and *Tonight.*

The network faced a crisis in the fall of 1977 when it appeared in grave danger of settling into third place, but the CBS programs rallied, as those of NBC went into a tailspin, and CBS entered the 80s as the leading network once again.

CBS operates from a sense of being "the Tiffany of television." The corporation's fastidiousness about graphics, decor and good form inevitably has carried over to the picture on the screen, and the company's obsession with being First and Best has yielded an unbroken string of television gems, from *Playhouse 90* and *Studio One* through *I Love Lucy, The Honeymooners, The Defenders, East Side, West Side, The Dick Van Dyke Show, All In the Family, The Mary Tyler Moore Show* and *Kojak;* and from *See It Now* to *CBS Reports* and *60 Minutes.*

CBS's supremacy began with the raids by chairman William S. Paley in the late 40s on NBC's biggest attractions: Jack Benny, Red Skelton, Amos 'n' Andy and Edgar Bergen & Charlie McCarthy. It proceeded from there to the building and maintenance of a regular star roster unmatched by the rival networks, including Lucille Ball, Jackie Gleason, Arthur Godfrey, Garry Moore, Ed Sullivan, Jim Arness, Danny Kaye and Andy Griffith. CBS also spawned the most distinguished journalist in broadcasting, Edward R. Murrow, and the one most trusted by the public in the 70s, Walter Cronkite.

Contributing to the TV network's success was the effort made during the early years to secure the Channel 2 position for the owned stations and, wherever possible, with affiliates. Controlling the first spot on the dial proved especially advantageous because the Channel 2 signal carries farther than that of the other channels. This gave CBS better penetration into the countryside than either NBC or ABC and therefore a larger potential audience. During the 60s, CBS augmented its overall ratings by playing continually to its "rural advantage" with such shows as *The Beverly Hillbillies, Green Acres, Petticoat Junction, Mayberry R.F.D., Gomer Pyle* and *Hee-Haw.*

Eventually this began to reflect adversely on the CBS reputation for sophistication and quality, and it became also a demerit in an advertising marketplace that was growing increasingly conscious of demographics. The network shed its bucolic programs in 1970 and replaced them with shows more pertinent to contemporary life, thus leading the way to the mature fare that characterized prime-time programming in the early 70s.

Previously known as the Columbia Broadcasting System, the company changed its name in 1974 to CBS Inc. to reflect its diversification in other fields, such as publishing, musical instrument manufacturing, toys and recordings.

History—CBS began in 1927 as United Independent Broadcasters, founded by a small group headed by Arthur Judson, a talent agent and packager who determined to set up his own radio network when he was unable to do business with NBC. But the company was under-financed and before going on the air took in a partner, the Columbia Phonograph and Records Co. The network finally went into operation on Sept. 18, 1927 as the Columbia Phonograph Broadcasting System, providing 10 hours a week of service to 16 affiliates, with Judson's company supplying the programs.

Discouraged by the mounting losses, Columbia Phonograph withdrew, and the controlling interest was purchased by Jerome Louchheim, a wealthy Philadelphia builder and friend of Ike and Leon Levy, owners of Philadelphia station WCAU, the first affiliate of the network. The Levys also bought shares. Louchheim changed the name to the Columbia Broadcasting System, but just as the network was beginning to make progress with advertisers he was sidelined by a hip injury.

It was then that William S. Paley entered the picture. An advertiser on the network (for his father's company, Congress Cigar Co., whose most popular brand was La Palina) and brother-in-law of Leon Levy, he purchased control of CBS on Jan. 8, 1929 for less than $400,000 and moved to New York to operate the company.

The network then had 22 affiliates and 16 employees. Largely through devising a contract offering more favorable terms to affiliates, Paley succeeded in expanding the network, and by 1932 CBS realized a profit of more than $3 million, notwithstanding the Great Depression. In 1939, CBS bought up one of its original owners, the Columbia Phonograph and Records Co., and built Columbia Records into a leader in the field.

Frank Stanton, who was to exert a great influence on CBS, second only to that of Paley, joined the company as a researcher in 1934 and became its president in 1946. As CBS prospered, it acquired stations of its own in major cities and later built their TV counterparts, which were to become hugely profitable.

For all the ways in which it was successful, CBS had a share of failures over the years. One was its venture in TV-set manufacturing, which began in 1951 when it acquired Hytron Radio and Electronics for $17.7 million and ended a few years later with a total loss of $50 million. Another was its development of a spinning-disc color TV system that lost out to RCA's system in the early 50s because it was not compatible with the existing black and white sets. There were also failures with the New York Yankees, acquired in the late 60s and divested several years later, and with EVR, its venture into video-cassettes with a film system that was soon obsoleted by various other tape systems.

Since the mid-60s, CBS has been preparing for a future without Paley. Initially, the plan was to merge with a larger company, and when the brokers advised that the management cadre was too old for CBS to be attractive for an acquisition, CBS embarked on a youth program. But shortly after the heads of the divisions were replaced by younger men, the plan was changed. CBS decided instead to diversify in a way that would make broadcasting only one facet of a conglomerate rather than a small company's primary business.

To accomplish the growth and to prepare for the succession, CBS hired away Charles T. Ireland from ITT as its new president in 1972. But Ireland died of a heart attack a few months later, and in July 1972 CBS tapped Arthur R. Taylor, a 37-year-old executive of the International Paper Co., to be Stanton's successor, heir-apparent to Paley and engineer of the acquisitions. Paley dismissed Taylor in the fall of 1976 and appointed John D. Backe, head of the publishing division, president. At the same time, Paley announced that he would give up his title as chief executive officer in April 1977 but would retain that of chairman. But Backe was fired by Paley in 1980 and Thomas Wyman brought in to replace him.

The fall of 1977 brought the most drastic and extensive reorganization in the company's history. John F. Schneider was ousted as president of the CBS Broadcast Group and succeeded by Gene F. Jankowski. Robert Wussler was demoted from president of CBS-TV to head of CBS Sports and was to leave the company during the following year. James Rosenfield became president of CBS-TV, while the program department broke off from the network as a division unto itself, CBS Entertainment, headed by Robert A. Daly. Meanwhile, Richard S. Salant was a lame duck as president of CBS News, facing mandatory retirement at 65 in the spring of 1979. Eventually he was replaced by Bill Leonard, who himself was only about two years away from mandatory retirement. The group flourished under Jankowski's leadership, despite the loss of Salant, Daly, and Weissler. CBS, as it proved, had excellent reserves.

Structure—Below the corporate tier of chairman and president are four separate groups, each headed by a president, those for the businesses of broadcasting, manufacturing, publishing and technical research. The president of the CBS Broadcast Group, Gene F. Jankowski, oversees the presidents of CBS Entertainment, CBS-TV, the TV owned stations, CBS News, CBS Cable, the radio network and the radio stations.

CBS CABLE—see Cable Networks.

CBS FOX—cable and video partnership formed between CBS and Twentieth Century-Fox in February 1982 for ventures in cable and video disc and video cassette markets. In effect, the arrangement merged CBS's cable activities with Fox's home video operations. The new company, CBS Fox, was scheduled to begin late in 1982 independent of the two owners; it would be run by a six-person board and have a fully empowered president.

"CBS MORNING NEWS"—See *Morning.*

CBS STUDIO CENTER—film lot of CBS, formerly the Republic Pictures lot before CBS purchased it in February 1967 for $9.5 million. It occupies 70 acres in Los Angeles, contains 17 sound stages and is used primarily for the production of TV series, whether or not they are on CBS.

CBS TELEVISION STATIONS DIVISION—one of the most powerful station groups, with properties in five major markets. The division was created in 1958 when the management of the CBS o&os was separated from the management of the TV network. At one time there were seven stations in the CBS group, two of them on UHF—WXIX Milwaukee, and WHCT Hartford, Conn.—but given up by CBS in the late 50s. The TV stations are WCBS New York, KNXT Los Angeles, WBBM Chicago, KMOX St. Louis, and WCAU Philadelphia. Thomas Leahy has been president since 1977.

CEASE AND DESIST ORDER—an order by the FCC to a station or person to refrain from actions found to be in violation of the Communications Act, the United States Code of the conditions of the license that had been granted. Authorized by Section 312 of the Communications Act of 1934, a Cease and Desist order implies serious federal penalties for noncompliance.

CEEFAX—data transmission system developed by the BBC which, with a decoder attachment, allows the viewer to select news bulletins, stock quotations, sports results and other reading matter which he may call forth on the screen by pushing buttons. A similar system, Oracle, was developed at about the same time by Britain's Independent Broadcasting Authority. The two systems have become standardized and late in 1974 were activated for an experimental run of two years, partly to determine whether there would be enough public and institutional interest to prompt the necessary investment from set manufacturers to produce a new generation of color receivers with built-in decoders.

The systems utilize two unused lines (at the very top of the screen) out of the 625-line PAL television picture in the U.K. By pushing buttons on the decoder, the viewer can choose from around 100 pages of printed news and information that he may wish to have appear on his screen. See also Frame-Grabber.

"CENTENNIAL"—26-hour adaptation of James A. Michener's best-selling novel of that title, presented on NBC in the fall of 1978 mainly in Sunday-night time periods. With moderate success in the ratings, it spanned half a season and held the distinction of being the longest movie—actually miniseries—ever made.

An epic on the evolution of the Old West, it was filmed in various American locations by Universal TV. The large cast included Robert Conrad, Richard Chamberlain, Raymond Burr, Sally Kellerman, Clint Walker, Michael Ansara, Barbara Carerra, Chad Everett, Stephanie Zimbalist and Donald Pleasence. John Wilder was executive producer and wrote the screenplay for the first five hours. Virgil Vogel directed.

CHAMBERLIN, WARD B., JR.—president and chief executive officer of the Greater Washington Educational Telecommunications Assn. (GWETA), parent and licensee of public TV station WETA in the capital. Before assuming that post in 1975, Chamberlin had been senior v.p. of PBS, executive v.p. of WNET New York, and senior v.p. and general manager of the Corporation for Public Broadcasting, participating in CPB's formation in 1967.

As head of GWETA, Chamberlin was also top official of the production organization providing news and public affairs programming from Washington for the PBS stations, the organization that previously had been known as NPACT (National Public Affairs Center for Television). NPACT had been merged with WETA before Chamberlin became president, and he joined the two entities under the single WETA banner.

CHAMBERS, EVERETT—series and TV movie producer, whose long list of credits is headed by *Columbo*.

CHANCELLOR, JOHN—anchorman for the *NBC Nightly News* (1971–82). He stepped down as anchorman for the role of commentator on the *Nightly News* and was succeeded in the spring of 1982 by the team of Tom Brokaw and Roger Mudd.

Prior to becoming an anchorman on NBC, Chancellor had gained prominence as a correspondent, head of Voice of America (1965–67) and, for a brief time, host of NBC's *Today Show* (1961). Chancellor was one of NBC's "four horsemen" noted for aggressively covering the floor of the political conventions of the 60s (the others were Edwin Newman, Sander Vanocur and Frank McGee). Between 1958 and 1965 he was NBC correspondent in Vienna, London, Moscow and Brussels, and he returned to the network from VOA as national affairs correspondent in Washington.

Chancellor joined NBC News in Chicago, at WMAQ-TV in 1950 after having been a reporter for the *Chicago Sun-Times*. Several years later, he was transferred to New York.

John Chancellor

CHANDLER, ROBERT—v.p. and director of public affairs broadcasts for CBS News since September 1977, with overall responsibility for such programs as *60 Minutes, CBS Reports*, CBS news specials, *Magazine* and children's news and religious news broadcasts. For two years previously, he had been v.p. of administration and assistant to the president of CBS News.

Chandler began at CBS News in 1963 as head of information services, following a brief stint with MGM-TV and a considerably longer one with *Variety*, as a TV-radio reporter. During his rise at CBS, he was producer of various special events and documentaries and a supervisor of the network's election units.

CHANNEL—in the U.S., six megacycles of spectrum space in the television bands which provide a single path for a station's transmissions. TV channels are designated by number from 2 to 83, although the 13 highest-numbered have been reallocated to land-mobile use.

CHANNEL ONE—44–50 MC, deleted from VHF allocations by the FCC in 1948 and assigned to land-mobile and two-way radio service.

"CHANNELS OF COMMUNICATIONS"—bimonthly magazine of serious thought about television, concentrating on the new video technologies and their influence on society. *Channels* covers television as a force rather than as a field, with a view to the public's stake in the burgeoning electronic environment. The magazine is directed at practitioners, policymakers, and leaders, as well as avid followers of media and public affairs, generally.

Published by the nonprofit Media Commentary Council Inc., on funding by the John and Mary R. Markle Foundation, *Channels* made its debut in April, 1981. It editor is Les Brown, former TV-radio correspondent for *The New York Times*. George Dillehay, who developed the *Sesame Street* and *Electric Company* magazines for CTW, is publisher.

CHANNEL TELEVISION—smallest of Britain's regional independents governed by ITV, broadcasting to the eight inhabited but sparsely populated Channel Islands. The broadcasting problems are complicated by the fact that the islands differ in language, traditions and government, which means that program service has to be divided between French and English fare and that news and public affairs must serve each of the islands separately.

"CHANNING"—series of campus dramas focusing on the relationship between a dean and a professor. It was produced by Universal TV for ABC (1963) and featured Henry Jones and Jason Evers. It ran 26 weeks.

"CHARLIE BROWN" SPECIALS—perennial CBS animated specials, principally for children, featuring characters of Charles M. Schultz's popular comic strip, "Peanuts." Drawing high ratings from the start, the programs seem to gain viewers with repetition. Most are tied to holidays or to the seasons. The specials are via Lee Mendelson-Bill Melendez Productions, in cooperation with United Features Syndicate, with Mendelson as executive producer, Melendez in charge of animation, and Schultz the writer.

Of the 16 specials produced up to 1977, the first and most frequently played was *A Charlie Brown Christmas,* which had its debut on Dec. 9, 1965 and has been shown annually ever since. It won both Emmy and Peabody Awards. Others that are aired almost annually are *It's the Great Pumpkin, Charlie Brown* (original air date, Oct. 27, 1966); *A Charlie Brown Thanksgiving* (OAD Nov. 20, 1973); and *You're In Love, Charlie Brown* (OAD Nov. 12, 1967).

The other titles are *It Was a Short Summer, Charlie Brown; Play It Again, Charlie Brown; You're Not Elected, Charlie Brown; There's No Time for Love, Charlie Brown; It's the Easter Beagle, Charlie Brown; Be My Valentine, Charlie Brown; It's Arbor Day, Charlie Brown; You're a Good Sport, Charlie Brown; Charlie Brown's All Stars;* and *It's a Mystery, Charlie Brown.*

Two 90-minute theatrical movies, *A Boy Named Charlie Brown* and *Snoopy, Come Home,* also were telecast as specials. CBS also mounted a special on Jan. 9, 1975, to mark the 25th anniversary of the comic strip.

The voice of Charlie Brown has variously been played by Todd Barbee, Chad Webber and Peter Robbins; the voice of Lucy by Robin Kohn, Pamelyn Ferdin, Sally Dryer and Tracy Stratford; and that of Linus by Stephen Shea and Chris Shea.

"CHARLIE'S ANGELS"—ABC crime series introduced in September 1976 concerning three young and sexy female private detectives working for the fictive Charlie Townsend, whose voice is heard but who is never seen. The series was a runaway hit when it began. It featured Kate Jackson, Farrah Fawcett-Majors (wife of Lee Majors of *The Six Million Dollar Man*) and Jaclyn Smith. A Spelling-Goldberg production, it was produced by Rick Husky.

Within months, Farrah Fawcett-Majors was a household word, a new glamour queen featured on scores of magazine covers. She left the program before the second season's episodes went into production and was replaced by Cheryl Ladd. In 1979 Kate Jackson's contract was not renewed. She was replaced by Shelley Hack, a model who had been prominent in Revlon's "Charlie" perfume commercials. Miss Hack, in turn was replaced by Tanya Roberts in 1980; and the series itself was canceled after the 1980–81 season.

CHARMOLI, TONY—director and choreographer. Credits include the Mitzi Gaynor and Perry Como specials.

CHARNIN, MARTIN—producer-director responsible for *Annie, the Woman in the Life of a Man,* a tour-de-force for Anne Bancroft which aired on CBS in 1970; *George M!,* with Joel Grey, on NBC in 1971; *Dames at Sea* (1971); *'S Wonderful, 'S Marvelous, 'S Gershwin* on NBC (1972); and *Get Happy* (1973) with Jack Lemmon on NBC.

CHARREN, PEGGY—president of Action for Children's Television (ACT) and one of its founders. She was its moving force as it grew from a small group of Boston-area parents in 1968 to a forceful national organization instrumental in effecting changes in children's programming and commercials. Largely through Mrs. Charren's energies, ACT became a major component of the broadcast reform movement, receiving most of its support from foundations and national membership. As ACT's representative, she testified before congressional committees and the regulatory agencies whenever issues concerning children's television arose. See also ACT, Broadcast Reform Movement.

Chevy Chase on *Saturday Night*

CHASE, CHEVY—youthful comedian who attained TV stardom in 1975 on NBC's *Saturday Night*, the 90-minute live "underground" comedy show airing at 11:30 P.M., and left the following year to pursue a broader career. As a member of the program's repertory company and one of its several writers, Chase caught the public fancy with his regular spot, a satirical newscast, and with his impressions of a clumsy President Ford. Chevy Chase is his real name.

Since his departure from *Saturday Night* he has appeared in several television specials and motion pictures.

CHASEMAN, JOEL—president of the Post-Newsweek Stations since 1973 and previously president of Group W Television. During his years with the Group W organization, beginning in 1957, he had been variously national program manager for the radio stations; v.p. and general manager of the newly formed syndication division; executive producer of the syndicated *Steve Allen Show;* and v.p. and general manager of WINS New York, a station he converted into an all-news service in 1965. He became president of the radio division and eventually of the television division, shortly before being hired away by Post-Newsweek.

CHAYEFSKY, PADDY (d. 1981)—one of TV's foremost dramatists in the "Golden Age" of television in the 50s, responsible for such classics of the medium as *Marty, The Bachelor Party* and *Middle of the Night,* all of which were later

made into movies. Others by the Bronx-born playwright were *The Catered Affair, Printer's Measure* and *Holiday Song.* When TV drama gave way to filmed melodramas and quiz shows, Chayefsky turned to writing motion pictures. In 1976 his devastating satire of the TV industry, *Network,* was one of the big box office films of the year.

CHECKBOOK JOURNALISM—the competitive practice of paying for exclusive news stories or interviews. It is considered dangerous by most journalists because it can restrict the flow of news, pervert the newsgathering process and ultimately undermine public confidence in journalism.

Although the networks profess to have policies against outbidding each other for news exclusives, CBS News paid for interviews with H.R. Haldeman and G. Gordon Liddy, and earlier Sirhan B. Sirhan. In each case, CBS News justified its action by claiming to have purchased "electronic memoirs," but Richard S. Salant, president of the division, conceded at the height of the controversy over payment of $100,000 to Haldeman for a two-part interview that it is difficult to draw the line between news and memoir and called the decision a mistake. Nevertheless, months later CBS lost $10,000 to a bogus informer who claimed to know the whereabouts of the body of James Hoffa.

"CHECKERS SPEECH"—the historic political telecast in which Sen. Richard M. Nixon, on Sept. 23, 1952, used the airways to defend himself in a scandal that threatened his candidacy for the Vice-Presidency on the ticket headed by Gen. Dwight D. Eisenhower. It came to be known as the "Checkers Speech" because of the candidate's emotional digression into a tale of how his children came to possess a dog they loved named Checkers. The speech turned public opinion in Nixon's favor and, many believe, preserved his political career.

In midcampaign, Nixon was charged with having accepted from wealthy friends, an $18,000 slush fund for his personal use, the suggestion being that the money was given in return for political favors. When Gen. Eisenhower reacted by saying he would seek the Presidency only with a running mate who was "clean," there were strong recommendations by Republican leaders that Nixon withdraw. But Eisenhower accepted the view of other party members that Nixon be permitted to present his case to the public.

On $75,000 raised by the party, time was purchased on CBS-TV, NBC-TV and the Mutual radio network for Nixon's speech. He dealt with the charge at once, denying that any part of the $18,000 went to him for personal use but insisted instead that it was given and used for campaign purposes. Seated nearby and occasionally sharing the screen was his wife Pat, whose composure and forbearance was demonstrated that night and was to become legendary during the Nixon Presidency, some two decades later.

Having issued his denial, Nixon then discussed his childhood, his early poverty, his Quaker mother and his courtship of Pat. To close, he made a confession, as follows:

"One thing I should probably tell you, because if I don't they'll probably be saying this about me, too. We did get something—a gift—after the election. A man down in Texas heard Pat on the radio mention the fact that our two daughters would like to have a dog. And believe it or not, the day before we left on this campaign trip we got a message from Union Station saying they had a package for us. We went down to get it. You know what it was? It was a little cocker spaniel in a crate that he had sent all the way from Texas. Black and white spotted. And our little girl—(Tricia, the six-year-old—named it Checkers. And you know, the kids love that dog, and I just want to say this right now, that regardless of what they say about it, we're going to keep it."

In its time the speech was a winner, and it remains a television classic.

"CHECKMATE"—series about an organization of special investigators in San Francisco; it had been created by mystery writer Eric Ambler and played on CBS (1959–62). Featuring Sebastian Cabot, Anthony George and Doug McClure, it was by J&M Productions and yielded 70 episodes.

Sonny & Cher

CHER—pop singer of the recording team of Sonny & Cher who became a CBS star in the 70s. She was noted for her wry quips, bold wardrobe and rock-generation insouciance. With her partner and husband, Sonny Bono, she launched the successful *Sonny & Cher Hour* in 1971. Divorce snuffed out the program in 1974, however. ABC then signed Bono for a Sunday night series, *The Sonny Comedy Revue*, but it

flopped and was quickly canceled in the fall of 1974. At midseason, Cher (Cherilyn Sarkisian) returned on CBS as a solo in a Sunday night slot and scored so well in her new series, *Cher,* that she became one of the most publicized talents in the CBS stable. When the series began to falter in the first half of the 1975–76 season, she was teamed once again with Bono, and the series, retitled *The Sonny & Cher Hour,* regained its old footing for a while. But it skidded again and was canceled in 1977.

CHERMAK, CY—series producer. He was executive producer of *Ironside* (except for its last season), "The Doctors" element of *The Bold Ones,* the "Amy Prentiss" element of *Sunday Mystery Movie* and *Barbary Coast.* He took over as producer of *Kolchak—The Night Stalker* midway in the 1974–75 season. He took over as executive producer of *CHiPs* during its first year and continues in that post.

CHERTOK, JACK—producer of TV series in the 50s and 60s through his own Jack Chertok Productions, notably *The Lone Ranger, Sky King, Private Secretary, My Favorite Martian,* and *My Living Doll.* Earlier in his career, as head of short subjects for MGM, he produced the Robert Benchley and Pete Smith series and *The Passing Parade.*

CHESTER, GIRAUD—executive v.p. of Goodson-Todman Productions since 1964 and previously a program executive for NBC, ABC and Ted Bates Advertising. Earning a PhD from the U. of Wisconsin, he first taught speech at Cornell and Queens College, then joined NBC's program department in 1954, working with Pat Weaver. He became ABC's v.p. of daytime programming (1958–62) after a stint with the Bates agency and returned to NBC as v.p. in charge of program administration. He nourished his academic interests with several books, including the textbook, *Television and Radio* (1963).

"CHEYENNE"—a classic TV western with historical significance in the medium. It was the show that started the stampede of "adult" oaters in prime time during the late 50s, and as one of the first shows produced for TV by Warner Bros. it helped to pave the way for the other major studios to enter television. *Cheyenne,* which featured Clint Walker as a frontier scout, had an eight-year run on ABC (1955–62). See also Westerns.

CHICAGO SCHOOL—appellation for the style, ingenuity and resourcefulness that characterized the network television programs emanating from Chicago in the early years of commercial TV. Having been an originating center for radio, principally with soap opera, Chicago quite naturally became a hub of television production when the new medium began, and it contributed to the national scene such programs as *Garroway-at-Large,* with Dave Garroway; *Kukla, Fran & Ollie; Studs' Place,* with Studs Terkel; *Zoo Parade,* with Marlin Perkins; *Super Circus,* with Mary Hartline and Claude Kirshner; *Ding Dong School,* with Dr. Frances Horwich; and a number of soap operas, among other programs.

Chicago programs were marked by originality, and suitability to the medium, as if in compensation for a lack of star glamour. Principally active was NBC, which maintained network production facilities that were known as the Central Division until 1958. In charge was Jules Herbuveaux, one of the showman-broadcasters of the period, who had had an earlier career as a band leader.

Even before film and video tape made live television obsolete, the networks began to concentrate their production bases in New York and Los Angeles, and by the late 50s Chicago was reduced to a television market, making little more than a profit contribution to the networks.

Chico and the Man: Freddie Prinze & Jack Albertson

"CHICO AND THE MAN"—popular comedy series on NBC (1974–78) centering on the relationship between the crotchety owner of an automobile garage and the high-spirited young Chicano who worked with him. With Jack Albertson and Freddie Prinze in the leads, it became one of NBC's highest-rated shows. But Prinze died, a suicide, in February 1977. NBC decided to renew the show the following fall incorporating other characters but not recasting Prinze's role. The series could not survive the loss.

James Komack was executive producer for The Wolper Organization in association with The Komack Co.

CHILDREN'S ADVERTISING—commercials directed specifically at children. These became a highly controversial aspect of television, raising questions on the morality of subjecting children to sophisticated advertising techniques. In the 70s, consumer groups began protesting the differing commercial standards for children and adults, as well as other allegedly abusive practices, among them promoting nutritionally inadequate foods, using program hosts as salesmen, tempting purchases by offering premiums, and advertising expensive toys in a deceptive manner. The efforts by the consumer groups led to reforms by the industry.

Most of the deplorable practices began in the 60s when advertisers perceived, through demographic research, that the juvenile viewer could be reached separately and sold specialized products, from Froot Loops to Barbie Dolls. At the same time the networks began to concentrate children's programs on Saturday mornings where a potential audience of 16 million youngsters was found to exist. In 1974 the three networks realized around $80 million in gross advertising revenues from Saturday daytime alone.

Until 1973, the NAB code permitted 16 minutes of nonprogram material per hour in children's weekend television in contrast to a maximum of 9.5 minutes in adult prime time. Network commercial rates for children's time ranged up to $22,000 a minute, depending on the program ratings and the time of year. Saturday morning had become a tremendous profit center.

According to a study by the Council on Children, Media and Merchandising, approximately 50% of the ads in children's programming (from 1965 to 1975) were for food, primarily sugared cereals, cookies, candies and soft drinks; 30% were for toys and 10% for vitamins, often in novelty form.

In 1970, Action for Children's Television filed a petition with the FCC asking for the elimination of all commercials on children's programs and for a requirement that all TV stations provide at least 14 hours a week of children's programming geared to the needs of the various age groups. Four years later, the commission responded by issuing a set of guidelines for children's programs.

Pressure from the consumer groups prompted drug companies to withdraw vitamin advertising from children's shows as potentially dangerous to the young. Also in response to the criticism, the NAB Code Authority and the networks agreed to reduce the amount of nonprogram material (chiefly commercials) in children's weekend television from 16 minutes an hour to 12 minutes an hour, effective Jan. 1, 1973, and proposed further to cut back to 9 or 9.5 minutes an hour, the approximate level in adult prime time.

Revisions in the code also did away with "tie-ins," the mention of products in a program context, and with the use of program hosts or cartoon characters from the shows as the commercial spielers. See also ACT, Broadcast Reform Movement.

"CHILDREN'S FILM FESTIVAL"—a CBS weekend series since 1967 which presents outstanding films produced for children in other countries, many of them winners of international competitions. One of the purposes is to acquaint children with life in other countries. The one-hour programs are hosted by Kukla, Fran & Ollie.

CHILDREN'S PROGRAMMING—material designed expressly for the juvenile audience, which generally is defined as ages 2 to 11. During the early years of the medium, the appeal of television to children was exploited for the sale of TV sets; later, when most households had but a single TV set, programs were aimed at children on the theory that the young controlled the family viewing until bedtime; and even now, independent stations—particularly those on UHF—schedule blocks of children's programs, because the young help the stations to be discovered.

Although children helped to build circulation for stations, and good will for the new television medium, they were not initially perceived as a major marketing group for products. Television was considered too high-priced for child-oriented products in the 50s and early 60s when the single or dual sponsorship of programs was the rule. Prime-time programs were particularly expensive, considering that they reached a large proportion of adults who were not targets of the advertisers. But a number of factors converged around 1965 to make children's programs a major profit center of networks: first, the proliferation of multiset households, which broke up family viewing and loosened the child's control over the program his or her parents would watch; second, the drift to participation advertising as opposed to full sponsorships, which encouraged more advertisers to use the medium; and third, the discovery that a relatively "pure" audience of children could be corraled on Saturday mornings (and to a lesser extent on Sundays) where air time was cheaper, advertising quotas were wide open and children could be reached by the devices used years before by the comic books.

By the late 60s television programming aimed at children was confined, with few exceptions, to Saturday mornings in the form of animated cartoons. Moreover, the animation studios developed a form of limited animation for the undiscriminating youngsters, involving fewer movements per second, which was cheaper than standard animation. Recognizing that children enjoy the familiar, the networks played each episode of a series six times over two years, substantially reducing costs. And while prime-time programs, under the Television Code, permitted 9.5 commercial minutes per hour, Saturday morning children's shows carried as many as 16 commercial minutes per hour. Citizens groups did not become aroused, however, until the networks began to deal excessively—in their competitive zeal—with monsters, grotesque superheroes and gratuitous violence to win the attention of youngsters. Advertisers, by then, were making the most of the gullibility of children by pitching sugar-coated cereals, candy-coated vitamins and expensive toys (some retailing for as much as $50) in shrewdly made commercials that often verged on outright deception.

Such patent abuses of the child market—while Saturday morning grew into one of television's largest profit centers—prompted the formation of watchdog groups such as Action for Children's Television, whose pleas for reforms could hardly go unheeded by Congress or the FCC. In response to such pressure groups, the industry reduced advertising quotas in children's shows to 9.5 minutes per hour, toughened its code for advertising and reduced the violence content.

When public television introduced *Sesame Street* in 1969 and proved that it, and others such as *Mister Rogers' Neighborhood* and *The Electric Company,* could teach while they entertained, the commercial networks then strived to develop shows and inserts that promoted prosocial values (*Fat Albert and the Cosby Kids*), learning concepts (*Pop Ups, Go, Multiplication Rock*) or even current events, via CBS News. From syndication came such educational series as *The Big Blue Marble* and *Call It Macaroni.* The efforts of ACT and other citizens' groups led the FCC to issue guidelines for children's programming in 1974. This spurred local production of educational shows for the young on weekdays, in recognition of the fact that children are a daily audience.

But the only regularly scheduled weekday program on the networks remained *Captain Kangaroo,* an early morning show for pre-schoolers which began in 1955 on CBS and which has been exemplary in its treatment of the young viewer.

Among the earliest hit shows for children in the medium were *Kukla, Fran and Ollie, Howdy Doody, Ding Dong School, Mr. I. Magination, Superman* and *Hopalong Cassidy.* Some were uplifting and some simply entertaining, but all were relatively harmless and contributed to television's popularity. In 1949 around half the children's programs on the networks were sustaining—that is, presented without commercials.

In the second wave came *Rootie Kazootie, Time for Beany, Zoo Parade, Super Circus, Pinky Lee, Big Top, Watch Mr. Wizard, Rin Tin Tin, Captain Video, Lucky Pup* and *Juvenile Jury.*

And then, *Lassie, Disneyland, The Shari Lewis Show* and *Captain Kangaroo.*

Meanwhile, at local stations, the stock children's show featured a host designated as Sheriff Sam or Fireman Fred whose chatter inevitably led to the unspooling of old moviehouse cartoons. These were the "wrap-around" hosts, who sometimes brought in a live audience of children or worked with puppets. The syndication hits were the movie shorts of

The Three Stooges and *Our Gang* (repackaged for television as *The Little Rascals*).

More pretentiously produced were the prime-time children's series, such as *Fury, National Velvet, My Friend Flicka, Dennis the Menace, Leave It to Beaver, Flipper, Daktari* and *International Showtime,* among others. Animated situation comedies, such as *The Flintstones* and *The Jetsons,* became popular both with children and adults, and *The Wonderful World of Disney* on NBC, which grew out of ABC's *Disneyland,* became the "perfect" all-family series, although it has always been aimed primarily at children. *Batman,* too, spanned the generations in its initial network run as a campy entry, but the syndicated reruns have fallen distinctly into the children's program blocks.

To keep their profitable Saturday morning enterprises going the networks have sought to appease the pressure groups by eliminating *Spiderman, Gigantor* and other grotesques of comic book fiction and striking a balance between live-action shows and cartoons.

Among the dozens of animated programs for children scheduled by the networks on Saturday or Sunday mornings have been the following:

Program	Network	Rating
Smurfs II	NBC	21.6
*Scooby Doo Classics	ABC	17.7
Smurfs 1	NBC	15.8
Richie Rich/Scooby Doo II	ABC	15.6
Bugs Bunny/Road Runner Show IV	CBS	15.6
Laverne & Shirley	ABC	15.5
The Kid Super Power Hour I	NBC	15.4
The Kid Super Power Hour II	NBC	15.0
Richie Rich/Scooby Doo	ABC	14.2
Bugs Bunny/Road Runner Show II	CBS	13.5
Bugs Bunny Road Runner Show III	CBS	13.4
Spider Man and Amazing Friends	NBC	13.2
Blackstar	CBS	12.9
Fonz and The Happy Day Gang	ABC	12.3
Tom and Jerry Comedy Show	CBS	11.7
Space Stars I	NBC	10.7
Goldie Gold/Action Jack	ABC	10.6
Bugs Bunny/Road Runner Show I	CBS	10.4
Space Stars II	NBC	10.3
Heathcliff/Marmaduke	ABC	9.7

*Program cancelled or in hiatus

The newly developing cable-TV-satellite networks made their first foray into children's programming in 1978. UA-Columbia Cablevision added a 90-minute weekly children's series to its Madison Square Garden sports channel (now the USA Network), entitled *Calliope;* the series carries historical dramatizations, educational programs and entertainment developed by the Learning Corporation of America. Early in 1979, Warner Cable Corporation initiated the first full channel dedicated to children's programming under the title Nickelodeon, offering a 14-hour-per-day, noncommercial schedule of programs developed at its Qube system in Columbus, Ohio.

CHILDREN'S TELEVISION POLICY STATEMENT— an FCC report issued in 1974 which offered guidelines for children's programming, emphasizing the obligation of broadcasters to serve children because of their immaturity and special needs. The guidelines came in response to a petition by Action for Children's Television (ACT), a consumer group, for rules banning advertisements on children's programs and requiring certain amounts of children's programming at specific times.

The guidelines state that all television stations must provide a reasonable amount of programming for children and that a significant part of it should be educational. The report pointed out that the lack of weekday children's programs is recent. In the early 50s the three networks broadcast 20–30 hours of children's programs during the week, but during the late 50s and early 60s many popular shows such as *Howdy Doody, Mickey Mouse Club* and *Kukla, Fran and Ollie* disappeared, leaving *Captain Kangaroo* the only weekday show regularly presented by a network. The report said that children's programming should not be shown only on weekends and that hosts should not serve as salesmen.

Although critics of the FCC were disappointed that rules were denied, the commission defended its decision saying that issuing rules would not be consistent with its court-sanctioned role of imposing only general affirmative duties on broadcasters in return for their right to use the public airwaves.

In offering the policy statement, the commission also accepted the code worked out by the National Assn. of Broadcasters which reduced advertising minutes on children's weekend programming from 12 to 9.5 minutes per hour beginning Jan. 1, 1976. See also Television Code, Action for Children's Television.

CHILDREN'S TELEVISION WORKSHOP (CTW)—a nonprofit organization created in 1967 for the purpose of producing the experimental program, *Sesame Street,* which when it debuted in 1969 revolutionized ideas about children's programming. The "workshop," headed by Joan Ganz Cooney, attracted $8 million in original funding from the Federal Government through HEW and from private foundations, mainly Carnegie and Ford, and the Corporation for Public Broadcasting.

CTW went on to produce *The Electric Company* for public television and foreign-language adaptations of both shows. To free itself of financial dependency, CTW went into production of related materials in other media, such as books, magazines, teaching guides, posters, puzzles and games based on the content of *Sesame* and *Electric.* This generated some funds but still left CTW highly dependent on grants. In 1974 the workshop began a third series, *Feelin' Good,* to convey health information to adults, and produced an NBC special with the Muppets, its first for the commercial medium. In 1977, CTW produced for PBS an adult history series, *The Best of Families.* Three years later it produced a new series *3–2–1 Contact,* which ran for one season on PBS.

The CTW idea began in 1966 with Mrs. Cooney, a TV producer and foundation consultant, who recognized television's untapped ability to teach. With Lloyd Morrisett, a foundation executive, she developed the proposal which led to CTW and *Sesame Street.*

CHILE, TELEVISION IN—with ownership shared (half and half) by the government and the university, the system broadcasts to about 600,000 sets. It does not carry color but utilizes the South Atlantic satellite and a 525-line screen.

"CHiPs"—hour-long NBC police drama (1977–), which was one of the few hits to emerge from dozens of new series to be introduced by the network from 1976 to 1979. The title is an acronym for California Highway Patrol, of which the two leading characters, both bachelor motorcyclists, are members. Larry Wilcox and Erik Estrada are the principals.

CHOATE, ROBERT B., JR.—consumer advocate and nutrition expert who founded and is chairman of the Council on Children, Media and Merchandising, a Washington-based watchdog group that has affected regulatory standards for advertising in children's shows. Born into a wealthy family and with a background in engineering, he took up the cause of advertising addressed to children and made it his vocation. He received nationwide attention in 1970 when, in testifying before a Senate committee, he asserted that most children's breakfast cereals were not nutritious and contained what he called "empty calories." Choate, who has allied himself with ACT and other broadcast reform groups on children's advertising issues, has fought extravagant nutritional claims in TV food advertising and has sought to restrict drug advertising on TV.

CHOMSKY, MARVIN—director who divides his time between motion pictures and TV movies, the latter including *Attack on Terror: The FBI vs. the Ku Klux Klan* and *The FBI vs. Alvin Karpis, Public Enemy Number One.*

CHRISTIAN BROADCASTING NETWORK—broadcast group of commercial UHF and FM stations, more than 50% of whose programs deal with fundamentalist religion. Founded by Pat Robertson when he purchased WYAH-TV in Portsmouth, Va., in 1960, the group grew to include WANX-TV Atlanta, WXNE-TV Boston and KXTX-TV Dallas, along with a number of FM stations. It also has syndicated several of its programs domestically and abroad, principally *The 700 Club,* a Christian-oriented talk and variety show hosted by Robertson. The "network" aims to become a vehicle for international evangelism, toward which it has begun construction of an International Communications Center in Virginia Beach, Va. It became an actual national network in April 1977 by leasing the RCA satellite for distribution to cable systems. Two years later it was available to some 5 million households in 49 states. By that time two other evangelical religious networks had joined it on the satellite, the PTL Television Network and the Trinity Broadcasting Network. See also Religious Television; Cable-Satellite Network.

CHRISTIANSEN, ROBERT W., AND ROSENBERG, RICK—producing team for Tomorrow Entertainment, until the company folded in 1974. They were executive producers of *The Autobiography of Miss Jane Pittman, Queen Of the Stardust Ballroom, The Man Who Could Talk to Kids* and *Born Innocent,* among other filmed dramas.

"CHRISTMAS IN KOREA"—CBS News documentary (Dec. 29, 1953) in the *See It Now* series, unusual for the fact that it did not present events or probe issues but simply recorded the faces, voices and sights of Korea to reveal the impact of the war upon that country and its people. Edward R. Murrow, Ed Scott and Joe Wershba were the reporters, and Murrow produced with Fred W. Friendly.

CHROMA-KEY—a special effect which permits the electronic superimposition of an object or person in front of a background by using a saturated color (usually blue) to form a "hole" in the picture, so that a second video source (camera, VTR or film) is keyed into the "hole.

"CHRONOSCOPE"—CBS News interview series (1951–54) in 15-minute format, sponsored by Longines-Wittnauer Watch Co. Guests for some of the notable programs included Sen. Joseph McCarthy, Sen. Everett Dirksen, Socialist Party Leader Norman Thomas and black leader Rep. Adam Clayton Powell.

CHUBBUCK, CHRIS (d. 1974)—anchorwoman of a local program on WXLT-TV Sarasota, Fla., who, on July 15, 1974, put a gun to her head and committed suicide on camera. The incident took place a few minutes after the start of the morning program, *Seacoast Digest*. Miss Chubbuck reportedly said, before producing the gun: "In keeping with Channel 40's policy of bringing you the latest in blood and guts in living color, you're going to see another first—an attempt at suicide."

"CIMARRON CITY"—western series on NBC (1958) which featured George Montgomery as the mayor of a growing town. Other regulars were Audrey Totter and John Smith. It was produced by Mont Productions in association with NBC.

"CIMARRON STRIP"—series about a western marshal produced by CBS (1967) and starring Stuart Whitman. Featured were Randy Boone, Percy Herbert and Jill Townsend.

CINADER, ROBERT A.—an executive producer for Jack Webb's Mark VII Ltd. production company, whose credits in that capacity include *Emergency, Chase* and *Sierra*.

"CINDERELLA"—a Rodgers & Hammerstein musical created expressly for television in 1957. The 90-minute show starred Julie Andrews as Cinderella and featured Jon Cypher, Howard Lindsay, Dorothy Stickney, Ilka Chase, Kaye Ballard and Alice Ghostley. It was directed by Ralph Nelson.

CINEAMERICA—see Cable Networks.

CINEMAX—see Cable Networks.

CIRCULAR POLARIZATION—a transmission technique designed to improve reception, particularly in areas subjected to "ghosting." Broadcast antennas in the U.S. normally emit signals on a horizontal plane, but circular polarization transmits both vertically and horizontally. CP tests have been conducted by WLS-TV Chicago and KLOC-TV Modesto, Calif.

If CP is authorized by the FCC, stations electing to transmit in this mode will require new antennas and, in most cases, larger transmitters. Viewers will require newly designed antennas to realize the full benefits of CP, but reception is immediately improved for many viewers using indoor antennas. CP has been in widespread use by FM stations since 1964.

"CISCO KID, THE"—popular syndicated western (1951–56) based on a character created by O. Henry. Duncan Renaldo played the title role and Leo Carillo his sidekick. The series was one of the leaders in the Ziv TV stable, and the reruns continue to play in syndication, the films having been reconditioned and reprocessed in 1965.

CITIZENS' AGREEMENTS—formal agreements between stations and local citizens' groups, entered into mostly at renewal time, in which the stations promise to be responsive to the needs of the concerned groups to head off a petition to deny. Frequently this takes the form of an agreement not to play certain programs, such as movies offensive to a minority group or syndicated programs deemed by the complainants too violent or otherwise unsuitable for children. About such agreements, the FCC has been emphatic that the licensee not assign away to outside parties his responsibilities to make broadcast judgments. See also Broadcast Reform Movement, KTTV Agreement.

CITIZENS' COMMUNICATIONS CENTER—public-interest law firm in Washington, D.C., founded in 1969 to provide legal assistance to individuals and groups with valid complaints against local broadcasters and cable operators. The firm has helped negotiate dozens of citizen-broadcaster agreements and has aided groups in filing comments in rulemaking proceedings. It has also been active in fairness complaints, petitions to deny and minority-access situations.

Civilisation's Lord Kenneth Clark

"CIVILISATION"—13-part BBC cultural series (1970) which created a sensation on PBS for its scope, informational content and viewing appeal. With it, Lord Kenneth Clark, who created, wrote and hosted the series, broke ground for a raft of other series examining aspects of life from a historical perspective, notably Jacob Bronowski's *The Ascent of Man* and Alistair Cooke's *America.*

Civilisation, which had several repeats on PBS, traced the development of Western man from the 7th to the 19th centuries through paintings, sculpture, architecture and music. Apart from his scholarly achievement, Lord Clark was appreciated for his effective TV projection, which was neither priggish nor condescending. Testifying to the success of the series was the fact that a Harper & Row book based on the scripts was on the best-seller list for eight months and sold 230,000 copies at $15 each.

CLAMPETT, BOB—creator of classic cartoon characters, his roster including Bugs Bunny, Tweety and Porky Pig for Warner Bros. and Beany & Cecil for his own company. He not only produces, directs and writes the episodes but also performs some of the voices and composes the music.

Dick Clark

CLARK, DICK—TV show host and head of Dick Clark Productions, a company specializing in youth-oriented programs and rock artist tours. Best known for his teenage pop music series *American Bandstand,* which has run continuously on ABC since 1957, Clark branched out in the 70s to hosting daytime game shows such as *The $10,000 Pyramid* on ABC. He began his career as a radio disc jockey and in the 50s hosted a local TV record hop show for WFIL-TV Philadelphia, enjoying uncommon success with a common local program form. In 1957, the show, *American Bandstand,*

began to be carried by ABC, and Clark became a national TV personality.

With considerable fanfare, NBC in 1978 signed Clark to a contract for series and specials, but his first weekly variety series for the network that fall, *Dick Clark's Live Wednesday,* failed and was canceled before midseason.

CLARK, MICHELE (d. 1972)—young CBS News correspondent whose highly promising career ended with her death in a plane crash. In a time when the networks were under pressure to provide opportunities for minorities and females, Miss Clark, who was black, attractive and an able TV journalist, moved to the CBS News forefront almost immediately after her graduation from Columbia's Graduate School of Journalism in 1970. Columbia's broadcast fellowship program for minorities, funded by CBS, NBC and the Ford Foundation, was renamed the Michele Clark Fellowship Program in her honor.

CLAVER, BOB—series producer. Credits as executive producer include *The Interns, The Partridge Family* and *The Girl with Something Extra.*

CLAXTON, WILLIAM F.—producer-director, his credits including *High Chaparral, Bonanza, The Cowboys, The Rookies* and *The Little House on the Prairie.*

CLEARANCE—the acceptance of a network show by a station. Since stations have the right to reject a program or series from the network, they govern the program's potential distribution. A series that receives inadequate clearance, such as *Calucci's Department* on CBS in 1973, is usually doomed before it goes on the air for lack of access to the full national audience. In some instances, however, new programs with low clearances were so popular in the cities where they were carried that the other stations hastily cleared the time for them. Network o&os rarely deny clearance to a network show.

CLIO AWARDS (AMERICAN TV COMMERCIALS FESTIVAL AWARDS)—annual competition by the advertising and commercials-production industries for the best American-made radio and TV advertisements. Awards in television are presented for 34 categories of product campaigns and for 15 craft categories (cinematography, editing, writing, etc.). Radio has 15 product awards and seven craft categories.

CLIPP, ROGER W. (d. 1979)—pioneer broadcaster who built and headed the TV and radio stations group for Walter Annenberg's Triangle Publications, expanding from the initial property, WFIL Philadelphia. After working as an engineer for NBC in New York, Clipp moved to Philadelphia in 1935 as business manager of WFIL. Later he became general manager and, in 1943, president. When Annenberg's *Philadelphia Inquirer* purchased the station in 1946, Clipp embarked on acquiring additional stations as head of Triangle's broadcast division. Clipp retired in 1968, and two years later the group was dismantled—some stations sold to Capital Cities and some to Gateway—when Annenberg became President Nixon's Envoy to the Court of St. James's.

CLOSED-CIRCUIT (CCTV)—a television installation in which the signal is not broadcast but transmitted to a limited number of receivers or monitors by cable or microwave. Closed-circuit systems are widely used in industry, commerce and education. North American Philips Corp., a leading source of closed-circuit television equipment, estimates the CCTV market at $175 million for 1975, rising to a $300-million annual rate by 1980.

CLUTTER—the congestion of nonprogram matter between shows, including commercials, program credits, previews of next week's episode, promotion for other programs, public service announcements, station identification, network identification, billboards and program titles. All constitute messages and are recognized by the television industry as a source of viewer irritation and by the advertising industry as vitiating the impact of the commercial. Yet, while the clutter menace has been discussed for years, little apparently can be done about it. Station and network identification are necessary, and the craft unions require that credits be given. Moreover, it has been established that on-the-air promotion helps to build audience for other television shows, and public service spots have to be accommodated.

CNN (Cable News Network)—see Cable Networks.

COALITION FOR BETTER TELEVISION, THE— organization formed in February 1981 for the purpose of driving out the kinds of televison programs its members considered morally offensive. Led by the Rev. Donald E. Wildmon, a United Methodist minister from Tupelo, Miss., the Coalition—claiming a membership of 300 fundamentalist and conservative groups—announced that it would begin monitoring prime-time programs for three months to determine by consensus which were the most abusive of traditionally wholesome values. Once the list was produced, the organization would identify the advertisers in those programs and urge its members to boycott their products.

The boycott might not have been taken seriously were it not for the fact that a key member of the coalition, and its largest financial contributor, was the Moral Majority, led by the charismatic and politically influential Reverend Jerry Falwell. The Moral Majority, formed in 1979, achieved its visibility initially by attacking television's excesses; it then branched into politics. The principal representative of the Religious New Right, it proved a powerful force in the 1980 elections. Falwell's personal popularity and power was boosted by the election of President Reagan, and the stature of the organization itself rose when Reagan selected as his religious-affairs advisor the Reverend Robert Billings executive director of the Moral Majority.

In saturating the media to discuss the boycott—arguing that the Coalition was not practicing censorship but free speech—Falwell conveyed a sense of a snowballing movement in the country; and the organization's initiative began to be taken quite seriously. The president of Procter & Gamble supported the Coalition's views in a widely publicized speech, and a number of network advertisers began sending representatives to Tupelo for meetings with Wildmon. These visits convinced the minister that reforms were at hand, and he called off the 1981 boycott.

A year later, however, he revived the threat and in March declared a boycott against NBC, its parent RCA, and all the RCA subsidiaries, including Hertz auto rentals. Saying that NBC had done the least to clean up its programming, Wildmon cited *Saturday Night Live* and *Hill Street Blues* as shows that were notably objectionable.

But this time, Falwell and the Moral Majority were not supporting the effort, and the boycott was perceived as arbitrary and opportunistic. NBC was considered to have been selected as victim because, as the weakest and most financially troubled of the networks, it would feel the effects of a boycott more than the others and would be most likely to respond to the censorial demands of the Coalition. In some critical circles, Wildmon lost sympathy and credibility.

COAXIAL CABLE (COAX)—a transmission line consisting of a tube of conducting material surrounding a central conductor held in place by insulation, widely used to carry television signals. The first inter-city television networking was via coaxial cable of the American Telephone & Telegraph Co., but these interconnections now have been largely replaced by microwave relays, and in some cases by satellite, both of which have higher signal-carrying capacity and are more economical.

Coaxial cable is still widely used in cable-TV systems, master antenna (MATV) distribution systems in apartment houses and is increasingly employed in home color television installations because it is less susceptible to interference than other types of antenna-to-set connections.

COBB, GROVER C. (d. 1975)—a leading industry figure who served as executive v.p. of NAB from 1973 until his death. He was also v.p. of broadcasting for the Gannett Co. since 1969 and before that part owner and general manager of KVGB Radio in Great Bend, Kan.

COCA, IMOGENE—comedienne whose career held great promise, but actually reached its height when she was teamed with Sid Caesar on NBC's memorable *Your Show of Shows* in the early 50s. Neither she nor Caesar enjoyed much success in TV after their split in 1954. They reunited for a series in 1958 but it failed, as did Miss Coca's situation comedy *Grindl* (1963), in which she played a housemaid. See also Caesar, Sid.

COE, FRED (d. 1979)—noted producer of TV plays when drama was in flower in the medium and, in the early years, manager of new program development for NBC. He was executive producer of *Mr. Peepers* (1952–53), producer-director of *Television Playhouse* (1948–53) and producer of *Producers Showcase* (1954–55) and *Playwrights '56,* among others.

COLE, NAT (KING) (d. 1965)—a top recording artist who became the first black to have his own network variety series (1957), but it was to be the shame of network television. In a period before the civil rights movement, and in a time when TV shows needed sponsorship to survive, no national advertiser would buy the Cole series for fear of a boycott in the South. Many performers rallied to its support and offered to appear free, but it was to no avail. The NBC show was canceled early in 1958.

COLIN, RALPH—outside counsel to CBS for more than 40 years, a member of its board of directors and personal lawyer to chairman William S. Paley. After a dispute with Paley in 1969, reportedly on matters concerning the Museum of Modern Art, of which both men were trustees, Colin was abruptly dismissed as attorney, and he resigned from the CBS board in anticipation of Paley's next action. Colin had been with CBS even before Paley, having worked for Jerome H. Louchheim, whose interests in the young network Paley had purchased.

COLLERAN, WILLIAM A.—producer-director who was executive producer of *The Judy Garland Show* for CBS in the

60s and earlier director of *The Lucky Strike Hit Parade*. He also produced or directed shows with Dean Martin, Frank Sinatra, Bing Crosby and Debbie Reynolds.

Charles Collingwood

COLLINGWOOD, CHARLES—correspondent for CBS News since 1941 who worked under Edward R. Murrow during World War II. Later he became the network's first United Nations correspondent and in 1964 chief European correspondent based in London. His annual *Vietnam Perspective* telecasts (1965–69) won several journalism awards in the field of foreign affairs. In 1968 he became the first U.S. network newsman admitted into North Vietnam, which resulted in two specials: *Charles Collingwood's Report From Hanoi* and *Hanoi: A Report by Charles Collingwood.* He also anchored special telecasts on the Middle East conflict and in 1970 and 1971 was moderator of the CBS year-end reviews featuring its correspondents.

During the 50s, while based in the U.S., he covered the political conventions, served as White House correspondent, and took a two-year leave of absence to become special assistant to Averell Harriman, then Director for Mutual Security in Washington. He returned to CBS News in 1957 as London bureau chief but was called back in 1959 to succeed Murrow as host of the *Person To Person* series. He was also host of such public affairs series as *Conquest, Adventure* and *WCBS-TV Views the Press* and moderated the first *CBS Town Meeting of the Air* to be transmitted by satellite, which involved a discussion among government officials in New York, Washington, London and Paris. He was also the host for *A Tour of the White House With Mrs. John F. Kennedy* (Feb. 14, 1962).

COLLINS, HAL—comedy writer of the 50s, who wrote for Milton Berle's *Texaco Star Theatre* (1948–52) and the Red

Buttons shows (1952–55). Later he was associate producer of the Berle specials and attempted a comeback on ABC with *Jackpot Bowling* (1960–61).

COLLINS, LeROY (GOV.)—former Florida governor who became president of the National Assn. of Broadcasters in 1961, serving almost four years. He became unpopular within the industry when he mistook what was essentially a lobbying job for an inspirational one and continually exhorted the industry to aspire to higher goals than the pursuit of profit. His calls for more public service in the public interest did not sit well with broadcasters who felt the industry needed a defender and not another critic. Nor did he endear himself to the NAB membership when he advocated the adoption of industry restrictions on cigarette advertising aimed at youth. Collins resigned from the NAB in 1964, after the board voted 22–19 to retain him, to head a new civil rights division in the Justice Department.

COLLYER, BUD (d. 1979)—indomitable game-show host whose chief credits, among many, include *Beat the Clock,* which premiered in 1950, and *To Tell the Truth,* the big CBS hit which began in 1956.

COLOMBIA, TELEVISION IN—a hybrid of nationalization and private enterprise in which a government organization, Intravision, controls the system, giving out three-year contracts to various independent producers. Under Intravision, and on two nationally owned channels, Colombian TV employs a 525-line black and white system which broadcasts via ground satellite to approximately 1.3 million sets.

Three production companies—RTI, Punch and Caracol—control 75% of all programming and pool their efforts at the Gravi Studios in Bogota. The three pay costs of production to Intravision. Each company owns its programming, the bulk being soap operas, musical shows, features and teleseries.

"COLONEL FLACK"—early comedy-dramatic series on CBS (1953) about a likeable con-man and his aide, played by Alan Mowbray and Frank Jenks. Produced by Jody Pam Productions with Stark-Layton.

COLOR TELEVISION—a technology that was in progress but not yet perfected when the television boom occurred in the late 40s, and which led to a bitter fight between RCA and CBS for the color system that would prevail. The arrival of colorcasting in the U.S. is marked as Dec. 17, 1953, when a modified version of RCA's "compatible," all-electronic (dot

sequential) system was approved by the FCC—in a reversal of its decision three years earlier to authorize the CBS "spinning disk" (field sequential) system.

CBS had been the first to perfect color transmission, but although its system offered a superior picture the programs broadcast in color could not be received on existing unmodified black & white receivers. There were, at the time, almost 20 million black & white TV sets in homes and millions more in manufacture. RCA challenged the FCC's approval of the CBS system, arguing that its engineers were rapidly developing a color system that would be compatible with the black & white sets; and although the courts denied the RCA appeals to set aside the FCC decision in CBS's favor, the time was gained for RCA and others working within the industrywide National Television System Committee (NTSC) to develop a compatible system which ultimately was approved by the FCC in place of the CBS system. DuMont and Chromatic TV Inc. (CTI) also vied for the commission's approval with their systems.

RCA's successful petition for the American color system stated that the corporation had spent $21 million in research and promised that it would expedite equipment production in its manufacturing division and would promptly begin colorcasting over its network, NBC.

During 1954 only 1% of U.S. homes had color sets, but both NBC and CBS began televising some of their programs in color. The fact that colorcasting raised costs that could not be passed on to the advertiser, since he derived no significant benefits from it, deterred CBS. But NBC steadily increased its color output to help its parent company sell color TV sets. In 1956 NBC's Chicago station WNBQ (now WMAQ-TV) became the first in the world to originate locally all-color signals, as a prototype for the company. Approximately 2% of the homes in Chicago could receive color at the time. Gen. David Sarnoff, chairman of RCA, predicted then that "by 1963, all of America would be blanketed by color, and each and every home will be receiving its entertainment in full color."

The prophesy was not realized on schedule. The high cost of conversion to color production and transmission, coupled with a slowness in the manufacture of color TV sets, found NBC alone offering a regular schedule of color programs through 1964. But by 1965 the penetration of color sets in the U.S. was sufficient to be a significant factor in program ratings, and in the spring of that year CBS made a sudden announcement that it would present virtually all its fall schedule in color, which meant that it would have to reshoot several of its pilots that had been made in black & white. Prompting the CBS action, in part, was the fact that it had just replaced James T. Aubrey with John A. Schneider as president of CBS-TV, and the dramatic move to color underscored the start of a new era. Not to be left behind, ABC immediately tooled up and became a full-color network the following year.

By 1977, more than 75% of TV-equipped homes were able to receive color on one or more sets.

"COLT '45"—western series on ABC (1957–59) produced by Warner Bros. and featuring Wayde Preston and Donald May.

COLUMBIA PICTURES TV (CPT)—See Screen Gems, Outlet Company.

Peter Falk as Columbo

"COLUMBO"—popular detective series airing approximately once a month in a rotation with three other detective series in the NBC showcase *Sunday Mystery Movie.* A vehicle for Peter Falk, who portrayed a shambling and seemingly ineffectual super sleuth, it premiered in the fall of 1971 and was by far the most successful of the rotating elements. *Columbo* was created by Richard Levinson and William Link, who modeled the character after Petrovitch, the detective in Dostoevski's *Crime and Punishment.* Before it became a series the concept was introduced in two "World Premiere" movies, *Prescription: Murder* (1968) and *Ransom for a Dead Man* (1971). Depending on the network's scheduling needs, the series episodes ran either 90 minutes or two hours. Roland Kibbee and Dean Hargrove were executive producers for Universal TV. The series ended in 1977 because Falk tired of it.

NBC attempted to revive the premise, if not the show, in the spring of 1979 by trying out a limited series entitled *Mrs. Columbo,* in which Kate Mulgrew starred as the wife of Columbo who busies herself solving crimes. The program did well enough to make the fall schedule under the new title, *Kate Columbo,* retitled *Kate Loves a Mystery,* before its cancellation at the season's end.

"COMBAT!"—hour-long adventure series on the exploits of an Army unit during World War II that featured Rick Jason, Vic Morrow and guest stars. It premiered on ABC in October 1962 and ran until the spring of 1967 via Selmur Productions.

COMBINED COMMUNICATIONS CORP.—broadcast group based in Phoenix, which was absorbed in June 1979 by the Gannett Co. in a $370 million transaction. It represented the largest merger in broadcast history up to that time but was likely to be exceeded by the half-billion dollar General Electric acquisition of Cox Broadcasting Co. that was awaiting FCC approval. The company born of the Gannett-Combined merger was a communications giant that owned 80 daily newspapers plus a number of weeklies, seven TV stations, six AM and six FM radio stations, outdoor advertising companies in the U.S. and Canada, and the polling concern of Louis Harris & Associates.

The Gannett-Combined TV stations are KTAR-TV Phoenix, KBTV Denver, KOCO-TV Oklahoma City, WXIA Atlanta, KARK-TV Little Rock, WLKY-TV Louisville and WPTA Roanoke-Ft. Wayne, Ind.

Karl Eller, who was founder and president of Combined, became president and chief executive officer of the Combined Communications subsidiary of Gannett. Alvin G. Flanagan, who previously headed Combined's broadcast division, reports to Eller as president of the Gannett Broadcasting Group.

COMMERCIAL TV IN EUROPE—Nearly all television systems in Europe are supported wholly or primarily by taxation in the form of annual set license fees. At the same time, most systems augment their income by the sale of commercial spots. This is true even of the east European socialist bloc, with the notable exception of the Soviet Union.

Among other exceptions to commercial taint, in its direct form of paid advertising, are the British Broadcasting Corp. and the two-channel, state-controlled service in Sweden.

On systems where spots are sold, a common practice is to run them in nightly commercial "ghettoes" of 15- or 20-minute duration. The spot format nearest to the American model is the pattern adopted by the British independent stations, the one service in Europe that does not benefit from a tax subsidy. The U.K. independents are permitted up to six commercial minutes an hour, usually bunched at convenient breaks in two-minute groupings.

But even in the case of "pure" public broadcasting, such as the BBC, a good deal of indirect commercialism creeps in, chiefly in the form of company-sponsored (cigarettes, beers, etc.) sports events. The promotional auspices are dutifully acknowledged on the air, and visual plugs usually are inescapable as well. Also a good deal of billboard advertising is picked up by the cameras covering football (i.e., soccer) games, the biggest mass audience sport in Europe.

COMMON CARRIER—designating a medium that delivers messages prepared by others for a fee and that is required by law to sell or give access to all who desire it, at posted rates, on a nondiscriminatory basis. Notable examples are telephone, telegraph and certain communications satellites.

The Communications Act of 1934 established that broadcasting would not be a common carrier, but there are many pursuing such a designation for cable-TV, believing that it would be in the public interest. To a certain degree, cable-TV already performs as a common carrier—as on its leased-access channels. But most cable operators are opposed to being given full common carrier status (prohibiting them from operating company channels for program origination) because that would give the FCC the right to regulate profits. They reason that, under those circumstances, investment capital for the expansion of cable would probably dry up.

COMMUNICATIONS ACT OF 1934—the embodiment of national policy for communications in the U.S.; its basic concept is one of private ownership of broadcasting, telephone and telegraph systems, with an administrative agency overseeing these private interests and regulating them in the public interest.

The goal of the Act was to make available to all people a rapid, efficient, national and worldwide wire and radio communications service with adequate facilities at reasonable charge. To accomplish this goal Congress created the FCC.

While the agency's predecessor, the Federal Radio Commission, had done much to alleviate the chaotic conditions that had existed in radio, the FRC had no authority to regulate telephone and telegraph companies. The growing awareness that these industries were highly interrelated and that the country was in need of a coherent national policy for communications led to the establishment of the Federal Communications Act in 1934.

The FCC's authority under the Act is clearly limited to interstate communications; those activities that remain local in nature are the responsibility of the states. While states have the right to regulate the rates and services for such communications as local telephone calls, the courts have given an expansive interpretation to this provision. For instance, they have determined that radio and television signals—while they remain strictly within one state's territory—may be regulated by the FCC because those signals may affect or cause interference with other radio and TV

signals. By this reasoning, the FCC is also enabled to regulate significant aspects of the cable-TV industry.

The Act structured the telephone, telegraph and broadcast industries in entirely different ways. Recognizing that telephone and telegraph were national monopolies, the Act designated them common carriers—that is, carriers for hire which must furnish service on request, at reasonable prices as defined by the commission. Broadcasting, however, was conceived as a private ownership system based on a concept of competition and operation in the public interest.

The theory behind the Act is that since broadcasters are allowed to use the public airwaves they must agree to operate as public trustees, serving the public interest, convenience and necessity. The law set up a concept of short-term licenses with no vesting of any rights in the private owner of the frequencies allocated. In fact, the Act specifically notes that the granting of any license is not to be construed as creating any right beyond the terms, conditions and periods of the licenses.

With respect to broadcasting, the FCC's authority is to classify stations, prescribe the nature of the service to be rendered, determine what power and type of facilities shall be used and establish times of operation and the areas to be served.

The most important function of the commission is that of allocating portions of the electromagnetic spectrum to various classes of broadcasting service and of assigning frequencies for station operations. In doing so, the FCC is directed by the Act to make a fair, efficient and equitable distribution of broadcast service among the various states and communities.

The commission thus is empowered with broad licensing authority. In considering whether an applicant will operate a broadcast facility in the public interest, the commission may take into account the character of the applicant and his financial and technical qualifications. The commission has the power to deny an application but is required to give the applicant a public hearing before the decision can be final.

If the commission determines that a licensee fails to operate substantially as required by his license or fails to observe or violates any provisions of the Act or regulations, the FCC may issue Cease and Desist orders and, in cases of willful or repeated failures, may revoke a license. While this provision of the Act is quite explicit, the FCC has been reluctant to revoke licenses and has done so only in rare and exceptional circumstances.

A chief reason for that is an ambiguity of the Act which makes it difficult for the FCC to take program service into account in determining whether a station is serving the public interest. While the Act requires the commission to hold the licensee accountable, it also specifically forbids the commission from censoring broadcast programs or from interfering with the right of free speech.

Since networks are not licensed, the FCC does not have any direct authority over them. But the Act does empower the commission to make special regulations applicable to broadcast stations engaged in network operations. The authority granted to the FCC under this provision gives it a form of indirect control over the networks.

The Act has been the target of increasing criticism in Congress and among some segments of the communications industry in recent years. House Communications Subcommittee chairman Lionel Van Deerlin (D-Calif.) had proposed a "basement to attic" rewrite of the law, which he believed stood as a major barrier between new technologies and the consumer. Van Deerlin began the process of rewriting the Act in 1977, proposing new bills in 1978 and 1979, but he abandoned hope for an extensive rewrite in July 1979.

See also Federal Communications Commission, WHDH Case, WLBT Case.

COMMUNICATIONS TECHNOLOGY SATELLITE (CTS)—most powerful communications satellite ever orbited, launched in January 1976. In stationary orbit west of South America it provides experimental educational, medical and data communications, community interaction and broadcasting services to remote areas. Designed for use with small earth-station antennas, it is operated jointly by the United States and Canada.

COMMUNITY SERVICE GRANT—mode by which the Corporation for Public Broadcasting passes federal funds directly to PTV stations. Under the "partnership agreement" between CPB and PBS in 1973, a sizable percentage of all monies received by CPB from the Federal Government must be distributed to the stations in the system for their unrestricted use (except for construction or the purchase of equipment). Qualified public TV and radio stations apply for the grant, and the amount of the awards is determined by a complex formula, the key to which is the size of the station and its market. Generally, the grants are used for local production or the purchase of national programming in the PBS markets known as SPC and SAM.

COMO, PERRY—long-popular TV performer, whose easy manner and casual singing style were ideally suited to the intimate medium. Switching back and forth between NBC and CBS, he was one of the few early stars to remain active and popular in TV through the medium's first quarter century. He made his TV debut Dec. 24, 1948 in *The Chesterfield Supper Club,* a 15-minute, thrice-weekly entry on NBC that was a continuation of his four-year-old radio series. In October 1950 he moved to CBS in a weekly variety series, *The Perry Como Show,* which ran five seasons. For the next eight years he was back on NBC, first in a Saturday night variety show whose ratings overwhelmed those of Jackie Gleason, his competitor on CBS, and then (starting in 1959) in a Tuesday night series, *The Kraft Music Hall.*

Como gave up the grind of a weekly series in 1963 to do three to six specials a year, and again he switched between CBS and NBC. The supporting company for most of his variety shows consisted of Frank Gallop as announcer, Goodman Ace as supervisor of scripts and the Mitch Ayres orchestra. Since the late 70s he has confined his TV work to two holiday specials a year, some of them on ABC.

COMPARATIVE HEARINGS—hearings before the FCC in instances when two or more parties apply for the same license, their purpose being to determine which of the applicants is most qualified for the award. An FCC administrative law judge presides over the hearings and issues an initial decision which may or may not be accepted by the commission in its review of the proceedings.

Over the years the commission has developed a number of comparative criteria to use as a standard in evaluating the applicants and in 1965 issued its definitive criteria for comparison. The commission said it primarily considers programming proposals and the applicant's ability to effectuate the proposals. Other points of comparison are ownership (whether local or remote), ascertainment efforts, staffing and equipment plans, integration of ownership and management and management's past broadcast record and broadcast experience if it should be outstanding or very poor. A critical factor in awarding licenses is diversity of ownership. See also License Renewal, Multiple Ownership Rules.

COMPLAINTS AND COMPLIANCES—a division of the FCC established around 1960 as the commission's own investigative arm and as a watchdog over the stations. It was created when the House Legislative Oversight Committee, in its 3.5-year investigation of regulatory agencies during the late 50s, uncovered abuses and shady practices in broadcasting that had eluded notice by the FCC.

COMPOSITE WEEK LOG—a representative sample of a station's programming efforts submitted to the FCC as part of a station's annual program report. The composite week is devised each year by the FCC, which selects three days at random from the first part of the year, three from the last part and one from the summer months for a representative seven-day period. Days are not selected from weeks with holidays. For the programming report, known as Form 303A, broadcasters are required to present their programming logs for the specified days. See also Ascertainment of Community Needs.

COMPUSERVE—a form of videotext which uses telephone lines for transmission and therefore became available to owners of home computers while most other systems using broadcast and cable transmission remained in the test phase. A subsidiary of H&R Block Co., CompuServe began as a company providing business data, then hitched itself to the growing home computer market by offering a range of video games, information services and news from the Associated Press and a number of newspapers, including the *Washington Post, Atlanta Constitution* and *Columbus Dispatch*. The CompuServe adapter, marketed by the Radio Shack chain, not only connects the consumer's television set to the telephone but also includes a memory bank and keyboard to convert the TV set into an interactive computer terminal.

COMSAT—Communications Satellite Corporation, a privately owned United States communications common carrier company operating under a congressional mandate, for the purpose of establishing and operating commercial communications satellite services. Comsat is the U.S. representative in Intelsat, which owns the worldwide international communications satellite system. It has a 50% interest in all seven of the U.S. earth stations operating in the global system and operates six of them. Comsat carries on a research and development program for itself and Intelsat, and, through a wholly owned subsidiary, Comsat General Corporation, engages in programs to establish domestic satellite services and assists foreign countries in telecommunications development programs. It also owns and operates Marisat Communications Systems for ships at sea. Comsat has approximately 100,000 stockholders and had operating revenues of $153.6 million and net income of $38.3 million in 1976. Joseph V. Charyk is president and Joseph H. McConnell chairman of Comsat. See also Satellites.

COMSTAR—AT&T's domestic satellite launched in May 1976 that marked the beginning of the Bell System's satellite era. The 1,800-pound craft, built by Hughes Aircraft at a cost of around $20 million, was one of four Comstars, each with 24 transponders.

CONNELL, DAVID D.—vice-president of production for the Children's Television Workshop who joined in 1968 as producer of *Sesame Street*. He was involved in the planning of that show and of *The Electric Company*. Connell had previously been executive producer of *Captain Kangaroo* on CBS-TV.

CONRAD, WILLIAM—actor who became a producer-director and then returned to acting full-time, as star of *Cannon* in the early 70s. Conrad had been the star of the radio

version of *Gunsmoke* but lacked physical attributes for the role when the series moved to TV. He found work in television as a director and became producer of the *Klondike* series in 1960, before giving up the medium for motion picture production. His portly build and unheroic appearance were right for *Cannon,* which enabled him to resume his acting career 20 years after the visual medium had brought it to a halt. In the 1980–81 season, he starred in *Nero Wolfe* on NBC.

"CONTINENTAL CLASSROOM"—early morning strip on NBC (6:30 A.M.) that began as an experiment in 1958 in the use of commercial channels to teach university-level academic subjects. The project was funded for a three-year period by $1.7 million in grants from the Ford Foundation, the Fund for the Advancement of Education and several business corporations. The first year's program, a refresher course in physics primarily designed for high school teachers, was estimated to have had an audience of around 500,000 regular viewers, including 5,000 teachers who received credit for the course at 270 cooperating institutions. Subsequent courses were in chemistry and contemporary mathematics. See also *Sunrise Semester.*

CONTROVERSIAL ISSUES—a critical factor in Fairness Doctrine complaints, since the FCC was explicit in stating that fairness obligations obtain only in matters involving "controversial issues of public importance." Numerous appeals for time under the Fairness Doctrine, to answer commercials, were denied by the FCC on the ground that the issue in dispute was not one of public controversy, among them: *Green* v. *FCC* [447 F2d 323 (D.C. Cir. 1971)], a case in which public interest petitioners sought time to present opposing views to military recruitment announcements sponsored by the U.S. Armed Forces. The FCC held that recruitment announcements raised no controversial issue, and its decision was affirmed by the D.C. Court of Appeals and the Ninth Circuit Court of Appeals.

Neckritz v. *FCC* [502 F2d 411 (D.C. Cir. 1974)], a case in which a public interest advocate sought time to present opposing views to commercials by Chevron which urged the use of a gasoline, F-310, that Chevron claimed combatted air pollution. The D.C. Court of Appeals affirmed the FCC's judgment that the use of F-310 was not a controversial issue of public importance.

There have been, however, notable instances when the FCC held a fairness complaint to be valid for advertisements, among them: *Wilderness Society and Friends of the Earth Concerning Fairness Doctrine re National Broadcasting Company* [30 FCC 2d 643 (1971)], which grew out of a commercial message by Esso asserting the need for developing Alaska oil reserves quickly. The commercial stressed the capability of the oil companies to retrieve and distribute Alaskan oil without environmental damage. On a petition by public interest groups, the FCC held that the commercial was subject to fairness obligations, and it directed the station (WNBC New York) to inform the commission of the additional material it intended to broadcast to satisfy its Fairness Doctrine obligations. See also Banzhaf Case.

"CONVERSATION WITH DR. J. ROBERT OPPENHEIMER, A"—a 30-minute CBS News documentary which aired on Jan. 4, 1955, probing the mind and heart of the controversial physicist, who had lost his security clearance as a consultant to the Atomic Energy Commission because he had opposed the building of the hydrogen bomb. The program, which had started out as a feature story on the intellectual resources at Princeton's Institute for Advance Study, settled on the man who headed it because of Edward R. Murrow's fascination with Oppenheimer's genius and humanity. The program was notable as a profile of a brilliant thinker but created problems for CBS News in the McCarthy era because Red-baiters had doubted Oppenheimer's patriotism. Murrow was the reporter and co-producer with Fred W. Friendly.

CONWAY, TIM—a leading TV sketch comedian in the late 60s and 70s who appeared destined to play the second banana. He came to notice first as a regular in the situation comedy *McHale's Navy* (1962–66) and then did numerous skits as a guest on variety shows. In 1970 he was given a variety series of his own on CBS but it ran only 13 weeks. He became a regular featured player on *The Carol Burnett Show* in 1975. When that show folded at the end of the 1977–78 season, CBS built another variety series around him, *The Tim Conway Show*, but it fared poorly. Another hour-long series was essayed in the spring of 1980, then was cut back to a half-hour series without musical acts, and in that form ran the entire 1980–812 season. (Z5COOK, FIELDER—noted director of quality drama during the "golden age," who later became executive producer of *DuPont Show of the Week* in the early 60s. His more recent directing credits included *The Homecoming* (from which *The Waltons* was derived) and the pilot for *Beacon Hill*.

COOKE, ALISTAIR—British journalist and U.S. television host whose charm and intelligence contributed to the success of *Omnibus* (1952–60). In the early 70s he created for NBC a series entitled, *America: A Personal History of the United States,* which he also wrote and narrated. A naturalized U.S. citizen, he served until 1972 as the American correspondent for the British paper *The Guardian,* and simultaneously broadcast a weekly radio series for the BBC,

Letter From America, which continues to run. He also is host of the British-produced series on PBS, *Masterpiece Theatre.*

COOKE, JACK KENT—sports promoter and major stockholder in Teleprompter Corp. In 1971 he led a group of insurgents in a major proxy fight for control of the cable concern. Later that year the conflicting groups reached a compromise, and in 1973 Cooke became chairman and chief executive officer of the firm. He owns the Los Angeles Lakers basketball team, the Los Angeles Kings hockey team, the Washington Redskins football team and the Los Angeles Forum sports arena.

COONEY, JOAN GANZ—president of Children's Television Workshop and originator of the proposals leading to the production by CTW of *Sesame Street* and *The Electric Company* for public TV.

Trained as a teacher, she worked for a time as a reporter on the *Arizona Republic,* then as a publicist for NBC and *The U.S. Steel Hour,* and later as a producer of documentaries for public television. All this served as basic training for what was to become a significant career in television, one that was responsible for a major development in education through entertainment.

In 1966 she and Lloyd Morrisett designed the experimental research project for *Sesame Street* and raised $8 million in foundation grants for it. Formed by *Sesame,* CTW later produced *The Electric Company, Feelin' Good* and several commercial television specials. Mrs. Cooney won wide recognition as both an innovator and outstanding executive. See also Children's Television Workshop.

COOPER, JACKIE—actor, producer and studio executive. A former child movie star, he came into prominence in TV as a star and co-producer of a successful situation comedy in the late 50s, *Hennessy,* which led to his appointment as head of production for Screen Gems in that studio's most prolific years. In the 70s he teamed with Bob Finkel in Cooper-Finkel Co., which created a number of situation comedy pilots that did not sell (Finkel producing, Cooper directing). Returning to acting, he had the starring role in *Mobile One,* a short-lived series in 1975. Since then he has concentrated on directing, and won an Emmy for developing the *White Shadow* pilot.

COOPERMAN, ALVIN—program executive who served three hitches with NBC-TV and, in the intervals, was variously executive director of the Shubert Theaters, president of Madison Square Garden Productions, chairman of Athena Communications Corp. (a cable-TV company), managing director for the Harkness Theater and a producer of TV specials (*The Bolshoi Ballet: Romeo and Juliet*). He also conceived and produced the format for the 1972 Republican National Convention in Miami Beach.

Cooperman's second return to NBC, in 1976, was as program officer for the series of weekly special attractions, *The Big Event,* with the title of consultant, but he remained for less than a season. In 1967–68, he had returned to become v.p. of special programs for the network. In the mid-50s, after working as unit manager for the Milton Berle show and later as an associate producer of *Wide Wide World,* he became manager of program sales for NBC and later executive producer of *Producers' Showcase.*

COORS, JOSEPH—ultra-conservative Colorado businessman (executive v.p. of Adolph Coors Brewing Co.) whose flirtation with TV was frustrated on two fronts in 1975 when the Senate Commerce Committee effectively rejected his nomination by President Ford to the CPB board and when the syndicated news company he financed, TVN, folded after losing $5 million in two years.

Coors created TVN (with Robert Pauley, formerly of ABC Radio and Mutual, as chairman) because he believed network news had a liberal bias. The company in 1974 absorbed a competing news service, UPITN, but there were not enough subscribing stations between them to justify the high cost of news gathering and daily transmission over AT&T long lines. Coors withdrew his financial support in October 1975 and TVN was dissolved.

Ironically, only a few weeks earlier Coors had been asked by the Senate Committee to resign his seat on the TVN board to make him a more acceptable nominee to the Corporation for Public Broadcasting, but Coors refused to do so. The committee's majority disqualified Coors on the conflict of interest issue, but there were other contributing factors.

While his nomination was pending, Coors had written several ill-advised letters to Henry Loomis, president of CPB. One seemed an attempt to influence CPB on selecting a company to construct its proposed satellite earth stations; another expressed displeasure, on behalf of a mortician friend, with a PBS documentary on the high cost of dying. Coors told Loomis he hoped to steer public TV away from programs critical of business, indicating a misapprehension on his part of the board's function.

COPYRIGHT LAW REVISION—[Public Law 94—553 (90 Stat. 2541)] bill enacted in 1976 which substantially revised and superseded the 1909 United States Copyright Law; the earlier law had become obsolete, particularly in its inability to remain relevant to the emerging technologies.

The underlying concept of the new law is a single nationwide system of statutory protection for all copyrightable works, whether published or unpublished. For works already under copyright protection, the new law retains the present term of copyright of 28 years from first publication, but it increases the length of the second term to 47 years. For works created after Jan. 1, 1978, the law provides a term lasting for the author's life, plus an additional 50 years after the author's death.

The new law adds a provision to the statute specifically recognizing the principal of "fair use" as a limitation on the exclusive rights of copyright owners, and it indicates factors to be considered in determining whether particular uses fall within this category. In addition to the provision for "fair use," the act lists certain circumstances under which the making or distribution of single copies of works by libraries and archives for noncommercial purposes do not constitute a copyright infringement.

The new law creates a Copyright Royalty Tribunal whose purpose is to determine whether copyright royalty rates, in certain categories where such rates are established in the law, are reasonable and, if not, to adjust them.

In addition, the Copyright Law retains provisions added in 1972 to the old copyright law which accords protection against the unauthorized duplication of sound recordings but does not create a performance right for sound recordings as such.

Passage of the new copyright act was delayed by the difficulties in resolving a number of complex policy issues, some of which related to the broadcasting industry. One snag concerned public TV. Under the old law, there was a general exemption for public performance of nondramatic literary and musical works when the performance was not "for profit." This was generally interpreted to include a public television broadcast.

The new law removes the general exemption. Instead, it provides several specific exemptions for certain types of nonprofit uses, including performances in classrooms and in instructional broadcasting. Noncommercial transmissions by public broadcasters of published musical and graphic works would be subject to a compulsory license. Copyright owners and public broadcasting entities who do not reach voluntary agreement will be subject to the terms and rates prescribed by the Copyright Royalty Tribunal.

Broadcasting organizations in general are given a limited privilege of making "ephemeral recordings" of their broadcasts under the new law.

Where cable-television is concerned, the new law provides for the payment—under a system of compulsory licensing—of certain royalties for the secondary transmission of copyrighted works on cable systems. The amounts are to be paid to the Register of Copyrights for later distribution to the copyright owners by the Copyright Royalty Tribunal. Although these provisions took effect only in 1978, they have already come under attack from program producers and sports interests, who maintain that the compulsory license deprives them of control over their product and that the levels of compensation set in the Act are far too low. They have proposed that Congress repeal the cable-TV provisions and substitute a direct negotiation system in a new Federal communications act.

Also to be paid in this manner are annual royalty fees by jukebox owners. The new law removes the exemption for performances of copyrighted music on jukeboxes and substitutes a system of compulsory licenses.

As a mandatory condition of copyright protection, the old law required that the published copies of a work bear a copyright notice. Under the new act, notice is required on published copies, but omission or errors will not immediately result in forfeiture of the copyright and can be corrected within certain time limits. Innocent infringers misled by the omission or error will be shielded from liability.

As under the old law, registration is not a condition of copyright protection but is a prerequisite to an infringement suit. Subject to certain exceptions, the remedies of the statutory damages and attorney fees will not be available for infringements occurring before registration.

Copies of phonograph records or works published with the notice of copyright that are not registered are required to be deposited for the collection of the Library of Congress, not as a condition of copyright protection but under the provisions of the law making the copyright owners subject to certain penalties for the failure to deposit after a demand by the Register of Copyright.

"CORONATION STREET"—prime-time British serial on the commercial network which was the No. 1 show in the United Kingdom for many years but literally could not be given away free in the U.S. Frustrated at the indifference to the series in the States, Lord Cecil Bernstein, head of Granada Television which produced the show, offered it free in the early 70s to any American station that would guarantee the serial a substantial run. There were no takers. The program's handicap was not just that the accents were British but that they were in a northern working-class dialect difficult even for some Londoners to understand.

During its 12th year on the air (it had begun in Britain in 1960) *Coronation Street* finally received U.S. exposure on public TV, but the serial did not develop a strong following among American viewers, and it ran only a year. Among those featured in the large cast were Violet Carson, Margot Bryant, Jack Howarth, William Roache, Patricia Phoenix and William Lucas.

"CORONET BLUE"—series about an amnesiac portrayed by Frank Converse that was purchased by CBS and then rejected after 13 episodes had been shot. When CBS decided not to air the series, although the episodes were already paid for, producer Herbert Brodkin demanded they be aired; and the show was slotted as a summer replacement in 1967. Frank Converse, Joe Silver and Brian Bedford were featured, and it was by Plautus Productions.

CORPORATION FOR PUBLIC BROADCASTING — nonprofit corporation established by the Public Broadcasting Act of 1967, responsible chiefly for administering the federal funds for the system, promoting its growth and keeping it free from political influence. CPB also obtains grants from private sources, disburses funds for the development and production of national programming, arranges for the interconnection of the stations (both TV and radio) and conducts training programs and research for the system.

An attempt was made to restructure the corporation in 1979 after the Carnegie Commission on the Future of Public Broadcasting recommended that CPB be abolished. The House bill to revise the Communications Act similarly proposed its elimination.

Although it is nongovernmental in character and charged with providing the insulation between government and broadcasters, CPB has a built-in political coloration that, from the beginning, has made it suspect even to the industry it serves. Appointments to its 15-member board are made by the President, on advice and consent of the Senate, and the law specifies that no more than eight may belong to the same political party. The suggestion of partisan ties has raised doubts that a CPB so constituted would work to safeguard public television from becoming an instrument of government policy.

Indeed, memoranda from the Nixon Administration files, made public in 1979, reveal that appointments to the board were made with a view to having the White House control it, and through it the system. The papers also reveal that several of the board members and Henry Loomis, president of CPB at the time, cooperated with, and sometimes collaborated with, the White House inner circle.

Appointments to the board are for six-year terms, on a staggered basis so that there is a turnover of five members every two years. The board annually elects its own chairman and vice-chairman and employs the CPB president and other officers. Board members receive no salary but are paid $100 a day while attending meetings and conducting CPB business. They are also allowed travel expenses and per diem.

Born of a recommendation by the Carnegie Commission on Educational Television, CPB was organized early in 1968 with an initial federal appropriation of $9 million. President Johnson, who under the law could appoint the first chairman, named Frank Pace, Jr., onetime Secretary of the Army and Director of the Budget. John Macy, Jr., former chairman of the U.S. Civil Service Commission, was CPB's first president.

Subsequent chairmen were Dr. James R. Killian, Robert Benjamin, Dr. W. Allen Wallis and Lillie Herndon. Macy was succeeded as president by Henry Loomis and he in 1979 by Robben W. Fleming. Edward Pfister, former president of KERA Dallas, succeeded Fleming in 1980.

CPB created the Public Broadcasting Service in 1969 as an agency to operate the national interconnection of the PTV stations and to select much of the programming for it. When it began operations in the fall of 1970, sending out 12 hours of programming each week, PBS performed the role of a network. But soon after, it became the representative body of the individual stations in the system which was to challenge CPB as the central authority.

CPB never attained the parental leadership that had been ordained for it. Its powers hampered by the meager funds appropriated by the Government, it also stood meekly by when Clay T. Whitehead, director of the White House Office of Telecommunications Policy, told the industry on Oct. 20, 1971 what in effect the Nixon Administration required of it if it were to recommend a bill for long-range PTV financing. Moreover, there were several episodes during both the Johnson and Nixon Administrations which hinted of a willingness by CPB to comply with certain programming wishes of the Government.

These occurrences fostered a union of the stations' managements, through PBS, to oppose CPB and diminish its influence. In 1973 the two organizations resolved their differences in what was termed *the partnership agreement,* the essence of which was a guarantee from CPB that 40% of the federal funds would flow directly to the TV stations, increasing to 50 and 60% as appropriations grew.

Conceding those funds, CPB conceded half its power to the organization of stations. Its activities since then have consisted of lending major financial support to several PTV program series each year, seeding new program projects, conducting research and helping to develop National Public Radio. See also Carnegie Commission on Educational Television, Community Service Grants, Partnership Agreement, PBS, Public Television.

CORWIN, NORMAN—noted radio playwright whose occasional work for TV included an adaptation of his best-known work for radio, *On a Note of Triumph.* He also wrote, directed and produced an anthology series for CBS, *26 by Corwin,* in the early years and, in 1971, hosted and directed a prime-access syndicated series, *Norman Corwin Presents,* produced by Group W.

COSBY, BILL (WILLIAM H., JR.)—one of the first black performers to have a lead role in a weekly network TV series—he co-starred with Robert Culp in the adventure-drama series *I Spy* (1965–68)—but otherwise was known as a

comedian. He had a situation comedy on NBC, *The Bill Cosby Show* (1969–71), and then a successful Saturday morning animated series on CBS based on his comedy routines, *Fat Albert and the Cosby Kids.* During the early 70s he was a regular on the PBS educational series, *The Electric Company,* and proved so adept at working with and entertaining youngsters that ABC gave him a prime-time hour variety series oriented to children in the fall of 1976 entitled *Cos.* It was canceled after a few months, however.

Howard Cosell

COSELL, HOWARD—TV personality who enjoyed a vogue in the 70s as the most controversial of sportscasters, favored by some viewers for his outspokenness and annoying to others for his abrasive personality and stentorian speech. A former lawyer, he became a sportscaster on ABC Radio in 1953 and he has been with ABC ever since. He came into prominence during the controversy over the boxer Muhammed Ali (then known as Cassius Clay) who had been denied the title of heavyweight champion in the late 60s because he had sought an exemption from the draft on religious grounds. Hostile to most other broadcasters, Ali was friendly to Cosell and granted him exclusive interviews. Cosell later became a commentator for ABC's popular *Monday Night Football,* a contributor to Wide World of Sports and host of *Howard Cosell's Sports Magazine.* In 1975, ABC made Cosell host of a weekly prime-time variety show, *Saturday Night Live with Howard Cosell,* but it was not well received and was canceled at midseason. In 1981, he began a weekly half-hour series, *Sportsbeat,* on Sunday afternoon on ABC.

"COSMOS"—13-part PBS series on astronomy, space exploration, and humanity's place in the universe, conceived and hosted by Dr. Carl Sagan, which began its run in the fall of 1980. For a time it ranked as the highest-rated regular series in public television history; it also spun off a best-selling book, *Cosmos*, published by Random House, which sold more than three-quarters of a million copies in the U.S. alone. *Cosmos* was perhaps the most successful effort by American public television in the educational documentary genre developed by the British Broadcasting Corp. with *Civilisation* and *The Ascent of Man*.

Carl Sagan on *Cosmos*

Sagan, an astronomer at Cornell University, was principal writer of the series and head of the independent company which co-produced with KCET Los Angeles. The funding came from the Corp. for Public Broadcasting, the Arthur Vining Davis Foundation, and the Atlantic Richfield Co. Adrian Malone, who produced *Ascent of Man* for the BBC and now works in the U.S. as an independent, was executive producer and director of *Cosmos*. The writers, along with Sagan, were Ann Druyan and Dr. Steven Soter. Senior producers were Geoffrey Haines-Stiles and David Kennard, and KCET program executive Gregory Andorfer was producer.

Cosmos aimed to reveal the interconnectedness of all things, from the origin of matter to the collision of continents, and dealt with such mysteries as black holes, alternative universes, life on other worlds, time travel and the ultimate fate of the universe.

COSTIGAN, JAMES—writer of quality dramas such as *Little Moon of Alban, War of Children* and *Love Among the Ruins*.

COST PER THOUSAND (CPM)—the ratio of the cost of a commercial spot to the size of the audience reached reported in thousands. CPM is the advertiser's index of how efficiently he has spent his money. Although a 30-second network announcement may cost as much as $45,000, it may reach such a vast audience that in terms of the cost of

reaching 1,000 households it is likely to be cheaper than other forms of advertising. Further, there are CPM's for general audiences and for specific audiences. An advertiser would pay, for example, $4.50 per thousand to reach viewers of either sex and all age groups but perhaps $10 or $12 per thousand to reach either women or men in the age group of 18 to 35. Although the target audiences (teenagers, college graduates, etc.) may matter to some advertisers, the most frequent usage is in cost-per-thousand households per commercial minute.

COTT, TED (d. 1973)—broadcast executive who in a 40-year career headed several New York stations: WNEW-AM (1943–50); WRCA-AM-TV (now WNBC-AM-TV) (1950–55); the DuMont-owned stations (1955–57); and WNTA (now public station WNET) (1957–60). Earlier, he had been a producer-director at CBS Radio and at WNYC-AM, the New York municipal station. In later years, he became an independent producer and consultant.

COUNCIL FOR UHF BROADCASTING —organization of commercial and public broadcasters formed in 1974 to spur the growth of the UHF band, whose priority status with the FCC and Congress had slipped away with the growing interest in cable and other new technologies. The coalition of public and commercial forces was arranged by Hartford N. Gunn, then president of PBS, and Richard C. Block, then president of Kaiser Broadcasting. Gunn's interest was that almost 65% of PBS stations were on UHF, Block's was that his seven independent TV stations were all on UHF.

In 1975, CUB made recommendations to the FCC that were formalized in "The Action Plan for Further UHF Development," which was sponsored also by CPB, PBS, NAB, NAEB, INTV and AMST. The plan asked the FCC to require improvements in the manufacture of UHF tuners, indoor and outdoor antennas and lead-in wire, and to adopt measures to reduce UHF receiver noise and increase the efficiency of transmitters.

COUNCIL ON CHILDREN, MEDIA AND MERCHANDISING—See Choate, Robert.

COUNTER COMMERCIALS—TV spots that dispute or rebut the claims of paid commercials. They may be granted under the Fairness Doctrine when the issue involved is deemed a controversial one of public importance. In 1967, responding to a petition by public-interest attorney John F. Banzhaf III, the FCC ruled that cigarette commercials constituted such an issue, and stations broadcasting such advertisements were required to provide time for announcements to discourage smoking. But the FCC said this was a unique case, and it would not apply the Fairness Doctrine to other product commercials that raised a controversy, because the result would "undermine" the commercial system of broadcasting.

The FCC also rejected a proposal by the FTC that broadcasters allow citizens to respond to commercials that explicitly or implicitly raise controversial issues of public importance, that make claims based on scientific evidence which is in dispute, or that do not point up the negative aspects of a product. However, it maintained that editorial advertising—time purchased for a commentary on an important public issue—would be subject to counter commercials under the Fairness Doctrine. See also Banzhaf, John F., III, Fairness Doctrine.

COUNTER-PROGRAMMING—the tactic of scheduling programs opposite those of rival stations or networks in a manner that would win audience away. If two networks are competing with dramatic series in a time period, the third might *counter* with a comedy or variety show. Sometimes, counter-programming is achieved through *bridging,* that is, by starting a 60- or 90-minute program a half-hour earlier than the competing hit. If the program proves popular, most viewers will not likely switch away while it is in progress.

"COUNTRY MATTERS"—series of five one-hour adaptations of British short stories, produced by Granada TV of England and carried by PBS in *Masterpiece Theatre* in 1975. Derek Granger was executive producer.

"COURT MARTIAL"—series on ABC (1966) about officers-lawyers working in the Judge Advocate General's Department in Europe during World War II. Produced by Roncom Films, ITC and MCA, it featured Bradford Dillman and Peter Graves.

"COURTSHIP OF EDDIE'S FATHER, THE"—a "heart" situation comedy about a widower, well-to-do and not lacking women, and his attempts to raise a young son. It had a fair run on ABC (1969–72). In the leads were Bill Bixby and Brandon Cruz, and the series featured Myoshi Umeki, James Komack (also executive producer) and Kristina Holland. It was via MGM-TV.

COVERAGE CONTOURS—the official and uniform rough guide to a television station's physical coverage. Upon grant of a station construction permit, the FCC requires each grantee to submit a map showing its predicted coverage based on the Commission's engineering formulas. The three grades of predicted contours required by the FCC are: Grade B service: the quality of picture expected to be satisfactory to the median observer at least 90% of the time for at least 50% of the receiving locations within the contour in the absence of interfering co-channel and adjacent-channel signals.

Grade A service: satisfactory service expected at least 90% of the time for at least 70% of the receiving locations.

Principal city service: satisfactory service expected at least 90% of the time for at least 90% of the receiving locations.

COWAN, LOUIS G. (d. 1976)—president of CBS-TV from March 1958 to December 1959, and before that a noted producer of programs, among them, *Stop the Music, The Quiz Kids, Kay Kyser's Kollege of Musical Knowledge* and *Conversation.* Cowan had created *The $64,000 Question* and was president of the network when the quiz-show scandals erupted. Although he no longer had an interest in any show, he was dismissed by CBS at the height of the scandals in an act many believe was sacrificial. It had become *pro forma* to serve up an executive in a bad moment.

Cowan went on to found book and music publishing companies, and he directed seminars in broadcast communications at the Columbia University Graduate School of Journalism, while serving as publisher of the Columbia Journalism Review.

After a career as an independent producer of radio and TV shows, he joined CBS in 1955 as a producer and became v.p. of creative service in 1956. He was succeeded as network president by James T. Aubrey, whose expertise was in Hollywood films while Cowan's was in live game shows and variety.

He and his wife died in a fire in their Manhattan penthouse said to have been caused by faulty wiring in their TV set.

"COWBOY IN AFRICA"—short-lived ABC series (1967–68) which featured Chuck Connors as an American cowboy teaching African ranchers the ropes. It was by Ivan Tors Productions.

COWDEN, JOHN P.—long-time CBS executive, principally in charge of information services (advertising, promotion, research and press information) but also a key aide and advisor to each of the presidents of the network from the mid-60s until he retired in 1978. A former child actor in radio, he joined the CBS promotion department in 1938, became an executive in 1951 and a vice-president in 1958. In

1972, he was named v.p. and assistant to the president of CBS-TV.

COWGILL, BRYAN—veteran production and programming executive with the British Broadcasting Corp., who left in 1977 to become chief executive of Thames Television, a British commercial station. Cowgill joined Thames shortly after being named head of news and public affairs for the BBC. For the three previous years he had been controller (chief executive) of BBC-1, the company's mass-audience TV channel.

Cowgill was formerly a producer with, and later the head of, BBC-TV's sports department. Subsequently, he was additionally responsible for all remote programs.

COX BROADCASTING CORP.—major broadcast group which was an outgrowth of the Cox newspapers founded by former Ohio Governor James M. Cox. In 1978, it announced a $560 million merger with General Electric Broadcasting, but changed its mind in June 1980 after the FCC had already approved the deal.

Based in Atlanta, Ga., the company owns WSB Atlanta, an NBC affiliate; WHIO Dayton, CBS; WSOC Charlotte, N.C., NBC; KTVU San Francisco-Oakland, independent; and WIIC Pittsburgh, NBC.

COX BROADCASTING CORP. v. COHN [420 U.S. 469 (1975)]—invasion of privacy case in which the Supreme Court held that a broadcaster could not be held liable for accurately reporting the name of a rape victim when that information was part of a public court record.

A Georgia statute prohibited anyone from giving the name of any rape victim. A Ms. Cohn had been raped, and when her assailant was tried for the attack, information surrounding the incident became part of the court record. Cox Broadcasting Co., in covering the trial, reported the facts of the case, including the name of the victim. The girl's father brought suit against the broadcaster for violating his daughter's right of privacy, and he won a substantial verdict in the state court.

Cox Broadcasting appealed the case to the U.S. Supreme Court, which reversed. The court held that the "public interest in a vigorous press" precluded recovery in this case, although the court refused to decide whether truthful publications could ever be subjected to civil or criminal liability, or whether the state could ever define and protect an area of privacy free from unwanted publicity in the press.

COX, KENNETH A.—scholarly FCC commissioner (1963–70) widely respected by the industry for his independence and intelligence, although he frequently voted for reform measures that were unpopular with broadcasters. Cox, who had been chief of the FCC's Broadcast Bureau before he was appointed to the commission by President Kennedy, was instrumental in forging the Prime Time Access-Rule and the rules and primer for community ascertainment. He and fellow commissioner Nicholas Johnson collaborated on studies of the records of promise and performance by broadcasters in Oklahoma, New York and Washington, D. C., and they were joined on the issue of diversifying media ownership.

A Seattle lawyer, Cox went to Washington in 1956 as special counsel to the Senate Commerce Committee, whose chairman, Sen. Warren G. Magnuson, was from his home state. He directed hearings for the committee that touched on network practices, TV allocations, pay-TV, cable and ratings. When Newton Minow became chairman of the FCC, Cox joined the commission as broadcast bureau chief.

When his term on the FCC expired, and he was not reappointed by President Nixon, Cox joined the Washington law firm of Haley, Bader and Potts and simultaneously served as v.p. of MCI Communications Inc., a common-carrier microwave firm offering intercity communications services for business.

COX, WALLY (d. 1973)—bespectacled, schoolboyish comedian who found his niche in TV with the classic series, *Mr. Peepers* (1952–55). But his popularity was never to grow, although he starred in another series, *The Adventures of Hiram Holliday* (1956), and was featured in such specials as *Heidi* and *Babes In Toyland* in 1955.

COY, A. WAYNE (d. 1957)—chairman of the FCC (1947–52), appointed by President Truman while he was director of the *Washington Post* stations. He was chairman during four critical years on the commission, when decisions were made regarding allocations, color TV, editorializing (containing the seed of the Fairness Doctrine), the reorganization of the commission and the lifting of the freeze on stations. Upon his resignation, he became consultant to Time Inc. and led that company into buying broadcast stations, the first acquisition being KOB-AM-TV, Albuquerque, N.M., for $600,000. He was succeeded as FCC chairman by Paul A. Walker.

CRAMER, DOUGLAS S.—producer for Aaron Spelling Productions. He has shared all executive producer credit with Spelling on all ASP shows since joining Spelling,, including the *Dynasty* and *Strike Force* series.

A former ABC program executive and later head of production for Paramount TV, he became an independent producer in the 70s and was executive producer of such situation comedies as *Bridget Loves Bernie* and *Joe & Sons.* Among his made-for-TV movies are *QB VII, The Sex Symbol, Cage Without a Key* and *Search for the Gods.*

CRANE, LES—brash young host of ABC's first attempt to compete with NBC's *Tonight* show. He had a short-lived TV career. A former disc jockey, he began *The Les Crane Show* in 1965, pinning his hopes on controversy and an abrasive style, but despite ample publicity he failed to build a strong following and was canceled a few months after his premiere.

CRAVEN, T.A.M.—twice an FCC commissioner, serving a term from 1937–44, leaving to establish his own engineering consulting firm in Washington and returning for a second term, 1956–63. He opposed any proposals by the commission that smacked of government control over programming or over business practices, and he held fast to the view that decisions on programming in the public interest were best left to the judgment of broadcasters.

CREAN, ROBERT (d. 1974)—TV playwright whose credits include *A Time to Laugh* (1962), *The Defenders* (1964) and *My Father and My Mother,* among a number of other dramas for *CBS Playhouse* and TV religious series.

CRONKITE, WALTER—TV's premier newscaster for two decades, quitting his post at CBS in 1981 while still the leader at his craft. He relinquished the anchor post to Dan Rather in 1981, but stayed on with CBS as special correspondent. The following year he was host of the short-lived CBS series *Universe.* He also conceived and coproduced for public

television a news-education series, *Why in the World.* It began on PBS in October 1981 as a weekly program and the following January went to a twice-weekly schedule.

His avuncular manner and journalistic integrity won him the distinction, in the 70s, of the most trusted man in America. He became the CBS anchorman on the evening news in 1962 and weathered the popularity of NBC's Huntley & Brinkley to emerge as the medium's No. 1 news star and the CBS "iron man" who anchored, from start to finish, the political conventions (except one in 1964, when he was relieved briefly by the team of Robert Trout and Roger Mudd), every space mission, numerous special reports and the day-long CBS observance of the Bicentennial on July 4, 1976.

A former UPI foreign correspondent who later had his own syndicated radio news program, he joined CBS as a correspondent in 1950 and became managing editor as well as newscaster on the first half-hour evening newscast a dozen years later, *The CBS Evening News with Walter Cronkite.* A warm TV personality, "as comfortable as old slippers" according to one description, he occasionally allowed human emotions to interfere with journalistic objectivity, as when he wept on camera during the coverage of President Kennedy's assassination.

Cronkite conducted scores of historic interviews, such as those with Archibald Cox and Leon Jaworski during Watergate, and others with such newsmakers as President Johnson, President Ford, Anwar El-Sadat and Alexandr Solzhenitsyn. He accompanied President Nixon on his summit visits to Peking and Moscow and anchored numerous documentaries and special reports. Winner of a vast number of TV and journalism awards, Cronkite was peculiarly the consummate newsman and popular TV personality.

His familiar sign-off on the newscast, "And that's the way it was," served for a trademark. He was noted for his expertise, fairness and restraint rather than for a spry wit or theatrical style.

CROSBY, BING (d. 1978)—radio, recording and motion picture star who was host of numerous TV specials but who —except for a short-lived entry in 1964—eschewed a series of his own. He made his TV debut on Feb. 27, 1951 singing on *The Red Cross Program* on CBS. Later he was host of such specials as *Bing Crosby and His Friends* and *The Bing Crosby Golf Tournament* and made guest appearances on scores of other shows. His only series, a domestic situation comedy for ABC entitled *The Bing Crosby Show,* was unsuccessful and ran only one season in 1964.

He founded Bing Crosby Productions in the 50s, a company that had a fair degree of success producing network and syndicated programs; Crosby sold it in the 60s to Cox Broadcasting Corp., which retained the Crosby name.

Crosby had been one of the great performing names in radio for nearly two decades. He had been an obscure singer with the Tommy Dorsey orchestra when William S. Paley, chairman of CBS, heard him on a recording and, impressed with his crooning style, signed him for the network. As was the fashion among radio stars, he developed a mock feud with Bob Hope, and their friendly rivalry inspired several motion pictures in which they co-starred, notably the *Road* series (*Road to Morocco,* etc.).

Bing Crosby

CROSS, JOHN S. (d. 1976)—FCC commissioner (1958–62) named to succeed Richard Mack when he was forced to resign in a scandal. Cross had previously been chief of telecommunications in the State Department. Cross's statement at his confirmation hearing during the height of the ex parte scandals, that he was "as clean as a hound's tooth," became an industry byword. He became an engineering consultant on leaving the FCC.

CROSS-OWNERSHIP—any ownership interest in two or more kinds of local communications outlets by an individual or business concern, such as a newspaper's ownership of a television station or of a cable-TV system. Concern in latter years over media monopolies has prompted the FCC to prohibit TV stations and telephone companies from owning cable systems in their service areas, TV networks from owning cable systems at all and existing newspapers from acquiring existing stations in the same cities. While the present rules "grandfather" the existing cross-ownership arrangement, they do not permit the sale of a newspaper and a television station to the same party, and co-owned radio and TV stations must find separate purchasers when sold if the license transfers are to be approved.

In 1977, the D.C. Court of Appeals ruled that the FCC was wrong in reasoning that the existing cross-ownership situations could continue unless tangible harm to the public interest were demonstrated. The court held that exactly the opposite standard should apply— that divestiture should be required except in those cases where the evidence clearly shows cross-ownership to be in the public interest.

The implications of that decision were that all cross-ownerships would be broken up, even the newspaper-radio combinations of 40- or 50-year standing. While segments of the industry prepared to mount legal challenges to the court's ruling, several affected owners began a swap of stations while others studied the possibilities for trading, either of stations or of newspapers.

CRTC (CANADIAN RADIO-TELEVISION COMMISSION)—regulatory agency for Canadian broadcasting created in 1968 to take the place of the Board of Broadcast Governors, which had been established in 1958. The CRTC was conceived as "proxy-holder" for the citizens as owners of the public airwaves, with the mission of administering the Canadian Broadcasting Act.

Unlike the FCC in the U.S., the CRTC does exercise limited controls on programming, chiefly to assure that the specific aims of the Broadcast Act are being met. The Act provides, among other things, that the broadcast systems should be owned and controlled by Canadians, that the programming be varied, comprehensive and permit differing views and that broadcast services be provided for both English- and French-speaking Canadians.

The CRTC consists of five full-time members who form the executive committee and ten part-time commissioners selected to represent the various geographic regions.

CRYSTAL, LESTER—president of NBC News (1977–79), after having been v.p. of documentaries (1976–77) and for the three previous years executive producer of the *Nightly News*. He was named to head the division by Herbert Schlosser, after the ouster of Richard C. Wald; in turn, he was ousted in the Fred Silverman administration. He stayed on as senior executive producer of political and special news broadcasts. Crystal had joined NBC News in Chicago in 1963 as producer of the local newscast on WMAQ-TV, after having worked in the newsrooms of other stations in Chicago, Philadelphia and Altoona, Pa. He later was associated with the *Huntley-Brinkley Report,* predecessor to the *Nightly News,* as a regional manager, field producer, London producer, news editor and producer.

C-SPAN (CABLE-SATELLITE PUBLIC AFFAIRS NETWORK)—nonprofit and noncommercial enterprise founded in 1978 by Brian Lamb to provide cable systems with a public affairs and news service. Initial programming, which presumably is to remain the heart of the service, is daily gavel-to-gavel coverage of proceedings of the House of Representatives, using the House's own feed and distributing it live by satellite to subscribing cable systems. Transmission began early in 1979. Systems carrying the C-SPAN channel pay one cent a month per subscriber to receive it. See Cable Networks.

CTV TELEVISION NETWORK—Canada's commercial network owned by the 11 independent stations that made up the network in 1966. CTV actually began operation on Oct. 1, 1961, with eight interconnected stations which covered approximately 70% of the English-speaking households. By the mid-70s, it had a total of 15 affiliates, three supplementary affiliates and 162 rebroadcast transmitters, giving the network coverage in 93% of the English-speaking homes.

From its origination center in Toronto, CTV feeds approximately 6612 hours of programing a week, about 60% of which is natively produced and the remainder imported from other countries, principally the U.S. American distributors customarily grant CTV the right to play programs at least a week in advance of the U.S. networks.

Keystones of CTV's news and information programming are *W5*, *Canada AM*, *Maclear*, *CTV News/Backgrounder*, *Human Journey*, *Inquiry* and *Window on the World*. Its sports fare includes major international and Canadian sporting events, principally hockey. President of the network since the mid-60s has been Murray Chercover.

"CUBA: BAY OF PIGS"—NBC News documentary (Feb. 4, 1964) analyzing the events leading to the disastrous attempt at an invasion of Cuba by an American-assisted military force of exiles in 1961. Irving Gitlin was executive producer of the hour, Fred Freed the producer-writer, Len Giovannitti associate producer and Chet Huntley reporter.

"CUBA, THE PEOPLE"—first news documentary produced on portapak equipment to be presented nationally on television. In 1974 an independent video group in Manhattan, the Downtown Community Television Center headed by John Alpert, was admitted into Cuba and permitted to tour the country to record on 12-inch tape the living and working conditions of the people; thus the group became the first TV journalists from the U.S. to work in Cuba after the Revolution of 1959. Brought up to broadcast standards by a time-base corrector, the tape documentary was broadcast on PBS in 1974. A second documentary, *Cuba, The People, Part II*, was produced by the group during a second visit in

December of 1975 but was shown in exhibitions and not on TV.

CUBA, TELEVISION IN—a single-channel operation in the total service of the state. Everything aired, including entertainment fare, has a "message." The result is such shows as *Frederick's Papers*, a comedy series which promotes security consciousness at all times, and *The Traitor*, an action series in which revolutionaries assault the pre-Castro regime.

The service operates on 525-line NTSC (monochrome only), and there are about half a million sets in the country. Most of them, made in the Soviet Union, are used in community centers but a few are in private hands.

CUKOR, GEORGE—famed movie director whose rare TV assignments included the film drama, *Love Among the Ruins* (1975), which won him an Emmy.

Bill Cullen

CULLEN, BILL—one of the durable emcees of the game-show field, having hosted since the early 50s such shows as *Where Was I?*, *Place the Face*, *Quick As a Flash*, *Hit the Jackpot*, *Give and Take*, *Down You Go*, *The Price Is Right* and *Eye Guess*, but his most successful association was with *I've Got a Secret*, the long-running CBS prime-time show on which he was a panelist. Cullen began as a radio announcer in Pittsburgh and then became a staff announcer for CBS in 1944. His career took a fateful turn when CBS made him emcee of the radio series, *Winner Take All*, in 1946.

CULLMAN PRINCIPLE (Fairness Doctrine) [Cullman Broadcasting Co./40 FCC 516 (1963)]—FCC policy requiring access where there is no money to pay for time. The

Cullman Principle was developed by the FCC to deal with the Fairness Doctrine in situations in which the broadcaster has sold time to one side to present its views but has not presented, or made plans to present, nonpaid contrasting viewpoints. The FCC determined that the licensee could not properly insist upon payment because of the right of the public to be informed.

CUME—shorthand for cumulative audience, the number of different or unduplicated households that viewed a specific program, time period or station at least once over a specified period of time. Cume ratings are a valuable index for programs that are scheduled daily, since they tell not how many people watch any single installment but how many different people are reached by some part of the program over the course of a week or month. Cumes are also widely used by independents and public TV stations as an indication of their ability to reach different households in the market.

CURRAN, CHARLES (SIR) (d. 1980)—former president of the European Broadcasting Union and, since September 1977, chief executive of Visnews, London-based international newsfilm agency. From 1969 until 1977, Curran had been director-general of the British Broadcasting Corp.

Born in Dublin and educated in England, he joined the BBC in 1947 as a radio producer, then left for newspaper reporting before returning to the BBC in 1951. He held a series of administrative positions and for a time headed BBC's shortwave services before succeeding Sir Hugh Greene as director-general.

CURRLIN, LEE—NBC program executive after having been CBS program chief (1975–76); he had been a sales administration executive when network president Robert D. Wood tapped him to succeed Fred Silverman, who had shifted his alliance to ABC. Currlin presided over the fall schedule devised by Silverman and could not be held accountable for its disappointing results; but when ABC took the ratings lead in the spring of 1976 and NBC answered by bringing in two veteran program executives (Irwin Segelstein and Paul Klein) to mount a new challenge, CBS replaced the inexperienced Currlin with a more seasoned programmer, Bud Grant. Currlin received a new position in the sales department and then became head of programs for the CBS

o&o's. When Silverman became president of NBC, he promptly hired Currlin away as a program executive.

CURTIS, DAN—producer specializing in terror and action-adventure TV movies and series pilots such as *The Night Stalker, Melvin Purvis, G-Man, The Night Strangler, Trilogy in Terror, The Great Ice Rip-Off, Dracula* and *The Norliss Tapes.* He began his independent production career with the successful ABC gothic daytime serial, *Dark Shadows,* after having been a network program executive. Curtis was also producer of the highly expensive series *Supertrain,* the 1979 fiasco rushed into production by the new president Fred Silverman in hopes that it would be the breakthrough show to turn the network's ratings around. When the program opened to poor ratings and reviews, Silverman fired Curtis, but the change in creative personnel was to no avail.

CURTIS, THOMAS B.—chairman of the Corporation for Public Broadcasting from September 1972 to April 1973. A former Republican congressman from Missouri, he was appointed chairman at a time when many suspected the Nixon Administration of attempting a takeover of public television. After repeatedly issuing assurances that the Administration was not involved in CPB affairs, Curtis suddenly resigned during a dispute with PBS, amid charges that members of the Administration had made contact with members of the CPB board and were "tampering." He was succeeded by the vice-chairman, Dr. James R. Killian.

"CUSTER"—attempt by ABC and 20th Century-Fox to create an hour-long western series purporting to be based on the early, flamboyant career of Gen. George Custer. The series, which featured Wayne Maunder, Slim Pickens and Michael Dante, had a brief run in 1967.

CZECHOSLOVAKIA, TELEVISION IN—state-controlled system, operated by Czekoslovenska Televize based in Prague. It has a network of satellite stations that also produce regionally oriented shows. The service is supported by a combination of state subsidies, license fees and spot advertising. The advertising is informative rather than competitive as would be expected in a socialist system. Czech TV is regarded as one of the most creative in Europe. There are about 3.3 million sets on 625-line SECAM.

D

DAGMAR—buxom blonde comedienne who gained some notoriety for her low-cut and off-the-shoulder gowns on NBC's first late-night series, *Broadway Open House* (1950–51). The flaunting of her bust on TV was considered racy at the time. A relative unknown who had played straight-woman for a number of comedians under the stage name of Jenny Lewis (real name, Virginia Ruth Egnor), she was signed for the NBC show to portray a sexy, not-very-bright blonde and was given the name Dagmar by the comedy writers. It became her permanent stage name.

"DAKTARI"—an hour-long color adventure series based on the popular motion picture *Clarence, the Cross-Eyed Lion;* it premiered on CBS Jan. 11, 1966, and remained for 89 episodes before being placed into syndication. Starred were Marshall Thompson, Cheryl Miller, Hari Rhodes, Hedly Mattingly, Bart Jason, Erin Moran, Clarence (the lion) and Judy (a chimpanzee.) The series was produced by Ivan Tors in association with MGM-TV.

"DALLAS"—hour-long CBS dramatic series, with elements of soap opera, that made the grade in its 1978 debut and was renewed for the 1979–80 season. It went on to become a national sensation, regularly the highest-rated show in television. The producers, Lorimar Productions, also pulled off one of the greatest television stunts of all time by ending the 1979–80 season with the shooting of *Dallas'* principal character, J.R. Ewing, and building up to the fall premiere by promoting the mystery of who shot J.R and

whether he would survive. The shooting was written into the script because Larry Hagman, who played J.R., was in the midst of a contract dispute and might not have returned in the fall.

The episode that opened the 1980–81 season, a masterpiece of hype, was the highest-rated individual show in history, scoring a 53.3 rating and 76 share audience.

The series centers on a fabulously wealthy family, in contemporary times, that built an empire on cattle and oil. Prominent in the large cast are Larry Hagman, Barbara Bel Geddes, Patrick Duffy, Victoria Principal and Linda Gray.

The *Dallas* success provided a windfall of spinoffs for Lorimar Productions—*Knots Landing, Flamingo Road, Falcon Crest,* and *King's Crossing.*

DALY, JOHN CHARLES—newsman and panel-show personality during the 50s and 60s who doubled as the suave moderator of the *What's My Line?* series on CBS while serving as v.p. in charge of news, special events and public affairs for ABC. He held the ABC post from 1953 to 1960, resigning in a clash with corporate management over policy matters. His tenure with the panel show spanned 17 years.

Born in South Africa, Daly worked first for NBC as a news correspondent, then for CBS (1937–49), variously as White House correspondent and on foreign assignments. In addition to *What's My Line?*, he hosted such programs as *We Take Your Word* and *News of the Week.* He served briefly as director of Voice of America (1967–68), leaving in a publicized dispute over personnel changes made without his advice. And he made news again in 1969 when he resigned as host of a public TV panel show, *Critique*, because a remark made by one of the panelists, which he considered obscene, was not removed from the video tape.

Robert Daly

DALY, ROBERT A.—president of CBS Entertainment since October 1977 when, in a major reorganization of the company, the TV network's program department was designated a division. Daly thus became head of programming with scant experience in that field. He had joined the network in 1955 in the accounting department, later shifting to business affairs where he rose to become v.p. in charge. In that post, he demonstrated skill in negotiating program contracts. When Robert Wussler became network president in April 1976, he elevated Daly to executive v.p. of CBS-TV. Then came the move to CBS Entertainment, which involved his transfer to the West Coast.

DAMM, WALTER J. (d. 1962)—head of the *Milwaukee Journal*'s radio and TV stations, the WTMJs, and a leading figure in the industry as president of the Television Broadcasters Assn. until it merged with the NAB in 1951. Later he became president of NAB and also helped to organize the NBC television affiliates organization in 1953.

"DAMON RUNYON THEATRE"—anthology based on the works of the noted writer, with different stars weekly. Produced by Screen Gems, the half-hour series premiered in 1955 on CBS with Donald Woods as host and ran one year.

"DAN AUGUST"—hour-long police-action series that was a failure in its first presentation on ABC (1970) but was a success in 1973 when CBS carried the reruns as a summer

replacement. In that three-year interval the series' star, Burt Reynolds, had climbed from relative obscurity to fame, largely through his appearances on TV talk shows and through the notoriety he acquired from posing nude for a national women's magazine. The series, by Quinn Martin Productions, featured Norman Fell, Richard Anderson, Ned Romero and Ena Hartman.

DANCY, JOHN—senior White House correspondent for NBC News since July 1978. For the five previous years, he had foreign assignments and covered three wars—the Middle East (1973), Cyprus (1974) and Lebanon (1975). Later he became the network's Moscow correspondent. From 1968–73, he was a national correspondent.

D'ANGELO, BILL—executive producer of *Room 222* during its last seasons and producer of *Run, Joe, Run,* NBC Saturday morning children's show.

"DANIEL BOONE"—frontier adventure series which starred Fess Parker and featured Patricia Blair, Darby Hinton and Dallas McKennan. Ed Ames co-starred as Boone's Indian friend during the first four seasons. The series debuted on NBC Sept. 24, 1964, and remained six seasons. It was produced by Arcola-Fesspar in association with 20th Century-Fox.

Michael H. Dann

DANN, MICHAEL H.—colorful head of programs for CBS from 1963 to 1970 and one of the thoroughgoing professionals who could provide quality or pap, as higher management required. An embattled executive who survived five changes of presidents during his tenure as v.p. of programs, Dann lost out finally to Robert D. Wood when he took a stand against the wholesale cancellation of such hit shows as those of Red Skelton and Jackie Gleason and *The Beverly Hillbillies, Green Acres, Petticoat Junction* and *Hee-Haw.* Dann believed in preserving the hits; Wood was intent on creating a new youth- and urban-oriented image for CBS.

On leaving, Dann joined the Children's Television Workshop to work on the foreign sales of *Sesame Street* and became a program consultant to a number of large companies, principally IBM and Warner Communications. He left CTW in 1976 and concentrated on helping to develop Warner's two-way commercial cable system in Columbus, Ohio. In 1981, he joined ABC Video as a programming executive.

He started in broadcasting in the late 40s as a comedy writer, then joined NBC's publicity department as head of trade and business news. Later he joined NBC's program department, where he worked in a number of capacities, and in 1958 he switched to CBS as v.p. of New York programs. In 1963 he became head of programming and in 1966 was named senior v.p. of programs. He built a strong department and was succeeded on his resignation by a protege, Fred Silverman.

D'ANTONI, PHILIP—producer whose successful TV specials such as *Sophia Loren In Rome* led to motion pictures (*Bullitt, The French Connection*). In the 70s he placed the series *Movin' On* with NBC.

"DARK SHADOWS"—Gothic soap opera on ABC, dealing with witchcraft, werewolves and vampires and popular for a time in its late afternoon berth where it was accessible to teenagers home from high school. Rare among the medium's soaps, its reruns had an afterlife in syndication.

The story was built around a young woman (Alexandra Moltke) who becomes governess to a young boy in a house of mystery, that of the Collinses. The stories switch between the contemporary family and their ancestors of the 19th century. Joan Bennett was star of the series, but Jonathan Frid, as Barnabas, the vampire, became its best-known character among youthful viewers.

Others in the cast were Mitchell Ryan, Louis Edmonds, Mark Allen, Nancy Barrett, Grayson Hall and Lara Parker. *Dark Shadows* was produced by Dan Curtis and Robert Costello.

"DAVEY AND GOLIATH"—long-running 15-minute children's series usually dealing with moral questions. It is produced by the Lutheran Church in America and offered to stations free for public service use. Since it began in 1961, the puppet animation series has aired on more than 350 TV stations in the U.S. and Canada, usually early in the morn-

ing before the commercial schedules begin, and is also carried widely abroad.

The title characters are a 10-year-old boy and a talking dog, Goliath, whom only Davey can understand. The dog is both friend and conscience to the boy. The shows are produced in three-dimensional stop motion technique, using free-standing figures with full backgrounds and props in miniature. As of 1976 a total of 65 regular episodes and 10 half-hour seasonal specials had been produced.

"DAVID FROST SHOW, THE"—syndicated talk-variety series via Group W Productions (1969–72) which was that company's successor to *The Merv Griffin Show* when that established entry shifted to CBS for the late-night derby. Frost, the noted English interviewer and humorist, was equal to managing a daily talkfest (which ran 60 minutes in some markets and 90 minutes in others) but lacked acceptance in many parts of the country and never amassed the number of markets necessary to make the show a success.

DAVIS, BILL—music-variety show director whose extensive credits include *Hee-Haw, The Julie Andrews Hour, Cher* and numerous specials.

DAVIS, DAVID, AND MUSIC, LORENZO—line producers of *The Mary Tyler Moore Show*, executive producers of *The Bob Newhart Show*, and producers of *Rhoda*. In 1976 Music and his wife starred in a syndicated series, *The Lorenzo and Henrietta Music Show*, but it was unsuccessful and had only a brief run.

DAVIS, ELIAS & POLLOCK, DAVID—comedy writing team serving as story consultants on *Chico and the Man* during its first season and as executive producers of *That's My Mama* during its second.

DAVIS, ELMER (d. 1958)—highly respected broadcast newsman known for his dry, unemotional delivery and penetrating analyses of news events. A former *New York Times* reporter, he joined CBS in 1939 and quickly became one of radio's most popular commentators. He left CBS in 1942 to become director of the Office of War Information, and when World War II ended he joined ABC. Davis conducted a nightly newscast on TV until illness forced him to leave it in 1953. He returned a year later with a weekly commentary.

DAVIS, JERRY—producer of *The Odd Couple* its first two seasons; executive producer of *Funny Face;* creator and executive producer of *The Cop and The Kid.*

DAVIS, PETER—documentary producer and writer for CBS News (1965–72), whose credits have included *The Selling of the Pentagon, The Battle of East St. Louis, Hunger in America* and *The Heritage of Slavery.* On leaving CBS he produced the controversial motion picture documentary on the Vietnam war, *Hearts and Minds* (1974), and in 1976 signed with WNET New York to produce a public TV series, *Middletown,* on life in Muncie, Ind. It aired in the spring of 1982 as a six-part documentary. Before joining CBS, he was an associate producer of NBC News and a writer for ABC

DAVIS, SID—Washington news director for NBC News (1977–82); for nine years previously he was Washington bureau chief for Westinghouse Broadcasting, after having been White House correspondent. He was one of the three reporters to witness the swearing-in of President Johnson aboard Air Force One in Dallas after the assassination of President Kennedy in November 1963. In 1979, he became NBC news bureau chief in the capitol.

DAVIS, BILL—director specializing in variety shows. His credits include *The Julie Andrews Hour* and *The Lennon Sisters Show* on ABC; *The Smothers Brothers Comedy Hour, The Jonathan Winters Show* and *Hee-Haw* on CBS; *The Lily Tomlin Special;* Marlo Thomas's *Free to Be You and Me;* and other specials with Herb Alpert and John Denver.

DAWSON, THOMAS H.—president of CBS-TV, Dec. 15, 1966 to Feb. 17, 1969, having risen through the sales ranks of both CBS radio and television. Dawson entered broadcasting in 1938 as a salesman for WCCO Radio, the CBS affiliate in Minneapolis. He moved to New York with CBS Radio Spot Sales in 1948 and switched to TV spot sales several years later. In 1957 he became sales v.p. for the TV network and was made senior v.p. of the network in July 1966, a few months before his promotion to president.

After leaving CBS, he became for a time director of radio and TV in the Office of the Commissioner of baseball.

DAY, JAMES—a leading force in public television, as president of NET and previously general manager of KQED San Francisco until he left the executive ranks in 1973. He then created the interview series *Day at Night,* which ran for

two years on a number of PBS stations, and in 1975 became a professor of communications at Brooklyn College.

Day was named president of National Educational Television in August 1969 after managing KQED for 16 years. When NET merged with WNDT New York to form WNET in 1970, Day became president of the new entity. Under Day's leadership, the station took on a vitality it had lacked, while the production center turned out such national series as *The Great American Dream Machine, Black Journal, Soul!, Playhouse New York* and the Frederick Wiseman documentaries. Day was also instrumental in acquiring the British series, *The Forsyte Saga,* and Kenneth Clark's *Civilisation* for the public TV system.

At KQED particularly, Day combined performing with his executive duties. For 14 years he hosted the local interview program *Kaleidoscope* and conducted nationally televised interviews with Eric Hoffer and Arnold Toynbee on *Conversation.*

James Day

"DAYS OF OUR LIVES"—NBC daytime serial which premiered Nov. 19, 1965 and later became the first soap opera to incorporate musical numbers. Members of the cast have included Macdonald Carey, Frances Reid, Brenda Benet, Jed Allen, Melinda Fee, Meg Wyllie, John Clarke and Robert Clary.

DAYTIME—see Cable Networks.

"DEAN MARTIN COMEDY HOUR, THE"—high-rated NBC variety series (1965–74) in which Martin starred with at least four guests each week and a resident chorus line, The Goldiggers. It was by Claude Productions in association with Greg Garrison Productions.

"DEATH OF A PRINCESS"— two-hour docudrama that created a diplomatic flap, and a domestic one as well, prior to its airing May 21, 1980 in the PBS series, *World*. The film attempted to illuminate the Middle East for Westerners through piecing together the story of a 19-year-old Saudi Arabian princess who in 1977 was executed, along with her lover, for committing adultery. Under Islamic law, adultery is a capital crime.

The furor began when the film aired in Britain and the Netherlands, a few weeks before the U.S. showing. The Saudis voiced objections to the British government, saying the film misrepresented the social, religious and judicial systems of the country and was insulting, besides, to the heritage of Islam. The Saudi ambassador then called upon the U.S. State Department to block the showing in this country. Implicit were threats of breaking off diplomatic relations and suspending oil exports at the height of the energy crisis.

Pressures on PBS began to mount, from Congressmen, Mobil Oil (a chief benefactor of public television) and at least one unidentified philanthropic foundationall arguing that in showing the film PBS would be going against the best interests of the United States. But PBS stood its ground, refusing to be censored, especially by a foreign government. Nevertheless, several public stations refused to carry the program, including KUHT Houston, a city where oil interests are concentrated, and eight stations in South Carolina, home state of the Ambassador to Saudi Arabia, John C. West.

The episode received wide attention and debate in the press, a fact which contributed greatly to the large audience for the broadcast. After the airing, the issues quickly evaporated. Mobil did not withdraw its underwriting of public television shows, the Saudis did not suspend oil shipments, and there was no move in Congress to end Federal support of the public broadcasting system.

The film was a coproduction of WGBH Boston, ATV in England and Telepictures Corp. It was written by British journalist and filmmaker Anthony Thomas, in collaboration with David Fanning, the executive producer of *World* for WGBH. The script was taken from interviews conducted by Thomas during 1978. Featured in the program were Paul Freeman, portraying a British journalist (apparently modeled on Thomas) trying to dig out the story of the young princess and why she died, and Suzanne Abou Taleb, an Egyptian actress playing the princess. *World* was funded by the Ford Foundation and the German Marshall Fund.

After the U.S. showing, the film was aired on Israeli TV, whose signal reaches across borders to some 8 million Arabs.

"DEATH VALLEY DAYS"—long-running western anthology series owned and syndicated by U.S. Borax for its 20 Mule Team products. The sponsor started the series on radio in 1930 and carried it into TV in 1952, usually placing it in around 120 markets. Host for the first 12 years was Stanley Andrews, who was represented as the Old Ranger. Subsequent host-narrators included Ronald Reagan, Robert Taylor and Merle Haggard. Production continued beyond 1975.

"DEBBIE REYNOLDS SHOW, THE"—an attempt by NBC (1969) to establish a successor to Lucille Ball in a situation comedy that followed the wacky lines of the various *Lucy* shows on CBS. Although Miss Reynolds had been given a two-year contract, the series ended a failure after 17 episodes. Via Harmon Productions and Filmways, it featured Don Chastain, Patricia Smith, Tom Bosley and Bobby Riha.

"DECEMBER BRIDE"—popular CBS situation comedy (1954–59) which served as a vehicle for Spring Byington, who portrayed an attractive widow living with her daughter and son-in-law. Featured were Frances Rafferty, Dean Miller, Harry Morgan and Verna Felton. It produced a successful spin-off, *Pete and Gladys*. It was via Desilu.

deCORDOVA, FRED—TV and motion picture producer and director who since 1970 has been producer of Johnny Carson's *Tonight* show on NBC. In the 40s, deCordova was a film director at Warner Bros. and Universal; in the 50s he moved into television, winning one Emmy as producer of *The Jack Benny Show* and another as director of *The Burns and Allen Show*. He also directed *My Three Sons* for four years and such other series as *The George Gobel Show*, *Mr. Adams and Eve* and *December Bride*. In all, before joining *Tonight* as successor to Rudy Tellez, he directed some 500 TV programs.

"DEC. 6, 1971: A DAY IN THE LIFE OF THE PRESIDENCY"—An NBC News special covering President Nixon in a 15-hour working day, reported by John Chancellor. It was broadcast in December 1971 shortly after the filming. Although NBC considered it a news coup, others found the timing of the flattering piece uncomfortably close to the election season of 1972.

DEF (DELAYED ELECTRONIC FEED)—Also known as ABC/DEF, the newsfilm service syndicated by ABC News to subscribing affiliates for use in their local newscasts, at their discretion. All the networks provide such a service to their affiliates for a nominal charge. The DEF material is sent out over the lines in the manner of a closed-circuit broadcast and is taped by the stations, allowing them to select the items suitable for their newscasts. The DEF feed consists of top national and foreign stories, as well as some that might be classified as overset from the network newscasts, and sports and news items with special regional appeals.

"DEFENDERS, THE"

"DEFENDERS, THE"—highly respected dramatic series which starred E.G. Marshall and Robert Reed as father-son lawyers and featured noted Broadway and Hollywood guest performers. Produced by Herbert Brodkin's Plautus Productions and Defender Productions, it premiered on CBS in September 1961 and continued until 1965. The series outshone most others on the networks for the quality of writing and acting and for its straightforward treatment of serious themes.

DEFICIT FINANCING—term frequently used by production studios to describe the networks' financial arrangements for shows, referring specifically to the fact that the networks usually license programs for substantially less than they cost to produce. A network may put a ceiling of $800,000 on made-for-TV movies, although the producer's cost is likely to exceed $1 million. The network may also hold the line at $320,000 an episode for a series that costs $350,000 an episode to produce. Studios are expected to make up the difference in foreign and domestic syndication, but shows that fare poorly on the networks rarely are in sufficient demand in secondary markets to recover their costs.

The fact that selling one or more shows to the networks can result in ruinous losses, when seemingly the achievement should guarantee profits, has deterred a number of companies from producing for television.

DE FOREST, LEE (DR.) (d. 1961)—Inventor of the three-element audion tube in 1906, vital to the development of radio and television and for which he became known as "the father of radio." He was credited with more than 300 inventions including, in 1948, a device for transmitting color television. But he was disappointed with the use to which his inventions were put and in later years bitterly criticized the broadcasting industry for what he felt was its excessive commercial orientation.

DEINTERMIXED MARKET—an all-UHF market. In its 1952 table of assignments, the FCC put all stations on UHF in certain markets that previously had been "intermixed" with some VHF and some UHF stations. Fresno, Calif., for example, originally had a Channel 12. That station was reassigned to Channel 30 when the market was deintermixed.

DELLA-CIOPPA, GUY—producer and CBS executive, whose career with the network ran from the late 30s through the late 60s, principally as a vice-president in Hollywood. He served as executive producer of the *Red Skelton Hour* during its last several years on the network.

DELTA— see Cable Networks.

"DELVECCHIO"—hour-long CBS police drama series (1976–77) about a brash but brilliant cop with a law-school education. The series was created as a vehicle for Judd Hirsch, who came into TV prominence in a 1975 NBC miniseries, *The Law.* Sam Rolfe and Joseph Polizzi created *Delvecchio,* and it was produced by Crescendo Productions and Universal TV. William Sackheim was executive producer and Steven Bochco and Michael Rhodes the producers.

DEMOGRAPHICS—rating data descriptive of *who* is watching, breaking down the audience by age group, sex, income levels, education and race. For most buyers of TV time desirous of reaching a target audience, demographic information is vital. For example, money may be considered misspent when advertising, intended for women, occurs in a program that has a large proportion of male viewers.

In rating reports at both the national and local levels, at least 30 demographic breakouts are given. The age breakouts generally occur in these groupings: 2–11, 12–17, 18–49 and 50–plus. On the premise that young married couples with young children are the prime prospects for most of the products and services advertised on television, the advertising industry consensus gives the highest priority to reaching viewers 18–34 and 18–49. Programs reaching those age groups predominantly command higher rates per thousand than programs more successful with younger or older viewers.

The purchasing of time according to demographics, which became widespread practice in the mid-60s, led networks and stations to design virtually all their programming for the most desirable age groups, in the knowledge that older and younger viewers would watch television in any case. A corollary effect is that programs with favorable demographics are often retained in the schedule even if their total audience falls below the required one-third share, while those with unfavorable demographics may be canceled despite substantial ratings.

Demographic data are often referred to as "persons" or "people" ratings.

DENIAL OF CONSTRUCTION PERMIT—[Young People's Association for the Propagation of the Gospel/6 FCC

178 (1938)]—instance in which the FCC denied an application for a construction permit primarily because of the applicant's policy of refusing to allow its broadcast facilities to be used by persons or organizations wishing to present any viewpoints different from its own.

DENMARK, TELEVISION IN—operates on a single channel under the Danish State Radio & Television Service, supported by license fees and spots. Color standard is 625-line PAL, and the country has around 1.6 million sets. TV households pay the equivalent of $140 to license color TV.

"DENNIS THE MENACE"—successful CBS situation comedy (1959–63) about a likeable but mischievous child. Inspired by the Hank Ketcham comic strip, it was produced by Darriel Productions and Screen Gems. It featured Jay North in the title role and Herbert Anderson, Gloria Henry and Gale Gordon (whose role was played the first three years by Joseph Kearns). NBC began carrying the reruns in daytime in 1961.

DENNY, CHARLES R. JR.—FCC chairman (1946–47) who resigned to join NBC as v.p. and general counsel. He became executive v.p. of operations for NBC in 1956 and two years later was made a v.p. of RCA. A lawyer, he had joined the FCC in 1942 as assistant general counsel and was appointed a commissioner in 1945, before being named chairman.

DEREGULATION—the reduction or elimination of existing regulation, discussed principally in connection with radio and cable-TV. It was born of the real or imagined mood of the FCC under the chairmanship of Richard E. Wiley during the 70s.

Deregulation fever became a Washington epidemic during the Carter Administration, born of desires from liberals and conservatives alike to diminish the government's involvement with business and to reduce the cost of government by eliminating unnecessary paperwork and scaling down the size of Federal agencies. Under Charles Ferris, the FCC undertook a number of initiatives to cut back the regulation of cable-TV and remove outmoded rules for television. But its major effort in this sphere was to propose an experimental, partial deregulation of radio. The spirit of deregulation was also evident in the various bills to rewrite the Communications Act, the House bill going so far as to propose replacing the FCC with a new and smaller agency.

Under Mark Fowler, in the Reagan Administration, deregulation went into high gear with the elimination of ascertainment procedures and initiatives to eliminate the Fairness Doctrine and Equal Time rules. Fowler declared himself not for deregulation, but rather what he termed "unregulation."

See also Cable-TV Rules; Communications Act; Federal Communications Commission; Ferris, Charles D.; Rewrite of Communications Act; Inquiry into the Economics Relationship between Television Broadcasting and Cable Television; Van Deerlin, Lionel.

DESILU—independent Hollywood production company created to produce *I Love Lucy* in 1950 and which went on to provide numerous other series to the networks, such as *December Bride, The Untouchables* and *Mannix,* until it was purchased by Gulf and Western (Paramount TV) in 1967 for about $20 million. The company's name melded the first names of the stars of *I Love Lucy,* Desi Arnaz and Lucille Ball, who were real-life husband and wife until their divorce in 1960. Miss Ball then became president of Desilu Productions and Desilu Studios, the old RKO lot which the company had purchased. Over the years Desilu also produced such series as *Our Miss Brooks, The Greatest Show on Earth, Glynis, Fair Exchange* and *Whirlybirds.*

"DESILU PLAYHOUSE"—dramatic anthology hosted by Desi Arnaz which ran on CBS from 1958 to 1961. It produced 54 films that later were put into syndication by Desilu Productions.

"DETECTIVES, THE"—police series starring Robert Taylor which premiered on ABC in 1959 as a half-hour show and in 1961 was expanded to a full hour, terminating in the spring of 1962. Produced by Four Star and Hastings Productions, it featured Tige Andrews, Adam West and Mark Goddard and added Ursula Thiess when it went to an hour.

"DIANA"—situation comedy on NBC designed for British actress Diana Rigg *(The Avengers).* It ran 13 weeks in 1973, was produced by Talent Associates/Norton Simon Inc. and featured David Sheiner, Richard B. Shull and Barbara Barrie.

"DICK VAN DYKE SHOW, THE"—a classic situation comedy that overcame initial viewer indifference to run five seasons on CBS, starting in 1961. It was a domestic comedy with show biz wrinkles, since Van Dyke's fictive office life was as a writer of a television show. Mary Tyler Moore, who played his wife, was propelled to TV stardom by the series. Rose Marie, Morey Amsterdam and Richard Deacon portrayed Van Dyke's colleagues, and Carl Reiner, who created the show and was its producer (also writing and directing some episodes), made occasional appearances as the star of the TV show within the show. Sheldon Leonard was executive producer. It was via Calvada Productions and T&L productions.

Mary Tyler Moore & Dick Van Dyke

DICKERSON, NANCY— first female correspondent for CBS News (1960-63) and the first woman on television to report from the floor of a national convention. She moved to NBC in 1963, where she was given a daytime news program of her own, from Washington. She became a favorite of Presidents and a television celebrity, matters which became difficult to reconcile with her work as a news correspondent. She left NBC at the end of the decade after a contract dispute.

Later she represented PBS as one of four network correspondents in a live one-hour interview with President Nixon. Since then she has participated in special news programs and an independent production company, Television Corporation of America.

She began her career as Nancy Hanschman (before marriage), starting as an associate producer for *Face the Nation* during the 50s. Later she became a producer of special events programs for CBS News, before being named a correspondent in 1960.

DIEKHAUS, GRACE M.—since January 1982 she has been producer of *60 Minutes*. Previously she was senior producer of *CBS Reports* in addition to being executive producer of *Magazine*, CBS News's monthly daytime informational series, and also of *Your Turn: Letters to CBS News*, a bimonthly letters-to-the-editor broadcast. Earlier she was senior producer of *Who's Who* (1976–77) and a line producer of *60 Minutes*.

"DIFF'RENT STROKES"—NBC situation comedy, introduced late in 1978, that became one of that network's few hits in the 1978–79 season. Produced by Norman Lear's T.A.T. Productions, the show developed as a vehicle for a precocious 12-year-old comedian, Gary Coleman. The premise is that two black youngsters from Harlem become residents of Park Avenue and are introduced to upper-middle-class life, when they are adopted by a wealthy white gentleman, played by Conrad Bain.

DIGITAL SIGNALS—information transmitted in discrete pulses rather than as continuous signals. Digital signals on a cable system are used principally to transmit data and messages, rather than television programs.

DILDAY, WILLIAM H., JR.—first black general manager of a TV station, WLBT-TV in Jackson, Miss. He was named to head the station by a nonprofit caretaker group that had been given temporary custody of WLBT when the original owners, Lamar Life Insurance, lost the license in 1969 on race-discrimination grounds. The station, an NBC affiliate, prospered under Dilday's stewardship, increasing its share of the market. Previously, Dilday had been an executive with WHDH-TV in Boston.

DILL, CLARENCE C. (d. 1977)—former Democratic U.S. Senator from the State of Washington, known as the "Father of the Communications Act." Dill served two terms in the U.S. House of Representatives and in 1922 was elected to the U.S. Senate, where he also served two terms. He was co-author of the Radio Act of 1927 that established the Federal Radio Commission, the predecessor of the FCC. In 1933, Dill was named chairman of the Senate Interstate Commerce Committee and helped draft the Communications Act of 1934, which created the FCC and gave the seven-member Commission authority to regulate all interstate and foreign communication by wire and radio.

DILLE, JOHN F., JR.—prominent broadcast figure in the 50s and 60s as head of an Indiana group with stations in South Bend, Elkhart and Fort Wayne; chairman of the ABC-TV Affiliates board of governors; and chairman of the NAB's joint board of directors. Dille also headed the National Newspaper Syndicate and was president of the *Elkhart Truth*, a daily newspaper.

DILLER, BARRY—Chairman of Paramount Pictures who had been a program v.p. for ABC when he was tapped for the post in 1974. The appointment of a middle-echelon TV executive to head a major movie studio startled the industry, but Diller had unique qualifications for the job: As executive in charge of the ABC Tuesday and Wednesday *Movie of the Week* showcases, he was responsible for more movies a year than any Hollywood studio released theatrically. Moreover, the ABC *Movie of the Week* series were successful, and several of the individual films were outstanding. Diller had demonstrated an ability to judge movie properties, deal with a variety of producers and bring the films in on budget.

He had joined the ABC program staff in 1966 and soon became its specialist in feature films. When his *Movie of the Week* series carried ABC through an otherwise lackluster season, he was elevated to v.p. in charge of prime-time programs, his position when Paramount hired him away.

DIMBLEBY, RICHARD (d. 1965)—journalist whose unique career over nearly three decades established him as the best-known broadcast personality in Great Britain. His fame crested as the semiofficial voice of the BBC on state occasions ranging from coronations to royal weddings and funerals. He died of cancer in December 1965 at the age of 52.

Born of a newspaper family in the London suburb of Richmond, in Surrey, he began his career in print journalism. He joined the BBC news department in 1936, became its first foreign correspondent and, in 1939, its first war correspondent. During the war he reported from 14 countries, culminating with his entry into Berlin with British troops in 1945. A year later he went from BBC staff status to freelance, launching a career as an "actuality" anchorman-commentator that saw him cover everything from the coronation of Queen Elizabeth II to state visits by foreign dignitaries, royal weddings, and the funerals of presidents, prime ministers and popes.

His career benchmarks included participation in the first live Eurovision broadcast, the first pickup by Eurovision from the Soviet Union and the first trans-Atlantic telecast by satellite.

Over the years he also appeared on British television in a variety of specials, as a quiz show panelist, and as a public affairs anchorman, notably on BBC-TV's long-run prime weekly documentary hour, *Panorama.*

His famous name is perpetuated in British television by his two sons, David and Jonathan, both prominent as on-camera newsmen.

"DINAH!"—syndicated talk-variety series featuring Dinah Shore and guests. Sold in both 60- and 90-minute versions, the daily programs began in September 1974 with the CBS o&os as its base. The series was produced by Henry Jaffe Enterprises and syndicated by 20th Century-Fox TV.

"DING DONG SCHOOL"—an early NBC program for preschoolers (1952–56) conducted by Dr. Frances Horwich, who became widely known to viewers as "Miss Frances." The network series originated in Chicago and was continued (1957–59) on the independent station there, WGN-TV. Later it was syndicated on tape.

The program was celebrated at first for its educational value but later drew some criticism for "talking down" to children. Dr. Horwich, who had been a professor of education at Roosevelt University, had a slow, deliberate way of speaking and was very much in the schoolmarm mold. Billed as "the nursery school of the air," the program was essentially instructional. As TV developed more exciting techniques of presentation, *Ding Dong School* seemed ever more talky and heavy, and when cartoon shows for children were placed in competition *Ding Dong* was done.

DIRECT BROADCAST SATELLITES (DBS)—technology of transmitting broadcast-quality signals from a satellite directly to home antennas, eliminating the TV station as intermediary. Japan and Canada began practical experiments with DBS in the late 70s, using low-powered satellites, as a means of reaching remote rural areas that cannot otherwise receive TV service. In Europe, France, West Germany and the Nordic countries have been considering DBS as an alternative to national television by terrestrial lines, chiefly because it would be the most cost-effective way to reach the last 1% of the population. In the experiments underway, parabolic dishes of 1-meter diameter suffice for home receiving equipment.

DBS is a source of concern to the American broadcast lobby for the obvious reason that it could obviate the need for TV stations. It also poses competition to cable systems. By 1982, there were 13 applications before the FCC for orbital slots, including CBS and RCA. The FCC's action on the bids was deferred until the 1983 Regional Administrative Radio Conference, which would determine how much spectrum would be allocated to the U.

See also Satellite, Communications.

"DIRECTIONS"—ABC's principal religion series since 1965 which deals—through drama, documentary and other forms—with history, heroes and social issues pertinent to each faith. Begun by the network as *Directions '65,* the series rotates among the Jewish, Catholic and Protestant faiths with production assistance from each, through the Jewish Theological Seminary, the National Council of Catholic Men, the National Council of Churches. Sid Darion is the executive producer.

DIRECTORS GUILD OF AMERICA —union representing directors, assistant directors and unit managers in TV-radio and films, organized in 1960 from a merger of the Screen Directors and the Radio-TV Directors guilds. Later it absorbed unions for a.d.'s and stage managers and in 1965 also was joined by the Screen Directors International Guild. The power of numbers—more than 3,000 members—gave DGA the leverage to bargain effectively with networks and individual production companies.

In addition to providing for minimum wage scales, working conditions, residual payments and screen credits in its contracts, DGA in its 1964 negotiations with the Assn. of Motion Picture and Television Producers established a "Director's Bill of Rights," which in spelling out the director's creative responsibility also guaranteed certain artistic rights. Those included his right to screen the *dailies,* to receive a *director's out* of the completed film and to participate in the dubbing and scoring process.

DISC RECORDER—see Instant Replay.

"DISNEYLAND"—Walt Disney's first show for television which was probably also ABC's first prime-time smash hit, causing CBS and NBC affiliates in two-station markets to scramble for a secondary affiliation with ABC. The show, which premiered Oct. 27, 1954, was presented in the early evening and included action-adventure segments, animation with the familiar Disney characters and nature sequences. The success of *Disneyland* altered industry thinking about programs designed for children and contributed to raising the overall quality of productions for children.

The following fall Disney added *The Mickey Mouse Club* on ABC, another huge success. *Disneyland,* meanwhile, not only helped the financially ailing network but served to promote the Disneyland amusement park in Southern California, which opened almost concurrently with the TV series. *Disneyland* ran on ABC until the fall of 1958, when it moved to NBC as *Walt Disney's Wonderful World of Color,* with essentially the same format. Becoming a Sunday night fixture, it was later retitled *The Wonderful World of Disney.* NBC gave up the series in 1981 because it had gone into a ratings decline. CBS then entered into a pact with Disney for an occasional series.

DISNEY CHANNEL— see Cable Networks.

DISNEY, WALT (d. 1966)—founder of the multimillion dollar Walt Disney empire—movies, television, amusement parks, recordings and merchandise—that grew out of the garage studio where he created his famous cartoon characters Mickey Mouse and Donald Duck. His Anaheim, Calif., animation studios expanded from shorts to full-length features *(Snow White and the Seven Dwarfs),* television series, and amusement parks (Disneyland in California and Disney World in Florida).

As one of the first movie producers to move into TV production, Disney produced such programs as *The Mickey Mouse Club, Zorro* and the weekly series running under various titles such as *Walt Disney's Wonderful World of Color* and *The Wonderful World of Disney,* which became the anchor of NBC's Sunday night schedule. Disney refused to release most of his movies to television, contending that every seven years there would be a new theatrical audience for them. However, in 1982, with pay-cable blossoming into a new market, the entire Disney library was being readied for the new premium service, The Disney Channel, scheduled to begin in 1983.

DISTANT SIGNAL—a station from another market imported by a cable system and carried locally by it. The right to bring in distant signals on a limited basis has been a selling point for cable-TV, especially in areas served by fewer than three networks or in those having no independent stations. Because of their coverage of sporting events and their heavy reliance on vintage movies, the independents have tended to be a desired distant signal for cable systems in the smaller towns and have thereby extended their reach. See also Leapfrogging; Inquiry into the Economic Relationship between Television Broadcasting and Cable Television.

"DIVORCE COURT"—a successful low-budget syndicated series, produced on tape, which recreated actual divorce trials in the setting of Los Angeles Domestic Relations Court. The cases were argued by professional lawyers, with actors portraying the plaintiffs, defendants and witnesses. Produced from 1958 through 1961, the series, owned by Storer Programs Inc., amassed a library of 130 programs. Syndication sales were handled by NTA.

DIXON, PAUL (d. 1975)—popular local TV personality in Cincinnati whose *Paul Dixon Show,* a live morning variety strip, ran 20 years on WLWT and in its later years on three other Avco TV stations in Ohio.

DJNS (DOW JONES NEWS/RETRIEVAL SERVICE)—Dow Jones Co.'s powerful early entry in the videotext field. During 1981, DJNS quickly soared past its chief rivals and predecessors in the "online data base" or videotext field (The Source and CompuServe), gaining more than 30,000 subscribers by year's end. This was accomplished by offering a strong mix of financial services and business news of the sort that Dow Jones's print products—*The Wall Street Journal, Barron's*—had long found profitable. Interestingly, DJNS also offered an online transcript of Louis Rukeyser's popular PBS series *Wall Street Week*.

DMA (DESIGNATED MARKET AREA)—Nielsen Station Index term used in local market reports for a geographic market design that defines the market, exclusive of any other. A DMA includes all counties, or portions thereof, in which home market stations command the preponderance of viewing. There is no geographic overlap. See also ADI.

"DOC"—CBS situation comedy introduced in the fall of 1975, centering on an aging general practitioner and his friends and patients in an urban neighborhood. It had a satisfactory initial season on Saturday nights and won renewal for a second year but was canceled in 1976. Produced by MTM Enterprises, it featured Barnard Hughes, Irwin Corey, Elizabeth Wilson and Mary Wickes.

"DOC ELLIOTT"—short-lived ABC series about a modern rural doctor that premiered in October 1973 and lasted 15 weeks. It starred James Franciscus, featured Neva Patterson, Noah Beery and Bo Hopkins, and was produced by Lorimar.

Richard Chamberlain

"DOCTOR KILDARE"—a 1961 re-creation of the MGM movie series of two decades earlier. Produced for NBC in response to ABC's hit doctor series *Ben Casey, Kildare* starred Richard Chamberlain and Raymond Massey. It received good ratings and ran through 1966, totaling 142 episodes.

Produced by Arena Productions in association with MGM-TV, the series made a TV star of Chamberlain.

"DOCTORS, THE"—popular NBC daytime serial that premiered April 1, 1963, and whose stories are built on the personal lives of the medical staff of Hope Memorial Hospital. Members of the cast have included Lydia Bruce, Jim Pritchett, Maria Danziger, Larry Riley, Amy Ingersoll, Jill Kasoff, David O'Brien, Meg Mundy and Elizabeth Hubbard.

DOCU-DRAMA—TV dramas based on actual historical occurrences and real people, blending, as it were, the elements of documentary and theater. With the rise of the miniseries and made-for-TV movies, the form grew increasingly popular during the 70s, but it created an uneasiness among journalists and historians for the distortions of history and truth that could result from invented dialogue, minor plot fabrications, the rearranging of facts and other uses of literary license. Shakespeare's histories, of course, were a form of docu-drama, and critics of the trend in television cited them as examples of how those imaginative and often careless renderings of history have been widely accepted as true and definitive accounts.

Most of the fact-based dramas on TV were drawn from contemporary history. Among the dozens presented on the networks in the 70s were *The Missiles of October, Pueblo, King, Ike, Eleanor and Franklin, Blind Ambition, Friendly Fire, Tail Gunner Joe, The Ordeal of Patty Hearst, Fear on Trial, Clarence Darrow, A Man Called Intrepid, Backstairs at the White House, The Trial of Lee Harvey Oswald, The Amazing Howard Hughes, Holocaust* and *Roots I* and *II*.

DODD, THOMAS J. (SEN.) (d. 1971)—leader of an intense and widely publicized 1961 crusade against TV violence as chairman of the Senate Subcommittee on Juvenile Delinquency. Formal hearings on the subject, presided over by the Connecticut senator in June and July of that year, produced evidence that network officials continually ordered programers and film studios to increase the violence content in programs for the sake of ratings. Dr. Albert Bandura, professor of psychology at Stanford, who appeared as a witness, cited studies showing that after watching violent films children played more aggressively than before and tended to imitate the behavior they saw.

But Sen. Dodd suddenly lost his fervor, drew back from the investigation and let the crusade peter out. Disillusioned members of his staff attributed his change of attitude to friendships he had made in the broadcast industry.

DOERFER, JOHN C.—chairman of the FCC from 1957 to 1960 whose resignation came as a result of charges of official misconduct during an investigation of the commission by the House Subcommittee on Legislative Oversight.

Doerfer was accused by Dr. Bernard Schwartz, chief counsel for the subcommittee, of undue fraternization with the broadcasting industry, of taking trips paid for by organizations regulated by the FCC and of receiving honoraria while being reimbursed by the Government for his expenses.

Answering the charges in public hearings on Feb. 3, 1959, Doerfer argued that a commissioner is not to be likened precisely to a judge, since he is also engaged in legislation and rule-making, and therefore must move outside the FCC for information on the problems and conditions of the industry. He noted also in his defense that, under the Communications Act, commissioners were permitted to receive honoraria for the publication or delivery of papers.

Since there was no evidence that any of Doerfer's decisions in official matters had been affected by *ex parte* influences, the subcommittee took no punitive action against him. But the tempest raised by the charges made it difficult for Doerfer to continue to serve on the commission and he resigned on March 10, 1960. He was promptly hired as a vice-president and counsel by Storer Broadcasting Co., the company prominently cited in the investigation as having entertained him lavishly.

DOERFER PLAN—a tacit bargain struck with the networks by FCC chairman John C. Doerfer in 1960 that each would present a weekly one-hour public affairs series in prime time in atonement for the quiz show scandals. Moreover, Doerfer asked the networks to agree not to place those programs in competition with each other. When it was argued that such an agreement would be in violation of the antitrust laws, Doerfer produced a written opinion from the Justice Department permitting collusion by the networks for such a worthy purpose. The Doerfer plan resulted in a boon for the news documentary that lasted about three years.

The precedent made it possible for future FCC chairmen to suggest that the networks rotate a quality program at the same time each day or that they rotate a noncommercial children's show. Neither of those suggestions was acted upon, but the Doerfer plan did break the ground for network agreements to rotate news coverage of such protracted events as the Senate Watergate hearings, with the provision that any network at any time could duplicate part of the coverage on its own news judgment.

DOMSAT—colloquial abbreviation for domestic satellite, relaying communications within a single country: Canada's Anik, the U.S.'s Westar and Satcom and the Soviet Union's Molniya system. See also Satellites, Communications.

DONAHUE, PHIL—host of a long-running syndicated talk show emanating from the Midwest. It began on WLWD-TV Dayton, in 1967 as *The Phil Donahue Show* and initially was fed to sister Avco stations but soon extended its syndication nationally. After six years the program was retitled *Donahue* and moved its originating base to WGN-TV Chicago. In March 1976 Avco, which had been disbanding its broadcast group, sold the program to Multimedia.

An hour-long daytime program, it featured interviews and discussions with celebrities and notable figures in a broad range of fields but emphasized topical matters.

The program grew to become one of the most successful in syndication, carried on 160 stations, and in 1979 it added a half-hour evening edition—an edited-down version of one of the morning shows—for prime-access slotting. Meanwhile, in the spring of 1979, Donahue began doubling on NBC, his interview features becoming a thrice-weekly element in the *Today* program.

DONALDSON, SAM—Capitol Hill correspondent for ABC News who joined the Washington bureau in 1967 after six years as a news producer, moderator and sometime anchorman for WTOP-TV Washington. He began his career in 1959 with KRLD-TV Dallas.

Sam Donaldson

"DON KIRSHNER'S ROCK CONCERT"—90-minute syndicated series showcasing leading rock music artists, slotted by most stations in a late Friday or Saturday night time period and hosted by Kirshner, a music publisher and record producer. Produced by Kirshner Entertainment in association with Viacom, the distributor, it premiered in the fall of 1973.

"DONNA REED SHOW, THE"—wholesome, domestic situation comedy that had a long run on ABC (1958–66) and whose reruns began to be stripped in daytime in 1964. Produced by Todon-Briskin in association with Screen Gems, it starred movie actress Donna Reed and featured Carl Betz, Paul Petersen and Shelley Fabares.

"DONNY AND MARIE"—popular weekly variety hour (1975–78) built around the youthful brother-sister team of Donny and Marie Osmond. It featured the Osmond Brothers vocal group and Shipstad and Johnson's Ice Follies. By Osmond Productions and Sid & Marty Krofft Productions, it was guided by Raymond Katz as executive producer, the Kroffts as co-producers and Art Fisher as director. See also Osmonds, The.

The Osmonds

DORFSMAN, LOUIS—top art director and designer for CBS, serving as v.p. of advertising and design for the Broadcast Group since 1968 and previously as director of design for CBS Inc. He was responsible for much of the interior decor at the CBS headquarters and for the graphics and design of all printed or packaged matter representing the company. He joined CBS in 1946 as a staff designer.

"DORIS DAY SHOW, THE"—a "heart" comedy designed for movie actress Doris Day, in which she portrayed a widow working as a magazine writer and raising two sons. Produced by Arwin Productions–Terry Melcher for CBS (1968–72), it featured Rose-Marie, Denver Pyle, John Dehner, Jackie Joseph and in the latter seasons Peter Lawford.

DORSEY, JOHN—director associated with Chuck Barris Productions whose work included *The Newlywed Game, The Dating Game, The Game Game, The New Treasure Hunt* and *Dream Girl of '67*. Earlier, as a producer-director for NBC, he directed sports and variety programs.

DORSO, DICK—executive of United Artists TV during the early 60s who was close to CBS-TV president Jim Aubrey. He was a generally successful supplier of programs to the network, but in 1965, after five of his pilots were rejected by CBS at a cost of more than $1 million, UA began its withdrawal from network production and Dorso left the company and the industry.

DORTORT, DAVID—producer specializing in western series who was executive producer of *Bonanza* and *High Chaparral*.

DOUBLE BILLING—fraudulent station practice of sending duplicate bills to a local advertiser participating in co-op advertising with a national company. Under the procedure, the invoice with the higher rate is sent to the national advertiser, who compensates the local merchant by paying 50% or more of the total under their cooperative agreement. But the local advertiser will in fact, by prearrangement, pay the station by the invoice with the lower rate. Occasionally the national advertiser's payment will be more than the actual cost of the spot, so that the local merchant makes a profit from the advertisement. Stations caught at double billing are subjected to heavy fines from the FCC, and the practice also carries the threat of revocation of license. The penalties are severe because the FCC views the fraudulent procedures as indicative of poor character for public trustees.

Variations on the double-billing ruse are the "upcutting" of a network program to insert local commercials (or the replacement of network commercials with local spots) and the failure of stations to advise advertisers that they have earned a discount for frequency.

DOUGLAS, MIKE—uniquely successful syndication performer whose daily talk-variety program, *The Mike Douglas Show*, via Group W Productions, is in its second decade and was valued by stations for late afternoon or early evening time periods. Douglas hasn't a single major credit on the networks and yet is one of the highest-paid performers in TV. In 1972, when other syndicators tried to woo him away from Group W, he negotiated a new contract paying him $2 million a year. In 1980, Group W dropped Douglas for a younger host, John Davidson. Douglas decided to keep his show going by producing it himself with a different distributor, but he threw in the towel in late 1981. The following year he became host of a daily hour interview show on the Cable News Network.

Originally a singer, Douglas acquired experience as host of a celebrity talk-show on radio in Chicago. His 90-minute daily TV show began at the Westinghouse (Group W) station in Cleveland in the early 60s, and when the FCC ordered Westinghouse and NBC to swap their Cleveland and Philadelphia stations, Douglas' base was moved to KYW-TV in Philadelphia. His show emanated from there until 1978, except for occasional tours about the country, and was such an effective showcase for talent that performers gratefully traveled from New York and other cities to appear on it. More recently, the show has been produced in Hollywood and Las Vegas.

DOWNS, HUGH—TV personality who began as a network announcer in Chicago. He became known to the mass audience as Jack Paar's sidekick and foil during Paar's reign on NBC's *Tonight* show and then won the assignment as host of *Today* (1962–71). For many of his years with *Today,* he doubled as host of *Concentration*. In semi-retirement, he continued to do commercials, specials and the syndicated series *Not For Women Only* as alternate with Barbara Walters. He became host of ABC's newsmagazine *20/20* in 1978. and is considered to have been the key to one program's rise in the ratings.

In his Chicago years he was associated with such series as *Kukla, Fran & Ollie* and *Hawkins Falls,* and later worked on NBC's *Home, The Sid Caesar Show, The Arlene Francis Show* and the radio series, *Monitor.*

Hugh Downs

DOZIER, WILLIAM—CBS program executive (1951–64) assigned to the West Coast for most of that period. He then

formed an independent company, Greenway Productions, which produced *Batman* and *The Green Hornet,* among other series.

DRAFT NOTICE—an FCC document issued in provisional form to permit discussion before the final version is written.

"DRAGNET"—hit police series based on actual case histories on NBC carried over from radio in 1952 for a seven-year run and then revived again successfully in 1967 for three more seasons. Starring and directed by Jack Webb, the half-hour series was noted for its style, which anticipated cinema verite even on radio, and made a household word of Sgt. Joe Friday, the character played by Webb. It also made famous Friday's workaday line of dialogue, "All we want are the facts, ma'am." The first version featured Ben Alexander as Webb's sidekick, the second, Harry Morgan. Both shows were produced by Webb's own firm, Mark VII Ltd., but the revival was in association with Universal TV. Reruns went into syndication with the title of *Badge 714.*

DRAMA ON TV—see "Golden Age."

DRASNIN, IRV—producer, director and writer for *CBS Reports* who also narrated some of his programs. His work included *The Guns of Autumn, New Voices From the South, Voices From the Russian Underground, Health In America, Cuba: 10 Years of Castro* and *Football: 100 Years Old and Still Kicking.* He also made documentaries for *The Twentieth Century* series.

DROP-INS—additional VHF channels that could be "dropped in" to the existing FCC table of allocations without causing interference, according to an engineering study by the White House Office of Telecommunications Policy in October 1973. The OTP study said that 62 VHF channels could be added in the top 41 markets; a subsequent FCC study put the figure at 30 stations in 27 cities. The study argued that the FCC's standards for its table of allocations in 1952 was too conservative for the present state of the art, in requiring a spacing of 170 miles for stations assigned to the same channel in the northeast, 190 miles for those in the west and 220 miles for those in the south. The FCC requirement was also that stations on adjacent channels be a minimum of 60 miles apart. The OTP said that if the distance criteria were reduced 10% to 15%, a substantial number of VHF channels could be added in most major cities.

Because the proposed drop-ins would afford an opportunity for minority ownership of stations, the Office of

Communication, United Church of Christ, petitioned the FCC for a rulemaking on the feasibility of the new channels in 1975. The organization noted that of the 608 licensed VHF stations, only two were owned by blacks, and both stations were outside the continental U.S. (in the Virgin Islands).

In the FCC inquiry on the proposal in 1976, established broadcast interests argued that the proposed drop-ins would cause signal interference with existing stations and that if the FCC adopted the plan it would be a severe setback for UHF and would constitute a "breach of trust" by the commission, which had pledged to help develop the UHF band. The proponents contended that the drop-ins would improve competition and make possible a fourth commercial network.

Drop-ins were recommended for such cities as Miami, Atlanta, Kansas City, Milwaukee, Little Rock, Dayton, Spokane, Dallas-Ft. Worth, Houston, Seattle, San Francisco, Memphis, Nashville, Des Moines and Birmingham, among others.

DUBIN, CHARLES S.—producer-director whose work ranged from productions of Rodgers & Hammerstein's *Cinderella,* the Bolshoi Ballet and segments for *Omnibus* to episodes of *Kojak, Toma, Kung-Fu* and *Room 222.* He also directed the *Sanford & Son* pilot.

James E. Duffy

DUFFY, JAMES E.—president of the ABC television network since March 23, 1970, succeeding Elton H. Rule when he became president of the parent corporation. Duffy rose from the ranks, having joined ABC in 1949 as a publicity writer for its Chicago stations upon his graduation from Beloit College. He moved into sales for the radio network in

1953 and ten years later became vice-president in charge of sales for ABC-TV.

As network president Duffy gained industry recognition for the upgrading of program and advertising policies in television directed to children.

Under reorganization of the ABC television division in 1972, Duffy's authority as network president was substantially diminished, although his title remained in force. The new organizational structure established a higher new post, president of ABC Television, and at the same time removed the program department from network supervision. The areas left reporting to Duffy were sales and affiliate relations.

"DUKES OF HAZZARD, THE"—a "Southern," relative to a Western, specializing in automobile chases at high speeds. A one-hour CBS series set in the South, about a pair of car-jockeying brothers who delight in outwitting criminals and harassing the local police, it entered the schedule as a midseason replacement in 1979 and immediately caught on. Tom Wopat and John Schneider play the Duke brothers, Luke and Bo. The series was renewed became the ideal lead-in to *Dallas* during the 1979–80 season and their popularity was one of the main reasons CBS returned to prime-time dominance that season.

DuMONT, ALLEN B. (d. 1965)—pioneer television broadcaster, picture tube and TV set manufacturer. An outspoken, brilliant inventor, DuMont transformed the cathode ray tube from a fragile, short-lived device to a reliable piece of equipment around which practical TV receivers could be built. Before World War II his Allen B. DuMont Laboratories was manufacturing picture tubes and television sets, and after the war it took the leadership in developing larger and larger direct-view tubes. The DuMont Television Network, although it did not survive television's early days, bred much of the industry's programming and business talent. DuMont appeared frequently before the FCC particularly to oppose the CBS field-sequential color system and the intermixture of VHF and UHF channels in the same areas.

DuMONT TELEVISION NETWORK—a venture into nationwide broadcasting by the Allen B. DuMont Laboratories in 1946. It failed in competition with ABC for third-network position and went out of business as the fourth network in 1955. Its owned stations were later to form the nucleus of a major independent group, Metromedia Television.

While it was in operation, the DuMont Network primarily offered low-budget quiz, variety and sports shows, and on occasion it shared special programs with the other networks. It was DuMont that covered the historic Army-McCarthy hearings, but its leading programs otherwise were those of Bishop Fulton J. Sheen, Jackie Gleason and the Monday night boxing matches.

The network grew out of DuMont's ownership of commercial stations WABD in New York (originally W3XWT) and WDTV in Pittsburgh. Paramount Pictures, which had purchased a half-interest in the parent company in 1938, separately owned two West Coast stations.

But the network was unable to put together fully a competitive lineup of affiliated stations, chiefly because it lacked the radio station relationships that the other networks were able to transfer to television. Nor was its unglamorous array of programs an incentive for uncommitted stations, and even Paramount's own KTLA in Los Angeles declined to clear for the DuMont shows.

The merger of ABC with United Paramount Theaters in 1951 strengthened its position as the third network, leaving DuMont a weak fourth, making it difficult to justify the cost of transcontinental lines. Although its affiliations were variously reported as between 80 and 178 stations, the DuMont Network was a losing operation, and less than two years after building a $5 million studio in New York—the largest production center in the industry at the time—DuMont Labs terminated the network and separated itself from the broadcasting division. The owned stations were operated as the Metropolitan Broadcasting Corp. in 1955, and Paramount's holdings were purchased in 1959 by John Kluge, who later changed the company's name to Metromedia.

DuMont's collapse was a chief reason for a 68% rise in billings for ABC during 1955.

"DUMPLINGS, THE"—one of Norman Lear's few failures, an NBC situation comedy (1976) about a loving fat couple who run a small restaurant. It featured James Coco, Geraldine Brooks, George S. Irving and Mort Marshall and was by T.A.T. Communications and NRW Productions.

DuPONT-COLUMBIA AWARDS—citations for outstanding broadcast journalism conferred by the Alfred I. DuPont Awards Foundation and the Columbia University Graduate School of Journalism. The awards are made by a blue-ribbon jury and carry great prestige largely because they are the by-product of a comprehensive nationwide survey of local, national and syndicated broadcast journalism conducted throughout the year. Each survey since 1968 has resulted in the publication of a book, *Survey of Broadcast Journalism (year),* edited by Marvin Barrett, director of the DuPont-Columbia Survey and Awards. When the commercial networks gave up the special telecasts for the news and documentary Emmy Awards, PBS moved to fill the vacuum

by televising the DuPont-Columbia Awards. The first of those telecasts was in 1977.

DUNHAM, CORYDON B. JR.—NBC v.p. and general counsel since 1968 who supervises the legal staff in New York, Hollywood, Chicago and Washington. In 1977 he was elevated to executive v.p. and given charge of broadcast standards as well as the legal department. He joined the network in 1965 as an assistant general attorney from the prestigious New York law firm of Cahill, Gordon, Reindel & Ohl.

DUOPOLY—the FCC rule limiting ownership to one of each type of broadcast service—TV, AM and FM—in a community. The rule exempts public television, whose licensees are permitted to operate two TV stations in the same market, one on VHF and one on UHF. The duopoly question is raised whenever the coverage of stations under the same ownership overlap. Waivers are granted in most cases if the overlap is minimal. If the overlap existed before the present rules, "grandfathering" is applied.

Public TV duopolies exist in Chicago, Philadelphia, Boston, San Francisco, Pittsburgh, Minneapolis, Miami, Milwaukee and Richmond, Va. No new public TV duopolies were created between 1971 and 1976, but New York's Educational Television Corp. (WNET) hoped to establish a new one if it could secure New York City's divested municipal station, WNYC-TV on Channel 31.

Jimmy Durante

DURANTE, JIMMY (d. 1980)—comedian with background in nightclubs, radio and films, who became one of TV's preeminent personalities in the 50s. His trademarks were a substantial nose, earning him the affectionate nickname of "Schnozzola;" a gravelly and tuneless voice which he was able to use effectively to put over novelty tunes, such as his theme song, "Inka Dinka Doo"; a weather-beaten fedora and the closing line, "Good Night Mrs. Calabash, wherever you are." His exit on TV was always made on a darkened stage through three circular spots of light issued from above.

Don Durgin

Occasionally his TV shows featured the strutting Eddie Jackson, surviving partner in the famous nightclub act of Clayton, Jackson and Durante, formed in 1920, which started his career as a comic. Durante made his TV debut in *Four Star Revue* on NBC, in which he appeared on a rotating basis from November 1950 to May 1951. That gave way in the fall of that year to *All Star Revue,* on which he appeared until the spring of 1953.

In October 1953 he starred in *The Colgate Comedy Hour,* and the following fall headlined his own show, *Texaco Star Theater,* which ran a year. A loveable lowbrow, he reached some of his highest TV moments clowning and singing duets with Ethel Barrymore, Helen Traubel and Margaret

Truman. When his series ended Durante became a frequent variety show guest and always scored with his old nightclub finale, which involved the passionate destruction of a piano.

DURGIN, DON—president of NBC-TV (1966–73) and champion of the TV special, the 90-minute series, the rotating series and made-for-TV movies. He came into the presidency after holding a number of executive positions with ABC and NBC in the fields of promotion, research, sales and radio. For seven years before his appointment, he had been v.p. in charge of sales for NBC-TV.

When his administration ended he became corporate executive v.p. of NBC and a year later left to become president of Caffery-McCall Advertising. He resigned from the agency in 1976 and then became senior v.p. for Dun & Bradstreet, overseeing the broadcast subsidiary, Corinthian Broadcasting.

DURR, CLIFFORD J. (d. 1975)—FCC commissioner (19448) noted for his vigorous support of the commission's controversial Blue Book in 1946, the document that attempted to define the public interest for purposes of license renewals. Durr also was concerned with reserving TV channels for education. A political liberal from Alabama, he turned down reappointment to the FCC saying he was unable to administer the Government's loyalty program in good conscience. He returned to the practice of law in Washington and then moved to Montgomery, Ala., to handle civil rights cases. One of his clients was the late Martin Luther King, Jr.

gaining more than 30,000 subscribers by year's end. This was accomplished by offering a strong mix of financial services and business news of the sort that Dow Jones's print products*The Wall Street Journal, Barron's*had long found profitable. Interestingly, DJNS also offered an online transcript of Louis Rukeyser's popular PBS series *Wall Street Week.*

E

EARLY BIRD—Intelsat I, launched April 6, 1965, the world's first commercial communications satellite and the first synchronous communications satellite. Stationed over the Atlantic, it provided 240 telephone circuits or one television channel, increased trans-Atlantic communications capability by 50% and made live commercial television across an ocean possible for the first time. Although designed for a life of 18 months, it provided continuous full-time service for more than 3½ years. It was built by Hughes Aircraft Co. and launched by NASA. See also Satellites, Communications.

EARTH STATION—installation for transmitting and/or receiving electronic communications (such as television) between the earth and a space satellite. The prominent feature of the earth station is a parabolic antenna aimed at the satellite. The station also contains specially designed transmitters and/or receivers and amplification equipment. Earth stations began to proliferate at cable systems around the country in 1977 after the FCC approved the use of 5-meter dishes, smaller and considerably less costly than the previously authorized models.

"EAST SIDE, WEST SIDE"—hour-long dramatic series centering on a social welfare agency in New York which starred George C. Scott and featured Cicely Tyson, both relatively unknown at the time. Premiering on CBS in the fall of 1963, it was respected by discriminating viewers but was only moderately successful in the ratings—a fact attributed by many in the trade to its serious content and depressing themes. The series was produced by Talent Associates-Paramount in association with United Artists Television.

An RCA satellite antenna

EAST GERMANY, TELEVISION IN—a state-controlled service on two channels on the 625-line PAL standard. It is supported by government subsidies (the German Democratic Republic), license fees and spot advertising.

EBSEN, BUDDY—a former novelty dancer who became star of *The Beverly Hillbillies* (1962–71), a bucolic situation comedy and then star of *Barnaby Jones* (1973–80), an urban private-eye series. In films and variety revues, before TV, he was known for his pretzel-like soft-shoe turns. Before *Hillbillies* he was featured in an adventure series, *Northwest Passage* (1958). See also *Barnaby Jones, Beverly Hillbillies.*

Buddy Ebsen

ECKSTEIN, GEORGE—contract producer at Universal TV, who served as producer of the *Sunshine* and *Name of the Game* series and executive producer of *Banacek*. He also produced a number of TV movies, among them *Duel.*

EDELMAN, LOUIS F. (d. 1976)—Hollywood motion picture producer who began producing for TV in 1954. His hits on the networks included such series as *Make Room For Daddy, The Big Valley, Wyatt Earp* and various shows with Danny Thomas, including the 1952 remake of the movie, *The Jazz Singer.* Among his films were *White Heat, A Song To Remember* and *Hotel Berlin.* He was president of the Screen Producers Guild, 1965–67.

"EDGE OF NIGHT, THE"—soap opera which began its run on CBS on April 2, 1956, and switched to ABC in 1975. The cast has included Ann Flood, Forrest Compton, Beah Ayers, Frank Gorshin, Lois Kibbee, Joel Crothers, Sharon Gabet and Meg Miles.

EDITORIAL ADVERTISING—commercials intended to promote a point of view or an editorial position on an issue, rather than to stimulate the sale of a product or service. Standards & Practices policies of the networks reject such advertising (although the print media and some TV stations accept it) essentially because it may subject them to claims for response time under the Fairness Doctrine. The Supreme Court has ruled that broadcasters may not be forced to sell time for discussion of controversial public issues.

The networks contend, in defending their policies, that to sell time for editorial commercials ultimately would give those with the most money the loudest voices on the issues. They argue also that the public is better served when the points of view are expressed in news programs or information forums than when the message is wrapped in the manipulative techniques of advertising. ABC, however, experimentally began acceptiing issue advertising in late night slots in 1981.

EDUCATIONAL TELEVISION AND RADIO CENTER (ETRC)—early "networking" center for ETV established in Ann Arbor, Mich., in 1952 on a grant of $1.5 million from the Ford Foundation's Fund for Adult Education. Headed by Harry K. Newburn, former president of the University of Oregon, the center stored old educational programs and contracted for new material to serve neophyte educational TV stations. The filmed transcriptions were also rented to schools and private groups. In 1958 Newburn resigned and was replaced by John F. White, who had operated the Pittsburgh ETV station, WQED. White moved the production arm to New York and renamed it the National Educational Television and Radio Center (NETRC), which in 1962 became NET, the acknowledged national production center for ETV, supported by annual Ford Foundation grants of $6 million. NET broke with the classroom approach to ETV and provided the system with a broad range of cultural, public affairs, documentary and children's programming, breaking the ground for a network service and for the industry's eventual redesignation as Public TV. See also Public Television.

"EDWARD THE KING"—British historical series of 13 half-hours, airing in the U.K. as *The Royal Victorians* but retitled by Mobil Corp. for the presentation on 49 commercial TV stations in the U.S. in January 1979. The series, which spans the reigns of the austere Victoria and her sybaritic son, Edward VII, was originally purchased by CBS in 1974 for $2.5 million, but it sat in storage for nearly five years while the network struggled to keep its ratings up. Eventually Mobil, which had tried to sponsor the series on CBS, bought the rights for $1.8 million and put together an ad hoc network that was to create discomfort for CBS. As it turned out, 19 of the 49 stations assembled by Mobil were major-market CBS affiliates, which virtually wrecked the network's Wednesday night prospects for three months. Mobil required all stations to carry the program at 8 P.M. on Wednesdays. Of the remaining 30 stations in the Mobil lineup, 22 were independents, seven NBC affiliates and one an ABC outlet.

The program was both a critical and ratings success, beating on some occasions its network competition. The opening program ranked third in New York, second in Los Angeles and first in Washington.

Timothy West played Edward; Annette Crosbie, Queen Victoria; John Gielgud, Disraeli; Michael Hordern, Gladstone; and Helen Ryan, Alexandra.

EDWARDS, BLAKE—producer and writer who left television for movies after making his mark in the 50s as creator of such series as *Peter Gunn, Mr. Lucky* and *Dante's Inferno*. He became a topflight film producer with the success of *A Shot in the Dark* and the Pink Panther movies.

EDWARDS, DOUGLAS—CBS newscaster who delivered the network's 15-minute early evening report *Douglas Edwards with the News* from 1948 until Walter Cronkite replaced him in 1962. Edwards, who had begun his career in radio, remained with CBS News and continued to do daytime newscasts on television into the 70s.

EDWARDS, RALPH—game-show producer and host of *This Is Your Life,* popular series in the late 50s. That show and Edwards' *Truth or Consequences* were both revived for first-run syndication in the 70s.

EEN (EASTERN EDUCATIONAL NETWORK)—originally a hookup of around a dozen PTV stations in the northeast for purposes of sharing programs of common interest, but now a misnomer since the member stations extend to Texas and California. Since the late 60s, EEN has been a consortium of major public stations jointly purchasing programs not otherwise available through PBS. In the main, the EEN stations are the boldest and most sophisticated in the system.

EFFECTIVE RADIATED POWER (ERP)—the power of the signal radiated from the transmitting antenna, a function of transmitter power and antenna gain (or loss). The FCC generally permits maximum ERP of 100 kilowatts (100,000 watts) for stations on Channels 2 through 6, 360 kilowatts for 7 through 13, and 5 megawatts (5,000,000 watts) for UHF stations. If the antennas are above specified heights, reduced ERP is required.

EGER, JOHN—acting director of the White House Office of Telecommunications Policy (1974–76), after serving briefly as deputy to the OTP's first director, Clay T. Whitehead. Although he served responsibly and well in the post, he failed to receive President Ford's nomination as permanent director; some believe it was because he had alerted Congress and—through leaks—the press to the fact that the Office of Management and Budget was laying plans early in 1975 to abolish OTP as a White House agency. The controversy that ensued saved the OTP but cost Eger the position as its head. When Thomas E. Houser was named to the office in the summer of 1976, Eger resigned to practice law. In 1981 he signed on with CBS as an executive in its newly formed international division.

Before joining OTP Eger had been a legal aide to FCC chairman Dean Burch.

"EIGHT IS ENOUGH"—hit ABC comedy-drama, in one-hour format, which premiered in March 1977 and became a Wednesday night mainstay. The series centers on a family with eight children ranging in age from 8 to 23. Dick Van Patten plays the father, who works as a successful syndicated columnist, and until her death Diana Hyland was the mother. Van Patten ran the family as a single parent for a time, then took on a new wife, Betty Buckley.

The children in the series are portrayed by Grant Goodeve, Lani O'Grady, Lauri Walters, Susan Richardson, Dianne Kay, Connie Newton, Willie Aames and Adam Rich. The series was drawn from a book by Tom Braden. Lee Rich and Phil Capice are executive producers, Robert Jacks producer and William Blinn, who created the show, executive consultant. The series is by Lorimar Productions.

"87th PRECINCT"—police series set in New York and carried by NBC (1961–62). It featured Robert Lansing, Gregory Walcott, Ron Harper, Norman Fell and Gena Rowlands. It was by Hubbell Robinson Productions and owned by MCA-TV Ltd.

"EISENHOWER PRESS CONFERENCE" (Jan. 19, 1955) —the first presidential news conference simultaneously covered by the print press, newsreels and television, and at which candid photographs and direct quotations were authorized for broadcast, as recorded, by the White House.

EISNER, MICHAEL D.—president and chief operating officer of Paramount since October 1976, after 10 years as a program executive for ABC. He resigned from the network for the Paramount post several months after he had been named senior v.p. of ABC Entertainment in charge of prime-time production and program development in Hollywood. Eisner and Paramount's chairman, Barry Diller, had been colleagues in the ABC program department during the late 60s and early 70s.

Eisner had come to ABC after a brief stint with the CBS program department, and he served variously as v.p. of daytime programming, v.p. of program development and manager of specials and talent. He was instrumental in developing a number of series that figured in ABC's banner 1975–76 season, among them *Rich Man, Poor Man, Welcome Back, Kotter, Laverne & Shirley* and *Bionic Woman.*

Edward Herrmann & Jane Alexander as Franklin & Eleanor Roosevelt

"ELEANOR AND FRANKLIN"—five-part cycle by ABC on the relationship of Eleanor and Franklin D. Roosevelt through the various stages of their careers. It began as a four-hour adaptation of the Joseph P. Lash book, *Eleanor and Franklin,* presented in two parts on a Sunday and Monday night, Jan. 11 and 12, 1976. The ratings and generally favorable response prompted ABC to order two sequels, one covering the Roosevelts during their years in the White House to be presented in 1977 and a final sequel based on another book by Lash, *Eleanor: The Years Alone,* for 1978. Jane Alexander and Ed Herrmann portrayed the Roosevelts. Executive producer was David Susskind, producer Harry Sherman, director Daniel Petry and writer James Costigan.

ELDER, LONNE, 3d—black playwright who wrote the script adaptation of his *Ceremonies in Dark Old Men* for *ABC Theatre* and also the teleplay for the *Sounder* pilot.

"ELECTRIC COMPANY, THE"—a PTV series intended to help teach basic reading skills to slow readers, as a derivation of *Sesame Street.* Produced by the same company,

CTW, it premiered as a daily hour in October 1971, two years after the resounding success of *Sesame,* and was itself an educational/entertainment success. According to PBS estimates it reached 6.5 million youngsters regularly, both in school and in the home.

The program was planned to reinforce school reading curricula, using the winning entertainment techniques of television—comic vignettes, animation, electronic effects and music. CTW also produced a variety of support materials, such as teaching guides, books and puzzles, for use in schools and day-care centers.

Produced at a cost of approximately $33,000 per show, it was funded initially by foundations and later was sustained by payments from the Station Program Cooperative and outside grants. See also *Sesame Street,* Children's Television Workshop.

ELIASBERG, JAY—head of research for CBS-TV since 1957, receiving the title of v.p. of TV network research for the CBS Broadcast Group in 1974. Earlier he was a research executive for several ad agencies and ABC Radio.

"ELIZABETH R"—BBC six-part miniseries that played on PBS (1973) tracing the life of Elizabeth I from the ages of 15 to 69. It starred Glenda Jackson and aired in Britain in 1971.

"ELLERY QUEEN"—twice-produced series, first during the 50s and again by NBC in 1975. The original (1950–59) was successful, with the title role passing from Lee Bowman to Hugh Marlowe and then to George Nader and Lee Phillips (it was syndicated under the title, *Mystery Is My Business).* The 1975 edition, with the late Jim Hutton and David Wayne, was canceled after one season. It was by Universal TV, with Richard Levinson and William Link as executive producers.

EMBASSY PRODUCTIONS—company formed in 1982 when Norman Lear and his associates acquired Embassy Pictures. Embassy absorbed T.A.T. and Tandem Productions, the companies through which Lear and Bud Yorkin produced their most popular series, and became the umbrella for those enterprises and others in which the two were partnered with Jerry Perenchio.

"EMERGENCY!"—NBC adventure series (1972–77) about the heroics of the paramedic unit of the Los Angeles County Fire Department. The program's chief distinction was that it drew a sizable audience in its first two seasons against TV's No. 1 show at the time, *All In the Family* . An animated version, *Emergency Plus 4,* was spun off by NBC for Saturday mornings. Featured in the original version were Robert Fuller, Julie London, Bobby Troup, Randolph Mantooth and Kevin Tighe. It was produced by Jack Webb's Mark VII Productions with Universal TV, in association with NBC-TV, with Robert A. Cinader as executive producer and Ed Self as producer.

Faye Emerson

EMERSON, FAYE—a former Broadway actress who became one of early TV's most popular hostesses. Somewhat controversial for her decolletage, she worked steadily at both CBS and NBC, with *The Faye Emerson Show* running three seasons at the former (1949–52) while she was a panelist on *Leave It to the Girls* and frequent guest on *Who Said That?* on the latter.

Her other shows were *Fifteen with Faye* and *Faye andSkitch* (Henderson, her husband for a time). She also had dramatic roles in *Studio One, Ford Theatre* and *Goodyear Playhouse,* among others, and was a panelist on *I've Got a Secret*. At her popularity peak, she substituted for such personalities as Edward R. Murrow, Garry Moore, Dave Garroway and Arlene Francis on their shows.

EMMY AWARDS—annual awards conferred in recognition of outstanding achievement by the National Academy of Television Arts and Sciences, an organization of professionals in the industry. A rift between the Hollywood chapter and the rest of the National Academy caused the 1977 Emmy Awards telecast to be postponed, and a dispute ensued over the secessionist group's claim to ownership of the rights to the statuette and its name. The 1977 awards were conferred in the fall rather than the spring and were made by the Hollywood group rather than the national organization. The ceremonies were televised by NBC.

The first Emmy awards were conferred on Jan. 25, 1949 at the Hollywood Athletic Club, with the name "Emmy" adopted for the trophy as a variation on "Immy," a nickname for the image orthicon tube. The name was suggested by Harry Lubcke, a pioneer television engineer who served as president of the Academy from 1949–50. The statuette was designed by Louis McManus.

The awards ceremonies had their first national(telecast on March 7, 1955 and thereafter became annual, star-studded TV events rotated among the three networks. Local chapters of the Academy also make awards for TV achievement in their cities, and those ceremonies are sometimes televised. See also NATAS.

ENG—electronic news-gathering, the use of all-electronic means for television news coverage that eliminates film as an intermediate step. Virtually from the start of television, electronic news-gathering techniques have been used on special occasions, when there was opportunity to install portable microwave links from the scene of the news event to the studio, transmitter or networking point.

Electronic news-gathering received its biggest impetus after 1973 with the development of the time-base corrector (TBC), which made possible the use of lightweight portable (and usually inexpensive) helical-scan videotape recorders as a substitute for film cameras. The TBC converts the output of the helical-scan recorder into a picture with sufficient stability to be broadcast.

The major ENG components are a miniature color camera (minicam), small VTR and a power supply. The power supply, usually battery-operated, can be carried on the cameraman's back or worn on a special belt. Some VTRs are also equipped for backpacking. An additional ENG accessory is a portable microwave transmission station for use in cases where it is desirable, and possible, to send the live picture directly to the television studio, thus eliminating the necessity for the videotape recorder.

In most cases, however, the VTR is an integral part of the ENG outfit. When the news event has been recorded, the tape is taken to the station and usually rerecorded through the time-base corrector into a studio quadruplex recorder and then edited for airing. The editing may be done first on a helical-scan editing system. The helical-scan recorders usually employed in ENG are videocassette units.

Among the principal advantages of ENG over film coverage of news are: speed, through elimination of film processing; (2) saving in manpower, since electronic news-gathering usually can be handled by two persons, in place of a normal camera crew of three or more; economy, since the tape can be erased and reused; mobility through the use of an ordinary station wagon or even a sedan, as compared to a light truck often employed for sound film work. See also Minicam, Time Base Corrector.

"ENGELBERT HUMPERDINCK SHOW, THE"—one of several attempts by the British concern, ATV, to market a variety show in the U.S., and, like the others, unsuccessful. Humperdinck, a popular singer on both sides of the Atlantic, was the focus, with international performers as guests. ABC carried the series in 1970 for 18 weeks.

ENGLANDER, ROGER—director associated with artistic and cultural programs. He directed the CBS *Young People's Concerts, Vladimir Horowitz at Carnegie Hall, S. Hurok Presents, The Bell Telephone Hour* and episodes for *Omnibus, The Great American Dream Machine* and *The Performing Arts*.

"ENSIGN O'TOOLE"—one-season series on NBC (1962) about the comedy adventures of a naval officer and his shipmates, based on the William J. Lederer novel, *All the Ships At Sea*. The half-hour episodes starred Dean Jones and featured Jay C. Flippen, Harvey Lembeck, Jack Albertson, Jack Mullaney and Beau Bridges. Reruns of the 26 shows were later carried on ABC (1964). It was by Four Star and Lederer Productions.

"ENTERTAINERS, THE"—26 hour-long revues variously featuring Carol Burnett, Bob Newhart and Caterina Valente, with a permanent repertory company. Produced by Bob Banner Associates in association with CBS, it aired in 1964.

EQUAL TIME LAW—the provision under Section 315 of the Communications Act that requires broadcast licensees who permit their facilities to be utilized by a legally qualified candidate to provide equal opportunities to opposing candidates if such time is requested. The statute was intended to guarantee that broadcasting will be responsive to the fact that politicians are extraordinarily dependent upon the mass media's portrayal of their candidacy.

Section (a) of 315 states that if any licensee shall permit a legally qualified candidate to use a broadcasting station, he shall afford "equal opportunities" to all other candidates for that office. Thus, if one candidate buys time on a station, his opponents may buy equivalent time, but the station is not required to give free time in response to purchased time. The Equal Time Law differs in this respect from the Fairness Doctrine, with which it is often confused.

A legally qualified candidate is defined as one who has publicly announced his candidacy and who meets the qualifications prescribed by law for the indicated office (McCarthy *v.* FCC). The broadcast licensee has no power of censorship over the material broadcast by a candidate (Farmers Educational and Cooperative Union *v.* WDAY).

As a general rule, any use of broadcast facilities by a legally qualified candidate imposes an equal time obligation on the broadcaster for all other candidates for the office. In 1959, however, Section 315 was amended to exclude certain kinds of programs, namely, bona fide newscasts, bona fide news interviews (e.g., *Meet the Press*), bona fide news documentaries in which the appearance of the candidate is incidental to the presentation of the subject, and on-the-spot coverage of bona fide news events.

To establish whether a program is "bona fide," the FCC looks to a number of elements, such as the nature of the format and content, whether the program is regularly scheduled, when the program was initiated and who initiated, produced and controlled it. The commission has held that a news program scheduled to begin 11 weeks before the start of an election was not exempt.

In an important 1975 ruling, which made possible the televised presidential and vice-presidential debates of 1976, the FCC determined that live debates between candidates were bona fide news events exempt from equal time obligations to minor candidates if the debates were under the direction and control of an independent organization. This exemption bars the involvement of the broadcaster in all respects but that of covering the event live with cameras and microphones.

In the same reinterpretation of the Equal Time Law, the commission held that press conferences arranged by a candidate were also exempt if they were considered newsworthy by the broadcaster. However, the exemptions for both the debates and press conferences specified that they be covered live and in their entirety. (The FCC did grant waivers to independent stations to carry delayed telecasts of the 1976 debates.)

Section (b) of 315 requires that charges made for the use of a broadcasting station cannot—during the 45-day period preceding a primary election or during the 60 days preceding a general or special election—exceed the lowest unit charge of the station for the same class of time. At any other time, the charges made to candidates would be comparable to those for other users.

In 1971 Congress passed the Campaign Communications Reform Act which added a new sub-section to Section 312 of the Communications Act. The new sub-section specifies that a station license may be revoked for willful and repeated failure to allow reasonable access to candidates for federal elective office or to permit them to purchase reasonable amounts of time on the station. The FCC ruled that these provisions applied as well to noncommercial stations, and public TV stations were challenged to comply by Senator James Buckley of New York during his reelection campaign in 1976.

In 1970 the commission adopted a concept called "quasi-equal opportunity" or the "Zapple Doctrine," which specifies that when a station sells time to supporters of a candidate during an election, or to his/her spokespersons, then the licensee must afford comparable time to supporters or spokespersons for all opposing candidates. See also Fairness Doctrine, Lar Daly Decision, Political Advertising.

ERLICHT, LEWIS—ABC executive who shifted from station management to network programming and in the spring of 1979 became v.p. and assistant to the president of ABC Entertainment, Anthony D. Thomopoulos. In a career with ABC that began in 1969 in the sales department of WABC-TV New York, he held a succession of posts that included manager of research for ABC Television Spot Sales; v.p. and general manager of WLS-TV Chicago (1974–77); v.p. of programs, east coast, for ABC Entertainment; and in July 1978, v.p. and general manager of ABC Entertainment, based in Hollywood.

ESCAPADE—see Cable Networks.

ESPN (Entertainment & Sports Programming Network)—see Cable Networks.

ESTES v. TEXAS [381 U. S. 532 (1965)]—case in which the Supreme Court held that television broadcasting of a criminal trial, either live or on tape or film, denies a defendant his right to due process of law as guaranteed by the Fourteenth Amendment.

The case was brought by Billy Sol Estes after his conviction in a state court in Texas on charges of swindling. Estes had asked that TV cameras and broadcast microphones be barred from covering the trial, but the court denied his motion. However, the court ordered all photographers and cameramen to remain in a constricted press booth which was totally enclosed except for a door and a narrow slit barely wide enough for lenses to fit through. The judge also restricted coverage to certain parts of the trial.

After his conviction, Estes appealed to the Texas Court of Criminal Appeals, arguing that the presence of cameramen and photographers deprived him of his right to a fair trial. His conviction was affirmed, and he then appealed to the U. S. Supreme Court for review.

The Supreme Court reversed his conviction because of the prejudice introduced by the cameras. At the outset the Court noted that, while Canon 35 of the Judicial Canons of the American Bar Assn. prohibits cameras in courtrooms, the Judicial Canons were not the law. But the Court held that due process of law was violated because of the substan-

tial possibility that the cameras interfered with the court's attempt to determine if the defendant committed a particular crime.

The cameras, the Court suggested, might well have an impact upon the jurors, the witnesses, the judge and the defendant. Since the cameras might well have an adverse effect upon the participants of the trial, and since they in no way aided the trials court in its role as fact-finder, the Supreme Court held that Estes had been denied due process.

"THE ETERNAL LIGHT"—religion series on NBC which has presented a variety of interview and discussion programs by Protestant, Catholic and Jewish clergymen on ethics in our time. After more than a quarter century on television it was canceled in 1978.

ETV (Educational Television)—the name for noncommercial television broadcasting until the Public Broadcasting Act of 1967 which, on the recommendation of the Carnegie Commission, changed "educational" to "public" to broaden the scope and promise of the system. The original noncommercial stations were built in the 50s with an instructional purpose, but their inability to attract a broad audience was attributed partly to the designation "educational," which connoted dry and didactic television. The term is still occasionally used, and well into the 1970s there were organizations such as the National Assn. of Educational Broadcasters, but officially ETV has been superceded by PTV.

EUROPEAN BROADCASTING UNION (EBU)—launched in 1950 by 23 nations at a founding conference in Torquay, England. It is essentially a system of international broadcasting cooperation—radio and television—(with 34 active and 64 associate members. Among the latter are the major U.S. networks.

With administrative and legal headquarters at Geneva, and a technical center at Brussels, the EBU has been instrumental in the drafting of treaties and conventions covering international broadcasting standards and copyrights. Through its Eurovision interconnect, which now links up 23 countries, the organization coordinates coverage of major actuality specials—coronations, space shots, sports championships, etc. Out of its Brussels technical center come, among other things, twice-daily news-film feeds by and for constituent TV systems. EBU also sponsors symposia and workshops as well as regular program trade markets.

EUROPE'S TELEVISION FESTIVALS AND MARKETS—competitive annual events treating television as an art form and conferring international awards for excellence. Festivals abound in Europe, some of them prestigious and consequential, others little more than chamber of commerce-type promotions. The principal ones annually are the Prix Italia, the Golden Rose of Montreux, the International Television Festival of Monte Carlo and the Prix Jeunesse International in Munich.

What had once been a major event, the annual Cannes festival for news and documentary programs, has been merged into the Monte Carlo Festival, where the categories otherwise spotlight drama and children's programs. The categories at Montreux, in Switzerland, are limited to light entertainment entries, while the Prix Italia is for dramas and documentaries. As the name suggests, the Munich event is for children's shows.

Other European festivals are at Kielce in Poland, for films and TV shows with musical themes; Knokke in Belgium for variety shows; Leipzig in East Germany for short films and documentaries; Prague in Czechoslovakia for dramas and documentaries; and Nordring in Norway for music shows.

The Prix Danube at Bratislava in Czechoslovakia is for children's shows; the Prix Futura in Berlin for programs themed to the world of tomorrow; the festival at Annecy in France for animation; and the Common Market-sponsored festival in Brussels for programs on European affairs.

Still others are sponsored by religious organizations for programs on themes such as peace and justice. Among these are Monte Carlo's UNDA festival for the International Catholic Federation of TV and Radio Awards; the International Christian TV Festival, a skip-year event promoted by the World Association for Christian Communication and the International Catholic Federation of TV and Radio; the Christian Unity TV Awards in Geneva; and the International Religious Broadcasting Festival at Seville, Spain.

Two major events for the international television trade are not festivals but program markets at which networks, syndicators and independent producers meet to buy and/or sell programs. They are the MIP-TV at Cannes in April and the twice-yearly (spring and fall) MIFED events in Milan.

The European Broadcasting Union (of which the three major American networks are associate members) also holds twice-yearly screenings of new products by its constituent broadcasting organizations.

EUROVISION SONG CONTEST—a once-a-year live TV special via the Eurovision interconnect in which European Broadcasting Union member countries compete with nominated songs and singers. Started in 1955, the event annually plays to vast television audiences from Ireland to Israel and from Finland to Portugal.

EVANS, BERGEN—erudite TV host and panelist who was also a professor of English. During the 50s he moderated, from Chicago, the word game *Down You Go,* and later *The Last Word.* He also appeared on *Superghost, Of Many Things* and *It's About Time,* and he supervised the questions for the big quizzes, *The $64,000 Question* and *The $64,000 Challenge.*

Maurice Evans

EVANS, MAURICE—noted Shakespearean actor from Britain who provided U.S. television with numerous distinguished performances of the classics during the 50s and 60s, most of them for NBC's *Hallmark Hall of Fame.* He also directed some of the early Hallmark productions and helped set the lasting standard for the series.

Evans made his U.S. TV debut in *Hamlet* in 1953 on the Hallmark series, a production which NBC claimed had a larger audience than all the theatrical productions of the play since its Elizabethan premiere. There was, of course, no way to substantiate the claim but it remains part of TV lore.

Evans performed *Macbeth* live on Hallmark in 1954 and again on film in 1960; *Richard II* in 1954; *The Taming of the Shrew* in 1956; *Twelfth Night* in 1957. He also performed Shaw's *Man and Superman, The Devil's Disciple* and *Caesar and Cleopatra,* the latter for *G.E. Theatre.*

"EVENING AT POPS/EVENING AT SYMPHONY"— companion PBS series produced by WGBH-TV Boston, the first with the Boston Pops orchestra, the second with the Boston Symphony, each running about three months of the year. *Pops,* hosted and conducted by Arthur Fiedler until his death in the summer of 1979, generally played the warm-weather months; *Symphony,* headlined by Seiji Ozawa and guest conductors, the fall-winter season. Both were an hours' length weekly and premiered in 1974.

Arthur Fiedler conducting the Boston Pops Orchestra

"EVENING/PM MAGAZINE"—innovative prime time-
-access program syndicated five nights a week by Group W
Productions, after it was successfully launched as a local
nightly magazine on Group W's five stations. On the originating stations the half-hour series is entitled *Evening Magazine;* in syndication it goes by the name of *PM Magazine.*

Evening was developed in 1976 at KPIX San Francisco,
and a year later spawned local counterparts at the four other
stations. Since some of the six- or seven-minute segments in
the telecasts were not distinctly local in nature, they were
bicycled among the Group W stations to reduce the burden
and the cost of producing a nightly half-hour. These "national" segments then were offered in a syndicated package
to other stations, which created their own magazines by
producing around 50% of the material locally. The concept
was successful, and by the fall of 1979 there were 47 stations
in the fold.

The original program was created by producer B. Ziggy
Stone and KPIX program manager William Hillier, who
became executive producer for all *Evening/PM* shows. Hillier
left to start his own company in 1979 and was succeeded by
Richard Crew.

EVENT TELEVISION—the medium engaged in the live
coverage of something happening in real time, whether a
news or sports event, an awards presentation or a beauty
pageant. Considered by critics to be television at its purest,
the idea was broadened by some networks in the late 60s to
include specials and movies as "events" created by television.

EVR—Electronic Video Recording, now obsolete, but first
demonstrated by CBS Inc. in 1967, a system whereby recorded video material may be played back through a television receiver. The recording medium was cartridged *black-
and-white* photographic film, on which picture information,
color and sound were electronically coded. With tape dominating the videocassette field, CBS abandoned EVR in 1971.

EXECUTIVE STORY CONSULTANT—common designation in the credits for a script "doctor," an independent
writer engaged to improve a faulty TV play or playlet.

"EXECUTIVE SUITE"—hour-long CBS prime-time dramatic serial (1976–77) about the internal struggles within a
large corporation and the intertwining private lives of its
employees. Loosely based on Cameron Hawley's novel of that
title, it was one of a number of expensive nighttime soap
operas introduced by the networks in 1976 following the
success that spring of *Rich Man, Poor Man* and *Family.*
Featured were Mitchell Ryan, Stephen Elliott, Sharon Acker
and Leigh McCloskey. Norman Felton and Stanley Rubin
were executive producers, Den Brinkley producer and
Charles Dubin and Joseph Pevney directors.

"EYEWITNESS NEWS"—title widely used by local stations for newscasts whose format permits reporters to deliver
their own stories on the air. The concept was revolutionary in
1968; before then virtually every newscast had a star newscaster who read the copy written by reporters or rewritten
from the wires by staff members.

The concept probably had its origins at the public TV
station in San Francisco, KQED, during a newspaper strike
in 1968. KQED created a program on which newspaper
reporters presented the stories that would have been in the
papers and conversed with each other on details. But for
practical purposes it began at KYW-TV Philadelphia later
that year. When the station's news director, Al Primo,
learned that the KYW reporters were members of AFTRA,
the performing union, and that no extra fees were involved
in putting them on the air, he broke with commercial
broadcast tradition and let the news-gatherers report the
news under their own bylines. This contributed to a looser,
more conversational presentation, which a significant number of viewers found more appealing than the standard
formal newscast.

Modifications of the basic idea led to the station's newsmen assuming roles as members of a happy-go-lucky team,
who exchanged quips between news items. In its most base
form—where the joshing stopped just short of spitball
throwing—*Eyewitness* became known to journalism critics as
"happytalk news."

Contributing to its spread across the country was its
demonstrable ability to break the established news-viewing

habits. In the years when all newscasts took the formal, single-anchorman approach, it was extremely difficult for any challenger to overtake the news leader in the market. The axiom of the time was that the news habit was the hardest of all to change. But *Eyewitness News*—with its comedy, chitchat and conviviality—won new viewers almost overnight and devastated the traditional newscasts that had ruled their markets for years.

The *Eyewitness* concept, which in some markets took other names such as *Action News* or *The Scene News,* was promulgated in many instances by news consultants like Frank Magid Associates and McHugh-Hoffman, who helped their clients break the news grip on the market held by more orthodox competitors.

Primo went on to develop successfully the *Eyewitness News* techniques at the ABC-owned television stations in the early 70s, and in 1976 he left to become a consultant to local stations on the implementation of the format he invented.

F

"F TROOP"—ABC military situation comedy (1965–67) set in the post-Civil War West, with Forrest Tucker, Larry Storch, and Ken Berry as inept cavalrymen, and Edward Everett Horton playing an Indian. Melody Patterson was also featured. The series was by Warner Bros. TV.

"FACE THE NATION"—CBS news-making series, scheduled Sundays at noon, that probes government officials, world leaders and others in the news on current issues. It premiered Nov. 19, 1954. The CBS counterpart of NBC's *Meet the Press* and ABC's *Issues and Answers* uses the format of a permanent correspondent and two guest journalists putting questions to the news-maker. In latter years George Herman has been the moderator and regular reporter, and Sylvia Westerman and Mary O. Yates the producers.

FACSIMILE—the electronic transmission of written or still pictorial or graphic material. The first recorded facsimile transmission system was patented in 1843—it used telegraph wires. Facsimile has often been proposed as a consumer service to deliver newspapers and other reading material via television channels during the time when no programs are being transmitted or during the vertical interval between pictures. Although many broadcast facsimile tests have been conducted successfully—most notably RCA's "Homefax" system—the technique has failed to capture the public imagination. Facsimile transmission today is confined largely to telephone wires as a business tool.

FAIRNESS DOCTRINE—a content-regulation concept that underlies the entire structure of broadcasting in the U.S. and distinguishes the broadcast media from all other journalistic endeavors relative to the guarantees of the First Amendment. The Fairness Doctrine requires a two-fold duty from the broadcaster, namely: (1) that he devote a reasonable amount of time to the discussion of controversial issues of public importance, and (2) that he do so fairly, by affording reasonable opportunity for the opposing viewpoints to be heard.

While these concepts appear relatively simple, they have presented American broadcasters with a tortured and complex series of regulations, legislation and litigation which many people, both within and outside the system, maintain undermine the journalistic integrity of broadcasting.

The Fairness Doctrine stems directly from the basic scheme Congress set forth for broadcasting in the Radio Act of 1927 and the Communications Act of 1934. Before 1927, the allocation of frequencies was left entirely to the private sector, and the result was chaos. The 1927 act originated because of the need for government to allocate radio frequencies among applicants to prevent problems of interference among transmissions.

The statutory scheme that was chosen by Congress was one of short-term licensing with the licensee obligated to operate in the public interest.

Very shortly thereafter, the Federal Radio Commission expressed the view that the public interest required ample play for the free and fair competition of opposing views; it also determined that the principle should apply to all decisions or issues of importance to the public. This tradition was carried on through the development of the FCC in 1934, and the Doctrine was applied through the denial of license renewals and construction permits.

For a period the licensee was obliged not only to cover, and to cover fairly, the views of others but to also refrain from expressing his own personal views. After much confusion about the noneditorial rule, the commission published its Editorializing Report in 1949, which was the first formal articulation of the Fairness Doctrine.

Ten years after the Editorializing Report, Congress amended Section 315 of the Communications Act and passed language which seemed to codify the Fairness Doctrine. After the 1959 amendment, the commission continued to enforce the principle in individual cases. Originally the commission considered fairness complaints only at renewal time. Its procedure was to refer appropriate complaints to the station at the time received, obtain the station's response and then consider the matter definitively at renewal time.

In 1962, the commission changed its policies and began to resolve fairness matters as they arose. If it determined that the station was found to have violated the Doctrine, the station was directed to advise the commission within 20 days of the steps it had taken to remedy the situation.

In addition the FCC began to remind broadcasters of their special obligation with regard to personal attacks, first brought to broadcasters' attention in the Report on Editorializing in 1949. In 1967 the FCC promulgated rules on personal attack.

These rules and their application—and by implication the entire Fairness Doctrine—were eventually challenged in the Supreme Court in the Red Lion case, and their constitutionality was upheld. Later attempts to turn the Fairness Doctrine into a concept for enforced access (the BEM case) were defeated in the Supreme Court.

The FCC has continued to apply the Fairness Doctrine on a case by case basis, but in the 70s it proposed an experiment that would exempt radio stations in the ten largest markets from fairness obligations, since in those situations there was no longer a scarcity of outlets for the full range of views. The proposal cheered broadcasters, who have sought freedom under the First Amendment equal to that of the printed press, but was condemned by broadcast reform groups which hold the Fairness Doctrine to be the cornerstone of regulation.

See also Red Lion Decision, BEM Case, Mayflower Decision, Equal Time, Report on Editorializing.

"FAITH FOR TODAY"—syndicated series by Seventh Day Adventists which presents a family view, with practical rather than theological emphasis. The program is carried on close to 200 local stations.

FALSE OR LIBELOUS STATEMENTS BY CANDIDATES [Farmers Educational and Cooperative Union of America v WDAY, Inc./360 U.S. 525 (1959)]—case resolved by the Supreme Court which established that broadcasters cannot be held liable for material that they are statutorily unable to control.

In October 1956 a candidate for the U.S. Senate in North Dakota demanded and received equal time on WDAY-TV Fargo. He then charged that communists controlled the North Dakota Farmers Union. The Farmers Union brought a damage suit against the candidate and the station. The North Dakota Supreme Court ruled that radio and TV stations were not accountable for false or libelous statements made over their facilities by political candidates because Section 315 (a) of the Communications Act prohibits stations from censoring the remarks of candidates.

The case was ultimately appealed to the U.S. Supreme Court which in a 4–3 decision affirmed the North Dakota Supreme Court and found no liability for the station.

FALWELL, JERRY—fundamentalist religious broadcaster from Lynchburg, Va., who gained nation-wide attention during the 1980 national elections as head of The Moral Majority, a quasi-religious organization that campaigned for Ronald Reagan and other ultra-conservative candidates. A television preacher with a nationally syndicated program, *The Old-Time Gospel Hour*, he came into prominence with his off-camera attacks on immorality in television. In 1981, he was a moving force in the Coalition for Better Television, which sought to purge the medium of sex and violence by boycotting advertisers in programs the organization deemed unwholesome. Falwell and the Moral Majority dropped out of the Coalition the following year.

Falwell founded the Thomas Road Baptist Church in Lynchburg in 1956 and a week later started a radio program. Within six months he was on television with a local program. This grew into his syndicated series. The broadcast exposure led to the growth of the church from 35 original members to a congregation of 17,000, one of the largest in the country. See also Coalition for Better Television; Religious Television.

"FAMILY"—ABC weekly prime-time serial introduced during the spring of 1976 as a miniseries and brought back in the fall as a continuing weekly entry. Centering on the lives of a contemporary middle-class family, it was created by Jay Presson Allen and was the first TV production by noted stage and film director Mike Nichols, in association with Spelling-Goldberg Productions. Sada Thompson portrayed the mother, James Broderick her husband and Jane Actman, Gary Frank and Kristy McNichol the children. Meredith Baxter-Birney was a late addition to the cast in 1976 and

Quinn Cummings was added in 1978 as an adopted daughter. The program went on hiatus in 1979 and was not in the fall starting schedule but returned at mid-season before finally giving up for good. Spelling and Goldberg were executive producers; Nigel and Carol Evan McKeand were the producers.

The Lawrence Family: Gary Frank, James Broderick & Sada Thompson

"FAMILY AFFAIR"—hit situation comedy on CBS (1966–71) about a wealthy and worldly bachelor who becomes the foster parent of his orphaned nieces and nephew— the quintessential "heart" comedy. Much of the humor derived from the maternal role that was incumbent upon the British manservant as portrayed by Sebastian Cabot. Produced by Don Fedderson Productions, it starred Brian Keith and featured child actors Anissa Jones, Johnnie Whitaker and Kathy Garver.

"FAMILY AT WAR"—British serial drama, popular on the independent channel there via Granada TV. It played on U.S. public television's Eastern Educational Network in the fall of 1974. The series focused on the lives of a Liverpool family and their friends during World War II.

"FAMILY HOLVAK, THE"—unsuccessful attempt by NBC in 1975 to put over a series like CBS's *The Waltons*. Despite a cast headed by Glenn Ford and Julie Harris, the Universal TV series about a family in the Tennessee backwoods during the 30s fared poorly and was canceled at midseason. Featured were Elizabeth Cheshire and Lance Kerwin. CBS carried the rerun in the summer of 1977.

FAMILY VIEWING TIME—an industry-wide policy adopted in 1975 designating the first two hours of prime time (7–9 P.M.) for programs that would be suitable to all age groups. The policy, which became part of the television code, was later declared illegal by U.S. District Court Judge Warren J. Ferguson, who ruled that FCC chairman Richard C. Wiley had coerced the industry to adopt such a plan in violation of the First Amendment. The networks said, nevertheless, that of their own choice they would continue the practice of keeping the early evening programming free of excessive sex and violence. They also said they would appeal the court's decision.

The Family Viewing concept had evolved from discussions between network officials and the FCC during a time, late in 1974, when Wiley was under pressure from three congressional committees to take some regulatory steps to protect children from the moral liberties assumed by television. Wiley, a staunch believer in industry self-regulation, chose not to issue rules but instead called for meetings with the network chiefs and sternly made suggestions for procedures they might voluntarily adopt. Some at the networks accused him of arm-twisting and "jawboning."

But during Christmas week, Arthur R. Taylor, then president of CBS, issued a statement that his network would, in the fall of 1975, consider the first hour of network prime time (8–9 P.M. EST) a family hour and, in addition, would post warnings on the screen for shows that were either intended for adults or that required parental guidance. NBC, and then ABC, soon afterwards made similar pledges.

Wiley then persuaded the NAB to add the Family Viewing concept to the television code, making it effective for the hour preceding network time, as well. Then he met with managers of independent stations that did not subscribe to the code and persuaded them to honor the plan. When it was unanimous, Wiley then praised the industry publicly for its achievement in self-regulation. "Family Viewing" had no precise definition, and what it meant besides the elimination of gratuitous violence was left to the broadcasters' judgment.

The policy affected numerous programs that were in development at the time for the new fall schedule and caused some, such as *Fay*, intended as a sophisticated comedy about a middle-aged divorcee, to undergo severe script editing. Series such as *All in the Family* and *The Rookies* were ousted from their accustomed 8 o'clock time periods. All this proved upsetting to Hollywood producers and writers, several of whom challenged the Family Viewing policy in a suit filed in the Federal Court for the Central District of California.

On Nov. 4, 1976, Judge Ferguson, in a lengthy opinion, accused Wiley of overstepping his authority by unconstitutionally pressing upon the networks and stations "a programming policy they did not wish to adopt." He said Wiley's actions, implying a threat if the networks did not respond to his urgings, had primarily intended "to alter the content of entertainment programming in the early evening hours," thereby violating the broadcasters' constitutional guarantee of free speech. See also Television Code; Wiley, Richard C.

"FAMOUS JURY TRIALS"—series which appeared in two versions: first in a 1949 production by DuMont and then as a syndicated entry designed for strip presentation dramatizing actual trials and serializing them over a period of weeks. The series featured Donnelly Rhodes, Tim Henry, Allen Doreumus and Joanna Noyes. It was produced in 1970 and 1971 by 20th Century-Fox in association with Talent Associates.

"FANTASY ISLAND"—moderately successful ABC series which entered the schedule early in 1978 and remained the following season as a Saturday night parlay with *Love Boat*. Ricardo Montalban stars as the mysterious ruler of an island who has the power to make the dreams of his guests come true. His assistant, a dwarf, is played by Herve Villechaize. Wendy Schaal, as Montalban's daughter, was added at the start of the 1981–82 season. The series is a Spelling-Goldberg production.

"FARMER'S DAUGHTER, THE"—ABC situation comedy (1963–66) based on a movie concerning a farm girl who marries a widowed congressman for whom she worked as governess to his two sons. It starred Inger Stevens and featured William Windom and Cathleen Nesbitt. Screen Gems produced in association with ABC.

FARNSWORTH, PHILO T. (d. 1971)—one of the two inventors of modern all-electronic television. A contemporary of Vladimir K. Zworykin, he independently demonstrated in 1927 a device similar to Zworykin's iconoscope—the "dissector tube" or orthicon, capable of dividing an image into parts whose light values could be restored to form a reproduction of the original picture. He was founder and research director of the Farnsworth Television and Radio Corp. in 1938. It later became part of International Telephone and Telegraph Co.

"FAT ALBERT AND THE COSBY KIDS"—(animated CBS children's series (1972–77) acclaimed for attempting to promote prosocial values through juvenile situation comedies. It was also a landmark show in minority programming, since its principal characters were black. Hosted by comedian Bill Cosby, the series was based on his monologue routines about the kids he grew up with, one of whom was

known as Fat Albert. Created in response to mounting criticism over exploitative programs for children, *Fat Albert was designed for affective rather than cognitive learning and dealt with themes on human feelings, personal relationships and ethics and values. It proved to be very popular with children of all races, and CBS social research found that the prosocial messages usually got through. Fat Albert* was produced by Lou Scheimer and Norm Prescott of Filmation Associates.

Philo T. Farnsworth

FATES, GIL—executive producer for Goodson-Todman of *What's My Line?, To Tell the Truth* and *I've Got a Secret* during the 50s. Earlier he produced *Stop the Music* and *The Faye Emerson Show.*

"FATHER KNOWS BEST"—a classic family situation comedy of the 50s starring Robert Young, a noted film actor, as the sensible head of a purportedly normal middle-class American family. It played on all three networks, premiering on CBS Oct. 3, 1954, then switching the following year to NBC, returning to CBS from 1958 to 1962, and continuing in reruns on ABC (1962–67). The films subsequently went into syndication.

NBC attempted to capitalize on the years of popularity of the series in December of 1977 by presenting a 90-minute movie, *The Father Knows Best Reunion*. In the intervening years the family had gone the way of the purportedly normal "middle-class" American family of the 70's. While the parents were suffering from "empty nest" syndrome, Betty (Princess) was a widow with two children, a career, and marriage on her mind; Bud was in a troubled marriage; and Kathy (Kitten) was in a semirebellious relationship with a much older divorced man.

Jane Wyatt portrayed Young's wife, and the children were played by Elinor Donahue, Billy Gray and Lauren

Chapin. It was produced by Young's own company, Rodney-Young Productions.

"FATHER OF THE BRIDE"—CBS situation comedy (1961–62) based on the movie and the book of that title, which featured Leon Ames, Ruth Warwick, Myrna Fahey and Bert Metcalfe. It was by MGM-TV.

FAULK, JOHN HENRY—folk raconteur and television personality victimized by the anticommunist crusaders of the 50s. His successful suit for libel contributed to ending the blacklist practices and purges for subversives in the entertainment industries.

In 1956 Faulk was elected a vice-president of the New York chapter of AFTRA, the performers' union, on a platform which deplored the activities of Aware Inc., one of the organizations leading the purge in broadcasting. Aware then issued one of its bulletins, citing Faulk for "communist activities." Although he had been enjoying popularity on both radio and TV, Faulk immediately began losing sponsor support and soon was released by CBS-TV from his panel show. He sued Aware and its principals, Vincent W. Hartnett and Lawrence Johnson, in June 1956. With a contribution of $7,500 from Edward R. Murrow (who called it "an investment in America"), he hired the famed Louis Nizer as his lawyer.

For six years, until the case was decided, Faulk found he was unemployable. Proving in court that the charges made against him by Aware had been either false or based distortedly on half-truths—and with witnesses from the industry testifying to the threats and fear tactics used by the organization to achieve the boycott of an entertainer—Faulk, in June 1962, was awarded damages of $3.5-million. It was the largest judgment ever returned in a libel suit, although it was subsequently reduced by the courts.

In 1976, CBS presented a 90-minute dramatization of Faulk's book, *Fear on Trial,* with George C. Scott as Nizer.

"FAY"—short-lived NBC situation comedy (1975) created as a vehicle for Lee Grant; it stirred a controversy when the network issued a cancellation notice after the fourth episode because its ratings were inadequate. On the *Tonight* show, Miss Grant protested both the hasty decision and the fact that NBC scheduled the program in family hour, forcing the sophisticated dialogue to be tempered. *Fay* concerned a fortyish divorcee determined to make a new life for herself, while her ex-husband continues to bob into the picture. The following summer NBC slotted the previously unexposed episodes outside of family time with a favorable lead-in, but *Fay* failed again to find an audience.

Featured were Audra Lindley, Joe Silver, Stewart Moss and Norman Alden. The series was by Danny Thomas Productions and Universal TV.

Efrem Zimbalist, Jr.

"FBI, THE"—series dramatizing FBI investigations of crime and subversion, based to some extent on cases from the agency's files. It became a Sunday night mainstay on ABC for eight seasons after its premiere in 1965. Initially, the program was sponsored by Ford, to offset the successful Chevrolet-sponsored Sunday night show *Bonanza.* Automobiles were used abundantly in the series, and most of them were Ford models. Additionally, since the series concerned the workings of a sacred federal agency, the violence employed in the stories was held by some in Congress to be above reproach. A total of 234 episodes were produced by QM Productions and Warner Bros., with a cast headed by Efrem Zimbalist, Jr., William Reynolds and Philip Abbott.

A new series, *Today's FBI,* starring Mike Connors, was introduced by ABC in 1981 in the same Sunday night time period; but it was not really a revival of the old series and was via David Gerber instead of QM.

FCBA (FEDERAL COMMUNICATIONS BAR ASSN.)—organization of lawyers specializing in communications law, many of them alumni of the FCC. It functions both as a trade association and as the auspices for forums on major communications issues. Founded in 1936, two years after the FRC became the FCC, it has today around 800 members, all of whom practice before the commission. On occasion the association has assisted the FCC, such as in providing ideas for the streamlining of the commission's adjudicatory proceedings.

FCC FEES—a schedule of charges set by the commission in 1970 so that the industries it regulates would cover the costs of operating the agency. But in 1976, the U.S. Court of Appeals for the District of Columbia ruled that the fees were improperly levied and that the basis on which they were collected amounted to taxation. Acting on suits by the NAB, Capital Cities Broadcasting, NCTA and a group of common carriers, the court cited the Supreme Court's ruling in March 1974 that the FCC had been collecting fees in an illegal manner when it based their apportionment on re-covering all its budget. Both courts held that the fees charged had to reflect only the direct benefits realized by those receiving them.

After the Supreme Court's ruling, the FCC issued a new schedule in 1975 with expectations of bringing in slightly more than one-third of the commission's annual budget of approximately $45 million. Television stations were charged annually 4.25 times their highest 30-second sport rate, or a minimum of $100, for regulatory services.

The Court of Appeals struck that down, saying that the FCC should base its charges on itemized services and that it should demonstrate the actual costs of those services. The commission was left in a quandary on how to process refunds for the fees collected since 1970. After the court's ruling, it suspended all license fees, including those for the use of citizen's band radio.

FCC PROGRAM POLICY STATEMENT, 1960— guidelines for programming in the public interest issued by the FCC in July 1960, four years after the controversy over the *Blue Book* report, which had attempted in broad strokes to define program standards that would bear on license renewals. The 1960 policy statement raised no controversy because the commission made it plain that it would not use the program areas outlined as a definition of service in the public interest for license-renewal purposes.

In fact, the Policy Statement was welcomed by broadcasters, not only because it waffled on the question of a licensee's responsibility but because it challenged the Blue Book by declaring that sustaining programs did not constitute, per se, a better service in the public interest than sponsored programs. Without specifying percentages or suggesting that these should be elements of a complete programming plan, the commission listed the following as usually necessary to meet the public interest: Programs that provide opportunity for self-expression, programs for children, religious programs, educational programs, public affairs programs, editorials, political broadcasts, news programs, agricultural programs, weather and market reports, sports programs, entertainment programs, service to minority groups and, overall, the development and employment of local talent. See also Blue Book.

FCC v. WOKO [329 U.S. 223 (1946)]—case in which the Supreme Court held that a licensee that misrepresents its ownership to the FCC may have its license revoked despite satisfactory broadcasting over a period of years.

WOKO Inc. satisfactorily operated a radio station in Albany, N.Y., for 12 years. About one-fourth of its capital stock was owned by a Mr. Pickard who was a vice-president of CBS. Pickard secured a CBS affiliation for WOKO but wished his ownership of the stock to remain a secret. In each application to the FCC, in which it was required to list its shareholders, WOKO concealed the participation of Pickard.

But the commission eventually learned of it, and with the next license-renewal application held that the misrepresentation disqualified WOKO from holding a broadcast license. The station appealed, and the D.C. Court of Appeals reversed, saying that the denial of renewal was too drastic a penalty to enforce upon innocent shareholders.

The FCC appealed to the Supreme Court, which reversed the Court of Appeals, noting that the Communications Act and the FCC required all applicants to provide information about ownership under oath or information. Nowhere could the Court find any indication that the statute required misrepresentation to be material. "The fact of concealment may be more significant than the facts concealed," the Court stated.

FEDDERSON, DON—one of the most successful packagers of family situation comedies through the 50s and 60s. He was producer of *Do You Trust Your Wife* and *The Millionaire,* then had a string of sitcoms for which he was executive producer, including *My Three Sons, To Rome with Love, Family Affair* and *The Smith Family.* He was also consultant to the *Lawrence Welk Show* and distributed that series since it went into syndication.

FEDERAL COMMUNICATIONS COMMISSION (FCC)—an independent government agency administered by seven commissioners and reporting to Congress, which is responsible for regulating interstate and foreign communications by radio, television, wire, cable and newer technologies. Operating as a unit, the seven commissioners are charged with interpreting the public interest in broadcasting under a national policy which makes the public interest paramount while providing for private operation of stations without government intrusion.

The FCC has the authority to award broadcast licenses and the power to revoke them when broadcasters have demonstrably failed to serve the public interest, convenience or necessity, or when they are found guilty of serious infractions of commission regulations or of the U.S. Criminal Code.

As the link between the public and the industry—by nature of having selected the trustees for the airwaves, which are in the public domain—the FCC may create rules, conduct hearings and issue guidelines and policy statements to affect broadcast performance in the public interest, but it is forbidden by the Communications Act of 1934 to engage in any form of censorship. This prohibition has kept the commission wary of interference in matters of programming, and the agency has adhered to a policy of trusting the licensee to determine what is best for his community.

While the FCC has no direct regulatory authority over the networks, since they are independent program services and not licensed entities (anyone, in theory, may start a network), historically it has dealt with the networks through the licenses of their owned and affiliated stations and through the regulations for chain broadcasting.

Created by the Communications Act of 1934, the FCC began operating on July 11, 1934, superseding the five-member Federal Radio Commission. The FRC had been formed by the Radio Act of 1927 to undo the chaos that prevailed on the radio band, where signals collided with each other as stations increased power, switched frequencies and extended their broadcast hours as they chose.

The FCC was established when a study of electrical communications, made at the request of President Roosevelt in 1933, recommended that a single agency be created to unify the regulations of all wire and radio communication—telephone and telegraph, as well as broadcast, all of which were engaged in interstate commerce and were to some degree interdependent. Eugene Octave Sykes, a Democrat who had been an original member of the FRC, became the first chairman of the FCC.

Amendments to the Act over the years gave the FCC regulatory responsibility for cable-TV, pay-television and communications satellites, and its duties were augmented by a Presidential Executive order in 1963 to ready the communications services under its jurisdiction for possible national emergency situations.

Under its mandate, the FCC allocates TV channels and assigns frequencies with a view to "fair and equitable distribution" of service from market to market, at the same time avoiding the collision of signals. It may also determine the power and types of technical facilities licensees may use and the hours during which they may operate. Licenses or construction permits cannot be transferred or sold without FCC permission, and when licenses are challenged the agency, after holding comparative hearings, must judge which is the most deserving applicant.

Members of the commission are appointed by the President for seven-year terms, with approval of the Senate. Anyone who is nominated to fill a vacancy is appointed only for the unexpired term of his predecessor. The President also designates the chairman. No more than four commissioners may be of the same political party and none may own securities of any corporation over which the FCC has jurisdiction. Since 1967 the annual salary for commissioners has been $50,000 and for the chairman $52,000.

The chairman presides over meetings, coordinates the work of the commission, represents the agency in all legislative matters and usually is the moving force who imparts a style and working spirit to the commission. Four commissioners constitute a quorum.

Each commissioner has two professional assistants plus three secretaries, while the chairman may have as many as six assistants. Much of the daily work of the agency is delegated to five bureaus and several staff offices. The FCC employs more than 2,000 persons, including its field staff in major cities.

The organization consists of the Broadcast Bureau, Cable Television Bureau, Common Carrier Bureau, Personal Services Bureau and Field Operations Bureau. Also there are the offices of the executive director, general counsel, chief scientist, administrative law judges (formerly, hearing examiners), review board, opinions and review, secretary, information. Growing in importance is the office of Plans and Policy.

The Broadcast Bureau, whose responsibilities include processing applications for stations and setting requirements for broadcast equipment, is made up of seven divisions: rules and standards, renewal and transfer, complaints and compliances, research and education, broadcast facilities, license, hearing. It also embraces the office of network study.

The Cable Bureau administers and enforces cable-TV rules, advises the commission on cable matters, and licenses private microwave facilities used to relay TV signals to cable systems.

Each of the bureaus is responsible for considering complaints, conducting investigations and taking part in commission hearing proceedings, among its other duties.

Administrative law judges, whose appointments are subject to Civil Service laws, conduct adjudicatory proceedings assigned to them by offices of the agency and issue initial decisions. As judges they operate independently and may not be supervised or directed by FCC officials in their investigative work or in the preparation of their opinions.

Most initial decisions are subject to review by the five-member Review Board, a permanent body made up of senior commission employees. Initial decisions may also be reviewed by one or more commissioners designated by the commission. In such cases the board or commissioner issues a final decision, subject to review by the full commission. On occasion the initial decision is reviewed directly by the commission.

The Office of Opinions and Review assists and advises the commission in the review of initial decisions and in drafting final decisions.

The FCC's contradictory mandates—that of looking after the public interest and that of refraining from any involvement with programming which might constitute censorship have kept the agency under constant criticism from Congress and citizens for failing to exert stricter program controls and for in effect "rubber stamping" license renewals. It has been called weak, bureaucratic and overly protective of the industry it regulates. On several occasions, it has been under formal investigation by Congress.

The FCC proceedings follow in these steps:

A Notice of inquiry is issued by the commission when it is asking for information on a broad subject or trying to generate ideas or ascertain points of view.

A Notice of Proposed Rule Making is issued when the commission is ready to introduce a new rule or take steps to change one.

A Memorandum Opinion and Order is issued to deny a petition for rulemaking, conclude an inquiry, modify a decision, or deny a petition asking for reconsideration of a decision.

A Report and Order is issued to state a new or amended rule or to state that the FCC rule in question will not be changed.

FEDERAL TRADE COMMISSION (FTC)—an independent administrative agency of the Government, consisting of a chairman and four commissioners, that is intended to promote free competition by preventing unfair methods of, or deceptive practices in, commerce. Its basic connection with the broadcasting industry is its concern with false and misleading advertising.

Section 5 of the Federal Trade Commission Act, which was passed in 1914, makes it unlawful to broadcast any false advertising relative to commodities that move in interstate commerce. Section 12 of the Act declares that false advertising with regard to foods, drugs, devices or cosmetics are unlawful whether or not they move in interstate commerce. Section 15 of the Act states that any advertisement is false which is misleading in a material respect.

In looking at what is false, the commission can take into account not only what is represented by the ad but also the extent to which the ad fails to reveal material facts with respect to the consequences that may result from the use of the product.

While the FTC has limited authority to punish, it may hold hearings on complaints charging violation of the statutes administered by the commission. If the charges are found to be fact, the FTC issues a cease and desist order requiring discontinuance of the practice. Formal litigation would follow if the order were ignored. Typically, however, voluntary compliance is brought about through advisory opinions by the commission, trade regulation rules, and the issuance of guidelines delineating legal requirements as to particular business practices.

In addition to its concern with unfair advertisements, the FTC aims to restrict discrimination in pricing and exclusive dealing, as well as corporate mergers and joint ventures when they may lessen competition or tend toward monopoly. The agency regulates packaging and labeling within the purview of the Fair Packaging and Labeling Act, supervises the operations of associations of American exporters and protects consumers against circulation of inaccurate credit reports.

The FTC has issued cease and desist orders on product demonstrations staged to deceive the TV viewer or that employ devices to enhance the appearance of the product. Such demonstrations have included the substitution of oil for coffee, to make the "coffee" in the commercial look richer and darker than it actually could be, and the placing of marbles in a bowl of vegetable soup forcing the vegetables to the top and making the soup appear more substantial than it was.

The commission also ordered a halt to a TV commercial that represented a brand of shaving cream as capable of shaving sandpaper, after having found that a compound had been added to the sandpaper that allowed a razor to shave it in a stroke. In another case, the FTC held misleading a commercial for a product advertised as affording relief of vitamin and iron deficiency anemia, because the commercial did not disclose the fact that a majority of the people who experience the symptoms noted in the advertisement do not have vitamin or iron deficiency.

In February 1978, chairman Michael Pertschuk announced the commission's investigation into children's television advertising. Public hearings on FTC staff recommendations opened in San Francisco on Jan. 15, 1978, with FTC administrative law judge Morton Needleman presiding. The staff recommendations, which stirred furious debate among the interested parties, included: a total ban on TV advertising directed to children under 8; a ban on all advertising of highly sugared foods to children 8–12; and corrective advertising (nutritional messages) to counterbalance commercials for sugared foods. Testimony was heard from more than 80 witnesses representing public interest groups, advertising and broadcast industries, as well as the fields of child development, education and medicine.

The investigation proceeded in spite of several serious setbacks. In the summer of 1978, the Senate Appropriations Committee, led by Sen. Lowell Weicker (R-Conn.) threatened to cut off all FTC funding if the commission did not abandon its inquiry. The budget was passed by Congress but not without a strong reprimand to the commission. The second blow occurred when manufacturing and advertising trade associations successfully petitioned the U.S. District Court in Washington to disqualify Pertschuk from par-

"had conclusively prejudged" the issues relating to the case. The inquiry report has not been released.

Members of the commission are appointed by the President, subject to approval by the Senate. The five commissioners are appointed for staggered seven-year terms. Not more than three commissioners may be members of the same political party, and no commissioner may engage in any other business or employment. The chairman is vested with the administrative responsibility for the agency, headquartered in Washington, and its eleven field offices.

"FEELIN' GOOD"—the first adult series created by the Children's Television Workshop (after its success with *Sesame Street* and *The Electric Company*). It sought to convey health care information on public television in the format of a weekly comedy-variety series. A noble idea, researched and tested over a three-year period, it premiered on PBS Nov. 20, 1974, and was withdrawn after 11 episodes for a substantial revision by the Workshop, which admitted that the program was misconceived.

Initially a one-hour program employing songs, comedy sketches, parodies and production numbers, held together by a situation comedy ensemble, it was poorly received by the key critics and was low-rated, even for public television. The program returned on PBS April 2, 1975, by previous arrangement, in a 30-minute format with a more direct approach to health-care information and less concern with entertainment. The situation comedy cast was dismissed in favor of a host, Dick Cavett. It ran 13 weeks in the new form.

Budgeted at $6.1 million, the series was created on grants from Exxon, Aetna Life & Casualty, the Corporation for Public Broadcasting, and a number of other sources.

Its executive producer was William Kobin, and regular members of the original cast were Rex Everhart, Priscilla Lopez, Ethel Shutta, Ben Slack, Marjorie Barnes and Joe Morton.

FEIN, IRVING—Jack Benny's manager, who also served as executive producer of the comedian's shows from the mid-60s on and has done the same for George Burns since his revival as a TV star.

FELDMAN, EDWARD H.—series and TV movie producer whose most durable entry was *Hogan's Heroes*, for which he was executive producer.

FELLOWS, HAROLD E. (HAL) (d. 1960)—president of NAB since 1951 who led the industry in numerous battles including, in the year of his death, its fight against the FCC's attempts to intrude in programming.

"FELONY SQUAD, THE"—half-hour police series on ABC (1966–69) featuring Howard Duff, Ben Alexander and Dennis Cole. It was via 20th Century-Fox TV.

FELTON, NORMAN—producer whose credits include *Robert Montgomery Presents* in the mid-50s, *Studio One, CBS Workshop* in the early 60s and later such series as *Dr. Kildare, Mr. Novak* and *Hawkins*.

FENNELLY, VINCENT M.—producer of *Trackdown, Wanted Dead or Alive, The David Niven Show, Richard Diamond, Dick Powell Theatre, Target: The Corruptors* and *Rawhide*.

FENTON, THOMAS TRAIL—senior European correspondent for CBS assigned to the network news bureau in London since 1979. Fenton joined the network in September 1970 as reporter-producer in the Rome bureau and since that time has been based in Tel Aviv (1973–77) and Paris (1977–79). Named a correspondent in 1971, he distinguished himself as a combat correspondent, covering the India-Pakistan War in 1971, the Arab-Israeli War in 1973, and the war in Cyprus in 1974.

FERBER, MEL—producer-director whose credits range from *Studio One, Seven Lively Arts* and *That Was the Week That Was* to episodes of *The Mary Tyler Moore Show, The Odd Couple* and *Alias Smith and Jones*. He directed the pilot for *Happy Days*, was executive producer of the 1972 Democratic national convention and in 1975 became exec producer of ABC's *Good Morning, America.*

FERRIS, CHARLES D.—chairman of the FCC (1977–81), as President Carter's nominee. A research physicist who decided to become a lawyer, graduating from Boston University Law School in 1961, he allied himself with the Democratic party and gained prominence on Capitol Hill first as aide to Senate Majority Leader Mike Mansfield (1964–77) and then as general counsel to Speaker of the House Thomas P. (Tip) O'Neill. Ferris's political backing was such as to make insignificant his lack of experience with communications matters.

In recruiting his key assistants from prominent public-interest organizations, Ferris signaled that his would be an activist administration. It proved not to be, however, at least during the first two years of his chairmanship—not, at any rate, activist in the sense that citizen groups had expected.

As the first Democrat to head the FCC since 1966, and as the liberal successor to the conservative Richard E. Wiley, Ferris endured a difficult two-year period with an agency predominating in Republican appointees. But even when the political balance finally shifted in 1979, he remained unadored by the citizen action groups as well as by the regulated industries. Largely this was an effect of the paradox of a liberal intention to achieve some degree of deregulation. Efforts to deregulate had begun in the Wiley administration, but they were spurred in Ferris's time by mandates from the White House and the Congressional oversight committees.

Ferris adopted a policy of what he called zero-based regulation, which meant examining the existing rules and moving for the elimination of those that were outmoded, unjustly restrictive or otherwise superfluous. Among the more significant regulations to be discarded was the certification procedure for cable-television, which permitted cable to spread more rapidly than before. At the same time, Ferris attempted to steer a course that favored neither the broadcaster nor the cable operator, and his addresses to both industries typically were laced with sharp criticism of their unwillingness to aspire to fulfilling their potential. Over all, he positioned himself as an advocate of open competition among the emerging technologies and conventional broadcasting, but an upholder of the public-interest standard for the licensed media.

In the perception of public interest groups, Ferris tilted more toward the doctrine of diversity in the industry than to citizens' causes. His principal innovation at the FCC was to base decisions for deregulation on economic criteria, and to that effect he created staff positions for economists. A study conducted by economist Nina Cornell indicated that most radio stations were broadcasting more news and public affairs than the FCC required of them and that most were operating well within the commission's commercial limits. This led to a commission proposal in 1979 for a general deregulation of radio, at least to the extent of eliminating news percentages, commercial ceilings and the community ascertainment requirement. The initiative did not endear Ferris to the aroused citizen groups, which saw it as an erosion of the people's rights in broadcasting.

"FIBBER McGEE AND MOLLY"—unsuccessful attempt by NBC (1959–60) to adapt the hit radio series to TV with Bob Sweeney and Cathy Lewis in the title roles. Featured were Hal Peary and Addison Richards. The show was produced by NBC.

FIBER OPTICS—the technique of using very thin flexible fibers of glass or plastic or other transparent materials to carry light. Because the interior of the fiber is reflective, it can convey light around corners. Used with a laser beam or light-emitting diode as the light source, fiber optics can carry wide bands of frequencies—many television channels

—and therefore has been proposed as a substitute for coaxial cable.

By 1977 it was in experimental use in both telephone and cable-TV installations. Because the signal carrier is in the light spectrum rather than in the radio frequencies, a tiny strand of optical fiber can carry as much information as a thick bundle of coaxial cable. See also Laser.

FIELD—one half of a complete television picture, constituting every other line. In the NTSC 60-cycle system, a field is transmitted in 1/60 of a second. See also Frame, Interlace, Scanning Line.

FIELD INSPECTION—an examination of a station's broadcast facilities by a representative of the FCC who has the right to make such an inspection "at any reasonable hour." During inspections, stations must make available their technical and program logs, on request. The FCC requires access to station logs for a period of two years after the date of broadcast.

FIELD-SEQUENTIAL COLOR—color television system developed by CBS Laboratories under Dr. Peter C. Goldmark and approved for broadcast use in a controversial 1950 FCC decision. In field-sequential broadcasting, the red elements of an entire picture were transmitted, followed by blue elements and then by green. At the receiver, a revolving disc containing red, blue and green filters was mounted over a black and white picture tube and synchronized with similar filters in front of the camera tube.

Although the system had the advantage of simplicity, it suffered from the drawback of incompatibility—that is, the 20 million black and white television sets in use at the time the system was approved would not have been able to receive a field-sequential color broadcast in black and white without a special adapter. After the development of the compatible NTSC system, the FCC reversed its color decision and approved NTSC for color broadcasting in 1954.

The field-sequential system is still in use, however, in certain closed-circuit applications, and was used in transmitting color pictures from the moon and from the Apollo space vehicles. The field-sequential signal was converted into standard NTSC color before being broadcast to the public. See also Color Television.

FILM CHAIN—see Telecine Chain.

FILM PACKAGE—a group of feature films assembled by a distributor and marketed at a single price. Packaging is more efficient than selling movies one at a time, and it permits the distributor to dispose of inferior films in a mix with a number of highly desirable titles. When movies first were sold to TV, packages consisted of vast libraries of pre-1948 titles from the major studios; more recently, they have been marketed in groups of from 5 to 30 features, with made-for-TV movies often included among theatrical releases.

Federal law requires that all pictures in a package be priced separately, but as a practical matter stations generally purchase the pictures at the package price and average the costs over the entire group of films.

FILMWAYS—one of Hollywood's leading independent TV suppliers, which came into prominence during the 60s when it placed with CBS, in rapid succession, such hits as *The Beverly Hillbillies, Green Acres* and *Petticoat Junction.* Headed by Martin Ransohoff, the company began as a producer of TV commercials and, after a huge success, expanded into series production. Filmways also produced such series as *Mister Ed, The Addams Family, My Sister Eileen, The Phyllis Diller Show* and *Trials of O'Brien.*

FINKEL, BOB—producer-director specializing in variety programs; his credits include *The Colgate Comedy Hour, The Perry Como Show, The Tennessee Ernie Ford Show, The Eddie Fisher Show, The Dinah Shore Show* and the Emmy Awards telecasts of 1960 and 1963. He was under exclusive contract to NBC in the late 60s.

FINLAND, TELEVISION IN—a two-channel system operated by the state-controlled Suomen Televisio on 625-line PAL. The service, reaching an estimated 1.3 million sets, is funded by license fees and sold time. This latter is through the device of selling "dark hours" to a privately owned brokerage, Oy Mainos Beklam, which as subcontractor fills with products acquired from foreign syndicators, in which it sells spot time to advertisers. Finland's first cable-TV station began operations in 1975 with a starting subscription of 1,500 homes in a neighborhood of Helsinki. Its principal fare was old movies.

The license fees amount to $42.75 a year per household for black & white sets and $78.66 for color TV.

FIORENTINO, IMERO—TV lighting expert whose independent company has designed the lighting for numerous specials and special events, including the first transmission by the Telstar satellite and the 1976 Presidential debates. After the first of the Great Debates in 1960, when Nixon's advisors felt he had been victimized by the electronic me-

dium's cosmetics, Fiorentino personally was retained to supervise the lighting of Nixon in subsequent debates and in his other TV appearances.

Fiorentino had been a lighting designer for ABC during the network's early years, then left in 1960 to start his own consultancy, Imero Fiorentino Associates. The company employs experts in a number of staging and production services, besides lighting, and maintains offices in New York, Las Vegas and Hollywood.

William F. Buckley, Jr.

"FIRING LINE"—interview and discussion series on current issues hosted by William F. Buckley Jr., the politically conservative publisher and columnist. The program began in 1966 on WOR-TV New York, which offered it in commercial syndication; in 1971 it shifted to PBS while public TV was under fire from the Nixon Administration for favoring the liberal view. Although Buckley resided in New York, the series was produced by SECA, the Southern Educational Communications Assn. in South Carolina. Warren Steibel was producer. Annoyed with the poor time periods the program received from many PBS stations, Buckley in 1975 switched the syndication of the program to a combination of commercial and PTV stations, with the RKO General stations as its base. In the fall of 1977 it became exclusively a PBS series again.

FIRST AMENDMENT AND DEFAMATION—the First Amendment to the United States Constitution provides that "Congress shall make no law abridging the freedom of speech, or of the press..." Although this prohibition was written in seemingly unambiguous terms, it was never considered a bar to civil liability for a defamatory falsehood by a newspaper or broadcaster [Chaplinsky *v.* New Hampshire, 315 U.S. 568 (1942) and Beauharnais *v.*

Illinois, 343 U.S. 250 (1952)]. This assumption was dispelled in New York Times *v.* Sullivan [376 U.S. 254 (1964)], when the Court determined that one may not be held liable for the "publication" (or broadcasting) of defamatory falsehoods about "public officials" as long as the publisher did not publish with malice. Malice was defined as either having actual knowledge of the falsehood or acting in "reckless disregard" of whether the defamatory statement was true or false.

The case arose in the context of an advertisement published in the *Times* which allegedly defamed the police commissioner of Montgomery, Alabama. He sued the *Times* and collected a judgment of $500,000 in the Alabama courts. The *Times* appealed to the Supreme Court of the United States, which reversed the judgment because, it wrote, the First Amendment required protection for those criticizing "public officials" acting in the course of their public duties.

The *Times* rule was subsequently broadened to include "public figures" [Curtis Publishing Co. *v.* Butts, athletic director at the University of Georgia; and Associated Press *v.* Walker, retired Army General, 388 U.S. 130 (1967)], and there were indications that it would be used to protect publishers of defamatory falsehoods whenever the subject matters under discussion were "issues of public importance" [Rosenbloom *v.* Metromedia, 403 U.S. 29 (1971), plurality opinion of Justice Brennan]. However, this expansion of protection, which at the same time prevented those who were defamed from compensation for their injuries, came to a halt in Gertz *v.* Robert Welch Inc. [418 U.S. 323 (1974)]. In Gertz, the plaintiff was a prominent lawyer who had been defamed in Welch's publication *American Opinion* because of his participation in an unconnected civil damage suit against a policeman. As a threshold matter, the Court reasserted its prior position that only "public officials" and "public figures" triggered the rule announced in Times *v.* Sullivan protecting the publisher of a defamatory falsehood. Gertz, though, was not a public figure despite his prominence among those in the legal profession.

The Court noted that the limited protection for defamations of public figures was in part due to their "significantly greater access" to the media, affording them an opportunity to reply. More importantly, "public figures" generally have "thrust" themselves to the forefront of particular controversies in order to influence the resolution of "issues involved." As a result, the Court indicated that the communications media are entitled to assume that public figures "have voluntarily exposed themselves to increased risk of injury from defamatory falsehoods." Gertz did not have any meaningful access to the media, nor had he "thrust himself" in the arena of public controversy. Therefore, the *Times* privilege did not apply to the defamation, and the *American Opinion* would be liable if Gertz were able to show that the publisher did not act reasonably.

The Gertz rule was reaffirmed most recently by Firestone *v.* Time, Inc. [424 U.S. 448 (1976)]. The Court held that Mrs. Firestone, a Palm Beach socialite involved in a well-publicized divorce action, was not a public figure because she had not "thrust herself in the forefront of any particular public controversy in order to influence the resolution of the issues involved in it." The Court was particularly anxious to reject the contention that Mrs. Firestone was a "public figure" merely because she was involved in a divorce which was of interest to the public.

FIRST COMMERCIAL TELECASTS—inaugurated by WNBT New York (now WNBC-TV) on July 1, 1941, the day its experimental call letters were changed from W2XBS. The station had been operating experimentally under RCA ownership since 1928 and actually began a schedule of regular service under NBC in 1939, two years before it embarked on commercial service. There were about 4,000 TV sets in the New York area for the first commercial telecast.

The first advertiser was Bulova Watch, which paid $9 for a 10-second Bulova Watch Time announcement superimposed on the test pattern at 2:29:50. The charge was broken down to $4 for the time and $5 for facilities. At 2:30, the telecast began from Ebbetts Field in Brooklyn, with Ray Forrest doing the play-by-play of a baseball game between the Dodgers and the Phillies. The station went off the air at 6:30 and returned 15 minutes later for a simulcast of the Sunoco newscast with Lowell Thomas. Dark again for two hours, it resumed again with a USO program, then *Uncle Jim's Question Bee* sponsored by Lever Brothers, then a Fort Monmouth Signal Corps show followed by Ralph Edwards's *Truth or Consequences* sponsored by Procter & Gamble. Ed Herlihy delivered the Ivory Soap "dishpan hands" commercial. The station signed off at 10:57:19 P.M. with the national anthem.

During the first week of commercial broadcasts, WNBT presented 19 hours of programming, 15 of which were devoted to sports—boxing, baseball and tennis. There were also films: *Death from a Distance, Where the Golden Grapefruit Grows* and *Julien Bryan: Photographer-Lecturer.*

With its commercial development hobbled by World War II, WNBT spent the war years broadcasting about four hours a week, mostly offering instructional programs for air raid and fire wardens on sets installed in New York's police precinct stations. It also presented occasional feature films and live coverage of sports events watched mainly by hospitalized veterans on sets provided by the broadcast industry. But on V-E Day, May 8, 1945, WNBT presented 15 hours of programming on the end of the war in Europe, all of it relayed to WRGB Schenectady, and WPTZ Philadelphia, on the first TV "network." Similar coverage was provided on V-J Day.

But on V-E Day, May 8, 1945, WNBT presented 15 hours of programming on the end of the war in Europe, all of it relayed to WRGB Schenectady, and WPTZ Philadelphia, on the first TV "network." Similar coverage was provided on V-J Day.

In December 1945 WNBT went to a six-day-a-week operation, with no service on Tuesdays. Early in 1946 Washington joined the network. The first network sponsor on the four-city hookup was Gillette, which underwrote coverage of the Joe Louis-Billy Conn fight on June 19, 1946. This was a five-camera pickup, with Ben Grauer announcing. That same year Standard Brands sponsored the first hour-long variety show, *Hour Glass,* featuring Edgar Bergen, Edward Everett Horton, Joe Besser and a chorus line.

WNBT went off the air from March 1 to May 9, 1946 to switch from Channel 1 to Channel 4. When it returned, it added three days of daytime programming. On June 8, 1946, the historic Milton Berle weekly series began. In 1947 the station carried the World Series between the Yankees and Dodgers. *Howdy Doody* started on Dec. 27, 1947. In 1948, Tuesday joined the schedule, and the programming expanded, with WNBT originating for the network.

FIRSTRUN PROGRAMS—programs or series episodes being presented for the first time on television, without having had previous network or local broadcast exposure. Motion pictures that have played in theaters are considered firstrun in their initial television presentations. Existing network or resurrected series are considered firstrun syndication fare if the episodes offered to local stations have never before been aired.

Through most of the 50s and part of the 60s the accepted practice at the networks was to schedule 39 firstrun episodes of a series and 13 repeats over the lower-viewing summer months. Later, for economic reasons and because rerun programing was found to be acceptable even in high-viewing seasons, the networks changed the formula to 26 firstruns and 26 repeats. By the early 70s, partly to accommodate specials, the pattern had been reduced to 22 firstruns for most series.

FIRSTRUN SYNDICATION—programs created expressly for the syndication market, or imported from abroad, which have not had previous exposure on the U.S. networks. Their opposite number in syndication is off-network reruns.

"FIRSTS"—TV broadcasts that were the first of their kind. When TV was new, each "first" carried the sense of history in the making. A list of claimed "firsts" would be endless and many of the claims open to dispute. The following are among the more significant landmarks:

• Inaugural of regular television service: by NBC on April 30, 1939, with President Roosevelt appearing in a telecast of the opening of the New York World's Fair.

• First professional boxing telecast: Max Baer vs. Lou Nova, June 1, 1939.

• First one-hour TV production: Gilbert & Sullivan's *The Pirates of Penzance,* June 20, 1939.

• First major league baseball telecast: Brooklyn Dodgers vs. Cincinnati Reds, Aug. 26, 1939, a doubleheader.

• First coverage of political conventions and first telecast of presidential election returns: 1940.

• First network broadcast: the hookup of WNBT-TV New York, and WRGB-TV Schenectady, N.Y., Jan. 12, 1940.

• First TV network sponsor: Gillette, with the telecast of the Joe Louis vs. Billy Conn boxing match, June 19, 1946.

• First opera telecast from the stage: Verdi's *Otello* from the Metropolitan Opera House by ABC, December 1948, with Texaco as sponsor.

• First telecast of a presidential inauguration: President Truman in January 1949.

• First telecast from the White House: President Truman's address to the nation on food conservation, October 1949. (This was before there was national interconnection.)

• First presidential speech carried coast to coast: President Truman's address at the Japanese peace treaty conference in San Francisco, Sept. 4, 1951. It was pooled by the networks to inaugurate AT&T's coast-to-coast network facilities.

• First presidential news conference to be televised: President Eisenhower, January 1955.

• First network editorial: delivered by Dr. Frank Stanton, president of CBS Inc., in August 1954. It asked that TV and radio be given the right to cover congressional hearings.

• First use of videotape on television: the West Coast feed of *Douglas Edwards with the News* by CBS, Nov. 30, 1956.

FISHER, ART—director associated with comedy-variety shows, pageants and occasionally sports. His credits include *The Sonny and Cher Comedy Hour, The Andy Williams Show, America's Junior Miss Pageant, Miss Teenage America* and specials with Ann-Margret, Bing Crosby, Dionne Warwick and the Jackson Five. He was also producer-director of the coverage of the Ali-Frazier championship fight.

FISHER, EDDIE—a popular singer of the 50s who starred in *Coke Time with Eddie Fisher,* a 15-minute variety show which aired twice a week on NBC from April 1953 to April 1957. The following September he began an hour-long prime-time series, *The Eddie Fisher Show,* which ran alternate weeks with *The George Gobel Show* for two seasons.

FLASHCASTER—device for superimposing written news bulletins over the television picture, usually in a horizontal strip moving across the bottom of the screen.

FLATT, ERNEST—director and choreographer. His credits include *Julie Andrews and Carol Burnett at Carnegie Hall, The Carol Burnett Show, The Lucky Strike Hit Parade, Annie Get Your Gun* and *Kiss Me, Kate.*

FLEISCHMAN, STEPHEN—director, writer and producer of news documentaries, initially with CBS (1957–63) and thereafter with ABC. His credits with the latter include *Close Up: Life, Liberty and the Pursuit of Coal, Assault on Privacy, The Long Childhood of Timmy, Anatomy of Pop—The Music Explosion* and *Close Up: Oil.*

FLEMING, ROBBEN W.—president of the Corporation for Public Broadcasting (1979–81) and for the previous 12 years president of the University of Michigan. His selection to succeed Henry Loomis at CPB came after a long search, but he arrived while public television was in turmoil and at just the time the Carnegie Commission on the Future of Public Broadcasting recommended that CPB be disbanded and replaced by a new organization. Fleming's first acts were to mend fences with PBS and to design a reorganization of CPB that might obviate the need for the legislative changes proposed by Carnegie. He was succeeded at CPB by Edward Pfister.

FLICKER, THEODORE J.—director and writer; credits include *Dick Van Dyke Show, Andy Griffith Show, The Rogues, I Dream of Jeannie, Night Gallery* and *Man From U.N.C.L.E.*

"FLINTSTONES, THE"—a Hanna-Barbera animated cartoon series of the 60s that utilized the situation comedy form in a fanciful portrayal of Stone Age domesticity. The first animated sitcom to be presented in prime time, its humor derived chiefly from the anachronism of a prehistoric setting for activities and concerns that were distinctly of modern suburbia.

In the relationships between the two married couples involved, and in their blue-collar attitudes and styles of speech, *The Flintstones* owed a large debt to Jackie Gleason's *The Honeymooners.* Also modeled on that series was Hanna-Barbera's *The Jetsons,* the reverse image of *Flintstones* in that it presented modern situations in futuristic times. *The Flintstones* premiered on ABC-TV in the fall of 1960 in an early evening slot; *The Jetsons* followed two years later. The reruns of both have been in continuous syndication.

Hanna-Barbera skirted the high cost of producing animation, with both series by developing an assembly-line process which reduced the number of lip movements to the exclusion of consonants. It sufficed. *The Flintstones* contributed to the English language Fred Flintstone's all-purpose cry: "Yabba dabba doo.

"FLIPPER"—hour juvenile interest adventure series on NBC (1964-67) built around a dolphin befriended by two young sons of a marine preserve ranger. Featured were Porter Ricks as the ranger and Luke Halpin and Tommy Norden as his sons. Filmed in Florida and the Bahamas, the series made a specialty of underwater scenes in which Flipper assisted in the capture of malefactors. It was by MGM-TV and Ivan Tors Films and based on a movie produced by Tors.

"FLIP WILSON SHOW, THE"—first comedy-variety hit starring a black performer, which ran on NBC from 1970 to 1974 and during the first three seasons was among the leaders in the Nielsen ratings. Flip Wilson, the star, based much of his comedy on satirical characterizations of Harlem types, such as the chatterbox Geraldine Jones and the Reverend Leroy of the Church of What's Happening Now. The Thursday night hour from Clerow Productions in association with Bob Henry Productions and NBC was bested in the ratings finally by *The Waltons* on CBS. Wilson continued to appear in specials.

Flip Wilson & Steve Lawrence

FLY, JAMES LAWRENCE (d. 1966)—outspoken chairman of the FCC (1939–44) who made enemies in the broadcast industry when he charged the networks with seeking a monopoly and called the NAB a "stooge organization." Before he became chairman, he was largely responsible for the commission order in 1941 forcing NBC to divest itself of one of its two radio networks, the Red and the Blue. It was the Blue that was sold off, and it became the foundation of the third major network, ABC.

A famous clash with the industry occurred at the NAB convention in 1941, where Mark Ethridge, former NAB president who was then general manager of the *Louisville Courier and Journal*, accused Fly of meddling and of trying to take over the FCC. Fly told newsmen the next day that the management of radio reminded him "of a dead mackerel in the moonlight, which both shines and stinks."

Appointed by President Roosevelt in 1939, Fly, when he became chairman, improved the efficiency of the agency. He was critical of radio soap operas and lotteries and contended that some programs were contributing to juvenile crime. He resigned from the FCC in 1944 to head the Muzak Corporation, from which he retired 10 years later.

"FLYING NUN, THE"—ABC fantasy-comedy (1967–69) about an American novitiate in a Puerto Rican convent who has the gift of being able to fly through the air. Via Screen Gems, it featured Sally Field in the title role and Alejandro Rey, Marge Redmond and Madeleine Sherwood.

FOGARTY, JOSEPH R.—FCC commissioner since September 1976, appointed by President Ford for a seven-year term to fill the non-Republican vacancy of Glen O. Robinson. Fogarty had previously been counsel to the Senate Communications Subcommittee while it was headed by Sen. John O. Pastore (D.-R.I.) and for 13 years a staff member of the Senate Commerce Committee. His work on the Commerce Committee had involved transportation, coastal fishing rights and East-West trade but never broadcasting until Sen. Pastore appointed him in 1975 to succeed Nicholas Zapple as chief counsel of the Communications Subcommittee.

FOLSOM, FRANK M. (d. 1970)—RCA executive who succeeded Gen. David Sarnoff as president in 1949, while Sarnoff retained the title of chairman. Folsom later became chairman of the executive committee board (1957–66).

FOOTBALL ON TV—the quintessential TV program: fast-moving, melodramatic, frequently violent—a familiar program form with an uncertain outcome. Football's popularity on TV, combined with the negotiating prowess of Pete

Rozelle, commissioner of the National Football League, has made it also the most expensive of sports attractions for broadcasters (the quadrennial Olympic Games excepted).

The rights for the Super Bowl game alone carried a price of $3 million in the mid-70s, which would have been huge for a prime-time event, let alone for a game played during the daytime hours. The entire NFL package with the networks was worth $54.6 million a year under four-year contracts with the networks that began in 1974. That amount contrasts strikingly with CBS's exclusive two-year contract with NFL for $9.3 million in 1961. Two years later, the rights rose to $14 million for two years, which was then considered a staggering amount.

Billy Kilmer & the Washington Redskins

Rozelle not only succeeded in boosting the rate substantially with each new contract but maximized revenues from TV by creating a separate package for each network. He was aided by the merger of the American Football League with the NFL in 1966; this was after NBC in January 1964 guaranteed the AFL $42 million for five years of TV rights, giving the young league a bankroll to bid for star players, such as Joe Namath.

With the merger, NFL had two network contracts to negotiate, in addition to the local broadcast pacts for each team. In 1969, ABC joined the fold by buying up a new piece of NFL rights with *Monday Night Football*, paying a reported $8 million a year. Rozelle's willingness to schedule night games and Sunday double-headers, and otherwise to set the starting times of games for TV's convenience, made it easier for the networks to meet his terms. By 1974 each of the 26 NFL teams received more than $2 million a year from broadcast rights.

Of the $54.6 million realized each year by the NFL from 1974–78, CBS pays a reported $23.5 million a year for a package of 83 games, NBC $16.6 million for 89 games of the American Football Conference and ABC $11.5 million for 14 prime-time games from September through December. The network carrying the Super Bowl pays an additional $3 million. CBS sweetened the pot with an extra $500,000 in 1977 in order to have the game played in the evening.

ABC, which has had a hold on the NCAA Saturday afternoon football games since 1965, paid about $15 million a year for the college rights in the mid-70s. The initial contract with ABC a decade earlier had set the rights at $8 million a year.

FORD FOUNDATION AND PUBLIC TV—a relationship that began in 1951 and ended in 1977, except for individual program projects. In that 26-year period, Ford will have pumped $289 million into the ETV and PTV systems for the development of stations, the building of production centers, the financing of program production, the interconnection of the stations, the perfecting of fundraising techniques and numerous internal projects.

If Ford is not actually the parent of public TV in the U.S., it is at the very least the godfather. For most of the years educational (later called "public") television has existed, Ford was the single largest source of support for the system. It announced its final four-year phase-out of institutional support in 1973 when it determined that long-range federal funding was in view, that mechanisms had been developed to insulate the system from government interference and that the system had a nucleus of contributing subscribers sufficient for it to stand on its own.

Ford's first activity in broadcasting was the creation in 1951 of the Radio-Television Workshop, conceived as an agency to improve the cultural use of TV and radio in the commercial systems. The Workshop brought forth *Omnibus,* a celebrated weekly series which ran four years on CBS and one on ABC, with its losses covered by the foundation. But as early as 1951, Ford had also contributed to the efforts to secure allocations from the FCC for noncommercial stations that would be dedicated to education.

Ford created the Fund for Adult Education, which in turn financed the National Citizens Committee for Educational Television and made grants to the various organizations working for ETV. Between 1952 and 1961, the Fund made grants of more than $3.5 million to assist 37 new educational stations that were going on the air, and it financed NAEB seminars, workshops and technical consultation services for the new stations. It also provided the backing to establish the Educational Television and Radio Center (forerunner of NET) in Ann Arbor, Mich., as a program supplier and distribution service. From 1953 to 1963, Ford's grants to the center amounted to around $30 million.

A similar amount was distributed during that period for individual experiments with TV as part of formal education, both at the university and public school levels. Meanwhile, the foundation underwrote the *Continental Classroom* early

morning series on NBC and made substantial grants to support experiments with the distribution of courses to several states at once from a transmitter in an airplane. By 1963, the total Ford investment in ETV was $80.7 million.

That year, Ford increased its support of NET (which by then had moved to New York) to $6 million a year so that it might provide national programming of a higher quality than before. Next it began to work at interconnecting the stations and allocated $10 million to demonstrate what public TV might achieve with adequate programing funds and a national hookup. This resulted in the creation of the Public Broadcasting Laboratory and its two-hour Sunday evening series, *PBL* (1967–69). Later, when a 24-hour rate was successfully negotiated with AT&T for interconnection by long lines, Ford made a grant to cover some of the expenses and then contributed to the start-up costs of PBS as caretaker of the interconnection system.

Ford contributed to such programs as *San Francisco Mix, The Advocates, Hollywood Television Theatre, Sounds of Summer, Soul!, NET Opera, Black Journal, The Great American Dream Machine, An American Family, V.D. Blues,* Elizabeth Drew's interview series *Thirty Minutes With…, Evening at the Pops* and *Zoom.*

Also *Sesame Street, The Electric Company, Visions, The Mac-Neil-Lehrer Report* and scores of other national and local programs. Ford also helped to finance national affairs programming through NPACT, created in 1972; the foundation's grants were reflected in such programs as *Washington Week in Review, Washington Connection* and the live coverage of the Senate Watergate hearings and the House Judiciary Committee hearings on impeachment.

Much of Ford's final $40 million grant went toward the support of the Station Program Cooperative, the Station Independence Program, the principal production centers and the plan to interconnect the stations by domestic satellite. See also Friendly, Fred W.; Public Television; WNET.

FORD, FREDERICK W.—FCC chairman for one year (1960) of the seven he served on the commission (1957–64). Although a Republican appointed by President Eisenhower, he held the liberal view that the commission should set up program guidelines for broadcasters in the public interest. After 30 months as a commissioner, following service as an attorney for the FCC, Ford was named chairman by Eisenhower to succeed John C. Doerfer, who resigned by request after the scandals raised by the House Legislative Oversight Committee. When President Kennedy won the election in 1960 and was privileged to appoint his own FCC chairman, Newton N. Minow, Ford returned to his old status as a commissioner. He resigned in 1964 to become president of NCTA at $50,000 a year.

FORD, TENNESSEE ERNIE (ERNEST J. FORD)—one of few country-music artists to become a major TV personality. His popularity spanned the years from 1955 to 1965 and was enhanced by his recording of "Sixteen Tons," one of the top song hits of the 60s. Calling himself the "ol' pea-picker," Ford camouflaged his theatrical polish with down-home tales, epigrams and jokes delivered in a becoming country accent.

He came to notice on TV as host of *The Kollege of Musical Knowledge* on NBC upon the retirement of Kay Kyser in the summer of 1954. The following year he began a daytime variety program, *The Tennessee Ernie Ford Show*, featuring country music and comedy. In the fall of 1956 he added a weekly evening series, *The Ford Show*, which was more sophisticated and which ran to the early 60s. Ford concluded each of his programs with a hymn.

Eric Porter & Margaret Tyzack

Tennessee Ernie Ford

FORREST, ARTHUR—director; credits include *The Dick Cavett Show,* Jerry Lewis Telethons, David Frost specials and daytime game shows.

"FORSYTE SAGA, THE"—BBC adaptation of the John Galsworthy novels of 19th-century England, completed in 26 episodes. Its popularity on public television here inspired the commercial networks to experiment with the miniseries form, especially in the adaptation of modern popular novels (*QBVII, The Blue Knight*). The series was produced by the BBC in 1967 and played on National Educational Television (before it became PBS) in the 1969–70 season. MGM-TV, which had been associated with the production, then made it available for commercial syndication.

Featured were Kenneth More, Eric Porter, Nyree Dawn Porter, Ursula Howells, Jenny Laird, Joseph O'Connor, June Barry and John Bennett.

FORTE, CHET—producer-director specializing in sports coverage; his credits include the Olympic Games on ABC since 1964, events for *Wide World of Sports,* NBC basketball and special events.

FORTNIGHTLY DECISION [Fortnightly Corp. v. United Artists Television Inc./392 U. S. 390 (1968)]—landmark ruling of the Supreme Court in a cable copyright case which established that cable systems do not violate a picture company's copyrights when they carry movies broadcast on distant TV stations.

Fortnightly Corp. owned and operated CATV systems in two small West Virginia towns, Clarksburg and Fairmont. When the systems were introduced there were no local TV stations, although two subsequently began to broadcast before the suit by United Artists TV was filed. In the classic use of cable-TV, Fortnightly brought in signals from stations in Pittsburgh, Steubenville, Ohio, and Wheeling, W. Va.

United Artists TV, having licensed some of its movies to the stations which Fortnightly imported to the communities it served, brought suit in the mid-60s to prohibit Fortnightly from showing its copyrighted movies without a license and asked for damages, as well. The film company was successful in the District Court and Court of Appeals, but Fortnightly carried the case to the Supreme Court.

The Supreme Court stated that the copyright laws give the holder exclusive right only to "perform in public for profit." It then proceeded to analyze the position of a CATV system to determine if it "performed" the movies for profit. It noted that the broadcaster "selects and procures the program to be viewed" and sends out the broadcast signal. The viewer does not perform; he merely provides the receiving equipment.

The Court felt that a CATV was more like a viewer than like a station, since it did not select programs but merely enhanced "the viewer's capacity to receive the broadcast signals.

Therefore, the CATV presentations were not a performance, the Court said, and Fortnightly did not violate United Artists' copyrights in movies by importing signals from distant stations for its subscribers in Clarksburg and Fairmont. See also Teleprompter Corp. v. CBS.

FOSTER, DAVID H.—president of National Cable Television Assn. from 1972–75, resigning after policy disputes with the association's board during a period of general frustration for the cable industry. He later joined the Natural Gas Supply Committee as executive director.

FOSTER, NORMAN—director; credits include *It Takes a Thief, Batman, The Loner, Loretta Young Show, Davy Crockett, Zorro* and *Hans Brinker.*

FOUHY, ED—v.p. of CBS News in Washington since 1978 and previously Washington-based producer for the *CBS Evening News with Walter Cronkite.* For five years (1969–74), he produced hard news coverage from the capital for the evening newscast, then went to NBC News where he was first producer of the *Nightly News* and then director of news. He returned to CBS News in 1977.

"FOUR-IN-ONE"—an attempt by NBC in 1970 to rotate four diverse hour-long dramatic series, all produced by Universal, in a single timeslot. The project was unsuccessful but led to the *Sunday Mystery Movie,* which did well. The rotating series, each of which contributed 6 episodes to *Four-In-One,* were *McCloud* with Dennis Weaver, *The Psychiatrist* with Roy Thinnes, *Night Gallery* with Rod Serling and *San Francisco International* with Lloyd Bridges. McCloud went on to become a *Mystery Movie,* and *Night Gallery* became a weekly series on its own.

FOURTH CHANNEL—new British network (in addition to BBC-1, BBC-2, and ITV) to debut in mid-eighties. The result of a lengthy debate during the late seventies, the Fourth Channel will immediately increase employment opportunities for U.K. TV/filmmakers by some 33%.

FOWLER, MARK S.—chairman of the FCC since May 18, 1981, designated for the post by President Reagan. A staunch upholder of Reagan's conservative policies, Fowler moved aggressively to accelerate the deregulation of broadcasting. He was also outspokenly sympathetic to the business concerns of broadcasters and quickly became the industry's all-time favorite FCC chairman.

Fowler called himself an advocate of "unregulation" and said he believed the public interest would be better served by market forces than by bureaucrats in the Federal agency. His commission, dominated by conservatives, did away with many of the procedures that were adopted to hold broadcasters accountable to the public and reduced much of the paperwork. The Fowler commission went so far in its first year to recommend the abolition of the fairness doctrine and equal time law but were thwarted by liberals in the Congress.

Fowler had worked for Reagan's election. As a senior partner in the Washington communications law firm of Fowler & Meyers, formed in 1975, he served as legal counsel for communications to the various committees working on the Reagan campaign. Fowler also served on the Reagan transition team.

Before entering the field of law, Fowler worked in broadcasting for about 10 years, chiefly as an announcer and salesman for small radio stations in Florida. Most of these were part-time jobs held during his college years.

Sonny Fox

FOX, IRWIN (SONNY)—one of the few to pass from TV performer to TV executive. He became v.p. of children's programs for NBC in 1976 after having come to prominence as a host of children's programs. He was the star of *Let's Take a Trip* on CBS (1955–58) and during the same period host of the prime-time quiz show, *The $64,000 Challenge.* During the 60s, he did such shows as *On Your Mark* and *Wonderama.*

He was fired by NBC in 1977 when his attempt at mounting a quality children's schedule was undone by the poor ratings for his shows. Moving to the Coast, he became associated with Alan Landsburg Productions and then formed a production company of his own.

FOX, MATTHEW (MATTY) (d. 1964)—motion picture executive and promoter who became a pay-TV pioneer as president of Skiatron, which he acquired in 1954. Fox's promotional skills caused the industry and the investment community to take Skiatron seriously at first, but the company foundered in the FCC's long delay in giving pay-TV permission to proceed. Earlier, Fox had been an executive with United Artists, Universal-International Pictures and United World Films.

"FRACTURED FLICKERS"—syndicated comedy series (1961) built around scenes from silent movie classics with humorous narration delivered by Hans Conreid. It was produced by Jay Ward and Bill Scott.

FRAIBERG, LAWRENCE P.—broadcaster and independent producer, who served two substantial hitches with Metromedia Television, rising the second time to president of the company (1977–79). He left, after a philosophical dispute with higher management, for a private venture.

In 1980, he became president of the Group W television stations. As a broadcaster, he became noted as an innovator. For example, while he was v.p. and general manager of WNEW-TV New York, the Metromedia flagship (1965–69 and again from 1971–77), he elevated the station's public image by instituting a successful one-hour newscast in prime time as an alternative to network fare and created campaigns of "total station involvement" for concentrated periods of time with particular issues of public affairs. Later, as president of Metromedia Television, he arranged for annual satellite telecasts of the Royal Ballet.

Fraiberg joined Metromedia in 1959 to assist in the acquisition of additional stations, after having spent the previous decade with KPIX-TV San Francisco in sales and program production. He became v.p. and general manager of Metromedia's Washington station WTTG in 1963 and two years later took over management of WNEW-TV. He left in 1969 to start his own company, Parallel Productions, Inc., then returned in 1971 for another eight-year stretch.

FRAME—a complete television picture, consisting of two interlaced fields, 525 lines in the NTSC system, 625 lines in the CCIR system. Thirty frames are transmitted every second in the NTSC system, 25 in the CCIR system.

FRAME-GRABBER—an adjunct of cable-TV technology which permits the subscriber, using a modified receiver, to call up still pictures or frames of print at will from a local information bank. The subscriber, by means of a selector dial and using a reference guide, may select the set of frames he wishes to view in a wide range of subject areas. The system has a number of applications beyond information retrieval, including programed instruction and student examinations or quizzes.

FRAMER, WALT—creator and producer of such shows as *Strike It Rich, The Big Payoff, Million Dollar Family, Lady Luck, It's in the Cards* and *Meet the People* during the 50s. He started in broadcasting as a performer on KDKA Pittsburgh.

FRANCE, TELEVISION IN—what used to be a monolithic and dull television service, operated under tight hegemony by a government administrative agency, ORTF, has since undergone a drastic overhaul, the result of both political and public pressure. The three TV channels still operate, but under separate and autonomous companies, among them providing a competitive and much more mettlesome service than formerly existed. The medium is no longer the official "Voice of France," as Charles de Gaulle and Georges Pompidou had insisted it be.

The first channel, TF-1, relies heavily on popular entertainment shows to attract the largest mass audience. Its chief competitor, A-2, puts more of an emphasis on cultural programs and documentaries. The youngest of the networks, FR-3, emphasizes regional programs and is also known as the cinema channel because it airs an average of five feature films a week.

Taxation (license fees) remains the primary source of revenue for the medium, though TF-1 and A-2 supplement their income by selling spots, the revenue from which may add up to 25% of their total income. The license fees come to $72 a year.

The French-developed 625-line SECAM scan is, of course, the system in use for color transmission. Black-and-white reception in France is on an 819-line scan.

FRANCIS, ARLENE—one of the leading female TV personalities of the 50s, tapped by NBC as hostess of the *Home* show (it was no reflection on her that it was the only one of the *Today, Home, Tonight* trio that did not succeed in the ratings). She was best known as a regular panelist on *What's My Line* but also, for a time, conducted the *Arlene Francis Show* on NBC; a brief entry, *Talent Patrol;* and on radio, *Arlene Francis at Sardi's.* An actress, she also had film and stage credits, including *Once More with Feeling.* Since 1981, she has been co-hosting *The Prime of Your Life* on WNBC-TV, New York.

FRANK, BARRY—v.p. in charge of CBS Sports (1976–78), picked for the post when it was vacated by Robert Wussler's promotion to network president. When Wussler returned to CBS Sports as president in the fall of 1977, Frank became senior v.p. Like Wussler, he left CBS in 1978 in the wake of the "winner take all" tennis scandal. For five years previously, Frank had been senior v.p. of Trans World International, an independent packager of sports programs. There Frank had developed such sports shows as *The Challenge of the Sexes, The Heavyweight Championship of Tennis, The Superstars* and *The World Invitational Tennis Classic.*

Earlier, he had been with ABC as v.p. of sports planning and with J. Walter Thompson as broadcast supervisor. He first worked for CBS in 1957 as assistant to the v.p. of operations.

FRANK, REUVEN—twice president of NBC News, named the second time in March 1982 after the resignation of Bill Small. His first tour (1968–73) ended when Frank asked to return to producing. He subsequently became executive producer of the magazine-style program *Weekend.* Before becoming a news executive in the mid-60s, Frank had produced more than a dozen documentaries, including *The Tunnel, The Road to Spandau, A Country Called Europe* and *The Daughters of Orange.* He had also been producer of *The Huntley-Brinkley Report* when it began in 1962 and then became its executive producer (1963–65).

Frank joined NBC News as a writer in 1950 after working as a reporter for the *Newark Evening News.* In 1954 he developed such weekly half-hour news programs as *Background, Outlook* and *Chet Huntley Reporting.* His career accelerated when Huntley and David Brinkley were teamed for the first time at the 1956 political conventions, a triumphal event with which Frank was closely associated.

FRANK, SANDY—syndication executive heading his own firm, one of the companies that became prominent distributing shows for prime time access periods. In 1975 Frank petitioned the FCC to bar strip programming from the access slots, a move against the low-budget game shows that were able to play nightly; after hearings the FCC denied the petition in 1976. A court appeal did not avail.

FRANKENHEIMER, JOHN—one of the outstanding TV directors of the 50s who, like many another who developed in the medium, abandoned it for movies. With CBS he was director of *Mama* and *You Are There* and of TV dramas

for *Danger, Climax, Studio One, Playhouse 90, DuPont Show of the Month* and other anthologies.

FRANKLIN, JOE—local New York television personality who has conducted a daily program of interviews, chiefly with book- and show-pluggers, almost continuously since 1953. His program *Down Memory Lane* switched stations over the years but settled in at WOR-TV.

Pauline Frederick

FREDERICK, PAULINE—correspondent for NBC News for 22 years and probably best known of the female journalists during the decades in which there were few in broadcasting. She became prominent as United Nations correspondent for NBC, an assignment she held until her retirement in 1975, but had served earlier as a foreign correspondent, both for NBC and ABC. She joined ABC in the late 40s after a period of freelancing for newspapers and moved to NBC in 1953.

FREEDOMS FOUNDATION AWARDS—citations presented each Washington's birthday for TV and radio programs contributing to an understanding of America in the judgment of Freedoms Foundation at Valley Forge (Pa.), source of the awards. Along with programs on patriotic themes, broadcasts on education, ecology, drug abuse and other contemporary topics have qualified for the citation.

FREED, FRED (d. 1974)—noted producer of documentaries for NBC News, principally those under the rubric of *American White Paper.* He received three George Foster Peabody Awards, two duPont-Columbia Awards and seven Emmys for his programs.

Three of the Freed *White Paper* productions spanned an entire evening of prime-time programming: *NBC Reports: The Energy Crisis* (1973); *Organized Crime In the United States* (1966); and *United States Foreign Policy* (1965). Two others were presented in two parts: *And Now the War Is Over—The American Military in the '70s* (1973) and(*Vietnam Hindsight* (1971).

Specializing in major current issues, Freed produced a series of *White Paper* reports examining the urban and environmental crises; these resulted in three programs entitled *The Ordeal of the American City* (1968ND69), and two others, *Who Killed Lake Erie?* (1969) and *Pollution Is a Matter of Choice* (1970).

Freed began as a magazine editor and writer, entered broadcasting in 1949 and joined NBC in 1955 as managing editor of the daytime *Home* show. He left to become a documentary producer for CBS, then rejoined NBC in 1961 as producer of *Today* before being assigned exclusively to documentary production.

FREEDMAN, LEWIS—TV drama producer working between PBS and CBS between 1956 and 1974. In 1980, he became head of the Program Fund, a newly established division of the Corporation for Public Broadcasting.

He began as a producer on *Camera Three* for CBS (1956–58), then became vice-president of programing for WNDT (now WNET) where he produced *New York Television Theatre* and then joined the Public Broadcasting Laboratory as director of cultural programming for its Sunday night "network" series *PBL*. In 1970 he became producer of *Hollywood Television Theatre*, a PBS series that originated at KCET Los Angeles. There he produced such plays as *The Andersonville Trial, Awake and Sing, Monserrat, Big Fish, Little Fish* and *Poet Game*. CBS hired him away in 1972 to spearhead its new thrust in drama and the mini-series. After producing several full-length plays, he resigned to live in Europe for a few years.

FREEMAN, LEONARD (d. 1973)—executive producer of *Hawaii Five-O, Storefront Lawyers* and such TV movies as *Cry Rape* as head of Leonard Freeman Productions.

"FREEZE" OF 1948—a halting of licensing and transmitter construction by the FCC when field studies showed that haphazard channel assignments were creating chaos on the airwaves. Instituted in October 1948, the freeze held until July 1952 while the FCC developed a blueprint for "a fair, efficient and equitable" distribution of service among the states and communities, under its mandate from Congress. Affecting the freeze, along with the need to sort out frequencies and establish engineering standards, was the uncer-

tainty over which of three color TV systems proposed would be authorized for the U.S.

The freeze was lifted with the FCC Sixth Report and Order, issued in April 1952 but effective in July, which presented the commission's frequency allocation plan. That table of assignments, which has guided the regulatory process ever since, envisioned 2,053 stations in 1,291 cities, including 242 noncommercial outlets for ETV. There were to be 617 stations on VHF and 1,436 on UHF, with Channel 1 designated for land-mobile and two-way radio service. The plan also established three geographic zones with different mileage-separation and antenna-height regulations for each. See Table of Assignments, Zones.

The French Chef, Julia Child

"FRENCH CHEF, THE"—long-running public TV series (1963–73) featuring the kitchen artistry and humor of Julia Child, a middle-aged holder of the cordon bleu who succeeded in taking the pomposity out of French cookery. A genial, plain-mannered woman who did not mind taking a swig of the cooking wine after adding the soupcon, she became public television's most popular personality in the 60s, enjoyed for herself as much as for her ability to instruct. The series emanated from WGBH-TV Boston and went off the air when corporate funding evaporated.

In 1978, Julia Child returned to public television with a new series and simultaneous book, both titled *Julia and Company*. Featuring menu themes rather than individual dishes, the program enjoyed the popularity of her first series.

FRENCH, VICTOR—actor and director, who both appeared in and directed episodes of *Gunsmoke, Little House on the Prairie,* and *Carter Country.*

FRIEDLAND, LOUIS N.—chairman of MCA-TV, the syndication arm of Universal, since 1978, after having been president for five years. A creative executive who was a key figure in MCA syndication for two decades, he was adept at devising new marketing patterns to maintain his company's leadership. One of his innovations was the selling of "futures" to local stations—the reruns of network hits years before they were spent on the networks, with the understanding that they would not become available to stations until their network runs ended. By selling the shows while they were hot, MCA commanded higher prices for them than it could get for a network program whose ratings had already deteriorated. This and other Friedland innovations were quickly adopted by other syndicators.

FRIEDMAN, PAUL—veteran producer at NBC News who switched to ABC News in the Spring of 1982. He had been executive producer of NBC's weekly newsmagazine, *Prime Time Sunday,* (1979–81). As executive producer of *Today* (1976—79), he revamped the program's format and added new features and contributors. He was executive producer and one of the principal architects of *NewsCenter 4,* the first two-hour local newscast in New York. On two different occasions, before and after he joined *News-Center 4,* Friedman was producer of the *NBC Nightly News.*

Fred W. Friendly

FRIENDLY, ED—independent TV producer since 1967, after having been v.p. of specials in the NBC program department. Initially he teamed with George Schlatter to produce a number of specials, which led to their becoming executive producers of *Rowan and Martin's Laugh-In.* In 1974, after acquiring the rights to the Laura Ingalls Wilder series of "Little House" children's books, he produced the two-hour made-for-TV movie, *Little House on the Prairie,*

which became a successful series. He also produced the miniseries *Backstairs At The White House.*

"FRIENDLY FIRE"—three-hour TV movie on ABC, airing April 22, 1979, which was based on C.D.B. Bryan's journalistic book of that title on a conservative and patriotic Iowa farm couple who become radical antiwar activists after learning that their son was killed in Vietnam by the "friendly fire" of American forces. Despite the grim theme, and the bitterness expressed toward the American government by the principal characters, the program attracted a huge audience.

The film received considerable advance press attention because it starred Carol Burnett, the famed comedienne, in a heavy dramatic role, which she carried off brilliantly. Ned Beatty also gave a strong performance as her husband. The supporting cast included Sam Waterston, Dennis Erdman, Timothy Hutton, Fanny Spies and Sherry Hursey.

Martin Starger was executive producer for Marble Arch Productions and Philip Barry coproducer with Fay Kanin, who wrote the screenplay. David Greene directed.

FRIENDLY, FRED W.—one of the larger-than-life figures in broadcasting who proceeded from a distinguished career as a news producer and partner of the famed Edward R. Murrow, to the executive echelons of CBS as president of CBS News and then on to becoming a pervasive influence in U.S. public television as consultant to its chief benefactor, the Ford Foundation. Beyond that, as Edward R. Murrow Professor of Broadcast Communications at the Columbia University Graduate School of Journalism, and as author of several books (notably, *Due to Circumstances Beyond Our Control*), Friendly impressed his values on a new generation of broadcast journalists.

Friendly was president of CBS News only two years (1964–66), but his tenure was marked with an activism that included publicly discussed clashes with the network for greater news access to the air. Indeed, his departure from CBS ostensibly was over one such battle—the denial of air time for live daytime coverage of the Senate hearings on Vietnam that would have meant the preemption of an old rerun of *I Love Lucy.* But some at CBS maintain that the denial was not really the issue. A layer of management, represented by John A. Schneider, president of the newly formed CBS Broadcast Group, had been placed between the news division and top corporate officials. Friendly, it was said, attempted to regain direct access to corporate president Frank Stanton by threatening to resign, and his resignation was accepted.

Soon afterwards, he became TV consultant to the Ford Foundation and professor at Columbia. At Ford he immediately set into motion a plan to develop a public TV news organization that would be superior to those of the networks in that it would be free of the commercial constraints. This involved first the interconnection of the stations to form a national network (they had not been hooked up at the time) and the creation of the Public Broadcasting Laboratory to develop a prototype program series on current affairs. On Oct. 31, 1967, the two-hour series, entitled *PBL,* made its debut as a Sunday night magazine of the air.

Friendly also developed a proposal to the FCC for a domestic satellite system operated by a nonprofit corporation which would not only provide interconnection for public TV free of charge but also would dedicate a portion of its revenues to the funding of public TV. Friendly's proposal, which became known as his "people's dividend" plan, was not adopted by the FCC, but it did spur the full-time interconnection of stations by land lines.

Friendly's efforts in public TV were resented by many public broadcasters, who considered him a presumptuous outsider determined to reduce the stations to mere carriers of a network. Nor were they as ob-sessed as he with the idea of creating a great news organization. Their objections forced Friendly to recede into the background, and Ford adopted a policy of noninterference in the affairs of public TV.

Friendly's remarkable career in national broadcasting began with an idea he had conceived for a spoken-history record album in 1948. He was, at the time, an obscure producer at a radio station in Providence, R.I. The project appealed to Edward R. Murrow of CBS, and the album they produced together, *I Can Hear It Now,* was a huge success. It began the legendary 12-year partnership of Murrow and Friendly, through the radio series, *Hear It Now,* and the historic TV version, *See It Now.* He retired from the Ford Foundation in 1980.

FRIENDS OF THE EARTH CASE [Friends of the Earth and Garie A. Soucie v. FCC and United States of America/449 F2d 1164 (1971)]—Court of Appeals case which expanded the application of the Fairness Doctrine to commercial advertisements after the cigarette commercial decision.

In 1971 Friends of the Earth , an organization dedicated to the protection and preservation of the environment, wrote a letter to WNBC-TV New York asserting that the station had broadcast a number of ads for automobiles and gasolines, which it maintained were heavy contributors to air pollution in New York City. FOE asked WNBC to make time available to the organization to inform the public of its side of the controversy. WNBC turned down FOE's request, claiming that the Banzhaf decision did not impose any Fairness Doctrine obligation on broadcasters with respect to product advertising other than cigarette commercials.

FOE complained formally to the FCC. The commission ruled that the air pollution problem was a complex issue and

that Congress had not urged people to stop using automobiles, as it had with cigarettes. The commission refused to extend the Banzhaf Case to other products.

FOE appealed to the D.C. Court of Appeals, which reversed the commission, holding that there was no difference between gasoline and cigarette advertising, since both had built-in health hazards. The court said that although it had indicated the Banzhaf ruling was not to be applied to product advertising generally, the FCC was being too restrained.

Holding that the FOE complaint was not distinguishable from the Banzhaf cigarette complaint, the court remanded the case to the FCC to determine whether WNBC had adequately discharged its public service obligations. After some negotiations, WNBC agreed to broadcast a series of one-minute antipollution announcements.

In a subsequent Fairness Doctrine ruling, in July 1974, the commission closed the door on fairness claims with regard to advertising by in effect stating that the cigarette commercial decision was in error. The FCC stated that the interpretation by the Court of Appeals in the FOE Case would in effect destroy the concept of the American commercial system of broadcasting. The commission went on to say that in the future it would apply the Fairness Doctrine only to those commercials that are devoted in an obvious and meaningful way to the discussion of public affairs. See also Banzhaf Decision, Fairness Doctrine.

FRIES, CHARLES W.—independent producer who was in charge of production for Metromedia Producers Corp. until that company curtailed its film producing activities in 1974. He has been one of the major suppliers of made-for-TV movies.

FRINGE TIME—time periods outside prime time but near to it, as *early fringe* or *late fringe.* Generally considered to be the hours between 5 and 7 P.M. and 11 to just past midnight.

FRITO BANDITO—commercial campaign for Frito corn chips which was pulled off the air in 1970 because of complaints from Mexican-American groups. The commercials featured an animated character representing the stock Mexican bandit speaking in heavily accented English. The groups charged that the commercials were promoting a comic stereotype that was damaging to Mexican-Americans. Their efforts also succeeded in keeping off the air in many markets the film classic, *Treasure of Sierra Madre,* because it involved a similar stereotype portraying the villain.

"FRONTIERS OF FAITH"—weekly religion series on NBC which features respected scholars lecturing on various aspects of the Bible. The program is produced in cooperation with the National Council of Churches.

David Frost

FROST, DAVID—British TV personality and humorist who became known to U.S. audiences in 1964 when the popular British series he helped to create, *That Was the Week That Was* (*TWTWTW*), spawned an American version on NBC. The program of satire and topical humor was well received by many but fell short in mass appeal; nevertheless, it launched a career for Frost on this side of the Atlantic.

In 1969, when Group W lost the Merv Griffin syndicated talk show to CBS, Frost became host of a talk and interview show for that company with moderate success; his British accent and Cambridge manner were said to have been resisted in mid-America, although the program did well enough in major cities. During the run, Frost developed the reputation of ocean-hopper, doing shows here, in England and in Australia. He became part of the ownership group of London Weekend Television when it was franchised, and he operated his own production company, Paradine Productions, based in London.

Frost made headlines in 1975 when, after the U.S. networks rejected an offer of an exclusive series of talks with former President Nixon for a reported fee of $1 million, he secured the rights for $600,000. The financing, in the main, came from conservative West Coast businessmen who believed that Nixon had been wronged in the Watergate scandal and who wanted him to have a forum to tell his side of the story.

Through Syndicast Services, the Frost-Nixon interviews —four 90-minute programs taped with the former President at his home in San Clemente, Calif.—were syndicated in more than 150 markets on a barter basis. Initially there

were to have been six minutes of national commercials in each program, but the resistance of advertisers to an association with a controversial figure resulted in the stations getting an extra minute (a seventh minute) to sell locally. The interviews aired over four weeks in May 1977. Public TV stations carried the broadcasts in certain key markets—such as Buffalo, N.Y., and South Bend, Ind.—where commercial clearances weren't achieved. The first of the telecasts, airing on May 4, attracted the largest audience ever for a news interview program. See "Nixon Interviews with David Frost, The."

"FROSTY THE SNOWMAN"—Christmas season perennial on CBS since 1969, produced by Rankin-Bass as an hour animated special. The children's story of a snowman who came to life for one day was narrated by Jimmy Durante with character voices by Billy DeWolfe, Jackie Vernon, June Foray and Paul Frees.

FRUCHTMAN, MILTON A.—widely traveled director based in New York who has done work on every continent and who set up a worldwide TV network for the Eichmann trial in Jerusalem. Credits include *Verdict for Tomorrow*, episodes for *High Adventure* series on CBS, *Assignment Southeast Asia* and *Lost Men of Malaya*.

David Janssen as Dr. Richard Kimball

"FUGITIVE, THE"—hour-long ABC adventure series which enjoyed great popularity throughout the world and inspired a cycle of man-on-the-run shows (*Run for Your Life, Run, Buddy, Run,* etc.). Running from 1963 to 1967, it starred David Janssen as a doctor wrongly accused of murder who is relentlessly pursued by a police inspector (Barry Morse) as he himself pursues the actual murderer, a mysterious one-armed man. The final episode of the 120 that were produced resolved the story, and it scored enormous ratings here and in every country where the series was carried. After its prime-time run, it was repeated on ABC as a daytime strip and went into syndication in 1970. QM Productions (Quinn Martin) produced it.

"FUNNY SIDE, THE"—a short-lived NBC series (1971) which, with a repertory company of five couples of different ages, attempted to bridge the variety show and the situation comedy. In songs and sketches, each episode attempted to highlight the universal stresses between the sexes. Produced by Bill Persky and Sam Denoff, it yielded 13 films. Gene Kelly was host, and Teresa Graves, John Amos, Warren Berlinger and Pat Finley were among the regulars.

FUNT, ALLEN—see "Candid Camera."

FURNESS, BETTY—TV performer who was one of the best-known commercial spielers in the 50s and a leading consumer advocate in the 70s. A former stage and screen actress *(My Sister Eileen, Doughgirls),* she earned a degree of fame in 1949 as TV spokeswoman for Westinghouse appliances. She became the talk of the nation one evening when, in a live commercial for refrigerators, she struggled in vain with a stuck door.

In the meantime she hosted a number of local shows in New York, among them *At Your Beck and Call* on WNTA-

TV and *Answering Service* on WABC-TV, as well as such radio programs as *Ask Betty Furness* and *Dimension of a Woman's World.* Active in Democratic party politics, she was appointed by President Johnson in 1967 to be his special assistant for consumer affairs. In 1970 she became head of New York State's Consumer Protection board but resigned a year later. Soon after she joined WNBC-TV New York as consumer reporter and occasional contributor to network news specials and the *Today* program. When Barbara Walters left *Today* early in 1976, Miss Furness became her temporary replacement through most of the year. Later she became regular consumer reporter for the local newscasts on WNBC-TV New York. She is married to Leslie G. Midgley, a news executive at NBC.

Betty Furness

G

GALAVISION—see Cable Networks.

"GALE STORM SHOW, THE" (also "OH! SUSANNA")
—series about the wacky antics of a social director for a
luxury liner, carried by CBS (1956–59), and starring Gale
Storm, with ZaSu Pitts, Roy Roberts and James Fairfax. A
hot property in its day, it was involved in several multimillion-dollar transactions. After 99 episodes, ITC bought
the films from Roach Studios for a reported $2 million and
then produced 26 more, this time for ABC. That network
purchased the reruns for its daytime schedule for 3 years,
plus one season's worth of new nighttime episodes, for $5
million.

"GALLOPING GOURMET, THE"—half-hour syndicated daily cookery series (1969–73) featuring an amusing
and good-looking Australian chef, Graham Kerr. The series
proved ideal for the barter form of syndication and was
placed in more than 130 U.S. markets by Young & Rubicam
for its clients, Hunt-Wesson Foods, American Cyanamid
and American Can Co. An inexpensive program to begin
with, it was produced in Canada, to shave the costs even
more. It was also sold in Canada, the U.K., Australia and
other English-speaking countries. Fremantle produced and
handled foreign distribution.

"GALLANT MEN"—1962 ABC series about two war correspondents traveling with the infantry during World War

II. Produced by Warner Bros. and featuring Robert Mc-
Queeney and William Reynolds, it lasted one season.

GALLO, LILLIAN—independent producer whose TV
movie credits include *Hustling, What Are Best Friends For*
and *Stranger Who Looks Like Me.* She had been executive in
charge of ABC's *Movie of the Week* before venturing out on her
own.

GARAGIOLA, JOE—TV personality whose quick wit
and amiable manner propelled him from local sportscasting
to a regular spot on NBC's *Today* show and baseball *Game of
the Week.* For a time he was also host of the game show, *He
Said, She Said,* and did commercials for national advertisers.
A former major league catcher, he started his broadcasting
career in 1955 as a sportscaster for the St. Louis Cardinals and
then was play-by-play announcer for the New York Yankees
before joining NBC.

GARLAND, JUDY (d. 1969)—one of the great motion
picture stars for whom television was a bitter experience,
partly a function of her temperament. A weekly series that
was to have begun in 1957 was scuttled after a round of
quarrels with CBS programmers. In 1963 a series finally was
produced, *The Judy Garland Show,* but it was plagued with
format problems and with a turnover of producers. Worse, it

was scheduled opposite the No. 1 show on TV, *Bonanza,* which demolished the Garland songfest in the ratings. The series lasted one season.

The famed singer had fared considerably better in specials, giving a memorable performance in a 90-minute spectacular in 1955 which drew a large TV audience. Before she undertook her own series, Miss Garland had another successful outing in a special with Frank Sinatra and Dean Martin. Her TV appearances were scant after the flop of the series, but ironically she will remain one of the medium's perennial favorites for generations to come through her timeless and inexhaustibly popular movie, *The Wizard of Oz.*

"GARRISON'S GORILLAS"—World War II action series which premiered on ABC in September 1967 and ran 26 weeks. Produced by Selmur Productions, it featured Ron Harper, Cesare Danova, Brendon Boone, Rudy Salari and Christopher Cary as an Army guerilla band, each fictive member on leave from a U.S. prison.

GARRISON, GREG—producer-director and long-time associate of Dean Martin in his series and specials. Other productions include *Golddiggers* and *Country Music,* a summer series.

GARNETT, TAY—director working in movies and TV during the 50s and 60s. His TV credits include *The Loretta Young Show, Four Star Theatre, Wagon Train, The Untouchables, Naked City, Gunsmoke, Rawhide, Death Valley Days* and *Bonanza.*

GARROWAY, DAVE—one of the early television "personalities," and probably the prime representative of the "Chicago School," whose intelligence and low-pressure style were in marked contrast to the aggressive pitchmen and manic show hosts of the 50s. A former radio disk jockey for WMAQ, the NBC station in Chicago, he established himself as a refreshingly unconventional TV personality with his first effort for NBC-TV, *Garroway-At-Large* (1949–51). The variety show was startling in forsaking the frills of production; instead of painted backdrops it used the bare studio stage, with a stagehand's ladder as the main prop.

Garroway's success prompted NBC to move him to New York as first host of *Today,* the two-hour morning show, which premiered Jan. 14, 1952. During the 1953–54 season, he doubled as emcee of a night time NBC variety series, *The Dave Garroway Show.* Later he hosted *Wide, Wide World* (1955–58) on that network. He also made numerous guest appearances. The strain of his heavy work schedule, combined with a personal tragedy, the death of his wife, led to

his retirement from *Today* and NBC in 1961. Later he hosted a series for NET, *Exploring the Universe,* and an entertainment show for CBS. In 1969, he began a syndicated talk show from Boston, *Tempo,* but it was short-lived.

Dave Garroway

"G.E. COLLEGE BOWL"—long-running Sunday afternoon series on NBC (1958–72) on which two teams of college undergraduates competed in a quiz on liberal arts subjects. Each week's winning team of four members received scholarship grants. Robert Earle was moderator for most of the years; Allen Ludden had conducted the show initially.

GEISEL, THEODOR—writer of "Dr. Seuss" children's books who also wrote the scripts for all the TV adaptations of his stories, along with original teleplays. His TV credits include *How the Grinch Stole Christmas, Cat in the Hat, Horton Sees a Who* and *The Lorax.*

GELBART, LARRY—veteran comedy writer and playwright who in TV worked variously for Jack Paar, Bob Hope, Sid Caesar, Red Buttons, Art Carney and Jack Carson, and in radio for *Duffy's Tavern.* In the 70s he created the *M*A*S*H* and *Roll Out* series, serving also as co-producer of the former. His plays for the stage included *A Funny Thing Happened on the Way to the Forum,* and *Sly Fox.*

To lure the "dean" of comedy writers back to television, NBC's Fred Silverman gave Gelbart full control of a situation comedy project, letting him write and produce it without network supervision. The program, *United States,* was given a lavish build-up as an adult series combining comedy and drama that would set a new direction for programming. But when *United States* finally went on the air in 1980, it fell far short of its billing and lasted less than two months.

GELLER, BRUCE—producer of action-adventure series, among them *Rawhide, Mannix, Mission: Impossible* and *Bronk* (as executive producer of the last three).

GELLER, HENRY—chief advisor to President Carter on telecommunications policy in his capacity as director of the National Telecommunications and Information Administration in the Commerce Department. With the election of President Reagan, Geller founded and became director of the Washington Center for Public Research.

A lawyer and communications savant who had worked 16 years at the FCC, mainly as general counsel, he joined NTIA as acting director when it was created in 1977 from the ashes of the White House Office of Telecommunications Policy. He received his appointment as director, and coincidentally as Assistant Secretary of Commerce, in 1978.

On leaving the FCC in 1974, Geller became a consultant and one-man think tank on a host of issues. In various capacities, as attorney, *amicus curiae* or intervenor, he participated in court proceedings on numerous matters concerning the FCC. During this period, he also worked on a range of telecommunications projects supported by foundation grants. Meanwhile, his personal philosophy underwent change. He became increasingly enthusiastic for First Amendment principles, and this led him away from his previous staunch belief in the Fairness Doctrine and the public trustee concept for broadcasting. In his NTIA post, he became an outspoken advocate of broadcast deregulation in the age of cable and home video, maintaining that there would be larger public benefits if all forms of television were allowed to be governed by market forces. Many of his ideas, with some modifications, found their way into Rep. Lionel Van Deerlin's 1978 and 1979 bills to rewrite the Communications Act.

Geller was consultant to the House Communications Subcommittee for its controversial 1976 policy report on cable-TV; one of the authors of the American Bar Assn.'s 1976 report on TV journalism, the First Amendment and the Fairness Doctrine; and a principal figure in the Aspen Institute's program on communications and society, funded by the Ford and Markle Foundations. For two years after leaving the commission he worked on projects for the Rand Corp.

He served three hitches with the commission, 1949–50, 1952–55 and 1961–73. He began the latter term as associate general counsel, then became general counsel in 1964 and became special assistant to chairman Burch in 1970.

Among his other contributions on the commission, Geller helped write the 1972 cable rules and the definitive explication of the Fairness Doctrine. See also Federal Communications Commission, National Telecommunications and Information Administration, Retransmission Consent, and Office of Telecommunications Policy.

"GENE AUTRY SHOW, THE"—early Western series on CBS (1950–53) starring Autry and featuring his horse Champion and comedian Pat Buttram. It was via Flying A Productions.

GENERAL ELECTRIC BROADCASTING—broadcast group founded by General Electric Co. in 1922 when it put WGY Schenectady, N.Y., on the air. GE was to grow from a minor group to a major one when, in the summer of 1978, it announced an agreement to acquire Cox Broadcasting for $560 million. The transaction, which was to have included Cox Cable, received FCC approval in June 1980 but never was consummated. At the last minute Cox pulled out. This created some havoc because GE had already begun divesting itself of stations, in some cases conditionally, so as not to exceed the lawful limit.

Between them, the two groups owned eight TV stations, eight AM radio stations and 12 FMs. This meant that GE, under the FCC's multiple ownership rules, had to dispose of three TV stations, one AM and five FMs for the merger to be completed, assuming FCC approval.

At the time it entered into the agreement with Cox, GE's television properties were WRGB Schenectady, WNGE Nashville, Tenn., and KOA-TV Denver; only the latter was to remain in the new group. In preparation for the merger, GE tentatively sold off WRGB to Group Six Broadcasting and WNGE to a subsidiary of a black-controlled insurance company. All three stations are currently owned by General Electric. All thrree stations are owned by General Electric.

"GENERAL ELECTRIC THEATER"—long-running weekly anthology series on CBS (1953–61) in which top Hollywood film stars appeared, including many who otherwise were holdouts to TV. It began as a one-hour series but in 1955 settled into the 30-minute form, with Ronald Reagan as host. Reagan also starred in some of the playlets. Charles Laughton, Myrna Loy and James Stewart were among those who took part. The series was by Revue Productions.

In 1973 General Electric began a series of occasional dramatic specials, ranging from an hour to 90 minutes in length, under the umbrella title, *GE Theater*. Produced by Tomorrow Entertainment and airing on CBS the series included such fine filmed teleplays as *In This House of Brede, Things in Their Season, Larry, I Heard the Owl Call My Name* and *Tell Me Where It Hurts*.

"GENERAL HOSPITAL"—ABC daytime serial that premiered April 1, 1963 and began to soar in popularity during the 70s, principally because it achieved something new for the genre, a powerful attraction to teenagers. This shot it to the top of the daytime ratings. More than any other daytime program, *General Hospital* is credited with inspiring the remarkable interest in daytime soaps that started in the late 70s; it was manifested in daily newspaper columns carrying the storylines, fan magazines devoted to the daytime serial colony, and scholarly analyses of the soap world.

The program did well enough during the 60s with a cast that included John Beradino, Emily McLaughlin, Denise Alexander, Rachel Ames and Peter Hansen, but it was the storylines involving Leslie Charleson, Robin Mattson, Chris Robinson, and rock star Rick Springfield that caught the fancy of teenage viewers. However, nothing quite matched the offbeat romance between Anthony Geary and Genie Francis (Luke and Laura) for interest; their wedding in an episode during the fall of 1981 became something of a national event, with Elizabeth Taylor making several guest-star appearances in segments leading up to the nuptials. Miss Francis left the series soon after to pursue other acting opportunities, but Geary remained with the series and is regarded as the first superstar created by exposure only in daytime drama.

GENESSEE RADIO CASE [Genessee Radio Corp./ 5 FCC 183 (1938)]—first major case in which the FCC determined that it would not grant a second broadcast license to an existing licensee in the same area unless the two stations were to be in real competition with each other. This was a harbinger of the antiduopoly rule and an early case of regulatory concern about the concentration of media control.

Genessee Radio Corp., which operated one of seven existing radio stations in Flint, Mich., applied to the FCC for a license to operate a second station that would be similar to the existing one. It would have the same manager, program format and network affiliation as the original station but would be run by a separate staff and have different advertising rates. The FCC denied the application on the ground that no additional service would be provided to the listeners of Flint. Additionally, the commission suggested that this concentration of media control might prevent the entry of a truly competitive station. There was no appeal.

"GENTLE BEN"—CBS prime-time adventure series for children (1967–69) featuring a 650-pound bear. It was produced by Ivan Tors Films. In the regular cast were Dennis Weaver, Clint Howard and Beth Brickell.

GENUS, KARL—director associated with cultural programs; his credits include *Studio One, DuPont Show of the Month, Robert Herridge Theater, The World of Mark Twain, Sibelius: Symphony for Finland, Duet for Two Hands, I, Don Quixote* and *New Orleans Jazz.*

Burns & Allen

"GEORGE BURNS AND GRACIE ALLEN SHOW, THE"—a half-hour CBS domestic comedy series (1950–58) built upon the established, daffy vaudeville routines of Burns & Allen, who moved gracefully into TV from radio. The 239 syndicated reruns of the series later became a staple of local programming.

Burns was the consummate straight-man to the dizzy observations and antics of Gracie, his fictive (and real-life) wife. The episodes opened with a monologue by Burns, a champion of the one-liner, which led into the stories. Regularly featured were Bea Benaderet, Fred Clark and Harry Von Zell. The series ended Sept. 22, 1958, with the retirement of Gracie from show business.

Burns then began a situation comedy of his own, *The George Burns Show,* on NBC, which ran a single season, Oct., 1958-April, 1959. Gracie Allen died in 1964. Admired by most professional comedians for his superb timing and dry style, Burns has remained active with TV guest shots.

GEORGE, PHYLLIS—CBS sportscaster (1972–78), the first female sportscaster on the network's regular staff. She joined CBS after having been Miss America of 1971. In addition to covering football and other sports, she served as co-host of several Miss America telecasts. She became a prime-time performer as host of the *People* series (based on *People* magazine) in September of 1978, but only briefly. The series was canceled after two months. She married John Y.

Brown, chairman of the Kentucky Fried Chicken chain, who later became governor of Kentucky.

GEORGE POLK MEMORIAL AWARD—bronze plaques given for special achievement in reporting, writing, editing, photography and production in either the print or electronic media. They are given in memory of the 34-year-old CBS correspondent who was murdered in 1948 in northern Greece covering the civil war. The circumstances surrounding his death are a matter of controversy with a charge by two freelance journalists that the Greek government and the C.I.A. were involved in a cover-up of his murder.

"GERALD FORD'S AMERICA"—four-part alternate media view of Washington politics presented on public TV (1974–75), notable for having been produced on half-inch video tape. It was by TVTV and the Television Workshop of WNET New York, on grants from the Ford Foundation and Rockefeller Foundation.

GERBER, DAVID—one of the leading independent producers of TV series in the 70s, having previously been a studio executive, first with 20th Century Fox-TV and then with Columbia Pictures TV. He was executive producer of *Nanny and the Professor, The Ghost and Mrs. Muir, Cade's County, Police Story, Police Woman, Needles and Pins, Born Free, Joe Forrester, The Quest, Gibbsville,* and *Today's FBI.*

Before joining 20th Century-Fox as v.p. of TV sales in 1965, he had been a packaging agent with General Artists Corp. and Famous Artists Corp. He became an independent producer in 1972, working mostly in association with Columbia Pictures TV, where he held the title of executive v.p. for worldwide production.

GERBNER'S VIOLENCE PROFILE—annual study of the extent and nature of violence in network programs conducted since 1967 by Dr. George Gerbner, dean of the Annenberg School of Communications, U. of Pennsylvania, and Dr. Larry Gross. The continuing research, funded by the National Institute of Mental Health, is based on observations of a team of trained analysts coding videotaped samples of the "violence content" of each season's programming. A second part of the study, examining the effects of TV violence on viewers, began in 1973 and is based on surveys of child and adult viewers.

A tabulation of violent acts in the programming results in a Violence Index, which reveals the extent to which each network deals in violence and has comparative value with indices of previous years. Among the study's various findings over the years is that heavy television viewers are more apprehensive of becoming victims of violence themselves and more distrustful of other people than are light viewers.

Using the Gerbner Index, The National Citizens Committee for Broadcasting, in its war on TV violence, identified the advertisers whose commercials most often appeared in violence-oriented programs. Partly in response to the NCCB campaign, many TV advertisers eschewed participation in violent programs.

"GET CHRISTIE LOVE"—hour-long series about a beautiful and able female police detective. It premiered on ABC in September 1974 and was canceled in its first season. The series starred Teresa Graves and was produced by Wolper Productions.

Don Adams & Barbara Feldon as Agents 86 & 99

"GET SMART"—successful NBC comedy series (1965–69) spoofing the movies and TV shows concerned with international espionage. Comedian Don Adams played bumbling Secret Agent Maxwell Smart, and Barbara Feldon and Ed Platt were featured. When NBC canceled, CBS picked up the series for an additional season (1969–70). It was by Talent Associates and Heyday Productions.

GHETTO—a period of the week where programs of a single general type tend to be segregated. Mornings and afternoons on Sundays, which historically have had low viewing levels, have come to be the religious, public affairs and cultural ghettos. The period before 7 A.M. on weekdays, where *Sunrise Semester* and *Continental Classroom* became moored, developed into an education ghetto in commercial TV.

Saturday mornings formed a children's ghetto in the mid-60s when advertisers found it possible to isolate that target group—and economically desirable, since the advertising rates were low for time periods where adult viewing was practically negligible.

"GHOST AND MRS. MUIR, THE"—fantasy-comedy about a beautiful widow in love with the handsome ghost who haunts her house, based on the novel and motion picture of that title. It featured Hope Lange and Edward Mulhare, with Reta Shaw, Charles Nelson Reilly, Harlan Carraher and Kellie Flanagan. Produced by 20th Century-Fox TV, it played one season on NBC (1968–1969) and was picked up the next by ABC (1969–70). Although well done, it failed both times.

"GHOST STORY"—hour adventure series dealing with the supernatural which aired on NBC (1972) and in midseason changed its title to *Circle of Fear.* It starred Sebastian Cabot and was by Screen Gems.

"GIBBSVILLE"—hour-long serial drama based on a number of semiautobiographical John O'Hara stories chronicling the life and the people of a fictional Pennsylvania mining town in the late 40s. Although scheduled to premiere on NBC in the fall of 1976, it was withdrawn for alterations (along with the situation comedy, *Snip*) before the season began. It thus became, technically, a series canceled before it aired. *Gibbsville* finally entered the NBC lineup as a midseason replacement that year but only ran a few weeks. The pilot had aired the previous season as a 90-minute movie.

Featured were Gig Young, John Savage, Biff McGuire and Peggy McCay. It was by David Gerber Productions in association with Columbia Pictures TV.

"GIDGET"—situation comedy based on the spirited teenager established in a series of motion pictures; it ran for one season on ABC (1965–66). Produced by Screen Gems, it featured Sally Field and Don Porter.

GIFFORD, FRANK—ABC sportscaster most prominent as play-by-play announcer on the *NFL Monday Night Football* telecasts and as a contributor to *Wide World of Sports.* A star football player with the New York Giants for 12 years, he began his broadcasting career part-time with CBS stations in New York well before his retirement from the sport in 1965. After several years as a full-time CBS sportscaster, he was hired away by ABC.

GILL, MICHAEL—producer-director for BBC, best known in the U.S. as producer of *Civilisation* and producer-

director of Alistair Cooke's *America* series. He left the BBC in 1978 to produce independently in the States and set up partnership with a fellow alumnus, Adrian Malone.

Frank Gifford

GILLASPY, RICHARD—director first associated with NBC (*Tonight, Home, America After Dark, Mr. Wizard*) and later v.p. of Ivan Tors Studios. Other credits include *The Chevy Show, Arthur Murray Dance Party, Ernie Kovacs Show* and *Open Mind.*

Bob Denver

"GILLIGAN'S ISLAND"—CBS situation comedy (1964–67) about a group of diverse buffoon types shipwrecked on an island and forced to develop their own society. Created and produced by Sherwood Schwartz for Gladasya Productions, in association with United Artists TV, the series featured Bob Denver, Alan Hale Jr., Jim Backus,

Natalie Schafer and Tina Louise. Strongly appealing to children for its low comedy, the series enjoyed a sustained sale in rerun syndication.

"GIRL FROM U.N.C.L.E., THE"—1966 spin-off of the popular *Man From U.N.C.L.E.* on NBC; it featured Stefanie Powers, Noel Harrison and Leo G. Carroll as secret agents. The hour series from Norman Felton's Arena Productions and MGM-TV lasted one season.

"GIRL WITH SOMETHING EXTRA, THE"—NBC situation comedy (1973–74) about a young wife whose marriage is complicated by her ability to read other people's minds. Produced by Columbia, it featured Sally Field, John Davidson, Zohra Lampert and Jack Sheldon.

GITLIN, IRVING (d. 1967)—news executive for CBS in the 50s and NBC in the 60s who developed and headed outstanding documentary production units for both networks. Gitlin's organization was overshadowed at CBS by Edward R. Murrow's *See It Now* unit, and in 1960 NBC president Robert Kintner hired Gitlin away to establish a creative projects unit for NBC News. Moving to NBC with Gitlin was Albert Wasserman, who had produced the stunning *Out of Darkness* on a psychiatrist's progress with a catatonic patient; his first effort for NBC was *The U-2 Affair* (Nov. 29, 1960). Gitlin swiftly put together a team of documentarians and launched the *NBC White Paper* series of specials. He was responsible for such efforts as *Sit-In* and *Angola: Journey to War.*

Gitlin had been a producer and writer in radio and entered TV as a producer of public affairs programs, among them *The Search, Conquest* and *The 20th Century Woman.*

"GLASS MENAGERIE, THE"—TV production of the Tennessee Williams play on ABC (Dec. 16, 1973) which starred Katharine Hepburn, one of her rare appearances in the medium. It was produced by David Susskind and Talent Associates, was directed by Anthony Harvey and featured Joanna Miles.

GLEASON, JACKIE—one of TV's great comedy stars of the 50s and 60s whose series, *The Honeymooners,* with Art Carney featured, ranks as one of the medium's classics. Portly, brash and enamoured of the ambiance of nightclubs and the glitter of chorus lines, Gleason billed himself as "The Great One" and made popular phrases of his patented expressions, "How sweet it is" and "Away we go."

He came into television in 1949, from nightclubs, radio and Broadway, as the original Chester Riley in the NBC situation comedy *The Life of Riley* (which later became the vehicle for William Bendix). After a season, he gave it up for *Cavalcade of Stars* on the DuMont Network. Then came *The Jackie Gleason Show,* a weekly comedy-variety hour in which he developed his repertoire of comic characters, Reggie Van Gleason, The Poor Soul, Joe the Bartender, Charlie the Loudmouth, and Ralph Kramden, the boastful but ineffectual bus driver of a recurring skit, "The Honeymooners." Played against the slow-witted sewer worker portrayed by Carney, Kramden was by far Gleason's most successful creation, and when the variety series closed after three seasons in 1955, the logical next step was for *The Honeymooners* to be transformed from skit to series. The filmed series began Oct. 1, 1955, with Audrey Meadows playing Gleason's wife (as she did in the skits) and ran until Sept. 22, 1956. Gleason went back to the hour live program that fall, but it closed after one season. In 1958 he essayed a new live half-hour series with Buddy Hackett, which ran but a few months.

Jackie Gleason

In the late 60s, CBS attempted a revival of *The Honeymooners* as a one-hour program, with Sheila MacRae and Jane Kean as the new "wives." It was canceled in 1970, but Gleason received a large annual sum from CBS for several years afterwards, under his contract, which prevented him from signing with another network. In the mid-70s, he was proposed for other series, but none came to fruition.

Although Gleason's production tastes ran to the grand and garish, it was noteworthy that *The Honeymooners* achieved its popularity on what was possibly the most drab and depressing of the standing sets in the history of television—a kitchen-living room with an ancient refrigerator, a table in the center and a window looking out on the bricks of the next building.

Gleason played his stardom to the limit. One of his contractual demands was that CBS build for him a one-bedroom circular building in Peekskill, N.Y., at a reported cost of $350,000. Late in the 60s he determined that he would live in Miami and do his shows from there, although there were no network production facilities in Miami. CBS, which was given to coddling its stars, agreed to build television studios adjacent Gleason's favorite golf course and to conduct an annual press junket to the site, by train, so that a party could be held going and coming.

"GLEN CAMPBELL GOODTIME HOUR, THE"—variety show on CBS (1969–72) starring country and popular music singer Glen Campbell, with weekly guests. It was produced on tape by Glenco Productions.

GLICKSMAN, FRANK—series producer whose credits include *12 O'Clock High* and *Medical Center.*

Tom Conti as Adam Morris

"GLITTERING PRIZES, THE"—much lauded BBC novel for television, written by Frederick Raphael, which aired on U.S. public TV via the Eastern Educational Network in January 1978 and was repeated the following year. The six 80-minute episodes trace what happens to a group of promising graduates of Cambridge who go after the glittering prizes of the professional world in the 1950s. With Tom Conti as star, it featured John Gregg, Dinsdale Landen, Barbara Kellerman, Natasha Morgan, David Robb and others. Mark Shivas was producer and Warren Hussein and Robert Knights directors.

GLOBO NETWORK—Brazil's dominant radio-TV broadcaster, which bills itself as the world's fourth largest television network, based on viewing that reaches levels of

32 million. The network excels in the production of novellas, or soap operas running 20 to 80 episodes, and hopes to syndicate them in the U.S. with a new process for dubbing the programs into English.

GLUCKSMAN, ERNEST D.—producer, director and writer associated with comedy-variety programs, principally those starring Jerry Lewis. His credits also include *The Chevy Show, The Saturday Night Revue* and specials with Donald O'Connor, Betty Hutton and Ethel Merman.

"GO"—NBC Saturday nonfiction series for children which made early practical use of portable video cameras (minicams) in programs designed to "go anywhere" for interesting subject matter, whether into the cockpit of a plane or to a recording studio. The brainchild of George Heinemann, NBC v.p. of children's programs, *Go* premiered Sept. 8, 1973 in a time when the networks were becoming responsive to public complaints about the low state of children's television. In 1976, for the Bicentennial, the series took a new title of *Go-USA*. Heinemann served as executive producer.

GOBEL, GEORGE—a comedy sensation in the mid-50s who became part of NBC's star roster, first in a half-hour series that premiered in 1954, which featured Jeff Donnell as his stage wife and Peggy King as singer, then in *The George Gobel Show* (1957–59), an hour variety program that alternated with *The Eddie Fisher Show.*

Gobel had been an obscure Chicago comic playing in small nightclubs until he did a guestshot on a major TV spectacular. His comedy persona—that of a bewildered innocent, a born loser, gamely coping with the world—and his boyish low-key style won the immediate enthusiasm of the television audience. His trademark exclamation, "Well, I'll be a dirty bird," quickly entered the popular language.

Gobel's overnight success typified how TV made sudden stars of journeyman entertainers; it was part of television's second wave, occurring after the move-over of established radio stars. Gobel's popularity faded in the 60s, and his TV work mainly consisted of guest shots and occasional dramatic roles until he became a regular on *Hollywood Squares* in the late 70s. This led to a starring role in the sitcom *Harper Valley.*

"GODFATHER, THE" (full title: "Mario Puzo's 'The Godfather': The Complete Novel for Television")—nine-hour serialized version on NBC of Francis Ford Coppola's two smash-hit movies, *The Godfather* and *The Godfather, Part II,* reconstructed and interwoven as a single drama about a gangster clan. The TV version, incorporating approximately one hour of material that had been trimmed out of the theatrical presentations, was assembled by Coppola and film editor Barry Malkin. NBC presented it on four consecutive nights in November 1977, but the ratings were somewhat disappointing considering the cost of the project. This may have been because the original movies had already been aired a few years before.

Arthur Godfrey

GODFREY, ARTHUR—one of TV's most successful personalities, who during most of the 50s conducted two weekly prime-time series for CBS as well as a daily radio show. A master commercial pitchman, with a deep-voiced style, who managed to blend folksiness with sophistication, he was reported by *Variety* to have been responsible for $150 million in advertising billings for CBS in 1959.

Godfrey's first TV venture, *Arthur Godfrey's Talent Scouts*, began Dec. 8, 1948 and ran nearly 10 years. The show presented young professional talent in the traditional amateur show manner, but each was introduced by a celebrity who professed to be the "discoverer." An applause meter determined who won. A month after the premiere, he began a second weekly series, the hour variety show, *Arthur Godfrey and His Friends* (changed in 1956 to *The Arthur Godfrey Show*). This involved a resident cast which at various times included Janet Davis, Julius LaRosa, Marion Marlowe, LuAnn Sims, The Chordettes, Haleloke, Frank Parker, The Mariners, Carmel Quinn, Pat Boone and the McGuire Sisters. Tony Marvin was his announcer. His trademarks were a ukulele and the chucklesome greeting, "Howa'ya, Howa'ya, Howa'ya."

Godfrey's firing of LaRosa on the air, with the charge that he lacked humility, was national news, as was his "buzzing" of a New Jersey airport in his private plane. Later, his successful battle with cancer was widely publicized.

In 1959 Godfrey was forced to give up his TV shows because of ill health, although he continued on radio. By the early 60s the increasing sophistication of TV entertainment had replaced the loose informality of television in the 50s, and performers such as Godfrey were no longer in demand.

When he was at his height with two prime-time shows, his weekly audience was estimated at 82 million viewers, and in 1954 his combined broadcasts were reported to have accounted for 12% of CBS's total revenues.

GODFREY, KEITH (d. 1976)—leading syndication figure for 20 years, all of them with MCA-TV. He joined in 1955 as a salesman in Houston and retired as executive v.p. in 1975 while ailing with cancer.

"GOING MY WAY"—sentimental dramatic series based on the movie of that title, produced by Revue and Kerry Productions for ABC in 1962. Gene Kelly and Leo G. Carroll starred, and the series ran for 39 episodes.

"GOLDBERGS, THE"—early situation comedy about a Jewish family in the Bronx. It had run 17 years on radio and had inspired a Broadway play when CBS brought it to TV on Jan. 17, 1949. The show was cut down at the height of its popularity during the McCarthy era because a member of the cast, Philip Loeb, who portrayed the father, Jake, was listed in *Red Channels*. Gertrude Berg, who was the star as well as the creator and writer of the half-hour series, refused to fire Loeb, and when advertising support evaporated CBS canceled. NBC then picked up the series, but not for long. The original TV run ended on June 25, 1951.

Loeb, harassed by the Red-baiters and blacklisted in show business, committed suicide in 1955.

The Goldbergs was revived in a 1956 syndicated version, by Guild Films, with Robert H. Harris in Loeb's role and with the original cast virtually intact, Miss Berg as Molly, Arlene McQuade as Rosalie and Eli Mintz as Uncle David. In the brief revival, Tom Taylor played Sammy, the role that had been originated by Larry Robinson. See also *Red Channels*.

GOLDEN AGE (of Television Drama)—appellation commonly used for TV during the 50s when live studio drama, in the theater tradition, was part of the medium's main nightly fare. Most of the plays were produced in New York under somewhat primitive conditions, but the opportunity to be produced and "discovered" overnight drew scores of young playwrights to television. In yielding a new body of literature, the drama anthologies gave the new medium respectability and prestige, heightened when stage adaptations (and then movies) were made of such TV plays as Reginald Rose's *Twelve Angry Men* (1954), Gore Vidal's *Visit to a Small Planet* (1955), William Gibson's *The Miracle Worker* (1957) and Mac Hyman's *No Time for Sergeants* (1955). Then films were made of Paddy Chayefvsky's *Marty* (1953) and *The Bachelor Party* (1955), Rod Serling's *Patterns* (1955) and J P Miller's *The Days of Wine and Roses* (1958), among others.

The era opened with the premiere of the *Kraft Television Theatre* on May 7, 1947, and closed with the final presentation of *Playhouse 90* as a weekly series ten years later. It reached its height between 1953 and 1955, when as many as a dozen original plays were offered by the networks almost every week. As drama flourished, Kraft expanded to two *Theatres*, one on NBC Wednesday nights, the other on ABC Thursdays. *Playhouse 90*, which began on CBS Oct. 4, 1956, was perhaps the most ambitious venture of all, calling for a major 90-minute production every week.

There were also the *Philco Playhouse* (alternating with the *Goodyear Playhouse*), *Studio One, U.S. Steel Hour, Robert Montgomery Presents, Omnibus, General Electric Theatre, Motorola TV Hour, Lux Video Theatre, Ford Theatre, Ford Startime, Elgin Hour, Alcoa Theatre, Kaiser Aluminum Hour, Medallion Theatre, Pulitzer Prize Playhouse, Schlitz Playhouse of Stars, Sunday Showcase, Armstrong Circle Theatre, Four Star Playhouse, Four Star Jubilee, Climax!, Producers' Showcase, Matinee Theatre, Revlon Theatre, Breck Golden Showcase, Front Row Center, Playwrights 56, Camera Three, Actors Studio, Hallmark Hall of Fame, DuPont Show of the Month, Desilu Playhouse, Special Tonight* and others.

Out of these showcases came such writers as those mentioned and Robert Alan Aurthur, Robert Anderson, A. E. Hotchner, Tad Mosel, Horton Foote, Calder Willingham, N. Richard Nash, David Shaw, Sumner Locke Elliott, Paul Monash and S. Lee Pogostin. Chayefvsky had been writing sketches for nightclub comics, Aurthur was part owner of a record company, Mosel an airlines clerk, Serling employed by a Cincinnati radio station, Nash a teacher, Foote an actor, Miller a salesman and Rose an advertising copywriter. Although writers were usually paid less than $2,500 a script, the wide-open TV market opened new career vistas for each of them.

Directors who came into prominence included Delbert Mann, Arthur Penn, Sidney Lumet, John Frankenheimer, Fielder Cook, George Roy Hill, Franklin Schaffner, Alex Segal, Dan Petrie, Fletcher Markle, and Ralph Nelson.

And the era spawned such producers as Fred Coe, Martin Manulis, Worthington Miner, George Schaefer, Paul Gregory, Robert Saudek, Robert Herridge, Herbert Brodkin, Albert McCleery, John Houseman, Norman Felton, David Susskind and Gordon Duff.

Along with the established names at the time, the acting talent included George C. Scott, James Dean, Kim Stanley, Julie Harris, Eva Marie Saint, Paul Newman, Sidney Poitier, Grace Kelly, Lee Remick, E. G. Marshall, Jack Palance, Jack Lemmon, John Cassavetes, Eli Wallach, Rod Steiger, Charlton Heston, Sal Mineo, Dina Merrill, Lee Marvin, Keenan Wynn, Piper Laurie, Rip Torn, Lee Grant, Jack Warden and Lee J. Cobb.

Although the shows were produced live, some of the more successful ones, such as *Patterns* and *A Night to Remember* (1956), which used 107 actors and 31 sets, were repeated weeks after the original telecasts.

But by the late 50s, virtually all the drama series were gone, having given way to filmed series and quiz shows, and most of the artists who had emerged in the Golden Age fled to movies and the theatre. A few, however, adapted to the new requirements of television and remained; those included Serling, Susskind, Monash, Felton and many of the actors.

Hallmark Cards continued *Hall of Fame* through the 70s as seasonal specials, and CBS revived the *Playhouse 90* concept in the late 60s for two or three plays a year. Studio drama, on videotape, has had sporadic revivals, but never on the scale of the Golden Age, and not with the conviction that it was a natural form for television.

No one can say for certain what caused the wave of drama to pass, but there are several theories. According to one, drama was practical in the years when the wealthier and better-educated families owned most of the television sets, but impractical when sets proliferated to virtually every home in the country, defining a new mass audience. Another holds that advertisers, dealing as they do in their commercials with instant solutions to problems, found it inconsistent with their purposes to sponsor serious plays on human conflicts, which revealed that in real life there are no easy solutions.

Studio drama is costly to produce and inevitably varies in quality from program to program. Networks and advertisers can never be sure how large an audience an original play will attract; episodic series are more predictable. Although many of the more recent drama specials have been sponsored, network officials have indicated that drama is less profitable than other forms of programming because it permits fewer commercial breaks and is usually limited to advertising that is artistically in keeping.

GOLDEN, BILL (d. 1959)—head of advertising for CBS in the 50s noted for his creativity, impeccable taste and for designing the "CBS eye," the network's distinctive logo. He also played a large role in projecting a classy image for CBS, which wanted to be thought of as the "Tiffany of the networks."

GOLDEN NYMPHS—see Monte Carlo International TV Festival.

GOLDEN ROSE OF MONTREUX—international TV awards competition for light entertainment programming, held annually in Switzerland since 1961.

Leonard H. Goldenson

GOLDENSON, LEONARD H.—top officer of the American Broadcasting Companies Inc. since its formation in 1953 from the merger of ABC and United Paramount Theatres; he held the title of president until January 1972 when he was elected chairman. Goldenson, who engineered the merger, had been president of UPT since its divorcement in 1950 from Paramount Pictures by government order. He had joined Paramount soon after receiving his degree from Harvard Law School; by 1938, at age 32, he was executive in charge of the company's 1,700 movie theaters.

Under Goldenson, ABC not only expanded rapidly in TV and radio but also diversified its activities by adding a phonograph record and distribution company, a network of scenic and amusement parks and publishing subsidiaries. Goldenson steered ABC through a number of financially shaky periods, fending off takeover bids by Howard Hughes and Norton Simon, and finally saw the company attain a secure footing when, in the early 70s, the ABC network achieved parity with CBS and NBC.

GOLDMARK, DR. PETER C. (d. 1977)—a pioneer in the development of new communications technology and, until his retirement in 1971, CBS's resident inventor as head of engineering research and development and later as president of CBS Laboratories. On leaving CBS he formed his own company, Goldmark Communications Corp., concentrating on the social uses of broadband communications.

Joining CBS in 1936, not long after his arrival in the U.S. from his native Hungary, he became involved with more than 160 inventions and is personally credited with

developing the 33⅓ r.p.m. long-playing record, which revolutionized the recording industry. Goldmark also helped create the first successful color TV system for CBS—the so-called "color wheel"—but because that system was not compatible with black-and-white sets, the FCC authorized instead the system developed by RCA for domestic use.

Dr. Peter Goldmark

Goldmark also developed EVR (Electronic Video Recording), a form of videocassette using 8.75 mm film instead of tape. CBS created a new division around the invention, in hopes of dominating what seemed to be an emerging home video recording industry, but closed it down when it was clear that the tape systems would prevail.

Frank Sutton & Jim Nabors

"GOMER PYLE, USMC"—smash hit situation comedy on CBS (1964–68) about a sweet, naive yokel in the U.S. Marine Corps. The countrified antics and dialect of Jim Nabors, in the title role, were the chief interest, and he was nicely foiled by the late Frank Sutton, as his tough sergeant. It was created by Aaron Ruben and produced by Ashland Productions and T&L Productions.

"GONE WITH THE WIND"—David O. Selznick's 1939 movie classic which, when finally released to TV in 1976 for a single showing, scored the highest rating in history for any television entertainment program. It had only a brief stay at the top, however, being surpassed by *Roots* in February 1977.

NBC paid MGM a reported $5 million for the rights to a single national airing. It was played in two parts on successive nights, Nov. 7 and Nov. 8, 1976. The telecasts scored ratings of 47.7 and 47.4, respectively, reaching an estimated average households of 33,960,000 the first night and 33,750,000 the second. *GWTW* retains the distinction as the highest-rated movie on TV of all time. CBS later bought the TV rights for 20 years.

"GONG SHOW, THE"—bizarre talent show out of the Chuck Barris stable in which oddball and freak acts perform before a celebrity panel that dismisses the most outrageous of them in mid-performance by striking a gong. For its sheer crudeness, it was rewarded with a measure of popularity, both as a daytime show on NBC (1976–78) and as a prime time–access evening entry (1976–80). The network version began in the summer, the added syndicated edition in the fall of the same year. Gary Owens was host the first year, but Barris himself took over the role thereafter, bringing to it a style of manic philistinism that apparently struck just the right note of tastelessness. Among the recurring panelists have been Rex Reed, Jaye P. Morgan, Arte Johnson and Jamie Farr. John Dorsey directed. Firestone Program Syndication, headed by Len Firestone, flourished with the show, selling the off-network library of reruns for stripping, as well as the freshly produced prime-access segments. In 1979, the prime-access version expanded from one a week to two a week.

"GOOD GUYS, THE"—CBS situation comedy (1968–70) about a hash-slinger and a cabbie that was an attempt by the network to establish a blue-collar comedy like Jackie Gleason's *The Honeymooners*. The effort failed. By Talent Associates, it featured Bob Denver, Herb Edelman and Joyce Van Patten.

GOODMAN, JULIAN—president of NBC from April 1, 1966 until April 1, 1974 when he became chairman. He was an unusual choice for the corporate post—in a time when broadcast leaders were typically drawn from the sales ranks—because his entire training had been with NBC News, where he worked 20 years, latterly as executive v.p. for administration. But while Goodman's background suggested that NBC would strive for even greater achievement in news, in fact news received a lower priority under Goodman than it had had with his predecessor Robert E. Kintner.

When Fred Silverman and Jane Cahill Pfeiffer were brought into NBC in 1978 to form a new administration, Goodman became chairman of the executive committee. He chose early retirement in 1979.

Julian Goodman

Goodman, a Kentuckian, joined NBC in Washington as a news writer for David Brinkley in 1945. He later became director of news and public affairs, v.p. of NBC News and then executive v.p. He produced such programs as *Comment*, *Ask Washington* and *Report From Alabama* and supervised the *JFK* series that reported periodically on the Kennedy Administration.

"GOOD MORNING, AMERICA"—two-hour early morning show on ABC (7–9 A.M.) that in November 1975 superseded *A.M. America* with a new cast and a largely new creative team. The changes paid off, and the program promptly began to make inroads into the audience for the long-established *Today* on NBC. Within a year NBC had begun to revamp *Today* to keep from losing more ground to the ABC entry.

A key change was the installation of actor David Hartman (who had starred in *Lucas Tanner* and other prime-time series) in the role of host, in place of Bill Beutel and other newscast personalities. Woody Frasey was brought in as executive producer. In general format the program resembled *Today* but with a stronger accent on features. A roster of outside personalities—Jack Anderson, Rona Barrett, Jonathan Winters, Erma Bombeck and John Lindsay—contributed five-minute pieces several times a week. Regulars in the cast were Nancy Dussault, Steve Bell, Margaret

Osmer and Geraldo Rivera. Sandy Hill succeeded Miss Dussault in the spring of 1977.

David Hartman & Nancy Dussault: first season of Good Morning, America

GOODSON-TODMAN PRODUCTIONS—company formed in 1946 by Mark Goodson, a former radio announcer, and Bill Todman, former radio writer. It hit the jackpot in the game-show field with a succession of big hits: *What's My Line?, I've Got a Secret, Stop the Music, Beat the Clock, To Tell the Truth, The Price Is Right, Password, The Name's the Same, Two for the Money, The Match Game* and others. Not only did the shows enjoy long and prosperous runs in the 50s and 60s, many of them in prime time, but virtually all were revived in the 70s either for prime-access syndication or by the networks for their daytime schedules. Goodson-Todman ventured also into the field of episodic filmed series with *The Rebel, Jefferson Drum, The Richard Boone Show, Philip Marlowe* and *The Don Rickles Show.*

"GOOD TIMES"—half-hour comedy series concerning an urban black family, which began on CBS in February, 1974 as a spin-off of *Maude* by Yorkin & Lear's Tandem Productions. Its debt to *Maude,* however, was chiefly in the appropriation of Esther Rolle, who had portrayed Florida, the maid, in that series. Miss Rolle headed a cast that included John Amos, Ja'net DuBois, Ralph Carter, Jimmie Walker and BernNadette Stanis. The series made Jimmie Walker instantly popular as a young black comedian and led to lucrative personal appearances and TV commercials. Amos left in 1976 to take a role in *Roots.* The series was canceled during the 1978–79 season.

GORGEOUS GEORGE—symbol of the theatricality of TV wrestling in the early 50s, as an effeminately coiffed and ostentatiously dressed grappler.

GOULD, JACK—radio and TV critic of *The New York Times* (1944–72), one of the first of the daily journalists to recognize that TV required full-time coverage, just as any other major news beat. As reporter, commentator and critic, who at one time headed a staff of eight, he became known as "the conscience of the industry" because his critical observations frequently influenced those who wielded power in broadcasting. At the top of his field in 1957, he received a special George Foster Peabody Award—normally restricted to TV practitioners—for his "fairness, objectivity and authority." Gould joined *The Times* in 1937 after five years with the *New York Herald-Tribune.* He worked initially in the drama department but shifted to radio in 1942 and became critic in 1944. Early in the 60s he left the newspaper to join CBS in a corporate post but found it an unhappy experience and shortly returned to *The Times.* He retired in 1972 to live in California.

"GOVERNOR AND J.J., THE"—political situation comedy about the beautiful daughter of an elected official who sometimes jeopardized, but always saved, his career. It featured Dan Dailey, Julie Sommars, Neva Patterson, James Callahan and Nora Harlowe. Carried by CBS (1969–71) and then rebroadcast in the summer of 1972, it was by Talent Associates-Norton Simon Inc.

GOWDY, CURT—veteran sportscaster who, in a career that has spanned four decades, probably broadcast more major events than any other TV announcer. He ended a 15-year stint with NBC in April 1979 to sign on with CBS Sports on a three-year contract to cover, among other assignments, the NFL football games. During his years with NBC, he freelanced as host of ABC's *The American Sportsman* and conducted a daily five-minute show for the Mutual radio network.

Over the years his credits include the play-by-play for World Series and baseball All-Star games, college and pro football, college and pro basketball and the Olympic Games. After broadcasting sports in Cheyenne, Wyo., and Oklahoma City during the 40s, he became co-announcer in 1949 with Mel Allen for the New York Yankees. Three years later he became announcer for the Boston Red Sox and held the position for 15 years. Since then, he has covered a wide range of sports for all three networks and was host of the PBS (1975–77) series *The Way It Was.*

GRADE A SIGNAL—denoting the areas in the coverage pattern of a station where the transmissions should be received best and most reliably. The Grade B Signal denotes the area of secondary service, farther from the transmitter, where reliable service can usually be received with an outdoor antenna.

Grade A conforms to specific electronic criteria established by the FCC: the Grade A perimeter equates to the point at which the field density is 68 dBu for channels 2–6, 71 dBu for channels 7–13 and 74 dBu for channels 14–83.

A station's Grade A contour, usually defined by a circle, is the geographic sphere of coverage where satisfactory service can be expected at least 90% of the time. In the Grade B contour, the quality of the picture is expected to be satisfactory at least 50% of the time.

Lord Lew Grade

GRADE, LEW (LORD)—British television entrepreneur who became a leading exporter of programs in the 60s, making numerous sales to the U.S. networks, principally ABC-TV. Moreover, he promoted them here with a flamboyancy that was not characteristically British.

As chairman and chief executive of Associated Television Corp., he headed the independent commercial station, ATV, based in Birmingham, Eng., as well as a major film production and international distribution company, ITC. (In the western hemisphere, ITC stands for Independent Television Corp., in the eastern for Incorporated Television Co.). Other subsidiaries of the parent company were a theater chain, the Pye record company and a music publishing firm, Northern Songs.

Russian-born (1906), Grade started in show business as a Charleston dancer in England, later operated a talent agency with his brother Leslie and entered the TV field in 1956 with the advent of commercial television in the U. K. His background as an agent gave him a familiarity with American

show business and served him in tailoring programs for both sides of the Atlantic. He also engineered major coproductions with foreign broadcast systems, such as Italy's RAI-TV, for programs that could be sold internationally. Among the notable ones were *Moses, The Lawgiver,* which starred Burt Lancaster, and *Jesus of Nazareth,* directed by Franco Zeffirelli.

Under Grade, ITC released in the U.S. such series as *Interpol Calling, Whiplash, The Saint, Gideon's Way, The Tom Jones Show, Shirley's World* (with Shirley MacLaine), *Secret Agent, The Persuaders, The Julie Andrews Show, The Muppet Show,* and others. Grade's organization also financed and produced a number of theatrical movies.

GRAHAM, FRED—lawyer-journalist who has been CBS News' law correspondent since 1972. Based in Washington, he covers the activities of the Supreme Court, Justice Department, FBI, and the legal profession. He serves regularly as substitute moderator on *Face the Nation.* Graham came to CBS News from *The New York Times,* where he had been Supreme Court correspondent (1965–72). Prior to that he held government positions.

GRAHAM, KATHARINE M.—board chairman and chief executive officer of the Washington Post Co. since the death of her husband, Philip L. Graham, in 1963. In addition to the *Post, Newsweek* magazine and other print properties, she heads Post-Newsweek Stations, a group that includes WTOP-AM-TV Washington, D. C.; WPLG-TV Miami; WJXT-TV Jacksonville, Fla.; WFSB-TV Hartford, Conn.; and WCKY-AM Cincinnati. Her control over the media empire earned her the reputation as one of the most powerful women in America.

GRALNICK, JEFF—v.p., executive producer and director of special events at ABC News since January 1979. In five prior years at ABC News he produced *ABC Evening News with Harry Reasoner and Barbara Walters, Weekend News,* and *ABC News Special Events.* He spent 12 years in the special events unit and had been a producer and on-air reporter in Vietnam.

GRAMPIAN TELEVISION—regional commercial station in the U.K. licensed to serve northeast Scotland, with headquarters in Aberdeen.

GRANADA TELEVISION—British commercial independent licensed to the Lancashire region, with studios in Manchester, the first of which was built in 1956. Among its notable programs are the long-running and high-rated (in the U.K.) *Coronation Street,* and the daily public affairs series, *Granada Reports.* Granada won praise in the United States and internationally for *Brideshead Revisited, Hard Times, World in Action, The Collection, Lawrence Olivier Presents* and *Clouds of Glory.* See also *Coronation Street.*

GRANATH, HERBERT A.—president of ABC Video Enterprises, a division created by ABC Inc. in 1979 to move the company into programming for cable, video disks and video recorders. Previously, for three years, he had been v.p. and assistant to Elton H. Rule, president of ABC Inc., after having left ABC for a year to be senior v.p. of Transworld International, a sports packaging company. He was a sales executive for ABC Sports (1970–75) after 10 years in sales with ABC Radio. Earlier, he was with NBC in sales.

GRANIK, THEODORE (d. 1970)—lawyer who also produced and moderated numerous TV shows, among them *Youth Wants to Know* and *All America Wants to Know* in the 50s. He also owned WGSP-TV in Washington.

GRANT, B. DONALD (BUD)—president of CBS Entertainment, the network's programming arm, since November 1980. In 1982, he signed a five-year contract to continue in that post, with oversight responsibilities also for the new Theatrical Films Division headed by Bill Self.

Grant was recruited by CBS in 1972 from NBC-TV, where he had given a good account of himself as head of daytime programs for five years. He was credited with introducing the serial, *Days of Our Lives,* that became the pivotal show in the NBC daytime schedule. Grant began with NBC in its executive training program in 1958.

At CBS, he served four years as v.p. in charge of daytime programs, jockeying the schedule into a strong first place and developing a successful youth-oriented soap opera, *The Young and the Restless.* This earned him the promotion to CBS program chief at the start of the Robert Wussler administration in April 1976. However, when the CBS Entertainment division was created in the October 1977 management realignment, Grant yielded the top spot in programming to Robert A. Daly and stayed on as v.p. When Daly resigned in 1980 to become head of Warner Bros., Grant was named president of the division.

"GREAT AMERICAN DREAM MACHINE, THE"—unusual, wry-spirited and sometimes remarkable magazine series on PBS (Jan. 1971–Jan. 1972). The one-hour episodes successfully combined such disparate disciplines as satire and serious documentary, and there were also dramatic and musical segments. A regular in the series was comedian

Marshall Efron. *Dream Machine* disbanded after a single year for lack of funding. It was produced at WNET New York with A.H. Perlmutter and Jack Willis as executive producers.

John Kennedy & Richard Nixon in 1960

"GREAT DEBATES"—series of four one-hour face-to-face encounters on national TV and radio by the 1960 presidential candidates, John F. Kennedy and Richard M. Nixon, made possible when Congress suspended the equal time law just for the presidential race that year. The debates served to erase Nixon's advantage as the better-known candidate and probably contributed to Kennedy's razor-thin victory that November.

With the three networks carrying the debates simultaneously in prime time, the four programs drew the largest TV audience ever, up to that time—75 million viewers for the first, 61 million for the second, 70 million for the third and 63 million for the final one, according to ARB estimates. The number of different viewers reached by the debates was 101 million. Most analysts attributed the record voter turnout in the national elections of 1960 to the *Great Debates.*

The debates were the fruit of a broadcast industry campaign, led by CBS president Frank Stanton, for the repeal of Sec. 315 of the Communications Act (the equal time rule) because of its stifling effect on political coverage by TV and radio. After the 1960 political conventions, Stanton secured acceptances from Kennedy and Nixon for a televised debate if Congress agreed to remove the obstacle of equal time for the 14 other legally qualified candidates. With the two candidates agreeing to meet on TV, there seemed a chance for the repeal of Sec. 315. But Congress voted only to suspend the rule, on a test basis, and President Eisenhower signed the bill on Aug. 24.

A drawing of lots gave CBS the responsibility for producing the first debate telecast, which was to be limited to the discussion of domestic issues. The event originated at the WBBM-TV studios in Chicago, with Howard K. Smith as moderator and Don Hewitt as producer-director. The panel of news correspondents posing questions to the candidates following their delivery of 8-minute opening statements consisted of Bob Fleming, ABC; Stuart Novins, CBS; Sander Vanocur, NBC; and Charles Warren, Mutual.

The first debate proved to be the most controversial of the four and also the most consequential with respect to its impact on the campaigns. Nixon physically showed the effects of strenuous campaigning, and the loss of weight from a recent illness had made his shirt collar too large. In striking contrast, Kennedy was sun-tanned and full of vigor. When Kennedy refused TV makeup before the broadcast, Nixon did also, although he was pale and needed it. Hewitt's occasional reaction shots during the program found Kennedy relaxed and Nixon perspiring. Overall, Kennedy's TV projection was far superior to that of Nixon, for whom the debate was a catastrophe.

For the subsequent debates, Nixon's entourage included a lighting consultant and a makeup artist, but he could not undo the negative impression he made on the first telecast. The second debate was held at WRC-TV in Washington, Oct. 7, and was produced by NBC with Frank McGee as moderator. The panelists were Edward P. Morgan, ABC; Paul Niven, CBS; Alvin Spivak, UPI; and Harold Levy, *Newsday.*

The third confrontation, produced by ABC on Oct. 13, had Nixon in the Los Angeles studios and Kennedy in the New York studios meeting on the air electronically. Bill Shadel of ABC was moderator and the reporters were Frank McGee of NBC; Charles Von Fremd, CBS; Douglass Cater of the *Reporter* and Roscoe Drummond, *N.Y. Herald-Tribune.*

ABC also originated the fourth debate, in its New York studios, on Oct. 21. Devoted entirely to foreign policy, it followed the general format of the first confrontation. Quincy Howe of ABC moderated. The news panel consisted of Walter Cronkite, CBS; John Chancellor, NBC; John Edwards, ABC; and Frank Singiser, Mutual.

J. Leonard Reinsch, representing Kennedy, and Fred Scribner, representing Nixon, had with their aides met more than a dozen times in negotiating sessions to set the procedures for the first debate. The vice-presidential candidates, Lyndon Johnson and Henry Cabot Lodge, had been asked to debate but declined.

The Kennedy-Nixon debates were the last of their kind until the campaigns of 1976, when a change in the FCC's interpretation of the equal time rule permitted TV coverage of debates by major candidates if they were legitimate news events. They were considered to be legitimate news if the debates were conducted by independent organizations, with no direct involvement by the networks. The 1976 debates

were held in rented halls, under auspices of the League of Women's Voters, and the TV networks were invited to cover them—which, of course, they did. See also Carter-Ford Debates, Equal Time Rule.

"GREAT GILDERSLEEVE, THE"—episodic comedy series brought over from radio in 1955–56, with Willard Waterman in the title role and featuring Ronald Keith and Stephanie Grin. It was produced by Robert S. Finkel and Matthew Rapf.

"GREAT PERFORMANCES"—umbrella title for rotating performing arts series on PBS (1973–) that included *Theater in America, Dance in America, Live from Lincoln Center* and *Music in America,* all produced at WNET New York and underwritten by Exxon on a 39-week basis. Jac Venza is executive producer. See *Theater in America.*

"GREEN ACRES"—hit CBS situation comedy (1965–71) about a chic city couple with no aptitude for country living who choose the bucolic life, a straight reversal of the *Beverly Hillbillies* premise while that series was red hot. The two series were a parlay for their creator, Paul Henning, and for Filmways TV. The cast was headed by Eddie Albert, Eva Gabor and Pat Buttram and featured Alvy Moore and Frank Cudy.

GREENBERG, PAUL W.—executive producer of special broadcasts for NBC News since June 1978, having been recruited from CBS News where he worked 14 years. His particular responsibilities at NBC are in political programs and so-called instant specials—those covering breaking stories of national and international importance. At CBS, where he was a writer and producer, he worked on *The CBS Morning News,* the *Weekend News* and *The CBS Evening News with Walter Cronkite.*

GREENBERG, STANLEY R.—screenwriter noted in television for his work in the docu-drama form. He wrote the scripts for *The Missiles of October, Pueblo* and *Blind Ambition,* all dramatic re-creations of modern historical events. Movie credits include *Soylent Green* and *Skyjacked.*

"GREEN HORNET"—ABC series based on a radio and comic book hero that was introduced in 1966 on the heels of the network's success with *Batman.* The series failed to last the season, however. The leads were Van Williams and Bruce

Lee, and it was produced by Greenway Productions in association with 20th Century-Fox TV.

GREENSPAN, CAPPY AND BUD—independent producing team whose credits include the series, *The Olympiad,* for PBS; the drama *Wilma* for NBC; three one-hour sports documentaries for James Michener's PBS series and a raft of special pieces, *Olympic Moments,* for NBC in connection with the 1980 Moscow Olympics.

GREGORY, PAUL—stage and movie producer who became active in TV in the late 50s, producing such CBS specials as *Three for Tonight* (1955), an adaptation of his Broadway show, with Marge and Gower Champion; *The Day Lincoln Was Shot* (1956) and *Crescendo* (1957), a musical.

GRIES, TOM (d. 1976)—writer who graduated to director and producer of motion pictures. For TV he directed Truman Capote's *The Glass House* and was producer-director of *The Migrants.* He directed *QBVII,* wrote and created *The Rat Patrol* series for ABC and wrote and directed episodes for *East Side, West Side.*

GRIFFIN, MERV—host of one of TV's most successful syndicated talk shows, which also had a run as CBS's late-night entry (1969–72). *The Merv Griffin Show* began in 1965 as a Group W syndication stablemate to *The Mike Douglas Show.* More suitable for evening stripping than the Douglas show, and somewhat more sophisticated, it played on a lineup of more than 170 stations by 1968, the year CBS became interested in Griffin as a possible competitor to the Johnny Carson *Tonight* show on NBC. Griffin joined CBS but was unable to dent the Carson ratings. A move from New York to the West Coast midway in the run did not avail and CBS canceled the show.

Group W bid for Griffin's services again, but he signed instead with MPC, the Metromedia syndication arm, and returned to the prime-time spot on the Metromedia independents that he had held before going to CBS. In 1981, Metromedia began to distribute the show by satellite to stations equipped with satellite-receiving equipment.

Griffin had begun his career as a singer and actor and in 1956 was the vocalist on the *CBS Morning Show* and *The Robert Q. Lewis Show.* Later he hosted a raft of game shows, including *Play Your Hunch* and *Keep Talking* on ABC.

"GRINDL"—situation comedy vehicle for Imogene Coca, the former partner of Sid Caesar in the classic comedy series, *Your Show of Shows.* NBC introduced it in 1963 but it was

unsuccessful. It was by Screen Gems, in association with David Swift Productions.

Lawrence K. Grossman

GROSSMAN, LAWRENCE K.—energetic president of Public Broadcasting Service (1976–   ), recruited from outside the system when there was a pronounced need for new leadership. He succeeded Hartford Gunn, who assumed the new title of vice-chairman. In taking the post, Grossman gave up his own advertising agency, whose clients had included PBS and WNET, and he resigned as president of Forum Communications Inc., the group that had challenged the license of WPIX New York. Before creating his own agency, Grossman had been v.p. of advertising for NBC and earlier had worked in that area for CBS.

GROUP—(1) a company which owns and operates a number of broadcast stations. Under FCC ownership limits, companies may own or have financial interest in up to seven TV stations (with no more than five of them on VHF) and seven AM and seven FM radio stations. Group stations generally are supervised by a central administration, which establishes policy, lends sales support and frequently makes program decisions.

Network affiliations may vary within a group, or they may be concentrated with a single network, as in the case of the Corinthian Stations and CBS. After the networks, the major groups are the most influential forces in television. Some of the more prominent ones are Group W (Westinghouse), Capitol Cities, Taft, Storer, Golden West, Metromedia, WGN Continental, Cox, Post Newsweek, Scripps-Howard and Kaiser.

(2) a level of management at CBS between the corporate tier and the operating line. When CBS became a diversified company in the late 60s, acquiring publishing and musical

instrument manufacturing firms, among others, it reorganized the company into groups such as the CBS Records Group, the CBS Publishing Group, the CBS Broadcast Group, etc., the presidents of which report to the president of the corporation.

The president of the Broadcast Group oversees and coordinates the various TV and radio divisions: CBS-TV, CBS Radio, CBS News, the company-owned television stations and the radio station group, and their respective spot sales organizations, along with other related enterprises.

GROUP W—corporate sobriquet for Westinghouse Broadcasting Co. adopted in the mid-60s to emphasize the *group,* or multiple-station owner, as a force in broadcasting separate from the networks and individual stations. The new name was sought also because the company initials, WBC, had the ring of station call letters. Under the leadership of Donald H. McGannon, its chairman and president (1955–82), Group W became both prosperous and a power in the industry, generally recognized as the leading broadcast organization after the three networks.

The power derives partly from the fact that Westinghouse TV stations are concentrated in major markets—the best VHF concentration outside the networks—and from their affiliation, variously, with all three networks. This not only gives Group W a voice in the policies of each network but makes it privy to the philosophies and secrets of each.

Moreover, McGannon, who earlier had been an executive with the DuMont Network, was a model broadcaster and earned the status of an industry statesman. McGannon gained credibility in Washington for having run a profitable business, while at the same time championing news and public affairs programming. For several years his company conducted annual industry seminars in public affairs and—partly to support the Group W all-news radio stations—was the first station group to establish its own world-wide news organization.

Group W was also a successful producer and distributor of programs for general syndication and was the source of such talk shows as *The Mike Douglas Show, The Merv Griffin Show, The David Frost Show* and, earlier, *The Steve Allen Show.* Since McGannon was a leading figure in persuading the FCC to adopt the prime time–access rule, limiting the networks to three hours of programming in the peak viewing period, his Group W Productions initially produced syndicated programs for the prime-access period but retreated after a disastrous first year.

Variety dubbed Group W "the Harvard Business School of Television" because so many of its executives went on to head other broadcast groups or to assume high positions at the networks.

Group W dropped cigarette advertising on its stations six months before those commercials were ordered off the air.

It also led the opposition of affiliated stations to the networks' proposals to expand their evening newscasts to an hour and it filed a petition on Sept. 3, 1976 asking the FCC to institute a comprehensive inquiry and rule-making procedure into the increasing control of the medium by the three networks.

McGannon, retaining the title of chairman, yielded the titles of president and chief executive officer to Daniel Ritchie in 1979. Ritchie's mandate was to harness the emerging communications technologies for the company's expansion. He moved the company into the new age by acquiring Teleprompter Corp., the second largest cable company; creating a cable satellite network division, which entered into partnership with ABC in establishing two cable news networks; and leasing a large number of transponders on new satellites well before their launching.

Group W owns and operates two CBS affiliates, KDKA Pittsburgh, and KPIX San Francisco; three NBC affiliates, WBZ Boston, KYW Philadelphia, and WPCQ Charlotte; and one ABC affiliate, WJZ Baltimore.

GUDE, JOHN—one-time CBS publicist who became an agent in the 40s representing such leading news figures as Edward R. Murrow, Raymond Gram Swing and Elmer Davis. It was Gude who brought Murrow into collaboration with Fred W. Friendly, then an obscure radio producer in Rhode Island, for the spoken-history record album, *I Can Hear It Now.* That grew into the long Murrow-Friendly partnership on radio's *Hear It Now* and TV's *See It Now.*

"GUESTWARD HO!"—ABC situation comedy (1960) about a couple who run an inn in the southwest, in which J. Carrol Naish portrayed an Indian. Produced by Desilu, the series featured Joanne Dru and Mark Miller.

Don Stewart & Charita Bauer

"GUIDING LIGHT, THE"—one of TV's oldest soap operas, premiering on CBS June 30, 1952, as the continuation of a popular radio serial. Owned by Procter & Gamble, the program kept its 15-minute format until the late 60s, although the half-hour form had long before become standard for the genre. Heading the cast were Millette Alexander, Charita Bauer, Christopher Bernau, Jane Elliot, William Roerick, Lisa Brown, Don Stewart and Robert Milli.

GUNN, HARTFORD N., JR.—vice-chairman of PBS (1976–79) after having been its first president (1970–76). Earlier he had been general manager of WGBH, the Boston public station. As PBS president Gunn devised the basic plan for the Station Program Cooperative, the apparatus by which member stations choose the national programming. He also spearheaded the plan for the interconnection of PBS stations by domestic satellite. In 1979, he became the architect of PBS's reorganization as a network, devising the plan by which three program services would be fed out simultaneously over the satellite: one for entertainment and cultural programs, another for educational and instructional programs and a third for special-interest and regional programs.

"GUNS OF AUTUMN"—90-minute CBS News documentary on the "sport" of hunting (Sept. 5, 1975) that drew letters of condemnation and praise even before the telecast. It aired with virtually no advertising because of a campaign by gun enthusiasts upon the scheduled advertisers. Only Block Drugs, among the advertisers, would not be intimidated. The letters of protest, which came from the National Rifle Assn. and other pro-gun organizations, expressed certainty in advance of the telecast that the program would be unfavorable to their interests. Although CBS professed to cover the issue of hunting dispassionately and in a balanced manner, several scenes of sheer brutality sufficed for a powerful indictment.

The program was produced, directed and written by Irv Drasnin and narrated by Dan Rather. Perry Wolff was executive producer.

"GUNS OF WILL SONNETT, THE"—half-hour western series starring Walter Brennan and featuring Norman Rambo. Produced by Thomas/Spelling, it ran two seasons on ABC (1967–68).

"GUNSMOKE"—classic western built around the fictional Matt Dillon, Marshal of Dodge City, which had a 20-year

run on CBS (1955–75) after having established itself as a Saturday night staple on radio. Beginning as a Saturday night half hour, it was expanded in 1961 to a full hour. The series survived a planned cancellation in the late 60s. William S. Paley, board chairman of CBS, overruled the network's program council and ordered the program reinstated on Monday nights at 7:30. There it was reborn as a hit, placing among the top 3 series in popularity for several seasons. James Arness starred as the marshal, whose deputy in the early years was Dennis Weaver. Amanda Blake was the love interest for 19 seasons, and Milburn Stone portrayed the running character, Doc. Other regulars were Kent Curtis, Buck Taylor and Glenn Strange. The series was owned and produced by CBS.

Milburn Stone & James Arness

GURNEE, HAL—director whose credits include *Tonight, The Jack Paar Show, That Was the Week That Was, The Garry Moore Show, The Jimmy Dean Show, The Joey Bishop Show* and the Emmy Awards telecast.

GUY LOMBARDO AND NEW YEAR'S EVE—a radio and TV tradition for ushering in the New Year, which was passed among the networks over the years and became a live syndicated entry by Worldvision in the 70s. The program featured the dance music of Lombardo and his orchestra, the Royal Canadians, from the ballroom of the Waldorf-Astoria in New York, with shots of the New Year's Eve revelers in Times Square and a countdown of the seconds to midnight with the globe atop the Allied Chemical Tower (originally called Times Tower).

The first of Lombardo's New Year's Eve broadcasts ushered in the year 1930 from the Hotel Roosevelt, the site of many of his later radio broadcasts. In later years, the program emanated from Grand Central Station before settling into the Waldorf. Lombardo reputedly started the tradition of associating *Auld Lang Syne* with New Year's Eve.

H

HAGEN, CHET—director, producer and executive of NBC News for 18 years; then, in the early 70s, head of an independent company, The Colin Group. He produced the country music awards shows on *Kraft Music Hall, The Johnny Cash Show* on CBS (1976) and documentaries, and he worked freelance on segments of NBC's four-hour 50th anniversary special in 1976. He remains active as a producer of network country music specials originating from Nashville.

HAGERTY, JAMES C. (d. 1981)—ABC executive who joined the network as v.p. in charge of news and public affairs in 1961 following eight years of service as President Eisenhower's press secretary, a job he handled with distinction and in which he was popular with the press corps. In 1963 he was elevated by ABC to v.p. of corporate relations and served as a liaison to the Nixon White House; Hagerty had come to know Richard Nixon well while he was Eisenhower's vice-president. Illness forced Hagerty into retirement in the early 70s.

HALEY, JACK, JR.—president of 20th Century-Fox TV since January 1975, succeeding Bill Self. The son of actor-comedian Jack Haley (*The Wizard of Oz*) he had been a producer, director and writer of movies and TV programs before joining 20th. Haley worked for Wolper Productions (1959–70), Paramount Pictures, Columbia Pictures and MGM. In TV, he directed a notable Nancy Sinatra special, *Movin' With Nancy* (1967), and in the same year a National Geographic special on insects, *The Hidden World*. His big-

gest movie was *That's Entertainment* (1974), a knitting together of old MGM movie scenes, which was a boxoffice smash.

HAITI INVASION PLOT—1966 episode for which CBS was censured following a 1970 House Investigations Subcommittee probe. CBS News had been charged with financing a plot by Caribbean exiles to invade Haiti for the sake of exclusive news coverage of the invasion. CBS denied the charge but admitted paying $1,000 for the maintenance of a three-man news crew on the invasion boat. The network was censured for encouraging the illegal invasion scheme and was charged by the subcommittee with staging scenes for use in a documentary.

HALL, MONTY—TV personality known particularly for hosting the hit game show *Let's Make a Deal*. A native of Canada, he began his television career there as emcee and producer of the long-running *Who Am I?*. After coming to the U.S. he worked on such shows as *Strike It Rich*, NBC Radio's *Monitor*, *Video Village* and *Your First Impression*. His popularity through *Let's Make a Deal* led to a Las Vegas nightclub engagement and specials for ABC's *Wide World of Entertainment*. See also *Let's Make a Deal*.

"HALLMARK HALL OF FAME"—long-running series of drama specials on NBC (since 1952) that over its course has presented some of the medium's most distinguished programs. Its productions—mostly in 90-minute form and offered at the rate of five or six a year—have ranged from Shakespeare, Shaw and Gilbert & Sullivan to modern adaptations from the stage and original plays, the casts made up of leading actors of the American and British stages.

The series ended its exclusive relationship with NBC after the 1978–79 season by mutual agreement. NBC was anxious to be rid of the prestigious series because its low ratings were hampering the network's overall prime-time strategies; Hallmark, meanwhile, was unhappy with the time periods it was being assigned by NBC.

Genevieve Bujold & Sir Alec Guiness in Shaw's Caesar and Cleopatra

In the 1979–80 season, CBS began airing the *Hall of Fame* specials at the rate of two a season, and scored well with them. Nearly all that played on CBS ranked among the top specials that season: *All Quiet On the Western Front* and *Aunt Mary* in 1979–80, *Gideon's Trumpet* and *A Tale of Two Cities* in 1980–81, and *The Hunchback of Notre Dame* in 1981–82.

The presentations generally are timed for holiday seasons, which are peak marketing times for the sponsor, Hallmark Cards Inc. Although ratings for the plays frequently fall below 17.0, which represents the base line of success in prime time, the *Hall of Fame* series has proved an immensely effective vehicle for the sponsor. Ratings aside, the company is frequently praised in the press reviews of the shows—a large publicity dividend that TV's scatter-plan advertisers do not receive. Moreover, the quality of the shows reflect on the company and suggest a quality product.

Because of such benefits it scarcely mattered that the productions for the 1961–62 season (*Macbeth, Victoria Regina, Arsenic and Old Lace* and *Give Us Barabbas*) averaged only a 14.5 rating and that those from 1975–76 (*Eric, Valley Forge, The Rivalry, Caesar and Cleopatra* and *Truman at Potsdam*) averaged 14.2. Some of the seasons between were considerably better, averaging as high as 19.0, and some considerably worse, at 11.5, but Hallmark continually re-affirmed its commitment to the series.

Hallmark's beginnings in TV were an extension of its sponsorship of *Hallmark Playhouse* on radio. In January 1952 Sara Churchill hosted *Hallmark Television Playhouse*, a weekly series of half-hour shows. Just before that, however, on Christmas Eve of 1951, Hallmark sponsored NBC's initial production of Gian Carlo Menotti's *Amahl and the Night Visitor*, the first opera to be commissioned for TV, and that signaled the direction it would take in the medium. By 1953 Hallmark had shifted to a schedule of ambitious productions under the *Hall of Fame* banner.

Over the years Hallmark has presented 10 productions of six works by Shakespeare—*Hamlet* (the first in 1953 with Maurice Evans, the last in 1970 with Richard Chamberlain), *Macbeth, Richard II, The Taming of the Shrew, Twelfth Night,* and *The Tempest.* There were also six by G.B. Shaw: *The Devil's Disciple, Man and Superman, Captain Brassbound's Conversion, Pygmalion, Saint Joan* and *Caesar and Cleopatra.*

Among Broadway shows it has offered *Kiss Me Kate, Ah! Wilderness, Arsenic and Old Lace, The Fantasticks, Inherit the Wind, Teahouse of the August Moon, Harvey, The Price, All the Way Home* and *The Green Pastures.*

Outstanding original productions included Rod Serling's *A Storm in Summer*, Allan Slone's *Teacher, Teacher*, James Costigan's *Little Moon of Alban*, John Nuefeld's *Lisa Bright and Dark* and Sherman Yellen's *Beauty and the Beast.*

Mildred Freed Alberg was executive producer of the series for many years and Maurice Evans a frequent director; together they set a standard that was maintained by George Schaefer and others who later assumed charge.

Hallmark switched networks in 1979 for the first time in 27 years. Unable to secure desirable time periods and suitable properties at NBC, in a time when the network was concerned with boosting its ratings, the sponsor bought the three-hour remake on CBS of *All Quiet on the Western Front* for a fall air date. Hallmark explained that it was not severing ties with NBC but would henceforth deal with all three networks to fulfill its advertising needs.

"HALLS OF IVY"—syndicated vehicle (1954 — 1955) for Ronald Colman, about the president of an Ivy League college. Produced by T.P.A. Hall Productions, it featured Benita Hume.

HAMILTON, JOE—Carol Burnett's husband and executive producer of her series. Among his other credits were *The Garry Moore Show* in the early 60s and the *Smothers Brothers* series (1974).

HAMNER, EARL—writer of serious dramas, whose *The Homecoming* gave forth *The Waltons* series, for which he served as narrator and executive story consultant. He also created the series, *Apple's Way*, and *Falcon Crest*, serving as co-executive producer of the latter. He also wrote the script for *A Dream For Christmas*.

HANGEN, WELLES—foreign correspondent for NBC News, and previously for *The New York Times*. Reported missing in Vietnam in 1970 while on assignment for the network, he is believed dead. Unconfirmed reports had said that he, an NBC soundman and a CBS cameraman had been killed in an ambush set by the North Vietnamese. Hangen joined NBC in 1956 and had worked out of Hong Kong since 1966.

HANNA-BARBERA PRODUCTIONS—principal producer of TV animated cartoons headed by Joseph Barbera and William Hanna, who established the California firm in 1957. By 1974 Hanna-Barbera had produced over 100 half-hour series and some 20 TV specials—syndicated in 80 countries—the most widely circulated being *The Flintstones* and *Yogi Bear*. The company is now a subsidiary of Taft Broadcasting, which acquired it in 1966 for $26.4 million.

At a time when soaring costs of full animation forced the closing of most cartoon-producing studios in Hollywood, Hanna-Barbera developed a cheaper method which helped it to dominate the animation market. The system reduced lip movements to simple cycles—vowel to vowel, ignoring consonants entirely—and held body movements to a minimum. This animation stressed plot action and ignored the time-consuming and expensive detail that was superfluous on the small TV screen.

For 20 years prior to the incorporation of Hanna-Barbera Productions, the pair had created and produced the noted *Tom and Jerry* theatrical series of cartoons for MGM. H-B's first TV effort on their own, *Ruff and Reddy*, was followed quickly by *Huckleberry Hound*, which became an immediate success. These led to *The Flintstones*, the "stone-age" situation comedy, and then *The Jetsons*, both of which were big hits in prime time. Other H-B series were *The Pebbles and Bam Bam Show*, *Yogi's Gang*, *The New Adventures of Huckleberry Finn*, *The Banana Splits Adventure Hour*, *Scooby Doo*, *Jonny Quest*, *Top Cat*, *Magilla Gorilla* and *Quick Draw McGraw*.

Barbera, now president of Hanna-Barbera Productions, was at first a New York banker, but also a doodler and dreamer who started submitting gag drawings to leading magazines and eventually became a regular contributor. At the first sign of success, he quit banking and mastered animation and then moved to Hollywood to work with MGM where William Hanna happened to be employed.

Hanna, now executive v.p., was a structural engineer until he found himself unemployed during the Depression and turned to cartooning. He went to work for Leon Schlessinger's cartoon company in Hollywood and in 1937

was hired away by MGM to be a director and story man in the cartoon department. Barbera was employed at the same studio as an animator and writer.

Hanna and Barbera began their collaboration in 1938, first in the production of a single, six-minute animated short. Then they created a cartoon entitled *Puss Gets the Boot*, involving a cat and mouse conflict. This subsequently became the Tom and Jerry series.

Henry Winkler & Ron Howard

"HAPPY DAYS"—ABC situation comedy introduced in January 1974 that initially traded on youthful nostalgia for teenage life in the 1950s. Two years later it exploded into TV's No. 1 hit, with the focus shifting from a middle-class family to a street-wise tough named Fonzie. The series at first centered on the Cunningham family in Milwaukee, but particularly on the teenaged son, played by Ron Howard, and his friends (Anson Williams and Danny Most). Tom Bosley and Marion Ross played the parents.

During the program's second season, a supporting character—Fonzie (played by Henry Winkler)—caught the fancy of youthful viewers, and the series began to be built around him.

It was a Miller-Milkis production in association with Paramount TV, with Thomas L. Miller, Edward K. Milkis and Garry Marshall as executive producers and Tony Marshall and Jerry Paris as producers. Early in 1976 the series spun off a companion hit, *Laverne & Shirley*, which on occasion was cross-pollinated with Fonzie. In 1982, *Joanie Loves Cachi* was spun off the series.

The reruns of the series, offered in syndication on a "futures" basis while the program was still on ABC, drew record prices. WPIX New York agreed to pay $35,000 an episode, KHBK-TV San Francisco $20,000, and WTTG Washington $15,000.

"HAPPY TALK" NEWS—derisive characterization of the style of local newscasts, popular in the 70s, in which the reporters exchange banter between news items. Apart from permitting scarce news time to be consumed by small talk and clowning, the style lent itself to extreme bad taste, as when reports on tragedies were followed by light-hearted repartee.

The informal, ensemble approach to newscasting—which in extreme instances involved fun-loving teams dressed alike in blazers, jolly comrades-in-news—swept the country because it allowed last-place stations to overtake the long-established news leaders almost overnight. In self-defense, stations that had prided themselves on serious newscasting and on their journalistic credibility gave in to the trend. The practitioners preferred to call it by other names, such as "eyewitness" or "action" news. See also "Action News"; Primo, Al.

HARD DOCUMENTARY/SOFT DOCUMENTARY—"hard" includes the probing and muckraking type on major issues of public importance, as well as in-depth examinations of significant news events or trends; "soft" is without urgency and deals with customs, fashions, sports, culture, history or natural phenomena. Hard documentaries usually provoke controversy; soft documentaries frequently are closer to general nonfiction than to journalism.

HARD NEWS—daily reporting, the occurrences and topical matters covered factually by journalists as items for the regular newscasts. This is as opposed to feature material, opinion pieces (editorials, commentary and reviews), essays, trend stories and human interest items not directly related to international, national or local affairs, which may also appear in a newscast.

"HARDY BOYS, THE"—ABC adventure series for children (1977–79), never better than a marginal performer in the ratings, which featured Shaun Cassidy and Parker Stevenson. The series was based on the numerous Hardy Boys adventure books turned out by Franklin W. Dixon five generations earlier. ABC scheduled it in the early Sunday evening hour, the period designated by the FCC's prime time–access rule for children's shows or public affairs, but it proved no match for CBS's *60 Minutes*.

Initially, the series alternated with *The Nancy Drew Mysteries* and was itself entitled *The Hardy Boys Mysteries*. They premiered in January 1977 as midseason replacements and were renewed for the fall. *Nancy Drew*, which starred Pamela Sue Martin and William Schallert, was not renewed for the 1978–79 season, and *Mysteries* was trimmed from *The Hardy*

Boys title. To no avail. The show was a midseason casualty and ended its run in January 1979.

The series were via Glen Larson Productions, in association with Universal, MCA. Larson was executive producer of both. Michael Sloan was producer, with Arlene Sidaris and Joyce Brotman credited as coproducers.

HARGROVE, DEAN—writer turned producer in the Universal TV stable, with initial credits on *The Name of the Game, McCloud* and *Columbo*. He became executive producer with Roland Kibbee on *Columbo* in 1974–75 and the following season on *The Family Holvak* and *McCoy*. Alone, he was executive producer of *Madigan*.

HARLEY, WILLIAM G.—first president of the National Assn. of Educational Broadcasters (NAEB), who established its headquarters in Washington in 1960 and was chief executive of the association until he retired in 1975. He was succeeded by James A. Fellows, former executive director of NAEB.

HARRINGTON, JOHN B., JR. (d. 1973)—co-founder of Harrington, Righter & Parsons (HRP), first station representative firm to handle only TV stations. He and his partners, James O. Parsons, Jr., and Volney Righter, had been with Edward Petry & Co. before starting their own sales representative company in 1949. Harrington retired in 1970.

HARRIS, HARRY—director with extensive credits on prime-time series, including *The Waltons, Kung-Fu, Marcus Welby, Gunsmoke, Perry Mason, Naked City* and *Apple's Way*. (A namesake, Harry Harris, is long-time TV critic for the *Philadelphia Inquirer.*

HARRIS, STAN—producer-director; credits include *The Dick Van Dyke Show, The Smothers Brothers Comedy Hour, That's Life, The Melba Moore Show, The Midnight Special, Music Scene* and *Duke Ellington—We Love You, Madly.*

"HARRY O"—hour-long police series on ABC (1974–75) set in San Diego and featuring David Janssen, Henry Darrow and Anthony Zerbe. The pilot played as a TV movie entitled *Smile, Jenny, You're Dead*. It was by Warner Bros.

HARTFORD PAY-TV EXPERIMENT—one of the major tests of pay-TV in the U.S. prior to the FCC's authorization of full-time subscription TV operations (December 1968). The experiment was conducted by Zenith Radio Corp., with its Phonevision system, and RKO General on its Hartford, Conn., UHF station, WHCT, between June 1962 and January 1969. RKO paid the expenses for the test.

The fare consisted of movies at a charge of 50 to $1.50 per film; sports, cultural events and variety performances at $1 to $3; and educational courses at 25 to $1. Approximately 2% of the TV homes in the area subscribed, and the average subscriber spent only around $1.25 a week for pay programs. As a business venture the Hartford test was a failure with so little penetration into the potential audience, but as a test it revealed that movies would be the dominant programming for pay-TV with sports next in importance.

HARTMAN, DAVID—TV actor who became a talk-show host when ABC tapped him to anchor its two-hour morning strip, *Good Morning, America,* on Nov. 3, 1975. The renovation of the program (previously called *A.M. America*), which included Hartman's services, boosted its lagging ratings. Hartman has been featured in several prime time series and played the lead in NBC's *Lucas Tanner* during the 1974 season.

HARVEY, PAUL—conservative news commentator based in Chicago, who, since 1968, has syndicated a successful 5-minute TV strip, *Paul Harvey Comments.* In the 70s the program was carried in around 100 markets and had a devoted audience in certain parts of the country. Harvey continued his association with ABC Radio, which he had joined in 1944, as news analyst and commentator. In addition, he has written a syndicated newspaper column since 1954.

HARTZ, JIM—NBC newsman who for two years (1974–76) served as co-host of *Today,* then became a roving reporter for the program. In 1977, he became co-anchor with Sue Simmons of the local newscasts on WRC-TV Washington, an NBC o&o. Before joining *Today,* Hartz had worked chiefly in local news, initially as a reporter (starting in 1964) and later as anchorman for the newscasts on WNBC-TV New York.

"HARVEST OF SHAME"—outstanding investigative hour-long documentary on CBS (Nov. 25, 1960) that revealed the suffering of, and inhumanities against, the agricultural migrant workers in the U.S. and their impoverished families. Edward R. Murrow was the reporter, David Lowe the producer and Fred W. Friendly executive producer.

HAUSER, GUSTAVE M.—chairman and chief operating officer of Warner Amex Cable Corp. since 1973. A Harvard lawyer, he previously held executive positions with Western Union Intl. and GT & E Intl., and earlier was international affairs counsel for the U.S. Defense Department.

Hauser was the moving force behind Qube, the futuristic interactive cable system that began its commercial test in Columbus, Ohio, in December 1977. It was he who authorized the $20 million test and who shepherded the project almost daily to determine which aspects of it could be turned into profitable businesses. The national attention gained for Qube helped Warner Amex with franchises in Cincinnati, Pittsburgh, Dallas, Houston, and New York.

"HAVE GUN, WILL TRAVEL"—western hit produced by CBS (1957–62) that rose above the rash of sagebrush series on the networks because of its unusual protagonist, Paladin, an erudite loner who hires out his shooting skills and carries business cards. The inscrutable character was played with finesse by Richard Boone.

The series, however, brought a successful plagiarism suit from an arena performer who not only billed himself as Paladin and distributed "Have Gun, Will Travel" cards but also bore a strong resemblance to Boone. The courts awarded him damages in 1974.

Jack Lord

"HAWAII FIVE-O"—hard-action police series filmed in Honolulu that has been a steady but unspectacular winner for CBS from 1968–80. Produced by Leonard Freeman Productions in association with CBS, it stars Jack Lord and features Kam Fong, Richard Denning, James MacArthur, Al Eben, Peggy Ryan and Linda Ryan in regular roles.

"HAWAIIAN EYE"—Warner Bros. adventure series about private investigators working from a swank Hawaiian hotel; it had a successful run on ABC (1959–63). The leads were Robert Conrad and Anthony Eisley (later replaced by Grant Williams). It featured Connie Stevens, Troy Donahue and Poncie Ponce.

"HAWKINS"—90-minute series for CBS (1973) designed to rotate with *Shaft* as a vehicle for James Stewart. It cast him as a country lawyer, renowned as an expert in criminal cases. Produced by Arena Productions in association with MGM-TV, *Hawkins* yielded eight episodes.

HAYDEN, JEFFREY—director whose credits range from *Philco-Goodyear Playhouse, Omnibus* and the Max Liebman specials in the 50s to *Mannix, Ironside* and *Alias Smith and Jones* in the 70s.

HAYES, JONATHAN—president of Group W Satellite Communications, the cable networking arm of Westinghouse Broadcasting Co., since its formation in May 1981. Hayes previously was v.p. and general manager of the company's Pittsburgh station, KDKA-TV (1975–81) and had been sales manager of KYW-TV Philadelphia and WJZ-TV Baltimore. He joined Group W in 1964 as a trainee.

HAYWARD, LELAND—noted Broadway producer who produced lavish specials for television and served as executive producer of *That Was the Week That Was.*

HAZAM, LOU—documentary producer for NBC, known for the high quality of his work, chiefly in the humanities. He produced *Shakespeare: Soul of an Age, Vincent Van Gogh, The Way of the Cross* and *Michelangelo: The Last Giant.* He traveled extensively for another genre of documentary, which pre-dated the National Geographic specials. Those included *The River Nile, The Sahara, Japan: East Is West* and *Ganges—Sacred River.*

He also worked extensively in the field of medical documentaries, producing the first programs on TV to visit inside a mental hospital and show the birth of a baby; he also produced the first live telecast of a surgical operation.

He began his career in 1933 as a writer of commercials for J. Walter Thompson and then spent a dozen years as a freelance writer for radio and the Federal government before joining NBC as a radio writer. He entered TV in 1952 as associate producer of the political conventions. Hazam took a year's leave of absence from NBC in 1966 and retired from the network several years later.

"HAZEL"—popular situation comedy derived from the *Saturday Evening Post* cartoons of Ted Key about a maid who runs the family she works for. A vehicle for Shirley Booth, it featured Don DeFore and Whitney Blake and was via Screen Gems. NBC carried it 1961–64, and it shifted to CBS in 1965.

Robert Klein at Yale from HBO's On Location

HBO (HOME BOX OFFICE)—first pay-TV network, which started as a pay-cable service in November 1972 on a cable system in Wilkes Barre, Pa., spread by microwave to systems in nearby states and began national distribution by domestic satellite in 1975. A wholly owned subsidiary of Time Inc., it was conceived as a channel of special programs —primarily current movies and sports events—offered to subscribers on affiliated cable systems for a monthly fee ranging up to $10, which was added to the fee for basic cable service. In effect, it served as a "pot luck" station, offering viewers a program assortment generally not available on commercial television.

By 1976 it had 300,000 subscribers nationally and operated approximately 12 hours a day. Its fare included 8 movies of recent release each month, all repeated several times for viewer convenience; live sports not covered by commercial TV; and a number of special programs, including children's shows, comedy performances from nightclubs and British TV series. The popularity of HBO spurred the growth of cable in the late 70s, especially in the larger cities. And as the leading pay-television network, HBO became one of Time Inc.'s most profitable enterprises in the 80s.

HBO arranges all program rights, schedules the programs and distributes them to the individual cable systems from a central studio and control center in Manhattan. It also publishes a monthly guide for subscribers and provides affiliated systems with marketing, technical, legal and administrative support. The cable operators are responsible for the delivery to subscriber homes, the marketing and the billing. Under the standard affiliate arrangement, HBO receives $3.50 of the first $6 of monthly subscriber revenue and shares monies above that amount equally with the cable system. See also Levin, Gerald; Pay-Cable; Cable Networks.

HDTV (High Definition Television)—technology already in development that provides a TV picture comparable in image quality to motion pictures. The high resolution, however, requires an ultra-wide bandwidth, available to over-the-air broadcasting only in the 12 gigahertz spectrum which the FCC has designated for direct-broadcast satellites. Since it may be achieved by media that do not use spectrum, such as cable and video cassettes, conventional television could enter the next century with the present, inferior 525-line resolution while its electronic rivals are able to provide pictures with as many as 1125 scanning lines. Therefore, CBS and certain other companies have been lobbying at the FCC to hold back on direct-broadcast satellites and reserve the newly available spectrum for HDTV.

High resolution television has been in development in Japan for years and was first demonstrated in the U.S. in 1981 and then more widely the following year. It has caught the fancy of such filmmakers as Francis Ford Coppola and George Lucas, who envision producing motion pictures in HDTV instead of film and transmitting them to theaters by satellite.

HDTV has an aspect ratio of 5:3, which resembles that of wide-screen movies. It carries stereophonic sound and lends itself best to projection television screens, although it also may be received on consoles. The development of HDTV would render all present-day TV sets obsolete, which is the main reason it is considered a technology for the next century.

HEAD END—a signal-processing facility, or the electronic control center of a cable-television system, often located at the antenna tower. The term usually refers to the land, the building and the electronic processing equipment normally associated with the starting point of the cable operation.

This center contains all the equipment able to receive off-the-air and microwave signals, to change frequencies and to maintain interface between signals and the cable distribution system.

Master Head End and Sub Head End are included in some cable networks, since more than one head end is required in large urban cable systems.

"HEADMASTER"—an attempt by CBS in 1970 to find a socially "relevant" vehicle for Andy Griffith, once one of its top stars. He was cast as headmaster of a school, with Jerry Van Dyke, Claudette Nevins and Parker Fennelly in support. The series, produced by ADA Productions, lasted only 13 weeks.

"HE AND SHE"—sophisticated situation comedy about a modern young couple which won favorable critical notice on CBS (1967) but not an audience. Produced by Talent Associates, it featured Paula Prentiss, Richard Benjamin and Jack Cassidy.

"HEART" COMEDY—a genre of sentimental situation comedy which enjoyed a vogue in the late 60s and usually involved a bachelor parent (i.e., widowed, not divorced) and children who in one moment tugged at the heart and in the next triggered the laugh-track. Typical were *Family Affair, The Courtship of Eddie's Father, To Rome, With Love,* and *The Doris Day Show.*

HEATH, LAURENCE—executive producer of *The Magician* and creator and producer of the short-lived *Khan* series.

"HEE HAW!"—free-form country-western comedy show, a cornball version of *Laugh-In,* which enjoyed popularity on CBS (1969–70) but was canceled with high ratings in the network's purge in 1970 of its rural-oriented shows. Produced by Yongestreet Productions originally for summer fare in 1969, it did well enough to be inserted in the regular lineup as a replacement in December. Yongestreet continued to produce the series for first-run syndication when the network dropped it, finding a reasonable market for it when the FCC's prime time–access rule went into effect in 1971.

"HEIDI" INCIDENT—a clash of television events on NBC on a Sunday night in November 1968 that distressed millions of viewers and held the network up to criticism for yielding to the tyranny of the clock. With only a few seconds remaining in a football game between the New York Jets and Oakland Raiders, and the Jets leading 32–29, NBC cut away to present, on time, the scheduled children's special, *Heidi,* in a new TV adaptation. But the Raiders, meanwhile, scored two touchdowns in the final 9 seconds of the game, a thrilling and incredible rally that was lost to viewers in the eastern time zone. So numerous were the calls to the network that the Circle 7 exchange in New York was knocked out of service. As a result of the incident, NBC adopted a policy of televising all football games to the end.

That policy backfired on Nov. 24, 1975, when the coverage of a game between the Raiders and the Washington Redskins ran 45 minutes overtime, disappointing millions of children who had tuned in at 7 P.M. for the movie, "Willie Wonka and the Chocolate Factory." In the eastern time zone, the movie was nearly half over when it came on the screen. NBC had considered showing the film from the beginning in the east but abandoned the idea because it would have kept children up almost until 10 P.M., and would have delayed the evening's other programs.

HEINEMANN, GEORGE—specialist in children's programming whose entire career has been with NBC. In 1970 he was named v.p. of children's programs for the network, the first person to receive a vice-presidency in that field; in 1974 he became v.p. of special children's programming, concentrating on experimental and educational fare.

His creations include *Go, Go-USA, Take a Giant Step, NBC. Children's Theatre* and *An Evening at Tanglewood*. In 1952, while program manager of WNBQ (now WMAQ-TV) Chicago, he developed *Ding Dong School,* which soon after went on the network. Heinemann joined the New York station, WRCA-TV (now WNBC-TV), in 1956 and gradually moved up to the network, initially as director of public affairs.

HELLER, JOEL—executive producer of children's broadcasts for CBS News since 1971, including *30 Minutes,* a magazine broadcast focusing on subjects of primary interest and concern to teenagers, which premiered in 1978; *Razzmatazz,* a television magazine for young viewers originally presented on a preemptive basis in 1977–78 and then monthly in 1979; and *In the News,* current events broadcasts presented in 2½-minute segments weekend mornings since 1971. See also *In the News; 30 Minutes.*

"HELLO, LARRY"—moderately successful NBC sitcom, produced by T.A.T. Productions as a vehicle for McLean Stevenson. It debuted in January 1979, in a near disastrous season for the network, but was kept afloat in the lineup as the program scheduled immediately behind *Diff'rent Strokes,* one of NBC's few hits. Stevenson was cast as a radio talk-show host who glibly dispensed personal advice but had trouble on the home front trying, as a single parent, to raise two young daughters. The girls were played by Kim Richards and Donna Wilkes, with Joanna Gleason and George Memmoli as other regulars. Perry Grant and Dick Bensfield were executive producers and George Tibbles producer. One series was canceled in April 1980.

"HELTER SKELTER"—four-hour TV adaptation of the best-selling book on the Charles Manson murders. It was the highest-rated telecast of the 1975–76 season despite the reluctance of some affiliates to carry it in prime time, and despite a temporary blackout of the film in the No. 2 market, Los Angeles.

Presented as a two-part made-for-TV movie (April 1 and 2, 1976), the docu-drama on the mass murders committed by the Manson "family" made many broadcasters uneasy because the subject suggested an exploitation of tabloid headlines in a period when TV was under fire for excesses in violence. Many affiliates delayed it to the late evening and some backed off completely and allowed an unaffiliated UHF station to carry the show. The Los Angeles o&o, KNXT, decided to delay the telecast several weeks because Vincent Bugliosi, the author of the book on which the show was based, and who was also the protagonist in the story, was a candidate for local public office at the time. (He lost.)

Most critics praised the script by J P Miller for downplaying the violence and moderating the sensational in examining the strange life of the group and the bizarre motives for the crimes. The program featured George Di-Cenzo, Steve Railsback, Nancy Wolfe, Marilyn Burnes, Christina Hart and Skip Homeier. Lee Rich and Philip Capice were co-producers for Lorimar Productions and Tom Fries the producer-director.

HEMION, DWIGHT—a leading music-variety director who has won more Emmy Awards than any other director in that classification and whose credits include specials by practically every top-name performer in show business. Hemion works with Gary Smith, and they are co-producers of any show or series he directs. Series credits include *Kraft Music Hall* and *The KopyKats.* In the early 70s Hemion and Smith were under contract to Britain's ATV-ITC production company, for which they turned out *Tony & Lena,* the Burt Bacharach specials and numerous others.

HEMION, MAC—director for ABC Sports whose work covered the full range of athletic events. He also directed music-variety latenight shows for ABC and various awards telecasts, including the Emmy Awards.

"HENNESEY"—series in the wave of service situation comedies featuring Jackie Cooper as a Navy doctor. Running on CBS (1959–62) via Hennessey Co. and Jackie Cooper Productions, it featured Abby Dalton, Roscoe Karns, Henry Kulky and James Komack.

HENNING, PAUL—writer of comedy series, dating to radio, who struck a mother lode with bucolic situation comedies as creator, writer and producer of *The Beverly Hillbillies* (1962) and *Petticoat Junction* (1963) and as executive producer of *Green Acres* (1965). Earlier he had written for the *Burns and Allen Show* (1942–52), passing with them from radio into TV; and the Ray Bolger, Bob Cummings and Dennis Day TV series. In radio he wrote for Rudy Vallee, Fibber McGee and Molly and Joe E. Brown.

HENNOCK, FRIEDA B. (d. 1960)—first woman appointed to the FCC, a New York lawyer who proved to be an activist, reformer and frequent dissenter. She served (1948–55) spiritedly and flamboyantly and worked particularly at reserving TV channels for education. She had been appointed to a Democratic seat by President Truman and was replaced during the Eisenhower Administration by Richard Mack, who left the commission in disgrace, charged with selling his vote in a Miami license contest. Often Mrs. Hennock was the lone dissenter on the commission, as when it approved the merger of ABC and United Paramount Theatres.

HENRY, BOB—producer, director and writer specializing in variety programs. Credits include *The Flip Wilson Show, The Mac Davis Show, Norman Rockwell's America* and specials with Perry Como, Lena Horne, The Young Americans, Andy Williams and Jose Feliciano.

HENRY, E. WILLIAM—chairman of the FCC (1963–66) who adopted the reformist posture of his predecessor, Newton N. Minow, but was never as controversial. Appointed a commissioner by President Kennedy in 1962 and named chairman a year later, Henry was critical of the industry's slavish devotion to ratings and its practice of raising the sound levels for commercials, but his administration in general was moderate. He resigned in 1966 to work in a political campaign in his native state of Tennessee and then became a principal in a closed-circuit TV company.

HENSON, JIM—creator and director of The Muppets, described as a cross between marionettes and puppets, which were featured on numerous network shows during the 60s and achieved great popularity as regulars on *Sesame Street.*

Organic to the celebrated PBS children's program, the principals of the Muppet cast were Big Bird, a 7-foot canary; Oscar, the misanthrope who dwells in a garbage can; Ernie and Bert; Kermit the Frog and the Cookie Monster. Many were specially designed for the show by Henson, who also directed their segments, assisted by Don Sahlin and Kermit Love.

Henson, who was also a painter, sculptor and animator, began puppeteering while a student at the University of Maryland. He introduced the Muppets on a late-night mu-

sic-comedy show in Washington, D.C.. It won them a local series, *Sam and Friends,* which ran for six years.

The Muppets later appeared on *The Ed Sullivan Show* and were featured regularly on ABC's *Jimmy Dean Show* and, for a time, on NBC's *Today.* They were also the stars of several specials, among them *Hey, Cinderella* and *The Frog Prince.* Over the years, Henson made scores of TV commercials and a number of industrial and experimental films.

The Muppet Show, a 30-minute weekly variety series, was produced by ITC for prime-access syndication in 1976 and won a staunch following both in England and the U.S. In 1979, Henson scored in another medium, theatrical films, with *The Muppet Movie.*

See also *Muppet Show, The.*

HERBERT, DON—former radio actor and writer who became known on TV as "Mr. Wizard" for the educational science series he created and performed in for NBC (1951–65). The series emanated from Chicago.

HERBERT LIBEL CASE—extremely controversial $44 million libel case, with far-reaching First Amendment implications, brought by former Army Lieut. Col. Anthony Herbert, who charged that journalists connected with CBS's *60 Minutes* defamed his character in order to discredit his report of American military atrocities in Vietnam.

The case is far from being resolved, but a legal decision made during initial proceedings already promises to have an enormous impact on freedom of the press. On April 18, 1979, the Supreme Court rejected the arguments of CBS News and ruled that the lawyers for Herbert could inquire during the course of the trial into "the thoughts, opinions, and conclusions" of journalists at CBS News for evidence of malice against him.

This case within a case developed at the pretrial "discovery" proceedings when producer Barry Lando and reporter Mike Wallace refused to answer specific questions about the editorial process involved in their *60 Minutes* segment of Feb. 4, 1973 on Herbert. Because in a libel case the burden is placed on the public figure to prove not just that falsehoods were published but that "actual malice" was intended against him, lawyers for Herbert argued that it was necessary to discover the basis and motivations behind their editorial decisions.

Initially a federal district court judge had ordered Lando and Wallace to answer the plaintiff's questions, but that decision was reversed by a 2–1 vote of a panel of judges from the Second Circuit Court of Appeals on Nov. 7, 1977. In a majority opinion written by Justice Byron White, the Supreme Court overturned the Second Circuit Court reversal and agreed that Herbert's lawyers had the right to obtain "direct evidence" concerning the editorial "state of mind" about Herbert at CBS News.

The Supreme Court's decision came on the heels of its decision in the *Stanford Daily* case that journalists could not protect their files from searches by law enforcement officials. Many members of the journalistic community direly predicted that these decisions at the least were intruding on press freedom and would have a "chilling" effect on the exchange of information and opinion among reporters, editors and sources.

HERBUVEAUX, JULES—v.p. of NBC's central division (Chicago) during the years of the "Chicago school" of television. He was the executive who made possible such shows as *Garroway-At-Large, Ding Dong School, Studs' Place, Zoo Parade* and *Kukla, Fran and Ollie.* He also helped to develop such talents as Hugh Downs, Dave Garroway, Bob Banner and Don Meier. He retired in 1962, after NBC had shut down the central division as a production entity, and became a broadcast consultant.

"HERE COME THE BRIDES"—adventure-comedy series based on the film, *Seven Brides for Seven Brothers,* about the importation of women to Seattle in the rugged pioneering days. The hour-long series, produced by Screen Gems, premiered on ABC in 1968 and ran two seasons. It featured Joan Blondell, Bobby Sherman, Robert Brown, David Soul, Bridget Hanley and Mark Lenard.

HERTZ—unit of frequency equal to one cycle per second. In the U.S. electrical current is 60 hertz. Megahertz equals a million cycles per second, and gigahertz a billion cycles per second. American VHF television stations are allocated on a band from 54 to 216 megahertz (MHz), UHF stations from 470 to 890 MHz.

The term is named for Heinrich Hertz, a 19th-century German physicist.

HEWITT, DON—one of TV's premier news and documentary producers who helped develop CBS's *60 Minutes* in 1968 and *Who's Who* in 1977 and served as executive producer of both. He had been producer-director of *Douglas Edwards with the News* (1948–62) and then exec producer of its successor, *CBS Evening News with Walter Cronkite* (1963–64). During the years before *60 Minutes,* Hewitt produced and directed documentaries and special reports.

He was producer-director of the first of the *Great Debates* of 1960, directed for CBS the three-network *Conversations with the President* (Kennedy in 1962, Johnson in 1964), and produced the CBS coverage of the assassination and funeral

of the Rev. Dr. Martin Luther King, Jr. His credits include the documentaries, *Hunger in America* (for *CBS Reports*), and Lord Snowdon's *Don't Count the Candles.* Hewitt also directed the CBS coverage of the national political conventions, from 1948 through 1960. Before joining CBS News in 1948 he had worked for newspapers and news services.

Don Hewitt

"HIGH CHAPPARAL, THE"—hour-long western dramatic series by the producer of *Bonanza,* David Dortort, in association with NBC. Airing from 1967 to 1970, it starred Leif Erickson and Cameron Mitchell and featured Mark Slade, Linda Cristal, Henry Darrow and Frank Silvera.

"HIGHWAY PATROL"—action series produced for syndication by Ziv Television (1955–59) which proved to be one of the highest-rated nonnetwork shows and a precursor to the filmed police series that became standard network fare. It starred Broderick Crawford.

HIKEN, NAT (d. 1968)—one of the outstanding comedy producers of the 50s whose masterpiece was *You'll Never Get Rich* (more generally known as the Sgt. Bilko show during its run from 1955 to 1959). Hiken also created and was producer-director of *The Phil Silvers Show* and *Car 54, Where Are You?,* among others.

HILL, PAM—v.p. and executive producer of documentaries for ABC News since 1978 and responsible for a resurgence of the form in bringing forth a succession of mettlesome reports in the *ABC Close-Up* series, in styles that departed from the conventional. The work she supervised, which was often controversial, contributed to the growing credibility of ABC News in the late 70s.

She joined ABC in 1973 after eight years with NBC, where she was a director on the *White Paper* series and producer of *Comment,* a half-hour series with Edwin Newman. Earlier, she had been a foreign affairs analyst for Henry A. Kissinger when he was consultant to New York Governor Nelson Rockefeller. She is married to Tom Wicker, columnist and associate editor of *The New York Times.*

At ABC, she became a documentary producer, and her 1977 film, *Sex for Sale: The Urban Battleground,* proved the highest-rated ABC documentary up to that time. She was named executive producer of documentaries in January 1978, when Marlene Sanders went to CBS, and she was named a v.p. several months later.

See also *ABC Close-Up.*

"HILL STREET BLUES"—new style of police-drama series, introduced by NBC in the fall of 1980, which won the praise of critics and a record number of Emmy awards but little encouragement from viewers its first season on the air. NBC renewed the series despite its poor ratings and was vindicated; the show caught on the second year.

Hill Street Blues is a phenonmenon of genre splicing, merging elements of soap opera with police-action melodrama. It also owes a certain debt to a retired MTM Enterprises stablemate, *The Mary Tyler Moore Show,* in focusing on the private lives and interaction of people in a workplace—in this case, a police precinct house. Realism is the series' idiom, reflected in its themes and storylines as well as in its format; individual episodes are never neatly resolved, as in conventional action shows.

The cast includes Daniel J. Travanti, Michael Conrad, Michael Warren, Bruce Weitz, James B. Sikking, Joe Spano, Rene Enriquez, Kiel Martin, Taurean Blacque, Barbara Bosson, Betty Thomas, Ed Marinaro, Charles Haid, and Veronica Hamel. Steven Bochco is executive producer, Gregory Hoblit supervising producer, and David Anspaugh and Anthony Yerkovich episode producers.

By odd coincidence, the president of MTM Enterprises, Grant Tinker, resigned after the first season of *Hill Street Blues* to become chairman of NBC; thus he remained a beneficiary of its success the second season.

HIRSCHMAN, HERBERT—producer whose credits include *Perry Mason, The Man from Shiloh* and *The Zoo Gang* and (as executive producer) *The Doctors, Planet of the Apes* and the film drama, *Larry.*

HITCHCOCK, ALFRED (d. 1980)—famed master of motion picture suspense who became a TV personality, pro-

ducer and director in the late 50s with his highly successful anthology, *Alfred Hitchcock Presents.* See also *Alfred Hitchcock Presents.*

HOBIN, BILL—veteran producer-director concentrating on variety shows. His credits range from *Garroway-At-Large, Your Show of Shows* and *Your Hit Parade* in the 50s; to *Sing Along with Mitch, The Andy Williams Show* and *The Judy Garland Show* in the 60s; and the Tim Conway and David Steinberg series in the 70s.

HODGE, AL (d. 1979)—actor who played the title role in the *Captain Video* series on the DuMont Network in early TV, after having portrayed *The Green Hornet* on radio in the 40s. One of television's first big stars, enjoying six years of celebrity, he became a victim of typecasting and was unable to find other work in the medium when *Captain Video* folded with DuMont's demise as a network in 1955. Driven to working at odd jobs, such as bank guard and store clerk, while striving continually for a comeback both in Hollywood and New York, he became a television tragedy, living his final years alone, impoverished and suffering from severe emphysema. He was found dead in a downtown New York hotel on March 19, 1979.

See also *Captain Video.*

"HOGAN'S HEROES"—one of the more successful series in the wave of military situation comedies in the 60s. It concerned a brash group of Allied soldiers in a German POW camp during World War II who make a luxurious life of imprisonment by constantly outwitting the buffoonish officers in charge. (Similarity to the play, *Stalag 17,* invited a plagiarism suit.) The series, by Bing Crosby Productions, ran on CBS (1965–71) and featured Bob Crane as Hogan and Larry Hovis, Robert Clary, Richard Dawson, Kenneth Washington and Ivan Dixon as the other prisoners. The nutty Nazis were John Banner and Werner Klemperer.

"HOLLYWOOD SQUARES, THE"—popular comedy game show on NBC (premiering in 1966) that features a regular celebrity cast and guest stars, each of whom sits in a square on a giant tic-tac-toe board. Although the game involves a competition for cash prizes between two studio contestants, the show's basic appeal has been the quips and byplay of the personalities whose answers to questions from host Peter Marshall are germane to the contest. In addition to the five-a-week daytime strip on NBC, the show spawned a prime time– access edition for syndication. It's a Heatter-Quigley Production, with Merrill Heatter and Robert Quigley as executive producers.

The program served as a showcase for the late comedians Paul Lynde and Charlie Weaver (Cliff Arquette), who were the regulars, as well as for such recurring participants as Rose-Marie, Jan Murray, Vincent Price, Karen Valentine and Nanette Fabray.

"HOLLYWOOD PALACE"—ABC Saturday night variety hour (1964–70) mounted as a modern version of vaudeville; it was staged at a refurbished Los Angeles movie house, the El Capitan, which was renamed The Hollywood Palace in honor of the New York theater that was vaudeville's top showplace. Each program featured seven or eight acts and a guest host of star stature. Bing Crosby was host for the premiere show (Jan. 4, 1964); others included Maurice Chevalier, Judy Garland and George Burns. Nick Vanoff was executive producer and William Harbach producer.

"HOLLYWOOD TELEVISION THEATRE"—PBS series of drama productions from KCET Los Angeles; it used major actors in adaptations of noted stage plays. Norman Lloyd was executive producer for the entire series. Productions have included *Ladies of the Corridor,* with Cloris Leachman and Neva Patterson; *The Lady's Not for Burning,* with Richard Chamberlain; *Six Characters in Search of an Author,* with John Houseman and Andy Griffith; and *The Chicago Conspiracy Trial,* with Morris Carnovsky, Ronny Cox and Al Freeman Jr.

HOLLENBECK, DON (d. 1954)—a CBS newsman associated with Edward R. Murrow. He died, by suicide, while under incessant attack by pro-McCarthy columnists for alleged leftist leanings. He delivered the late evening news on WCBS-TV in New York, worked occasionally with Murrow on *See It Now* and conducted a weekly radio series, *CBS Views the Press,* on which he was frequently critical of Hearst journalism. His own chief critic and accuser was Hearst television columnist Jack O'Brian.

Frail and suffering from ulcers, and said by his colleagues to have been deeply distressed and fatigued by the attacks upon him, Hollenbeck took his life by gas on June 22, 1954.

At the time, Murrow's *See It Now* had contracted for newsfilm with an outside organization, News of the Day, which was jointly owned by Hearst and MGM. On Hollenbeck's death, Murrow and his producer, Fred W. Friendly, decided to end the Hearst relationship and received permission from CBS to hire away all the technicians from News of the Day as the network's own newsreel unit.

HOLLYWOOD FILM COUNCIL—an organization concerned with the common interests of member craft guilds and related unions in the TV and motion picture fields. The Council has lent the support of its numbers to a host of rule-makings and legislative proceedings, usually in the interest of increasing employment. It played an active role, for example, in the Balmuth petition to the FCC to limit the number of network reruns and in the Sandy Frank petition to bar strip programming in the prime time–access rule time periods.

The New York equivalent of the Hollywood Council is the National Conference of Motion Picture and TV Unions (NACOMPTU).

"HOLMES AND YOYO"—ABC situation comedy (1976) about an accident-prone police detective partnered with a humanized but not-quite-perfected robot. It featured Richard B. Shull and John Shuck and was created by Jack Sher and Lee Hewitt. The series was by Heyday Productions in association with Universal TV, with Leonard Stern as executive producer and Arne Sultan as producer.

Meryl Streep as Inga Weiss

"HOLOCAUST"—powerful 9½-hour miniseries on the horrors of the extermination of Jews in Nazi Germany, presented by NBC on four consecutive nights in 1978, April 16–20, and repeated in September 1979. In the U.S., the initial broadcast reached a total audience of 107 million; worldwide, in the 50 other countries in which it aired—including West Germany, where it provoked a continuing controversy—it was seen by an estimated 220 million viewers. The story, written by Gerald Green, author of *The Last Angry Man,* covers a ten-year period of persecution, centering on one Jewish family and one Nazi family in Berlin. The production won eight Emmy awards, a George Foster Peabody award and twoscore other citations.

Some critics faulted *Holocaust* for trivializing one of history's darkest episodes with a dramatic scaffolding that too often lent itself to soap opera, but in the main it was hailed as a television landmark, a program that not only brought the reality of Nazi terror to young persons in America but also to many in West Germany who had never learned what had happened before their lifetimes.

Among those top-featured were Michael Moriarty, Meryl Streep, Joseph Bottoms and Rosemary Harris. Others in the large cast included Blanche Baker, Tom Bell, Tovah Feldsuh, Marius Goring, Anthony Haygarth, Ian Holm, Lee Montague, Deborah Norton, George Rose, Robert Stephens, Sam Wanamaker, David Warner, Fritz Weaver and James Woods.

Filmed entirely in Germany and Austria, *Holocaust* was produced by Herbert Brodkin's Titus Productions in association with NBC Entertainment. Brodkin was executive producer, Robert (Buzz) Berger producer and Marvin Chomsky director.

HOLOGRAPHY—a type of three-dimensional photography in which laser light is used to make a picture on an ordinary black-and-white photographic plate. What actually appears on the plate is a light-wave interference pattern, which, when illuminated by a laser beam, produces a three-dimensional image that can be viewed without special glasses. It is often predicted that holography will some day make possible three-dimensional television, but there are still a number of practical obstacles in the way of such a development.

HOME VIDEO—generic for a range of nonbroadcast television technologies—video recorders, video discs, video games, home video studios, portapacks, computer television and print retrieval systems such as teletext and viewdata —any or all of which are expected to blossom in the 80s and make an impact on television usage and ultimately on television programming.

Taking an optimistic view, commercial broadcasters predict that home video devices will appeal primarily to selective viewers and thus will cause only a slight erosion of their audiences. Indeed, they foresee a greater use of television than now exists because of the new and varied ways to use the medium and because of the increases in software for specialized tastes. Some speculate, however, on the emergence of TV stations that would rely on video discs for programming in the way that radio stations rely on phonograph records.

See also Teletext, Video, Videocassette, Video Disc.

"HOMECOMING, THE"—two-hour CBS dramatic special (Dec. 19, 1971) subtitled "A Christmas Story," which became the basis for *The Waltons* series although it was not intended as a pilot. The teleplay was an adaptation by Earl Hamner, Jr., of his own novel and was by Lorimar Productions, with Lee Rich as executive producer. The program has been repeated several times as a Christmas special. Patricia Neal, Cleavon Little and Andrew Duggan were featured, as was Richard Thomas who continued the role of John-Boy in *The Waltons*.

HOMES PASSED—cable-TV term for residences within reach of a cable trunk line and which could be wired if the household chose to subscribe. Cable penetration is defined by the number of subscribers in relation to the number of dwelling units passed (i.e., homes capable of receiving cable).

"HONEYMOONERS, THE"—See Gleason, Jackie.

"HONEY WEST"—ABC half-hour private-eye series (1965–66) about a female investigator, both seductive and expert in judo. Anne Francis had the title role and John Ericson and Irene Hervey were featured. It was by Four Star.

HONG KONG, TELEVISION IN—a system consisting of three television stations. Two of them, HKTVB and RTV, are bilingual (English and Chinese) operations owned by commercial firms; the third and newest, CTV, broadcasts in Chinese only and has a fixed commitment to educational programming in association with Hong Kong Chinese University for part of its broadcast day. The British crown colony has around 800,000 sets (625-line PAL), more than 120,000 of them color receivers.

HOOKS, BENJAMIN L.—first black FCC commissioner, appointed to the agency by President Nixon in 1972 in a Democratic seat. An ordained Baptist minister who was pastor of a church in Memphis during the mid-50s, he was at the same time a lawyer in private practice who eventually became a judge in the Shelby County Criminal Court (1965–68). He was also a v.p. of a savings and loan association and a producer, host and panelist on TV public affairs programs in Memphis.

Just before the start of the Carter Administration, Hooks accepted the position as head of the National Assn. for the Advancement of Colored People but stayed on with the FCC until July 1977 in order to qualify for a government pension.

He was considered most likely to be named chairman of the FCC if he had remained on the commission.

"HOPALONG CASSIDY"—early western series, and one of the medium's first financial bonanzas, drawn from the character created by William Boyd in low-budget movies of the 30s and 40s. The shows were enormously popular with children, and "Hoppy," as he was known to them, was extensively merchandised.

Hopalong was introduced on TV in 1948 through the old movies that had been purchased by Boyd. They were so successful that in 1951 and 1952 Boyd produced 52 low-budget half-hour *Hopalong* films for the series; they are still in syndication.

Bob Hope

HOPE, BOB—not so much a comedian as an American institution: the master dispenser of topical jokes, the people's emissary to the troops abroad in time of war, emcee *par excellence* for black-tie occasions, champion of the American Way of Life threading a patriotic spirit through a web of one-liners, chairman of the board of the entertainment fraternity. Hope's enormous popularity in radio and films carried into television and, if anything, increased with the years over a quarter century.

During the 70s his specials for NBC—four or five a season—continued to rank among the top shows of the year, not far behind the Super Bowl and Academy Awards. During the Vietnam War, the annual *Bob Hope Christmas Show*—which actually aired in January or February—was fairly consistently the highest-rated special of the year; this was the 90-minute film of his Christmas season tour of military bases with a troupe of entertainers, a Hope tradition that began in World War II.

Like Will Rogers, Hope drew on politics and current events for his monologues, but where Rogers' style was folksy Hope's was urbane. Rogers looked the western country boy, Hope the middle-America businessman.

Hope's TV programs held to the simple format of his radio shows, a monologue followed by skits with guest stars, interspersed with musical performances. He began *The Bob Hope Show* for NBC on Oct. 12, 1952 on a monthly basis, and it continued through 1955. Then came *The Chevy Show Starring Bob Hope* with episodes filmed in Iceland, England, Korea and Japan.

Thereafter, he contracted for a number of specials for NBC each season and made a tradition of hosting the Academy Awards telecast.

HORNER, VIVIAN —v.p. of program development for Warner Cable, a title conferred in 1979 after she mounted Nickelodeon, the company's cable network for children, and Pinwheel, a channel of violence-free programming for pre-schoolers for Warner's interactive system, Qube. Dr. Horner, who earned a Ph.D. in psycholinguistics from the University of Rochester, joined Warner Cable in 1976 from the Children's Television Workshop, where she was director of research for the children's series, *The Electric Company*.

See also Nickelodeon.

HORWITZ, HOWIE (d. 1976)—producer with Warner Bros. in the 50s when he was active on such series as *77 Sunset Strip, Hawaiian Eye* and *Surfside 6*. Later, with 20th Century-Fox, he was producer of *Batman,* and with Universal, *Banacek* and several made-for-TV movies.

HOTELVISION—a form of pay-cable which supplies a small selection of movies to hotel and motel guests over the TV receivers in their rooms for a fee (typically $2 or $3) that, like room service, is added to the bill. An industry with only a handful of companies involved as pay-TV suppliers, it began in 1972, and the concept was adopted rapidly by many of the large hotel chains. In some installations the movies are provided over the master antenna systems from videocassettes supplied to each hotel. Others transmit the movies from a central source to a number of hotels by coaxial cable or microwave link. The pioneer companies were Computer Cinema, an affiliate of Time Inc.; Trans-World Communications, a subsidiary of Columbia Pictures; Creative Cine-Tel; and Athena Communications.

"HOT L BALTIMORE"—ABC situation comedy which lasted six months in 1975, the first flop by Norman Lear's T.A.T. Productions. Based on the off-Broadway play of that title by Lanford Wilson, it concerned decadent characters in a decaying hotel and was found objectionable by several of the network's affiliates. The cast included Al Freeman, Jr., Charlotte Rae, Conchata Ferrell and James Cromwell.

HOTTELET, RICHARD C.—scholarly United Nations correspondent for CBS News, holding the assignment since 1960, after having been a foreign and war correspondent for the network since 1944. Before joining CBS, he was a foreign correspondent covering World War II for United Press.

HOUSE INVESTIGATIONS SUBCOMMITTEE—most powerful congressional investigative body for issues related to the communications media, although it has no legislative power. Known formally as the Special Subcommittee of Investigations of the House Interstate and Foreign Commerce Committee, it examines FCC policies and in 1972 held hearings on charges of staged TV news.

HOUSEMAN, JOHN—noted stage director, producer and writer who occasionally has worked in TV. In the 50s he produced the Emmy-winning CBS series, *The Seven Lively Arts,* and a number of dramas for *Playhouse 90.* In 1976 he had one of the acting leads in a PBS production of Pirandello's *Six Characters in Search of an Author* in the *Hollywood Television Theatre* series, and in 1977 was featured in the ABC serial, *Washington: Behind Closed Doors.* But he became best known in television as the star of the CBS series *The Paper Chase* (1978–79), which won the praise of critics but could not muster the support of viewers in its single season on the air. See also *Paper Chase, The.*

HOUSE TELECOMMUNICATIONS SUBCOMMITTEE—key subcommittee of the House of Representatives concerned with telecommunications issues and known formally as the Communications Subcommittee of the House Interstate and Foreign Commerce Committee. Having legislative power and oversight of the FCC and other communications agencies, the subcommittee has been involved with reconsidering the Fairness Doctrine, extending the broadcast license period, investigating TV violence and imposing blackouts on local sports. It has also been concerned with political advertising, cable-TV regulation, satellite policy and public broadcasting.

Under the chairmanship of Rep. Lionel Van Deerlin (D.-Calif.), the committee in 1976 proposed a "basement to attic" revision of the Communications Act of 1934 to make it more responsive to the emerging technologies. Van Deerlin introduced his first bill in 1978 and a revised bill in 1979, each with the support of the ranking Republican on the

subcommittee. Van Deerlin was succeeded as chairman in 1981 by Rep. Timothy Wirth (D.-Colo.).

See also Rewrite of the Communications Act; Senate Communications Subcommittee; and Van Deerlin, Lionel (Rep.).

"HOUSE WITHOUT A CHRISTMAS TREE, THE"—90-minute dramatic special on CBS (Dec. 3, 1972) based on a story by Gail Rock about a Nebraska family in the 40s. It led to such other CBS holiday sequels as *The Thanksgiving Treasure* and *The Easter Promise.* Jason Robards, Mildred Natwick and Lisa Lucas were featured; the script was by Eleanor Perry. Paul Bogart directed and Alan Shayne was producer.

HOUSER, THOMAS J.—director of the White House Office of Telecommunications Policy (July 1976–January 1977) and for nine months in 1971 a member of the FCC. In the years between he was a partner in the Chicago law firm of Sidley and Austin. Earlier he had been special counsel to Sen. Charles Percy (R.-Ill.), after having been his 1966 campaign manager, and then served as deputy director of the Peace Corps (1969–70).

HOVING COMMITTEE—a nationwide TV watchdog group (1967–70) nominally headed by Thomas P.F. Hoving, director of the Metropolitan Museum of Art, but actually run by its executive director, Ben Kubasik, now head of a public relations agency. The group was formed as the National Citizens' Committee for Public Broadcasting, coincident with the passage of the Public Broadcasting Act of 1967, on funding from the Carnegie Foundation.

Originally it was to have united leaders in the arts and humanities around the country into a body that would have served as an inspirational force for public TV, but it quickly evolved into an organization critical of commercial television. By 1968 it had assumed the name of National Citizens' Committee for Broadcasting (dropping the word "Public")—progenitor of the NCCB, now headed by Nicholas Johnson.

Kubasik left, and the organization came apart (until revived by Johnson) when the board disagreed on the question of whether to oppose Vice-President Agnew in his attack on network news in November 1969 or hold silence.

See also National Citizens' Committee for Broadcasting.

HOWARD, ROBERT T.—NBC executive who became president of NBC-TV on April 1, 1974, after having been for 7 years v.p. and general manager of KNBC, the company's station in Los Angeles. He was relieved as president in August 1977, replaced by Robert Mulholland, but instead of leaving the company as many deposed presidents do, he stayed on in the lesser job as v.p. and general manager of the New York station, WNBC-TV.

He left the company after a reorganization by Fred Silverman in 1980 and later became head of United Satellite Television, a new pay-television service using the Canadian ANIK-C satellite.

In an NBC tradition, Howard began his career as a page in the guest relations department in 1947 and gradually moved up to the company's research department, to NBC Radio Spot Sales and to Television Spot Sales. From 1959–63 he was national sales manager of WNBC-TV New York, later advancing to station manager, and in 1966 became general manager of KNBC.

Howdy & Buffalo Bob Smith

"HOWDY DOODY"—one of the earliest network children's series, lively, nonsensical and pretending to no educational value, which was wildly appealing to children. It was to become the symbolic show of the first generation nurtured on television.

The show featured Buffalo Bob Smith, a ventriloquist who was its host and creator; Howdy Doody, a four-foot puppet; and Clarabell, a clown whose voice was an auto horn. (For a time Clarabell was played by Bob Keeshan, who later became Captain Kangaroo.

Howdy Doody premiered live on NBC on Dec. 27, 1947, as a one-hour Saturday program and remained on the air until 1960. A measure of its success was the overwhelming demand for studio tickets. The waiting list was so long that expectant mothers were reported to have requested tickets for their unborn children. The program was unsuccessfully revived in syndication in 1976.

HTN (Home Theater Network)—see Cable Networks.

HTV—British commercial independent serving Wales and West of England, its contract area stretching from Anglesey to the borders of Devon and Wiltshire. Headed by Lord Harlech (and previously known as Harlech TV), the program company has its bases in Cardiff and Bristol and reaches a population of around 4 million. Its chief problems are a difficult terrain, which requires the use of 53 UHF and 12 VHF transmitters, and the fact that the programming must be divided for the English-speaking and Welsh-speaking viewer.

HUBBARD BROADCASTING INC.—broadcast group based in St. Paul, Minn., founded by Stanley E. Hubbard, with KSTP Minneapolis-St. Paul, an ABC affiliate, as flagship. Other stations are KOB Albuquerque, N.M., NBC, and WTOG St. Petersburg, Fla., independent. The owners are Stanley E. Hubbard Trust, Stanley S. Hubbard Trust, Hubbard Foundation, the estate of Thomas E. and Vera S. Bragg, Paulette Fownes and Stanley S. Hubbard and family. Stanley E. Hubbard is chairman and his son, Stanley S., president and general manager.

In 1978, Hubbard rocked NBC by switching its long-time and powerful Twin Cities affiliate, KSTP-TV, to ABC. The shift permitted KSTP to extend its signal with translators to areas not reached by ABC, and it gave ABC a strong news outlet in a major city where previously it had a relatively weak one.

HUBERT, DICK—independent producer of TV documentaries, after having been executive producer of Group W's Urban America Documentary Unit and a writer-producer for ABC News. When Group W disbanded the Urban American unit in 1973, Hubert formed Gateway Productions with Paul Galan and Morty Schwartz and produced *World Hunger: Who Will Survive?* and *It's Tough to Make it in This League,* a football documentary. Both aired on PBS. In 1979, for Capital Cities Broadcasting, he produced *Inflation: The Fire That Won't Go Out,* a special that was syndicated to 190 commercial stations.

HUBLEY, JOHN (d. 1977) AND FAITH—film animators known for their humanistic approach to the craft. A husband and wife team, each with a background in aspects of film production, they formed their own company in 1955, Storyboard. Their most ambitious film for TV was *Everybody Rides the Carousel* (1976), a 90-minute work adapted from the writing of psychoanalyst Erik H. Erikson and developed at Yale University. It aired on CBS, sponsored by Mobil Oil, with Cicely Tyson as host.

HUGGINS, ROY—action-adventure producer and writer who established his reputation as creator of *Maverick, Colt .45* and *77 Sunset Strip* in the mid-50s and added to his list *The Fugitive, Run for Your Life, The Rockford Files* and *City of Angels.*

Huggins has been executive producer of *Alias Smith and Jones, Cool Million, Toma, Baretta* and numerous other series and made-for-TV movies. His production company, Public Arts, was associated with Universal TV on such series as *Baretta* and *Rockford Files.*

HUGHES TELEVISION NETWORK—one of the first ad hoc networks which put together lineups of stations for special telecasts, usually of sports events. Most of the stations are regularly affiliated with the major networks but clear time for the HTN events.

Paramount Pictures purchased HTN in 1977 and established a division under Rich Frank to create original programs for the network. That plan was abandoned in 1979 and HTN was put under the wing of the Madison Square Garden Communications Network, with John A. Tagliaferro as general manager.

The company started as Sports Network in 1956 and was founded by Dick Bailey, a facilities expert for ABC who had an exceptional understanding of the economics of leasing AT&T long lines. Initially the company served as a facilities coordinator for sporting events, which usually involved arranging for the transmission of "away" games to a team's home city. But soon it began to distribute golf, college basketball, tennis and other sports around the country over a part-time network. In 1968 Sports Network was purchased outright by the reclusive billionaire, Howard Hughes, and given the name of Hughes Sports Network, later changed to Hughes Television Network.

In 1974 the network merged with a number of other businesses that Hughes owned personally, principally Las Vegas hotels (The Sands, Harold's Club, etc.) and real estate into a single company known as Hughes Television Network Inc. But Paramount purchased the network alone. Most of the sports telecasts are on weekend afternoons, but HTN also has distributed some entertainment programming in prime time, such as *Magic, Magic, Magic* (1976) and *Steve Allen's Laugh-Back* (1976). Since 1969 it has distributed the weekly series *Outdoors* in 70 markets for Liberty Mutual Insurance by means of a tape network.

In addition to its network projects, HTN has the TV facilities contract for Madison Square Garden, leases mobile equipment, produces commercials and provides dubbing services. HTN has been known to distribute 160 hours' worth of "network" programs a year, clearing 90% of the country for several of the events.

HULL, WARREN (d. 1974)—TV emcee, best known for *Strike It Rich,* a hit giveaway show in the early 50s.

"HULLABALOO"—one of the first rock 'n' roll variety series, presented by NBC (1964–66) and featuring a chorus line of wriggling dancers and guest performers. It was by Gary Smith Ltd.

HUNGARY, TELEVISION IN—one state-controlled channel which operates 45–50 hours a week and accepts advertising on a limited basis. There are more than two million sets in the country, and the technical standard is 625-line.

"HUNGER IN AMERICA"—CBS News documentary (May 21, 1968) produced by Martin Carr which became the subject of a controversy over an instance of misreporting. In the opening sequence, a baby receiving emergency medical treatment was said by the narrator to be dying of starvation. But the San Antonio hospital at which the sequence was filmed maintained that the baby actually died of premature birth. CBS News explained that the misinformation had been given by a hospital official, an account denied by the party in question. The FCC decided not to resolve the issue since it came down to a choice of whom to believe, the producer or the hospital official. The commission, at any rate, found no evidence of deliberate deception.

Chet Huntley

HUNTLEY, CHET (d. 1974)—anchorman on NBC's *Huntley-Brinkley Report,* the network's high-rated evening newscast which began Oct. 15, 1956 and ran until Huntley's retirement from NBC in July 1970. The format cut back and forth between Huntley in New York and David Brinkley in Washington, and the ritual closing—"Good Night, Chet" "Good Night, David"—entered the lore of television. In 1965 a consumer-research company found that Huntley and Brinkley were recognized by more adult Americans than were such stars as Cary Grant, James Stewart or The Beatles.

Huntley was the straight-man to the dry wit of Brinkley, and he projected sobriety and sincerity. He was selected for the newscasting post and was signed to a seven-year contract by NBC after the triumph over CBS by the Huntley-Brinkley tandem in covering the political conventions of that year. Huntley's penchant for speaking out on controversial issues, off camera, drew criticism from both the political left and right. A southern newspaper editor charged him with editorializing on-air with his eyebrows. In the 60s he raised a controversy for a radio commentary favorable to beef interests when it was learned that Huntley himself owned cattle in his native Montana. Other newsmen faulted him for becoming a spokesman in commercials after his retirement, calling it a betrayal of the news profession to lend his journalistic credence in support of an advertiser. But he was a respected and unflappable newsman in his time, and at the height of his career his salary with NBC was estimated at nearly $200,000 annually.

Huntley began his broadcasting career in 1934 at radio station KPCB in Seattle, as announcer, writer, disk jockey and salesman for $10 a month. Two years later he joined KHQ Spokane, then KGW Portland, Ore. In 1937 he was with KFI Los Angeles, and later joined CBS News in the West where he remained for 12 years. He moved to ABC in L.A. in 1951 and remained until NBC News hired him away, to New York, in 1955.

Huntley and Brinkley had what was perhaps their only public difference in 1967 when Huntley crossed AFTRA picket lines while Brinkley refused to do so. Huntley contended that newsmen did not belong in a union that represented "actors, singers and dancers."

He retired on Aug. 1, 1970 to pursue business interests in Montana, including the development of the Big Sky resort, which did not open until after his death. When he left NBC the evening newscast was retitled *The NBC Nightly News.*

HUT (Households Using Television)—a rating company's estimate of the unduplicated households tuned to television during a quarter-hour period. HUT levels vary through the day, from a low percentage of usage in the mornings to a high percentage in the peak viewing hours. The ratings for individual programs take on meaning when measured against the HUT; a high rating when the HUT is low is superb, a low rating when the HUT is high is a disaster.

The term supplanted *sets-in-use* with the proliferation of multiset households.

HYATT, DONALD B.—producer-director of public affairs programs for NBC whose most notable work was the distinguished *Project 20* series, begun in 1958. He became executive producer of special projects programs for the network in that year and added to his list of credits the *Wisdom* and *World of* series.

HYDE, ROSEL H.—long-time commissioner (1946–69) and twice chairman of the FCC, whose service to the agency as a staff member dated back to the Federal Radio Commission. In 1953 President Eisenhower elevated him to chairman for the specified period of one year; he then served several subsequent months as acting chairman until the President named George C. McConnaughey the new chairman in October 1954. The most significant action by the FCC during Hyde's term as chairman was to extend the license renewal for television period from one to three years.

Hyde, who had several reappointments to the commission, was named acting chairman by President Johnson for two months in 1966 when E. William Henry resigned and then received the full title of chairman a second time. He served as chairman for three years, until his retirement in 1969. During his second administration, the hard line began to be drawn between liberal and conservative factions on the FCC, partly because the radical reformer, Nicholas Johnson, had joined the commission.

HYLAND, WILLIAM H.—CBS sales executive of the 50s who exerted a powerful influence on TV program decisions in the 60s as senior v.p. and director of broadcasting for J. Walter Thompson Advertising. Hyland, who had joined CBS in 1937, became v.p. in charge of sales in 1953 and senior v.p. of the network in 1963, after which he left to join the ad agency.

HYPOING—the practice of scheduling stronger than usual programs and "blockbuster" movies by networks and stations during *sweep weeks,* when the rating services survey most of the markets for the three local rating reports they will receive during the year. The attempt to inflate a station's ratings during those weeks is widely recognized as an act of deliberately distorting the competitive picture in a market, and it has been condemned by both the FTC and FCC as fraud. But the age-old practice continues, because the sweep ratings are used by ad agencies in determining where to place their spot business and how much to pay for commercial time.

HYTRON (also CBS-HYTRON)—TV set manufacturing company purchased by CBS in 1951 for $17.7 million in stock—a venture which proved one of the corporation's most embarrassing mistakes. Not only had CBS misjudged the value of Hytron Radio and Electronics, the company produced an inferior set, under the brand name of Air King, that could not compete with those manufactured by the large electronic companies (RCA, Zenith, etc.). Moreover, CBS-Hytron placed its faith in the vacuum tube when other companies were switching to the newly developed transistor. CBS commissioned the noted designer Paul McCobb to give its sets a smart external appearance but to no avail. In 1959 CBS dropped the name Hytron and called the division CBS Electronics. In 1961 it dissolved the manufacturing arm after it had run up losses of about $50 million.

I

IATSE (INTERNATIONAL ALLIANCE OF THE-ATRICAL STAGE EMPLOYEES)—AFL-CIO union representing film craftsmen (cameramen, soundmen, editors, remote lighting crews) and certain other technicians at all three networks and at numerous TV stations. About 20% of the total international membership of 60,000 is involved in TV operations or production. The union also represents motion picture projectionists, makeup and wardrobe artists, set designers and screen cartoonists. Founded in 1893, it became international in 1902.

IBEW (INTERNATIONAL BROTHERHOOD OF ELECTRICAL WORKERS)—technical union formed in 1891, which in 1931 began to organize radio engineers and today represents around 12,000 technicians at CBS and at more than 170 local stations. Broadcast representation, however, makes up a small part of the diversified International, whose total membership from all industries is about 1 million. This is in contrast to NABET, the technical union at NBC and ABC, which is exclusively in broadcast.

The technicians' unions at the networks are protected from raids upon each other by being signatories to the AFL-CIO nonraiding agreement.

"I, CLAUDIUS"—highly acclaimed 13-hour BBC series, which for its scenes of violence and sex in ancient Rome, challenged the boundaries of what was acceptable on American television when it aired on PBS in 1978. The series, based on two novels by Robert Graves, *I, Claudius* and *Claudius the God,* covered the reign of the four emperors who followed Julius Caesar and preceded Nero. A political tale played against the background of the dissolute sexual affairs in the courts of the emperors, it chronicled the corruption of Roman life during the years 24 B.C. to 54 A.D.

Using Graves's literary device, the series had Claudius—the fourth of the 12 Caesars who presided over the decline and fall of Rome—narrate the story of his reign and those of his three predecessors: Augustus, Tiberius and Caligula. Adapted for television by Jack Pulman, the series skillfully wove the ingredients of domestic drama and comedy into the fabric of the historical tragedy, and it preserved the author's intended parallels to our times.

Derek Jacobi played Claudius, the presumed court idiot whose shrewd intelligence was camouflaged most of his life by a deformed body and severe stammer. Featured were Sian Phillips, John Hurt, Brian Blessed, George Baker and George Hart. Herbert Wise was the director.

The series made many public television stations uneasy for its beheadings, assassinations, gladiator games and episodes of incest, prostitution, adultery, rape, nymphomania, homosexuality, toplessness, sex orgies and sex tournaments. Nevertheless, all PBS member stations broadcast the series, under the *Masterpiece Theater* umbrella, via WGBH Boston and Mobil Corp.

ICM (INTERNATIONAL CREATIVE MANAGE-MENT)—a leading talent and packaging agency in the 70s, formed from mergers the previous decade of several agencies, principally International Famous, Creative Management

Associates and Marvin Josephson Associates. The agency represents actors, variety performers, writers, directors, producers, composers and other creative talent in their employment by movie and TV studios, networks, syndicators and other facets of show business. It also represents Washington writers and such former government officials as Henry A. Kissinger. ICM is a subsidiary of Marvin Josephson Associates.

"I DREAM OF JEANNIE"—fantasy situation comedy about an astronaut blessed with his own genie, who happens to be a luscious female. Running on NBC (1965–69), via Screen Gems, it featured Barbara Eden, Larry Hagman, Hayden Rorke and Bill Daily. The reruns proved very successful in syndication.

"IKE"—six-hour ABC miniseries on the life of General Dwight D. Eisenhowever during the World War II years when he became a military hero. It was televised over three nights, May 3, 4 and 6 in 1979 to very good ratings. Drawn partly from Kay Summersby Morgan's memoir, *Past Forgetting,* the serial focussed on Ike's close relationship with Miss Summersby while she was his personal driver in London and later an aide in his office, but it was careful not to link them romantically.

Robert Duvall portrayed Ike, Lee Remick was Kay and Bonnie Bartlett played Mamie Eisenhower. The cast included Dana Andrews as Gen. George C. Marshall, Wensley Pithey as Winston Churchill and Ian Richardson as Field Marshal Montgomery. Melville Shavelson was executive producer and writer, and he directed with Boris Sagal.

ILOTT, PAMELA—head of CBS religious broadcasts, a unit of CBS News, since 1957; she was named v.p. of religious and cultural broadcasts in September 1976. In that capacity she has been executive producer of such regular weekly series as *Lamp Unto My Feet* and *Look Up and Live* and of religious and arts specials. A native of England, she came to the U.S. in 1952 as an actress and supported herself by writing religious programs for a radio agency. She joined CBS News in 1954 as script editor for religious broadcasts, then became producer of *Lamp Unto My Feet* and eventually director of the religion department.

"I LOVE LUCY"—the supreme situation comedy, tops in the ratings for most of its years on CBS (1951–57) and popular in almost every country in the world; the reruns continue to be prized by local stations as daily fare. Apart from its significance in propelling Lucille Ball to stardom, the series was prototypical; its basic structure—the interac-

tion of two neighboring couples—served such other successful situation comedies as *The Honeymooners, The Flintstones* and *All In the Family.* See also Ball, Lucille.

ILSON, SAUL & CHAMBERS, ERNEST—comedy writing team, developed in Canadian television, who specialize in music-variety formats. Their extensive credits include producing the *Tony Orlando & Dawn* series.

In the late 70s, they split and went separate ways. Ilson joined NBC as executive in charge of variety and comedy development, and left in late 1981 to return to independent production. He produced *The Billy Crystal Comedy Hour* for NBC before signing an exclusive deal with Columbia Pictures TV. Chambers, as a freelance producer, became supervising producer for *Love, Sidney.*

"I MARRIED JOAN"—broadly played but popular situation comedy on NBC (1952–57) that featured Joan Davis as the zany wife of a judge, Jim Backus. It was by Volcano Productions.

"I'M DICKENS, HE'S FENSTER"—slapstick situation comedy about a pair of bumbling carpenters. It had a brief run on ABC in 1962 and featured John Astin and Marty Ingels. It was by Heydey Productions.

"INCREDIBLE HULK, THE"—hour-long CBS adventure drama on one of the Marvel Comics superheroes, represented in live-action. The role of The Incredible Hulk required two actors—one, Bill Bixby, to portray the normal human doctor; the other, weight-lifter Lou Ferrigno, to play the immense green monster into which he is transformed in situations of stress. *Hulk* was the third comic-book-inspired show on CBS when it premiered in March 1978, having been preceded by *The New Adventures of Wonder Woman* (salvaged by the network when ABC canceled it) and *Spiderman.* In preparing for the 1979–80 season, CBS discarded two of the shows, although they were both doing moderately well, because it did not want the identity of a comic book network. *Hulk* was the program retained, because it had developed a youthful cult following. Its network run ended in 1981–82 season.

"INCREDIBLE MACHINE, THE"—first of the National Geographic specials to play on PBS (Oct. 28, 1975) and the program attaining the highest rating in public TV history, a 24.8 rating and 36 share in New York, and shares above 25 in most major cities, even some with UHF outlets for PTV. The one-hour program, a documentary on the workings of the human body, produced, directed and written by Irwin Rosten, had been turned down by the commercial networks. When all rejected the four annual National Geographic specials, believing them to have exhausted their commercial popularity, WQED, the public TV station in Pittsburgh, secured underwriting from Gulf Oil and acquired the series for PBS. E.G. Marshall was narrator, Dennis B. Kane executive producer and Wolper Productions and the National Geographic Society the source.

INDEMNITY CLAUSES—standard provisions in program contracts under which the packagers assume responsibility for claims, liabilities and damages. They protect the networks and their affiliates in lawsuits that may occur over programs supplied by outside studios. There is also a moral turpitude clause allowing a network to void a contract if a principal performer should be involved in a scandal that may affect the value of the property.

INDEPENDENT BROADCASTING AUTHORITY (IBA)—the British regulatory agency that oversees independent advertiser-supported television and radio in the United Kingdom of England, Scotland, Wales, Northern Ireland and the Channel Islands.

Created by an Act of Parliament in 1954, it originally was called the Independent Television Authority, but underwent the change of name in 1972 with the advent of government-authorized commercial radio in the U.K.

Usually thought of as Britain's equivalent of the U.S. Federal Communications Commission, the IBA's scope and power are in fact much broader. The agency not only has the power of licensing independent stations but also ultimate control over, and responsibility for, all programming and advertising content. It also owns the transmitting towers used by the stations, which lease them for substantial annual fees.

Besides its regulatory functions, the IBA carries on research in broadcast technology and audience measurement.

INDEPENDENT NETWORK NEWS (INN)—successful national newscast for stations not affiliated with a major network, produced and distributed by WPIX New York and transmitted by the Westar II satellite. Growing out of the local newscast on WPIX, and utilizing newsfilm from Visnews for international and national coverage, INN began as a national service in June 1980. In less than two years, the lineup of 27 charter stations grew to 72 subscribers, many of which couple the half-hour national program with a local newscast for a full hour of local news in prime time. The evening edition, offered seven nights a week, is anchored by Pat Harper, Bill Jorgenson and Steve Bosch.

In October 1981 a midday broadcast was added, Mondays through Fridays, with Marvin Scott and Claire Carter. The INN service also includes a weekend news-interview program, *From the Editor's Desk*, hosted by Richard Heffner. The INN service, which raises the stature of independent stations, was conceived and developed by Leavitt J. Pope, president of WPIX, and John Corporon, vice president of news.

INDEPENDENT TELEVISION IN BRITAIN—commercial TV in the United Kingdom, otherwise known as the ITV Network, which arrived in 1956 as a rival to the monopoly of public broadcaster BBC. As with BBC, it operates 625-line PAL color.

ITV is not so much a network (at least not in the American sense) as an alliance of 15 independently owned-and-operated stations licensed by the Independent Broadcasting Authority, the regulatory agency which also has control over all program and advertising content. Station licenses are renewable every six years.

Two of the 15 stations cover metropolitan London—Thames TV, operating Monday to 7 P.M. Friday, and London Weekend TV, which operates for the balance of the week. The other stations and the regions they serve are Granada (Lancashire county in northwest England), Yorkshire, ATV (English midlands), Southern (south England coastal strip), HTV (west England and Wales), Westward (southwest England), Border (English-Scottish frontier and Isle of Man in the Irish Sea), Scottish Television (Central Scotland), Grampian (northeast Scotland), Anglia (east England), Tyne-Tees (northeast England), Ulster Television (Northern Ireland), and Channel Television (the islands in the English Channel).

As with BBC, the commercial stations are obliged to furnish classroom programming in the weekday morning time periods.

On a consortium basis, the stations hold proportional ownership of Independent Television News, the agency which provides them with networked news programming including a nightly prime-time half-hour strip called *News at Ten*.

Though independent of each other, the few stations that were operating in the first years of the commercial medium decided early on, for competitive reasons, to network certain of their programming, chiefly in the prime nighttime periods. Decisions as to which shows get network clearance are made by a committee of program executives from the five major stations—Thames, London Weekend, Granada, ATV and Yorkshire—and a representative of the IBA. A good deal of horse-trading is involved in these committee sessions.

License renewals are far from automatic in the U.K. The most dramatic instance of license revocation came in 1967, the year of the big shuffle. By the time the smoke had cleared, three licensees were canceled, both London franchises were in new hands, and some key regional coverage areas had been carved up and reassigned. Only nine stations came through the upheaval untouched by the IBA's traumatic rulings.

IBA's apparent rationale was twofold: to bring some "new blood" into the medium, and to extend the original concept of "regionalization" with its accent on programming, of whatever form, with the values of the region served by its licensee.

In practice, the degree to which those regional values pervade the schedules varies from station to station. And in the prime-time periods, with their networked comedy, variety and drama shows, adherence to the regional concept all but vanishes.

INDEPENDENT TELEVISION NEWS—the international news "department" for commercial television in Great Britain since the mid-50's. Headquartered in London with its own studios, ITN is consortium-owned by Britain's independent stations and supplies them with several networked news wrap-ups each day, as well as news specials. Its prime nightly feed is the half-hour *News at Ten* show. See *News at Ten*.

INQUIRY INTO THE ECONOMIC RELATIONSHIP BETWEEN TELEVISION BROADCASTING AND CABLE TELEVISION—two-year FCC study of its cable-TV programming rules concluded in April 1979. The commission found that increased competition from cable-TV would present no serious threat to the financial health or survival of local broadcast stations, and it initiated a rulemaking proceeding to eliminate its limits on distant-city television signals cable systems may carry and its "syndicated exclusivity" program blackout requirements, both established by the FCC's 1972 Report and Order on Cable Television.

The Economic Inquiry was the commission's first thorough assessment of the assumptions underlying its 15 years of cable-TV regulation. Broadcast interests had first sought protection from cable television in the mid-60s, when cable systems began to offer their first independent, competing service in the form of television stations brought from distant cities via microwave. Beginning in 1965, the Commission issued a series of regulations limiting the number of distant-city television signals cable systems could carry and requiring programming on those signals to be blacked out if it was under contract to local broadcast stations—regardless of whether the local station was showing the program or "warehousing" it for use sometime in the future.

The commission had based its regulations on an "intuitive model" of cable's potential impact on local broadcast stations. The increased viewing options available on cable-TV would cause a decline in local station viewing audience, the commission assumed, bringing on a serious decline in advertising revenue. Ultimately, the loss in revenue would cause a decline in local programming produced by stations, the commission believed. Beginning in 1977, the FCC set out to test each of these assumptions, seeking evidence from broadcast and cable-TV interests and commissioning its own independent studies from five of the nation's top television economists.

The commission concluded that each of the assumptions was groundless. Even in television markets heavily penetrated by cable, the studies showed, local stations had lost at most 1% of their audience; in the long run, that figure might rise to only about 10% even with tremendous cable growth. In fact, UHF stations were experiencing an average of 5.5% audience increase because of improved reception via cable. On advertising revenues, the commission found that the growth in demand for broadcast advertising continues to grow more than enough to offset minimal audience losses to cable-TV. Finally, the commission found that there is little relationship between a broadcast station's revenues and the amount of local or public affairs programming it produces.

The commission also found that its blackout rules, designed to protect the value of syndicated programs to local broadcast stations, were unnecessary and that syndicated programming carried on cable-TV has not had any impact on program suppliers' revenues from these productions.

"INSIDE NORTH VIETNAM"—documentary by Felix Greene, a British citizen residing in the U.S., which stirred a controversy when it was telecast on NET in 1968. The film, which depicted the Vietnam War from the enemy side, was denounced as Communist propaganda by a group of congressmen who had not seen it; they also charged NET with acting against the public interest for showing it.

Greene's film had actually been commissioned by CBS News when it learned that he would be going to North Vietnam for the *San Francisco Chronicle*. CBS used only a few scenes of the film for a report in its evening newscast, and the documentary was offered to NET. Greene swore that the footage had not been inspected or censored by the North Vietnamese and that none of the scenes were staged or recreated.

INSTANT ANALYSIS—the analytical reporting that usually follows a presidential address on the networks. Politicians and partisans have complained about it—particularly during the Nixon Administration (the term in fact was coined by Vice President Spiro T. Agnew in his famous speech denouncing network news delivered in Des Moines in November 1969)—because it appears to give broadcast pundits the last word over that of the President. But journalists point out that although newspaper analyses appear the next morning, they too are written immediately after the President's address.

Apparently in response to objections of the Nixon Administration, the CBS chairman William S. Paley banned instant analysis on his network in 1972 but retracted the order five months later.

INSTITUTIONAL OWNERSHIP—shares held in broadcast companies by banks, insurance companies and other financial institutions. The FCC considers an investor an "owner" when he holds more than 1% of the company's stock; he thereby is subject to the commission's multiple-ownership rules which limit broadcast holdings to seven stations each on AM, FM and TV (with no more than five on VHF). Institutions owning more than 1% of the stock in a broadcast company may not therefore exceed that benchmark with another company if the total number of stations involved exceeds the multiple-ownership limit.

INTELSAT—the International Telecommunications Satellite Organization, which manages a global system of communications satellites for 91 member countries. The Intelsat system consists of four fourth-generation satellites in synchronous orbit—two over the Atlantic, one over the Pacific and one over the Indian Ocean. There is also a spare, or backup, satellite over each ocean, used in instances of heavy traffic, such as during World Cup soccer matches. The system includes 88 earth stations (antennas that make possible the transmission and reception of signals) in 64 countries.

Comsat (Communications Satellite Corp.) in the U.S. manages the entire system under contract from Intelsat. It is also the largest owner with a 38.5% investment share. Telephone remains the primary use of the international satellites. In 1975 TV use accounted only for 3% of the volume.

Intelsat was established in August 1974. The satellites are launched from Cape Canaveral, Fla., by NASA, which is reimbursed by Intelsat.

INTERACTIVE SYSTEMS—See Two-Way Cable.

INTERCONNECTION—the hooking up, or linking, of TV stations or cable systems through microwave, cable relay or satellites so that they may simultaneously carry the same programs or exchange their services. Interconnection on a national scale is represented by the three TV networks, but in public TV and cable there are also statewide and regional

interconnections, and cable systems in different franchise areas of the same city maintain an interconnection in order to share certain programs. Interconnection is not synonymous with *network,* although as a physical capability it may imply a network.

INTERLACE—the method by which scanning is accomplished in alternate sets of lines on broadcast television. Instead of scanning each line consecutively, the television system scans alternate sets of lines. For example, in the NTSC system, 1, 3, 5, 7, etc., are scanned, down to line 525 in $\frac{1}{60}$ of a second; then the alternate lines—2, 4, 6, 8, etc.—are scanned, producing a complete picture in $\frac{1}{30}$ of a second. Each set of lines is a field. Two fields, comprising a full picture, comprise a frame. One major purpose of interlace is elimination of the flicker which would occur if all 525 lines, or 625 lines in the CCIR system, were scanned consecutively.

INTERNATIONAL CHRISTIAN TELEVISION FESTIVAL—sponsored by Protestant and Catholic broadcasting organizations such as the World Association for Christian Communication; this skip-year competitive festival is for television shows that best realize the gospel of Christian fellowship. Entries are accommodated in several categories, such as "children's, youth and family" and "documentaries on the theme of reconciliation." Originally a European affair—the very first was held in 1969 at Monte Carlo—the festival now draws participants from the U.S., Africa and the Orient as well.

INTERNATIONAL INSTITUTE OF COMMUNICATIONS—nonprofit organization based in London committed to promoting the exchange of communications ideas and concepts between broadcasters and policy-makers throughout the world. Known originally as the International Broadcast Institute, when it had its headquarters in Rome, IIC has a membership of more than 700 individuals representing 70 countries. In sponsoring research and conferences, the Institute seeks to improve the world's electronic media, to promote an understanding of new communications technologies and to facilitate free discussion of major problems common to the world's communications systems. It was founded as IBI in 1968.

"INTERNATIONAL SHOWTIME"—family series on NBC in which Don Ameche hosted performances by foreign circuses, aquacades, ice shows and other spectacles. The hour-long series premiered on Sept. 15, 1961 and ran through 1965.

INTERTEL—a step toward international communication taken in 1960 when broadcast organizations in four English-speaking countries agreed to produce a stream of documentaries in concert as the International Television Federation. The conditions were that the documentaries be presented on a bimonthly basis, be given prime time exposure and be distributed nationally in each country represented. Each group financed its own production. Although some brilliant work resulted, and talented documentarians received exposure outside their own countries, the arrangement was fraught with problems, and Intertel disbanded in late 1968.

The participants were Westinghouse Broadcasting and National Educational Television in the U.S.; Associated Rediffusion Ltd. of the U.K.; the Canadian Broadcasting Corp.; and the Australian Broadcasting Commission. Intertel's demise was brought on in part by the withdrawal of Westinghouse, which found it difficult to carry some of the foreign programs, and by Rediffusion's loss of its license when Britain revamped its broadcast assignments.

Although one of the ideals of Intertel was to promote international understanding, it developed that some countries took exception to the way they were represented by producers of other participating countries. A Canadian documentary on Castro's government, *Cuba, Si!,* was scheduled but never shown. Another CBC film, Douglas Leiterman's *One More River,* on racial problems in the U.S., was rejected by both Westinghouse and NET as sensationalized.

Britain had some problems with Michael Sklar's *Postscript to Empire,* an American view of the changes in English life and of the conservative-liberal polarity there, and neither Canada nor the U.S. was altogether comfortable with the Rediffusion entry, *Living with a Giant,* produced by Rollo Gamble, which pointed up Canada's economic, cultural and political dominance by the U.S.

There were a number of outstanding programs from each country, however, including such NET-Westinghouse offerings as Sklar's *A Question of Color* and Dan Klugherz's *Canada in Crisis* and *American Samoa: Paradise Lost?*

"IN THE NEWS"—brief CBS news broadcasts for children dispersed through the Saturday and Sunday morning entertainment lineups since September 1971. The 2½-minute pieces, presented 10 times on the weekends, strive to make the news comprehensible to school-age children. Joel Heller has been executive producer, Ken Witty the writer and Christopher Glenn the reporter.

INTV (ASSN. OF INDEPENDENT TELEVISION STATIONS)—organization formed in 1972 to represent the interests of independent (nonaffiliated) stations, particularly with respect to regulatory matters, and to promote sales for them. Its founder was Roger Rice, then v.p. and

general manager of KTVU Oakland, who since has become president of TvB. Herman Land, a broadcast consultant, became INTV's first executive director and later its president. In the mid-70s the membership consisted of 48 stations, 26 of them UHF broadcasters.

"INVADERS, THE"—adventure series about a man who knows the planet has been invaded, and what the invaders' plans are, but must work alone to foil them because no one will believe him. By QM Productions, with Roy Thinnes as star, it was carried by ABC (1967–68).

INVENTORY—the number of commercial positions in a broadcaster's schedule. Networks, more than stations, tend to have a fixed inventory, since they are permitted to sell no more than 6 minutes an hour in prime time (7 minutes in movies) and 12 minutes an hour in daytime shows. Under the Television Code local stations may sell up to 12 minutes an hour in the evenings and 18 minutes during the day, but some stations allow themselves fewer than the maximum number. Affiliated stations also receive minutes between the network shows, as well as station breaks within the shows, for their own sale.

The pricing of the inventory is largely a matter of supply and demand. Where the demand is great, owing to substantial ratings and a robust general economy, the commercial time will command premium rates. Where the demand is slight, the time will be sold below the posted rates or may even go unsold.

The Prime Time–Access Rule reduced the inventory of each network by three minutes a night and, in doing so, created a relative scarcity of prime time positions. The demand that resulted enabled the networks to raise their rates, so that they actually benefitted from the rule.

The networks expanded their inventories in 1970 when they allowed the 30-second spot to become the standard selling unit (previously, the minimum unit was 60 seconds). This, in effect, doubled the number of commercials without increasing actual commercial time.

IRAN, TELEVISION IN—two channels, both state-controlled, both operated commercially. The system is on the 625-line standard, and there are an estimated 300,000 sets in the country, largely concentrated in and around the capital city, Teheran.

IRELAND, CHARLES T., JR. (d. 1972)—president of CBS Inc., successor to Dr. Frank Stanton, for nine months. Three months after being hospitalized with a heart spasm and then returning to work, he died in his sleep. Ireland came to the company, after a long executive search by CBS, as a specialist in finance and acquisitions for ITT. His selection in October 1971 indicated the direction CBS intended to take preparatory to the eventual retirement of William S. Paley, its founder and chairman. Ireland had been senior v.p. of the ITT conglomerate and its moving force in the expansion with new companies. A consummate businessman, he was, however, an "outsider" to the show business activities of CBS—television, radio and recordings. Death came while he was still learning the ways of the company, and he was succeeded soon after by Arthur R. Taylor, a young executive previously with International Paper Corp. Taylor was fired by Paley in October 1976.

Charles T. Ireland

IRELAND, TELEVISION IN—see Radio Telefis Eireann.

"IRON HORSE, THE"—hour-long adventure series by Screen Gems about the building of the railroads in the western frontier in the 1870s. Starring Dale Robertson, it premiered on ABC in the fall of 1966 and was canceled in December 1967.

"IRONSIDE"—hit hour-long detective series on NBC (1967–74) whose principal character, a special consultant to the San Francisco Police Department, is confined to a wheelchair. Raymond Burr portrayed former detective chief Robert T. Ironside who becomes consultant, with his own crime-solving team, when an attempt on his life leaves him paralyzed from the waist down. Featured were Don Mitchell, Don Galloway and Elizabeth Baur (Barbara Anderson appeared in the first four seasons). It was by Harbour Productions and Universal TV in association with NBC-TV.

IRTS (International Radio & Television Society)—a grandiose name for New York City's version of the broadcast-advertising clubs common to many of the larger television markets. The Los Angeles equivalent is called the Hollywood Radio & Television Society and the Chicago counterpart the Chicago Broadcast Advertising Club. Although formed to bring together members of the advertising industry with executives of broadcasting, the IRTS membership also includes representatives of foreign systems and communications faculty from universities. The IRTS monthly luncheons, which traditionally begin in September with an address by the chairman of the FCC, serve as a forum for speakers on diverse broadcast subjects. One of the annual features of both IRTS and HRTS is a panel discussion with the program chiefs of the networks. Through its IRTS Foundation, IRTS also underwrites college-industry and faculty-industry seminars, as well as a number of in-service training programs for younger people in the broadcast and related communications fields.

John Jay Iselin

ISELIN, JOHN JAY—president since 1973 of WNET New York, largest of the public TV stations and the leading producer of national programs for the system. He succeeded James Day as president, after having served two years as general manager. With a background in publishing and journalism, and no previous experience in broadcasting, Iselin joined WNET in 1971 with the initial assignment of knitting together two PTV companies that had merged in 1970: the station and National Educational Television . As president, Iselin proved an effective fund-raiser, promoter and impresario, increasing the station's audience and expanding its financial base. Before coming to WNET he had been v.p. with Harper & Row and senior editor for national affairs with *Newsweek*.

Bill Cosby & Robert Culp

"I SPY"—hour-long espionage series, which featured Robert Culp and Bill Cosby as secret agents on international assignments posing as a top tennis player and his trainer companion. Notable as the first TV series to co-star a black actor, the series ran on NBC from 1965 to 1968. It was by Sheldon Leonard Enterprises.

ISRAEL, LARRY H.—broadcast executive who became president and chief operating officer of the Washington Post Co. in 1973 and resigned early in 1977, apparently after a dispute with the chairman, Katherine Graham. A businessman with a journalism background and a scholarly aspect, Israel earlier had risen to top positions with Group W and the Post-Newsweek Stations. On leaving the Washington *Post*, he proceeded to work at acquiring stations in forming a new broadcast group.

During the 50s, he had owned two struggling stations, WENS-TV Pittsburgh and KMGM-TV Minneapolis. After selling them, he joined Westinghouse Broadcasting (Group W) as v.p. and general manager of WJZ-TV Baltimore and moved rapidly upward. After organizing and becoming president of the Westinghouse station rep firm, TVAR, he next was named executive v.p. of the station group and, in 1964, president and chief operating officer of Group W. That post, however, left him in the shadow of Donald H. McGannon, chairman and the company's moving force.

In 1968, Israel accepted the presidency of the Post-Newsweek group. There he became an industry statesman in his own right, taking courageous positions on controversial broadcast issues (he dropped cigarette advertising about a year before it became outlawed because it reflected on the

credibility of the stations); fending off harassments by Nixon loyalists who tried to punish the *Washington Post* by challenging its station licenses; and rising to noble causes, such as supporting CBS in the controversy over *The Selling of the Pentagon*. He also expanded the group with the acquisition of stations—WTIC-TV Hartford, Conn.; WPLG-TV Miami and WCKY-AM Cincinnati. Meanwhile, he donated WTOP-FM (now WHUR) in Washington to Howard University.

On advancing to president of the Washington Post Co., he named a Group W protege, Joel Chaseman, to succeed him as president of the broadcast group.

"ISSUES AND ANSWERS"—ABC's Sunday afternoon news-interview series, that network's equivalent of NBC's *Meet the Press* and CBS's *Face the Nation;* like those programs, it frequently is the source of front-page news the following morning. The format usually involves a single newsmaker before a panel of ABC News correspondents, televised live from Washington, but on occasion—as when there is exclusive access to a foreign government official—the interviews have been one-on-one and on tape or film. Peggy Whedon is producer, and Bob Clark has been chief correspondent and permanent panelist since 1975.

ITALY, TELEVISION IN—a two-channel system, state owned, and operated by RAI, Radiotelevision Italiana. Though managed by a state holding company, RAI receives only a small direct subsidy from the government. The system is financed by combined revenues from commercials ($130 million a year) and license fees ($325 million). The owner of a black and white set pays $31 yearly, a color set owner $62. There are approximately 12.1 million black and white and 600,000 color sets in use.

After the parliamentary compromise with elements of the Leftist opposition in 1976, the ruling Christian Democratic party lost part of its 30-year hold on RAI-TV. The national program, or First Network, operating on VHF since 1954, remains under the management of the Christian Democrats. Programs usually avoid socially provocative issues which might offend its conservative Catholic viewers. The network attracts a total audience of about 18 million. The newer second program, reflecting the interests of the Communist and Socialist parties, has an audience of about 8 million.

Since 1971, an estimated 100 private commercial television stations have started operating without formal state approval, 30 of them in Rome alone. Operating on a rather shaky financial structure, most independent stations offer a standard fare of old movies and game shows. Independents in Milan and Turin have gained international notoriety by televising striptease acts by housewives in the early morning hours.

In 1979, RAI-TV produced three major programs in cooperation with American interests. The most ambitious, a dramatization of the life of Marco Polo, was scripted by England's Anthony Burgess and filmed in China, with financial backing from Procter & Gamble.

ITC (INDEPENDENT TELEVISION CORP.)—U.S.

film distribution arm of Associated Television Ltd. of Britain, which mainly syndicates the parent company's programs to local stations here. It was formed in 1958 by ATV, the Jack Wrather Organization and other firms as a general international distributor of programs, but later was bought out by ATV and became its American subsidiary. See also Grade, Lew.

ITFS (INSTRUCTIONAL TELEVISION FIXED SER-VICE)—omnidirectional microwave system utilizing channels in the 2500–2690 megahertz frequency range, which was established by the FCC in 1963 expressly for use by educational institutions. An ITFS station at a school district headquarters may transmit up to four programs simultaneously to its schools. "Fixed service" denotes reception by special receivers at the intended institutions. The Catholic Television Network, established in several major cities, broadcasts on ITFS to its parishes and schools.

In 1976, more than 150 of these point-to-point stations were in use. Operating on frequencies above the UHF band, with their power generally limited to 10 watts, the ITFS stations in most cases carry less than five miles. The successful utilization of ITFS made it possible for ETV to become public television, since the dependency on it for instructional programming was lessened.

ITNA (INDEPENDENT TELEVISION NEWS ASSN.)

—a cooperative formed by 10 major independent stations in the fall of 1975 to provide them with daily coverage of national and international news when the syndication company, TVN, collapsed. Using the Westar satellite to distribute a daily half-hour package of items and newsfilm to its members each evening, ITNA involves a sharing of news by the stations and their joint subscription to UPITN, which services Britain and European countries with world news and some coverage of Washington. Each ITNA member serves as a news bureau for the others, with WTTG the principal supplier as the independent station in the capital. The 10 charter stations include WPIX and WNEW New York; WGN-TV Chicago; KTLA, KHJ-TV and KTTV Los Angeles; KTVU Oakland-San Francisco; KPLR-TV St. Louis; and KTXL Sacramento, Calif.

Reese Schonfeld, ITNA's first managing director, resigned in 1979 to join Ted Turner in creating a proposed all-news national cable-TV channel. Charles R. Novitz, formerly of ABC News syndication, succeeded him. Novitz left in 1981.

"IT'S ACADEMIC"—weekly half-hour high school quiz program which has run continuously on WRC-TV Washington since 1961 and over the years spawned local versions at other NBC o&os. The Washington edition has been hosted by Mac McGarry from the start.

"IT TAKES A THIEF"—hour-long series taking tongue-in-cheek approach to counter-espionage, starring Robert Wagner as a master thief paroled to ply his trade solely for a government intelligence agency. Fred Astaire appeared in several episodes as his father, the greatest thief of them all. Produced by Universal TV, it played on ABC (1968–70).

"I'VE GOT A SECRET"—highly successful Goodson-Todman panel show which was a prime time staple for CBS (1952–64), with Garry Moore as moderator and a panel consisting primarily of Henry Morgan, Betsy Palmer, Bill Cullen and Jayne Meadows (although its composition changed somewhat over the years). The format involved the quizzing of guests for purposes of ascertaining a "secret" that had already been divulged to the viewers. It was revived for first-run syndication in 1972 with Steve Allen as moderator.

JACKER, CORRINE—playwright whose work in TV has included the American adaptation of Ingmar Bergman's *The Lie* and episodes for *Visions, The Adams Chronicles* and the CTW drama series on PBS. She also served as head writer and editor of *Bicentennial Minutes* on CBS.

JACKSON, KEITH—ABC sports commentator, known especially for announcing the NCAA football series and the Monday night baseball games. His other credits include the *Pro Bowlers Tour, ABC's Championship Auto Racing,* major league baseball and events for *ABC's Wide World of Sports.* Before joining ABC Jackson had announced sports events for the University of Washington for eight years, the football games of Washington State for four years and the games for several AFL teams in the west. He had also worked 10 years at KOMO-TV, the ABC affiliate in Seattle.

JAFFE, HENRY—producer who, through long association with Dinah Shore, has become primarily known as her guiding genius in television. He was in charge of the *Dinah Shore Chevy Show* in the late 50s, the *Dinah's Place* NBC daytime series in the 70s and *Dinah!,* the syndicated talk show that began in 1975.

JAGODA, BARRY—special assistant and chief television advisor to President Carter for the first two years of the administration. Previously, he had been an independent producer and for several years a producer in the CBS News special events unit.

JAMES, DENNIS—veteran emcee of daytime game shows who came into prominence in TV in the early 50s as commentator of the wrestling matches on the DuMont Network, adding sound effects and generally contributing to the comedy that was part of the grappling exhibitions in those times. He went on to host a flock of game shows over more than two decades, including *What's My Line, Chance of a Lifetime, High Finance, The Name's the Same, People Will Talk, Haggis Baggis* and *Judge For Yourself.* He also emceed a daytime variety show, *Club 60,* and early in his career was announcer for *Ted Mack's Original Amateur Hour.*

"JAMES AT 15"—NBC series written by novelist Dan Wakefield, as his first TV effort, about the growing-up process of a sensitive teenager, played by Lance Kerwin. The series entered the schedule in October 1977 to replace an NBC failure but was itself a marginal performer. After Wakefield resigned in a clash with NBC, the program's days were numbered, and it ran out its skein in the spring of 1978.

The dispute occurred after Paul Klein, then NBC program chief, offered an idea to perk up the ratings: James would turn 16 and lose his virginity on his birthday. Wakefield proceeded to write the script, but he quit the show when NBC Standards & Practices would not accept dialogue between the boy and girl on the matter of contraception, even when the euphemism "responsibility" was used for birth control. When the episode aired in February 1978, after Wakefield had left, the title of the series was changed to *James at 16*. Presumably, if it had continued to run, the title would have changed annually.

Featured in the cast were Linden Chiles, Lynn Carlin, David Hubbard and Susan Myers.

"JANE GOODALL AND THE WORLD OF ANIMAL BEHAVIOR"—umbrella title for an ABC series of nature documentaries featuring Miss Goodall, who came to national attention—through a National Geographic article and TV special—as a scientist who lived among the apes. Her first effort in the ABC series, narrated by Hal Holbrook, was *The Wild Dogs of Africa* (January 1973) and her second was *Baboons of Gombe* (February 1974). The two were produced by Marshall Flaum for Metromedia Producers Corp.

"JANE WYMAN THEATER"—half-hour dramatic anthology series on ABC (1955–58) with Jane Wyman as hostess and star. Some of the playlets were rerun on CBS in what was entitled *Jane Wyman's Summer Playhouse*. The series was via Lewman Productions.

Gene Jankowski

JANKOWSKI, GENE F.—president of the CBS Broadcast Group since October 1977, succeeding John F. Schneider, under whom he had previously served as executive v.p. Soon afterward, he was named a v.p. of CBS Inc. and a director of the corporation. His appointment as president was accompanied by the most extensive management reorganization in the company's history, with the creation of several new divisions and the appointment of five new divisional presidents, all reporting to Jankowski.

Jankowski rose in the company through sales, having started in 1961 as an account executive with the radio network, rising there to eastern sales manager. In 1969 he joined the CBS-TV sales staff, then became general sales manager of WCBS-TV, New York. He was named v.p. of sales for the owned stations division in 1973 and the following year became v.p. of finance and planning for the division. That proved the springboard to CBS Inc. Two years later he was elevated to v.p. and controller for the corporation and a year after that v.p. of administration.

Jankowski attracted notice in the company in the early 70s when, in the wake of the ban on cigarette advertising, he conducted a successful effort to bring new advertisers into television. Affable and businesslike, he maintained a low profile as president of the broadcast group and was not an autocratic leader, deeming management a team effort. Thus while the cult of personality flourished at NBC with Fred Silverman and at ABC with Frederick Pierce, Roone Arledge and Tony Thomopoulos, it was muted at CBS. Under Jankowski's stewardship, however, CBS regained first place in the prime-time ratings and made deft moves into the fields of cable and teletext.

JAPANESE BROADCASTING CORP.—otherwise known by the call letters NHK (for Nippon Hose Kyokai), it is Japan's behemoth public broadcaster sometimes thought of as the BBC of the Orient because of the sweep and range of its programming. Funded by license fees only (there are some 28 million sets in Japan, nearly half of them color sets), it is enormously rich and extravagant in programming its two TV and three radio channels. As the only nationwide service in Japan, it also operates 62 satellite stations around the country, each with its own production capability. The NHK Symphony is the oldest concert orchestra in the country. Japanese television operates on NTSC 525-line standard.

JARRIEL, TOM—White House correspondent for ABC News since 1969. He joined ABC's Atlanta bureau in 1965 from KPRC Houston and gained national recognition for his coverage of the civil rights movement in the South. In 1975 he was assigned to anchor the Sunday edition of the *ABC Weekend News* and later added the Saturday edition. He also, on occasion, substituted for Steve Bell on the news segments for *Good Morning America* and became one of the regular correspondents on *20/20*.

JARVIS, LUCY—documentary producer for NBC News (1960-76), whose notable achievements include filming the Kremlin, China's Forbidden City, Scotland Yard and The Louvre. Her work has spanned the political and cultural spheres, and her videography has an international flavor. She produced, among other TV specials and films, the following:

The Kremlin (1963), The Louvre (1964), Who Shall Live? (1965), Mary Martin: Hello Dolly! Around the World (1965), Dr. Barnard's Heart Transplant Operations (1968), Khrushchev in Exile: His Opinions and Revelations (1967), Bravo Picasso! (1967) and Trip to Nowhere (1970).

In 1976 she left NBC to produce independently, her first assignment being to produce prime time specials with Barbara Walters for ABC. That relationship ended after the first program. In the 1981–82 season she produced Family Reunion, a four-hour miniseries starring Bette Davis.

Sherman Hemsley & Isabel Sanford

"JEFFERSONS, THE"—Norman Lear situation comedy on CBS (1975–) centering on a middle-class black family in a luxury apartment building; its characters, in certain respects, parallel those of All In the Family, from which the series was spun off. Slotted immediately following Family and preceding Mary Tyler Moore, the series established itself at once and in 1976 took over the Saturday evening leadoff spot long held by the Bunkers. The Jeffersons broke new ground in TV by introducing a biracial couple as neighbors of the principals.

Featured were Sherman Helmsley, Isabel Sanford, Roxie Roker, Franklin Cover, Zara Cully and Berlinda Tolbert. It was by Lear's T.A.T. Communications with NRW Productions.

JENCKS, RICHARD W.—high-ranking executive of CBS during the early 70s who, for a time, was a leading candidate to succeed Frank Stanton as president of the corporation. But his star fell suddenly, and two years after attaining the post of president of the CBS Broadcast Group he was

assigned in 1972 to Washington as corporate v.p. He asked for early retirement in 1976.

Erudite, reserved and well-spoken, Jencks rose rapidly after rejoining CBS in 1965 as deputy general counsel. He shortly became general counsel, then executive v.p. of the network in 1968 and then head of the broadcast group. He had started with CBS as a lawyer in 1950, after two years with the NAB, then moved to Hollywood as West Coast general counsel and left in 1959 to become president of the Alliance of TV Film Producers, serving in that capacity six years.

JENKINS, CHARLES FRANCIS (d. 1934)—inventor and entrepreneur who in 1928 inaugurated the first regularly scheduled television broadcasts. Already known as the inventor of the basic motion picture theater projector, he regularly broadcast "radiomovies" from Washington, D.C., using a spinning disc which provided 1-inch-square pictures with 48 lines of resolution to eager hobbyists who modified their radio sets to receive the silhouette images. As early as 1923 he had received recognition for transmitting a picture of President Harding by wireless from Washington to Philadelphia. The Jenkins broadcasts continued until 1932. Jenkins Television Corp., which manufactured receivers, was taken over by De Forest Radio Co., which later was declared bankrupt. In 1929 Jenkins forecast: "The folks in California and Maine, and all the way in between, will be able to see the inauguration ceremonies of their President in Washington, the Army and Navy games on Franklin Field, and the struggle for supremacy in our national sport, baseball."

Barbara Parkins, Christopher Cazenove & Lee Remick as Jennie

"JENNIE: LADY RANDOLPH CHURCHILL"—seven-part miniseries on the life of Jennie Jerome, the wealthy

American who married Lord Randolph Churchill and became mother of Winston Churchill. Produced by Britain's Thames TV, and starring Lee Remick, it drew high ratings in its U.S. run on PBS in 1975. More than a biography, the serial marked the social changes that occurred from the late 19th to the early 20th centuries. The hour episodes were written by Julian Mitchell and directed by James Cellan Jones.

JENNINGS, PETER—ABC correspondent who twice had been an anchor of the network's evening newscast: first, as the solo anchorman when he joined ABC News in 1964, a stint that lasted about four years, and again in 1978 when ABC adopted a three-anchor format for *World News Tonight*. In the latter assignment, he headed the foreign desk from a base in London.

In the 10-year interval between his anchoring duties, he worked chiefly as a foreign correspondent in the Middle East, Europe and southern Africa. In 1974, he was Washington cohost of ABC's *A.M. America*. He came to ABC from the Canadian Television Network.

"JESUS OF NAZARETH"—6-hour and 37-minute film on the life of Jesus, directed by Franco Zeffirelli and coproduced by Britain's ATV and Italy's RAI-TV; it premiered in the U.S. in 1977 as a two-parter on NBC on Palm Sunday (April 3) and Easter (April 10), dominating prime time. The film, reported to have cost $18 million to produce, drew a 50 share both nights in its debut and had an estimated cumulative audience of 90 million viewers.

The U.S. rights had been purchased by General Motors for $3 million two years before the telecast, and G.M. had an additional investment of $1.5 million for air time on NBC when it decided to withdraw its sponsorship. The automotive company was reacting to a national campaign by a number of evangelical religious groups to block the telecast because they objected to the idea that Jesus was being presented as an ordinary human being in the film biography. Led by Bob Jones, III, president of Bob Jones University, the protestors—who had not screened the film—had formed their opinion from a statement Zeffirelli had made in a press interview. In literature that was widely circulated around the country to fundamentalist groups, those who opposed the film called for a boycott of General Motors products.

Procter and Gamble then purchased sponsorship of the premiere telecast at bargain rates, although G.M. retained the rights to repeat showings. By the time it went on the air, the film, after special screenings by NBC, had the endorsement and praise of religious leaders of the major faiths.

In the film, Robert Powell portrayed Jesus; Olivia Hussey, the Virgin Mary; Yorgo Voyagis, Joseph; Peter Ustinov, Herod; Isabel Mestres, Salome; Michael York, John the

Baptist; and James Farentino, Simon Peter. The international cast also included Claudia Cardinale, Anne Bancroft, Anthony Quinn, James Earl Jones, Donald Pleasence, Laurence Olivier, James Mason, Christopher Plummer, Stacy Keach, Rod Steiger, Ernest Borgnine, Ian Holm and Fernando Rey.

The program was repeated in April 1979 on NBC and scored an average rating of 21.2 with a 33 share.

Bernard J. Kinham was executive producer, Vincenzo Labella producer, and Anthony Burgess, Suso Cecchi d'Amico, and Zeffirelli the writers.

"JETSONS, THE"—animated situation comedy about a family in the 21st century whose concerns and spoken idiom were of the 1960s. Premiering on ABC in 1962 as an offshoot of Hanna-Barbera's *The Flintstones,* it enjoyed a successful run, was purchased by CBS in reruns and in 1971 was scheduled by NBC on Saturday mornings. The voices were by George O'Hanlon, Penny Singleton, Daws Butler and Janet Waldo.

JEWISON, NORMAN—producer-director who was active in New York network originations in the late 50s and early 60s specializing in music-comedy specials such as *Tonight with Belafonte* (1959) and *An Hour with Danny Kaye* (1960). He gained his initial reputation in the medium in his native Canada as a director for the CBC. Like many another accomplished TV director, he quit the medium for motion pictures. His film credits include *Fiddler on the Roof* and *Rollerball.*

JICTAR—acronym for Joint Industry Committee for Television Research, the British equivalent of America's A.C. Nielsen audience measurement service. Supported by major U.K. advertisers, their ad agencies and the independent stations, Jictar meter-samples some 2,600 British television homes for its weekly chart of top 20 shows as well as providing subscribers with demographic and related viewer data.

"JIMMY STEWART SHOW, THE"—half-hour situation comedy presented on NBC (1971–72) which failed to catch on despite the long movie stardom of James Stewart. In the series he portrayed a college professor faced with the problems of generation gap both on campus and at home. Featured were John McGiver, Jonathan Daly and Julie Adams. It was by Warner Bros. in association with AJK Abildon Productions.

"JIM NABORS HOUR"—hour-long comedy-variety show starring Jim Nabors and Frank Sutton, with Ronnie Schell and Karen Morrow. Produced by Naborly Productions for CBS (1969–71).

"JOEY BISHOP SHOW, THE"—situation comedy vehicle for Joey Bishop which underwent several changes of format, cast and network—all to no avail. Nevertheless, it ran from 1961 to 1965. Featured were Corbett Monica and Abby Dalton. Produced by Danny Thomas's Bellmar Productions, the series started on NBC and two seasons later moved to CBS.

"JOHN GUNTHER'S HIGH ROAD"—adventure-travel films scheduled by ABC in 1959 and produced by Blue J Productions.

"JOHNNY CASH SHOW, THE"—country-music variety series on ABC, produced in Nashville by Screen Gems. It had a successful run as a summer replacement in 1969 and was brought back as a midseason entry the following January. Although successful at first, it faltered the next season and was canceled.

JOHNSON, NICHOLAS—Enemy No. 1 to the broadcast industry for the more than seven years he served as an FCC commissioner (1966–73). He was a noisy reformer who campaigned for virtually everything the industry feared: counter-commercials, license challenges by citizens at renewal time, the break-up of media monopolies, an informed and activist FCC and access to the airwaves for minorities, political dissenters and representatives of the counter-culture.

He remained on the commission several months beyond his term of appointment because his successor had not yet been named; then he swung into action to oppose (unsuccessfully) the confirmation of the nominee, James Quello, because he was a former broadcaster.

On leaving the commission, Johnson—who had developed a following among youth—entered politics but was defeated in a bid for the Democratic nomination for congressman in his home district in Iowa. He then continued his work in broadcast reform as chairman of the National Citizens Committee for Broadcasting in Washington, publisher of the magazine, *Access,* and media commentator on the NPR program, *All Things Considered.* When Ralph Nader took over NCCB as chairman in 1978, Johnson became head of a new, related organization, the National Citizens Communications Lobby.

Johnson was the most unorthodox and flamboyant of commissioners, boldly taking his dissents to the press, writing magazine articles critical of the FCC and of the industry and making speeches to citizens groups, stirring their participation in license renewals and advising them of their rights to challenge broadcast licenses. He even wrote a book as a commissioner, *How to Talk Back to Your Television Set,* which detailed the public's rights in broadcasting. Unfazed by the industry's attacks upon him as a censor and a dictator, or by the anger he aroused in FCC chairman, Dean Burch, with his brashness and his ridiculing of the commission, Johnson pursued his causes in public, taking advantage of the print media's receptivity to any denunciation of broadcasting by a public official.

Proclaiming himself the public's advocate on the FCC, he accepted the appelation of "radical," and grew a mustache and wore his hair long as a message that he was not of the establishment. He was faulted, even by admirers, for being given to hyperbole, as when he called television "a child molester" in building a case for reforms in children's programming.

Although he was thought of as a dissenter, he voted with the majority on such critical issues as the denial of the WHDH license to the Boston *Herald-Traveler* and the adoption of the prime time-access rule.

Johnson was only 32 when he was appointed to the FCC by President Johnson. But by then he had already served as administrator of the U.S. Maritime Administration (1964–66), practiced law with the Washington firm of Covington and Burling, taught law for three years at the University of California at Berkeley and served as law clerk to the late Supreme Court Justice Hugo L. Black.

See also Broadcast Reform Movement, National Citizens Committee for Broadcasting, and National Citizens Communications Lobby.

JONES, ANNE P.—FCC commissioner since April 1979, appointed by President Carter to the female seat previously held by Margita White. She had been general counsel for the Federal Home Loan Bank Board and earlier, for 10 years, on the staff of the Security Exchange Commission. A Democrat from Boston, she had been recommended to the White House as a possible nominee by FCC chairman Ferris, with whom she had been a classmate at Boston College Law School. Despite that connection, she won confirmation easily on the judgment that she was a person of independent spirit who would not automatically cast her vote with the chairman. She became the third Boston Law School alumnus on the present commission, along with Ferris and Joseph Fogarty.

JONES, CHARLOTTE SCHIFF—v.p. of CBS Cable and former producer of the short-lived CBS series, *People*, based on *People* magazine. Before joining CBS Cable as one of its original staff when it was organized in 1980, she had been an executive of Time Inc., first with Manhattan Cable, then as assistant publisher of *People* at the time of the network series. Before joining Time, she was director of community programming for Teleprompter.

JONES, CHUCK—animation producer-director who for a time in the late 60s served as head of children's programs for ABC-TV. Most of his career had been spent with the cartoon division of Warner Bros. (1938–62), where he created the Roadrunner and Pepe LePew cartoon series and helped to develop Porky Pig and Daffy Duck. His association with ABC began while he was co-producer, writer and director of *The Bugs Bunny Show* on that network. He was producer of *Cricket on Times Square* and *A Very Merry Cricket* and was writer, director and producer of the *Jungle Books* animated specials.

JONES, MERLE S. (d. 1976)—long-time CBS executive who for 14 months was president of the network (1957–58) and then became president of the owned-stations division for 10 years until his retirement in 1968. His career with CBS spanned 32 years, beginning in 1936 as assistant to the g.m. of KMOX, the CBS-owned radio station in St. Louis. Except for three years with Cowles Broadcasting in Washington, D.C. (1944–47), his entire subsequent service was with CBS. He was g.m. of WCCO Minneapolis-St. Paul (1947), when it was owned by CBS; then headed KNX Los Angeles and the Columbia Pacific Network (1949); and then the TV counterpart, KTSL (now KNXT), before becoming v.p. in charge of the CBS-owned TV stations. In 1956 he was named executive v.p. of the network, and a year later became president. He was a director of CBS from 1957 to 1968.

JONESTOWN MASSACRE—ambush at the airport outside Jonestown, Guyana, on Nov. 18, 1978, which set in motion the events leading to the incredible mass suicides of the cult followers of the Rev. Jim Jones. Killed in the ambush, along with California Congressman Leo Ryan, who was investigating the cult, were two members of NBC News, Don Harris and Bob Brown. Also a victim was Greg Robinson, a photographer for the San Francisco Examiner.

Harris, a reporter based on the West Coast, had previously reported on Southeast Asia for NBC. Brown, a newsreel cameraman, had come on staff six months before the fatal episode, having previously worked as a free lance for CBS and other organizations. Brown filmed the ambush even as he was being shot, and that footage received wide exposure on network newscasts and special reports on Guyana. Harris's posthumous scoop was the final interview with the Rev. Jones, preserved on film. Bob Flick, an NBC News field producer who escaped the massacre, provided the eyewitness account of it for reporters.

JORDAN, GLENN S.—director of numerous dramatic specials for public television at NET and WNET in New York, who crossed over into commercial TV in the 70s to produce and direct the four Benjamin Franklin specials for CBS.

JOSEPHSON, MARVIN—major figure in the field of talent representation and program packaging; in a period of 14 years, through a series of mergers, he parlayed a small personal management business into the country's second largest talent agency, Marvin Josephson Associates (William Morris is the largest). The firm represents hundreds of performers, writers, producers, directors and production companies for the standard 10% commission and has been responsible for packaging scores of TV series. In 1973 Josephson negotiated the sale of the TV rights for the 1976 Montreal Olympics to ABC for a record sum of $25 million. In 1977, he negotiated the literary and TV contracts, said to total more than $5 million, for Henry A. Kissinger.

In 1955, after working briefly for CBS as an attorney, Josephson started a personal management office, Broadcast Management Inc., with Bob Keeshan (*Captain Kangaroo*) as his first major client. Six years later the rapidly growing company merged with a West Coast firm, Roenberg, Coryell Inc., and changed its name to Artists Agency Corp. In 1967 Josephson bought out his partners and renamed the agency Marvin Josephson Associates. Two years later he acquired the large Ashley-Famous Agency that earlier had been formed in a merger of the old International Famous Agency and the Ted Ashley organization. Ashley-Famous had been part of the Kinney conglomerate when Josephson acquired it. MJA then became the parent of two divisions: IFA, representing talent and packagers, and Bob Keeshan Associates, owner and producer of *Captain Kangaroo*.

"JUDD FOR THE DEFENSE"—dramatic series of a larger-than-life lawyer, represented as the most successful in the world, featuring Carl Betz, with Stephen Young as his junior partner. It was produced by 20th Century-Fox TV in association with Vanadas Productions for ABC (1967–69). The series was carried in 13 international markets.

"JUDGMENT" SERIES—series of docu-drama specials for ABC (1974–75) by Stanley Kramer in association with

David L. Wolper. The programs dramatized critical trials of modern history and consisted of *Judgment: The Court-Martial of the Tiger of Malaya—General Yamashita* (June 11, 1974); *Judgment: The Trial of Julius and Ethel Rosenberg* (Jan. 28, 1974); and *Judgment: The Trial of Lt. William Calley* (Jan. 12, 1975). All were 90-minute specials offered under the rubric of *ABC Theatre.*

JUDSON, ARTHUR (d. 1975)—prominent concert manager from the 20s through the 60s who was a founder of the small radio network in 1926 that was to become CBS. To create opportunities for the artists he managed, Judson purchased New York radio station WABC for $75,000 at a time when radio was a new and expanding field, and he organized a network of 16 stations by paying each $500 a week to carry his programs. With financing from the Columbia Phonograph Company, the network began operations on Sept. 19, 1926 as the Columbia Phonograph Broadcasting System. When William S. Paley acquired controlling interest in the failing company in 1928, "Phonograph" was dropped from the name. Judson remained the second largest stockholder in CBS.

He later founded and headed Columbia Concerts Corporation and its subsidiary, Columbia Artists Management, and in 1938 became sole owner of Columbia Records, which was eventually to become a division of CBS.

Diahann Carroll as Julia

"JULIA"—NBC situation comedy (1968–71) built around a black star, Diahann Carroll. Its success opened the way to other shows with black principals. Miss Carroll was not TV's first black star, but she was the first to carry a show on the same terms as a white star with good ratings and ample advertiser support.

Amos 'n' Andy, a syndicated TV series drawn from the radio program, had been an audience favorite during the 50s with an all-black cast, but it was driven off the air by black citizens groups as patronizing and embarrassing to the race. In 1965 Bill Cosby partially broke the color barrier with billing as co-star of the NBC action-adventure hour, *I Spy,* but in fact he was the second lead to Robert Culp, and the billing was really cosmetic. In 1957 Nat King Cole, like every other top recording artist, was given a show of his own on network TV, but that was before the civil rights movement; no sponsor would touch it, and some southern stations declined to carry it. *Julia,* in 1968, was the breakthrough program.

At that, it was widely criticized as unrealistic and unrepresentative of black life in America. Miss Carroll portrayed a beautiful widow raising a young son, she worked as a nurse and lived in an integrated housing project. On TV, however, her living quarters were idealized to the level of floor-wax commercials, and it was felt that the character could have been white but for the fact that a light-skinned black had been assigned to play it.

In the integrated supporting cast were Marc Copage as her son and Lloyd Nolan, Betty Beaird, Michael Link and Lurene Tuttle. It was produced by 20th Century-Fox TV.

The pilot for *Julia* had actually been rejected by NBC at first sight. But a half-hour slot was open on the network opposite *The Red Skelton Show,* a long-time hit on CBS, and several other situation comedies had been under consideration. NBC programmers gave themselves no chance of beating Skelton and chose *Julia* to salvage something from the loss—the appearance of having tried to do a program with a black lead. To their, and the industry's, surprise, *Julia*— with all its faults—turned out to be a hit.

"JULIE ANDREWS SHOW, THE"—musical-variety series on ABC (1972) lavishly produced by the British firm, ATV/ITC. It was canceled in its first season.

"JUNE ALLYSON SHOW, THE"—half-hour dramatic anthology on CBS (1959–61) with movie actress June Allyson as hostess and occasional star. Originally titled *DuPont Show with June Allyson,* it was by Four Star and Pamric Productions.

JUSTICE DEPARTMENT ANTITRUST SUIT—identical lawsuits filed April 14, 1972, by the Department of Justice against ABC, CBS and NBC charging them with illegally monopolizing prime time and with restraint of trade. The suits sought to bar the networks from securing financial interests in programs and from producing shows of their own.

In 1976, NBC agreed to a settlement, one whose terms appeared to have no serious economic implications for the network. Many of the restrictions, however, were conditional on the other networks reaching a similar settlement. But ABC and CBS declined to settle in the manner of NBC, and both continued to fight the suits, calling the charges unfounded.

The department's case, which had been prepared in 1970 but was withheld for two years, contended that the networks use their control over prime time to keep off the air programs in which they have no financial interests, that they thereby force producers to grant them part ownership of the shows and that they control the prices paid for programs and movies.

The networks struck back with the contention that the suit was politically motivated, brought against them by the Nixon Administration in retaliation for the news coverage of the government which President Nixon had felt was biased against him.

They cited as suspicious the fact that the suit might have been filed earlier but instead had been laid aside by previous Attorneys General and that it was entered even after a similar civil suit by Hollywood studios that was still pending. Moreover, they pointed out, the research that had been the basis for the suit was applicable to the year 1967 but not necessarily to 1972 since, in the interval, the FCC had already taken steps to drive the networks out of syndication and the ownership of programs for postnetwork sale.

In November 1974, motions to dismiss the Justice Department's lawsuits were entered in U.S. District Court in Los Angeles, after the networks were denied their petition for access to President Nixon's tapes, which they maintained would have substantiated their claim that the suit was politically motivated.

U.S. District Court Judge Robert J. Kelleher had dismissed the cases against the networks because the Ford Administration would not provide the tapes, since the question of whether they were owned by the government or by Nixon was still under debate. But the judge's dismissal was "without prejudice," which meant the Justice Department could file the suits again, under a new administration—and it did.

The networks then appealed to the Supreme Court to bar the department from prosecuting its case. In April 1975, the Court dismissed the appeal, permitting the case to proceed.

Under NBC's settlement agreement, the network would be limited in program ownership to 2½ hours' worth of entertainment programs a week in prime time; to no more than 11 hours' worth in fringe time and no more than 8 hours' worth in daytime (8 A.M. to 6 P.M.). The network's production of news programming would be exempt.

The agreement also limited to one year any contract between the network and a supplier that a program be produced at NBC's facilities. It limited the number of years over which a network could obtain exclusive rights to a program before it was developed. Further, the network was prohibited from retaining exclusive options on more than 35% of the shows presented to it that had not yet been selected for broadcast.

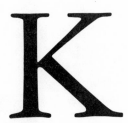

K

KAHN, IRVING B.—dominant figure in cable-TV and its leading visionary as president of Teleprompter Corp. during the 60s, until he was convicted of bribery and perjury in connection with an attempt to secure a cable franchise in New Jersey. In 1975, after serving a 30-month prison sentence, he formed his own company, Broadband Communications Inc., to secure films and other programs for pay-cable. He also began to work at developing fiber optics technology for cable-TV, and meanwhile built a large cable system in Camden, N.J. and surrounding communities. In 1980 he sold these franchises to the *New York Times* for $100 million, a record price, based on subscribers served. The system involved 55 franchised areas in Southeren New Jersey and pitched the *New York Times* into the cable business.

Under Kahn's leadership, Teleprompter grew to the largest and most powerful of the MSOs. Not long after his conviction and his total separation from the company, Teleprompter experienced a financial crisis when the expected cable boom in the cities did not materialize. Kahn became president and chairman of Teleprompter in 1951, after 16 years in show business advertising and publicity, chiefly with 20th Century-Fox.

KALB, BERNARD—widely traveled CBS News correspondent, specializing in foreign affairs who left in 1980 to join NBC News, following his old Boss, Bill Small there. Kalb joined CBS in 1962, after 15 years as foreign correspondent for *The New York Times,* as chief of the Southeast Asia and India bureau. Except for a year as Paris correspondent,

his base was Hong Kong until 1970, when he was reassigned to the Washington bureau. For nearly two years (1970–72), he was Washington anchor for the *CBS Morning News with John Hart.* With his brother, Marvin Kalb, CBS diplomatic correspondent, he wrote the political biography, *Kissinger.*

KALB, MARVIN—NBC News diplomatic correspondent, hired away from CBS News in 1980 after a distinguished 23-year career at the network. The Kalb brothers, Bernard and Marvin, defected to join Bill Small when he became president of NBC News, and they represented Small's prize catch. They were, however, resented at NBC News, which has proud traditions of its own. Small's ouster from NBC in 1982 was attributed in part to the morale problems resulting from what was perceived as his attempt to remake NBC News with old CBS colleagues.

Marvin Kalb's transfer to NBC was controversial also for another reason: reportedly, his new contract contained a guaranteed quota of air time. This raised the issue, in media circles, of star power superseding news judgment in television.

Kalb became diplomatic correspondent for CBS News in 1963, after extensive experience in foreign affairs dating back to the State Department post he had held at the American Embassy in Moscow. He joined CBS as a journalist in 1957. Fluent in Russian, Kalb is a specialist in Soviet-American relations and is the author of five books.

KANE, DENNIS—director with varied entertainment credits (Joyce Brothers's *Living Easy, Dark Shadows,*

Thanksgiving Day parades). He also directed network political specials for Republican candidates—Eisenhower, Goldwater, Nixon and others.

KANE, JOSH—assistant to the president of NBC Entertainment since 1981; he previously had been v.p. of East Coast programs for NBC since November 1977; He joined the program department in 1976, after having spent the previous 10 years in the network's press and public information department. He began at NBC as a page in 1965.

KANTER, HAL—comedy writer and producer who, after writing radio programs and screenplays for Bob Hope, made his mark in TV as producer-writer-director of *The George Gobel Show* in the 50s. He later became executive producer of *Julia* (1968–71) and in 1976 was named supervising producer of an established hit series, *Chico and the Man.* Kanter's credits also include *Valentine's Day, The Jimmy Stewart Show* and episodes of *All In the Family.*

KARAYN, JIM—news executive, chiefly with public TV. He was president of NPACT (1971–75) and before that Washington bureau chief and executive producer for NET. Earlier he had been with NBC News, KTLA and the *Los Angeles Examiner.* When NPACT merged with WETA Washington, Karayn went off on his own and became project director of the League of Women Voters' *Presidential Forums* in 1976, televised on PBS, and spearhead of a League-sponsored drive for the televised debates between the candidates for President and Vice-President. In 1977 he became general manager of WHYY, the Philadelphia public TV station licensed to Wilmington, Del.

KARP, RUSSELL H.—president and chief operating officer of Teleprompter Corp. since May 1974. An attorney, he had been v.p., general counsel and treasurer of Screen Gems and v.p. for corporate affairs of Columbia Pictures Industries. During the two years before he joined Teleprompter, Karp had his own financial and management consultancy in New York, with 20th Century-Fox and Children's Television Workshop among his clients.

KASS, JEROME—writer of dramatic specials and made-for-TV movies whose credits include the original teleplay for *Queen of the Stardust Ballroom.*

KATZ, OSCAR—a former program chief for CBS who returned in a lesser program capacity in 1973, after a period

in Hollywood as a studio executive and agent. With a background in broadcast research, dating to 1938, Katz changed the direction of his career in 1956 when he became v.p. in charge of daytime programs for CBS. His success there earned him the appointment, in 1959, as head of the network's program department. He left in 1963 to join Desilu Productions and three years later became an agent for General Artists Corp. He rejoined CBS in 1972 as a chief aide to his one-time protege in daytime television, Fred Silverman, who had become the network's program boss.

Danny Kaye

KAYE, DANNY—noted movie comedian who made TV appearances sparingly until 1963 when he began a weekly variety series on CBS that ran four seasons. The series did well enough in the ratings scheduled at 10 P.M. on Wednesdays, but many believed it would have been a smash at an earlier hour since Kaye was enormously appealing to children.

One of his memorable TV appearances was in a 1957 *See It Now* program, entitled *The Secret Life of Danny Kaye,* on which he appeared with children of various countries to demonstrate the work being done by UNICEF.

KEESHAN, BOB—creator and star of *Captain Kangaroo,* daily children's program on CBS which has entered its third decade. After working as a page for NBC, he became an assistant to Bob Smith on the *Howdy Doody* show and soon created the popular clown character, Clarabelle, which he played for five years. In the early 50s, he developed characters for his own TV series, *Time For Fun* and *Tinker's Workshop,* but on Oct. 3, 1955 he launched *Captain Kangaroo* and was on his way to becoming an institution.

In 1959, as an adjunct to the program, he conducted symphony orchestras in major cities for a series of classical concerts for children. In 1965, he added a Saturday morning program, *Mister Mayor,* which two years later gave way to a sixth day of *Kangaroo.* The series is frequently cited for its good taste and promulgation of positive social values.

See also *Captain Kangaroo.*

KENNEY, PETER B.—Washington v.p. and chief lobbyist for the National Broadcasting Co. since 1962, senior among the networks' government relations executives. Previously he had represented NBC International in Buenos Aires and had been station manager of the company's Washington station, WRC-TV. He became an NBC executive in 1956 when NBC bought UHF station WKNB-TV, and its radio counterparts, in New Britain, Conn. Kenney had been executive v.p. and general manager of those stations for the previous owner and stayed on as general manager with NBC's takeover. When NBC sold the stations three years later, Kenney remained with the company, taking a post with NBC International.

KENWITH, HERBERT—director with varied prime-time credits, including two seasons of *Here's Lucy, Good Times* and *One Day at a Time.* He also directed episodes of *Name of the Game, Marcus Welby, Star Trek, Daktari* and others, and had numerous stage credits.

KESTEN, PAUL W. (d. 1956)—a key management figure at CBS as head of promotion and public relations, until ill health forced his retirement in 1946. Apart from having the distinction of hiring Frank Stanton, who later became president of the corporation, and Dr. Peter Goldmark, who developed a color-TV system and the LP recording, Kesten was the father of the CBS "image," having determined that CBS could compete with NBC on a basis other than circulation. In the late 30s, years before commercial TV began, Kesten became the CBS officer in charge of developing color-TV, a project for which he had brought in Goldmark.

KEYES, PAUL W.—comedy writer-producer who wrote for Jack Paar in his late-night NBC show and later became producer and head-writer of *Laugh-In.* He was producer of The American Film Institute's *Salute to James Cagney* and the *Salute to John Wayne.*

KIBBEE, ROLAND—veteran producer-director who began as a radio writer for Fred Allen, Groucho Marx and Fanny Brice. He was executive producer of *Columbo,* co-exec producer of *The Family Holvak* and *McCoy,* and producer of *Barney Miller* for 1976–77 season. He was also producer-

writer for *It Takes a Thief* and *The Bob Newhart Show.* He returned to *Barney Miller* in the 1981–82 season as co-executive producer.

"KIDS ARE PEOPLE TOO"—ABC entertainment/information program for children which began in 1978 on Sunday mornings. Bob McAllister, former host of *Wonderama,* starred in the first 13 programs and Michael Young in each program thereafter. A variety format designed to introduce young people to interesting personalities from the worlds of art and entertainment, the program features certain network-produced segments but allows time for local stations to incorporate their own programming material. Produced by Lawrence Einhorn, the program introduced a 3½ minute segment of advice for young people titled *Alex and Annie,* which deals with youth-oriented problems.

KIHN, ALBERT (d. 1974)—news cameraman who, while in the employ of KRON-TV, San Francisco, filed a petition to deny against that station and KRON-FM, charging undue media concentration by the parent Chronicle Publishing Co. He was joined in the petition by an employee of the *San Francisco Chronicle,* Blanche Streeter. They charged that the company ordered newsmen to avoid some stories and to cover others to advance the Chronicle Company's private interests. The FCC resolved the various issues in the stations' favor and renewed the licenses in May 1973. Kihn and Mrs. Streeter pursued the case through court appeals, but before the case could be heard Kihn was killed in a plane crash in March 1974 with other members of a film crew while on a documentary assignment for Wolper Productions. His widow and Mrs. Streeter abandoned the appeal the following year, upon receiving a settlement of $150,000 from the *Chronicle* to cover their legal expenses.

KIKER, DOUGLAS—NBC News correspondent mainly assigned to Washington after having been White House correspondent for the *New York Herald Tribune* and director of information for the Peace Corps. He joined NBC News in 1966, went to Rome for two years in 1969 and, after his return, became prominent on the network in covering the Watergate developments. He was named Washington correspondent for the *Today* show in 1975.

A sometime novelist, he was the author of *The Southerner* and *Strangers on the Shore,* both published in 1959.

KILLIAN, JAMES R. (DR.)—called "the father of public television," because he had headed the Carnegie Commission on Educational Television, whose study and report led to the Public Television Act of 1967. An MIT professor, he served on the board of the Corporation for Public Broadcasting from its inception and was its chairman from May 1973 until he resigned in December 1974. He was succeeded by Robert Benjamin and was given the title of chairman emeritus.

James R. Killian

KINESCOPE RECORDING—a record of a television program made by filming it from a television monitor. Before the development of video tape, it was the means of preserving live television programs and news coverage. Although programs were sometimes repeated or syndicated by kinescope recording (or "kinnie"), the poor picture quality discouraged extensive use. The word *kinescope* is actually a synonym for "picture tube," but it became the accepted abbreviation for *kinescope recording*.

"KING"—a six-hour NBC docu-drama serial airing on three consecutive nights in February 1978 to very disappointing ratings. The series, tracing the life of Martin Luther King Jr. from 1953 to his murder in a Memphis hotel in 1968, was actually a portrait of the civil rights movement. Written and directed by Abby Mann, it starred Paul Winfield as King and Cicely Tyson as his wife, Coretta. Others in the cast were Howard Rollins, Kenneth McMillan, Dick Anthony Williams, Ossie Davis and Al Freeman Jr.

"KING FAMILY, THE"—ABC music-variety series (1964–66) which grew out of a guest appearance on *Hollywood Palace* of the six King Sisters and their husbands and

children—36 in all. They later produced a series of specials for syndication, each keyed to a holiday theme, including back-to-school—12 in all.

KINOY, ERNEST—prolific New York-based writer of plays for TV, radio, films and the Broadway stage who contributed to such acclaimed series as *Studio One, Playhouse 90, Naked City* and *DuPont Show of the Week.* In 1965 he won an Emmy for an episode of *The Defenders,* entitled "Blacklist," about a victim of the anticommunist campaigns of the McCarthy era. In the 50s he wrote a TV series, *The Marriage,* which starred Hume Cronyn and Jessica Tandy. His credits in the 70s include scripts for *The Story of Jacob & Joseph* and *Roots.*

Kinoy served for a time as president of the Writers Guild of America, East.

Robert E. Kintner

KINTNER, ROBERT E. (d. 1980)—forceful president of the National Broadcasting Co. (1958–65), noted for giving high priority to the network's news operations. During his administration, NBC achieved leadership in news and special events coverage. Earlier (1949–56), he had been president of ABC-TV.

Kintner was given the title of chairman and chief executive officer of NBC in 1966 but, owing to illness, never served in that capacity and went into retirement shortly afterward.

His shibboleth for NBC News was "CBS Plus 30," mandating 30 minutes more coverage of an event than CBS would provide. Kintner's emphasis on network news traced to his early career in the 30s as Washington correspondent for the *New York Herald Tribune* and later as writer of a nationally syndicated column with Joseph Alsop. With Alsop, he also wrote two best-selling books, *Men Around the President* and *Washington White Paper.*

In 1944, following military service, he joined ABC as v.p. over programming, public relations and advertising. Two years later he was promoted to executive v.p. and in 1949 to president. He resigned in 1956 and the following year joined NBC as executive v.p. responsible primarily for coordinating all color TV activities. In February 1958 he was put in charge of the television network and five months later became corporate president.

KIRSHNER, DON—rock music publisher, record producer and promoter who branched into TV in the early 70s and was responsible for *In Concert,* the ABC monthly 90-minute rockfest that began in 1973; the syndicated *Don Kirshner's New Rock Concert;* the annual *Rock Awards* telecasts; and the Saturday morning children's series, *The Kids From C.A.P.E.R.* Kirshner's first major work in TV was as music supervisor for the 1967 NBC series, *The Monkees,* a situation comedy about rock musicians that was designed to introduce a new number every week. Kirshner was head of Colgems Records at the time and reaped three number one singles from the series.

Kitchen Debate: Krushchev & Nixon

"KITCHEN DEBATE"—a spontaneous exchange between Soviet Premier Nikita Khrushchev and Vice-President Richard M. Nixon that took place at an exhibition hall in Moscow in July 1959, while RCA color cameras and Ampex video tapes were being demonstrated. The two men had dared each other to record a discussion that would be televised in their respective countries without editing.

As they walked through the American exhibits, trading taunts and criticism of each other's political ideology, with

the cameras following them, they stopped to argue before a display of American kitchen appliances. It was at this point that Nixon gesticulated in a manner in which he appeared to be poking a finger at Khrushchev's chest. The moment, preserved also in still photographs, suggested that he was standing up to the Soviet leader.

The kitchen exhibit gave the Cold War videotape event its name, and it served Nixon's future presidential campaigns.

"KKK: THE INVISIBLE EMPIRE"—noted CBS documentary broadcast Sept. 21, 1965, probing the history and workings of the Ku Klux Klan, the first TV film report to do so. It was produced by David Lowe under the *CBS Reports* banner, with Charles Kuralt as reporter.

KLAUBER, EDWARD (d. 1954)—executive v.p. of CBS during the 30s, considered by some to be "the father of broadcast journalism" because of his pioneering initiatives with radio network news. With Paul White as his chief aide, Klauber in 1933 organized the Columbia News Service (the forerunner of CBS News) when General Mills offered to sponsor daily news broadcasts. Klauber spurred the growth of the news organization, and by World War II it was one of the largest and most distinguished news services in the world.

Klauber joined CBS in 1930 as top assistant to the network's young president, William S. Paley, after having been night city editor of *The New York Times.* Intrigued with the possibilities of disseminating news by radio, he convinced Paley of its value to the network and then hired White to help build the staff. Klauber also served as the corporate force to insulate news judgments from advertiser interests.

When he was passed over for the position of wartime manager of CBS (Paley having gone to London to work for General Eisenhower), Klauber resigned from the network and joined the Office of War Information.

KLEIN, HERBERT G.—former director of communications for the Nixon Administration who, on resigning in July 1973, became v.p. of corporate relations for Metromedia Inc., based in Los Angeles. Klein—a longtime friend of Nixon, having helped him in every campaign since his first race for Congress in 1946—had been editor of the *San Diego Union* when he was named to the newly created post in the executive branch in 1968.

The job involved, among other things, receiving complaints from the news media and in turn forwarding the complaints of the Administration to the media. Mr. Klein's resignation after five years was prompted partly by Ron Ziegler's preemption of his role as confidant to the President.

KLEIN, PAUL L.—NBC audience research expert (1965–70), who returned to the network in 1976 as v.p. of programming. In the interval he founded and was chairman of Computer Television Inc., a company which pioneered *hotelvision* (movies for pay over hotel's master antenna systems). He also produced *Pop-Up,* educational spots for children carried by NBC and was periodically a consultant to PBS, the Ford Foundation, Lincoln Center and Janus Films. He sold his interest in CTI to Time Inc. in January 1976.

Klein came up short of hits as NBC programmer, and his gamble that a schedule of specials, movies and miniseries would prevail over stock weekly series was a chief reason for NBC's slide into third place. In two seasons, his only clear success among weekly series was *CHiPs*. Although his failures abounded, Fred Silverman kept him on as program chief when he came in as president of NBC, but Klein resigned in 1979 to become an independent producer.

Klein became widely known among media buffs for his "LOP Theory," which holds that most viewers do not turn on the set for a specific program but instead select what Klein termed "the Least Objectionable Program" after making a decision to watch television. The theory was first enunciated in an article for *New York Magazine.*

Isaac Kleinerman

KLEINERMAN, ISAAC—prolific producer-director of documentaries for CBS News (1957–76) and earlier for NBC (1951–57); he resigned from CBS in the summer of 1976 over differences with management and became an independent

producer. Kleinerman joined NBC from RKO-Pathe and became film editor of *Victory at Sea* and a producer in the *Project 20* unit *(The Great War, The Twisted Cross, The Jazz Age, Nightmare in Red).* He also produced episodes for the *Wisdom* series.

CBS hired him away in 1957 to help organize *The Twentieth Century* unit; he was to produce close to 200 films for that series, as well as others for *CBS Reports* and *The 21st Century.* Overall, in a 25-year career with the two networks, Kleinerman was responsible for more than 400 documentaries covering a broad spectrum from history and current events to the humanities and sports.

As an independent he produced *The Unknown War,* a syndicated series created from footage of World War II shot by Russians, which was distributed by Air Time in 1978–79.

KLUGE, JOHN W.—chairman and president of Metromedia Inc., diversified communications company that is parent to Metromedia Broadcasting, a leading group of largely independent stations. In 1959 he bought a controlling interest in the Metropolitan Broadcasting Corp., the remains of the recently expired DuMont Network, which essentially consisted of its stations. Kluge changed the company's name to Metromedia as it came to embrace outdoor advertising, publishing and other media ventures.

KOCH, HOWARD—producer of movies and series; his TV credits include *The Untouchables, Maverick, Cheyenne* and *Hawaiian Eye.*

Telly Savalas as Kojak

"KOJAK"—highly popular detective series, both in the U.S. and abroad, which began on CBS in the fall of 1973,

Although the series made extensive use of location footage, it was the Savalas persona—the stocky, totally bald, unglamorous rendition of hero—that gave the series its realistic edge. Savalas achieved true stardom with the series and was the source of its great appeal around the world.

Featured were Dan Frazer, Kevin Dobson and Demosthenes. Matthew Rapf was executive producer for Universal TV.

"KOLCHAK—THE NIGHT STALKER"—hour-long series on ABC (1974) about a fearless newspaper reporter who seeks out the bizarre and unworldly, played by Darren McGavin. It was by Universal TV.

KOMACK, JAMES—a leading producer of situation comedies in the 70s whose independent production company, The Komack Co., turned out *The Courtship of Eddie's Father, Chico & The Man, Welcome Back, Kotter,* and *Mister T & Tina.* A former nightclub comedian, he became an actor in situation comedies before turning to producing.

KOOP, THEODORE F.—long-time Washington v.p. for CBS (1961–72) responsible for government relations. While holding the post, he also served through the administrations of Kennedy, Johnson and Nixon as standby censor in the event of a national emergency; in that capacity, he was to have interdicted the dissemination of information that might aid the enemy. While the position was classified, for security reasons, the secret got out in 1962 during the Cuban missile crisis; Koop remained on standby, nevertheless. He was tapped by President Kennedy because he had served as deputy to Byron Price, director of censorship during World War II.

Koop had been a journalist before his corporate assignment. From 1948 to 1961 he was director of news and public affairs in the CBS Washington bureau. Before the war he had been with AP and on the editorial staff of the National Geographic Society. He retired from CBS in 1972.

KOPPEL, TED—Washington-based ABC News correspondent who spent some 17 years with the network before bursting into national prominence in November 1979 as anchor of the latenight series, *The Iran Crisis: America Held Hostage.* In March 1980, the crisis special became a regular nightly news program, *Nightline,* and Koppel was Johnny Carson's full-fledged competitor at 11:30 P.M. With *Nightline,* Koppel gained the reputation as one of television journalism's best interviewers; he also earned a co-anchor role in the network's coverage of such events as the inauguration of President Reagan, the assassination attempts on Reagan and Pope John Paul II, and the election coverage in 1980.

Before *Nightline,* Koppel had been ABC's chief diplomatic correspondent since 1971. Before that, he was Hong Kong bureau chief, Miami bureau chief covering Latin America, and a Vietnam correspondent. Born in England, but a graduate of Syracuse and Stanford, Koppel joined ABC News as a general assignment reporter in 1963 and went on to become anchor of a nightly newscast for ABC Radio, before receiving assignment in Vietnam. With Marvin Kalb of CBS News (now with NBC News), he collaborated on the novel, *In the Public Interest.*

Ted Koppel

"KOPYKATS, THE"—hour-long comedy-variety series on ABC (summer, 1972) on which noted impressionists enacted great stars. Produced by Gary Smith and Dwight Hemion for ITC, the summer series grew out of a special that seemed to have possibilities for an ongoing series, but it was never adopted for the regular season. Taking part were Rich Little, Frank Gorshin, George Kirby, Marilyn Michaels, Charlie Callas, Joe Baker and Fred Travalena. In January 1976 Little began a series of his own on NBC, but it was short-lived.

KORD INC. RENEWAL [21 R.R. 781 (1961)]—instance in which the FCC moved to hold a formal hearing on the renewal of a station's license because of a wide discrepancy between its programming proposals and its actual performance. The commission called off the hearing and granted the renewal when the licensee argued that similar discrepancies had previously been acceptable to the FCC in granting other renewals.

The station, KORD in Pasco, Wash., had in its original license proposal promised to broadcast 84% entertainment,

6% news and agricultural, educational and religious programs and local live talk shows in the remaining time. It also said it would carry 700 commercial announcements a week. When the license came up for renewal, however, the commission found the performance to have been 85.5% entertainment, 11.3% news, 0.6% religious and virtually nothing else. Moreover, KORD carried an average of 1,631 commercial announcements a week.

KORD maintained nevertheless that its operation was in substantial conformity with the original proposals. In granting renewal, the FCC asserted that a station must make a "good faith effort" to deliver on its proposals.

KORTY, JOHN—independent San Francisco area film producer and director (*Crazy Quilt, Funnyman, Riverrun*) who turned to TV as director of made-for-TV movies, including *The People, Go Ask Alice, The Autobiography of Miss Jane Pittman* and *Farewell to Manzanar.*

KOTLOWITZ, ROBERT—v.p. of programming for WNET New York, since April 1973, having previously been editorial director responsible for all national programming from the PTV station in the cultural and public affairs spheres. Before joining public TV, he had been managing editor of *Harper's* magazine (1967–71) after having served as senior editor for three years. Earlier he was senior editor of *Show* magazine. Kotlowitz is also a novelist (*Somewhere Else* and *The Boardwalk*).

Ernie Kovacs

KOVACS, ERNIE (d. 1962)—innovative comedian whose fanciful and often irreverent sight comedy made creative use of video effects more than a decade before the techniques became voguish with *Laugh-In.* His career was ended before

reaching full bloom by his death in an auto crash in Hollywood. He had been married to Edie Adams, a rising comedienne.

His first programs, on NBC, *It's Time For Ernie* (March 1951) and *Ernie in Kovacsland* (summer, 1951), emanated from Philadelphia. Though short-lived, *The Ernie Kovacs Show,* produced in New York for NBC (December 1955 to July 1956), established his reputation for zany originality and won him wide admiration, if not the general popularity required by television. His trademark became a cigar and mustache under a pair of horn-rimmed glasses.

PBS mounted a retrospective of his programs in the spring of 1977.

KOWALSKI, BERNARD L.—producer-director best known for his action-adventure work; he was producer of *Rawhide* in the mid-60s and executive producer of *Baretta* in the 1975–76 season.

KRAMER, ALBERT H.—one of the leaders of the broadcast reform movement who founded the Citizens Communication Center in 1969, a leading public-interest law firm for media reform activities. He helped to reestablish the National Citizens Committee for Broadcasting as a force in the movement, serving as its president and chief counsel (1974–75) and helped to organize the National Citizens Communications Lobby (1975). He was attorney for numerous public welfare cases in broadcasting that were brought before the FCC and into the courts. Early in 1976 he joined the prestigious Washington law firm of Arnold & Porter and in 1977 joined the FTC as chief of its consumer protection bureau.

KRIVIN, ALBERT—senior v.p. of Metromedia Inc. since 1974, in charge of all TV activities including MPC, the company's production and syndication subsidiary. For the previous eight years, he was president of Metromedia Television, credited with raising the stature and profitability of the group's independent stations.

During the 50s he was with Westinghouse Broadcasting, working in Philadelphia and Cleveland as a sales executive. In 1961 he joined Metromedia as v.p. and general manager of KMBC-TV Kansas City. He became president of the group in 1966. In tribute to Krivin, Metromedia chairman John Kluge gave his newest station acquisition, in Houston, the call letters of KRIV-TV.

KROFFT, SID AND MARTY—brother team of puppeteers who head a successful TV production company responsible for such children's shows as *H.R.Pufnstuf, The*

Bugaloos, Lidsville, Lost Saucer, Far Out Space Nuts and *The Krofft Supershow.* The Kroffts also produced *Donnie and Marie* and *Barbara Mandrell and the Mandrell Sisters* in prime time.

Born into a European family of puppeteers, the Kroffts made their mark at the New York and Seattle World's Fairs with their adult puppet show, "Les Poupees de Paris." In 1969, when the networks were responding to pressure for more wholesome Saturday morning children's shows, the Kroffts created *Pufnstuf,* an NBC fantasy series combining live actors and life-size puppets. Its success led to the other series. The Kroffts later created the first indoor, high-rise amusement park, "The World of Sid and Marty Krofft," in Atlanta's Omni International magnastructure; it opened in 1976.

KTBC—Austin, Tex., station owned by Mrs. Lyndon B. Johnson. Its history came to public light and caused eyebrows to raise when Mr. Johnson became President. The station, which was the source of the family fortune, appeared to have had unusual advantages, perhaps because of Johnson's power in the Senate, first as Democratic Whip and then as Majority Leader. KTBC-TV was one of the first 18 licenses to be awarded in 1952 after the FCC's four-year freeze, and it was the only VHF assigned to the market, giving it a virtual monopoly there against only UHF competition.

The Johnsons had acquired radio station KTBC in Mrs. Johnson's name in 1943 for $17,500. The station was a daytime-only operation with no network affiliation, therefore seriously handicapped. Soon, however, the FCC granted Mrs. Johnson's application for a power increase from 250 watts to 1,000 watts and for unlimited broadcasting hours. The station then acquired a CBS affiliation and became hugely profitable.

Mrs. Johnson gave up control of the stations in 1963, putting the stocks in trust while she was in the White House. She resumed control in 1969, after the Johnson presidency.

KTTV AGREEMENT—controversial pact negotiated in 1972 by Metromedia's KTTV Los Angeles and the National Assn. for Better Broadcasting, a citizens group, under which the station agreed to ban from the air a list of violent children's programs drawn up by NABB. The agreement was reached to call off a petition to deny by NABB, which argued that KTTV did not deserve to hold the license because, with its concentration on violence, it did not properly serve the child audience.

Among the 42 series KTTV agreed not to play were *Batman, Superman* and *Spiderman.* The station also said it would comply with NABB demands that it post warnings to parents for a list of 81 other shows—many of which it did not own—if they were played before 8:30 P.M.

In December 1975, the FCC shot down the agreement saying that "deals" could not be made between a citizens group and station, because under law a station could not delegate its program responsibilities. The commission was also troubled that the agreement sought to bind not only the parties involved but also future licensees and that the pact would generally have the effect of program censorship.

Kukla, Fran & Ollie with Burr Tillstrom

"KUKLA, FRAN AND OLLIE"—an artful puppet series, not strictly directed at children, that was highly popular in the early 50s for its endearing characters, topicality and gentle humor, sometimes verging on satire. A television classic of the "Chicago School," and uniquely suited to the medium, it has had sporadic revivals and during the 70s was used by CBS as the *wraparound* in its Children's Film Festival.

"KFO," as it was known in the trade, featured a live performer, Fran Allison, in the company of puppets created by Burr Tillstrom, and dealt with their foibles and concern for each other. Featured were the gnomish Kukla and the vain dragon, Ollie. Others in the Kuklapolitan Players included Beulah Witch, Delores Dragon, Madame Ogelpuss, Fletcher Rabbit and Cecil Bill. From time to time, they performed musicals, such as *St. George and the Dragon* and *The Mikado.*

Beginning as a local show on Chicago's WBKB on Oct. 13, 1947, it was so well received that it was carried on a midwestern network and then went to NBC, premiering there on Jan. 12, 1949. The show ended its NBC run on June 13, 1954, and the following September resumed on ABC, where it remained until Aug. 30, 1957. It was produced by Beulah Zachary until her death in a plane crash in 1959. An important contributor was the musical director and pianist, Joe Facianato.

Tillstrom had created Kukla in 1936 while he was performing in Chicago with the WPA Parks District Theatre. Miss Allison had been a Chicago radio actress when she teamed with the Kuklapolitans for the TV show. Tillstrom also perfected a hand ballet, which he performed occasionally on TV after the series was canceled.

KULIK, BUZZ—much-in-demand director of made-for-TV movie specials, whose recent credits include *Brian's Song* and *Cage Without a Key.*

Kung Fu: David Carradine

"KUNG FU"—hour-long adventure series on ABC (1973–75) about a Buddhist priest trained in the martial art of Kung Fu traveling the American frontier in the 1870s. David Carradine played the lead, and Keye Luke and Philip Ahn were featured. It was by Warner Bros. TV.

"KUP'S SHOW"—long-running syndicated program of topical discussion emanating from Chicago where its host, Irv Kupcinet, is a popular columnist for the *Sun-Times.* The weekly program assembled a guest group of celebrities, authors, political figures and academics for interviews and cross-talk on a variety of subjects. Originally entitled *At Random,* it began as an open-ended show that started Saturday at midnight and lasted until the conversation ran out;

later it was given normal time restrictions and has been syndicated in one-hour form.

As *At Random,* it premiered February 1958 on WBBM-TV as a local show. In the early 60s it switched to WMAQ-TV. In 1975 PBS began carrying the series through its Station Program Cooperative, and in 1976 the program switched its base to WTTW, the Chicago PTV station. Kupcinet has hosted the program since its inception, with Paul Frumkin as producer.

Charles Kuralt

KURALT, CHARLES—CBS News correspondent since the 50s, best known for his *On the Road* offbeat features which started in 1967. In 1979, he doubled as anchor of the new 90-minute newscast, *Sunday Morning,* which won such praise that Kuralt was later made ancohor of the revamped weekday CBS series, *Morning.* That program faltered, however, and Kuralt was relieved of the daily chore in March 1982. He was reassigned as special correspondent on the *Evening News* but retained the Sunday morning program.

Before beginning *On the Road,* he had been a political reporter, foreign correspondent and reporter on such documentaries as *Mayor Daley: A Study in Power, You and the Commercial, Misunderstanding China* and *Destination: North Pole* (the latter resulting in a 1968 book by Kuralt, *To the Top of the World*). He joined CBS as a news writer in 1956. Formerly he had been with the *Charlotte* (N.C.) *News.* In 1977, he became one of the regular correspondents for the CBS prime time series, *Who's Who.* See also *On the Road With Charles Kuralt.*

L

LACHMAN, MORT—head writer and producer for Bob Hope specials for more than 20 years. He was also executive producer of *One Day At a Time* (1975–76), *All In the Family* (1976–77) and other comedy series, the most recent being *Gimme A Break*.

Perry Lafferty

LAFFERTY, PERRY—veteran producer and director who became CBS v.p. of programs in Hollywood (1965–76) and then returned to production until NBC recruited him in the summer of 1979 as its v.p. of programs in Hollywood. Lafferty thus was reunited with Fred Silverman, new president of NBC, under whom he had worked for part of his tenure at CBS. Silverman chose Lafferty to strengthen the network's ties with the creative community on the Coast and to gain from Lafferty's extensive experience with TV programs.

Before taking the executive position he had been producer of *The Danny Kaye Show* for its first two seasons and earlier of such programs as *The Andy Williams Show, Your Hit Parade* and several Arthur Godfrey specials. He also directed dramas for *Studio One, The U.S. Steel Hour, Twilight Zone* and *Robert Montgomery-Presents* and episodes of *Rawhide*. Lafferty wrote and directed radio programs while studying music at Yale and began working in TV after graduation in 1947.

LA LANNE, JACK—host of a leading syndicated physical fitness series from the late 50s through the 70s. Assisted by his wife, Elaine, LaLanne demonstrated systematic exercises, or "trimnastics," for an early morning audience of housewives. The series was produced by Filmline Productions in association with the American Physical Fitness Institute. The program led to the creation of a national chain of Jack LaLanne health clubs.

LANDAU, ELY—syndication executive in the 50s who later became a film producer (*The Pawnbroker, The Fool Killer*). In 1953 he organized National Telefilm Associates , which for several years was a leading and innovative syndication concern. Landau, as chairman of NTA, created in 1956 the NTA Film Network, an attempt at bridging syndication and networking. He also was the moving force behind *Play of* *the Week,* a high-quality videotape series of two-hour stage dramas distributed by NTA and produced at the station the company owned, WNTA-TV New York (now WNET, the public TV station). Landau sold his interest in WNTA in 1961 to produce theatrical movies.

James C. Lehrer

LAIRD, JACK—producer of *Ben Casey* and *Rod Serling's Night Gallery,* and later supervising producer of *Kojak.* He was producer of *Doctor's Hospital* (1975–76) and *Switch* (1976–77).

"LAMP UNTO MY FEET"—long-running and prestigious CBS religion series which was canceled in January 1979 along with two other Sunday morning programs to make way for the news series, *Sunday Morning. Lamp* was doused because the vast majority of CBS affiliates had stopped carrying it, most of them choosing instead to play paid religious programs. CBS hoped to regain its clearances with the news program.

During more than a quarter-century on the air, *Lamp* aimed to illumine the common aspects of all religions. Employing theological discussions, programs on the arts and dramatic and documentary forms, the programs have featured representatives of the Syrian Orthodox, Quaker, Hindu and Buddhist religions, as well as the major faiths. Pamela Ilott was the producer. See also *Camera Three, Look Up and Live* and *Sunday Morning.*

"LANCER"—western on CBS (1968–69), produced by 20th Century-Fox TV and featuring James Stacy, Wayne Maunder, Andrew Duggan and Elizabeth Bauer.

"LAND OF THE GIANTS"—adventure series by Irwin Allen, master of special effects. It concerned the passengers and crew of a commercial rocket ship that had crashed on a planet inhabited by a race of giants. Produced by Allen in association with 20th Century-Fox TV, it was carried on ABC from 1968 to 1970 and featured Gary Conway and Kurt Kasznar.

LANDON, MICHAEL—star of *Little House on the Prairie* and also executive producer of that series, for which he wrote and directed a number of episodes each season. He also directed *Bonanza* episodes and the dramatic special, *It's Good to Be Alive.*

LANDSBURG, ALAN W.—head of Alan Landsburg Productions, specializing in documentaries. He was producer-director for the National Geographic specials, *Undersea World of Jacques Cousteau,* and *In Search of Ancient Astronauts,* among others. He has been executive producer of the *That's Incredible* and *Those Amazing Animals* in recent years, and his company produces *Gimme A Break.*

LANI BIRD—first Pacific satellite launched by Intelsat (Jan. 11, 1967) establishing satellite communication between the U.S. mainland and Hawaii and making live network TV transmissions possible there for the first time. Its formal name was Intelsat II.

LAR DALY DECISION—a 1959 FCC decision on a political equal time claim that caused Congress to amend the Communication Act exempting legitimate news coverage from the Sec. 315 requirements.

As a minor candidate (America First Party) for Mayor of Chicago that year, Lar Daly claimed equal time on television to the routine newscast coverage of several ceremonial acts performed by the incumbent, Mayor Richard J. Daley. The FCC, by a 4–3 vote, held that Daly was entitled to the time, a decision widely interpreted as meaning that elected officials could not receive legitimate news coverage during a campaign period without creating free time situations for all their rival candidates.

Congress moved then to amend the law, making newscasts, news interviews, news documentaries and spot coverage of news events exempt from equal time considerations. See also Equal Time Law.

"LARAMIE"—western on NBC (1959–62) with John Smith, Robert Fuller, Spring Byington and Hoagy Carmichael. It was produced by Revue Productions.

"LAREDO"—post-Civil War Texas Rangers series featuring Neville Brand, William Smith, Peter Brown and Philip Carey. Produced by Universal TV, the series aired on NBC (1965–66).

LA ROSA, JULIUS—pop singer of the 50s who earned a paragraph in TV history when he was fired on the air by Arthur Godfrey as a regular member of the weekly series, *Arthur Godfrey and His Friends,* in 1953. Bounced for an alleged lack of "humility," LaRosa had reason to be thankful for the firing, since for a brief time it made him a national folk hero and won him an RCA Victor contract. In 1955 he began his own 15-minute summertime TV show three times a week, as well as a half-hour series with Kitty Kallen, *TV's Top Tunes.* His popularity declined in the 60s and, by the end of the decade, he had become a disc jockey for WNEW New York. See also, Godfrey, Arthur.

Louise Lasser

LARSON, GLEN A.—an original member of The Four Preps singing group in the late 50s who moved into TV writing and then producing. He produced *Alias Smith & Jones* and *It Takes a Thief* and was executive producer of *McCloud, Six Million Dollar Man, Get Christie Love, Switch* and *Quincy.* He joined 20th Century-Fox Television in 1981 and produced *The Fall Guy* there.

LASER—acronym for "Light Amplification by Stimulated Emission of Radiation," a device which converts power into an intense, narrow beam of light. It has several potential uses in television: (1) because the laser beam can carry large bandwidths, it can be used in conjunction with fiber optics as a substitute for cable, (2) as a pickup device in optical videodisc player systems, (3) as a potential key component of three-dimensional television. See also Fiber Optics, Videodisc, Holography.

LASSER, LOUISE—comedy actress who became a national celebrity in 1976 in the title role of Norman Lear's satirical soap opera, *Mary Hartman, Mary Hartman.* The former wife of comedian Woody Allen, she had appeared in a number of his films and had roles in several TV and Broadway productions. In the 1981–82 season, she starred in the *Making A Living* series.

"LASSIE"—long-running weekly adventure series with particular appeal to children. Derived from the movie series about an intelligent and heroic collie, it was a Sunday night fixture on CBS from 1955 to 1970. A casualty of the FCC's prime time-access rule, which forced the networks to cut back their programming, *Lassie* continued in production three more years as a bartered prime-access entry, under Campbell Soup sponsorship. The series was owned and produced by the Wrather Corp.

The original cast featured Tommy Rettig as Jeff, the young owner of the collie, Jan Clayton as his mother and George Cleveland as his grandfather. The cast changed at various times in the show's long run, and over the course of it at least six different dogs, descended from the same line, played the title role. Although Lassie was supposed to be female, male collies were used.

The second *Lassie* family featured Jon Provost as the young boy now named Timmy, and Jon Shepodd and Cloris Leachman as his parents. In the third and probably best-known of the families, June Lockhart and Hugh Reilly were the parents and Provost continued as Timmy. Later episodes had Lassie traveling with an adult master, a ranger, played by Robert Bray.

Reruns of the early black-and-white version were syndicated as *Jeff's Collie.* The remainder were syndicated as *Lassie.*

"LAST KING OF AMERICA, THE"—an attempt by CBS News at reaching into history with the news interview; the effort, on June 6, 1973, fell short of a critical success. Eric Sevareid as a contemporary reporter, attempted an impromptu interview with King George III of England, who lost the colonies, portrayed by actor Peter Ustinov. It was filmed near Stratford-on-Avon, with Perry Wolff as producer for CBS News.

LAUGH-TRACK—laughter mechanically inserted on the soundtrack of a show by mixing "canned" or prerecorded audience responses into the final audio track of a show that had been filmed without audience. The device used is a portable console consisting of multiple cassette players containing a range of laughter from mild titters to uncontrollable guffaws. Working in an audio recording studio while the film or tape of the program is projected, the operator assesses scenes for the type of laughter indicated and activates the appropriate cassette to be mixed into the sound track. Laugh-tracks are mainly used for comedy, game and variety programs. See also Sweetening.

Cindy Williams & Penny Marshall as Shirley & Laverne

"LAVERNE & SHIRLEY"—ABC smash-hit situation comedy spinoff of *Happy Days*. It concerns two young female blue-collar workers in Milwaukee during the 50s played by Penny Marshall and Cindy Williams. Thomas L. Miller, Edward K. Milkis and Garry Marshall are co-executive producers and Mark Rothman, Lowell Ganz and Tony Marshall the producers. It is by Paramount TV and Miller-Milkis Productions.

"LAWMAN"—western series on ABC (1958–61) produced by Warner Bros. and featuring John Russell, Peter Brown and Peggie Castle.

"LAW AND MR. JONES, THE"—ABC series about an idealistic attorney (1960–62) featuring James Whitmore and Janet DeGore. It was created and produced by Sy Gomberg for Four Star-Naxon Productions.

LAWRENCE, BILL (d. 1972)—ABC News correspondent and national affairs editor, noted for his raspy style, wit and accurate prediction in 1966 that President Johnson would not run for reelection. He joined ABC in 1961 as White House correspondent after 20 years with *The New York Times*,

variously as a war, foreign and national correspondent. He became the network's national affairs editor in 1965 and in the years before his death broadened his activities to include some sports reporting, such as covering the Triple Crown of racing and the World Series. His only book, *Six Presidents, Too Many Wars*, was published shortly after his death.

Bill Lawrence

"LAWRENCE WELK SHOW, THE"—music-variety series that was a Saturday night fixture on ABC for 15 years (1955–70) and had a large enough following to continue in first run syndication as one of the more popular entries in prime-access time. Welk lost his ABC berth despite healthy ratings because he tended to attract older viewers.

Featuring traditional popular music and the continental but unsophisticated bandleader, Welk, the program made its debut on July 2, 1955 as the *Dodge Dancing Party* and three years later changed the title to *The Plymouth Show Starring Lawrence Welk*. The following year it became *The Lawrence Welk Show*.

Adhering to an old and simple format, the series presented Welk's 27-piece band, soloists and singers doing individual turns and Welk performing on an accordion or doing a fox trot with a cast member. Welk billed his style as "champagne music" and used "I'm Forever Blowing Bubbles" as his theme, but he was best known to TV audiences for his musical cue: "a-one, a-two!"

Among the featured performers over the years were Alice Lon, the Lennon Sisters, Joe Feeney, Larry Hooper, Tiny Little, Jr. and Jerry Burke. Miss Lon, who was the original "Champagne Lady" on TV, was fired by Welk for exposing her knees too often.

At his height, Welk had had a second TV series, *Top Tunes and New Talent*, which began in 1956 and lasted two seasons. He gave up making original programs in 1982, but specially selected reruns continued to be offered in syndication.

Lawrence Welk

LEACOCK, PHILIP—co-producer of *The New Land* and *Hawaii Five-0* the following season. He became executive producer of the latter in 1976.

LEAD-IN—the preceding program. In rating analysis, a program's performance is assessed with attention to the audience inherited from its lead-in. A show that overcomes a weak lead-in demonstrates that it has extraordinary appeal; one that cannot hold the audience of a strong lead-in is considered lacking in inherent appeal and is usually a candidate for cancellation.

LEAHY, THOMAS F.—executive vice president of CBS Broadcast Group, responsible for its new ventures and for the CBS television stations division, radio division, CBS Cable and CBS International. He was elevated to the Broadcast Group in February 1981, after serving four years as president of the owned-stations division.

He joined CBS in 1962 as salesman for the television network, rising to director of daytime sales in 1969. In 1971, he switched to the stations division as vice president and general manager of WCBS-TV, the flagship station, which served as his springboard to the presidency of the station division in 1977.

LEAR, NORMAN—TV's most successful producer in the 70s who, in pioneering the mature and frank-spoken situation comedy, led the medium into new directions of sophistication. Flouting the old broadcasting taboos and finding a huge audience highly receptive to his "adult" approach to TV comedy, Lear triggered the massive reappraisal of pro-

gram standards that led to the "new boldness" which characterized prime time in the early 70s.

Coming to TV from motion pictures with his partner Bud Yorkin (together they were Tandem Productions), he had an explosive hit in 1971 with *All In the Family,* which broke new ground in dealing with such subjects as bigotry, menopause, homosexuality and impotency, as well as with political issues. A string of successful spin-offs—*Maude, Good Times, The Jeffersons*—and another blockbuster, *Sanford and Son,* quickly made him the hottest independent program supplier to the networks and broadened his license to explore the outer boundaries of acceptability for the use of the language and subject matter. Usually, his ventures beyond the normally accepted limits were justified by well-crafted scripts that attempted to deal realistically and responsibly with contemporary society.

Under his own company, T.A.T. Productions, separate from his partnership with Yorkin, he also produced *One Day at a Time, Hot l Baltimore, The Dumplings, All's Fair, The Nancy Walker Show* and *Mary Hartman, Mary Hartman*—the latter making an impact in syndication in 1976 comparable to that of *All In the Family* five years earlier. A nightly soap opera both satirical and obsessed with sex, *Mary Hartman* too was a radical modification by Lear of an established program form, which was the essence of his genius. He followed in 1977 with another syndicated serial, *All That Glitters,* which fared poorly.

Norman Lear

Lear had been active in the early years of TV as a director-writer, principally with *The Martha Raye Show,* but like many others left the medium in the 60s to make pictures.

In 1978, after eight years and 16 television series, Lear took another extended leave from TV to concentrate on writing and directing movies, although his companies, T.A.T. and Tandem, remained active. Much as he tried to extricate himself from day-to-day involvements with his TV series, he lent a personal hand to *In the Beginning* and *Apple*

Pie, two sitcoms in the 1978 fall lineups that died swift deaths, and then became entranced with the syndication project, *The Baxters,* readied for the fall of 1979, for which he wrote scripts. He also became interested in over-the-air pay-television and with his partners, Jerry Perenchio, Bud Yorkin and Oak Industries, launched the first successful STV station in the country, KBSC-TV, Corona–Los Angeles. In 1981, they acquired Embassy Pictures and made it the umbrella for their separate production companies.

Lear began devoting much of his time in 1981 to counteracting the attempts of the Religious New Right to interfere with television programming. He created a national organization, People for the American Way, which sent out literature and produced television commercials to offset the media campaign of the Moral Majority and kindred groups. In February 1982 he staged a rally for liberty and patriotism in Hollywood which featured such politically diverse figures as Sen. Barry Goldwater and Jane Fonda. It was videotaped and telecast nationally a month later.

See also *All in the Family; Baxters, The; Mary Hartman, Mary Hartman;* and Pay-Television.

LEASED CHANNELS— channels made available on cable systems, usually at the behest of the franchising authority, for leasing to members of the public at posted rates, on a common carrier basis. As a form of paid public access, leased channels permit individuals or companies to present programs of their own with advertising sponsorship and with no interference from the cable operator except for reasons of obscenity.

In principle, the leased channel concept permits every citizen to become a broadcaster, or video publisher, at a relatively low fee—in marked contrast to over-the-air television which is a closed shop for a limited number of licensees. The FCC regulations require that leased cable channels be made available on a nondiscriminatory first-come, first-served basis according to posted rates applicable to all.

To prevent any party from monopolizing the leased channels, the rules provide that at least one channel must grant priority to part-time users. On New York's Manhattan Cable Television system, those who buy time on the leased-access Channel J for $50 an hour are limited to one year of cablecasting to avoid entrenchment by any producing entity. Those who have succeeded commercially and wish to continue are asked to shift to an unsubsidized leased channel at higher rates.

Because leased channels are open to free expression and are exempt from rules governing conventional broadcasting, they therefore may be used entirely for sales pitches or for the expression of political views without being subject to the Fairness Doctrine.

"LEAVE IT TO BEAVER"—family situation comedy of the late 50s whose reruns continue to play on local stations in the 80s. It premiered on CBS in October 1957 and moved to ABC a year later, where it ran until August 1963. Jerry Mathers played Beaver, a young boy, and Tony Dow his older brother. Hugh Beaumont and Barbara Billingsley were their parents. It was by Gomalco Productions and Universal.

LEE, H. REX—FCC commissioner (1968–73) who was appointed by President Johnson to a Democratic seat and resigned 18 months before the end of his term to benefit from an early retirement plan for government officials with long years of service. Lee entered government in 1936 as an economist with the Department of Agriculture. Later he served with the Department of Interior and in 1961 was appointed, by President Kennedy, governor of U.S. Samoa. On the FCC he was public television's leading advocate. He later became head of the Public Service Satellite Consortium.

LEE, JOANNA—writer whose script credits include *The Waltons' Thanksgiving Story, Babe* and *I Want to Keep My Baby.* She was also executive story consultant on *Room 222.* In the mid-70s she went into independent production of made-for-TV movies and has developed a reputation for quality productions in that area.

LEE, PINKY—burlesque comedian who was popular in the early years of TV. He starred with Martha Stewart (later replaced by Vivian Blaine) in the variety show, *Those Two,* from 1951 to 1954, which was followed by a successful children's series, *The Pinky Lee Show.*

LEE, ROBERT E.—FCC commissioner (1953–81) appointed to a Republican seat by President Eisenhower. He was reappointed by Presidents Eisenhower, Johnson and Nixon. Through all those years he was the commission's staunchest believer in the development of UHF and approached many issues with a view to their potential impact on UHF. Before joining the FCC, he was a special agent for the FBI, administrative assistant to J. Edgar Hoover at the FBI and director of surveys and investigations for the House Appropriations Committee. Lee served longer on the FCC than any previous commisioner. Before retiring, he served for one month as chairman of the Commission while Mark S. Fowler was preparing to take the office.

LEFFERTS, GEORGE—writer, producer and director; he worked in all three capacities on the *Purex Specials for Women* in the early 60s and earlier wrote for *Studio One, Kraft Theatre,*

High Adventure and *Ellery Queen*. He was also executive producer of *The Breaking Point*.

LEGALLY QUALIFIED CANDIDATES [Letter to Mr. Larry Seigle/FCC 72–924 (1972)]—vital to claims for equal time, determined by reference to whether the candidate can be voted for in the state or district in which the election is being held, and whether he is eligible to serve in the office in question.

The FCC made a key ruling in 1972 when it denied appeals by candidates for President and vice-president for the Socialist Workers Party for equal time to respond to broadcasts by the Democratic candidates. The appeals were by candidates who were 31 and 21 years old. The commission found, with one dissent, that since the Constitution states that a person under 35 years of age is ineligible to hold the office of President, neither candidate could serve and therefore neither was a "legally qualified candidate." See also Equal Time.

"LEGEND OF JESSE JAMES, THE"—short-lived western series on ABC (1965) about the notorious James brothers, produced by 20th Century-Fox and featuring Chris Jones and Allen Case.

"LEGEND OF MARILYN MONROE, THE"—re-creation of the life of the legendary actress through stills, newsreel and candid footage and clips from her films. The hour was produced in 1965 by Wolper Productions, with John Huston as narrator.

LEHRER, JAMES C.—Washington-based co-anchor of the successful PBS nightly news program, *The McNeil-Lehrer Report*. Previously he was correspondent for NPACT and moderator of many of its programs. He came to the Washington public affairs arm of PBS in 1972 from KERA, the PTV station in Dallas, where he was news director and executive producer of *KERA Newsroom*. With NPACT he held the title of public affairs coordinator. Lehrer and Robert MacNeil formed their own production company in 1982 for commercial as well as public television.

LEISER, ERNEST—vice president and assistant to the president of CBS News since Dec. 1981. Before that he had been a leading producer for CBS News for most of the years since 1953, when he joined. From 1964 to 1967 he was executive producer of *The CBS Evening News with Walter Cronkite* and then exec-producer of the CBS Special Reports Unit. He left in 1972 to become exec-producer of *The ABC*

Evening News and *The Reasoner Report* but returned to CBS after three years to work on special news projects, such as the all-day Bicentennial spectacular on July 4, 1976. He won Emmys for two documentaries, *Father and Sons* and *The World of Charlie Company,* and for producing the coverage of the assassination and funeral of Dr. Martin Luther King. In 1979, he was named CBS News v.p. of special events and political coverage

LEONARD, BERT (Herbert B.)—highly successful producer of adventure series in the 50s and 60s, whose credits include *Naked City* and *Route 66* (through his Shelle Productions), *Rescue 8* (through Wilbert Productions), *Tales of the Bengal Lancers* (through Lancer Productions) and, earlier, *Circus Boy* and *Rin Tin Tin* (Herbert B. Leonard Productions).

Bill Leonard

LEONARD, BILL (William Augustus)—longtime CBS News executive who in April 1979, after more than 30 years with the company, became president of the division, succeeding Richard S. Salant, forced to retire on reaching age 65. Leonard himself, when he assumed the post, was two years away from mandatory retirement. He was tapped by a relatively new CBS management that had not yet evaluated the longterm candidates for the presidency of the news division. Prior to his appointment, Leonard had been CBS Washington v.p. and chief liaison with the government and its agencies. He succeeded Richard W. Jencks in the post in 1975.

Leonard stayed on beyond his 65th birthday, under a waiver by the CBS board, and retired in March 1982, yielding his post to Van Gorden Sauter.

Starting in radio in 1946, with the CBS station in New York, Leonard conducted the *This Is New York* series that later shifted to WCBS-TV as *Eye on New York*. He moved on to the CBS News division in 1959 as producer-correspondent of *CBS Reports* and until 1962 produced, wrote and narrated those programs. He then formed the CBS News Election Unit and in 1964 became an executive with initial responsibility for editorial policies, production planning and special events coverage. In 1968, he was placed in charge of all special and documentary production and three years later became senior v.p. Among the *CBS Reports* documentaries produced and narrated by Leonard before 1962 were *Trujillo —Portrait of a Dictator, Thunder on the Right* and The Beat Majority.

Leonard, who married the former wife of CBS correspondent Mike Wallace, is the step-father of Chris Wallace, a rising star at NBC News.

LEONARD, SHELDON—character actor in movies and comedy actor in radio whose chief success in television has been as a producer and director. His credits in that sphere began in 1953 when he became executive producer and director of the Danny Thomas series during its first year. Later he and Thomas became partners in a production company. Leonard developed *The Andy Griffith Show* for CBS, then spun off *Gomer Pyle—USMC* and put together such other hits as *The Dick Van Dyke Show* and *I Spy,* all of which he supervised as exec producer. He was also exec producer of *My World and Welcome To It.*

In 1975, he returned to acting, playing his most familiar character role, that of a Runyonesque gambler in *Big Eddie,* a CBS situation comedy. The show flopped, however. During the 30s Leonard had been featured in a number of Broadway shows, including *Three Men on a Horse.* He went to Hollywood in 1939 and made more than 140 movies, frequently playing gangsters or gamblers. He also began making radio appearances and was a frequent guest on the comedy programs of Jack Benny, Phil Harris, Bob Hope and Judy Canova. He also dabbled in writing radio plays and in the 50s switched to television writing and directing.

LESCOULIE, JACK—TV personality who was a regular on NBC's *Today* from 1958 to the late 60s and earlier had been a TV sidekick to Jackie Gleason. He also hosted several daytime shows and in 1975 was anchorman for NBC's latenight experiment *America After Dark.* On leaving *Today* he moved to Cincinnati to work on television projects there.

"LESLIE UGGAMS SHOW, THE"—a CBS music-variety series (1969) which represented an early attempt to build a program of that type around a black star. But Miss Uggams, although fresh from a triumph on Broadway, lacked a national following, and her series was canceled after 13 weeks. It was produced by Ilson & Chambers.

LESTER, JERRY—rowdy comedian who made an impact on TV in 1950–51 as emcee of *Broadway Open House,* the pioneer latenight show on NBC. As a nightly show it drew heavily on Lester's broad comedy resources—and those of Dagmar, Wayne Howell and Milton Delugg, as well—but while it marked the height of his career, it also caused him to suffer from overexposure. Lester faded from prominence soon afterwards, while NBC learned that comedy in a lower key would sustain better on a nightly basis.

"LET ME SPEAK TO THE MANAGER"—innovative "Letters to the Editor" kind of program initiated in 1961 by Mike Shapiro, general manager of WFAA-TV Dallas to explain how and why television decisions are made. Originally a 15-minute prime-time program, airing on Sunday nights, it grew into a half-hour program and in recent years changed its title to *Inside Television.* Carried also by sister station KFDM-TV in Beaumont, Tex., the format has been copied by a handful of other local stations. Essentially, it involves Shapiro responding to gripes from viewers concerning programing on all networks and frequently includes interviews with experts in a pertinent field. The show is taped several days ahead of broadcast and is presented with program director James Pratt asking questions of Shapiro based on letters sent by viewers.

America's Top Trader, Monty Hall

"LET'S MAKE A DEAL"—highly successful game show involving substantial giveaways of merchandise and cash, which had a 12-year run as a weekday strip—first on NBC and then on ABC (1963–76)—and continued as a twice-weekly, syndicated prime-access entry. The format verged on the grotesque, and the sense of an orgy of greed was heightened by the practice of members of the audience dressing in bizarre costumes to attract notice for their selection as contestants. The essence of *Deal* was the heaping of temptation upon the players to risk their holdings for the prospect of still larger prizes. The pyramiding transactions were masterfully conducted by the host, Monty Hall.

Deal had been a successful daytime program for NBC for five years, but in 1968, when the network balked at raising the licensing fee, Hall moved the program to ABC. The effect of the switch was startling, since it had not previously been perceived that *Deal* was the pivotal show in NBC's daytime schedule. ABC's ratings immediately shot up for its entire daytime lineup while NBC's fell.

In 1971 Hall and his partner Stefan Hatos added a nighttime version of *Deal* for prime-access syndication and two years later expanded that venture to two a week. The series was by Hatos-Hall Productions, with Hatos as executive producer. See also Hall, Monty.

"LET'S TAKE A TRIP"—CBS series (1955) that was a travelogue styled for children. It was hosted by Sonny Fox.

David Letterman

LETTERMAN, DAVID—midwestern humorist (Indiana) who gained national recognition in 1979–80 as Johnny Carson's favorite guest host. Letterman was spoken of as the Carson successor, and was signed to a long-term NBC contract. His first opportunity came in the summer of 1980 with a morning talk show which was well reviewed—and cancelled after a few weeks. Early in 1982, "Late Night With David Letterman" replaced Tom Snyder's "Tomorrow" in the 12:30 A.M.–1:30 A.M. slot: an attempt at a hip talkshow for younger audiences.

LEVANT, OSCAR (d. 1972)—concert pianist, literary wit and raconteur who in the late 50s became host of a popular syndicated talk show emanating from the West Coast. But the program was considered in poor taste by many because it seemed to exploit Levant's neuroses and his eccentric behavior more than it made use of his conversational or musical talents. He guested frequently on other shows and for a time was panelist on *General Electric Guest House.*

LEVATHES, PETER G.—commercial TV and motion picture executive who became director of program development for the Corporation for Public Broadcasting (1976–77). He had previously been with 20th Century-Fox TV as executive v.p. of all production for the studio, movies as well as TV.

LEVIN, GERALD M.—vice president of the Time Inc. video group since 1979, previously chairman of Home Box Office, and dubbed by *Channels* magazine "the Lenin of video" and "the man who started Television II."

Levin became chairman of HBO, the pay-cable subsidiary of Time Inc., in 1973. He came to the post with a background in law, management and finance, after serving a year as v.p. of programs for Sterling Cable Network, the company that became HBO in 1972.

Levin built Home Box Office, a small, three-state operation when he took over, into the nation's largest pay-television service with over 2 million subscribing homes, and in the process sparked a revolution in the development of cable-TV. In the fall of 1975, Levin made the decision to distribute HBO by satellite, at a time when only two cable systems in the country were equipped with receiving earth stations. But the gamble paid off. The attractiveness of pay-cable services spurred systems to become part of the growing cable satellite network, and by the spring of 1982 more than 1,500 earth stations were in place, serving more than 8 million homes. Levin's move was a major leap forward for the medium, which had been primarily local, and a wide variety of new services have been stimulated by the availability of an inexpensive national distribution network. It was also highly profitable for HBO which controlled 60% of the pay-TV market by 1982.

LEVINSON, DAVID—producer of *The Senator* segments of *The Bold Ones* and also the *Sarge* series. He was executive producer of *Sons & Daughters* (1974–75) and produced *Sarah T,* a made-for-TV movie.

LEVINE, IRVING R.—veteran NBC News correspondent (since 1950) who became economic affairs correspondent in Washington in 1971 after covering foreign beats for nearly two decades. He was based in Rome for ten years, in

Moscow for four and in Tokyo for two. He also served a year in London.

LEVINSON-LINK—producer-writer team of Richard Levinson and William Link who created *Mannix, Columbo, Ellery Queen* and *Tenafly* and served as executive producers of the latter three. They were also the writers and exec producers for such made-for-TV movies as *The Execution of Private Slovik* and *The Gun.*

LEVITAN, PAUL (d. 1976)—director of special events for CBS (1961–72), responsible for coverage of such spectacles as the Miss America, Miss Universe and Miss Teenage America pageants and the Rose Bowl and Thanksgiving Day Parades. During the 50s he had been an executive producer for CBS News, principally on special events coverage, and before that worked with Fred Friendly on producing *See It Now.*

LEVY, DAVID—program v.p. for NBC from April 1, 1959 to Aug. 4, 1961, having joined from the broadcast department of Young & Rubicam. He later went to Goodson-Todman Productions on the West Coast and then to Filmways, serving as executive producer of such series as *The Double Life of Henry Phyfe, The Pruitts of Southhampton* and *The Addams Family.* In a novel, *The Chameleons,* published in 1964, Levy described with some acrimony the executive suite intrigues at a network.

LEVY, ISAAC (d. 1975) **AND LEON** (d. 1978)—pivotal figures in the evolution of CBS and later members of its board of directors; they were instrumental in William S. Paley's acquisition of the struggling radio network in 1928.

In 1922 Isaac bought Philadelphia's WCAU and Leon became chief administrator of the station. A few years later Leon married Blanche Paley, whose family owned Philadelphia's Congress Cigar Co., which soon became a major advertiser on WCAU. In 1927 Leon made the station an affiliate of the newly formed network, United Independent Broadcasters, which sought to compete with RCA's network. That same year the operating rights of UIB were purchased by the Columbia Phonograph Co. and the network was renamed the Columbia Phonograph Broadcasting System.

A year later the Levy brothers persuaded Leon's brother-in-law, William Paley, to buy the controlling interest in the network while they bought large shares in it themselves. Through the years they remained major shareholders of CBS. Ike Levy left the board of directors in 1951 but Leon continued to serve as a director and in 1975 his holdings in the company came to 330,756 shares.

LEVY, RALPH—producer-director of the 50s who guided the programs of Jack Benny, Ed Wynn and George Burns and Gracie Allen, among others. But his most significant credit was producing the pilot film for *I Love Lucy* in 1951.

LEWINE, ROBERT F.—former ABC, NBC and CBS program executive; later executive v.p. of the talent agency, CMA; and for many years an official of the Academy of Television Arts & Sciences, serving twice as national president. He was an ABC program exec from 1953 to 1957, rising to v.p. in charge, and then moved to NBC and became its program chief (1957–58). After a year in independent production he joined CBS and became the network's program v.p. in Hollywood. He joined CMA in 1964 and returned to NBC in 1977.

LEWIS, ROBERT Q.—ubiquitous TV personality during the 50s who hosted or was a panelist on numerous game shows and conducted daytime variety shows known as *Robert Q.'s Matinee* and *The Robert Q. Lewis Show.*

LIBERACE—pianist and entertainer who achieved an intensely devoted fan-following during the 50s with a 15-minute nationally syndicated program on which his ornate semiclassical playing style was abetted by his personal effusiveness, extravagant smile and outrageous wardrobe. His vast appeal, which from 1953 to 1955 made his program one of the most popular on television, was demonstrably with older women.

Although he was born and educated in Wisconsin, his persona was continental, an image that was also served by his trademark—a candelabra on his piano—and his use professionally of his surname only. His given names were Wladziu Valentino. Working with him always was his brother, George Liberace, as a violinist and conductor of the orchestra.

Liberace's excesses were frequently ridiculed, but he generally took the insults good-naturedly with a response that is now famous: "I cried all the way to the bank."

He did, however, sue the British columnist, Cassandra, for libel.

LICENSE CHALLENGE—procedure by which an individual or group may oppose the renewal of a license for a particular existing station and apply to receive it instead. A license challenge—also called a *strike application*—differs from a petition to deny essentially in the fact that it asks for the license while the petition to deny does not, although both contend that the incumbent is not worthy of the right to broadcast.

License challenges have been nerve-wracking for stations and costly to defend, but the FCC has shown a great reluctance to take licenses away from incumbent broadcasters and has generally been persuaded by the argument that the licensee under challenge has performed as well as the average broadcaster. The most serious strike applications—such as those against KHJ Los Angeles and WPIX New York—have taken years to adjudicate, with the incumbent prevailing ultimately, and the expense involved in pursuing such fruitless challenges has deterred the practice. See also Petition to Deny.

LICENSE RENEWAL—approval by the FCC of a broadcaster's privilege to operate for another five-year term upon the expiration of his license. Technically the licensee must demonstrate in his application for renewal that his continuance would serve the public interest. Licensees are required to give adequate notice to the public of the application for renewal, and if the incumbent's license should be challenged by a competing applicant at license-renewal time the FCC must hold a comparative hearing to determine who will get the license.

Historically, a proven—even though unexceptional—record of performance by an incumbent has been preferred, in comparative renewals, to the untried proposals of a challenger. Nevertheless, bills were introduced in Congress during the 70s to amend the Communications Act in ways that would establish that an incumbent is not to be treated as a new applicant. Legislation in 1981 lengthened the licensing period for television stations from three years to five.

In a notable comparative renewal case in the early 50s [Hearst Radio Inc. (WBAL), 15 FCC 1149 (1951)] the FCC favored an existing license over a challenger even though the incumbent's performance had not been outstanding. This decision of the commission so discouraged challengers that during the 1950s and 1960s, with one exception, no TV licensee in the regular renewal process faced a comparative hearing.

But in 1965 the FCC issued its policy statement on comparative hearings which said that past broadcast experience would not be an important factor unless it was unusually good or poor. The FCC also stressed that integration of management and diversity of ownership would be important considerations in assessing an incumbent's qualifications.

In the WHDH case, the standards enunciated in the 1965 policy statement were directly applied in a comparative renewal hearing, and although WHDH was far from a normal renewal situation the FCC's refusal to consider the incumbent's past performance caused great consternation in the industry. It also opened the floodgates to other license challenges.

Broadcasters immediately began lobbying Congress for some relief and in 1969 Sen. John O. Pastore, chairman of the Senate Communications Subcommittee, proposed a two-stage hearing for comparative renewals. Under this bill, the FCC would examine the incumbent's record to determine if the licensee had met the public interest standard; if it had, then the commission would authorize renewal. Competing applications would thus be considered only if the incumbent had failed to meet the standard.

After opposition to the bill stalled its progress, the FCC in 1970 issued the *Policy Statement Concerning Comparative Hearings Including Regular Renewal Applications*. Here the commission stated that if an incumbent could show that its program service had been "substantially" attuned to the needs and interests of its area, and that the operation of the station had not otherwise been seriously deficient, its application will be preferred over that of the newcomer.

In June 1971, the Citizens Communications Center *v.* FCC [477 F2d 1201 (D.C. Circuit 1971)], the D.C. Court of Appeals held the 1970 policy statement invalid because it operated to deprive the competing applicant of its statutory right to a full hearing. In addition, the court stated that the commission was attempting to establish by policy statement what Congress had not done by legislation.

The court suggested that "superior" service should be of major significance and that the commission in its rulemaking proceeding should try to clarify what constitutes "superior" service.

After numerous attempts to define an appropriate standard, the commission has struggled to avoid decisions such as WHDH, and it has proceeded as before in preferring the incumbent regardless of whether the licensee has rendered "superior" or "substantial" service. See also Alabama ETV Licenses, WLBT, WHDH.

LICENSE TRANSFER—the passing of a broadcast license from one owner to another, normally through the sale of the station, which cannot be consummated without FCC approval. To keep broadcasters from "trafficking" in licenses—that is, buying stations with a view to reselling them quickly at a profit—the commission has passed a rule that licenses must be held a minimum of three years before they may be transferred, except in cases of extreme financial hardship.

Section 310 of the Communications Act specifies that application must be made to the FCC for the transfer of licenses or construction permits and that the FCC should approve only if it finds that "the public interest, convenience and necessity will be served thereby." See also Trafficking.

LICENSING (OF PROGRAMS)—lease basis on which producing companies "sell" their programs to networks or stations. In ordering a program or series from a studio, a network merely secures the rights for two air plays under a standard agreement, but the property remains under the ownership and control of the producing concern. The studio may simultaneously license the rights to foreign systems but must wait until the network has completed its use of the property before releasing it in domestic syndication. There, individual stations may license the rights to the reruns for a specified time in their specific markets.

LIEBMAN, MAX (d. 1981)—a Broadway theater and film producer known principally in TV for his work as producer of the famed *Your Show of Shows* comedy series with Sid Caesar and Imogene Coca. In 1955 he produced a number of color "spectaculars" for NBC and in the late 60s was creative supervisor of *The Jackie Gleason Show.*

"LIE, THE"—drama written for Swedish TV by Ingmar Bergman which was translated and produced for the U.S. by CBS (April 4, 1973) as a 90-minute *CBS Playhouse 90* special. Lewis Freedman produced, and George Segal and Dean Jagger were the stars.

"LIEUTENANT, THE"—MGM-TV series about the adventures of a young Marine officer in peacetime, carried by NBC in 1963. It starred Gary Lockwood and featured Robert Vaughn.

"LIFE AND TIMES OF GRIZZLY ADAMS, THE"— NBC hour-long adventure series (1977–78) on life in the wilderness, filmed in Payson, Ariz., and in the Uinta Mountain Range in Utah. Dan Haggerty played the title role and Denver Pyle and Don Shanks were featured. Charles E. Sellier Jr. was creator and executive producer and Art Stolnitz producer. It was via Schick Sun Classic Productions.

"LIFE ON EARTH"—remarkable 13 episode British natural history series tracing the evolution of life on this planet over millions of years; it had its U.S. premiere on PBS in January 1982 and quickly became one of public television's most watched series ever. David Attenborough, former program director of the British Broadcasting Corporation, wrote and narrated the series.

In the tradition established at the BBC by *Civilisation* and *The Ascent of Man*, Attenborough's documentary series was meticulously researched and was filmed on locations in 30 countries, employing the most advanced photographic techniques. It took three years to produce. Christopher Parsons was executive producer and John Sparks and Richard

Brock co-producers. It was the biggest production ever undertaken by the BBC's Natural History Unit in Bristol, England.

Life on Earth was produced in association with Warner Bros. Television, which invested in the project, and was presented on PBS by WQLN, Erie, Pa., on funding by Mobil Corp. Attenborough's book based on the series was published in the U.S. by Little, Brown & Co. in 1981.

"LIFELINE"—weekly prime-time nonfiction series on NBC (1978), in the cinema verite style, focusing on the professional and private lives of actual doctors. The pilot had caught the fancy of Fred Silverman after he joined NBC as its new president in June 1978, and he altered the fall schedule that had already been set to make room for it.

An unusual program for prime time, it seemed to represent what Silverman promised NBC affiliates in his maiden speech to that body: innovative, high-quality fare—programming that would take the high road rather than pander to the youth market. NBC purchased 13 episodes from Tomorrow Entertainment and the Medcom Co., which produced it jointly, but was unable to spark the interest of the viewership.

The series was shifted to various time periods in hopes that it would find an audience somewhere, and one week, in a patented Silverman "stunt," it aired three times. But always, despite generally favorable notices, it landed at or near the bottom of the Nielsens. *Lifeline* was dropped at midseason. Thomas W. Moore and Dr. Robert E. Fuisz, chairman of Medcom and a practicing physician, were the executive producers.

"LIFE WITH FATHER"—a half-hour situation comedy based on the popular book and play by Clarence Day. It was aired on CBS from Nov. 22, 1953 until July 5, 1955, with Leon Ames as Clarence Day, Sr. (father) and Lurene Tuttle as his wife, Vinnie. The Day children were played by Ralph Reed, Freddie Leiston, Ronald Keith and Harvey Grant.

"LIFE WITH LUIGI"—popular early CBS situation comedy that moved over from radio in 1948. In it, J. Carrol Naish portrayed a newly arrived Italian immigrant. Alan Reed was featured.

"LIFE OF RILEY, THE"—the prototypical early situation comedy of the Bumbling Father school: it bloomed twice on NBC, first with Jackie Gleason and Rosemary DeCamp in the leads (1949–50) and then with William Bendix and Marjorie Reynolds (1953–58). The protagonist was a likeable but stupid working-class head of family who managed always to muddle through mild trouble. The show was popular and the repeats sold well in syndication.

"LIFE AND LEGEND OF WYATT EARP, THE"—hit western series on ABC (1955–60) about the marshal of Tombstone, which starred Hugh O'Brian. It was by Louis F. Edelman and Wyatt Earp Enterprises.

"LILLIE"—13-hour British series, produced by London Weekend Television, on the life of Lillie Langtry, the legendary British actress and professional beauty who became mistress of the Prince of Wales, later Edward VII. It began its U.S. run in March 1979 as an entry in the PBS *Masterpiece Theater* series. Based on James Brough's book, *The Prince and the Lily,* it was developed and largely written by David Butler and directed by John Gorrie. Francesca Annis starred in the title role, with Patrick Holt, Peggy Ann Wood, Anton Rogers and Simon Turner featured.

LIMITED SUBSCRIBER FEEDBACK—a feature in a cable-TV system that enables the subscriber to respond to incoming messages for polling purposes, ordering merchandise or audience participation. The feature is possible in two-way systems which permit viewers to send digital signals from their TV sets to the head end.

The most extensive use of this technology to date has been in Warner Communications' QUBE system in Columbus, Ohio, where viewers may use digital response for polling purposes, video games programmed by the system and to select individual programs on a pay-per-view basis.

LINDEMANN, CARL, JR.—v.p. of CBS Sports since 1978, after a long association with NBC where he was head of sports from 1963 to 1977. Lindemann's departure from NBC was sudden and never fully explained by the network, but it appeared to have been precipitated by a dispute over who would supervise coverage of the 1980 Moscow Olympics.

One of Lindemann's notable achievements at NBC was to gain the network a share of the professional football market at a time when CBS had an exclusive on the glamour package of sports, the National Football League. Lindemann persuaded NBC to put up $42 million for a five-year contract with the new American Football League in 1964, which was five times what ABC had been paying for the struggling second-class league. The money and TV exposure helped the AFL grow into a full-fledged competitor with NFL, which eventually prompted a merger of the two into separate conferences of a single league.

Lindemann also formulated an extensive college basketball schedule for NBC and secured the rights to the World Series, Rose Bowl, Orange Bowl, Senior Bowl, Wimbledon Tennis Tournament and, on alternating years, the baseball All Star game and the Super Bowl.

He began with NBC as a student engineer, became an associate producer of *The Kate Smith Hour* in 1953, unit manager of the *Home* show in 1954 and then business manager of the program department. In 1957 he was director of daytime programs.

Carl Lindemann, Jr.

"LINEUP, THE"—half-hour crime series on CBS (1954–59) and one of the first to make use of an authentic setting, in this case San Francisco. It featured Warner Anderson and Tom Tully as police detectives and was produced by Desilu. The syndicated reruns wisely gave billing to the colorful city, taking the title *San Francisco Beat.*

LINKE, RICHARD O.—producer long associated with Andy Griffith, serving as co-executive producer with Griffith of *Mayberry RFD* and as executive producer of *The Jim Nabors Show, Headmaster* and the made-for-TV movie, *Winter Kill.*

LINKLETTER, ART—daytime TV personality whose *Art Linkletter's House Party,* a program of interviews and features for housewives, ran almost 20 years on CBS after establishing itself on network radio. Linkletter, who entered TV in 1952, mined a rich lode from interviewing children. It not only made for amusing television but served for several books, including *Kids Say the Darnedest Things.* In 1954 he added to his TV exposure with *People Are Funny,* a popular

NBC entry in which contestants were asked to perform stunts.

"LITTLE ANNIE"—See Stanton-Lazarsfeld Program Analyzer.

Michael Landon & Karen Grassle

"LITTLE HOUSE ON THE PRAIRIE"—hour-long NBC dramatic series based on the *Little House* books of Laura Ingalls Wilder that recalled her life in the West in the late 19th century. The series began in September 1974 after its two-hour pilot drew an impressive rating in March of that year. Michael Landon, featured in the series as the father, served also as co–executive producer with Ed Friendly and occasional director. Friendly later dropped out after a series of "artistic differences" with Landon. Also in the regular cast were Karen Grassle, Melissa Gilbert and Melissa Sue Anderson.

The series became a Monday night fixture for NBC and was the network's highest-rated weekly program during 1978 and 1979, its only show to rank regularly in the Nielson Top 10. In 1981–82, supporting players Merlin Olsen and Moses Gunn were spun off to another series, *Father Murphy*, which was created and produced by Michael Landon. Both series were via NBC productions.

"LITTLE RASCALS"—TV title for the Our Gang movie shorts produced in the 30s. Repackaged with great success as a syndicated children's series, the old films continue to be purchased by stations for the new generations.

"LIVE FROM LINCOLN CENTER"—occasional series of PBS telecasts direct from Lincoln Center for the Performing Arts. It premiered on Jan. 30, 1976. Anxious to extend the audience for the performances of its constituent companies —and to increase their revenues, as well—Lincoln Center experimented for two years until it developed techniques that permitted telecasts from the stage that would neither alter the atmosphere of the theater nor compromise the integrity of the performances. The techniques involved the use of unobtrusively placed shotgun microphones and newly developed studio cameras able to provide a picture with normal stage lighting.

The telecasts—usually five or six a year—were generally carried in the PBS *Great Performances* showcase underwritten by Exxon. Financial support for the project came from the National Endowment for the Arts, CPB, PBS, the Charles A. Dana Foundation and Exxon. The premiere broadcast was a concert of the New York Philharmonic with Andre Previn conducting and pianist Van Cliburn as soloist. All of the Lincoln Center constituent companies, except the Metropolitan Opera, participated in the series.

The Met was a holdout because of its long radio relationship with Texaco, an oil company competing with Exxon. In 1977, Texaco underwrote a telecast of *La Boheme* with the Met, airing in the *Live from Lincoln Center* time period.

Ivan Nagy & Natalia Makarova

"LIVE FROM STUDIO 8H"—series of cultural specials begun at NBC during the Fred Silverman administration in an effort to revive the network's great traditions. Studio 8H in Rockefeller Center was the largest radio studio in the world when it was built in the 30s; it was converted permanently for television in 1950.

8H was the studio from which Arturo Toscanini conducted broadcasts of the NBC Symphony Orchestra for 17 years, and so, appropriately, the premiere of *Live From Studio 8H* on Jan. 9, 1980 was a 90-minute tribute to Toscanini featuring Zubin Mehta and the New York Philharmonic, with soloists Leontyne Price and Itzhak Perlman. Martin Bookspan was commentator and Judith DePaul and Alvin Cooperman producers.

The second in the series, airing July 2, 1980, presented five ballets and was entitled *An Evening with Jerome Robbins*. Subsequent programs, all running 90 minutes, included *Caruso Remembered* and *100 Years of America's Popular Music*.

LLOYD, FRANK—administrative assistant to FCC chairman Charles D. Ferris (1977–80), after having served briefly as consultant on public television to the White House Office of Telecommunications Policy in the Carter Administration. Previously, he was executive director of Citizens Communications Center (1973–77), a leading public-interest law firm concerned with broadcasting. A lawyer who had served in the Office of Economic Opportunity during the Johnson Administration, he worked for public television's NPACT in Washington before joining CCC. He returned to law practice at the end of the Carter Administration.

LLOYD, NORMAN—executive producer and guiding spirit of *Hollywood Television Theatre,* a PBS series of stage drama adaptations which premiered in 1970. A former stage and motion picture actor, he turned to producing and directing, compiling TV credits with *Movie of the Week, Alfred Hitchcock Presents* and *The Name of the Game,* among other series.

LOCAL LOOP—the physical connection of each local station to the coast-to-coast transmissions over AT&T's long lines. The loop, generally achieved by microwave or wire, is frequently likened to a cloverleaf exit off a main highway. The great expense is not in sending the signal cross-country from New York to San Francisco but in delivering it from the main highway, as it were, to the individual stations along the way.

Network affiliated stations usually contract for full-time service at $2,500 monthly. Occasional users of a loop, such as independent stations, are charged $1,500 by the telephone company for the basic connection, $80 per hour for its use and $9 for the physical process of switching from the incoming connection to that of the station. After the basic connection charge, maximum cost for the use of the loop is $1,000 a month.

Long-lines tariffs for commercial networks are 55¢ a mile per month. Occasional users are charged 75¢ a mile per hour.

LOEVINGER, LEE—scholarly FCC commissioner (1963–68) who, although a Democratic appointee, held to the conservative view that the commission may not undertake regulation of program content in any form in broadcasting. He subscribed to the belief that broadcasters give the public what it wants and on the whole serve the public well. He left the FCC to practice communications law. Before joining he had been an Associate Justice of the Minnesota Supreme Court and Assistant Attorney General in charge of the antitrust division of the Justice Department.

LOHR, LENOX RILEY (d. 1968)—president of NBC from 1936 until his resignation in July 1940. He had come to the network after serving as general manager of Chicago's successful Century of Progress exposition.

LONDON WEEKEND TELEVISION—commercial company licensed to serve London on weekends, from 7 P.M. Fridays to closedown on Sundays, while Thames TV has the franchise on weekdays. A major contributor to the ITV network and exporter of such hit series as *Upstairs, Downstairs* and the *Doctor* cycles, it operates from the South Bank Television Centre situated on the south bank of the Thames between Waterloo Bridge and Blackfriars Bridge.

"LONE RANGER"—half-hour filmed series that carried over into TV the classic radio series created by Fran Striker in 1933. ABC carried the first runs (1952–57). CBS picked up the reruns, and then NBC licensed them, giving the 182 films produced by the Wrather Corp. a network run until 1961. Clayton Moore played the lead in most episodes (John Hart also performed the role) with Jay Silverheels as Tonto.

A second version, in animated cartoon form, was introduced on CBS as a Saturday morning entry in 1966. Both versions continue to be sold in syndication by the Wrather Corp.

"LONG HOT SUMMER, THE"—hour-long series based on a movie of that title inspired by fictional works of William Faulkner. Produced by 20th Century-Fox TV, it starred Edmond O'Brien, Dan O'Herlihy, Nancy Malone, Roy Thinnes and Ruth Roman. It ran a single season on ABC in 1965.

"LONGSTREET"—ABC hour-long series about a blind insurance investigator who, with the aid of a guide dog and assistants, solved difficult cases. James Franciscus played the lead and Marlyn Mason and Peter Mark Richman were featured. Produced by Paramount TV, with Stirling Silliphant as executive producer, it ran a single season in 1971–72.

"LOOK UP AND LIVE"—CBS series on religion which incorporates sub-series by the major faiths, with production cooperation from each. Thus, the National Council of Churches assisted in the *Image of Man* series, which presented an overview of man as currently shown in stage, film, TV and musical productions; the National Council of Catholic Men cooperated in a two-program discussion on freedom within the Church and the relationship of Church to society; and the New York Board of Rabbis presented poetry and drawings by some of the doomed children of Terezin Concentration Camp in Czechoslovakia in the 40s.

It went off the air in January 1979, with *Lamp Unto My Feet* and *Camera Three,* to create a time block for the new CBS newscast, *Sunday Morning.* Essentially the series died for lack of station clearances.

LOOMIS, HENRY—president of the Corporation for Public Broadcasting (1972–79) after the departure of John W. Macy. Loomis, a physicist who spent most of his career in public service, had been a Deputy Commissioner of Education, director of the Voice of America, deputy director of the USIA and part of a group of policy advisors to President Nixon in 1968, before joining CPB. He was an advocate of the White House position to decentralize public television and spent virtually all his career on the CPB trying to preserve the legislated powers of the organization against the usurpation of most of them by PBS, the organization of public TV stations. But if Loomis failed to bring inspired leadership to the system, he played a key role in the effort to convert the system to satellite distribution from its dependency on AT&T long lines for interconnection.

James L. Loper

LOPER, JAMES L.—president and general manager of KCET, the Los Angeles public TV station and production center, since 1971. Loper had been assistant to the president when KCET went on the air in September 1964 and became v.p. and general manager in 1966.

Loper had been director of the educational television department at California State University at Los Angeles when he became involved in 1962 with a committee seeking to establish a noncommercial TV station in L.A. He subsequently was hired full-time to help put together the license application and became an officer of the licensee, Community Television of Southern California. Loper served as chairman of PBS from 1969–72 and later became a director on the PBS board of managers.

LORAIN JOURNAL DECISION—case in which the *Lorain Journal* of Lorain, Ohio, was found by the U.S. Appeals Court to be in violation of antitrust laws for refusing to sell space to local advertisers who also bought time on local radio. The suit, charging unfair competition, was brought by WEOL Radio in Elyria, Ohio. The Appeals Court decision was upheld by the Supreme Court in 1951.

LORD, WILLIAM E.—veteran executive of ABC News, who became v.p. and executive producer of *Nightline* in March 1980. Earlier he was v.p. in charge of TV news for ABC (1976–78) and later head of news for *Good Morning America.* Prior to that he had been v.p. in charge of ABC's Washington bureau and earlier had been Washington producer for *ABC Evening News* (1966–74). He joined ABC News as a reporter and writer in 1961.

"LORETTA YOUNG SHOW, THE"—long-running NBC anthology series (1953–60) with Loretta Young as hostess and star. The show became noted for Miss Young's swirling entrances in dazzling gowns. In its first year the series was entitled, *A Letter to Loretta.* The reruns went into syndication as *Loretta Young Theatre.* It was produced by NBC.

LORIMAR PRODUCTIONS—independent Hollywood company founded in 1968 by Lee Rich, former advertising executive, and Merv Adelson, financier and movie producer. The company has produced such series as *The Waltons, Doc Elliott, Apple's Way* and *The Blue Knight* and made-for-TV movies, including *Helter-Skelter, Sybil, Eric* and *The Widow.*

One of the leading independent producers in the 70s, Lorimar went into high gear in the 80s with the blockbuster, *Dallas,* and all its spinoffs—*Knots Landing, Flamingo Road, Falcon Crest,* and *King's Crossings.*

"LOST IN SPACE"—science-fiction series on CBS (1965–68), with plot characteristics of *Swiss Family Robinson*, produced by Irwin Allen Productions in association with Van Barnard Productions and 20th Century-Fox TV. Its regular cast featured Guy Williams, June Lockhart, Mark Goddard, Jonathan Harris, Marta Kristen, Billy Mumy and Angela Cartwright.

LOTTERY—either as a category of information or as a form of station promotion, barred from broadcasting under Section 1304 of the United States Code. The prohibition against the broadcast of lottery information was eased somewhat, however, when state-sponsored lotteries became prevalent. The courts held that the winning number in a state lottery was "news" and therefore protected by the First Amendment. Under a congressional amendment in 1975, it became possible for stations to televise the final drawings in state lotteries as a form of entertainment programming.

Section 1304 states that a licensee may not broadcast "any advertisement or information concerning any lottery." In several instances, promotional activities by stations were held to be violations of the law because they constituted lotteries in that they involved (1) a prize, (2) chance and (3) consideration. Court tests have determined that if one of the three elements is lacking, the activity is not a lottery under federal law.

The prohibition against the broadcast of lottery information was challenged early in the 70s in a petition by the New Jersey Lottery Commission, joined by New Hampshire and Pennsylvania. It called for a declaration that broadcasting the winning numbers from the weekly state drawings would not violate the statute. [New Jersey State Lottery Commission *v.* United States, 491 F2d 219 (Third Circuit 1974).]

The Court of Appeals for the Third Circuit held that the winning ticket number was of interest to nearly 3 million ticket holders, and therefore it was "news," protected both by the First Amendment and by Section 326 of the Communications Act which prohibits censorship. The court felt that Section 1304 should apply only to the "promotion of lotteries for which the station receives compensation."

While the New Jersey Lottery case was pending before the Supreme Court, Congress enacted Section 1307(a), which created certain exemptions to the blanket prohibition in Section 1304. These exemptions applied to lotteries that were conducted by a state and to broadcasts on stations in that state, or in an adjacent state which also conducts a lottery.

The Supreme Court thereupon remanded the case to the Court of Appeals to determine if the new statute would cover the New Jersey situation. [420 U.S. 371 1975]

On remand, New Hampshire argued that although it did conduct a state lottery, its western neighbor did not. Therefore, even under Section 1307 Vermont stations would be barred from broadcasting New Hampshire lottery results, thus hurting its sales. The Court of Appeals [34 RR2d 825 (Third Circuit 1975)] held that the question was not moot, and its original decision was reinstated. The Supreme Court has not yet decided whether to hear an appeal.

LOUDNESS GUIDELINES—advice issued by the FCC in a 1965 policy statement to curb attention-getting practices in behalf of advertisers (such as raising the volume level of commercials above that of the program, a tactic which had brought a stream of complaints from viewers). Deeming the practices "contrary to the public interest," the FCC advised stations to pre-screen commercials for objectionable loudness and to adopt control room procedures that would equalize the volume level of commercials and the program matter that precedes them. The policy guidelines also called for the avoidance of commercials presented in a "rapid-fire, loud and strident manner" and of those using sirens or other effects which might alarm.

"LOU GRANT"—much respected hour-long dramatic series on CBS, created as a vehicle for Ed Asner in the 1977–78 season after the demise of *The Mary Tyler Moore Show,* in which he played the newsroom boss, Lou Grant. Asner continues the role in the new series, but in a less comic fashion, as city editor of a Los Angeles newspaper. Nancy Marchand is featured as the publisher, and Mason Adams, Robert Walden and Linda Kelsey play Grant's colleagues in the newsroom. The series survived a shaky start in the ratings and has gradually built a loyal audience.

The series is produced by MTM Productions, with Gene Reynolds, James L. Brooks and Allan Burns as executive producers.

"LOVE, AMERICAN STYLE"—successful comedy anthology series on ABC (1969–73) which featured playlets, sketches, vignettes and blackouts, all on the general theme of love, with no regular performers. The series began in a one-hour format, was cut back to 30 minutes during 1970 and then was returned to its original length. It was produced by Paramount TV.

"LOVE BOAT, THE"—ABC hit series (1977–   ), an anthology of light romantic stories taking place on a luxurious cruise ship, which has become a Saturday night fixture. Several stories are interwoven in each episode, all of them featuring guest stars. The regular cast is the ship's crew: Gavin MacLeod, Lauren Tewes, Bernie Kopell, Fred Grandy and Ted Lange. The series frequently draws its guest stars from other ABC hits. Aaron Spelling and Douglas S. Cramer are executive producers for Aaron Spelling Productions.

"LOVE OF LIFE"—CBS soap opera which marked more than a quarter century on TV, after its radio run. It premiered Sept. 24, 1951 as a 15-minute program and was expanded to 25 minutes late in the 60s. Ron Tomme and Audrey Peters have played the romantic leads since 1959. Others in the cast have been Joanna Roos, Diane Rousseau, Jerry Lacy, Ray Wise and Tudi Wiggins.

"LOVE THAT BOB"—a high-rated situation comedy of the late 50s in which Bob Cummings (who had been Robert Cummings in movies) played a libidinous bachelor working as a commercial photographer. The series won a moment of fame for Ann B. Davis, who portrayed his homely secretary, Schultzy. Also featured were Rosemary DeCamp and Dwayne Hickman.

Premiering on NBC in January 1954, it moved to CBS for two more seasons, terminating in July 1959. ABC then picked up the reruns for its daytime schedule and carried them for two years. In reruns, the series was retitled *The Bob Cummings Show.* That was also the title of a new CBS series Cummings began in October 1961, which cast him as a trouble-shooting adventurer and featured Roberta Shore and Murvyn Vye. It ran only four months.

LOWE, DAVID (d. 1965)—producer of bold investigative documentaries for *CBS Reports,* who, in 1960 contributed one that is considered a classic, the disturbing *Harvest of Shame,* narrated by Edward R. Murrow. A few days after his death, he won an Emmy for *KKK: The Invisible Empire.* Other of his documentaries were reports on such matters as the racial strife in Birmingham, Ala., the threat of nuclear war, the right to bear arms, abortion and the law, the funeral industry in the U.S. and the racial situation in South Africa. He was married to columnist Harriet Van Horne.

LOWER, ELMER W.—retired TV news official who held posts at all three networks; he became v.p. of corporate affairs for the American Broadcasting Companies Inc. in 1974 after 11 years as president of ABC News. Lower was responsible for building the modern ABC News operation and for raising its stature from a routine service to one of the major news organizations in the world.

While he was with NBC News, Lower developed the modern electronic tabulating methods for election returns that became the foundation of NES (National Election Service), a co-operative of the three networks and the two U.S. wire services.

Lower joined ABC News in 1963 after a four-year stint with NBC News, during which he served first as Washington bureau chief and then as v.p. and general manager of the division in New York. Earlier (1953–59), Lower had been with CBS News as director of special projects in New York. He entered broadcasting after a career in print journalism, as foreign correspondent for *Life* magazine (1945–51) and previously as a reporter and editor for newspapers and wire services.

Elmer W. Lower

LOW-POWER TELEVISION—new kind of television station licensed to cover a limited geographical area in a radius of 10 to 15 miles and capable of being inserted into the VHF and UHF bands without interfering with an existing full-powered station. In New York, for example, an LPTV station might cover a single demographically cohesive area such as Bedford-Stuyvesant or Harlem.

The concept of low-power television was advanced by the FCC under Charles Ferris in the Carter Administration, both to increase over-the-air competition and to provide minorities with new opportunities to become broadcasters. Low-power stations, which are modeled on the translators (or repeater stations) that have existed for years in rural areas, are estimated to cost as little as $100,000 to construct. In short order, there were more than 6,500 applications for the 3,000 to 4,000 frequencies that might become available.

Many of the applicants were minorities, but NBC and ABC also put in bids for large numbers of stations, as did pay-television interests, Sears Roebuck, and trade unions. One company in the southwest sought a chain of low-power stations to create a network that would reflect small town values and interests. The first low-power station to be licensed was in Bemidji, Minn., which began operations in 1981; but a freeze was put on all other licensing because there were so many competing applications for the choicest frequencies.

The freeze was lifted in the Reagan Administration, in March 1982, when the FCC opened the way to licensing, promising to give preference to minority applicants. But it also envisioned competition for the major networks through the interconnection of low-power stations by satellite.

The LPTV stations would be subject to fewer restrictions and regulations than conventional television stations, but on the other hand they would not come under the FCC's "must carry" rules for cable television, which means that most of them would have to survive strictly on over-the-air transmissions and could not be received on sets hooked up to cable.

LOXTON, DAVID—director of the Television Laboratory, the experimental unit at WNET New York, since it was established in 1971. In that capacity he served as executive producer of the series, *VTR—Video and Television Review,* and supervised such alternate video documentaries as *Gerald Ford's America, Lord of the Universe, The Good Times Are Killing Me* and *Super Bowl.* Earlier (1966–71) he had been an associate producer and producer for National Educational Television. He made his commercial TV debut in 1976 as co-producer, with Frederick Barzyk, of a late-night NBC project, *People.*

"LOYAL OPPOSITION" SERIES—a plan by CBS president Frank Stanton in 1970 to grant 25 minutes of free TV time several times a year to permit the opposition political party to express its views on presidential policies. What was to have been a sporadic series was cancelled after the first installment because of the controversy it raised.

Stanton's purpose with the series was to offset, in a two-party system, the natural advantage enjoyed by a President to command air time for the promulgation of his views on issues. At the time, the networks were under pressure from several Democrats in Congress who opposed the war in Vietnam and wanted air time to challenge the Administration's policies; the networks were in the position of accessories to the prowar policies if they did not balance President Nixon's telecast with the views of the significant opposition.

CBS aired the first and only *Loyal Opposition* telecast on July 7, 1970, having left the use of the time to the discretion of the Democratic National Committee. DNC chairman Lawrence O'Brien chose to go on the air personally, and his presentation was sharply critical of President Nixon on a number of issues, beyond that of the war.

The 25-minute program was followed in many cities by a paid five-minute spot soliciting funds for the Democratic party.

The Republican National Committee immediately demanded time from CBS to reply to the O'Brien broadcast,

calling it an attack on the President rather than a discussion of the issues. When CBS refused to grant the time, the RNC complained to the FCC. After agonized deliberations, the commission held that under the Fairness Doctrine the RNC was entitled to the time. CBS then appealed to the courts, and the FCC decision was reversed. But the problems raised by *Loyal Opposition* caused CBS to abandon the plan for a series.

"LUCAS TANNER"—hour-long series for NBC (1974) about a former professional baseball player who becomes a free-thinking high school teacher. Produced by Universal-Groverton Productions, and starring David Hartman, it ran one season.

Allen Ludden

LUDDEN, ALLEN (d. 1981)—game-show host and moderator, whose principal credits include *Password* and *G.E. College Bowl,* although he has been ubiquitous in daytime TV since 1961. Earlier he had done multiple duty as program director for the CBS-owned radio stations, executive of CBS News and director of the network's creative services, while moderating the radio and then the TV versions of *College Bowl.* Ludden and his wife, actress Betty White, formed their own production company, Albets, which produced her syndicated series, *The Pet Set.*

LUMET, SIDNEY—noted film director who developed his skills with CBS during the 50s in such programs as *Mama, Danger, You Are There, Omnibus, The Best of Broadway* and *Alcoa-Goodyear Playhouse.* The plays he directed on TV included *Twelve Angry Men, All the King's Men, The Iceman Cometh, The Sacco and Vanzetti Story* and *The Dybbuk.* Like many others of TV's "Golden Age" who went into film, he never returned to television.

M

MACDONALD, TORBERT H. (REP.) (d. 1976)—chairman of the House Communications Subcommittee in the early 70s, responsible for a number of significant bills affecting TV and radio. In 1967 he sponsored the legislation in the House that led to establishing the Corp. for Public Broadcasting. Later he was a force behind the Federal Election Campaign Act of 1971, which limited media spending by candidates for federal office. He also introduced the bill in 1973 that banned TV blackouts of professional football home games and other sports events if they were sold out in advance. Before illness forced him to resign in the spring of 1976, his major project was to clear away many of cable-TV's regulatory obstructions so that it might grow. A Democrat from Massachusetts, Macdonald served 11 consecutive terms in the House. He had close political and personal ties to President Kennedy, having been his roommate at Harvard from 1936 to 1940.

MACK, CHARLES J. (d. 1976)—one of the chief cameramen for CBS News dating to the beginnings of the Edward R. Murrow-Fred W. Friendly *See It Now* unit. Mack, in fact, had been part of the Hearst *News of the Day* newsreel team that had been contracted for *See It Now* and later signed on with CBS when Murrow severed relations with Hearst. Mack remained a leading cameraman for the network until he retired in 1971. Among his credits was the documentary, *What Really Happened in Tonkin Gulf,* which won an Emmy for CBS News.

MACK, RICHARD A. (d. 1963)—an FCC commissioner who was forced to resign in March 1958 under charges that he had sold his vote to an applicant for Channel 10 in Miami. The House Legislative Oversight Committee, headed by Rep. Oren Harris (D.-Ark.), had produced checks made out to Commissioner Mack by Thurman Whiteside, who was alleged to be acting for National Airlines, the winning applicant for the station. The case was tried in federal court but resulted in a hung jury, and the retrial was never held. Whiteside committed suicide and Mack, who had become an alcoholic, died in a cheap rooming house.

In 1960 the FCC canceled its grant of the Channel 10 license to National Airlines and awarded it instead to L.B. Wilson Inc.

MACK, TED (d. 1976)—host of the *Original Amateur Hour* for 22 years, taking it from radio to TV. When CBS canceled the show in 1970, Mack tried to revive it in syndication, without success. Among the 10,000 amateurs who appeared on the program over the years, around 500 became professionals, among them Maria Callas, Frank Sinatra (as a member of a quartet known as the Hoboken Four), Robert Merrill, Ann-Margret, Pat Boone, Paul Winchell, Vera-Ellen, Regina Resnik, Mimi Benzell, Jerry Vale and Jack Carter.

Amateur Hour started as a radio program in 1934 and soon was taken over by Major Edward Bowes. Mack became its talent scout and Bowes's assistant, eventually taking over the show when Bowes died in 1946. Two years later he moved it into TV where initially it played on four stations of the

DuMont Network before becoming a fringe-time staple on CBS.

Mack began in show business as a clarinetist and saxophonist with a number of big bands and for a time headed a band of his own.

Robert MacNeil

Ted Mack

MacLEISH, ROD—news commentator for CBS since the fall of 1976, noted for his writing style and his perceptive essays on social change in the U.S. He joined the network after a 21-year association with Group W Broadcasting.

After serving as news director of the Westinghouse station in Boston, WBZ, MacLeish organized and headed the company's Washington bureau (1957–59) and then did the same with its foreign news service (1959–66). He became a commentator in 1966 but continued to go abroad from time to time on reporting assignments as events warranted. He is nephew of the poet, Archibald MacLeish.

"MacNEIL-LEHRER REPORT, THE"—nightly half-hour news program on PBS which premiered in 1976 as a co-production of WNET New York and WETA Washington, cutting between the studios of both. The program actually had begun the previous year as a local origination for the two cities but gradually gained other outlets and eventually was adopted by the whole system as its nightly news program.

Leaving to the commercial networks and stations the task of delivering the laundry list of news items, *MacNeil-Lehrer* concentrates on covering a single major story in depth each night, using guest experts for the discussions, including other journalists. Robert MacNeil is the New York anchor and Jim Lehrer the Washington. PBS scheduled the program at 7:30 ,P.,M., where it would follow the network news in most cities and compete with nothing more formidable than prime-access programming.

MacNEIL, ROBERT—TV journalist mainly with public TV in the U.S. but whose career also included stints with NBC, Britain's BBC and Canada's CBC. In August 1975 he left the BBC to become anchorman and executive editor of a nightly news and commentary program on WNET New York, *The MacNeil Report.* The program later was sent out over PBS as *The MacNeil-Lehrer Report.*

A native Canadian, he began his career with the CBC and in 1955 went to England for Reuters. He joined NBC News as a London correspondent in 1960 and three years later was transferred to the Washington bureau and then to New York. There he anchored WNBC's local evening newscast and co-anchored NBC's weekend *Scherer-MacNeil Report.* He left to join the BBC the first time in 1967, then returned to the U.S. to narrate and write a documentary for the Public Broadcasting Laboratory on the 1968 Democratic National Convention, *The Whole World Is Watching.*

In 1971 MacNeil became senior correspondent for public TV's NPACT, anchoring several of its shows including *Washington Week In Review* and the coverage of the Senate Watergate hearings. He rejoined the BBC in 1973 to make a series of documentaries about the U.S. and to participate in several co-productions of the British and American public systems.

His credits include numerous documentaries and a book on TV's influence on American politics, *The People Machine.*

"MacPHAIL, WILLIAM C.—head of sports for CBS from 1955 to 1974, receiving the title of v.p. in 1961 when the sports unit shifted from CBS News to become a separate department of CBS-TV. For nearly 10 years before joining CBS, he was a baseball executive, like his father, Col. Larry MacPhail and his brother Lee, both of whom were officials of the N.Y. Yankees. On leaving CBS he became a v.p. of the

Robert Wold Co. and, in 1979 joined Ted Turner's Cable News Network as chief sports executive.

MACY, JOHN W., JR.—first president of the Corporation for Public Broadcasting (1969–72) who resigned in frustration after a number of difficulties with the Nixon Administration, which disapproved of the direction public TV was taking, and after failing to win the support of PTV stations in trying to build CPB into a strong national organization. He went on to become president of the Development and Resources Corp.

Macy was appointed to the CPB by President Johnson when the organization was formed, after having been executive v.p. of Wesleyan University, chairman of the Civil Service Commission under Presidents Kennedy and Johnson and a recruiter for Johnson for government positions. He was driven from PTV when the Nixon Administration accused PBS (then an arm of CPB) of a liberal political bias, wanted federal funds to be funneled directly to the stations rather than to PBS and finally vetoed the $155 million authorization bill for CPB. On leaving, Macy said he believed the reelection of Nixon would be the death-blow to public TV.

"MADIGAN"—short-lived NBC series (1972), based on a motion picture, featuring Richard Widmark as a New York cop on special assignment. Four episodes were presented on a rotating basis. It was by Oden Productions in association with Universal TV.

"MAGAZINE"—occasional CBS daytime series in news-magazine format, which began in 1973 on a schedule of four-a-year and later expanded to six. With Sylvia Chase as anchorwoman, and with a predominantly female staff, the series concentrated on issues of specific interest to women, such as unnecessary hysterectomies and profiles of prominent women. The one-hour shows were presented as daytime specials, preempting regular soap operas and game shows.

MAGAZINE CONCEPT—commercial policy that permits advertisers to buy spots in programs the same way they buy pages in magazines, strictly on a circulation basis with no direct association with the program (or article) or its content. The opposite of program sponsorship, the magazine concept—or participating advertising—has prevailed in TV since the mid-60s.

A pioneer of the concept was Sylvester (Pat) Weaver who, while he was president of NBC in the early 50s, developed the *Today, Home* and *Tonight* shows to be sold in that manner. But the great impetus came from ABC around 1965 when, in desperation for prime-time sales, it began to sell inser-

tions in programs on scatter plans where it was unable to find sponsors.

The magazine concept served to give the networks full control over programs and overall has proved more profitable to them. At the same time it has afforded greater access to television for the smaller advertisers and has benefited the larger ones by eliminating the risks inherent in program sponsorship.

MAGAZINE FORMAT—a program made up of varied segments, long and short, on a variety of subjects or themes, just as magazines are made up of assorted articles, departments and graphic materials, some serious and some not. The application is usually to news or public affairs programming: *60 Minutes, Weekend, PBL* and *The Great American Dream Machine,* for example. But in the late 70s it was adopted for straight entertainment with *Real People, That's Incredible,* and *Those Amazing Animals.* The longest-running magazine show of all is the ABC anthology, *Wide World of Sports.*

Magazine programs with an entertainment accent began to proliferate in local TV during 1976, and several syndicated shows in the format came on the market for the 1977–78 season. At the local level the form was pioneered by such stations as KCRA Sacramento, KGW Portland, Ore., WMAL-TV Washington, and KPIX San Francisco. These efforts spawned Group W Productions' *PM Magazine,* one of the most potent syndicated programs of the 80s.

"MAGICIAN, THE"—hour-long adventure series for NBC (1973) about a performing magician who helps people in trouble through the skillful and timely application of the tricks of his trade. Starring Bill Bixby and produced by Paramount TV, it was canceled in its first season.

MAGID ASSOCIATES—market research firm specializing in consulting TV stations on improving the appeal of their newscasts. Based in suburban Cedar Rapids, Ia., and headed by Frank N. Magid, whose graduate school field was social psychology, the company became the industry's leading "news consultant" in the 70s with clients in more than 100 markets. Contributing to its growth was the discovery by the industry that stations which led their markets in news were usually first in overall ratings, regardless of their network affiliations.

The Magid influence was felt primarily in the explosion of *Action News* formats around the country and in the technique of presenting news items tersely, in among banter between members of a newscasting "team." The Magid recommendations extended to news content as well as to the manner of presentation and the selection of personnel.

MAHONY, SHEILA—lawyer who became executive in charge of franchising activites for Cablevision Corp., after having been executive director of the Carnegie Commission on the Future of Public Broadcasting (1977–79). For the previous two years she was executive director of the Cable Television Information Center in Washington, a nonprofit organization devoted to assisting state and local governments with cable television franchising, and before that she worked for the City of New York in cable television matters.

MAJOR MARKET—one of the 50 largest television markets, the designation having importance both in national spot sales and in the application of certain FCC regulations. Many spot advertisers specify only the major markets, which means that those below 50 will not be purchased. The FCC occasionally will establish rules that apply only to stations in the top 50 markets, the Prime Time–Access Rules for example. A market is defined by the effective reach of a station's television signal, and it is measured for size by the density of TV households.

"MAKE A WISH"—ABC half-hour informational children's series scheduled Sunday mornings (1971–77). Hosted by Tom Chapin, it was produced, directed and written by Lester Cooper of ABC Public Affairs.

"MAKE ME LAUGH"—comedy game show introduced on ABC as a prime-time series in 1958 and revived in syndication in 1979 by a consortium of independent stations calling themselves the Program Development Group. With Bobby Van as host, the game employs young comedians who work at making other participants break into laughter.

George Foster, who produced the original network show, also produced the syndicated version through Paramount Television.

"MAKE ROOM FOR DADDY"—early situation comedy vehicle for Danny Thomas on ABC in which he portrayed a family man working as a nightclub comedian, with Jean Hagen as his wife. When Miss Hagen left the cast in 1957 and Marjorie Lord replaced her, the series was retitled *The Danny Thomas Show.* The series moved to CBS the following year and ran until 1964. Reruns were purchased by NBC for daytime stripping in 1960 and the repeats then went into syndication. Angela Cartwright and Rusty Hamer portrayed Thomas's children.

In 1970 ABC made an attempt to revive the series as *Make Room For Grandaddy,* with the members of the previous cast a generation later, but it did not fare well and was canceled after a season.

The Making of the President 1964

"MAKING OF THE PRESIDENT, THE" (1960 and 1964)—90-minute CBS documentaries based on Theodore H. White's quadrennial best-selling books on the campaigns for the Presidency. The first, *The Making of the President— 1960,* centering on John F. Kennedy, was produced in 1961 by Metromedia Producers Corp. The second (1964), covering the political race between Lyndon B. Johnson and Barry Goldwater, was televised in October 1965 and was produced by Wolper Productions. Both were narrated by Martin Gabel.

The Making of the President—1972, documenting Richard M. Nixon's smashing victory over George McGovern, was produced by Metromedia but rejected by the networks, apparently because it was offered while the Watergate scandal was unfolding.

"MAMA"—popular sentimental comedy on CBS from July 1949 to March 1957, based on the play *I Remember Mama,* which had been drawn from the book, *Mama's Bank Account.* Thousands of letters from viewers brought the program back in December 1956, but the revival lasted only three months. Presented live, it played on Friday nights under sponsorship by Maxwell House coffee.

It starred Peggy Wood as the immigrant matriarch from Norway who, with her husband (played by Judson Laire), learned to adapt to American ways through their three children. Set in the early 1900s, it dealt not so much in humor as in the commonplace problems of youth that could be resolved by parental wisdom, and in the personal wounds that could be healed by family love. *Mama* brought on a wave of domestic comedies on TV.

Featured were Rosemary Rice as Katrin, Dick Van Patten as Nels, Robin Morgan as Dagmar and Ruth Gates as Aunt Jenny.

"MAN AGAINST CRIME"—early police series on CBS (1949–53) starring Ralph Bellamy and produced by MCA-TV. The reruns were syndicated as *Follow That Man*.

"MAN AND THE CITY, THE"—hour-long dramatic series starring Anthony Quinn as the mayor of a middle-sized southwestern city. Produced by Universal TV for ABC, the series lasted only 13 weeks in 1971.

"MAN CALLED SHENANDOAH"—post-Civil War western by MGM-TV which had a brief run on ABC (1965–66). Robert Horton was featured.

"MAN FROM U.N.C.L.E., THE"—popular NBC action-adventure series on international espionage (1964–68), which featured Robert Vaughn and David McCallum as secret agents and Leo G. Carroll as their superior officer. In 1966 the series spun off *The Girl from U.N.C.L.E.* with Stephanie Powers and Noel Harrison, which lasted only a single season. *Man's* demise was attributed by some to the frequent shifting of its time period by NBC. Norman Felton was executive producer of both *U.N.C.L.E.* series, which were by MGM-TV in association with Felton's Arena Productions.

MANDEL, LORING—New York-based dramatist who wrote original plays for the quality showcases of the 50s and 60s and more recently for *CBS Playhouse*, the Benjamin Franklin series, *Sandburg's Lincoln* and the CTW drama series, *The Best of Families*. He received an Emmy for *Do Not Go Gently into That Good Night*. He also has written screenplays and stage dramas, including the Broadway version of *Advise and Consent*.

"MANHUNT"—half-hour series based on the files of the San Diego Police Department, produced by Screen Gems for first-run syndication (1959–60). It featured Victor Jory and Pat McVey.

MANINGS, ALLAN—veteran comedy writer who, in his association with Norman Lear, became producer and then executive producer of *Good Times* and co-creator of *One Day at a Time*.

Abby Mann

MANN, ABBY—writer and producer, noted for such TV dramas as *Judgment at Nuremberg, The Marcus-Nelson Murders* (which led to the series, *Kojak*) and *A Child Is Waiting*, in addition to plays for *Studio One, Playhouse 90, Robert Montgomery Presents* and other showcases of the "golden age" of TV drama. He also created *Medical Story*, an anthology series for NBC (1975) which won praise but fell short in the ratings. Mann then set about working on a four-hour TV dramatic special for NBC, *King*, on the life of the Rev. Dr. Martin Luther King, Jr., for presentation in 1977 on the eighth anniversary of the civil rights leader's death. Mann was writer, director and executive producer of the film.

MANN, DELBERT—one of the group of motion picture directors who received their grounding in the live TV drama anthologies of the 50s. Mann directed both the TV and movie versions of such plays as *Marty, Bachelor Party* and *The Middle of the Night* and has concentrated on movies since the 60s. He had been staff director for NBC in the early years (1949–55) and worked on such series as *Lights Out, Philco-Goodyear Playhouse* and *Producer's Showcase;* later he did *Playhouse 90, Playwrights '56* and *Ford Star Jubilee*, among others. His TV directing credits include *The Day Lincoln Was Shot, Our Town, Heidi, The Man Without a Country* and a 1955 production of *The Petrified Forest* with Humphrey Bogart, Lauren Bacall and Henry Fonda. In 1980, he directed *Playing for Time*.

MANNING, GORDON—news executive, first with CBS as senior v.p. and director of news (1971–75) and then with NBC News (since May 1975) as executive producer of special broadcasts. For NBC he headed a new unit created to handle special events, from the Apollo-Soyuz space mission to the primaries, conventions and elections of 1976. In 1979 he became v.p. of politics and special programs for NBC News.

Manning joined CBS News in 1964 as a producer, having previously been back-of-the-book editor for *Newsweek* (1956–61) and managing editor of *Collier's* (1950–56).

"MANNIX"—long-running CBS private-eye series (1967–76) with Mike Connors in the title role and Gail Fisher as his secretary. It was via Paramount TV.

MANSFIELD, IRVING—producer of *Your Show of Shows* and *Talent Scouts,* two outstanding hit series of the 50s. Later he became business manager for his novelist-wife, the late Jacqueline Susann.

MANSFIELD JOURNAL CO. *v.* **FCC [180 F2d 28 (D.C. Cir. 1950)]**—case which strengthened the policy of the FCC to diversify control of the outlets of mass communications in a community.

The Mansfield Journal Co., which published the only newspaper in Mansfield, Ohio, refused as a competitive maneuver, to accept advertising from any firm that advertised on the town's only local radio station. When the Journal Company later applied for radio licenses in several areas outside its immediate community, the FCC denied all its applications because the newspaper's advertising policy bespoke an attempt to suppress competition and to secure a monopoly of mass advertising and news dissemination.

The Journal Company appealed to the courts, charging the commission with acting beyond the scope of its power by enforcing the antitrust laws. The D.C. Court of Appeals disagreed with the *Journal,* holding that a formal violation of the antitrust laws was irrelevant to the *Journal's* case because the commission never made such a determination. The only question had been the Mansfield *Journal's* ability to serve the public, and the commission had every right to consider whether the Journal's advertising policies, as an attempt to monopolize mass communications, were in the public interest. See also Cross-ownership.

MANTLEY, JOHN—executive producer of *Gunsmoke* and later of *How the West Was Won* (1976–77).

MANULIS, MARTIN—prominent producer-director of the 50s and early 60s associated with *Studio One, Playhouse 90, Suspense, Climax* and others, chiefly as a staff producer-director for CBS (1951–58). He later became head of produc-

tion for 20th Century-Fox TV and then formed his own motion picture company.

"MANY LOVES OF DOBIE GILLIS, THE"—CBS situation comedy (1959–63) created by humorist Max Shulman, which starred Dwayne Hickman and featured Bob Denver and Frank Faylen. It was via 20th Century-Fox TV.

MARCH, ALEX—producer and director of TV dramas, prominent during the 50s as producer of the *Studio One Summer Theater* on CBS. In 1960 he produced *The Sacco and Vanzetti Story* and directed *The Story of Margaret Bourke-White.* In 1982, he was named supervising producer for *Nurse.*

"MARCH OF TIME"—documentary series produced for syndication in 1966 by Wolper Productions as an attempt to revive, for TV, the series that had played movie houses in the 40s. Eight films were produced, all narrated by William Conrad. They included *And Away We Go,* on the proliferation of the auto; *Search For Vengeance,* on the international underground pursuing Nazi war criminals; and *Seven Days in the Life of the President,* on President Lyndon Johnson during a week of crisis.

MARCUS, ANN—writer of comedy serials who made her mark with Norman Lear's T.A.T. Productions in the mid-70s as cocreator of *Mary Hartman, Mary Hartman* and *All That Glitters,* then joined Columbia Pictures Television to write the NBC daytime serial, *Days of Our Lives.* Teamed with her husband Ellis, she created a new late-night syndicated strip in 1979, *Life and Times of Eddie Roberts,* known also by the acronym *L.A.T.E.R.* It was unsuccessful.

Robert Young as Marcus Welby

"MARCUS WELBY, M.D."—highly successful medical melodrama on ABC (1969–76) which served as a second TV vehicle for Robert Young almost a generation after the first, *Father Knows Best*. The hour-long series cast him as a general practitioner of the old school working with a young associate schooled in the new medicine who made house calls on a motorcycle. The series was a big hit from the start, but many attribute that to the fact of weakish competition; CBS had scheduled its newsmagazine *60 Minutes* in that Tuesday night timeslot and NBC its monthly newsmagazine, *First Tuesday*. James Brolin portrayed the younger doctor, and Elena Verdugo was featured. *Welby* was created by David Victor, who served as executive producer for Universal TV.

"MARK SABER MYSTERY THEATER"—half-hour detective series (1951–54) drawn from radio's *Mystery Theater* with Tom Conway as inspector Saber of the homicide squad and Paul Burke featured. The series was produced by Roland Reed Productions. A new version, *Saber of Scotland Yard*, was produced for syndication by Sterling Drug from 1957 to 1960, which turned Mark Saber into a one-armed British private eye. The role was played by Donald Gray, and the producers were Edward J. and Harry Lee Danziger.

MARKELL, ROBERT—drama producer who had started as a scene designer in the early days of TV, notching his first producing credits with *Playhouse 90*. He was producer of *The Defenders* in the 60s and in 1975 became executive producer of *CBS Playhouse* and the network's nightly *Bicentennial Minutes*. For *CBS Playhouse* he produced *20 Shades of Pink* and *The Tenth Level*, among other original plays. In 1981, he supervised the production of the CBS miniseries, *The Blue and The Gray*.

MARKLE, FLETCHER—writer, producer and director from Canada who was also prominent in the "golden age" of studio drama and directed such series as *Life with Father* (1953–55) and *Father of the Bride* (1961). He had been director for *Studio One* on CBS in 1947–48 and returned to that series as producer in 1952–53.

His initial broadcasting experience had been with CBC and BBC, chiefly in radio. Other U.S. credits include *Ford Theatre*, *Front Row Center*, *Mystery Theatre*, *Panic*, *Colgate Theatre*, *Lux Playhouse*, *Thriller* and *Hong Kong*.

MARKS, ALBERT A., JR.—executive producer of the *Miss America Beauty Pageant* telecast for many years.

MARSHALL, GARRY—prolific comedy writer who hit a jackpot in the mid-70s as creator and co-executive producer of *Happy Days*, *Laverne & Shirley*, *Mork and Mindy* and *Angie*. He also created and co-produced *The Odd Couple* series and *The Little People* (later retitled *The Brian Keith Show*).

Unlike other successful producers who started their own companies, Marshall continued to work through Paramount TV. His father, Tony Marshall, was on his staff as a producer and his sister, Penny, played Laverne in *Laverne and Shirley*. Another sister, Ronny, served as an associate producer.

Marshall began his TV career as a comedy writer for Jack Paar in 1960 and then became a writer for Joey Bishop's late-night show on ABC. For seven years he teamed with Jerry Belson in the writing of movies, specials and comedy series, one of which was the situation comedy adaptation of *The Odd Couple*. Marshall then became executive producer of that series for Paramount and in 1974 created *Happy Days*.

"MARTIN KANE, PRIVATE EYE"—progenitor of the long line of TV detective series, airing from 1949 to 1954 and revived by United Artists TV in 1957 as *The New Adventures of Martin Kane*, with William Gargan in the lead. In the earlier series the role was also played by Lloyd Nolan, Mark Stevens and Lee Tracy.

Martin & Lewis

MARTIN & LEWIS—red-hot comedy act of Dean Martin and Jerry Lewis which in 1950 began successful seasons in NBC's *Colgate Comedy Hour* rotation. The act was an outstanding modern vaudeville pairing of a suave and handsome singer (Martin) with a frantic buffoon (Lewis). The team split up in 1956. Martin went on to make films and had great success with his own NBC variety show (1965–74), which projected him as a likeable roue. Lewis was also successful as a solo but on a different scale. His movies appealed chiefly to youngsters and rural folk, and his TV triumphs were an animated Saturday morning children's show, which starred his likeness, and the annual Muscular Dystrophy Telethons, which he hosted. His venture in a weekly Sunday night variety series on ABC failed.

MARTIN, MARY—Broadway musical-comedy star whose occasional TV appearances resulted in some of the medium's memorable hours during the early years. One of her stunning contributions was on *Ford's 50th Anniversary Show* on June 15, 1953, carried on both CBS and NBC, when she and Ethel Merman sang a Broadway catalog, seated on stools on a bare stage. But even more potent was her performance in *Peter Pan,* in the title role, offered first on March 7, 1955 on NBC (before the video tape era) and repeated live on Jan. 9, 1956. Once taped, Mary Martin's *Peter Pan* was repeated on NBC for several years.

Later she co-starred with Helen Hayes in *The Skin Of Our Teeth* for an NBC *Sunday Spectacular* and starred in a TV version of *Annie Get Your Gun.* Her specials included *Magic with Mary Martin,* a program which traced her career in music, and *Mary Martin at Easter Time.*

MARTIN, QUINN—highly successful producer of adventure series, usually with police themes, whose QM Productions was one of the most active independent companies on network TV through the 60s and 70s. His first major hit, *The Untouchables,* was produced for Desilu Productions in 1959. On his own, he scored in 1963 with *The Fugitive* and in 1965 with *The F.B.I.* Then came *Cannon, Barnaby Jones, Streets of San Francisco* and *Bert D'Angelo/Superstar,* all of them running simultaneously in 1976 with Martin as executive producer. Among his lesser shows were *Banyon, The New Breed* and *Caribe.*

Earlier in his career he wrote for Four Star Playhouse and produced The Jane Wyman Show and Desilu Playhouse. The Untouchables emerged from the latter and started Martin on his way as a specialist in crime-adventure shows.

MARTIN, STEVE—comedian of the late seventies/early eighties who made his mark with the catchphrase "Excu-u-u-se me!," white suits, white hair, and a parodic wit. See Martin Mull.

MARX, GROUCHO (d. 1977)—the wise-cracking member of the four Marx Brothers of vaudeville and films, whose singular talents were harnessed for television in the half-hour quiz show, *You Bet Your Life,* which began Oct. 5, 1950 and ran for 10 seasons. The program, which offered modest prize money and often featured bizarre contestants, succeeded not

for the intrigues of the game but for the opportunities it presented for Marx's brash, ad lib humor in his interviews with the guests. Marx was noted for outrageously candid quips that verged on insult.

His comic trappings were a morning coat, an aggressive cigar and a broad, rectangular mustache. His show had a single extra gimmick which became famous: If a contestant chanced to speak a predetermined "magic word," an absurd stuffed duck descended to signify the winning of a bonus prize. *You Bet Your Life* was Marx's only TV series, although he made numerous guest shots. In 1974 video tapes and kinescopes of the old shows were syndicated to stations with a reasonable degree of success.

MARX, MARVIN (d. 1975)—TV comedy writer and producer, noted for writing the Jackie Gleason-Art Carney classic, *The Honeymooners.* His association with that series, which had periodic revivals, spanned 17 years. In 1968 Marx created a novel musical-comedy series for NBC, *That's Life,* which lasted a single season. He died in a nursing home in Miami at the age of 50.

Steve Martin

"MARY HARTMAN, MARY HARTMAN"—phenomenal syndicated strip that was part legitimate soap opera and part spoof. It enjoyed a vogue in around 100 markets during 1976 and would have had wider distribution but for its sex-oriented themes, which deterred many stations from buying it. Producer Norman Lear, who was the leading television independent with a string of situation comedy hits on the networks, had put the daily half-hour series into syndication after all three networks had turned it down (CBS financed the pilot). *MHMH* became one of the most talked-about shows in television, but because the 26-week charter contracts were for a low try-out fee, the series, paradoxically, was losing around $50,000 a week while it was scoring well in all the ratings, whether scheduled by stations in the afternoon or the late evening. Desirous of keeping it going, such groups as Metromedia and Kaiser, enjoying the benefits of its success, offered to renegotiate their contracts at higher fees.

The cast was headed by Louise Lasser, Phil Bruns, Dody Goodman, Debralee Scott, Mary Kay Place, Graham Jarvis, Greg Mullavey, Victor Killan and Claudia Lamb. Lear's T.A.T. Productions was its source, and it was distributed initially by Rhodes Productions. The series ended its run, voluntarily, in the summer of 1977 because of the strain of producing five shows a week. It gave way to a new fictional talk show, with characters from *Mary Hartman*, entitled *Fernwood 2-Night*.

"MARY TYLER MOORE SHOW, THE"—CBS situation comedy (1970–77) considered by many to have come closest of all sitcoms to perfection in the form, partly a function of superb casting. Miss Moore, best known previously as the wife and second lead in the original *Dick Van Dyke Show*, was presented here as a single girl rebounding from a romantic disappointment and trying to build a career as a functionary in the newsroom of a second-rate Minneapolis TV station. The programs relied on two principal sets, each with its own cast of characters: the newsroom and Miss Moore's bachelor apartment. The latter resulted in spin-off series for two neighbors, *Rhoda* and *Phyllis,* played respectively by Valerie Harper and Cloris Leachman.

But TV stars were spawned by the newsroom, as well. Edward Asner, a previously obscure character actor, won a batch of Emmys for his role as the crusty but kindly news director and became an actor in high demand. He went on to star in his own series, *Lou Grant.* Likewise, Gavin MacLeod got his own series, *The Love Boat.* Ted Knight, a virtual unknown until the series, became a household word as Ted Baxter, the vain and obtuse anchorman; and Georgia Engel and Betty White received strong career boosts from the show.

The series became the keystone of the powerful CBS Saturday night lineup and was terminated at its height of popularity at the will of the star. As for Miss Moore, symbol of the wholesome, determined and humane middle-American woman, she not only enjoyed great personal popularity from the series but built from it (with her husband, Grant Tinker) one of the leading independent production companies in Hollywood, MTM Enterprises.

James Brooks and Allan Burns, the creative nucleus, received continuing producer-writer credit.

"M*A*S*H" immensely popular CBS series based on Robert Altman's 1970 hit movie of that title concerning war-weary surgeons during the Korean War who labor to keep

their sanity with dark jokes, pranks, and irreverent wise-cracks. After a slow start in 1972 and several uncertain subsequent seasons, the half hour series caught fire in 1975 when it was given a Monday night timeslot. For the next seven years it ranked regularly in the Nielsen top ten, despite a succession of cast changes. But it has had even greater success in syndication, where it promises to earn more money than any program ever.

*M*A*S*H* belongs to a long line of war sitcoms (*You'll Never Get Rich, Hogan's Heroes, F Troop, McHale's Navy*) but differs from the others because it adheres to realism rather than contrived situations and because it is black comedy, with pain and death always on hand. Moreover, its philosophical roots are in the anti-establishment, anti-war movements of the 60s, making it an anachronism in the conservative climate of the 80s in which it flourishes. To a large extent, the program's appeal flows from the high quality of its scripts, the excellent production standard that is consistently maintained, and the superb ensemble playing of its cast. Among its other virtues *M*A*S*H* does not strain for laughs.

Loretta Swit, Alan Alda & Mike Farrell in *M*A*S*H*

What is most remarkable about the program, and helps to explain its phenomenal success in syndication, is that it appeals to all levels and age groups of the audience. Thus it pulls high ratings wherever it is scheduled by stations in the early evening, when children and the elderly are watching, or in late night time periods frequented by teenagers and young adults.

Reportedly, 20th Century-Fox Television reaped $25 million from the first round of syndication, and after the program's proven success on local stations asked five times the original price for the second round in 1982. As the star of the series with a sizeable profit participation, Alan Alda is expected to earn $30 million from syndication alone.

In addition to Alda the regular cast includes Loretta Swit, Mike Farrell, Harry Morgan, David Ogden Stiers,

Jamie Farr, and William Christopher. Gary Burghoff, Wayne Rogers, and McLean Stevenson, who were in the original cast, dropped out at various points during the run to pursue other performing opportunities.

Larry Gelbart was creator of the series and writer of its earliest episodes, and Gene Reynolds was original executive producer. Both left after several seasons. Burt Metcalfe, who was the program's first casting director, succeeded Reynolds as executive producer.

MASTER—an original video tape recording, from which copies are made.

MASTER ANTENNA (MATV)—a single receiving system serving multiple television receivers within the same building or group of buildings. MATV systems are widely employed in apartment houses and projects, hotels and motels, and office buildings. MATV differs from cable-TV (CATV) in that the latter connects a number of separate and distinct homes or buildings to a single antenna system.

"MASTERPIECE THEATRE"—year-long weekly PBS series of dramas and dramatic serials imported from England, which began in 1969 on underwriting by Mobil Oil Corp. The series had its base at WGBH-TV Boston, which was responsible for the selection of shows and the production of wraparounds, with Alistair Cooke as host.

Far the most popular entry in the series, and its only continuing element season after season, was *Upstairs, Downstairs,* an original serial from London Weekend Television. Otherwise, the series has presented such dramatizations as *Cakes and Ale, Notorious Woman, Madame Bovary, Sunset Song, Shoulder to Shoulder, I, Claudius, The Duchess of Duke Street* and *Lillie.*

MATER, GENE P.—key executive of the CBS Broadcast Group since 1970, whose areas of influence have expanded steadily through the administrations of Richard W. Jencks, John A. Schneider and Gene F. Jankowski. He served under all three with the title of assistant to the president and became a v.p. in 1972. While initially he concentrated on media relations, as a former news director for Radio Free Europe and onetime assistant news editor of the New York *World-Telegram & Sun,* he became increasingly engaged in policy matters and with oversight for such critical departments as Standards & Practices.

MATHESON, RICHARD—writer whose many credits include *The Night Stalker, The Night Strangler, The Morning After, The Stranger Within* and *Trilogy of Terror.*

"MATT LINCOLN"—hour-long series about a psychiatrist, a comeback attempt for Vince Edwards, who earlier had success in the doctor series, *Ben Casey.* Produced by Universal TV in association with Vincent Edwards Productions, it premiered on ABC in the fall of 1970 and was canceled after 16 weeks.

Bill Macy & Beatrice Arthur

"MAUDE"—highly successful and frequently controversial spin-off of *All In the Family* on CBS (1972–78) about an aggressive middle-aged woman in step with the times and committed to the liberal viewpoint. The series, which projected veteran actress Bea Arthur to stardom, once involved her confronting the explosive question of abortion and on another occasion being helped by modern psychiatry and medicine from manic-depressive illness. On yet another program, she was permitted to curse her husband as a "son of a bitch."

Bill Macy played her somewhat erratic husband, and other regulars were Conrad Bain, Adrienne Barbeau, Hermione Baddeley, and Brian Morrison. Esther Rolle had portrayed her maid for two seasons, until she was spun off in a new situation comedy, *Good Times.* It was produced by Norman Lear's Tandem Productions.

"MAVERICK"—highly successful offbeat western whose hero was cowardly, unskilled in the orthodox heroic arts and somewhat of a rogue. The perfect casting of James Garner in the lead shot him to stardom. He was later joined by Jack

Kelly and Roger Moore. Produced by Warner Bros., the series ran on ABC (1957–61).

The series was revived for ABC in the 1981—82 season under the title *Bret Maverick*, with Garner repeating the role. Darleen Carr, Richard Hamilton, Stuart Margoein, Ed Bruce, and John Shearin were featured.

Warner attempted to revive the series in 1979 with a made-for-television movie, *The New Mavericks,* also starring James Garner. Instead it resulted in a spin-off series without Garner. It featured Susan Blanchard and Charles Frank.

"MAYA"—adventure-drama of two boys and an elephant in the jungles of India. Jay North starred as an American boy searching for his father and Sajid Kahn as a Hindu orphan dedicated to freeing his pet elephant. Produced by MGM-TV in association with King Brothers for NBC (1967), the series lasted 18 weeks.

"MAYBERRY, R.F.D."—countrified situation comedy on CBS (1968–70) that was salvaged from the *Andy Griffith Show* when Griffith decided to leave. It was successful but fell to a CBS decision in 1970 to weed out the rural-oriented and demographically undesirable shows. The series starred Ken Berry and featured George Lindsay, Arlene Golonka, Jack Dodson and Paul Hartman. It was by RFD Productions and Paramount.

MAYFLOWER DECISION [Mayflower Broadcasting Corp./8 FCC 333 (1940)]—an FCC opinion that stood as a rule through most of the decade of the 40s which prohibited all licensees from taking positions of advocacy on the air. The decision was later reconsidered and reversed in the Report on Editorializing, issued on June 1, 1949.

The episode began when a radio station in Boston, owned by Mayflower Broadcasting Corp., was charged with making political endorsements and supporting partisan politics in public controversies, with no effort toward fairness and balance. After lecturing the licensee about its one-sidedness and bias, the FCC renewed the license on securing a promise from the licensee not to editorialize in the future. This stood as a rule for the industry. The FCC's opinion appeared to define public interest as the obligation to present all sides of important public questions fairly, objectively and without bias. That definition was laid aside along with the ban on editorializing in 1949. See also Report on Editorializing.

MCA—parent corporation of Universal Pictures, whose TV subsidiaries, Universal Television, the production company, and MCA-TV, the distribution company, are, respectively,

among the leading suppliers of network and syndicated programming. The initials had stood for Music Corp. of America, a giant talent agency formed in 1924 to handle orchestras and dance bands and which later added singers, comedians, actors, writers, directors and producers. Its founder was a Chicago optometrist, Jules Stein.

During the 50s and 60s the company grew to even greater power in the entertainment world, earning itself the nickname of "The Octopus," by acquiring, in order, Revue Productions, Universal Pictures and Decca Records. This put it in the business not only of selling talent but also of buying it. With television, MCA expanded into packaging, putting together the property, the actors and the creative unit. It also set up its principal clients in their own production companies, and then represented both the companies and the stars.

Ordered by the Government to divest itself of either the buying or the selling function, under the Sherman-Clayton Act, MCA elected to give up the talent agency. With the decision, it surrendered its original corporate name. As MCA Inc. it concentrates now on packaging, production, recordings and TV syndication.

McANDREW, WILLIAM R. (d. 1968)—head of NBC News from 1951 until his death, a respected executive under whose guidance the division grew both in stature and size. McAndrew assembled a topnotch roster of correspondents and producers as the division grew during his administration from 70 employees to nearly 1,000. Through much of the 60s NBC was widely regarded as the leading news network in journalistic initiative and achievement, and it led as well in the ratings. The organization's drive and esprit were largely attributed to McAndrew's almost fatherly leadership.

McAndrew began with the NBC Washington bureau in the late 30s, then went to work for *Broadcasting* magazine and later ABC News. He returned to NBC News in 1944 as director of the Washington bureau and five years later became station manager of the Washington o&os, WRC and WRC-TV. In 1951 he moved to New York as manager of news and special events for the radio and TV networks and received a succession of new titles in the post until in 1965 he was named president.

McAVITY, THOMAS A. (d. 1972)—v.p. in charge of programs for NBC and later general program executive until his retirement in 1971. In three stints with the network, the first dating to radio days, he helped develop numerous shows, among them, the Bob Hope specials and *Your Hit Parade*. In the 40s he left NBC to work in the advertising and talent agency fields, then took a position with CBS in 1950 and a year later rejoined NBC as director of talent and program procurement. He became v. p. of programs in 1954 and two years later head of both programs and sales. He left soon afterward to join McCann-Erickson Advertising, and then J. Walter Thompson, but returned to NBC again in 1963 as a general program executive.

McCARTHY v. FEDERAL COMMUNICATIONS COMMISSION [390 F2d 471 (D.C. Circuit 1968)]—a

Court of Appeals case which laid down the general proposition that a President is not a candidate for reelection until he announces his decision to run.

Following a practice that began in 1962 with a year-end interview with President Kennedy, the three TV networks in December 1967 carried a joint hour-long interview with President Johnson. Sen. Eugene McCarthy, who, prior to the broadcast had announced his own candidacy for the Democratic party nomination, requested equal time. The FCC denied the senator's request, saying that Section 315 only applied to legally qualified persons who had, among other things, publicly announced their candidacies. Since President Johnson had not officially announced his own candidacy, the commission felt there was no equal time requirement.

The Court of Appeals, while warning the commission not to adhere to arbitrary formulas, affirmed the commission because it felt the FCC's ruling was not unreasonable.

McCLATCHY BROADCASTING CO. v. FCC [239 F2d (D.C. Cir. 1956)/ rehearing denied, 239 F2d 19,/ cert. den., 353 U.S. 918 (1957)]—case which established that,

when other factors balanced out, the FCC may award a license to one applicant rather than another because it would lead to greater diversification of media ownership.

McClatchy and Sacramento Telecasters each submitted an application for a single TV license in Sacramento, Calif. The FCC's hearing examiner found that McClatchy was the licensee of several radio statons in central California and that it was a wholly owned subsidiary of McClatchy Newspapers, which published a number of newspapers in the same area. Telecasters, on the other hand, had no other media interests and would have been a newcomer in the field. The hearing examiner found, however, that since McClatchy had not used its concentration of media in a monopolistic way the fact of concentration should not be held against it.

But the commission itself disagreed. In finding only slight differences between the applicants in terms of their programming and staff proposals, the commission therefore considered the diversification question to be crucial. Since granting Telecasters the license would add a new media voice to Californians and since granting the license to McClatchy would not, the FCC awarded the license to the former because all other factors were virtually equal.

McClatchy appealed. The D.C. Court of Appeals held that the FCC had every right to consider diversification as the "decision" factor. The court reasoned that determination of the "public interest" required a series of ad hoc tests and that the commission should not be "imprisoned in a formula of general applications." The commission's obligation, the court said, was to avoid acting in an arbitrary fashion, and as long as it decided in a reasonable manner the court would not overrule its judgment. See also Cross-ownership.

McCONNAUGHEY, GEORGE C. (d. 1966)—chairman

of the FCC from 1954 to 1957, having been appointed to the commission in 1953 by President Eisenhower. He became embroiled in a scandal over alleged improper contacts between FCC officials and broadcasters, the so-called "$100 million lunch," which jeopardized the license of WHDH-TV Boston. The scandal resulted in the establishment by the Justice Dept. of a guideline for the commission that any applicant who met with an FCC commissioner outside of normal proceedings would automatically lose his case.

McConnaughey denied under oath charges made in House Subcommittee hearings that he had solicited bribes from some license applicants and that he often had lunched with applicants. Those charges were never substantiated, but McConnaughey did admit having private meetings with applicants. He left the commission when his term expired June 30, 1957 to return to practicing law.

McCONNELL, JOSEPH H.—president of NBC from

Oct. 7, 1949 to Dec. 31, 1952, a period of expansion and transition for the company. McConnell reorganized NBC into three major units—the radio network, the TV network and the broadcast stations owned and operated by NBC. He then proceeded to enlarge the television network.

McConnell had practiced corporation law in Washington and New York before joining the legal department of RCA, parent of NBC, in 1941. He rose to general counsel and executive v. p. of RCA before receiving the NBC assignment. McConnell resigned in 1952 to become president of Colgate-Palmolive-Peet Co.

McDERMOTT, THOMAS—production executive who

served as program director for RCA's videodisc subsidiary, SelectaVision (1970–76), after having been president of (and partner in) Four Star International, the TV production company. Earlier he had been v. p. for broadcasting of Benton & Bowles Advertising. He returned to TV production on leaving RCA.

McDONALD, EUGENE JR. (COMMDR.) (d. 1958)—board chairman of Zenith Radio Corp. and an early and ardent advocate of pay-TV. Under McDonald, Zenith developed the over-the-air pay system Phonevision (used in the 1962–68 experiment in Hartford, Conn.) and established a pay-TV subsidiary, Teco, to pursue the cause for subscription TV.

McEVEETY, BERNARD—director with numerous prime-time credits, principally in western series: *Death Valley Days, The Virginian, Rawhide, Laredo, Gunsmoke, Wild, Wild West* and *Dirty Sally,* among others.

McGANNON, DONALD H.—chairman and president of Group W (Westinghouse Broadcasting Co.) (1955–1982) and a forceful, independent figure in the industry whose views affected network and FCC policies. While building Group W into the most potent station group outside those owned by the networks, he also established Group W Productions as a major syndicator, producing and distributing such widely used fare as *The Mike Douglas Show, The Merv Griffin Show* (until 1970), *The David Frost Show, The Steve Allen Show, PM East and PM West, That Regis Philbin Show,* numerous documentaries and such notable children's series as *Call It Macaroni.*

Early in 1979, McGannon yielded the titles of president and chief executive officer to Daniel Ritchie and withdrew from the day-to-day operations of the company. He retained the chairmanship, however, until his retirement in 1982.

As an industry statesman, McGannon fell in and out of favor with his fellow broadcasters, depending on his latest cause. In 1964 he received the NAB's distinguished service award, then angered his fellow station operators by pushing for a rule that would limit network dominance over prime time. His efforts led to the FCC's adoption of the prime time–access rule in 1970, which at first made McGannon a pariah with fellow broadcasters and then a saint, when the PTAR arrangement proved more profitable to them than network service. For the first year of prime-access, Group W Productions, at McGannon's behest, produced such series as *Norman Corwin Presents, The David Frost Revue, The Tom Smothers, Space Ride* and other series which reportedly ran up a loss of $3 million for the company.

McGannon also fought the networks on increasing sex and violence in prime time and, in 1969, withdrew his stations from the TV Code because its provisions were not strict enough. Earlier, he successfully led the industry opposition to ABC's attempt to ramrod a fourth commercial minute in the half-hour prime-time series, *Batman.* During the early 60s he canceled all his radio stations' connections with the networks and made them all independent. To provide them with news, he built his own national and foreign news organization, and then adopted the all-news format for several of the Group W stations. He was also in the forefront of the fight in 1976 to block the networks from attempting to increase their newscasts from 30 minutes to an hour a night. Group W had TV stations affiliated with all three networks. A McGannon petition for a new FCC inquiry into network practices led to such a probe in May 1977.

A lawyer, McGannon made his initial mark in broadcasting as an executive of the old DuMont Network (1952–55) before moving on to Westinghouse. Under his stewardship, the company sponsored six annual public service conferences for the entire industry. McGannon also founded the Broadcast Skills Bank, an organization dedicated to discovering and recruiting able black personnel for the broadcast industry, which later became the Employment Clearing House under the NAB. He served also as president and trustee of the National Urban League, chairman of the Advertising Council and chairman of the Connecticut Commission for Higher Education.

Frank McGee

McGEE, FRANK (d. 1974)—NBC newsman who drew major domestic assignments during the 60s and, from 1971 until his death, served as host of *Today.* In a 17-year career with NBC, McGee had variously been a Washington correspondent, anchorman for the WNBC newscast in New York, Sunday anchorman for the network and (1970–71) co-anchor of the *Nightly News.* He also covered Apollo moonshots, presidential elections, political conventions and the assassinations. McGee was moderator of the second of the Kennedy-Nixon "Great Debates" in 1960. He came to the network's attention with his coverage of racial friction in Montgomery, Ala., while working as news director there for NBC affiliate, WSAF-TV, in 1957.

McGRAW-HILL BROADCASTING CO.—station group founded by McGraw-Hill Publishing in 1972 when it purchased most of the Time-Life broadcast stations. The group maintains headquarters in New York City, with Norman E. Walt (formerly of the CBS stations) as president. TV stations are KMGH Denver, a CBS affiliate; WRTV Indianapolis, NBC; KGTV San Diego, ABC; and KERO Bakersfield, Calif., NBC.

"McHALE'S NAVY"—one of the more potent items in the service comedy vogue of the 60s featuring Ernest Borgnine, Joe Flynn and Tim Conway as broadly inept officers of a PT boat. Produced by Sto-Rev, it ran on ABC from 1962–65 and then in syndication.

McKAY, JIM—ABC sports commentator and host of the network's keystone sports program, *ABC's Wide World of Sports*. He won recognition for his news reporting of the terrorist activities and killings that interrupted the 1972 Olympic Games in Munich. McKay entered television as a writer and production man at WMAR-TV Baltimore in 1947, and three years later moved to New York as host of a daily variety program. He became involved with sports reporting in the late 50s, covering college football and championship golf tournaments, initially for CBS. He appeared on the premiere program of *Wide World of Sports,* April 29, 1961, covering the Penn Relays in Philadelphia.

McLUHAN, MARSHALL (d.1981)—professor at the University of Toronto, author and avant-garde interpreter of the media and their cosmic meanings. He contends that societies have always been shaped more by the nature of the media by which men communicate than by the content of the communication. The now familiar phrase that summarizes his position, "the medium is the message," embodies the historic view that the means by which man communicates have always determined his actions.

In *Understanding Media: Extensions of Man* and his other books, McLuhan's underlying theme is that media—speech, printing, art, radio, telephone, television—function as extensions of the human organism to increase power and speed. He used the words "hot and cool" to describe the mode of impact of a particular medium on people's senses (television in his view is "cool"), and he observed that the mass media of today are turning the world into a "global village," shrinking the globe with respect to shared experience and the passage of news.

McNEELY, JERRY C.—writer and producer-director who created the *Lucas Tanner* series and was co-creator of *Owen*

Marshall. A former professor of communications at the University of Wisconsin, with a Ph.D. in drama, he served as executive story consultant to Universal Studios before branching into production.

MDS (MULTIPOINT DISTRIBUTION SERVICE)—a specialized private service in the superhigh frequency band (2,150–2,160 megaherz). MDS is a common-carrier service authorized to transmit special private television programming, data and facsimile to locations within a metropolitan area, on order from customers. MDS is widely used to transmit special television channels to hotels and has also been used to beam pay-TV programs to cable systems. In 1975 MDS began to be used for the transmission of pay-TV directly to homes in several major cities. MDS serves its subscribers by beaming its carrier from a high point in the area to specially supplied temporary or permanent parabolic antennas. Special converters reduce the frequency to match a vacant channel on a standard television receiver.

MEAD, ROBERT—TV adviser to President Ford, appointed at the start of the Administration but discharged by press secretary Ron Nessen in July 1976 when a live special on Queen Elizabeth's visit to the White House, carried by PBS, went badly. Mead had been a news producer for CBS at the White House when he was appointed. In addition to handling arrangements for the President's televised speeches and press conferences, and selecting the times of broadcast, Mead also coached the new President in advance of the telecasts to help him perform effectively.

MEADE, E. KIDDER, JR.—vice-president of corporate affairs for CBS Inc. (1957–1981). He joined the company after serving with its outside public relations counsel, Earl Newsom & Co. A graduate of West Point and a Lt. Colonel during World War II, he became a member of the staff of the Secretary of Defense and later (1950–53) a special assistant to the Under Secretary of State. He then became v.p. of Colonial Williamsburg Inc.

MEADOWS, AUDREY—actress and comedienne best known for her role as Jackie Gleason's wife (Alice Kramden) in *The Honeymooners*. She was also featured on the *Bob & Ray Show* in the early 50s and with Sid Caesar, Red Skelton and Jack Benny.

MEANEY, DONALD—an NBC News executive since 1960 and a v.p. since 1966. On July 1, 1974, he became v.p.

of news in Washington, after having been a principal of news operations in New York.

MEDIA ACCESS PROJECT—Washington-based public interest law firm affiliated with the Center for Law and Social Policy and concentrating on communications matters before the FCC, other regulatory agencies and the courts. Its aim is to improve coverage of significant news issues by advising local groups in their relations with broadcasters and by using the avenues afforded by the fairness doctrine and license renewal procedures. The organization is headed by Andrew Jay Schwartzman.

"MEDIC"—artful and meticulously researched medical-drama series which had been produced for syndication (1954–56), with Richard Boone as narrator and star of the first episode. In a fictional framework, the programs examined medical practices and problems and were filmed on location at Los Angeles hospitals. The 30-minute programs were written by James Moser and produced by Worthington Miner for Medic TV Productions.

"MEDICAL CENTER"—hour-long series of dramas set in a fictional medical center within a large university. It premiered on CBS in 1969 and performed well enough in the ratings to earn an annual renewal. Produced by MGM-TV in association with Alfra Productions, it starred Chad Everett and James Daly.

"MEET MILLIE"—a 1952–56 situation comedy on CBS about the business and romantic life of a Manhattan secretary. Elena Verdugo played the title character, Florence Halop her mother and Ross Ford the boss's son.

"MEET THE PRESS"—a weekly 30-minute, live press conference which began on NBC Nov. 6, 1947, the oldest series on network television. The format—essentially a 30-minute interview of a newsmaker by a panel of newspaper journalists—has been imitated by CBS with *Face the Nation* and ABC with *Issues and Answers*. But while all three regularly draw out statements from public officials that make front-page news the following day, *Meet the Press* remains the most prestigious of the TV forums.

The program was conceived by Lawrence E. Spivak, who served as producer and permanent panelist, and later as moderator, until he retired in November 1975. Martha Roundtree, the program's first moderator, was succeeded in the 50s by Ned Brooks. Initially, the program was scheduled after 10 P.M., for a while on Wednesday nights, then on

Mondays and then on Saturdays. In the mid-50s, it found its niche on Sunday afternoons.

Usually the format involves one newsmaker and four interviewers, but on occasion there have been multiple guests, and in unusual instances the program has run an hour and even 90 minutes.

Spivak created *Meet the Press* as a radio program in 1945 to promote *American Mercury* magazine, of which he was publisher and editor. After the program moved to TV, he sold the magazine and remained in broadcasting.

MEGAHERTZ —million cycles per second. See Hertz.

MEIER, DON—long-time producer-director of *Mutual of Omaha's Wild Kingdom,* both on NBC and in syndication, and president of the Chicago-based Don Meier Productions. Meier started his own company in the late 50s when NBC disbanded its Chicago central division program staff, which he had headed.

MELCHER, MARTIN—producer and talent agent who, after selling his agency to MCA in 1948, concentrated on representing his wife, actress Doris Day. Their own company, Arwin Productions, produced *The Doris Day Show* on CBS (1968–73), as well as some of her films.

MELENDEZ, BILL—producer (and voice of Snoopy) for the *Charlie Brown* animated specials on CBS-TV.

MELNICK, DANIEL—independent film producer. He had been a program executive for ABC (1959–63), a partner in David Susskind's Talent Associates and later production chief for MGM Pictures (1972–1976).

In 1977, while still active in independent production, he joined Columbia Pictures as head of worldwide production. He left in 1979.

"MEN IN WHITE" COMMERCIALS—advertisements which used actors to simulate doctors, suggesting that their patented medicines were recommended by medical authorities. The practice ended in 1958 when, in response to pressure, it was outlawed by the NAB Code.

"MEN WHO MADE THE MOVIES, THE"—public TV series presenting noted American movie directors with personal interviews and film clips of their most significant works. It premiered on PBS in 1974 and was produced by Richard Schickel with a grant from Eastman Kodak.

MENCHEL, DON—president of MCA-TV, Universal's syndication arm, since 1978. Earlier he was v.p. and director of sales for three years. He had previously been with Time-Life Films, Telcom Associates and ABC Films.

MENDELSON, LEE—independent producer based in the San Francisco Bay Area who specializes in animated and science-fact specials. He's best known as executive producer of the *Charlie Brown* animated specials, in which he is partnered with Bill Melendez.

MERCER, DONALD J.—head of affiliate relations for NBC since 1955 who became v.p. in 1967. He had joined the company in 1935 as a page and served in a number of positions with NBC and its parent, RCA, before being assigned to station relations.

MEREDITH, DON—former star football quarterback who became a star ABC sportscaster on *Monday Night Football* and in 1974 was hired away by NBC on a sports and entertainment contract. Meredith not only provided football expertise as color commentator for ABC but also displayed a lively, countrified wit. This prompted NBC to sign him for occasional acting roles and hosting assignments on *Tonight* as well as for its coverage of football and tennis. He returned to ABC sports in the summer of 1977.

MERRIMAN, JOHN (d. 1974)—editor, producer and reporter for CBS News since 1942 who served as news editor of the *CBS Evening News* from 1966 until his death eight years later in a plane crash. In 1973 he was president of Writers Guild-East.

MESTRE, GOAR—founder of two television networks in Cuba during the early 50s, each consisting of seven stations, which made him the first person in the world to create a national television system. When his broadcast properties were taken over by Fidel Castro's government, Mestre emigrated to Buenos Aires with his Argentine wife and there started a successful production company, Proartel, which created programs for the country's Channel 13. His wife, with financial support from CBS, was part owner of the station. *Variety*'s headline for Mestre's reemergence in television was "Mestre Rides Again."

CBS helped to sponsor Mestre's Argentina enterprise out of gratitude. Mestre, a graduate of Yale, had sold his Cuban properties to the American network and then called off the deal when he realized the stations were about to be confiscated by Castro. His gentlemanly act spared CBS a huge loss.

METROMEDIA—leading nonnetwork broadcasting group in the U.S., with six stations in the top 15 markets, only two of them having network affiliations. Metromedia also maintains a production subsidiary, MPC (Metromedia Producers Corp.), which produces programs for the networks as well as for syndication.

The group was founded in 1955 by John W. Kluge from stations that were sold off when the DuMont Network disbanded. Originally known as Metropolitan Broadcasting, the company changed its name in the early 60s to reflect its diversification in other advertising media, such as outdoor billboards and publications.

In July 1981, Metromedia agreed to buy WCVB-TV Boston, an ABC affiliate, from Boston Broadcasters Inc. for $220 million, a record price for a station. WCVB is widely regarded as the best local station in America and is heavily involved in production for syndication. Soon afterward it sold KMBC Kansas City to the Hearst Corporation for $79 million.

Metromedia owns WNEW New York, KTTV Los Angeles, WTTG Washington, D.C., KRIV Houston, and WXIX (UHF) Cincinnati, all independents, and WTCN Minneapolis-St. Paul, an NBC affiliate. Kluge is president and chief executive officer, Clemens M. Weber, executive vice-president, and Ross Barrett and Albert P. Krivin, senior vice-presidents. Tom Tilson is president of Metromedia Television, the operating division for the stations.

MEXICO, TELEVISION IN—sophisticated, privately owned Latin American system in which, by law, 12.5% of programming and two separate channels are controlled by the government. The system uses NTSC color, 525 scan lines, the South American satellite and beams to around 4 million sets.

The four private channels get most of their programs from Televisa, a Mexican production company which has cornered 90% of the Mexican market and exported 18,000 hours of programming in 1975. Televisa earmarks numerous programs for educational or public service purposes.

Each of Mexico's channels plays to a particular social or intellectual stratum: Channel 4 aims at the mass urban public around Mexico City; Channel 5 reaches about half the nation through the use of 20 repeater stations with programs for "youthful middle class," with documentaries, cartoons and American action-adventure or police shows; Channel 8 is the "highbrow" station for academics and establishment critics; Channel 13 offers feature films and sports; and Channel 11 is purely educational. In addition, two cable channels from Texas feature standard U.S. fare. See also Televisa.

MICHAELS, LORNE—youthful producer of the original NBC *Saturday Night Live*, credited with being most responsible for its success. When Michaels left the show in 1980 for other pursuits, virtually the entire cast and creative staff followed.

A Canadian, Michaels had honed his skills in Toronto (1967–73) writing, producing, and performing in a number of CBC comedy specials with a partner, Hart Pomerantz. He broke into American TV, while still in his 20s as a writer for *Laugh-in* and later for Lily Tomlin specials.

On leaving *Saturday Night Live*, he formed his own production company with a view to making movies and creating programming for cable TV as well as commercial TV. One of his first commissions was to develop a new comedy variety series for NBC.

MICHEL, WERNER—Viacom executive whose career spanned three decades, variously with networks, ad agencies and production companies. He previously was a general program executive for ABC Entertainment in Hollywood. Prior to 1975, he spent 18 years as v.p. in charge of buying network time for such ad agencies as SSC&B, Reach, McClinton and Kenyon & Eckhardt. In the mid-50s, he was executive producer for the DuMont Network, associated with such programs as *Captain Video* and the *Jackie Gleason Show,* and earlier (1946–51) a producer-director for the CBS documentary unit. Born and educated in France, and working in Vienna in the 30s as a conductor and composer, he came to the U.S. just before the Nazi Anschluss in 1938 and began writing for radio. In the 40s he was with the Office of War Information and the Voice of America.

MICKELSON, SIG—first to hold the title of president of CBS News, serving in that capacity from 1951–1961. He chaired the networks' joint committee which arranged and produced *The Great Debates* between John F. Kennedy and Richard M. Nixon in 1960 and was a founder and early president of the Radio-Television News Directors Assn.

Mickelson joined CBS as a reporter in 1941, after having taught journalism. On leaving the company, he became head of international broadcasting operations for Time Inc. and later moved to Chicago as v.p. of international operations and TV for Encyclopaedia Britannica Educational Corp., teaching journalism part-time at Northwestern University. Later he became president of Radio Free Europe and Radio Liberty.

"MICKEY MOUSE CLUB"—a popular early evening series on ABC in the late 50s which employed a cast of juvenile actors known as *Mouseketeers* and offered varied entertainment, both live and filmed, including Walt Disney cartoons and episodic series ("Spin and Marty," "The Hardy Boys").

Legions of children around the country became members of the club, wore its mouse-eared beanie and sang the club song. The program premiered Oct. 3, 1955, as a daily one-hour entry and ran through 1959. It not only provided ABC with a major show in a time when the youngest network was struggling for audience attention, but also was a windfall for Mickey Mouse merchandise and served to promote the newly opened Disneyland, Walt Disney's amusement park in Southern California.

Some of the reruns were syndicated for a few years during the early 60s. On Jan. 20, 1975—almost 20 years after it began—*Mickey Mouse Club* was brought back into syndication by SFM Media Service Corp. on the theory that its original audience would watch it nostalgically along with a new audience of children. Later a new series was mounted, but the revival was only moderately successful, and production ceased in 1977.

A movie career grew out of Annette Funicello's performance as a Mouseketeer. Others of the original cast were Bobby Burgess, Karen Pendelton, Cubbie O'Brien.

"MICKEY SPILLANE'S MIKE HAMMER"—private-eye series based on the character in Spillane's novels, portrayed by Darren McGavin. By Revue Studios, it played on CBS.

MICROWAVE, or MICROWAVE RELAY SYSTEM—a system of radio repeaters mounted on towers, each consisting of a receiving antenna and transmitter, spaced up to 50 miles apart. This is the principal means of interconnecting television stations as well as cable systems within continental boundaries. The microwave frequencies are 890 megaherz and above and give their name to this relay system. Microwaves are also used to connect studio to transmitter and for remote television origination equipment to the studio or transmitter.

MIDGLEY, LESLIE—CBS News executive principally responsible for producing the "instant" news specials until NBC News hired him away in 1979 as v.p. of documentaries and such programs as *Tomorrow* and *Prime Time Sunday*. He retired from NBC in 1982, following Bill Small's departure as president of the news division.

When CBS suspended all commercial programming following the assassination of President Kennedy in 1963, Midgley produced the network's nighttime schedule for four evenings. He also produced the four one-hour specials on the Warren Report in 1967, a number of programs on the energy crisis in 1973–74 and 10 hour-long special broadcasts on the unfolding Watergate story in 1973.

Midgley joined CBS News from *Look Magazine* in 1954 and became producer of the CBS Sunday News, then anchored by Eric Sevareid. In 1960 he produced the series, *Eyewitness to History* (later entitled, *Eyewitness*), and from 1967 to 1971 he was executive producer of *The CBS Evening News with Walter Cronkite*. Midgley married Betty Furness, NBC consumer affairs reporter.

"MIDNIGHT SPECIAL, THE"—90-minute contemporary music series on NBC scheduled at 1 A.M. Saturdays (following the Friday *Tonight Show*) as part of the network's exploration of postmidnight television. The series premiered Feb. 3, 1973, with Burt Sugarman as executive producer and Stan Harris as producer and director. Wolfman Jack, nationally known rock disk jockey, was regular announcer, and an array of pop music and comedy stars served as hosts. Each program presented approximately 10 acts from the various fields of pop music.

MIDWEST VIDEO CORP. CASE [United States *v.* Midwest Video Corp./406 U.S. 649 (1972)]—test in which the Supreme Court affirmed the authority of the FCC to create statutory policies for cable-TV, including requiring the systems to originate their own programming.

Rules issued by the FCC in 1969 mandated "significant amounts" of cablecasting from all systems with more than 3,500 subscribers. Midwest Video Corp., which operated a system of that size, challenged the rule in the Court of Appeals for the Eighth Circuit and won a decision that the FCC had not the authority to require cablecasting. Moreover, the court held that such a rule was not in the public interest.

On the FCC's appeal, the decision was reversed by the Supreme Court. No majority was assembled, but a four-justice plurality said its decision rested upon whether or not the rule was "reasonably ancillary to the effective performance" of the commission's regulation of television broadcasting. The plurality held that the commission's concern with cable-TV is not merely prohibitory by avoiding adverse effects but extends also "to requiring CATV affirmatively to further statutory policies." It maintained also that since cablecasting would assure more diversified programming, the FCC's rule furthers the goal of the Communications Act generally.

Chief Justice Burger concurred in the result, and four justices dissented, saying that congressional action was required to make cablecasting compulsory. See also Cable-TV, Cable-TV Rules.

MIDWEST VIDEO II [Federal Communications Commission *v.* Midwest Video Corporation et al. 47 U.S.L.W. 4335 (1979)]—case in which the Supreme Court struck down the FCC's cable-TV access rules and production facilities and channel capacity requirements. At issue were the FCC's 1972 rules requiring cable-TV systems with over 3,500 subscribers to offer access channels to the public, local governments and educational institutions, over which the cable system had no program control; to maintain production facilities for public use for a minimal fee; to provide a minimum of 20 channels by 1986.

Midwest Video Corp. challenged the FCC rules as beyond the commission's authority on statutory grounds and was joined by the NCTA, which sought a ruling that would have established broadcastlike First Amendment rights over cable programming for cable operators. The court upheld the ruling of the Eighth Circuit Court of Appeals, striking down the regulations, on statutory grounds alone, but specifically deferring judgment on any constitutional questions of a cable operator's right to control the programming on his system.

The court held that the FCC's authority over cable-TV derived from the 1934 Communications Act's grant of jurisdiction over broadcast television. However, the Act specifically forbids the commission to impose common carrier–like obligations—i.e. first-come, first-served access at federally regulated rates, like those prescribed for telephone companies—on broadcast entities. The court found the commission's cable-TV access rules to be common carrier–style regulations. Rules requiring maintenance of production facilities and prescribing future channel capacity were struck down because they were so intertwined with the access rules that they could not reasonably be separated, although the justices did not find them to be beyond the FCC's current statutory jurisdiction.

Because the court did not rule on constitutional questions but confined itself to judgment on the basis of the 1934 Communications Act, it left the door open for Congress to reimpose the access requirements through new legislation. In addition, state and local authorities may still require access channels through legislation or in new cable-TV franchises. In communities where current franchises already require access channels and facilities, cable operators must continue to honor the franchise provisions even though the federal requirement has been struck down.

"MIGRANT"—controversial NBC News documentary (1970) produced by Martin Carr as a followup ten years later to *Harvest of Shame*, the famous CBS documentary by Edward R. Murrow and David Lowe on the plight of migrant workers. Carr's one-hour film demonstrated that living and working conditions for the Florida agricultural migrants were still wretched and had scarcely improved at all in the decade

—a conclusion challenged acrimoniously by the Florida Fruit and Vegetable Assn., which asked NBC for reply time under the Fairness Doctrine.

Meanwhile, executives of Coca-Cola Foods—a company with vast interests in the Florida citrus industry through such orange juice brands as Minute Maid, Hi-C, Tropicana and Snowcrop—on learning in advance that theirs was one of the companies cited in the film as exploiting the migrant workers, applied pressure on NBC to make certain excisions. NBC did not, nor did it grant the association reply time. The program was televised without commercial support. Perhaps it was a coincidence, but the following quarter Coca-Cola shifted its scatter-plan billings from NBC to ABC and CBS.

In later hearings by the Senate Subcommittee on Migratory Labor, the program's revelations were upheld as valid, and Coca-Cola was the first company to announce a plan to transform the migratory work force into a stable year-round group with the same fringe benefits as were received by other Coca-Cola employees.

George Burns & Mike Douglas

"MIKE DOUGLAS SHOW, THE"—See Douglas, Mike.

MILEAGE SEPARATION—minimum distance formula used by the FCC as a guide for allocating TV channels in its 1952 Table of Assignments. The commission determined that stations on the same channel would have to be at least 170 miles apart, and it set the minimum separation for adjacent channel assignments (e.g., Ch. 2 and Ch. 3) at 60 miles.

MILLER, D. THOMAS—president of the CBS Television Stations Division (1970–77). Previously he had been v.p. of

the CBS Broadcast Group (1969–70) and v.p. of broadcast affairs for the New York Yankees when the team was owned by CBS (1967–69). Before joining CBS he was president of a syndication firm, North American Television Associates, and from 1963 to 1966, v.p. and general manager of WBKB-TV (now WLS-TV) Chicago. During the 50s he held sales positions in Chicago with CBS, ABC and the rep firm, Harrington, Righter & Parsons.

MILLER, HERMAN—writer specializing in the creation of series and the writing of pilot scripts. Among his credits are *Kung Fu* and *McCloud*.

MILLER, JP—one of TV's leading playwrights, perhaps best known for *The Days of Wine and Roses*. Active with *Philco-Goodyear Playhouse* and other drama showcases during the "golden age," he took to writing novels when TV drama went into its decline, although he has returned from time to time. He wrote the script for the two-part TV dramatization of the book, *Helter-Skelter* (1976), for which he won critical praise. The show ranked No. 1 in the Nielsens that season.

MILLER-MILKIS—producing partnership of Thomas L. Miller and Edward K. Milkis, both former production executives for Paramount TV. They were executive producers of *Petrocelli* and, in collaboration with Garry Marshall, of *Happy Days, Laverne & Shirley*, and *Mork and Mindy*. In 1981, Robert L. Boyett, another former Paramount production executive, joined their production company, which became Miller-Milkis-Boyett Productions. *Bosom Buddies* was their first network series.

MILLER, RON—executive producer of *The Wonderful World of Disney* series since the death of Walt Disney.

"MILLIONAIRE, THE"—big CBS hit (1954–59) whose stories involved the sudden gift of $1 million to ordinary persons by an anonymous donor, an eccentric billionaire. Marvin Miller was the sole continuing character, the dour presenter of the gift as secretary to the billionaire.

MILNE, ALISDAIR—chairman of the British Broadcasting Corp. and previously managing director for television (1977–81). Earlier he was its director of programs for several years.

Born in India of Scottish ancestry, he joined the BBC in 1954 as a public affairs producer. A high point in his programming career came as executive producer of *That Was the*

Week That Was, the topical satire series that later transferred to the U.K.

MINER, WORTHINGTON—a leading creative force in the early years of TV as writer, producer and executive. With a background in theatre, he became manager of TV program development for CBS in 1948 and was responsible for the creation of *Studio One, Toast of the Town (The Ed Sullivan Show), The Goldbergs* and *Mr. I. Magination.* Miner produced and wrote many of the early plays for *Studio One* and was also producer of the other three series. In 1952 he was hired away by NBC but later became a freelancer. He was executive producer of the syndicated series, *Medic,* then of *Frontier* and, in the later 50s, the syndicated drama series, *Play of the Week.* He later worked in motion pictures (*The Pawnbroker, The Fool Killer*).

MINISERIES—program series designed for limited runs, over several nights or several weeks, as opposed to those created in hopes of running indefinitely. The miniseries came into vogue in U.S. commercial television during the 70s after the success on public television of such short-term British series as *The Forsyte Saga, Elizabeth R, The Six Wives of Henry VIII* and *Civilisation.* Lending themselves particularly to short-term serialization were popular novels, and with the success on U.S. networks of the adaptations of Leon Uris's *QBVII* and Joseph Wambaugh's *The Blue Knight,* a raft of other best sellers were produced as TV miniseries.

The miniseries came into full flower with ABC's adaptation of Irwin Shaw's *Rich Man, Poor Man* in 1976. That inspired numerous others, and NBC even created a regular weekly series of mini-series under the title of *Best Sellers.* But the crowning achievement in the form, in terms of its ratings and public impact, was ABC's serialization of Alex Haley's *Roots* over eight consecutive nights in January 1977. It drew the largest audience for an entertainment program in the history of television.

MINOW, NEWTON N.—chairman of the FCC (1961–63) on appointment by President Kennedy who shook the industry with his "vast wasteland" speech at the NAB convention shortly after he took office. Minow, 34 at the time, and a former Chicago law partner of Adlai Stevenson, made it plain at once that the FCC in his administration would consider program performance as a condition for license renewal. While there were a number of fines, short-term renewals and a few radio license revocations administered, no major licenses were lost during Minow's term, but the climate he set nevertheless had broadcasters on edge. Later in his career he became involved in public television and was

elected chairman of PBS in 1978. He had been a member of the PBS board since 1973.

In his famous 1961 speech—which instantly won him high visibility with the public—Minow described the TV programming landscape as a "vast wasteland" and went on to say that he would not abide a "squandering of the public airwaves." He said broadcasters would be held to their program promises at license-renewal time, and he put the industry on notice that it was expected to do a better job in the public interest. From the industry view, his most teeth-rattling statement was, "There is nothing permanent or sacred about a broadcast license." FCC chairmen did not normally assume the role of TV critics, and because of the sensitivities toward government censorship they usually avoided making evaluations of the general programming.

Part of Minow's strength derived from his support by the press; unlike most FCC chairmen, he was able to make news at will. But he left the FCC without many lasting achievements to take a high-paying position with Encyclopaedia Britannica in Chicago. In 1965 he returned to law practice, with CBS as one of his clients; became chairman of the organization operating Chicago's public TV station, WTTW; and taught classes at Northwestern University.

Former host of the Miss America Pageant, Bert Parks

"MISS AMERICA PAGEANT"—premier national beauty competition televised live from Atlantic City since 1954 and consistently one of the highest-rated programs of the year. Occurring early in September, it has traditionally marked the opening of every new season. Among the event's own traditions has been the singing of the pageant's anthem, "There She Is, Miss America," at the finale by master of ceremonies Bert Parks. The song dates to the second telecast (1955), which was Parks's first as emcee.

A few of the Miss Americas went on to build modest careers as TV performers, among them Marilyn Van Derbur and Lee Ann Merriwether, winner of the first pageant to be televised. Bess Myerson has had the largest success in TV, ranging from panelist on *I've Got a Secret* to syndicated talk shows of her own, but she was crowned in 1945 and was not a TV Miss America.

The telecasts were carried first by ABC for three years, then by CBS (1957–65) and by NBC ever since. Although the event was popular with viewers from the start, it did not become blockbuster fare until 1958, when it scored a 49.2 rating (Total Audience), and 66 share, and reached 21.4 million homes over the course of the telecast. Since then it has typically been received in 25 to 28 million homes each year and in 1973 hit a high of more than 30 million homes. The ratings are remarkable for the fact that the telecasts occur after prime time and run past midnight.

In an effort to rejuvenate the event, the promoters fired the aging Bert Parks and made Ron Ely the host in 1979. After two seasons, Ely gave way to Gary Collins.

William Devane & Martin Sheen as John & Robert Kennedy

"MISSILES OF OCTOBER, THE"—landmark ABC documentary drama (Dec. 18, 1975) concerning the political maneuverings in the 1962 Cuban Missile crisis, with actors portraying the world figures involved. William Devane played President John F. Kennedy; Martin Sheen, Robert F. Kennedy; Howard da Silva, Khrushchev; and Ralph Bellamy, Adlai Stevenson.

The three-hour teleplay was written by Stanley R. Greenberg. Herbert Brodkin and Buzz Berger were co-producers and Anthony Page director. Irv Wilson was executive producer for Viacom.

"MISSION: IMPOSSIBLE"—hour-long series on CBS (1966–73) of foreign intrigue and intricately executed feats of espionage, produced by Bruce Geller for Paramount TV.

The original unit, for the first four seasons, featured Barbara Bain, Martin Landau, Peter Graves, Greg Morris and Peter Lupus. When Miss Bain and Landau dropped out, Leonard Nimoy and Leslie Warren took their places. Lynda Day George appeared for one season, and Steven Hill took part in the first 18 episodes. Dubbed in 15 other languages, the series has been sold in 71 countries.

"MISTER DUGAN"—unborn situation comedy with an unusual story of its own: it was canceled by its producers a few days before its scheduled premiere, although the network was eager to air it. The bizarre incident occurred in the spring of 1979 when Norman Lear and Alan Horn of T.A.T. Productions determined that the show, which was about a black freshman Congressman from Philadelphia, did not depict the lead character with sufficient dignity or as a "positive and accurate role model." A screening for the Congressional Black Caucus confirmed their own reservations, and they withdrew the program from CBS-TV and ordered production stopped after three episodes. CBS, which had no objection whatever to the portrayal of the Congressman by Cleavon Little, and which had promoted the show's May 11 debut in its Sunday lineup, was forced to fill the time with the second part of a two-part episode of the preceding show, *Alice*.

The setback was the third for the project that became *Mister Dugan*. The idea was born as a way to refurbish the faltering *Maude* series—Maude would become a freshman Congresswoman. But Bea Arthur, who portrayed Maude, elected not to continue for another season. The role was then adapted to a black Congressman and was to have been a vehicle for John Amos. But Amos withdrew before the shooting, and he was replaced by Little. Lear indicated, after his action, that the project was not dead but would have to be redesigned. It never was.

"MISTER ED"—one of the first of the fantasy situation comedies, a successful CBS entry (1961–66) about a talking horse. Alan Young, Connie Hines, Leon Ames and Florence MacMichael were featured (Larry Keating and Edna Skinner were regulars the first two seasons). It was by Filmways.

"MISTER ROGERS' NEIGHBORHOOD"—a low-key PTV children's series devoted to examining values, feelings and fears through the company of Mister Rogers (Fred Rogers) on a set representing his home. Rogers, a Presbyterian minister, employed an easy manner, gentle conversation and soothing songs in dealing with matters that were likely to concern children emotionally, such as nightfall, rejection, physical handicaps, going to the dentist, disappointment, death.

The program began locally on WQED Pittsburgh and then spread to other public stations until it became part of the regular weekday PBS children's fare, along with *Sesame Street* and *Electric Company*. In addition to performing in it, Rogers also wrote and produced the series.

A relatively inexpensive show, costing $6,000 per 30-minute episode when it began in February 1968, and just over $19,000 in 1974, it employed a small cast of regulars representing neighbors and friends of the host. Believing that after nearly eight years he had covered the full range of pertinent themes, and that he had created a library of programs—a total of 460—which were valid to new generations of young children, Rogers suspended production to package the best of his programs for repeat showings.

In 1979, he resumed production of 10 new *Neighborhood* programs, introducing some new themes and refreshing some of the old ones.

Fred Rogers

MNA RATINGS (MULTI-NETWORK AREA)—a regular audience survey report of the A.C. Nielsen Co. indicating the relative potency of the networks in the top 70 markets where the programs of all three are in direct competition. (Daytime MNA ratings are drawn from 63 of those markets.) Covering a Sunday through Saturday period and issued approximately seven days after the broadcast week, the MNA report contains demographic breakdowns by age and sex, along with household viewing estimates for time periods. The populations of the MNA markets represent 65.7% of the total national audience.

"MOD SQUAD"—successful hour-long series on a young trio of counterculture proclivities who channel their rebelliousness into useful police work. Produced by Thomas-Spelling Productions and playing on ABC (1968–73), it featured Peggy Lipton, Michael Cole, Clarence Williams, III, and Tige Andrews.

MOFFITT, JOHN—producer-director with credits chiefly in music-variety. Director of *The Ed Sullivan Show* for three years, he later directed numerous music specials and awards shows and was producer-director for the *Stand Up and Cheer* syndicated series in the 70s.

MOGER, STAN—president of SFM Entertainment and a founder and principal of the parent company, SFM Media Corp., which had a phenomenal growth in the 70s as a time-buying and program distribution company. Moger became a force in television syndication by setting up the *Mobil Showcase* network, creating the *SFM Holiday Network*, and reviving the *Mickey Mouse Club* and *The Adventures of Rin Tin Tin*. With the media planning and time-buying skills of SFM's other division, Moger broke the ground for the so-called "fourth network" movement, which brought forth a variety of occasional, or ad hoc, networks that cut into the programming of the major networks. He cleared the time on commercial television for such Mobil offerings as *Edward the King, Edward and Mrs. Simpson, Churchill and the Generals*, and *Ten Who Dared* and repackaged old family-audience movies into the *SFM Holiday Network*, which broke into the network schedules during the holiday seasons. In 1982, he purchased the old *March of Time* series for recycling of filmic history.

Earlier in his career, Moger had been a salesman for NBC Films, Storer Television, and George P. Hollingbery Co., station reps.

MOLINE RENEWAL [Moline Television Corp./31 FCC 2d 263 (1971)]—instance in which the FCC excused a licensee for failing to fulfill its license promise. In the comparative hearings in 1963 in which the Moline Television Corp. was awarded the license for Channel 8 in the quad-cities of Moline, East Moline and Rock Island, Ill., and Davenport, Ia., the company had proposed 12 locally produced public affairs series. By the conclusion of its initial three-year license period, the station, WQAD, had aired none of the promised programs. When Moline applied for renewal, a competing applicant, Community Telecasting Corp., also filed and the FCC scheduled a comparative hearing.

The commission, in considering Moline's application for renewal, excused the marked difference between promise and performance by noting that Moline's network, ABC, had substantially upgraded its public affairs programming during the three-year period. Additionally, the FCC found persuasive Moline's argument that its decision to spend money on its evening news program precluded financing the promised public affairs series.

In examining Moline's financial situation, the FCC found a substantial operating deficit, supporting the excuse for failure to produce the local series. The FCC held that these factors resulted in only a "slight demerit" against Moline. Since Moline integrated its ownership with its management very well, and because its owners were local residents, the commission awarded Moline a renewal of its license.

Commissioner Nicholas Johnson filed an angry dissent in which he chastised the majority for granting a renewal to an unqualified licensee when a qualified challenger stood ready to run the station. He felt that awarding a renewal to Moline in this instance indicated that the commission would never deny a renewal to an existing licensee, no matter how dismal his performance had been or how promising the challenger.

MOLNIYA (LIGHTNING)—Soviet Union's domestic satellite system, the first domsat system inaugurated by any country. Since 1965, multiple Molniya satellites, in random elliptical orbits, have been relaying television and other telecommunications within the Soviet Union's borders. See also Domsat, Satellites.

MOM & POP SYSTEMS—term used in the cable-TV industry for the early, small cable systems that continue to be operated locally by companies owning no other systems. Anachronisms in an industry dominated by MSOs (multiple system operators), they are likened to Mom & Pop grocery stores in an era of supermarkets and represent a contrasting mentality to the large companies eager for cable to spread. Single-ownership systems are gradually disappearing, however, through mergers and acquisitions.

MONASH, PAUL—drama producer whose TV credits include *Peyton Place,* an ABC series in the mid-60s, and *Judd For the Defense,* a legal drama series. In 1976, after producing a string of motion pictures, including *Butch Cassidy and the Sundance Kid* and *Slaughterhouse Five,* he joined CBS as a program v.p. in charge of movies for television and miniseries, but two years later he returned to independent production of TV movies.

"MONDAY NIGHT FOOTBALL"—see Football On Television.

"MONKEES, THE"—musical situation comedy on NBC (1966–68) about a wild rock quartet, not unlike the Beatles. A group of singer-musicians was assembled for the series and

their first recording promoted into a hit before the program premiered. The group then became, for several years, popular rock artists outside the fictive portrayal on TV. It consisted of Davey Jones, Peter Tork, Mickey Dolenz and Mike Nesmith. The series was produced by Raybert Productions and Screen Gems. CBS carried the reruns in 1969 and ABC in 1972. *The Monkees* went into syndication in 1975.

MONOCHROME—black-and-white television.

MONROE, BILL—executive producer and moderator of NBC's *Meet the Press* since November 1975, having previously been Washington editor for *Today.* He joined NBC News in 1961 as Washington bureau chief, after having been news director of WDSU-TV New Orleans for six years.

MONTANUS, EDWARD (d. 1981)—key executive of MGM-TV until his death, he was promoted to executive v.p. in 1976, and president in July 1977. He joined MGM-TV as director of syndication sales, then became head of network sales in 1970. Earlier he had been with ABC-TV sales and NBC Films in Chicago.

MONTE CARLO INTERNATIONAL TELEVISION FESTIVAL—TV awards competition organized in 1961 by the Monaco Principality, whose prizes (called Golden Nymphs) are awarded to writers in the fields of news and dramas, to directors and to actors and actresses. There are also other categories which vary from year to year.

MONTGOMERY, ROBERT (d.1981)—motion picture star who became prominent in TV in 1950 as host of a popular NBC drama anthology, *Robert Montgomery Presents,* which he also produced and in which he occasionally starred. He later became staff consultant to President Eisenhower for television—the first person to serve a President in that capacity—and a severe critic of the networks for what he considered their abuses of power.

Robert Montgomery Presents, slotted opposite *Studio One* on CBS, ran for six seasons with an assortment of one-hour adaptations and original TV plays. The series was to provide a springboard for the star's daughter, Elizabeth Montgomery, who was part of a stock company Montgomery had organized for the summer presentations in 1956.

Montgomery joined the White House at the urging of press secretary James Hagerty and functioned primarily as an acting coach to the President, who had felt uneasy before the TV cameras. Through Montgomery's instruction, Eisenhower gained the confidence to hold televised press conferences, and the coaching was also of benefit to the President's 1956 reelection campaign.

Montgomery had little to do with TV thereafter, except to denounce it, and his entry for *Who's Who in America* makes no mention of *Robert Montgomery Presents*. In 1969 he became president of the Lincoln Center Repertory Theater.

"MONTY NASH"—action half-hour starring Harry Guardino which was one of the first syndicated shows produced in response to the prime time–access rule (1971), via Almada Productions. It was not successful.

Monty Python's Flying Circus

"MONTY PYTHON'S FLYING CIRCUS"—BBC series of madcap and irreverent comedy which, after enormous success in England, was sold to the Eastern Educational Network in 1974 by Time-Life Films. Its popularity on public stations, especially among the young, prompted a few sales to commercial stations and then a sale to ABC for two late-night specials in 1975. The 30-minute programs were free-form, consisting of sketches and blackouts. In the regular cast were Eric Chapman, John Cleese, Terry Gilliam, Eric Idle, Terry Jones and Michael Palin; they were also the writers. Ian Macnaughton was producer.

MOORE, ELLIS O.—veteran network public relations executive, working first for NBC and then ABC; he has been v.p. in charge of public relations for ABC Inc. since 1972 and became v.p. of corporate information in 1979. A former reporter for newspapers in the midwest, he joined the NBC press department as a staff writer in 1952 and rose to v.p. in charge of the department in 1961. Leaving in 1963 for a public relations post with Standard Oil of New Jersey, he

returned to broadcasting three years later as v.p. of press relations for ABC in the Elton Rule administration.

MOORE, GARRY—amiable personality and program host who thrived during a period when the ability to maintain a friendly rapport with the audience mattered more than having musical, acting or comedy talent. For a period during the late 50s, when he was ubiquitous on CBS, Moore was reputed to have had the highest income of any TV performer.

His mainstay was *I've Got a Secret*, the Goodson-Todman panel show which he moderated from 1952 to 1964, but the high point of his career was the prime-time *Garry Moore Show*, a popular variety series on CBS which began in 1958 and which was otherwise notable for launching the comedienne, Carol Burnett.

One of the many who moved from network radio to TV, Moore made a number of efforts to establish a daytime variety show for CBS during the early 50s before finding his niche with *Secret*. In the late 60s an attempt was made to revive his Sunday night variety series; its failure marked his virtual retirement from the medium.

MOORE, MARY TYLER—star of the CBS situation comedy *The Mary Tyler Moore Show*, which became a fixture in the network's Saturday night lineup through most of the 70s. In her portrayal of Mary Richards in the series, she came to represent the quintessential clean-cut bachelor woman striving for a career and coping with life in the big city. The success of the series caused MTM Productions (the company in which she was partnered with her husband, Grant Tinker) to flourish as one of the leading independent suppliers to the networks. MTM was responsible also for *Rhoda, Phyllis* and *The Bob Newhart Show*, among other series.

Moore achieved her first prominence in TV as the wife in *The Dick Van Dyke Show*, which ran five years on CBS in the 60s. She went on to make several movies, including *Thoroughly Modern Millie*, and to attempt a Broadway show, but when none was successful she returned to television. She had been a professional dancer since the age of 17 and did several TV specials which allowed her to display her dancing talent.

Her first role in a TV series was as Sam, the secretary in the *Richard Diamond, Private Eye*. But it called for only her legs to be shown, and she asked for a release from the series after 13 weeks.

The Mary Tyler Moore Show ceased production in the spring of 1977, in a rare instance of a program canceling itself.

She attempted two comedy-variety series in the 1978–79 season, the first called *Mary* and the second *The Mary Tyler Moore Hour;* both failed.

MOORE, RICHARD A.—president of KTTV Los Angeles during the 50s and part of the 60s who became part of the White House inner circle during the Nixon Administration and thus a witness in the Watergate proceedings. During his broadcast career, Moore helped organize the Television Advertising Bureau, the sales promotion organization for the industry which later became TvB. After the election of Nixon he became first a consultant to Robert Finch, when he was Secretary of HEW, and then special assistant to Attorney General John N. Mitchell in 1970. Before joining KTTV, Moore had been an attorney for ABC.

MOORE, THOMAS W.—president of ABC-TV (1963–68) and later founder and head of Tomorrow Entertainment, a General Electric subsidiary involved in motion picture and TV production. When GE disbanded the company in 1975 because of dissatisfaction with its profits, despite the production of a number of distinguished shows, Moore retained the name and continued to operate as an independent producer. The company had been responsible for such acclaimed filmed dramas as *The Autobiography of Miss Jane Pittman, In This House of Brede, Larry* and *I Heard the Owl Call My Name.* In association with Medcom, it produces the *Body Human* series of occasional specials.

Moore had been president of ABC during the network's desperate years, when it perennially ran third, and he jockeyed ABC into near contention with the leaders in 1965. The former program chief of ABC (1957–63), he became president when Ollie Treyz was dismissed after congressional criticism of an overly violent episode of *Bus Stop.*

Moore was the loser in an internal power struggle that occurred in 1968 when the expected merger with ITT fell through. Although he had appeared to be receiving a promotion as head of all ABC broadcast operations, he was in fact shunted aside. He resigned in the summer of 1968. Before Tomorrow Entertainment was organized, he served for a brief time as head of Ticketron.

During the election campaigns of 1972, he was TV advisor to Vice-President Agnew and later was appointed by President Nixon to the board of the Corporation for Public Broadcasting. In 1976 he became vice-chairman of the board. His CPB term ended the following year.

MORAL MAJORITY—See Coalition for Better Television, The.

MORD, MARVIN S.—v.p. of research for ABC Television since 1975, in charge of the departments giving research support to the TV network and ABC News. He joined ABC in 1959 as a research analyst for the network and two years later moved into sales as an account executive. In 1963 he left to become a sales executive for the American Research Bureau, then joined Young & Rubicam Advertising. He returned to ABC in 1967 in the research department, working primarily in audience measurements. He moved steadily up the ranks to become ABC's research chief.

MORGAN, CHRISTOPHER—producer-son of actor Harry Morgan, whose credits in the mid-70s included *Police Story, Medical Story* and *The Hunter.*

MORGAN, EDWARD P.—ABC newsman, prominent in the 50s and 60s, who became known as the "voice of labor" through his 15-minute radio program of news and commentary sponsored by AFL-CIO because of his liberal views. As if to prove that his labor sponsors did not influence his editorial opinions, Morgan often was critical of organized labor in his commentary.

Before joining ABC in 1955, Morgan had been a correspondent for CBS News. In 1967 he took a two-year leave of absence from ABC to become senior correspondent of the Public Broadcasting Laboratory and anchorman of its TV news-magazine, *PBL.* A critic of commercial broadcasting even while working in it, Morgan had been a board member of the National Citizens Committee for Broadcasting. He retired from ABC in 1975.

Robin William & Pam Dawber as Mork & Mindy

"MORK AND MINDY"—the runaway hit of the 1978–79 season, bringing instant stardom to comedian Robin Williams, who plays an explorer from the planet Ork learning the ways of earthlings in Boulder, Colo. The character of Mork was introduced in an episode of *Happy Days* in the 1977–78 season, with such pleasing results—thanks to Williams's gift for sight comedy and verbal acrobatics—that a new series was immediately developed as a vehicle for him. Mindy, a young single woman, is played by Pam Dawber.

The show was so popular in its freshman season that it won ABC dominance of Thursday nights. Confident that *Mork's* audience was a loyal one, ABC programmers moved it to Sundays for the 1979–80 season, pitting it against *Archie Bunker's Place* (the continuation of *All in the Family*) in a showdown of sitcom giants.

Garry K. Marshall, creator of *Laverne & Shirley* and *Happy Days,* and Tony Marshall are the executive producers. Bruce Johnson and Dale McRaven are the producers. The series is via Miller-Milkis Productions and Henderson Production Co., in association with Paramount Television.

"MORNING"—CBS's hour-long daily newscast at 7 A.M., which was called *The CBS Morning News* until it was revised, under Robert (Shad) Northshield, with a Sunday edition added, in January 1979. *Morning,* which competes with NBC's *Today* and ABC's *Good Morning America,* eschews the author interviews and entertainment elements of those shows and concentrates on straight television journalism. Partly as a result of that, but also partly because it was teamed in the morning with the children's program, *Captain Kangaroo,* the CBS program consistently ran a distant third in the time period.

In the fall of 1981, CBS cut back *Kangaroo* to 30 minutes and expanded *Morning* to 90 minutes with Charles Kuralt as anchor. When this strategy failed to produce higher ratings, CBS moved *Kangaroo* out of the 7 to 9 A.M. period to let *Morning* go head-on against the competing network programs. In the spring of 1982, Kuralt was replaced by Bill Kurtis, a popular Chicago anchorman; and Northshield yielded to George Merlis, whom CBS hired away from *Good Morning, America* to redesign and produce the program. Diane Sawyer, who had been teamed with Kuralt, remained to co-anchor with Kurtis.

The *CBS Morning News* began in April 1957 as a 15-minute program and lasted two years. It returned in 1963 in a half-hour format, making it the first 30-minute news program on the networks, and was expanded to an hour in 1969. Over the years it has had a variety of anchormen, among them Joseph Benti, John Hart and Morton Dean, but in 1973 it underwent its most publicized change, with Hughes Rudd and Sally Quinn named dual anchors. Ms. Quinn, who did not make the transition well from newspapers to television, lasted only a few months, and Rudd did a solo until October 1977. He was replaced then by Richard Threlkeld and Lesley Stahl.

On Jan. 29, 1979, with the title changed to *Morning,* and with Norshield as the new executive producer, Bob Schieffer became the regular Monday–Friday anchor and Charles Kuralt anchor of the 90-minute Sunday edition. Threlkeld

and Ms. Stahl received featured assignments in *Sunday Morning.*

See also Quinn, Sally; Rudd, Hughes; and Northshield, Robert.

MORSE, ARTHUR D. (d. 1971)—former CBS News producer-writer who in 1967 became the first director of the International Broadcast Institute, a nonprofit organization, then based in Rome, which undertook studies of communications technologies. While serving in that capacity, Morse was killed in an automobile accident in Yugoslavia.

At CBS, he had worked for the *See It Now* series as a reporter and director during the 50s, his credits including *The Lost Class of 1959* and *Clinton and the Law,* a report on integration in Clinton, Tenn. In 1960 he was assigned to *CBS Reports* as a producer-writer and worked on such programs as *Who Speaks for the South?, The Other Face of Dixie* and *The Catholics and the Schools.* He resigned from CBS in 1964 because of what he felt was a lack of support from the network for a report he was preparing on the tobacco industry and the health hazards of cigarette smoking.

MORTON, GARY—former comedian who, on marrying Lucille Ball, became producer of her *Here's Lucy* series and her subsequent TV specials.

MOSEL, TAD—dramatist who in the 50s wrote for such series as *Omnibus* and *Playhouse 90,* his TV works including *Who Has Seen the Wind?, The Playroom, The Lawn Party* and *My Lost Saints.* In 1971 a TV version of his Pulitzer Prize-winning Broadway play, *All the Way Home,* was presented on *Hallmark Hall of Fame.*

"MOSES, THE LAWGIVER"—six-part dramatization of the Biblical episode carried by CBS in the summer of 1975. The series was produced by Italy's RAI-TV in association with Britain's ATV-ITC and was filmed in Israel with an international cast. Burt Lancaster played Moses; Anthony Quayle, Aaron; Ingrid Thulin, Miriam; Irene Papas, Zipporah and Laurent Terzieff, Pharaoh. Richard Johnson was narrator.

The script was by Anthony Burgess and Vittorio Bonicelli. Vincenzo Labella was producer and Gianfranco DeBosio director.

"MOST DEADLY GAME"—hour-long series on ABC (1970) in which Ralph Bellamy, George Maharis and Yvette Mimieux portrayed a sophisticated team specializing in solving difficult crimes. By Aaron Spelling Productions, it lasted 12 weeks.

"MOST WANTED"—hour-long ABC crime series (1976–77) centered on an elite unit of the L.A. Police Department and featuring Robert Stack. Shelly Novack and Jo An Harris in supporting roles. It was by Quinn Martin Productions, with Martin, John Wilder and Paul King as executive producers and Harold Gast as producer.

"MOTHERS-IN-LAW"—NBC situation comedy (1967–69) about suburban neighbors who become in-laws. A vehicle for Eve Arden and Kaye Ballard, it featured also Herb Rudley, Richard Deacon, Jerry Fogel and Deborah Walley and was via Desi Arnaz Productions.

MOTION PICTURE ASSN. OF AMERICA (MPAA)— organization formed in 1922 to represent the principal distributors and producers of movies in the U.S. to preserve and enlarge the global market for their films. Television is one such market, and MPAA, under its president Jack J. Valenti, has aggressively sought FCC regulations that would promote the growth of pay-cable as a potentially lucrative new market for feature films.

MOVIE CHANNEL, THE see Cable Networks.

"MOVIE OF THE WEEK"—series of 90-minute features made expressly for ABC under arrangements with independent producers and various studios. Originally scheduled one night a week, the anthology did so well it was expanded to two nights. Each showcase carried approximately 24 originals and repeats. *MOW*, as it was abbreviated in the trade, ran from 1969 to 1975. See also *World Premiere Movies*.

MOVIES ON TV—a program staple from the earliest days of commercial TV (to the extent that films were available before 1956) and since the mid-60s a staple of the network prime-time hours, as well. As late-vintage theatrical movies entered TV, with their mature themes and racy dialogue, television's program standards were forced to loosen; movies thus were a chief influence in the revolution in TV mores that took place in the early 70s.

Television, which in the beginning was viewed as a threat to movies, has instead prolonged the life of motion pictures and indeed has given some of them perpetual life. It has also become a second market for movies (and pay-cable a third), which for a time represented a virtual guarantee to producers

of at least $1 million in advance of production for even average pictures.

The Godfather: Al Pacino, Marlon Brando, James Caan & John Cazale

William Boyd's old *Hopalong Cassidy* films, and other "B" westerns that had been sold to TV by Republic Studios, proved as early as 1948 that movies were dynamite in the new medium. The major studios held back their libraries, however, for fear of helping the rival medium seal their doom.

Until 1956 movies trickled into TV from small companies and private owners. In 1955 ABC purchased 100 British films from J. Arthur Rank for an afternoon film festival. But the dam burst the following year when RKO Teleradio Pictures sold 740 movies and 1,000 shorts to C&C Super Corp., which then leased them to stations. Soon after, Columbia Pictures released a batch of pre-1948 titles through its syndication subsidiary, Screen Gems, and the other majors followed in rapid order—some packages going for as high as $50 million. All held the line at the theatrical release year of 1948 and were able to sell the films without payment of residuals.

Local stations fed heavily on the movies during the 50s and the early 60s, the feature-length films solving innumerable program problems for 90 minutes and two hours at a time. Independent stations used them to compete with the networks in prime time. WOR-TV New York developed the *Million Dollar Movie,* repeating the same film every night of the week. WGN-TV Chicago tailored a batch of old *Bomba, The Jungle Boy* movies into a successful one-hour prime-time series, and then assembled some odd-lot juvenile titles (*Tom Sawyer,* etc.) into a Friday evening series, *Family Classics,* which outrated some network shows.

The CBS o&os invested heavily in choice titles and created the parlay of *Early Show, Late Show* and *Late, Late Show* (late afternoon, post-prime time and post-midnight), which other stations adopted. Each film was rotated among the three showcases, then was rested and then repeated. The release of a library of Shirley Temple movies in 1957 proved a bonanza for the stations, and especially for the Ideal Toy Co., which sponsored.

When the major studios began selling their post-48 titles after 1960, the networks began buying. NBC started the tide with *Saturday Night at the Movies*, and by 1966 each network had staked out one or two nights for movies. But the prime-time movies were a Pandora's Box: They wiped out most TV series in competition, leaving each network the problem of scheduling against them; their success brought escalations in price (by 1967 films of no particular distinction were fetching $800,000 for two plays); and, worst of all, there was a looming shortage of supply, made even more critical by the fact that the theatrical movies then in vogue were of a sophistication that made them unplayable under TV's program standards at the time.

NBC prepared to beat the shortage by contracting with Universal TV in 1966 for a steady supply of movies that would be made expressly for television at approximately $800,000 per title. The long-range deal resulted in the two-hour *World Premiere* anthology. ABC also dabbled in made-for-TV pictures by ordering three from MGM before devising the highly successful *Movie of the Week*, which arranged for the production of 90-minute features from a variety of producers at half the price NBC was paying Universal. CBS, ABC and Westinghouse Broadcasting all started their own motion picture production companies, but none was a financial success.

On the local level, the movie shortage was being felt even before the networks began buying features. The dwindling supply gave rise to the 90-minute talk shows—principally those of Mike Douglas and Merv Griffin—to fill the long gaps in the schedules. A number of syndicators speculated with dubbed foreign films but lost money in a market that rejected them.

In the early 70s the movies that had previously been considered too mature for TV (*The Graduate, Love Story* and scores of others) found their way onto the home screens, some with a bit of editing. The networks rationalized their change of heart by citing the changing morality in the country and asserting their need to be in step with the times. But it was also true that the movie houses had been getting precisely the audience television and its advertisers were most eager to reach—young adults in the 18–34 age group. To win them back to TV, the networks would have to alter their standards. The new permissiveness then was extended to TV's own series to make it possible for them to compete with movies.

By 1974 an average sort of movie no longer sufficed for prime time and the networks grew selective with titles and trimmed down to fewer showcases.

The new mania was for blockbuster movies, preferably fresh out of theatrical release, to be shown as specials. ABC had provided the eye-opener in 1966 with *The Bridge on the River Kwai*, for which it had paid $2 million for two show-

ings. The first outing on Sept. 25, 1966, drew a tremendous audience of more than 60 million, a record at that time for a televised movie. Later it was exceeded by *The Birds* and a number of big theatrical hits purchased by the networks for astronomical amounts.

NBC paid $7 million for a single airing of *The Godfather*, $5 million for a single play of *Gone with the Wind*, and $3 million for the first TV rights to *Dr. Zhivago*. ABC paid $3.3 million for *The Poseidon Adventure* and charged advertisers $150,000 a commercial minute in that film. NBC charged $250,000 a minute for spots in *Gone with the Wind*.

Television's Top 25 prime-time movies, as of October 1981, follow:

Movie	Rating	Share	Network/Date
1. Gone With The Wind-Part 1	47.7	65	NBC, 11/76
2. Gone With The Wind-Part 2	47.4	64	NBC, 11/76
3. Airport	42.3	63	ABC, 11/73
4. Love Story	42.3	62	ABC, 11/72
5. The Godfather-Part 2	39.4	57	NBC, 11/74
6. Jaws	39.1	57	ABC, 11/79
7. Poseidon Adventure	39.0	62	ABC, 10/74
8. True Grit	38.9	63	ABC, 11/72
9. The Birds	38.9	59	NBC, 1/68
10. Patton	38.5	65	ABC 11/72
11. Bridge on the River Kwai	38.3	61	ABC, 9/66
12. Helter Skelter-Part 2*	37.5	60	CBS, 4/76
13. Jeremiah Johnson	37.5	56	ABC, 1/76
14. Ben-Hur	37.1	56	CBS, 2/71
15. Rocky	37.1	53	CBS, 2/79
16. The Godfather-Part 1	37.0	61	NBC, 11/74
17. Little Ladies of the Night	36.9	53	ABC, 1/77
18. Wizard of Oz(R)	36.5	58	CBS, 12/59
19. Wizard of Oz(R)	35.9	59	CBS, 1/64
20. Planet of the Apes	35.2	60	CBS, 9/73
21. Helter Skelter-Part 1	35.2	57	CBS, 4/76
22. Wizard of Oz(R)	34.7	49	CBS, 1/65
23. Born Free	34.2	53	CBS, 2/70
24. Wizard of Oz	33.9	53	CBS, 11/56
25. Sound of Music	33.6	49	ABC, 2/76

*Made-for-TV-Movie

MOVIES ON TV IN EUROPE—Unlike the U.S., where television is saturated with film fare, and where home audiences can see movies as early as 18 months after initial theatrical release, the showing of features around Europe is saddled with restrictions as to frequency and age. A key reason is the safeguarding of the movie exhibition business.

In Britain and France the TV clearance on movies is five years. At one time in the United Kingdom it was 10 years, but there have been recent signs that the limbo period may be lowered again, to two or three years.

In Italy, where the motion picture industry flourishes and some 10,000 cinemas of all descriptions operate, the two TV channels are not allowed to air films on weekends and each schedules only one film a week.

In France, features may not exceed 10% of total programming on the three channels, and none are permitted at the weekends. Also, of all movies played on TV there, at least 50% must be of domestic origin.

West German TV airs only around 150 movies a year. By contrast, around a thousand are carried on Britain's three channels annually, the bulk of them by two-channel BBC. The British independent stations are permitted to air no more than seven movies a week.

Outside of Britain, probably the most movie-intensive TV market in all Europe is Belgium. This is not so much because of what the two state-subsidized channels carry (one in French, the other Flemish) as it is the result of an extensive cable system that pulls in programming from all neighboring countries. Some movies thus pulled into the country were aired even before their Belgian theatrical release, a headache for the local film industry that has led to a court test. As a result Belgian cable can no longer carry movies prior to their theatrical playoff.

Until recently in Finland there was no official clearance factor with respect to movies. The time lag was left to individual negotiation. Such elasticity was removed under a 1975 agreement between Finnish TV and the Finnish Film Producers Assn., which ratified a formal three-year clearance.

Movies on TV, in Europe as in the U.S., frequently suffer cuts for reasons of running time or sensitivity of subject. But at least one company, Britain's BBC, practices an integrity with regard to movies that may well be unique. The BBC is not only scrupulous about playing features intact, but it also has a reputation for going to great lengths to restore footage distributors have chopped out of their theatrical prints. BBC is also in the habit of spending its own money on the restoration of prints of old classic films, such as Chaplin two-reelers.

"MOVIN' ON"—hour-long series on the adventures and relationship of two "gypsy" truckers of different generations, starring Claude Akins and Frank Converse. Produced by D'Antoni-Weitz Productions in association with NBC, the series premiered September 1974 and was canceled after its second season.

MOYERS, BILL D.—correspondent and commentator for CBS News since Nov. 1, 1981, after having been a mainstay of public television with his documentary series, *Bill Moyers' Journal*. It was his second stint at CBS. In 1976 he became chief correspondent for *CBS Reports* but returned to resume

his old show because he disliked the trend to magazine-format documentaries.

In his new role at CBS, Moyers assumes Eric Sevareid's tasks with commentary on the *Evening News*, in addition to producing several documentaries a year with Howard Stringer.

At PBS Moyers demonstrated that he was one of broadcast journalism's best interviewers. A former Texas newspaperman, Moyers became press secretary to President Johnson and then publisher of *Newsday*, the Long Island daily. In 1971 he entered public television as host-moderator of the weekly half-hour public affairs series, *This Week*, produced by WNET New York, which was to be his production base. The following year, he began his noted weekly documentary series, *Bill Moyers' Journal*. In 1976, before leaving for CBS, he served as principal reporter for the weekly election-year series, *USA: People and Politics*.

Bill Moyers

"MR. ADAMS AND EVE"—domestic situation comedy, sophisticated for its time, which starred Ida Lupino and Howard Duff as a husband-wife movie-star team. It premiered on CBS Jan. 4, 1957 and ended its run Sept. 23, 1958.

"MR. I. MAGINATION"—children's show which began on CBS in 1949, created by and featuring Paul Tripp. Its setting, Imagination Town, purported to be magical.

"MR. LUCKY"—CBS series (1959) about an honest gambler which had a short run. John Vivyan played the lead. Spartan Productions was the source.

"MR. NOVAK"—hour-long dramatic series on NBC (1963–65) about an idealistic young teacher. It broke the ground for numerous other teacher shows. James Franciscus appeared in the title role with chief support from Dean Jagger (replaced by Burgess Meredith in the final 13 episodes). MGM-TV produced it.

Mr. Peepers

"MR. PEEPERS"—a TV comedy classic despite its failure to win a vast audience in three seasons on NBC (1952–55). Commended for its casting, intelligent scripts and humor in a low key, it concerned a bookish science teacher and some of his quirky colleagues at a small town Junior High. Wally Cox starred, and Tony Randall, Marion Lorne and Patricia Benoit were featured.

Mr. Peepers began as a summer replacement in July 1952 and impressed NBC sufficiently to win a premium slot in October. The programs were presented live.

"MR. ROBERTS"—short-lived comedy-drama series on NBC (1965–66) based on the characters of the movie and play of that title (from Thomas Heggen's book) and inspired by the success of other military situation comedies in the 60s. Roger Smith had the title role, and Steve Harmon, George Ives and Richard X. Slattery were featured. It was by Warner Bros.

MSO (MULTIPLE SYSTEM OPERATOR)—a cable-TV company which owns and operates more than one cable system. The first MSO was Pioneer Valley Cablevision, formed in the merger of two Massachusetts systems in 1965, those in Nutler Falls and Shelburne Falls. Since then, large MSOs have proliferated throughout the country. The 10 largest MSOs (as of January 1982), according to *Television*

Digest, are:

• American TV & Communications Corp. (ATC), owned, by Time Inc., with 1,690,000 subscribers.
• Teleprompter Corp., owned by Westinghouse Broadcasting, with 1,556,021 subscribers.
• Tele-Communications Inc., with 1,423,000 subscribers.
• Cox Cable Communications Inc., owned by Cox Broadcasting, with 1,070,000 subscribers.
• Warner Amex Cable Communications Inc., with 820,000 subscribers, a figure not yet reflecting the new installations in major cities.
• Storer Cable Communications, owned by Storer Broadcasting, with 777,011 subscribers.
• Times-Mirror Cablevision, sister company of the *Los Angeles Times*, with 642,976 subscribers.
• Viacom Communications, with 509,375 subscribers.
• Rogers-UA Cablevision, with 468,000 subscribers.
• United Cable TV Corp. with 437,344 subscribers.

"M SQUAD"—high-rated NBC police series (1957–60) starring Lee Marvin as a Chicago plainclothesman. The half-hour series was via Latimer Productions and Revue Studios.

MSN (Modern Satellite Network)—see Cable Networks.

MT. MANSFIELD TELEVISION INC. *v.* FCC [442 F2d 470 (2nd Circuit 1971)]—Court of Appeals action upholding the constitutionality of the prime time–access rules. After the rules were introduced in 1970, the networks brought suit claiming that the rules violated the First Amendment. The U.S. Court of Appeals for the Second Circuit upheld the validity of the rules, claiming that the Red Lion case had held that the public's First Amendment rights took precedence over the broadcasters', and that under that implied theory of access the prime time–access rules appeared to be a reasonable attempt to fulfill that concept.

MTV (Music Television)—see Cable Networks.

MUDD AND TROUT—news-anchor team mounted by CBS in 1964 in response to the popularity of Huntley and Brinkley on NBC, to no avail. Roger Mudd and Robert Trout had a short run as a team. They replaced Walter Cronkite as anchor for the 1964 Democratic national convention, but Cronkite was back at his post for the Republican conclave that year.

MUDD, ROGER—co-anchor, with Tom Brokaw, of the *NBC Nightly News* after a long association with CBS as Washington correspondent. Mudd joined NBC News in 1981 after CBS selected Dan Rather, and not him, to succeed Walter Cronkite on its newscast. He was assured of becoming John Chancellor's successor on the NBC newscast. Mudd later agreed to team up with Brokaw to smooth the way for Brokaw's new contract with the network. They made their debut in April 1982.

He became a CBS News correspondent in 1961 and was prominent since the late 60s for political coverage, special news broadcasts and as substitute anchorman during Walter Cronkite's illnesses or vacations. Handsome, articulate and respected by his peers for his writing abilities, he was considered by many to be heir-apparent to Cronkite as the network's premier anchorman.

Mudd, who joined CBS from its Washington affiliate, WTOP-TV, covered all political conventions and elections since 1962 and served as anchorman or reporter for documentaries such as *The Selling of the Pentagon* (1971) and special reports in 1973–74 on the unfolding of the Watergate story, the resignation of Spiro Agnew, the appointment of Gerald Ford as Vice-President, the Senate Watergate hearings and the resignation of President Nixon.

With the primary assignment of covering Congress, Mudd was anchorman of the Saturday edition of *The CBS Evening News* (1966–73) and also the Sunday edition (1970–71).

Roger Mudd

MUELLER, MERRILL—broadcast journalist (1945–74), working first for NBC News and then, for six years, ABC News. Between the two networks, he was a floor reporter at all political conventions from 1952 to 1974. At NBC, he was variously news director for London, Asia and the Mediterranean and newscaster for the *Today* program, preceding Frank

Blair. From 1957 until he joined ABC in 1968, he was a senior correspondent in New York and worked on numerous major stories, including the assassination of President Kennedy. With ABC, he was a roving correspondent, based in New York and for a time in Los Angeles.

On leaving the network, he became a consultant to the Federal government in Washington for two years and then Newhouse Professor of broadcast and print journalism at Syracuse University.

MUGGS, J. FRED—chimpanzee who received a five-year contract from NBC for appearances on the *Today* show in the early 50s and who later had his own local TV show in New York for a brief period. In 1972 he joined Busch Gardens, a family entertainment park in Tampa, Fla., but continued to do occasional TV guest shots. His trainers were Bud Mennell and Roy Waldron, pet shop owners who had purchased him as a baby for $600.

Robert E. Mulholland

MULHOLLAND, ROBERT E.—president of the National Broadcasting Company since 1981, named to the post when Grant Tinker succeeded Fred Silverman as head of the company with the title of chairman. Tinker chose to continue living on the West Coast, and Mulholland was in day-to-day charge of the company's New York headquarters. Previously he was president of the television network (since 1977), after having briefly been executive v.p. in charge of the Moscow Olympics. Most of his career was spent as an NBC news executive.

Mulholland, who began as a news writer for WMAQ-TV, the NBC o&o in Chicago, had risen in 13 years to executive v.p. of NBC News with overall responsibility for all TV newsgathering operations, news documentaries and news at the owned stations. From 1963 to 1967, he acquired background as a field producer in Chicago and then in Europe, and for the next four years served as West Coast news director, in which capacity he launched the first two-hour local news program in a major market. This earned him a chance to become producer of the *NBC Nightly News* and then, in 1972, executive producer.

MULL, MARTIN—comedian of the late seventies/earlies eighties who first gained attention as the fictional host of the mock talkshow "Fernwood 2-nite," later as the real occasional guest host of "Tonight." See Steve Martin.

MULLER, ROMEO—writer working with Rankin-Bass on their animated programs and specials; he wrote *Frosty the Snowman, The Little Drummer Boy, Santa Claus Is Coming to Town* and others.

MULTIPLE OWNERSHIP RULES—limits on station ownership imposed by the FCC to prevent undue concentration of economic power and to promote a diversification of viewpoints. Under present rules, any single licensee may own a maximum of seven stations in each of three categories: AM, FM and TV (with no more than five of the latter on VHF). No commercial licensee may own more than one of each category of station in the same market, and no licensee may own the only newspaper in town and the only broadcast service of its kind (AM, FM or TV) in that community. Public broadcasters are exempt from the one-to-a-market rule to the extent that they may operate one VHF and one UHF in the same community.

The first multiple ownership rule was adopted in 1940. It pertained only to FM stations and forbade ownership by any one person of more than one station serving substantially the same area. The rule also provided that ownership of more than six stations would be a concentration of control inconsistent with the public interest.

In 1953 the commission adopted a revised multiple ownership rule which provided that no person may hold any interest, direct or indirect, in more than five TV, seven AM and seven FM stations. The validity of these rules was sustained by the Supreme Court (U.S. *v.* Storer Broadcasting Co.). In 1954 the FCC amended the rules to permit ownership of seven TV stations but restricted ownership of VHF stations to five. This was designed to promote the development of UHF.

While the FCC has never decided to bar newspapers as a class from ownership, it has consistently favored applicants without newspaper experience in selecting among competing applicants for a license. In addition the commission has also followed the consistent practice of considering whether an applicant already has other broadcast interests even

though such interests were permitted by multiple ownership rules.

In 1970 the commission proposed a controversial rule that would require divestiture within five years to reduce one party's media holdings in any market to one or more daily newspapers, or one TV station or one AM-FM combination. The rule would require that if a broadcast station were to purchase one or more daily newspapers in the same market, it would be required to dispose of any broadcast station. No grant for a broadcast station license would be made to a newspaper owner in the same market.

After much discussion of this rule, the FCC, in 1975, adopted a modified version which would require divestiture by 1980 only in instances where the degree of media concentration could be considered egregious. The FCC defined an egregious situation as one in which a party owns and operates the only newspaper published in a community, as well as the only TV or radio stations in that community. A newspaper could own the only radio station if there is a separately owned TV station.

In 1977 the D.C. Court of Appeals, in NCCB v. FCC, determined that the FCC has the authority to forbid future formation of jointly owned newspaper-broadcasting combinations located in the same communities. However, the court ruled that the FCC erred in its reasoning that divestiture should be ordered only when it can be demonstrated that the public interest is harmed by the combinations. The court concluded that exactly the opposite standard should apply—that divestiture should be required except in those cases where evidence clearly shows cross-ownership to be in the public interest. See also Cross-Ownership.

MULTIPOINT DISTRIBUTION SERVICE—See MDS.

MUNRO, J. RICHARD—president of *Time, Inc.* since 1980, after having been head of all video enterprises of Time Inc. since October 1975. The Video Group flourished under Munro's direction and rose from the smallest of Time's three main divisions to the largest, in terms of revenues and profits. Its extraordinary success—especially that of HBO—pointed the way to Time Inc.'s future and made it possible for Munro to leap over executives of the publishing division to the presidency when Andrew Heiskell retired.

Munro joined Time in 1957 in the magazine's circulation department, then moved to *Sports Illustrated* where he rose to become general manager and eventually publisher. He has been a director of Time Inc. since 1978.

"MUNSTERS, THE"—fantasy situation comedy about a family of monsters living ordinary domestic lives. The head of the family, Herman Munster (Fred Gwynne), was physically modeled on the Frankenstein monster. Other regulars were Yvonne DeCarlo and Al Lewis. Produced by Universal, it ran on CBS (1964–65) and had been created by Joe Connolly and Bob Mosher.

The Muppet Show: Kermit & Jim Henson

"MUPPET SHOW, THE"—syndicated variety series (1976–) featuring Jim Henson's large cast of Muppets (imaginative puppets) and a live guest star each episode. As a prime-access entry, it opened on around 160 stations domestically, including the CBS o&os, and in scores of foreign markets. It was produced in England by ITC and Henson Associates, with Gary Smith and Dwight Hemion as creative consultants and Jack Burns as producer-writer. It grew to be a tremendous hit in England and Europe, while in the U.S. it ranked consistently among the top-rated prime time–access series. By 1978, with its global spread having reached about 100 countries, it had become one of the most popular television shows in the world. In 1979, the TV series spun off a theatrical feature, *The Muppet Movie.* See also Henson, Jim.

MURRAY,(KEN—early variety-show emcee who had three seasons (1950–53) with *The Ken Murray Show.* Later he did guest shots on other programs showing his home movies of celebrities.

MURROW, EDWARD R. (d. 1965)—broadcasting's supreme journalist, whose verbal gifts and superb delivery were matched by his humaneness and high professional standards. He not only was radio's premier reporter in covering World War II for CBS—famous for his broadcasts from the rooftops of London in language one commentator called

"metallic poetry"—but also was responsible for assembling the crackerjack team of foreign correspondents that served as a cadre for CBS News for the next quarter century.

Murrow's greatest work, however, was on the domestic front, principally with the series *See It Now,* on which in 1954 he succeeded in exposing the ruthless tactics of, and injustices perpetrated by, Sen. Joseph McCarthy. It was undoubtedly the most courageous undertaking in TV history. Several years later, for *CBS Reports,* he contributed the memorable documentary expose of the inhumanities suffered by migrant workers in the U.S., *Harvest of Shame.*

Murrow also was host of a popular prime time series, *Person to Person,* on which he visited electronically the residences of the famous for live interviews. His two nighttime series ran concurrently, *See It Now* for eight seasons and *Person to Person* for five.

Fascinated with TV's ability to broadcast from several places at once, Murrow extended the *Person to Person* concept to other continents for a weekend discussion series, *Small World* (1958–59).

He had two trademarks: the closing line, "Good night...and good luck," and the cigarette. Often on the screen he was engulfed in a cloud of smoke. He died of lung cancer at just about the time the advertising of cigarettes on TV had grown into a controversial issue.

Murrow had joined CBS in 1935 as director of talks and education, and two years later became European director. His distinguished work during the war, both in reporting and managing the CBS forces, earned him a promotion in 1945 as v.p. and director of public affairs and member of the board of directors. He gave up the vice-presidency, however, to do *See It Now.*

Murrow left CBS in 1961 to become director of the U.S. Information Agency in the Kennedy Administration. By that time he had become disenchanted with commercial television and openly critical of its devotion to escapism.

The cancellation of *See It Now* in 1958, for competitive reasons, had had a damaging effect on his spirit.

MUSBURGER, BRENT—host and managing editor of CBS Sports' weekend series, *Sports Saturday* and *Sports Sunday.* He previously had been broadcast reporter with the distinction of holding two posts simultaneously on KNXT Los Angeles—those of principal news anchor and sportscaster—in addition to two regular sports assignments for the network. Formerly he was CBS sportscaster based at the network's Chicago station, WBBM-TV. He has covered NFL football and golf for CBS and in 1975 became host of the sports anthology, *The CBS Sports Spectacular* while continuing to do sport news for the principal WBBM-TV newscasts and sports commentaries for the all-news WBBM Radio.

MUSEUM OF BROADCASTING—repository of audio tapes, video tapes, books and scripts and other written materials, situated at 1 E. 53 St. in New York City and open to the public. Established under a personal grant from William S. Paley, chairman of CBS, it opened its doors in November 1976 with 718 shows from all three networks and independent sources. Paley has guaranteed the first five years of operation at a probable cost of $2 million. Representatives of all three networks, CPB and broadcast groups are on the board, and Robert Saudek, one-time producer of *Omnibus,* was president until 1981. Robert Batscha succeeded him.

"MY FAVORITE MARTIAN"—moderately successful CBS fantasy series (1963–66) in which Ray Walston portrayed a whimsical visitor from Mars who attaches himself to Bill Bixby. It featured Pamela Britton and was by Jack Chertok Productions.

"MY FRIEND FLICKA"—sentimental half-hour series which ran several times on all three networks, and then in syndication, although only one season's worth of episodes had been produced. CBS carried it initially in 1956, NBC played the reruns the following year, then ABC purchased the 39 episodes for the 1959 and 1960 seasons. The reruns went to CBS in 1961, then back to NBC in 1963 and back again to CBS for the next two years.

Flicka, concerning a boy's love for his horse, and, based on the book by Mary O'Hara and the movie of it, featured Johnny Washbrook as the boy and Gene Evans and Anita Louise. It was by 20th Century-Fox TV.

"MY FRIEND IRMA"—early situation comedy (1952–54) that featured Marie Wilson as a voluptuous and naive good samaritan. Others in the cast were Cathy Lewis, Mary Shipp and Hal March.

"MY LITTLE MARGIE"—father-daughter situation comedy (1952–55) produced for syndication by Hal Roach, Jr., and Roland Reed. It proved a durable item and was one of the first filmed shows whose reruns were "stripped," i.e., presented daily. It featured Charles Farrell and Gale Storm.

"MY THREE SONS"—highly successful situation comedy vehicle for film star Fred MacMurray that ran six seasons on ABC (1959–65) and then seven more on CBS. MacMurray portrayed a widower raising three sons with the help of a housekeeping uncle (William Demarest), who was preceded in the early seasons by a housekeeping grandfather (William Frawley). The sons were Tim Considine, Stanley Livingston and Don Grady. As the boys aged, the initial premise of the series changed. In the 10th season, Beverly Garland joined the cast as MacMurray's new wife. Meanwhile, marriage involving the sons brought other women into the cast, as well as children. The series was by Don Fedderson Productions.

N

"NAKED CITY"—realistic police series based on Mark Hellinger's motion picture which had a season's run in half-hour form on ABC (1958), then was revived successfully with a new cast in hour form (1960–63). The shorter version starred John McIntire and James Franciscus; the longer, Paul Burke, Horace McMahon and Nancy Malone. Both were by Herbert B. Leonard's Shelle Productions and Screen Gems.

"NAME OF THE GAME"—an attempt by NBC to create three 90-minute series (1968–70) in one by rotating three heroes of a single publishing empire. Gene Barry had every third episode as the publisher, Robert Stack portrayed the editor of a crime magazine and Tony Franciosa, the editor of a celebrity magazine. Regularly featured were Susan Saint James and Ben Murphy. The series was produced by Universal TV and grew out of a made-for-TV movie, *Fame Is the Name of the Game.*

"NAME THAT TUNE"—game show created by Harry Salter in 1954 which had a successful run on NBC with a succession of emcees: George deWitt, Bill Cullen and Red Benson.

The series was revived for NBC daytime in 1974 by Ralph Edwards Productions and simultaneously once weekly for prime-access syndication. The daytime version was canceled after five months but the syndicated edition continued.

"NANNY AND THE PROFESSOR"—ABC situation comedy (1970–71) about an attractive British governess who helps a widowed professor raise his three children. It featured Juliet Mills, Richard Long, David Doremus, Trent Lehman and Kim Richards and was produced by 20th Century-Fox TV.

NASHVILLE NETWORK, THE—see Cable Networks.

NATIONAL ACADEMY OF TELEVISION ARTS AND SCIENCES (NATAS)—national organization formed in 1957 to confer annual awards of excellence in TV production and performance in the manner of the older motion picture academy. Until a rift developed in 1976 between East and West Coast factions, NATAS consisted of 13 chapters in major cities and had a total membership of about 11,000, representing both the creative and technical sides of the medium. The chapters not only participated in the national voting for awards but also presented local awards.

The schism occurred when the presidency of NATAS shifted from Robert Lewine of the Hollywood chapter to John Cannon of New York who derived his support chiefly from the hinterland chapters. The Hollywood group rebelled and mounted a boycott of the 1977 awards, arguing that its members were the ones who performed in and produced, wrote and directed virtually all the shows that would be considered for awards. The Hollywood members maintained that those who belonged to the hinterland chapters were not really peers and should not have had the right to oust a president who, in their view, had performed outstandingly.

Also vexing to the dissidents was the fact that Hollywood, as the largest chapter with 4,800 members, had only 10 trustees and therefore could not control an organization in which New York and the regional chapters together had 20 trustees.

When all attempts to resolve the differences failed, the Hollywood unit withdrew from NATAS and created a separate organization called the Academy of Television Arts & Sciences. In addition, it sued to retain the exclusive use of the name "Emmy" for its awards, basing the claim in that the Hollywood unit had been presenting Emmy awards for several years before the national academy was formed.

All suits were dropped and the matter settled with a compromise. ATAS, the Hollywood organization, would have the rights to the Emmy awards for nighttime programs. The National Academy of Television Arts & Sciences, with headquarters in New York, would control the Emmy awards for daytime, sports, news and local programs. It would also continue to publish the *Television Quarterly*. See also Emmy Awards.

NATIONAL ASSN. FOR BETTER BROADCASTING (NABB)—oldest of the broadcast reform groups, founded in 1949 and dedicated to "the public's right to better broadcasting." Based in Los Angeles, the organization monitors programming for quality and antisocial content and participates in regulatory and congressional hearings. Its newsletter is called *Better Radio and Television*. In 1973 it negotiated a controversial agreement with KTTV Los Angeles to ban a list of violence-oriented children's shows. See also Broadcast Reform Movement, KTTV Agreement.

NATIONAL ASSN. OF BROADCAST EMPLOYEES AND TECHNICIANS (NABET)—union representing engineers and certain non-technical personnel at NBC, ABC and more than 50 local TV stations, with a total membership of around 8,500 employees. NABET grew out of the Assn. of Technical Employees, organized in 1933, which became the bargaining agent at NBC. When NBC was forced to split off its Blue Network, NABET found itself with built-in jurisdiction at the new company, which became ABC. An attempt in the early 50s to gain jurisdiction over CBS technical personnel, represented by IBEW, failed.

The union adopted the name NABET in 1940 when it began to extend its representation beyond technical workers to other job classifications, including news writers, publicity writers and clerical departments. NABET struck NBC for three weeks in 1959 and, in a contract dispute with both networks, struck ABC in the fall of 1967. A protracted strike

at ABC in 1977 chiefly concerned the extent to which the network would be allowed to use freelance crews for electronic newsgathering.

NATIONAL ASSN. OF BROADCASTERS (NAB)—the major trade association of the U.S. broadcast industry; headquartered in the capital, it represents the interests of the three networks, some 540 TV stations and more than 4,000 radio stations. The NAB establishes codes for acceptable industry practices, speaks for the industry in national and policy matters relating to legislation and governmental regulation, and provides legal, labor relations and research services for its members.

An arm of the association is the Television Information Office, based in New York, whose essential work is to put forth a favorable image for the industry. NAB's annual convention, generally devoted to management and engineering aspects of the industry, is usually held in March or April and is the largest broadcast conclave of the year.

NATIONAL ASSN. OF TELEVISION PROGRAM EXECUTIVES (NATPE)—an organization for TV program directors, formed in 1963, whose annual February conference serves also as a domestic program exposition for syndicators. Although its original purposes merely were to raise the status of the program director in the industry and to foster an exchange of local program ideas, the organization gained prominence through the FCC's prime time–access rule, for which it provided the industry's chief forum and program market. By the mid-70s NATPE's conferences, drawing general managers as well as program directors, rivaled in importance those of the National Assn. of Broadcasters for the discussion of industry issues. The organization's presidency passes each year to an active program director.

NATIONAL ASSN. OF THEATER OWNERS v. FCC [420 F2d 922 D. C. Cir. (1969), cert. den./397 U.S. 922 (1970)]—case in which the Court of Appeals held that the FCC had the power to regulate pay-TV and that pay-TV was a permissible medium.

In late 1968, after many years of consideration, the FCC issued a report which stated its authority to regulate subscription television and then established regulations for pay-broadcasting. The National Assn. of Theater Owners challenged both the regulations and the power of the commission to issue such regulations.

The Court of Appeals for the District of Columbia Circuit affirmed the FCC's regulations, citing statutory requirements that the FCC act to "encourage the larger and more effective use of radio in the public interest" and the Supreme

Court's holding that the FCC could regulate cable-TV. The regulations, the Court said, did not constitute a general assertion of control over the rate-making decisions of conventional TV stations, nor were the regulations arbitrary because the FCC was not required to employ measures "less drastic" than rate-making.

Finally, the Court rejected a claim that even if the regulations were within the power of the commission, they were unconstitutional. The court held that the regulations were not discriminatory against the poor or in violation of the equal protection clause of the Fourteenth Amendment, as had been claimed, because the FCC had promulgated a number of safeguards to protect free-TV. Nor, it said, did the program restrictions placed upon pay-TV violate the First Amendment as a prior restraint of free speech. The court indicated that the rules did not ban speech.

The Theater Owners sought review in the Supreme Court but the court denied certiorari.

An NBC studio control room

NATIONAL BLACK MEDIA COALITION (NBMC)—

an aggregate of nine former media reform groups working for fair representation of black people in all aspects of broadcasting. NBMC, which had its origins in a Washington group called Black Efforts for Soul in Television (BEST), works to insure full employment of blacks in the industry and fair treatment of them in televised material.

Established in 1973, NBMC now has 75 affiliated groups in 62 cities and 29 states, which are coordinated by the Washington office under the leadership of Pluria Marshall. See also Broadcast Reform Movement.

NATIONAL BROADCAST EDITORIAL ASSN.

(NBEA)—industry organization for TV and radio stations, writers and presenters of editorials. Meetings are held in Washington every other year for briefings on national issues and regulations pertaining to editorializing on the air.

NATIONAL BROADCASTING CO.—See NBC.

NATIONAL BROADCASTING CO. v. UNITED STATES [319 U. S. 190 (1943)]—case in which the Supreme Court affirmed the power of the FCC to issue rules restricting the scope of agreements between the networks and their affiliates. Suit was brought by all the networks following the FCC's issuance in 1941 of regulations which greatly restricted the future scope of the national radio networks and their respective affiliates.

The FCC's new rules prohibited exclusive affiliation between a local station and a network; barred local stations from restraining other stations in the area from broadcasting a network show which it chose not to air; prohibited the networks from requiring affiliates to air a particular program at a particular time; and established that networks could not regulate the local advertising rates of affiliates. The rules also set a maximum of two years for affiliate contracts.

In their suits the networks challenged the regulations as being beyond the scope of the commission (since the networks are not licensed) and unconstitutional. The Supreme Court, as a threshold matter, decided that the FCC had the power to issue such regulations. This power, it said, came from the general powers inherent in the commission's congressional mandate to act within the "public interest, convenience and necessity" and to promote the "more effective use of radio." Additionally, Congress gave the commission power to make regulations concerning chain broadcasting. The court concluded that the regulations comported with the Communications Act of 1934.

In dealing with the constitutional challenge, the court held that the "public interest" standard was not too vague to constitute an overt delegation of power from Congress to a regulatory commission. Nor did the regulations conflict with the First Amendment prohibition of speech abridgments. The court held that some regulation of radio was necessary to create an orderly system of radio broadcasting, and that the regulation need not be limited to technical considerations alone.

The court stated, bluntly, that the "right of free speech does not include...the right to use the facilities of radio without a license." Since the court rejected the assertions of the networks on these two major points, it affirmed the power of the FCC to issue guidelines relating to the practices of radio networks.

NATIONAL CABLE TELEVISION ASSN. (NCTA)—

trade association for the cable-TV industry representing its interests before Congress, the FCC and state regulatory agencies. As cable's equivalent of the NAB (although smaller and weaker), NCTA was successful in countering the broadcast industry's propaganda war on pay-cable in the mid-70s and marshaled its forces to fight for recognition of cable-TV as an independent medium in Congress' efforts to rewrite the nation's basic communication law.

NATIONAL CITIZENS' COMMITTEE FOR BROADCASTING (NCCB)—a nonprofit public interest action group based in Washington, D.C., that serves as a central service organization for the various citizens groups in the broadcast reform movement. Essentially, the aim of the movement is to make radio and TV more responsive to the needs and interests of the public.

NCCB had begun in the late 60s as a citizens' organization to support legislation for public broadcasting and was originally the National Citizens' Committee for Public Broadcasting. Dropping "public" from its name in 1968, and headed by Thomas P.F. Hoving with Ben Kubasik as executive director, it became a watchdog organization on commercial broadcasting practices. NCCB went into a decline after a few years but was reorganized in 1974 with Nicholas Johnson as its chairman, giving the former FCC commissioner a base from which to continue his potent criticism of the industry and the regulatory process.

In 1978, the NCCB became part of the Ralph Nader organization, with Nader becoming its chairman. Johnson left to head the new National Citizens' Communications Lobby, which derived funding from NCCB.

NATIONAL CITIZENS' COMMUNICATIONS LOBBY (NCCL)—public-interest organization headed by Nicholas Johnson that spun out of the National Citizens' Committee for Broadcasting when Ralph Nader became its chairman in 1978. With funding from the Nader group, and with a separate board of directors, NCCL became a membership organization devoted to research, education and political action on broadcast issues and to representing citizen interests before government agencies and Congress.

NATIONAL CONFERENCE OF MOTION PICTURE AND TV UNIONS (NACOMPTU)—East Coast organization concerned with the common interests of film, TV and related unions, comparable to the Hollywood Film Council. In legislative and regulatory matters, it represents the following unions: Theatrical Protective Union, Local 1, IATSE; Theatrical Stage Employees, Local 11, IATSE; NABET, Local 11; NABET, Film, Local 15; NABET, Local 16; Motion Picture Studio Mechanics, Local 52, IATSE; Script Supervisors, Local 161, IATSE; Motion Picture Machine Opera-

tors, Local 306, IATSE; International Photographers, Local 644, IATSE; Theatrical Wardrobe Attendants, Local 764, IATSE; Motion Picture Editors, Local 771, IATSE; American Federation of TV & Radio Artists (AFTRA), N.Y. local; Make-up and Hair Stylists, Local 798, IATSE; Brotherhood of Theatrical Teamsters, Local 817; United Scenic Artists, Local 829; Screen Cartoonists, Local 841, IATSE; Radio and Broadcast Engineers, Local 1212, IBEW; Screen Actors Guild (SAG), New York branch.

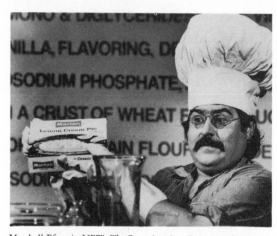

Marshall Efron in NET's *The Great American Dream Machine*

NATIONAL EDUCATIONAL TELEVISION (NET)—the principal supplier of national programming for the noncommercial system through all the phases of ETV and PTV. Until the Public Broadcasting Service began in 1970, many regarded NET as the public "network," although it had to deliver its programs by post since the stations were not interconnected. For economy reasons, it merged in April 1972 with New York station WNDT, which then changed its call letters to WNET.

NET grew out of NETRC, the National Educational Television and Radio Center created in December of 1952 by the Fund for Adult Education, and an original grant of more than $1 million. Its purpose initially was to facilitate an exchange of films and kinescopes between ETV stations, from a processing center at Ann Arbor, Mich. Later, for purposes of developing new programs, it expanded to New York, with the annual funding assumed largely by the Ford Foundation. In October 1962, it dropped the radio function and became simply NET headed by John F. White.

Ford regularly contributed $6 million toward NET's annual $8 million operating budget, and the organization supplied approximately half the general programming for the typical ETV station in the years before the Public Broadcasting Act of 1967. Exclusive of reruns, NET normally provided five hours of original evening programming a week–divided between cultural and public affairs–and 2.5 hours a week of children's programs. Most of the production was by staff or independent producers in New York, but some was contracted to local stations in various major cities. NET also imported programs from foreign systems, principally the BBC.

Although the programs were needed by all stations, many regarded NET as having a liberal bias on political issues and were wary of its documentaries. The establishment of PBS served to give individual stations a greater voice in what was being produced for the system nationally, but NET (as WNET) continued to be a major supplier of programs chiefly because it had the support of the Ford Foundation and access to the New York talent pool. See also Ford Foundation and Public Television (PTV).

NATIONAL GEOGRAPHIC SPECIALS—series of specials on natural history, anthropology and explorations, produced in association with the National Geographic Society and carried first on CBS (1965–73), then on ABC (1973–74) and later on PBS (1975–). The nonfiction hours were produced at the rate of four a year and were a surprising success when they began, since programs of their type did not usually draw high ratings in prime time.

CBS dropped the series after eight years when the ratings went into a decline, and after a lukewarm season on ABC the Geographics were no longer of interest to the commercial networks. Meanwhile, four new programs had already been produced. They eventually wound up on PBS on underwriting from Gulf Oil, through the Pittsburgh station, WQED. The first of the four, *The Incredible Machine,* a documentary on the workings of the human body, was an enormous hit, scoring the highest rating in the history of public television and in some markets outdrawing programs on the commercial networks. Thus the series was reborn, as a noncommercial entry.

The initial group of programs on CBS was produced by Wolper Productions. Metromedia Producers Corp. then took over for several years, and Wolper had the project again in 1973. The programs were narrated by Alexander Scourby, Orson Welles, Leslie Nielsen and Joseph Campanella, and they included such titles as *Grizzly!, The Amazon, Dr. Leakey and the Dawn of Man, Holland Against the Sea, Polynesian Adventure* and *Miss Goodall and the Wild Chimpanzees.*

The four that played on ABC, all by Wolper, were *The Big Cats, Bushmen of the Kalahari, Journey to the Outer Limits* and *Wind Raiders of the Sahara.*

The PBS group, produced independently by such talents as Irwin Rosten and Nicolas Noxon, featured E.G. Marshall as host and included *Treasure!, Voyage of the Hokule'a, The Volga, The New Indians* and *Search for the Great Apes.*

NATIONAL NEWS COUNCIL—an independent organization founded in August 1973 to receive, examine and report on specific complaints concerning the accuracy and fairness of news reporting by the press and broadcast media. The 15-member council has no power of enforcement over the media beyond the release of its findings. It was formed on the recommendation of a task force selected by the Twentieth Century Fund and has been supported by the Fund and other foundations.

NATIONAL PUBLIC AFFAIRS CENTER FOR TELEVISION (NPACT)—former designation for the Washington-based news and public affairs production arm for public television. The name NPACT was dropped in 1976 as a delayed effect of the merger of the organization with the Washington station, WETA. The NPACT programs— *Washington Week in Review, Washington Straight Talk* and live special events—continued to be produced, but under the WETA banner.

NPACT had been created as an independent production agency for the PTV system in July 1971 from what had been the Washington bureau of NET, on grants totaling $3 million from the Ford Foundation and CPB. It was a step taken toward diversifying the production sources for PBS. But shortly after NET was merged with WNDT New York in 1972 to conserve costs, NPACT was similarly merged with WETA, although it retained its own administration under James Karayn, the former Washington bureau chief for NET.

Karayn resigned in 1975 and WETA absorbed the organization totally. Wallace Westfeldt, former NBC News producer, in effect took Karayn's place but with the title of executive producer. He left in 1977 to join ABC News.

Frequently at the center of controversy during its early years, NPACT was attacked by aides to President Nixon for an alleged liberal bias, and it was a target of OTP director Clay T. Whitehead, who contended that it was improper for a system receiving federal funds to deal in news and public affairs programming.

A source of congressional indignation was the revelation that NPACT's principal anchormen, Sander Vanocur and Robert MacNeil, were being paid $85,000 and $65,000 a year, respectively. Although these fell below comparable salary levels in commercial TV, they were considered extravagant for a system that had been pleading with Congress for larger funds. This prompted the CPB board to put a ceiling of $36,000 on public TV salaries, which could be exceeded only with explicit approval of CPB.

When Vanocur and MacNeil left, they were succeeded by Jim Lehrer and Paul Duke, the latter, like his predecessors, an alumnus of NBC News. NPACT's other correspondents included Carolyn Lewis, Peter Kaye and Christopher Gaul, and the agency drew regularly from newspaper and magazine correspondents in the capital for on-air assignments.

NATIONAL TELECOMMUNICATIONS AND INFORMATION ADMINISTRATION (NTIA)—an

agency of the Commerce Department established in 1977 by the Carter Administration to assume the functions of the disbanded Office of Telecommunications Policy in the White House. NTIA is responsible for keeping abreast of developments in communications and the common-carrier field and for recommending policy and legislation. The OTP had lost its influence and was discredited when its first director, Clay T. Whitehead, used the office to advance the broadcast strategies of the Nixon White House.

Henry Geller, a former FCC general counsel and public interest lawyer, became acting director of OTP before the changeover and also of NTIA when the agency moved to Commerce. He was officially named Assistant Secretary of Commerce and director of NTIA early in 1978. In that capacity, he was also the President's chief advisor on telecommunications matters. Geller was succeeded in the Reagan administration by Bernard Wunder.

See also Geller, Henry; Office of Telecommunications Policy; Whitehead, Clay T.

NTSC—initials commonly used to designate the American-developed systems of both monochrome and color television. The letters stand for National Television System Committee, first established in 1936 to develop monochrome color television standards and reconstituted in 1950 to recommend a compatible color television system. Both groups consisted of engineers from a variety of receiving and transmitting equipment manufacturers and were formed under the aegis of the Radio Manufacturers' Association (now the Electronic Industries Association).

The first NTSC recommended the 525-line, 60-field-per-second black-and-white system currently in use. The second group developed a color television system compatible with these monochrome standards. Both recommendations, based largely on systems developed by RCA, were adopted by the FCC for American television. The NTSC monochrome system has a 6-megaherz channel bandwidth, amplitude-modulated picture frequency-modulated (FM) sound.

The major countries using NTSC monochrome and color standards, in addition to the United States, include Canada, most Western Hemisphere nations, Japan and other Far Eastern countries in the American sphere of influence. NTSC color broadcasting started in the United States on January 1, 1954. The other major television system is the European CCIR, which is used with two different color standards—PAL and SECAM.

"NATIONAL VELVET"—half-hour NBC series (1960–62) based on the movie from the Enid Bagnold novel about a young girl and the champion horse she raises. Lori Martin played the lead in the MGM-TV series.

"NAVY LOG"—series of documentary-like dramas produced for syndication (1955–57) by Sam Gallu Productions with the cooperation of the Department of the Navy. The 30-minute stories, each with a different cast, formed an anthology that enjoyed a wide sale overseas as well as domestically.

NBC (NATIONAL BROADCASTING CO.)—RCA's radio and television subsidiary, created to help sell receivers and other broadcast equipment manufactured by the parent company. Year by year, however, with the growth of television, NBC has become responsible for more and more of RCA's total profits.

Through most of the years in television competition NBC has been second in the ratings and second also in glamor. But if running second in this manner has been NBC's negative tradition, there is also a positive one: NBC has through the years been the most innovative of the networks, the one that brought TV to the masses with Milton Berle and a quarter century later opened new audience frontiers after midnight with *Tomorrow* and *Saturday Night Live.*

NBC introduced the desk & sofa talk show (*Today, Tonight*), the TV special, the perennial special (*Peter Pan, Amahl and the Night Visitor*), the prime time movie showcase, the made-for-TV movie, the Sunday press conference (*Meet the Press*), the free-form comedy show (*Laugh-In, Saturday Night Live*), the 90-minute series (*Virginian*), the rotating dramatic series (*Name of the Game, Sunday Mystery Movie*), the dual-anchor newscast (Huntley-Brinkley), the one-minute newscast (*Update*), the one-hour soap opera, the miniseries and the magazine concept for TV advertising.

Because of RCA's manufacturing interest, and because the RCA color system had beat out CBS's in the early 50s, NBC was the first network to push color and to present all its programs in color. CBS did not join until the penetration of color sets became significant in 1965, and ABC followed grudgingly. As with color, the other networks followed NBC's lead into early morning and latenight programming and adopted many of the forms it developed.

Indeed, CBS built its ratings supremacy on programs spirited away from NBC while they were at their peak of popularity during the declining days of network radio—*Amos 'n' Andy,* Jack Benny, Red Skelton and Edgar Bergen & Charlie McCarthy.

NBC withstood further raids, however, and in the early years of TV was able to field such stars as Bob Hope, Fred Allen, Eddie Cantor, Kate Smith, Groucho Marx, Sid Caesar, Jimmy Durante and Milton Berle. In pioneering latenight programming, NBC developed such stars as Jerry Lester, Steve Allen, Jack Paar and Johnny Carson.

NBC historically had scant success with situation comedy—the form that became the strength of CBS—but fared best with variety shows, movies and long-form adventure dramas. Its commitment to the special and the coverage of major events led it to seek identity as the network of "event television." A step taken to turn that phrase into a reputation was the installation in 1976 of a Sunday night block of time carrying no series but designated as *The Big Event.*

As it turned out, that move coincided with NBC's greatest crisis—its slide into 3rd place, behind CBS, as ABC began its emergence as the new power in network television. Third place was costly: By 1979, the profit differential between ABC and NBC was $100 million a year. Moreover, NBC became vulnerable to ABC's raids on its affiliates, and NBC lost important stations to its rival in Minneapolis, Indianapolis, Dayton, San Diego, Charlotte, N.C., and Jacksonville, Fla., among other cities.

Inevitably, these developments brought down the administration of Herbert S. Schlosser. Since programming was at the heart of the network's problem, RCA recruited the leading television program executive on the scene—Fred Silverman of ABC—at a reported salary of $1 million a year. Silverman became president and chief executive of the National Broadcasting Co. in June 1978. Several months later, Jane Cahill Pfeiffer was brought in as chairman, to assist Silverman in the management of the company, and Richard S. Salant joined from CBS as vice-chairman.

The Silverman era was chaotic and a travesty of the television business. In his desperate quest for a smash hit—on the theory that one or two would set everything right—Silverman continually revamped the schedule, almost week to week. His famed programming gifts failed him utterly, and he proved unequal to the task of running a large company. He hired and fired executives in the same frenzied manner with which he installed and discarded programs. By the time he left in 1981, NBC's programming was a shambles, both daytime and nighttime; some of the network's oldest and most prestigious affiliates had lost faith in NBC and joined a rival camp; and many of the company's most dedicated executives were gone. The debacle also brought down Edgar Griffiths, chairman of RCA Inc., and left NBC so deep in third place it actually became conceivable that RCA—which invented commercial networks—might sell off its broadcast operations

Thornton Bradshaw, president of Atlantic-Richfield and a member of the RCA board, was named to succeed Griffiths, and he in turn recruited Grant Tinker to head the National Broadcasting Company as chairman. Tinker, a

former NBC program vice president who became president of MTM Productions (named for his ex-wife, Mary Tyler Moore), was one of the most popular and respected television figures in Hollywood. He was expected to bring a more sober and dignified style to NBC and to upgrade the network's programming through his good judgment and Hollywood connections. Very little happened to change NBC's fortunes during Tinker's first year.

The character and stature of NBC as a company was molded by Gen. David Sarnoff, who headed RCA for nearly 40 years, and his son Robert, who moved up the executive ladder at NBC and then followed his father into the chairmanship of RCA. (He was ousted in a corporate shakeup in November 1975.) But the network's personality was forged by two other men—Sylvester Weaver and Robert E. Kintner. Both broke new ground: Weaver in entertainment, Kintner in news.

Weaver, the imaginative and venturesome NBC president in the 1950s, created the TV *spectacular* (later dubbed the *special*), the intimate talk show (*Today, Home, Tonight*) and the TV magazine (*Wide, Wide World*). Kintner, a former newspaper columnist who had become president of ABC-TV before NBC hired him away, countered the CBS leadership in entertainment by setting out to trounce that network in news.

Kintner gave news a priority it never enjoyed before or since in television, and through it he won a new kind of respect for the medium generally. His standing order to the news forces was "CBS Plus 30"—which meant that in all instances of common coverage NBC was to offer 30 minutes more than CBS. The Huntley-Brinkley team emerged during Kintner's administration, backed by a platoon of TV journalists who became star byliners—Frank McGee, John Chancellor, Edwin Newman, Sander Vanocur, Nancy Dickerson, Irving R. Levine, Pauline Frederick—and a battalion of crackerjack executives, writers and documentary producers. It was on Kintner's approval that Barbara Walters was allowed to develop as the female star of *Today.*

When Kintner left in 1966, NBC retreated from flamboyancy and one-man rule and experimented instead with a tripartite system of management and other forms considered more businesslike. Kintner was succeeded by a former news executive, Julian Goodman, but the company's dominant figure for the next decade was David C. Adams, an erudite attorney who functioned under various titles—chairman, vice-chairman, executive v.p.—but served in all as the grey eminence.

Adams, who always preferred to remain in the background personally, manifested a distaste for high visibility in management generally. But he yielded eventually to the peculiarities of the TV business, which has always thrived best on the cult of the personality. In 1975 Herbert S. Schlosser, a lawyer and former NBC program executive, was elevated to president of NBC with a clear mandate to provide aggressive personal leadership. Two years later, he became also chief executive officer. Goodman, as chairman, assumed the role of statesman. Adams directed the corporate staff. Goodman and Adams elected retirement, and Schlosser took a post with RCA, after the start of the Silverman regime.

Structure. Organizationally, the National Broadcasting Company has no exact equivalent at the other networks. It consists of "corporate" NBC and four divisions whose presidents report to the NBC president. NBC's management, in turn, reports directly to the president of RCA. NBC has its own board of directors. The RCA board always includes one officer of NBC. The NBC divisions follow:

- NBC-TV, the television network, which embraces engineering, operations, affiliate relations and sales;
- NBC Entertainment, the programming arm, responsible for programming, advertising and promotion;
- The Owned Television Stations, which includes the spot sales unit. NBC's stations are WNBC-TV New York, KNBC Los Angeles, WMAQ-TV Chicago, WRC-TV Washington, D.C., and WKYCD-TV Cleveland;
- NBC Radio, encompassing the network and owned stations;
- NBC Sports;
- NBC News, which serves both the radio and TV networks.

(For a brief time in the late 60s, RCA Victor, the recording division, came under the stewardship of NBC.)

The news division operates autonomously of the network and has certain fixed positions in the schedule—those for *Nightly News, Today, Weekend* and *Meet the Press.* Where preemptions are indicated for special reports or documentaries, the requests for the time are made of the company president.

History. Oldest of the networks, NBC was inaugurated on Nov. 15, 1926 with a special four-hour program originating from the ballroom of the Waldorf-Astoria Hotel in New York that included cutaways to other cities. The station lineup consisted of 21 charter affiliates and four other outlets, reaching as far west as Kansas City. CBS was organized the following year.

RCA had entered broadcasting in 1926 when it purchased WEAF (now WNBC), in New York from AT&T for $1 million plus an agreement to use AT&T for network interconnection. NBC was incorporated Sept. 9, 1926 by a consortium consisting of RCA, which owned 30% of the stock, Westinghouse which owned 20% and General Electric, 50%. At the time, an estimated 5 million homes were equipped with radio.

That same year, NBC acquired a second New York station, WJZ (actually licensed to Newark, N.J.), which had been owned by Westinghouse. Rather than duplicate the network programs on two New York stations, NBC created a second network. Stations hooked into the programs originating at WEAF became the Red Network; those carrying the WJZ feed, the Blue Network. The color designations resulted from the switchboard's use of red and blue to separate the networks. The Blue Network began Jan. 1, 1927 with six affiliates. (On FCC orders, it was divested by RCA in 1943 and became ABC.)

In 1930 RCA bought out the Westinghouse and GE interests and became sole owner of NBC, with Sarnoff doubling as its and RCA's president.

Radio not only developed the stars and program formats that were to be adopted by commercial television but also the entire modus operandi, including the economic system and the affiliate relationships.

NBC began its first TV transmission on Oct. 30, 1931 on experimental station W2XBS New York, and it inaugurated regular TV service on April 30, 1939 with a telecast of President Roosevelt at the opening of the New York World's Fair. The first "network" broadcast on Jan. 12, 1940 linked the New York station, WNBT (now WNBC-TV), with WRGB in Schenectady. World War II all but halted the development of TV, but the post-war spurt began around 1947, when *Kraft Television Theatre* and *Howdy Doody* started in the medium (along with other dramatic series on CBS and DuMont). But the medium's proliferation among working-class households was propelled in 1948 by the start of *The Texaco Star Theater* with Milton Berle. Then began the wholesale transfer of shows and stars from radio to TV—everything from *The Voice of Firestone* to the soap operas.

The network fanned out as AT&T facilities permitted and as new stations signed on the air when the FCC lifted its freeze on licenses in 1952. On Sept. 4, 1951 NBC and the other networks pooled coverage of the signing of the Japanese peace treaty in San Francisco, an event which inaugurated coast-to-coast service.

"NBC MAGAZINE"—weekly series begun in the fall of 1980 which represented the third try by NBC News to establish a weekly prime-time newsmagazine comparable to CBS's *60 Minutes* and ABC's *20/20*. These efforts started with *Weekend* in 1978. The following year it became *Prime Time Sunday,* with Tom Snyder as host-anchorman and with an emphasis on live explorations of news stories through interviews by satellite. When this failed after eighteen months, the series was revised and built around a new anchor, the veteran David Brinkley. It was slotted Fridays at 10 P.M. for the 1980–81 season and retitled *NBC Magazine with David Brinkley.*

The new format was less formal and more chatty than the previous ones. Between feature stories, Brinkley bantered with the principal correspondents–Jack Perkins, Garrick Utley, Dougls Kiker, and Betsy Aaron. The critical response to the program improved, but the ratings did not, and the show was shifted to Thursdays at 8 P.M.

When Brinkley suddenly left the network to join ABC News at the start of the 1981–82 season, the program proceeded without a host-anchor and with a shorter title, *NBC Magazine.*

NBC-WESTINGHOUSE SWAP—an exchange of stations between NBC and Westinghouse Broadcasting in 1955 which was held suspect for 10 years and finally led to an order by the FCC to reverse the deal. Thus, NBC became a sojourner in Philadelphia and Westinghouse in Cleveland.

The original exchange involved NBC giving up its Cleveland station WNBK-TV and $3 million in 1955 for Westinghouse's Philadelphia properties, WPTZ-TV and KYW-AM. The FCC approved the transaction but the agreement raised some eyebrows because it allowed NBC to trade up from the 6th largest market to the 4th, which was not a logical move for Westinghouse. Becoming concerned with possible antitrust implications, the Justice Department found a Westinghouse executive who said the trade was "forced" upon his company, and a federal grand jury probe into the alleged coercion was begun in 1956. Three years later the Supreme Court ruled that NBC would have to stand trial for the allegation. In 1964 the FCC said it would renew the licenses of the stations if the two companies would switch back, and the following year they did. NBC's Cleveland station then became WKYC-TV and Westinghouse's Philadelphia outlet, KYW-TV.

"NBC WHITE PAPER"—umbrella title for special documentary reports produced by NBC News, begun in 1960.

NEBRASKA GAG RULE—a controversial order by a Nebraska court restricting media coverage of the 1975 trial of a mass murderer, in order to preserve the defendant's right to a fair trial. The networks and RTNDA joined with members of the print press in opposing the prior restraint as a violation of the First Amendment and one that would establish a dangerous precedent. In 1976 the Supreme Court ruled the gag order unconstitutional.

The order by Judge Hugh Stuart, in the case in which Erwin Charles Simants was accused of murdering six members of a family and sexually assaulting two of them, barred the media from reporting certain matters that were discussed in pretrial proceedings (including confessions by the defendant) and aspects of the case discussed in open court. The case was brought to the Supreme Court by newspapers and broadcasters in Nebraska.

NELSON, GENE—veteran writer-director whose credits include made-for-TV movies, pilots (*I Dream of Jeannie*) and

episodes for such series as *Mod Squad, The FBI, Dan August* and *The Rookies.*

NELSON, OZZIE (d. 1975)—actor, producer and director who starred in ABC's long-running family situation comedy, *The Adventures of Ozzie and Harriet,* and also in the short-lived *Ozzie's Girls.* He also produced and directed both series and earlier had been producer-director of *Our Miss Brooks.* Before entering TV, he had been a popular bandleader on radio, with his wife, Harriet Hilliard, as vocalist.

NELSON, RALPH—one of the outstanding early directors in the medium, who directed the entire *Mama* series, the first production of Rodgers & Hammerstein's *Cinderella* (1957) and dramas for *Playhouse 90, Climax!* and other anthologies of the time. He received an Emmy for *Requiem for a Heavyweight* and also directed the movie version. Like many creative talents of the "golden age," he gave up TV for movies and directed such notable films as *Charly* and *Lilies of the Field.*

NESSEN, RON—NBC News correspondent who in 1970 was appointed press secretary to President Ford, the first TV newsman to hold that position. A former reporter for UPI, he had joined NBC in 1962 and was White House correspondent during the Johnson Administration.

NETHERLANDS, TELEVISION IN—probably the nearest thing to an access medium in that, structurally, it is geared to accommodate almost every significant political and religious persuasion in the land.

The single-channel system is governed by the NOS, the Netherlands Broadcasting Corp., which is also a consortium of a half dozen private programming companies. Each has a percentage of time to fill each week, but under the 1967 Broadcasting Act each must also collaborate on a sizable weekly percentage of neutral or "joint programs." In addition, the Act stipulates that a limited amount of time be allotted to institutions other than broadcast organizations, as a consequence of which Catholic, Protestant, Humanist and other groups also have regular access to the channel. Especially since the change to a more conservative government at the end of 1977, there has been growing pressure to make use of the regional cable stations in a more commercially oriented fashion.

The private NOS companies are supported out of annual set license fees (taxes), but are also permitted to sell a limited amount of advertising. About 12 hours a week on Dutch TV are in color, on 625-line PAL. The license fees for TV households are the equivalent of $58 a year for both color and black and white sets.

NETWORKS—chains of stations interconnected by microwave and coaxial cable for the efficient distribution of programs and advertising from a central source. Because the economies of scale permit more ambitious and expensive production than individual stations could normally afford, and because they tend to be highly promotable, network programs usually draw greater audiences than local shows in any TV market.

The major national networks in the U.S.—ABC-TV, CBS-TV and NBC-TV—are each affiliated with approximately 200 stations (ABC, with 195, has fewest) and cover virtually the entire country. Stations are compensated by the networks for the use of their local air time at a rate that makes it more profitable for them to be affiliated than to buy or produce, and then sell, their own programs. Because of this, the networks supply most of the programs that affiliated stations broadcast, and they have become the dominant force in American television, although the system was founded on the doctrine of localism.

There are regional, state and part-time national networks (such as the Hughes Television Network) devoted chiefly to sports, but the term usually denotes the big three, which largely are carryovers from radio. NBC, the first of the networks, was established by RCA in 1926 when it connected its flagship, WEAF in New York, to various stations halfway across the country. The following year it went coast to coast over AT&T lines. CBS started in 1928. ABC grew out of the Blue Network, one of two maintained by NBC (the Red and the Blue) until it was ordered to divest itself of one.

If there had been no networks in radio, they would undoubtedly have been created in television if only to build audiences for, and bring glamor to, the stations owned by the large companies in the largest markets. During the many years ABC-TV operated at a loss as a network, its continuation was justified by the fact that the five stations it owned became immensely profitable from the service.

Arrangements for station compensation vary and are subject to negotiation, but by a rule-of-thumb the network pays each station approximately 12% of its rate card charge for network commercials it carries. In addition, station breaks and other local commercial adjacencies are made more valuable by the program environment created by the potent network service. Networks relieve local stations from making massive program investments and from maintaining sizable production and creative staffs, while improving their popularity in the market and their ability to charge more for their time. This is why the license for a network-affiliated station is worth substantially more on the market than that for an independent.

When they broadcast live, the networks are unique in communications in that they produce, exhibit and distribute in a single process. With their stars and high-powered promotion, the networks are largely responsible for the vast penetration of television sets in the U.S. and for the fact that viewing in the average household exceeds seven hours a day. Among them, the three networks reach more than 100 million viewers every night.

But while prime-time is where they realize the largest audiences, in fact the greatest profit centers for the networks have at times been daytime, Saturday mornings and late-night—when they are successful—because the programs cost far less to produce. Annually, the networks divide close to $3 billion in advertising revenues.

Although they are not directly regulated by the FCC, the networks are held under control by the agency's authority over their owned stations and their affiliates. Network-affiliation contracts are limited by the commission to two years. The FCC has driven the networks out of the domestic syndication business and has barred them from seeking an equity in programs they buy from outside suppliers. To curb the network monopoly, the agency has restricted the networks to three hours of programming in prime time, and the Justice Department has filed suit to bar the networks from the production of entertainment shows.

There have been attempts to start a fourth commercial network with independent stations on both VHF and UHF, but the economic hazards and the inability to blanket the country in the manner of the established networks made each attempt futile except on an occasional and limited basis.

THREE NETWORK FINANCIAL SUMMARY
(1963–80):

Year	Revenues	Pre-Tax Profits	Employment
1963	$635,800,000	$56,400,000	9,732
1964	712,500,000	60,200,000	10,709
1965	788,600,000	59,400,000	11,012
1966	903,900,000	78,700,000	11,195
1967	953,300,000	55,800,000	11,538
1968	1,016,400,000	56,400,000	12,190
1969	1,144,100,000	92,700,000	12,913
1970	1,144,600,000	50,100,000	13,917
1971*	1,094,100,000	53,700,000	11,999
1972	1,271,300,000	110,900,000	12,410
1973	1,404,900,000	184,900,000	12,422
1974	1,545,900,000	225,100,000	13,158
1975	1,673,800,000	208,500,000	13,256
1976	2,117,000,000	295,600,000	13,802
1977	2,581,000,000	406,100,000	14,153
1978	2,964,000,000	373,500,000	14,542
1979	3,454,000,000	370,200,000	16,028
1980	3,865,000,000	325,600,000	16,590

*Cigarette advertising banned as of Jan. 1 and prime time-access rule begun in September.

Source: FCC.

Revenues are net, after commissions to agencies, representatives and brokers, after cash discounts and after deducting station compensation to o&os and affiliates.

NETWORK SYNDICATION—the resale of off-network series and specials by the syndication divisions of CBS, NBC and ABC, until the FCC forced the networks to end those operations in June 1972. The commission's purpose was to reduce network control over programs and to remove a potential for anticompetitive developments in the syndication market. The networks were given until June 1973 to cease distributing programs for nonnetwork exhibition.

CBS Films was spun off, along with the CBS cable-TV division, into a new company, Viacom. ABC Films was purchased for close to $10 million by a group of its executives and became Worldvision. The syndication properties of NBC Films (known during the 50s as California National Productions, or CalNat) were sold outright to NTA.

Until the FCC action, the networks had made a regular practice of sewing up the subsidiary rights to programs before putting them on the air, or at least of acquiring financial interests in the programs to enable them to share in the syndication revenues. Because the networks provided a steady stream of desirable programming, their syndication operations exerted a unique power in the syndication market; and stations were known to buy programs they did not want, to remain in the good graces of the network syndicators.

The FCC rules forcing divestiture were actually adopted in 1970, at the same time the prime time–access rule was hatched, but the measure was stayed pending an appeal by the networks to the U.S. Court of Appeals in Washington. In 1972 the court upheld the commission. Under the rules the networks may only engage in foreign syndication with programs they produce entirely themselves, which basically are news and documentaries.

NEUMAN, ALAN—veteran director who worked on such series of the 50s and 60s as *Person to Person, Wide, Wide World, This Is Your Life, Colgate Comedy Hour, NBC Opera, Wisdom* and *Producer's Showcase.*

NEUMAN, E. JACK—lawyer turned writer-producer who created such series as *Police Story, Petrocelli* and *Kate McShane.* He also served as executive producer of *Kate McShane* and of the *Law and Order* pilot, which he wrote. He works through his independent company, P.A. Productions.

"NEW ADVENTURES OF CHARLIE CHAN"—syndicated series produced in England by Vision Productions and ITC in 1957. It featured J. Carroll Naish and James Wong.

"NEW ADVENTURES OF HUCK FINN, THE"—fantasy series loosely based on the Mark Twain characters. It combined live performers and animated characters against a cartoon background. Produced by Hanna-Barbera and carried in prime time on NBC (1968), it featured Michael Shea, LuAnn Halsam, Kevin Schultz and Ted Cassidy.

"NEW ADVENTURES OF PERRY MASON, THE"—futile attempt by CBS to re-create, in 1973, its old and amazingly durable hit of the 50s. The second version featured Monte Markham as Erle Stanley Gardner's famous fictional attorney and was short-lived, running less than a season. The regular cast included Harry Guardino, Sharon Acker, Albert Stratton and Dane Clark. It was produced by 20th Century-Fox TV and Cornwall Jackson.

"NEW DICK VAN DYKE SHOW, THE"—unsuccessful attempt by CBS (1971–73) to equal the star's initial success in the situation comedy form. Produced in Arizona, the series co-starred Hope Lange and was created by Carl Reiner. It was produced by Cave Creek Enterprises.

NEWHOUSE BROADCASTING CORP.—station group founded by Samuel I. Newhouse, who with his family also owns the Newhouse newspaper chain. TV stations are WSYR Syracuse, an NBC affiliate; WTPA Harrisburg, Pa., ABC; WAPI Birmingham, Ala., NBC; KTVI St. Louis, ABC; and WYSE, Elmira, N.Y., NBC. E.R. Curley Vadeboncoeur is president. In 1979, the company announced plans to enter the cable-TV field, hiring a number of key Cox Cable Communications executives to establish and develop a new cable-TV subsidiary. In 1980, Times-Mirror Corp. purchased the Newhouse group for $82 million with FCC approval.

NEW JERSEY COALITION FOR FAIR BROADCASTING—citizens organization formed in 1971 to campaign for adequate VHF television service for New Jersey, one of two states without a commercial VHF assignment (Delaware is the other). Situated between two of the largest urban centers, New Jersey has been blanketed by the New York City stations in the north and the Philadelphia stations to the south. Because the state's residents constituted approximately 30% of the viewership in each market, the Coalition was formed by 18 organizations of the larger cities to demand from the stations commensurate coverage of New Jersey news events and state issues. Public officials argued that the lack of a television station not only created a news vacuum in the state but mitigated against a sense of state identity among Jerseyites.

Although it is a more populous state than Florida, Wisconsin, Minnesota and most others, New Jersey wound up without a major TV station by accident of geography and the FCC's allocation policies in 1952, when it lifted the four-year "freeze" on station licenses. Attending first to the major urban centers, the agency set up a schedule of allocations based on market size and suitable mileage separation for the signals.

New York was assigned channels 2, 4, 5, 9, and 11 and Philadelphia channels 3, 6, and 10—the frequencies fitted as interlocking fingers. Channel 8 was allocated to New Haven, Conn., Channel 12 to Wilmington, Del. and Channel 13 to Newark, N. J., as a commercial operation.

But in 1961, as educational television was spreading, Channel 13 was sold to the newly formed noncommercial Educational Broadcasting Corporation and has since served as New York area's public TV station (first as WNDT, later as WNET). Similarly, Wilmington's Channel 12 went educational and, as WHYY-TV, is generally identified as Philadelphia's public station.

On pressure from the Coalition, several of the New York stations assigned full-time news reporters to New Jersey, and others created weekly public affairs programs devoted to issues of the state. But the organization, regarding the efforts as token, petitioned the FCC in 1974 to reassign one of the existing New York stations to New Jersey or to determine whether a completely new station could be "dropped in" on the VHF band. Otherwise, it recommended that the agency "hyphenate" the New York and Philadelphia markets, designating them officially as the New York-northern New Jersey and Philadelphia-southern New Jersey, which would require that all stations establish studios in the state.

In 1976 the FCC denied all those requests but ruled the New York and Philadelphia stations would, in some manner, have to establish a "presence" in New Jersey.

"NEWLYWED GAME, THE"—successful prime-time game series on ABC (1966–70) which ran as a parlay with *The Dating Game,* both shows creations of Chuck Barris Productions. Later, *Newlywed* was converted into a daytime strip and ran through 1974. Bob Eubanks was host.

NEWMAN, EDWIN—NBC newsman with raspy voice and dry wit, considered one of the most literate members of the electronic journalism fraternity. He joined NBC News in London in 1952, later became bureau chief there (1956–57),

then in Rome (1957–58) and then in Paris (1958–61). Based in New York since 1961, he has been one of the regular reporting stars of the national political conventions and has reported on other special events, and has been narrator of documentaries. Newman has been a frequent substitute on *Today* and also hosted a weekly interview show, *Speaking Freely.* For a time he was drama critic for WNBC-TV. His book on the use of language in news media, *Strictly Speaking,* was a best seller in 1974.

Edwin Newman

NEWMAN, SYDNEY—Canadian broadcast executive who made his mark in British television as a drama producer and later as a program executive. For independent television there, he created and produced *Armchair Theatre,* a highly successful weekly showcase with a strong bias for gritty, realistic drama. As such it was a benchmark for the British medium and helped put many a young writer and director on the map. Newman subsequently shifted to BBC-TV as a program executive in charge of anthology and series drama. His major achievement in that capacity was to establish the *Wednesday Play* as a slot for contemporary drama. On his return to Canada in the late 60s, he took over the leadership of the National Film Board of Canada, and more recently has been serving as an independent film and television consultant.

NEWS—television's noblest service, the major source of its prestige and, in commercial respects, an increasingly lucrative form of programming, growing over the years from loss-leader to profit-center at most stations. It is a service with such impact on the society that television journalism itself has become a national issue, its entire procedure—news-gathering, editing and presentation—continually scrutinized for hidden motives and meanings.

Television's uniqueness among news media lies in its ability to get the story, distribute it and exhibit it all in a single process. With polls showing that nearly 65% of U.S. adults rely on TV as their principal source of information, the medium's importance in shaping public opinion has given TV broadcasters extraordinary power in their communities and has fostered concern over the integrity of the news product almost equal to the concern over integrity in Government.

Time devoted to news on television has expanded with each decade reflecting the appetite of the public for topical information and coverage of the real world. The audience for network news has multiplied from around 15 million viewers a night in the mid-50s to about 55 million in the mid-70s. The newscasts themselves increased in length from 15 minutes nightly to a half-hour nightly in 1963, and the first steps were taken in 1976 to expand the network evening news to an hour on weeknights.

As the TV news "hole" has expanded, so have the opportunities of the networks and stations to justify, and even profit from, their huge investments in maintaining news organizations—since obviously a one-hour newscast provides four times the revenue potential of a 15-minute newscast. It has been estimated that the networks collectively would gain $75 million a year in news revenues from increasing their evening newscasts from 30 minutes to an hour.

In pushing for the increase, the networks have maintained that 30 minutes is insufficient for covering the complex domestic and international developments of the times. A half-hour network newscast actually affords only 22 minutes for news, the remaining time being consumed by commercials, titles and introductory matter.

The network news operations are believed to carry an annual budget of around $100 million each. The half-hour evening newscasts alone, each staffed with about 50 people, cost approximately $200,000 a week, or $10 million a year to produce. In addition to those nightly programs, there are daytime and weekend newscasts, the early morning programs (*Today, CBS Morning News, Good Morning America*), weekend public affairs and religious shows, news-magazine series (*60 Minutes, Weekend*), monthly documentaries, news specials, special event coverage and radio news to absorb the work of the armies of TV reporters, producers and camera crews.

Local news received a great impetus in the late 60s from the general perception that the station which led its market in news usually led it in entertainment programming as well, regardless of network affiliation. When the CBS o&o in Chicago lost its news credibility in an extensive alteration of staff, the station's overall ratings dropped to fourth in the market (even below the independents) although CBS was the No. 1 network nationally.

The phenomenon of news ratings affecting a station's overall standing gave rise to a new force in television, the news consultant, an independent "show doctor" of sorts who, through the use of research, advised stations on the style, decor and pacing of their newscasts for purposes of improving their popularity. These consultants were, to a large degree, responsible for the widespread adoption of the "eyewitness news" concept—a relatively informal type of newscast, flavored with conviviality (often strained) and occasional joshing by members of the news "team.

During the 50s, when TV programs were sponsored rather than sold on a participating basis, local newscasts usually ran 15 minutes in length; the more ambitious ones in larger cities had a second 15-minute segment broken into separately sponsored programs for weather, sports and household tips. In the 70s the one-hour local newscast was not uncommon, and in some of the larger markets 90-minute and two-hour programs were operating successfully.

Like radio news of the 30s, TV news faced the hostility of the printed press (which reviewed and evaluated it), not only because it competed for the same audience and advertisers but because TV news had the capability of breaking stories almost instantaneously, beating print technology by hours. The development of miniature electronic cameras in the 70s allowed television to report the news faster than ever. Newsfilm had to be processed before it could be put on the air; that lag was eliminated by ENG (electronic news gathering), since the cameras produced tape that was immediately usable or could even send the picture directly to the master control room by microwave.

But for all its virtues, TV news had serious shortcomings recognized by its own journalists. Because broadcasters are competitive and always fearful of boring the viewer and sending him to another station, TV news has rarely been more than a headline service. Stories that may run 800 to 1,000 words in print are usually delivered in 20 or 30 seconds on TV. CBS News once set an entire newscast in columns of type and found that the words scarcely covered half the front page of *The New York Times*.

Pictorial stories almost always are preferred in TV to those requiring long verbal explanations, which was why economic news was ignored in the medium until the recession of the early 70s. Foreign news, because it is seldom as interesting to most viewers as national or regional news, tends to be given short shrift and usually becomes prominent in TV only for wars and disasters.

These sins born of commercial competition compounded the traditional problems of broadcast news stemming from the Fairness Doctrine, the Equal Time law and the endless charges of "slanting" from the political left and right. Because broadcast stations are licensed, because they use the public airwaves and because of the relative scarcity of frequencies, TV journalism has never enjoyed the same First Amendment guarantees of the printed press. Added to the restrictions imposed by the Communications Act and the FCC, the American Bar Assn.'s Canon 35 has barred broadcast gear from the courtrooms, forcing TV and radio to cover trials much in the manner of the printed press.

One of the essential differences between TV and print journalism is that TV shows the process of news-gathering while newspapers and magazines do not. On TV, the reporter is shown asking the questions that elicit news; in print, the questions are discarded and only the responses presented. Thus the TV reporter's foolish, abrasive or insensitive probing goes on record; the print reporter may have been more foolish or more abrasive, but his secret is safe.

TV network news began on CBS in 1948 with *Douglas Edwards and the News*. NBC followed soon after with *The Camel News Caravan* with John Cameron Swayze. Both were 15 minutes in length and essentially involved a recitation of the stories by a news reader. Edward R. Murrow was the CBS news star at the time but did not appear on the evening newscast, confining his work to the weekly *See It Now* series and CBS Radio. ABC entered later with John Daly but was hobbled by a short affiliate lineup and a comparatively small news organization.

Network newscasts doubled in length in the fall of 1963, fronted by Walter Cronkite on CBS and Chet Huntley and David Brinkley on NBC. ABC had a large turnover in anchormen—including Bob Young, Peter Jennings, Howard K. Smith and Frank Reynolds—until it hired away Harry Reasoner from CBS in 1970. In 1976 Reasoner was teamed with Barbara Walters, first female news anchor on the networks, who was hired away from NBC's *Today* show. By then, John Chancellor had become NBC's anchor, teamed with Brinkley. Cronkite remained CBS's ace.

Over the years, the ratings lead switched between NBC and CBS, while ABC made slow progress toward parity. NBC held the lead from 1953 to 1956, then lost it to CBS, which held it until 1960. That year, Huntley and Brinkley forged ahead and gave NBC the most popular newscast for five years. From 1965 to 1967, NBC and CBS ran neck and neck, until Cronkite moved in front, holding the lead for about nine years. In 1976, CBS and NBC were close again, each reaching around 20 million viewers a night to ABC's 15 million. ABC News, with Roone Arledge as its new president, began a build-up of staff in the summer of 1977 for a new assault on the news ratings of CBS and NBC. In 1979, ABC's *World News Tonight*, chiefly anchored by Frank Reynolds, closed in on NBC's *Nightly News* for second place and on several occasions passed it in the ratings

In 1979, ABC's *World News Tonight,* chiefly anchored by Frank Reynolds, closed in on NBC's *Nightly News* for second place and on several occasions passed it in the ratings. ABC's rise, and its attempts to raid the other networks for news stars, led to changes at CBS and NBC.

In 1981, Cronkite yielded the anchor post on the evening newscast to Dan Rather, who was joined within the year by

Bill Moyers, serving as commentator. Meanwhile, Roger Mudd moved to NBC News and in 1982 became co-anchor with Tom Brokaw of *Nightly News*. In the wave of personnel changes, David Brinkley ended a long career at NBC News and switched to ABC.

"NEWS AT TEN"—a cross-the-board 10 P.M. half-hour news program in Britain originated by Independent Television News and cleared by all independent stations in that country. Introduced in the late 60s, it has consistently won high audience shares and enjoyed a reputation for journalistic hustle, ingenuity, and showmanly presentation. Not long after its inception, the show went to a two-man anchor format, the first in Britain, admittedly modeled after NBC's prime nightly newscast. The program's graphics include an opening shot of "Big Ben" and the houses of Parliament.

NEWS OF THE DAY—newsreel company jointly owned by Hearst and MGM which had been contracted by the CBS *See It Now* unit in the 50s for all the film and post-production work. Edward R. Murrow canceled the arrangement, however, after the suicide of his protege, Don Hollenbeck, who had been under constant attack by Hearst columnists accusing him of slanting the news to the left. The same columnists had been critical of Murrow's devastating programs on Sen. Joseph McCarthy. When Murrow dropped the services of the Hearst newsreel, the News of the Day technicians resigned to join *See It Now.*

"NEWSROOM"—public TV program created by KQED San Francisco during a newspaper strike in the 60s. It was the precursor to the *Eyewitness News* format that spread throughout the commercial TV system. Central to the *Newsroom* concept is the gathering of reporters around a desk to discuss the day's news and its significance. Several public stations borrowed the format, but none was able to equal the bold, straightforward tone of the original.

NICHOLL, DON—writer associated with Norman Lear who was co-creator and co-producer, with Michael Ross and Bernie West, of *The Jeffersons* and *The Dumplings*. He was also executive producer of *All In the Family* for the 1974–75 season.

NICHOLS, MIKE—noted stage and motion picture director who made his entry into TV production as co-executive producer of *Family* in 1976. Earlier in his career he frequently appeared on TV as a comedian teamed with Elaine May.

NICKLEODEON—a cable channel of programming exclusively for children, introduced by Warner Communications and made available for national distribution to cable systems by satellite April 1, 1979. Large portions of the programming for the channel were developed in 1977–78 at Warner's experimental two-way cable installation in Columbus, Ohio, known as Qube. Packaged into a 13-hour-a-day service and carrying no advertising, the programming is assembled by an authority in children's education, Dr. Vivian Horner, who helped develop *The Electric Company* for the Children's Television Workshop. Supported by the various cable systems that carry the network, each paying ten cents a month for every household delivered, the channel was expected to attract new customers for the cable systems that carry it because of the unique service it provides. See Cable Networks.

NIELSEN, A.C., COMPANY—a company whose best-known service is the measurement of TV audiences and whose reports are recognized as the index to success or failure in TV programming. In television research, the company maintains two divisions: NTI (Nielsen Television Index), which produces the ratings for network shows, and NSI (Nielsen Station Index), which concentrates on local market reports. An additional service, NAC (Nielsen Audience Composition), produces demographic data.

The company's main business, however, since its founding in the 20s, has been the tabulating and reporting of consumer consumption of drug and food items from retail shelves. Nielsen conducts its retail surveys in 17 countries. Although Nielsen is best-known for its TV research, that activity accounts for only slightly more than 10% of corporate revenues.

It was Nielsen's success in pinpointing the flow of retail merchandise that prompted manufacturers, in the 30s, to ask Nielsen to evaluate radio advertising, in which they were beginning to invest heavily.

Nielsen purchased the rights to an audience-counting device, a black box attachment to radio sets known as the Audimeter, and then bought out the methodology of an existing ratings company. In 1942 it introduced an audience measurement service that could document which stations were being listened to by the slow-crawl film contained within the Audimeters. In the late 40s, with television's emergence, Nielsen adapted its methodology to the new medium.

The increase in the number of independent radio stations, coinciding with the expansion of TV, forced Nielsen out of radio measurement—the machines simply couldn't read the growing flow of signals. By 1963 Nielsen had ceased producing radio ratings and concentrated on network and local market audience data for television.

In local television, Nielsen has stiff competition from Arbitron. But in network television, Nielsen has the field virtually to itself. Thus, Nielsen is not only the scorekeeper in the TV game; its numbers are, in fact, the score. Its founder, A. C. Nielsen, Sr. died in 1980. See also Arbitron, Audimeter, NTI, NSI, Pocketpiece, Rating.

NIELSEN STATION INDEX (NSI)—Nielsen's local audience data (both size and composition) gathered for more than 220 geographic areas in the U.S. which it recognizes as individual markets. Unlike the national Nielsen ratings which utilize electronic meters, the NSI data primarily come from diaries placed in randomly selected households, with telephone contact as the major means of securing cooperation. Through the phone contact, Nielsen collects information on color TV, multi-set ownership and the penetration of UHF. The local market reports range in number from three to eight annually and are published in booklets entitled *Viewers in Profile (VIP)*.

The three largest markets—New York, Los Angeles and Chicago—are metered with SIA for fast overnight ratings. In these markets the meter data is augmented with diary information for the eight NSI reports to provide the demographic profiles for day parts, programs and time periods. See also Arbitron, DMA, VIP.

NIELSEN TELEVISION INDEX (NTI)—the basic national audience measurement service of the A.C. Nielsen Co. that publishes estimates of network programming, as opposed to the local market audience estimates . The 1,200 households used for the NTI metered sample are drawn from census tracts. See also Audimeter, Audilog, Pocketpiece, Ratings.

"NIGHT GALLERY"—anthology series of bizarre tales of the occult, with Rod Serling as host-narrator. It began on NBC in 1970 as one of four rotating one-hour series under the umbrella title of *Four in One*. The following season it became a full-time series under its own title. In 1972 it was reduced to half-hour form and canceled in 1973. Universal TV produced it.

"NIGHTLINE"– ABC's latenight news program, anchored by Ted Koppel, which grew out of the network's special nightly coverage of the Iranian crisis in the winter of 1979. The program represented a major expansion of network news and proved surprisingly successful as a competitor to NBC's *Tonight* with Johnny Carson and CBS's reruns of action-adventure shows in the 11:30 to midnight time period.

As a series *Nightline* was a fluke. ABC had been carrying rerun programming in prime time and ABC News, under the relatively new leadership of Roone Arledge, was struggling to be regarded seriously when the American hostages were taken at the embassy in Iran on Nov. 4, 1979. The story was of enormous interest to the American people, and Arledge persuaded the network to give up some of the latenight time to a daily series of updates, *The Iran Crisis: America Held Hostage.* It is doubtful that the commitment would have been made if anyone had imagined the hostages would be held for more than a few months: ABC found itself stuck covering a news story that had few developments day by day and that seemed to threaten its profits. But *America Held Hostage* proved surprisingly successful, and in short order Koppel, who had been with ABC News since 1963, became popular. By February 1980, ABC discovered it had a new regular latenight program and a month later dubbed it *Nightline.* William E. Lord, a longtime executive of ABC News, became its executive producer. The program reaped a number of awards in its first year, including an Emmy, a duPont Columbia, and a Christopher.

ABC had so little faith in the news program that it deliberately scheduled the broadcasts in 20-minute lengths, instead of the customary half-hour, to frustrate local stations that might have chosen to preempt the news show for a syndicated program. Since there are no 20-minute syndicated programs, the ABC affiliates stayed with *Nightline* so as not to lose out on the entertainment programs that followed. When it was clear to all that *Nightline* had won the viewers' acceptance and that it could compete with *Tonight*, ABC was able to expand it to a full half-hour without fear of losing affiliate clearances.

Nightline differs from the typical newscast in that it generally develops a single story in the news through interviews with guests and reports from correspondents. The program is structured to respond to late-breaking stories.

"90 BRISTOL COURT"—a trio of related situation comedies mounted by NBC in 1964 in an attempt to penetrate the CBS dominance of Monday night. The comedies, scheduled in successive half hours, were linked by the fact that their characters all lived at the same address. The three series—entitled *Karen, Tom, Dick & Mary* and *Harris Against the World*—all failed in their first season.

NHK (NIPPON HOSE KYOKAI)—Japan's public broadcasting network. See Japanese Broadcasting Corp.

NIXON, AGNES—packager and head writer of two long-running ABC soap operas, *One Life to Live* and *All My Children.* She became heiress to the title of "Queen of the Soaps" with the death of her former employer, Irna Phillips. In the 50s she became a dialogue writer for Miss Phillips and for several years wrote the daily episodes of *As the World Turns.* Later she became head writer for serials owned by Procter & Gamble. In 1968 she struck out on her own, creating and packaging *One Life to Live,* which premiered July 15, 1968. *All My Children* followed on Jan. 4, 1970. She wrote the story for *The Mansions of America* miniseries, which aired on ABC in 1981.

Agnes Nixon

"NIXON INTERVIEWS WITH DAVID FROST, THE"—series of four 90-minute telecasts in May 1977 in which the British talk-show personality David Frost interviewed former President Nixon over a special U.S. network of 155 TV stations. Representing Nixon's emergence from nearly three years of seclusion after resigning from office in disgrace, the broadcasts stirred up the press and national magazines, helping to build the programs a tremendous audience.

The first of the telecasts, airing on May 4 and devoted to issues surrounding the Watergate episode, scored shares of 47 in New York and 50 in Los Angeles—and an estimated national audience of 45 million people—making it the highest-rated news interview program in TV history and one of the most watched news programs ever (excepting events covered simultaneously by all three networks).

Under an intense prosecutorial barrage from Frost, Nixon admitted to having done some lying as Watergate was unfolding but continued to deny an involvement in the Watergate conspiracy and cover-up. Nixon confessed only to letting the American people down and said he impeached himself.

Frost secured the rights to the interviews in 1976 by offering Nixon a guarantee of $600,000 and 10% of the profits, after CBS News and ABC News turned away the ex-President's agent, Irving (Swifty) Lazar, who had asked $1 million for TV access to Nixon. NBC News made an offer of $400,000 for two telecasts with the ex-President and then learned, before it could raise its bid, that Frost had won the rights. For Frost the project was a chance for a comeback in the U.S., where his star had fallen after his syndicated shows for Westinghouse had been canceled three years before.

CBS and ABC both rejected the program on the principle of "checkbook journalism," and ABC was particularly averse to the idea of paying a former president for his accounting to the public.

Frost secured financial backing from a group of affluent Nixon followers in Southern California and engaged the syndication firm, Syndicast Services, to put together an ad hoc network. Meanwhile, his own company, Paradine Productions, sold the radio rights to the Mutual Broadcasting System, the 16-millimeter educational film rights to Universal Pictures and the foreign rights to 10 countries. The broadcasts were carried in Britain on the BBC on a one-day delay.

Using the barter incentive and scheduling the programs for a "sweep" month when the networks were expected to be given over largely to reruns, Syndicast lined up stations sufficient to cover 90% of the country. It also arranged for the Robert Wold Co., specialists in arranging the simultaneous distribution of programs for special networks, to handle the interconnection.

Initially, the commercial time was to have been divided evenly between national and local spots—6 minutes for each —but when advertisers resisted the network buy at $125,000 a minute, Syndicast sold one of the minutes of national time to the stations.

After a masterfully engineered publicity campaign at the 11th hour, which through selective leaks of the program's content gained Frost/Nixon the covers of *Time, Newsweek* and *TV Guide,* the first program was sold out by broadcast time.

With $2 million from "network" advertising sales, $1 million from foreign sales, and substantial other amounts from the sale of radio and educational rights, the entire project was in the black.

Moreover, Frost exercised his contractual right to create a fifth program from the outtakes. He had taped 28 hours with Nixon at the ex-President's home in San Clemente, Calif., and was privileged to use a total of 7 hours on the air. The four original broadcasts covered only 6 hours. The fifth program was put on the syndication market as a straight sale to stations for airing during the week of September 5, 1977.

To protect what he believed to be the rerun value of the program, Frost denied the networks excerpt rights for their newscasts. Estimates were that Nixon would realize at least $1 million overall from the telecasts.

Marvin Minoff of Paradine Productions was executive in charge of the production and Jorn Winther the director.

NOBLE, EDWARD J. (d. 1958)—founder of the American Broadcasting Company and its first board chairman. A multi-millionaire who made his fortune manufacturing Life Savers candies, Noble purchased the NBC Blue Network from RCA for $8 million in 1943, when the FCC ordered the company to give up one of its two radio networks. At the time the Blue Network had 155 affiliated stations. Noble's prior experience with broadcasting had been as owner, for two years, of WMCA in New York, a station he sold in 1943. Noble served as chairman of ABC until 1953 when it merged with United Paramount Pictures. He remained a director of the company until his death.

NON-DUPLICATION RULES—rules adopted by the FCC in 1965, and modified in 1966, prohibiting cable systems from importing distant stations broadcasting the same programs at the same time as local stations. In the initial rules, cable could not bring in programs carried within 15 days of the local broadcast. The amended rules in 1966 required cable systems to carry all local stations and reduced the duplication prohibition to the day of broadcast.

Robert Northshield

NORTHSHIELD, ROBERT (SHAD)—noted producer of news programs whose career has swung between CBS and NBC, with a brief stint at ABC. After developing and producing the successful 90 minute *Sunday Morning* news broadcast on CBS on 1978, Northshield was rewarded with responsiblility for revamping the weekday morning news program, retitled *Morning*, in January 1979. Three years later, when the weekday edition failed to rise to ratings parity with *Today* and *Good Morning, America*, Northshield's responsibilities were reduced to *Sunday Morning* and a new executive producer, George Merlis, was brought in for the daily shows.

Northshield had rejoined CBS in 1977 after a 17-year absence. He worked initially as an executive producer in sports and then as a documentary producer for CBS News, before receiving the *Sunday Morning* assignment. He had spent the previous years, since 1960, with NBC News. In the early 60s, he was producer of *Today* with occasional documentary assignments. He was made general manager of NBC News in 1964, supervising political coverage, and a year later became executive producer of the *Huntley-Brinkley Report.* He produced such documentaries as *Suffer the Little Children* (1972), *And Who Shall Feed This World?* (1974), and *The Navaho Way* (1974).

He began his TV career with CBS in 1953 and was producer of the *Adventure* series and of *Seven Lively Arts.* Later he became a columnist for the *Chicago Sun-Times* and then produced public affairs programs for ABC (1958-60) before joining NBC.

NORWAY, TELEVISION IN—a single state-controlled noncommercial channel operated by Norsk Rikskringskasting on 625-line PAL. It is on the air about 30 hours a week. The country has an estimated 900,000 sets. Individual households pay the equivalent of $63.27 in annual license fees for black & white sets and $82.08 for color.

NOTICE OF INQUIRY—an FCC procedural device preliminary to rule-making, which is its means of alerting the public to the prospective adoption or new rules or to possible changes in existing rules. The Notice of Inquiry is issued for information on a broad subject or when the commission is seeking ideas on a given topic.

"NOVA"—major PBS series begun in March 1974, noted for examining complex scientific questions in a manner comprehensible to laymen and in a relatively entertaining fashion. The series concerned itself with the effects on society of new developments in science. Produced by WGBH Boston, it was underwritten for two years by the Corporation for Public Broadcasting and then was sustained by the pooled funds of PBS stations and by outside grants.

NOXON, NICOLAS—producer-director-writer specializing in nonfiction programs who, with Irwin Rosten, headed the MGM documentary department in the early 70s, chiefly responsible for the *GE Monogram Series.* That unit yielded three programs a year for NBC. Earlier, at Wolper Productions, where he had met his partner, Noxon produced and wrote such shows as *Dr. Leakey and the Dawn of Man, The Voyage of the Brigantine* (both programs winners of Peabody Awards), *Epic of Flight,* seven half-hours of *Hollywood and the Stars* and 17 half-hours of the *Biography* series.

"NURSES, THE"—drama series on CBS (1962–64) set in a large hospital. Produced by Plautus Productions, it featured Shirl Conway, Zina Bethune, Michael Tolan and Joseph Campanella.

"N.Y.P.D."—half-hour ABC series (1967–68) produced on 16 mm film, drawing its stories from cases of the New York City Police Dept. It starred Jack Warden, Robert Hooks and Frank Converse and was produced by Talent Associates.

O&O—trade shorthand for an owned-and-operated station, usually in reference to those owned by the parent corporations of the networks but applying also to group-owned stations. WBBM-TV Chicago is a CBS o&o; WBZ-TV Boston a Group W o&o. See Owned Stations.

OAK INDUSTRIES—one of the companies that has flourished in the second age of television, both in manufacturing and services. Headed by Everitt A. (Nick) Carter, as chairman, president, and chief executive, Oak was restructured in 1978 to operate through two wholly-owned subsidiaries—Oak Communications Inc., providing over-the-air subscription television services, and Oak Technology Inc., a company producing switches and components for the electrical and electronics industries.

Oak Communications, based in Rancho Bernardo, Calif., operates National Subscription Television (NST) and manufactures equipment for cable and pay television, including tuners, decoders and scrambling systems. NST began broadcasting in Los Angeles in 1977 as an over-the-air pay service known as ON-TV. Sixteen months after inauguration, the service was in the black, with more than 100,000 subscribers. Oak also has STV stations in Phoenix, Dallas, Philadelphia, Chicago, Miami, and Minneapolis.

"'OBJECTIONABLE' MATERIAL [Pacifica Foundation/ 1 R.R. 2d 747 (1964)]—programs found offensive by some people but otherwise well within the public interest standard for broadcast. The FCC decided in the Pacifica case that broadcasters have a right to air "provocative" programs but suggested that they be confined to the later evening hours when children presumably are asleep.

The Pacifica Foundation, licensee of several noncommercial FM stations, frequently broadcasts avant-garde programs. When the foundation sought to renew its licenses, the FCC questioned it about five programs which had drawn complaints. They included two programs in which poets read from their works; a reading of Edward Albee's play *Zoo Story;* a discussion group in which homosexuals spoke of their problems and expressed their attitudes; and readings of an unfinished novel by its author.

Pacifica defended most of the broadcasts by stating that they fell well within the public interest standard. The commission recognized that while such programs may tend to offend some listeners, "this does not mean those offended have the right, through the commission's licensing power, to rule such programming off the air waves." The commission also noted that the programs had been broadcast at hours when most children had gone to sleep.

Pacifica did not defend the two programs of poetry, which it admitted contained some passages that did not conform to its own standards of acceptability. It said the programs were aired through errors in the screening process. The FCC found the explanation to be "credible" and decided that two isolated errors over a period of four years were hardly sufficient to question seriously a licensee's fitness to continue broadcasting. There was no appeal, and Pacifica's licenses were renewed.

OBSCENITY AND INDECENCY [in re-WUHF-FM
Eastern Education Radio/ 24 FCC 2d 408 (1970)]—issues
generally held to be within the FCC's censorial purview,
applicable to material that is found to be patently offensive
by contemporary community standards and totally without
redeeming social value. The critical case was that of WUHY-
FM, a noncommercial radio station in Philadelphia, which
in 1970 aired a taped interview with Jerry Garcia, a member
of the rock group known as The Grateful Dead, whose
comments were dotted with such expletives as "fuck" and
"shit."

The FCC, in its opinion, stated that it had the right to
prevent widespread use on broadcast stations of such expressions since they are patently offensive to millions of listeners.
The commission noted that unlike the purchase of books and
theater tickets, which involve a deliberate act by the consumer, broadcasts are disseminated generally to the public
and may be heard by anyone, including children, whose
families might prefer not to expose them to such language.

The FCC also noted that the U.S. Code 18 USC 1464
makes it a criminal offense to utter obscene, indecent or
profane language by means of radio communication, although the commission conceded that the words on WUHY
were not obscene under criteria laid down by the U.S.
Supreme Court.

WUHY-FM informed the commission that it had fired
the producer of the program for violating the station's internal standards, but in light of the sensitivity of the subject,
the FCC imposed a fine upon the station of $100. Although
two commissioners dissented, claiming that the commission
was not sensitive to two decades of First Amendment law and
concepts of vagueness and overbreadth, WUHY chose not to
appeal the FCC decision to the courts.

OBSCENITY CASES AND THE SUPREME COURT—
The Supreme Court has attempted to define First Amendment protection of allegedly obscene speech for many years
without success, owing primarily to the difficulty of defining "obscene" in the first instance. Originally it was assumed that obscenity, like defamation and "fighting words,"
was not "speech" within the meaning of the First Amendment protections [Chaplinsky v. New Hampshire, 315 U.S.
568 (1942)]. This was subsequently affirmed in Roth v.
United States, 354 U.S. 476 (1957), a landmark decision
because it was the first case in which the Supreme Court was
required to decide if obscenity was protected speech.

Roth, a New York publisher, was convicted of mailing
obscene advertising material and an obscene book in violation of Federal law. His conviction was appealed to the
Supreme Court, which affirmed. The court reasoned that the
First Amendment "was not intended to protect every utte-
rance," and that obscenity was "utterly without redeeming

social importance," and was not protected by the Constitution.

Subsequent cases focused sharply on the word "utterly,"
and the court found itself reviewing dozens of cases to decide
whether allegedly obscene matter was "utterly without redeeming social value" [Memoirs v. Massachusetts, 383 U.S.
413 (1966)]. Since a majority of the Justices were unable to
agree upon criteria which made material obscene under this
"definition," the court disposed of case after case by refusing
to grant certiorari, or reversing in per curiam opinion.

This ad hoc approach which continued to leave undefined
the parameters of protected communication was generally
unsatisfactory to the bench and bar alike because it required
the Supreme Court to sit in review of *factual* determinations
of state and Federal trial and appellate courts. The Supreme
Court attempted to rectify this situation in Miller v. California, 413 U.S. 15 (1973) and Paris Adult Theater I v. Slaton,
413 U.S. 49 (1973).

In the Miller case, defendant was convicted of violating a
California obscenity law by conducting a mass mailing
campaign to advertise "adult material." These advertisements explicitly depicted sexual activities, and included
prominent display of genitals. The Supreme Court, speaking through Chief Justice Burger, not only affirmed Miller's
conviction but established general guidelines for lower courts to follow in future obscenity cases. The guidelines required the application of "contemporary community
standards" to determine whether the material appealed to
"prurient" interest by a "patently offensive" display of sexual
conduct where the work, "taken as a whole, lacks serious
literary, artistic, political or scientific value."

In the Paris Theater case, the court held that a movie
theater owner could be held liable for showing obscene
movies despite warnings at the theater entrance that admis-
sion was limited to people over 21 because the films displayed explicit sexual acts. The individual, said the court,
had no fundamental privacy right to view obscene movies in
places of public accommodation.

If the court hoped by these two decisions to rid itself of
the chore of examining material to review lower court findings of obscenity, it was mistaken. In Jenkins v. Georgia,
418 U.S. 153 (1974), the Supreme Court was required to
apply its Miller test to the film *Carnal Knowledge,* which had
been found obscene in the Georgia courts. Jenkins, a theater
owner, was convicted by a jury of displaying an obscene film.
The Georgia Supreme Court affirmed the conviction because, it said, a jury's determination of obscenity precluded
all further review. The U.S. Supreme Court disagreed and
held that juries simply do not have "unbridled discretion in
determining what is patently offensive." Since the Justices
did not find that the film was "obscene," the court reversed
the conviction, reaffirming an earlier pronouncement of a
distinction between sex and obscenity.

Although majorities of Justices have consistently agreed that obscene speech is not entitled to the protection given to other kinds of speech by the First Amendment, the Supreme Court has not as yet successfully delineated the bounds of this limitation, and until it does the court will continue to review lower court decisions by substituting its own moral judgment.

Tony Randall & Jack Klugman, the Odd Couple

"ODD COUPLE, THE"—ABC situation comedy (1970–75) drawn from Neil Simon's Broadway comedy hit of that title, concerning a compulsively neat photographer sharing an apartment in New York with a slovenly sportswriter. The roles were played by Tony Randall and Jack Klugman, respectively, with Al Molinaro featured as a policeman-friend. The series was always only marginally successful, but the syndicated reruns did well. It was by Paramount TV, with Garry Marshall as executive producer and Tony Marshall as producer.

OFFICE OF NETWORK STUDY—department of the FCC responsible for keeping abreast of issues involving the TV, radio, sports and regional networks and for making policy recommendations to the commission. The office has been involved with the prime time-access rule and its modifications, the issue of restricting network reruns, the definition of "network" where regional or ad hoc hookups are concerned and the regulatory restrictions against the networks, such as those limiting their ownership of stations and prohibiting their ownership of cable systems.

OFFICE OF TELECOMMUNICATIONS POLICY —a White House agency established in 1971 as principal advisor to the President on national telecommunications matters, with the additional responsibility of formulating policies and coordinating cost-effective operations for government communications systems. OTP was dissolved in 1977 and its functions assumed by the new National Telecommunications and Information Administration in the Department of Commerce. OTP assigned the frequencies used by federal agencies and provided policy direction for the national communication system and emergency communications.

The office was politicized by its first director, Clay T. Whitehead, who used it to carry out the media strategies of the Nixon Administration. In his well-publicized speeches, he intimidated the networks and their affiliates with charges that network news dealt in "ideological plugola" and "elitist gossip," and he admonished public broadcasters against centralizing under a fourth network. He made it plain that the Administration would not support a long-range funding bill for public television if that industry did not adopt a system based on "grassroots localism." Whitehead resigned in 1975, after producing several important studies that redeemed the office, but the damage to the image of the agency during his tenure nearly caused it to be abolished during the Ford Administration.

With John Eger as acting director, OTP later made policy recommendations to the FCC regarding cable-TV, domestic satellites, land mobile radio service, VHF spectrum allocation and the economic impact of competition in certain telecommunications markets. Thomas J. Houser, a Chicago lawyer who once had served briefly on the FCC, was named director by President Ford in 1976. He resigned in January 1977 at the start of the Carter Administration. William Thaler then became acting director.

OTP maintained a staff of around 48 people and operates on an annual budget of about $8.5 million, although nearly three-fourths of that amount went to the Office of Telecommunications of the Commerce Department, from which OTP received support services. More than 100 employees of the Commerce OT were assigned to OTP on a full- or part-time basis in connection with spectrum management and general policy-making.

"OF LAND AND SEAS"—moderately successful syndicated series (1966–69) on which guests presented films of their travels and explorations and then discussed their adventures with the host, Col. John D. Craig. It was by Olas Productions.

O'HANLON, JAMES—writer who created *Maverick*. He also wrote for *My Favorite Husband, 77 Sunset Strip, Cheyenne* and *Going My Way*.

OHLMEYER, DON—production executive with extensive experience in covering the Olympic Games, and thus a prize catch for NBC Sports after it had acquired the rights to the 1980 Moscow Olympics for $100 million. Ohlmeyer joined NBC in 1977, at the age of 32, after having spent ten years with ABC Sports in a variety of production capacities.

He was director of the summer Olympics in Munich (1972) and Montreal (1976) and also producer-director of the 1976 Winter Olympic telecasts from Innsbruck. In 1968, only a year out of Notre Dame, he was associate director of the Summer Olympics from Mexico City. At ABC, he also produced the NFL Monday night football coverage, NCAA football, NBA basketball, coverage of several world heavyweight championship fights and segments for *Wide World of Sports*. Other credits include such prime-time specials as *Battle of the Network Stars, Superstars* and the Harlem Globetrotter specials. His work has netted him a passel of Emmy awards.

With such a background, he was given overall creative control of NBC's coverage of the 1980 Moscow event. In addition, he is executive producer of all NBC Sports programming and serves also as a program packager for the TV network.

OIRT—International Radio & Television Organization, an organization of Communist-bloc countries, whose initials are applied to the monochrome television standards used in most of those countries.

OLYMPICS, THE—quadrennial international sporting event and television extravaganza credited with lifting ABC to the top of the Nielsens in 1976 and expected, at the astronomical price of $100 million, by NBC executives to do the same for their network in 1980.

But to NBC's misfortune, the U.S. boycotted the 1980 Olympics in Moscow because of the Soviet invasion of Afganistan. The network recovered much of its prepaid investment from insurance, but the loss of the event was a severe blow to the network's strategists.

In 1979, ABC won the television rights to the 1984 Los Angeles Summer Olympics with a record bid of $225 million. By capturing an average of 50% of the American viewing audience each night over a two-week period, the Olympics have become the most sought-after and most expensive event in the history of television, and technically one of the most difficult to produce.

For half a month during the appointed summer, and to a lesser extent for a similar period during the winter, a global audience estimated at one billion focuses on a parade of nations as they display their finest amateur athletes in competition. From the inception of the modern Olympics in 1896, politics and nationalism have been interwoven in the games. In recent years, the worldwide attention provided by television has created an unrivaled forum for ideological statement. Thus the world watched in horror as members of the Israeli delegation to Munich in 1972 were tragically killed by Palestinian terrorists. In 1976, 25 nations shattered the dreams of their 697 athletes by pulling out of the games at the last minute because New Zealand had been allowed to participate after sending a rugby team to South Africa. For the athletes, years of training came to naught over matters of diplomacy that seemed to have no place in the Olympic arena.

Coverage of the games has grown commensurate with the cost of telecasting them. From Mexico City, ABC broadcast on a delayed basis 44 hours of the 1968 Summer Olympics. In 1972, ABC presented 67 hours of coverage from Munich after securing the American rights to the games for what now seems a paltry $13.5 million. Utilizing the 100 cameras and huge production personnel of the Deutsches Olympic Zentrum , a special joint effort by West Germany's two networks, ABC managed to broadcast superb coverage of the Olympics using only 13 additional cameras of its own and a crew of 175.

By 1976, the cost and logistics of televising the Montreal Summer Olympic Games had escalated dramatically. ABC paid $25 million for the U.S. rights and for the right to provide a feed to Latin America. An additional $10 million was earmarked for production, which included a staff of 470 to supplement the Canadian Broadcasting Corporation's truly Olympian-sized crew of 1,805 people.

Some 30 ABC cameras were added to CBC's 104, while 74-foot mobile trailer units broadcasted from 24 locations. Coordination came from a specially-built control center complete with two studios. Thirty sportscasters, among them Jim McKay, Howard Cosell, Chris Schenkel, and former Olympic athletes Cathy Rigby Mason, Mark Spitz and Bob Seagren, provided color and commentary through 80 hours of coverage.

For 1980, NBC engaged itself in a project of awesome proportions. In early 1977 NBC agreed to pay $35 million for the American television rights to the Moscow games, $50 million for technical equipment and facilities, and $15 million for production and travel costs, bringing the estimated total to $100 million. For that sum NBC planned to broadcast 150 hours of coverage, a total of 9 hours a day over two weeks.

Negotiations for the 1980 games covered six months, with the Soviets first appearing to have awarded the rights to Satra, an American company specializing in trade with Russia. The international Olympic Committee subsequently rejected Satra because its rules require that the rights go to actual broadcast companies and not to middlemen. Lothar P. Bock, a West German entrepreneur who had been representing CBS before the network dropped out of the competition, offered his services to NBC and quietly wrapped up the $100 million deal in January 1977 before ABC could make a move. For his efforts, Bock received a $1 million commission and a commitment from NBC for future programs.

"OMNIBUS"—one of commercial TV's most honored series (1951–56), created by the Ford Foundation's Radio-Television Workshop and underwritten by the foundation to demonstrate that a program of cultural and intellectual value could attract a grateful audience and sponsorship. Hosted by Alistair Cooke, and produced by Robert Saudek, former head of public affairs for ABC, the program took in $5.5 million in advertising revenues during its five years on the air, against $8.5 million in costs, Ford making up the difference.

The program began on CBS, which scheduled it Sunday afternoons, and went to ABC its final year in a Sunday evening berth. As Sundays became more lucrative to the networks, none felt it could afford any longer to carry a program like *Omnibus,* which had virtually no potential for profits. Ford withdrew from the series, giving the rights to any future production to Saudek, and turned its attention to building up educational (now public) TV.

The title and concept were reviewed for a limited series of specials in 1980 and 1981 on ABC. Produced by Martin Starger for Marble Arch Productions, with Hal Holbrook as host, the specials were neither critical nor ratings hits, and production was abandoned.

OMMERLE, HARRY G.(d. 1976)—a key CBS executive in the 50s who became v.p. of programs in 1958. He left a year later to head the radio-TV department of the ad agency, SSC&B, and in 1966 was named executive v.p. He retired in 1970. Before joining CBS he operated his own talent and program packaging firm.

"ONE DAY AT A TIME"—Norman Lear situation comedy on the trials of an independent, middle-aged divorcee raising two teenaged daughters; it was introduced by CBS as a midseason replacement in 1975. Featured were Bonnie Franklin, Mackenzie Phillips, Valerie Bertinelli, Richard Masur and Pat Harrington. The series was by Lear's T.A.T. Communications and Allwhit Productions, with Jack Elinson and Norman Paul as executive producers; Dick Bensfield and Perry Grant, producers; and Herbert Kenwith, director. Subsequently Bensfield and Grant became executive producers, along with director Alan Rafkin.

"ONE LIFE TO LIVE"—ABC soap opera created by Agnes Nixon which premiered July 15, 1968, and concerns residents of the town of Llanview. The cast has included Nat Polen, Ellen Holly, Michael Storm, Al Freeman Jr., Judith Light, Philip MacHale, Lee Patterson and Erika Slezak. In recent years the key cast members have included Anthony Call, Philip Carey, Anthony George, Sally Gracie, Chip Lucia, Clint Ritchie, Jeremy Slate, and Robin Strasser. Joe Stuart is producer and Sam Hill, Peggy O'Shea and S. Michael Schnessel, the chief writers.

Valerie Bertinelli, Pat Harrington & Bonnie Franklin in
One Day at a Time

"ONE STEP BEYOND"—dramatic anthology series based on investigated cases of ESP, hosted and produced by John Newland. It was on ABC (1959–61).

"ON THE ROAD WITH CHARLES KURALT"—innovative CBS News series, begun in 1967, in which correspondent Charles Kuralt reported on American life in towns and hamlets around the country. Kuralt's journeys on the backroads have mainly yielded periodic features for the *CBS Evening News,* but, approximately once a year, have also resulted in a prime-time special. Kuralt's features had such appeal to viewers that the other networks, and certain stations, developed a similar reporting beat. The series was discontinued in the late 70s when Kuralt became anchor of the new *Sunday Morning* program and later the weekday *Morning* programs, as well.

"OPEN END"—free-wheeling syndicated conversation show, hosted by television producer David Susskind, which broke new ground in the medium in 1958 by allowing its discussions with varied guests to be oblivious of the clock and to run indefinite lengths, usually for several hours. By the mid-60s, however, it was given a fixed one-hour format and eventually took the title of *The David Susskind Show.*

OPEN UNIVERSITY, THE—an independent, degree-awarding home instruction university operated in collaboration with the British Broadcasting Corp. It is the largest adult higher education system of its kind in the world, with a tuition-paying home enrollment of more than 50,000 in Great Britain.

The OU offers more than 80 courses, nearly all of which make use of radio and television programming specially produced by the BBC, whose production expenses are covered by government subsidies. BBC airtime, both radio and TV, devoted to the OU adds up to around 55 daytime hours per week for 34 weeks of the year.

Several American and other universities make use of OU study material, and a few countries have been inspired to adopt their own versions of the OU.

OPERATION PRIME TIME—a nebulous cooperative of independent stations formed in 1976 to finance the production of significant prime-time programs to compete with those of the networks. The effort was joined by MCA-TV, the syndication arm of Universal TV, and was spearheaded by Al Masini, president of Telerep, the station representative firm. Masini had suggested the idea at an INTV convention.

The project bore fruit in May 1977 with the presentation by 95 stations of a six-hour adaptation of the Taylor Caldwell novel, *Testimony of Two Men.* Twenty-two of the stations were independents and the rest network affiliates; all had put up money to finance the Universal production in exchange for the right to play the serial six times. Most were able to recoup their investment with the first showing.

Among the independents active in starting the cooperative were WPIX New York, WGN Chicago, KCOP Los Angeles and KTVU San Francisco. The linking of independents in this manner caused word to spread in the industry that a "fourth network" was being born, but that characterization of the effort was largely promotional hyperbole.

OPT was significant because it represented a new way to generate national programming, but as an ongoing venture it was limited by the scarcity of independents and the high risk of mounting programs on network-scale budgets.

Testimony had a production budget of $550,000 an hour. Its cast included David Birney, Steve Forrest, Inga Swenson, Barbara Parkins, William Shatner, Margaret O'Brien, Cameron Mitchell, Dan Dailey, Ralph Bellamy and Ray Milland. It was produced by Jack Laird and directed by Larry Yust and Leo Penn.

Subsequent presentations included serial adaptations of John Jakes's *The Bastard, The Rebels* and *The Seekers,* Irwin Shaw's *Evening in Byzantium* and Howard Fast's *The Immigrants.*

OPTION TIME—periods of a station's schedule that for many years were effectively controlled by a network, because affiliation contracts specified blocks of time that a network might claim for its use. Local programs were forced to evacuate those time periods whenever the networks exercised their options.

In 1963 the FCC banned "network option" clauses from affiliation contracts as a practice with "anti-competitive effects" which involved an abdication of the licensee's responsibility to program the station in what he considers to be the public interest.

ORACLE—system developed by Britain's ITA, almost concurrently with the BBC's Ceefax, which permits the viewer to call up printed news bulletins and other reading material on the screen by means of a decoder attachment. See also Ceefax, Teletext, Viewdata.

ORBEN, ROBERT—comedy writer whose credits include the shows of Red Skelton and Jack Paar. In 1975 he joined the speechwriting staff of President Ford and later became headwriter.

"ORIGINAL AMATEUR HOUR, THE"—a fringe-time variety show featuring amateur talent sought out from all parts of the country. It had a 21-year-run on TV (1949–70) as a carry-over from radio. Its host was Ted Mack. The program switched back and forth among the networks. It had its TV premiere on NBC Oct. 4, 1949, switched to ABC (1954–57), went back to NBC for a season and then went to CBS for a long run (1958–70). In its final years its title became *Ted Mack's Original Amateur Hour.* An attempt was made to keep the show going in syndication, but it was unsuccessful.

ORKIN, HARVEY (d. 1975)—writer of *You'll Never Get Rich,* the Phil Silvers hit situation comedy (1955–59); later he became head of the creative services department for Creative Management Associates, the talent agency.

ORR, WILLIAM T.—head of TV production for Warner Bros. whose deals with ABC for *Cheyenne* and other series in 1955 opened the way for the major Hollywood studios to produce films for television. A sometime actor who was also son-in-law to Jack L. Warner, Orr rose to executive assistant to Warner after his coup in opening an important new market for the company.

ORTF—See France, Television In.

"OSCAR LEVANT SHOW"—West Coast talk show which began in the late 50s contemporaneously with David Susskind's *Open End* in New York and Irv Kupcinet's *At Random* in Chicago. Levant—concert pianist, author, actor, wit and raconteur—invariably was more interesting than his interview guests and after a few years the program degenerated into a kind of psychiatry session in which the host discussed his own neuroses.

OSMONDS, THE—family of singers from Utah who received their first regular television exposure on *The Andy Williams Show* in the 60s and later had several regular series and specials of their own. Their biggest show, *Donny and Marie* (1975–78) on ABC, produced by Sid and Marty Krofft, gave way to *The Osmond Family Show,* produced by The Osmond Brothers and Dick Callister. ABC slotted it early Sunday evening, opposite *60 Minutes,* as a midseason replacement for *The Hardy Boys,* but without success. When the series was canceled, the Osmonds, through their own production company, kept it going in syndication.

See also *Donny and Marie.*

O'STEEN, SAM—director of drama specials and movies-for-TV, among whose credits are *Queen of the Stardust Ballroom* and *Rosemary's Baby II.*

"OUR MISS BROOKS"—a CBS situation comedy (1952–57) about a high school teacher that served as a vehicle for Eve Arden, a movie comedienne noted for her wisecrack delivery. Produced by Desilu, on film, the series enjoyed great popularity for several years and had a successful syndication run when it was retired by CBS. Gale Gordon was regularly featured as the school's principal, and Bob Rockwell played a fellow teacher.

"OUTER LIMITS, THE"—science fiction anthology series created by Leslie Stevens. Produced by Daystar-United Artists TV, the series ran on ABC (1963–65).

OUTLET BROADCASTING—Rhode Island-based station group owned by the Outlet Company, a retail chain, which put its first radio station on the air in the 20s. The TV group consists of WJAR Providence, an NBC affiliate; WDBO Orlando, Fla., CBS; KSAT San Antonio and KOVR Stockton (Sacramento), ABC; and WCMH Columbus, Ohio, NBC.

The outlet group was sold to Columbia Pictures in 1981 for $165 million. A year later, Columbia was acquired by Coca-Cola, with Outlet part of the transaction, pending FCC approval.

"OUT OF DARKNESS"—noted CBS documentary (1956) produced by Albert Wasserman which traced the progress of a psychiatrist with a catatonic patient in a mental hospital, leading to a powerful climactic scene when the woman begins to speak again. The commentary was by Orson Welles.

"OUTSIDER, THE"—hour-long NBC series (1968) about a nonconformist private eye played by Darren McGavin. Created by Roy Huggins, it was by Universal TV and Public Arts, Inc.

OVERMYER NETWORK (United Network)—a short-lived commercial TV network formed by UHF station owner Daniel H. Overmyer and one-time ABC-TV president Oliver Treyz, which broadcast for 11 nights in May 1967 before collapsing under the high cost of line charges. Overmyer and Treyz laid plans for the network in 1966, projecting a nighttime service that would be carried chiefly by independent stations and that would feature mainly a two-hour variety show from Las Vegas. While preparations were underway, a West Coast group of investors gained control of the project and renamed it the United Network. The two-hour *Las Vegas Show* premiered on 127 stations with support from 13 advertisers, but the audience response was slight, and the venture collapsed during its second week on the air.

Arthur Hill as Owen Marshall with Lee Majors

"OWEN MARSHALL: COUNSELOR AT LAW"—hour-long series on ABC (1971–74) about a brilliant lawyer portrayed by Arthur Hill. Lee Majors, who played his young law partner for two seasons, left the cast for his own series, *The Six Million Dollar Man,* and was replaced by Reni Santoni and David Soul. Featured were Christine Matchett and Joan Darling. David Victor, who produced *Marcus Welby, M.D.,* fashioned *Owen Marshall* as a legal counterpart, both for Universal TV.

P

PAAR, JACK—volatile TV personality who became a nightly institution as host of NBC's *Tonight* show, which he took over in 1957. He frequently made news with his capricious behavior on camera and his feuds with the press and other show business personalities. Paar was known to weep on camera, and on one occasion he walked off the show when the network censored one of his jokes which made reference to a water closet. Paar's hallmarks were his emotionality, his adeptness at conversation and his penchant for irritating both his guests and his audience.

Jack Paar

He grew so popular that after his first year on *Tonight* the network changed the title to *The Jack Paar Show.* He had done four network daytime shows previously, including a quiz program, *Bank on the Stars,* but had not caught the public fancy until he became the late-night host. He was selected

by NBC after making four impressive appearances as guest host. On the nightly program he developed a roster of regular guests—among them Zsa Zsa Gabor, Cliff Arquette, Genevieve, Dody Goodman, Hans Conried, Alexander King and Hermione Gingold—some of whom were faded celebrities reborn.

Tired of the grind, Paar gave up the show in 1962 and switched to a weekly variety series on NBC that failed. He purchased a TV station in Poland Springs, Me., and sold it several years later. Meanwhile, Johnny Carson took over *Tonight* and was even a bigger hit than Paar. In 1973 Paar signed with ABC to compete with Carson on a limited schedule of one week a month and failed to recapture his earlier glory.

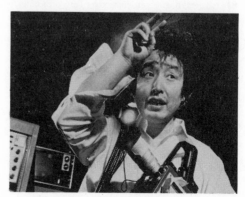

Nam June Paik

PAIK, NAM JUNE—Korean-born avant-garde video artist who fathered the video arts movement in the U.S., conducting experiments with abstract images on the screen as early as 1963 and producing the first non-professional works on half-inch portapak equipment in 1970. Working with old TV sets whose circuitry he modified, Paik developed the technique of creating fantastic abstractions by moving a magnet across the screen. Later he achieved complex kaleidoscopic effects through electronic "feedback," a circular electronic process which occurs when a TV camera is focused upon a TV monitor. Collaborating with Shuya Abe at the studios of WGBH Boston, he built the first video synthesizer, a device to add color to black-and-white images, which was yet another boon to video artists. Paik's own works are largely electronic collages flavored with humor and a sense of the absurd. See also Alternate Television.

Bruce Paisner

PAISNER, BRUCE—president of Time-Life Films (1973–81), engaged not only in marketing BBC programs in the U.S. but also in co-financing new programs with the BBC and developing domestic shows, including those based on the magazine properties of Time Inc. A graduate of Harvard Law School, Paisner served for a time as a reporter for *Life* magazine and as an assistant to Andrew Heiskell, chairman of Time Inc.

When Time-Life Films folded, he formed an independent company and then in 1981, joined King Features Entertainment as president.

PAL—Phase Alternate Line color television system, developed by Germany's Telefunken and widely in use in Western Europe. An offshoot of the American NTSC system, it is nonetheless incompatible with NTSC and with the other major color system, SECAM. The PAL system has been adopted by all Western European countries except France, and by Brazil and Yugoslavia, and is expected to be selected by the People's Republic of China. It was first broadcast in the United Kingdom in 1967.

William S. Paley

PALEY, WILLIAM S.—legendary patriarch of CBS for nearly five decades, and the company's largest stockholder (1.6 million shares). He purchased a floundering radio network with 16 affiliates in 1928 and built it into an enormously powerful and profitable broadcast organization. Paley's ability to deal with stars and keep them contented, his instinct for what would succeed in mass entertainment and his keen eye for the extraordinary executive were among the factors in the CBS rise to preeminence in a field that had been dominated by NBC and its parent, RCA.

More than chairman, Paley had been monarch of CBS, and under his demanding leadership the television network was first in ratings for 20 consecutive years, during which it billed itself as "the largest advertising medium in the world." Scheduled to retire in 1966 on reaching age 65, Paley waived the company's mandatory retirement rule for himself and, with the board's approval, stayed on to chart a line of succession and to put CBS on a new economic footing through business acquisitions that would reduce the corporation's dependency on profits from broadcasting.

Ten years later, just after his 75th birthday, Paley fired the executive who had been in line to succeed him—Arthur R. Taylor, recruited in 1972 from International Paper Co.—and named John D. Backe, head of the CBS Publishing Group, the new president of CBS Inc. Then in April 1977, Paley announced at the annual shareholders meeting that he would yield the title of chief executive officer to Backe but would retain indefinitely the chairmanship.

Although Backe technically was "boss" of the company, it was clear that as long as Paley was on the premises CBS was still in his hands. In 1980, he fired Backe and brought in Thomas Wyman as president and heir apparent.

Paley began in commercial broadcasting as a sponsor, purchasing air time in Philadelphia for La Palina and other brands of the Congress Cigar Co., his family's concern. Intrigued with radio's possibilities, he purchased for $300,000 the failing Columbia Phonograph Broadcasting Co., a small network formed earlier from what had been United Independent Broadcasters. On Jan. 8, 1929, he renamed the company the Columbia Broadcasting System and, from five floors of office space in Manhattan, began building his communications empire by purchasing stations, establishing affiliations with scores of other stations around the country and selling sponsors on the concept of network radio.

When CBS purchased the Columbia Phonograph and Records Co. for $700,000 in the 30s, Paley hired away the president of RCA Victor, Edward Wallerstein, to run it. Thus he acquired the most experienced executive in the recording business and with the same stroke eliminated his most formidable competitor. Under CBS ownership, Columbia Records grew to become the most successful recording company in the world.

With a $5-million bank loan, Paley executed the coup in 1948 for which he became famous and which was to propel CBS to the forefront of broadcasting. In startling succession, he raided NBC for some of its biggest stars, among them Jack Benny, Amos 'n' Andy, Red Skelton, Edgar Bergen and Charlie McCarthy and Frank Sinatra. As the audience followed the stars, the CBS penetration grew. *Variety* dubbed the phenomenon, "Paley's Comet."

World War II crystallized Paley's idea for the *CBS World News Roundup*, a daily program around which formed an army of stellar foreign correspondents organized by news director Paul White and headed by Edward R. Murrow. Among them were William L. Shirer, H.V. Kaltenborn, Raymond Gram Swing, Eric Sevareid, Robert Trout and Charles Collingwood, a group that formed the nucleus of what later became CBS News.

Paley's failures were chiefly in his attempts to move CBS into electronic manufacturing. His purchase of Hytron Radio and Electronic Corp. in the early 50s was a notable embarrassment, reported to have cost the company close to $50 million by the time it was dissolved in 1956. Contributing to its failure was the FCC rejection in 1953 of the color system devised by CBS, in favor of the RCA system which was compatible with black-and-white receivers. The CBS attempt in the 60s to pioneer the video-cassette with a division called EVR (Electronic Video Recording) also fizzled, resulting in a write-off of around $25 million. Despite such setbacks, the profits of CBS rose steadily each year.

Paley's influence on the company extended to styles of dress and office decor, which he chose to describe as "modern but conservative." Although he lived in a patrician manner,

moved in high social circles and was president of the Museum of Modern Art, he maintained throughout his career a mastery of the popular entertainment objectives of broadcasting and was the industry's foremost impresario.

In 1976, he founded the Museum of Broadcasting in New York. His autobiography, *As It Happened*, was published in 1979.

PALMER, BETSY—ubiquitous TV performer during the 50s and 60s who appeared in dramas as an actress, on game shows as a panelist and on *Today* as one of Dave Garroway's regular sidekicks. She was a regular panelist on *I've Got a Secret* and had dramatic roles in productions for *Studio One, Danger* and the *U.S. Steel Hour* and in the series *Martin Kane*.

"PAPER CHASE, THE"—widely praised CBS dramatic series (1978–79) that ran the full season despite anemic ratings but was not renewed beyond that. The series was drawn from the 20th Century-Fox movie of that title, which in turn was inspired by John Jay Osborn Jr.'s novel about the trials of first-year law students with an intimidating, autocratic professor, Kingsfield. John Houseman, who had played Kingsfield in the movie and won an Oscar for the performance, recreated the role for the TV series. Robert C. Thompson, producer of the film, was executive producer of the series for Fox.

The intelligent scripts and high level of acting endeared the series to many TV critics, but the subject matter was not the stuff on which television hits are usually made, and the series was handicapped further by its placement in the schedule. CBS assigned it the Tuesday night slot opposite two of ABC's top-rated shows, *Happy Days* and *Laverne & Shirley*. Later in the season the series was shifted experimentally to other time periods to no avail. The regular cast included Francine Tacker, Tom Fitzsimmons, Robert Ginty, James Keane, Jonathan Segal, James Stephens and Deka Beaudine.

Reruns of the series aired on PBS in 1981.

PAPP, JOSEPH—famed theater producer and head of the New York Shakespeare Festival who produced an embarrassing episode for CBS in 1973. In pursuing leadership among the networks in the sphere of serious contemporary drama, CBS had signed Papp in 1972 to produce 13 plays for the network over a period of four years. He delivered only two, canceling the contract after a widely publicized row with CBS when it postponed the showing of his production of David Rabe's antiwar drama, *Sticks and Bones*.

The two-hour TV adaptation of the prize-winning Off Broadway drama had been scheduled for broadcast March 9, 1973. But after a closed-circuit preview, dozens of affiliates indicated that they found it too rough to carry, or too difficult for viewers to comprehend. CBS itself became concerned over the timing: The program was airing the week that Vietnam POWs were being released and returning home, a week when patriotic sentiment was running high. Moreover, there was the worry that President Nixon, who was personally greeting the returning prisoners, might construe the showing of *Sticks and Bones* an impertinent attempt by CBS to undermine his effort to create a national celebration around the event.

When the network decided to put off the broadcast indefinitely, while as many as 69 affiliates were on record as refusing to carry it, Papp called CBS "cowardly" and charged the network with censorship. CBS finally did present the play, without commercials and in the light viewing month of August; more than 90 affiliates preempted it, some delaying it to a time period around midnight. But by then, the relationship with Papp had ended.

In November 1973 ABC announced that it had signed Papp to produce two dramas for prime time and to develop other shows for children and for late evening viewing. None has ever appeared, however.

Papp's first program for CBS was a three-hour TV adaptation of his stage hit, Shakespeare's *Much Ado About Nothing,* in turn-of-the-century dress. The program was charming and was generally well-received by the critics, but it drew anemic ratings. Papp later complained that in giving the show away free on TV he hurt the box office for the stage version, forcing it to close nine days after the telecast. The program aired on Feb. 2, 1973.

PARKER, EVERETT C. (The Rev. Dr.)—a leading crusader for the public's rights in broadcasting, as director, since 1954, of the Office of Communication of the United Church of Christ. With some notable success, his continuing effort has been to make broadcasters accountable to the public; to gain employment and access on the air for blacks and other minorities, and for women; and to oppose the broadcasting of extremist views when voice is not given to opposing opinions.

His office filed the petition to deny the license renewal of WLBT-TV Jackson, Miss., for discriminating against blacks, who constituted more than 40% of the local population. This led to two landmark decisions by the Court of Appeals for the District of Columbia Circuit: the first, granting members of the public the right to intervene in license-renewal proceedings; the second, revoking the WLBT license for disobeying the Communications Act requirement that it serve "the public interest, convenience or necessity.

Working with citizens groups in local communities, Parker's office brought pressure upon a number of stations to correct alleged abuses and in several instances obtained

agreements from the stations to perform certain specified services for minorities. Parker was also instrumental in causing the FCC to adopt rules requiring stations to maintain a "continuous dialogue" with members of the public so that dissatisfaction with the station's perception of community needs might be resolved locally.

See also Broadcast Reform Movement, United Church of Christ, WLBT Case.

PARKS, BERT—TV personality who, despite numerous daily programs, came to be best known for a single annual appearance as master of ceremonies for the Miss America Pageant. His singing of the ritual song, "There She Is, Miss America," had been one of the medium's constants through two decades. His ouster from the pagent for a younger host in 1979 became a national scandal.

Outgoing and exuberant, Parks was an ideal game-show host. He had a program of his own on NBC, *The Bert Parks Show* (1950–52), after which he hosted *Stop the Music* on ABC (1954–56) and later *County Fair, Masquerade Party* (in the original version) and *Break the Bank.*

PARIS, JERRY—actor-turned-director who first made his mark directing episodes of *The Dick Van Dyke Show* in the 60s, becoming thereafter a busy director of situation comedy episodes and of pilots. He has been director of *Happy Days* since 1975, adding the co-producer credit in 1976.

PARTNERSHIP AGREEMENT—a joint resolution adopted on May 13, 1973, by the Corporation for Public Broadcasting and the Public Broadcasting Service, the two governing bodies of public television, establishing ground rules for working together. Although CPB had been deeded leadership over the system in the Public Broadcasting Act of 1976, PBS asserted its primacy as the representative body of the licensed television stations. The PBS claim to equal power drew from the "bedrock of localism" principle affirmed in the agreement as "mutually desired.

A key provision of the agreement was that CPB distribute a substantial portion of the Federal funds it received directly to the PTV stations, as Community Service Grants, under the following formula: 30% at a $45 million funding level, 40% at a $60 million level, 45% at a $70 million level and 50% of funds exceeding $80 million.

The partnership was strained in the fall of 1976 when CPB acted on a number of projects that had not received PBS approval, among them the allocation of $1 million to the BBC to help finance the production of the complete dramatic works of Shakespeare. The conflict led to another round of negotiations between committees of both organizations. They resolved to reaffirm the agreement, with modifications that would give CPB discretion only in the financing of pilots but otherwise requiring joint approval for grants to programs that would actually be aired.

See also Community Service Grants, CPB, PBS, Public Television.

"PARTRIDGE FAMILY, THE"—ABC musical situation comedy (1970–74) about a family of children who become a professional rock group. Via Screen Gems, it starred Shirley Jones and featured David Cassidy, Susan Dey, Danny Bonaduce, Brian Forster, Suzanne Crough and Dave Madden.

PASETTA, MARTY—director specializing in musical-variety. He has directed, and sometimes produced, the Emmy, Oscar and Grammy Awards shows in the 70s and scores of star-centered specials, among them *Sandy in Disneyland* and *Bing Crosby & His Friends, Barry Manilow In Concert,* and *Ole Blue Eyes Is Back.* He was director of the weekly series with the Smothers Brothers, Andy Williams, and Glen Campbell.

Pasetta began his career in local television at KGO-TV, San Francisco, where he directed and created a wide range of shows. After 16 years with KGO-TV, he moved to Hollywood in 1968 and soon began receiving choice assignments.

"PASSWORD"—Goodson-Todman game show which had a vogue in the 60s, was revived in the ABC daytime lineup (1971–74). Allen Ludden hosted both the original and the revival, *Password Plus.* The game featured celebrities who team with studio contestants in attempting to guess the "password" from one-word clues.

PASTORE, JOHN O. (Sen.)—feisty Rhode Island senator who held great power over the broadcast industry during his 21 years as chairman of the Senate Communications Subcommittee as principal legislative force. Pastore, who retired from the Senate in 1976, was perhaps best known for his unrelenting war on excesses of sex and violence in TV, and the pressures he exerted played a part in the industry's adoption of *family viewing time* in 1975. But he figured importantly in much communications legislation over the two decades, concerning such matters as public broadcasting, communications satellites and the Fairness Doctrine. His power also derived from his committee's oversight of the FCC and its confirmation of appointees to communications-related posts in the Government. Not the least of his strength was his profound knowledge of the issues and his readiness to trade something the industry desired, such as longer licenses periods, for the reforms he sought.

Pastore was a vigorous defender of the Fairness Doctrine and successfully fought bills offered in 1975 by Sen. William Proxmire (D.-Wis.) and Sen. Roman Hruska (R.-Neb.) to abolish the doctrine. On the other hand, he opposed the equal-time law and was unsuccessful in his efforts to have it repealed.

It was Pastore's inquiry in the late 60s into the possible influence of TV violence on violent behavior—and his questioning of William H. Stewart, Surgeon General of the Public Health Service—that led to the $8 million, three-year government study of the subject, culminating in the publication in 1972 of the five-volume *Surgeon General's Scientific Advisory Report on Television and Social Behavior.*

In the midst of confusion about the conclusions of the report, caused partly by accounts in the press, Pastore managed to preserve the credibility of the work through hearings with members of the advisory committee that established a reasonable consensus on the social effects of TV violence.

See also Family Viewing Time, Fairness Doctrine, Equal Time Rule, Surgeon General's Report on Television and Social Behavior.

PATCHETT-TARSES—producing team of Tom Patchett and Jay Tarses, who were executive producers of *The Bob Newhart Show* on CBS (1975–78) and also producers of *The Tony Randall Show* on ABC (1976–77). They had been comedy performers, as a team, before turning to TV comedy writing, which led to producing. They created and produced *Open All Night* which debuted in the 1981–82 season.

"PATTY DUKE SHOW, THE"—ABC situation comedy (1963–66) in which Patty Duke played two roles, that of an American teen-aged girl and her look-alike British cousin. Featuring William Schallert, Jean Byron and Paul O'Keefe, it was by Chrislaw and United Artists TV.

PAUL, BYRON—long-time friend and associate of Dick Van Dyke, credited with having gotten him started in TV, who has been executive producer of *The New Dick Van Dyke Show* and *Van Dyke & Company* (Sheldon Leonard performed that function on the original Van Dyke series).

PAULEY, JANE—female member of the *Today* show cast, who joined in October 1976 after a five-month search by NBC News which began when Barbara Walters was hired away by ABC News. Miss Pauley had been with the NBC Chicago Station, WMAQ-TV, for a year when she was selected and had been out of college for only four years. She married the cartoonist Gary Trudeau and by 1982, when Tom Brokaw left to become co-anchor of the *Nightly News,* she was a senior member of the *Today* cast and represented the star power. From time to time she was assigned to anchor the weekend newscasts.

Jane Pauley

PAY-CABLE—pay-television by means of cable-TV, which results in the viewer paying first for basic cable service and then an additional monthly amount for the pay fare, the total ranging from $11 to $20, depending on the cable system. As in over-the-air pay-TV, the principle of pay-cable is to feed a scrambled channel that cannot be received without a special decoder.

Most pay-cable proponents perceive the ideal system to be one that charges viewers on a per-program basis—such as Warner Cable's Qube System—but the hardware for such a system has not been economically feasible, and the pay-cable operations that have developed are in the main those providing a channel of service, primarily movies and sports, for a monthly fee.

The success of pay-cable since its introduction in 1972 has fostered the belief in the cable industry that the pay-TV feature is the key to the expansion of cable-television into the major cities and in areas of the country where cable's retransmission of the existing TV signals is not a selling point.

Fearing the growth of cable, conventional TV broadcasters contributed more than $1 million in 1975 for a campaign administered by the NAB to keep the FCC from liberalizing the stringent rules imposed on pay-TV and pay-cable. Seeking the support of the public, the TV industry in its newspaper and magazine ads portrayed pay-cable as a menace to the elderly and the poor, and as a malignancy that would ultimately force all viewers to pay for what they were then receiving free. The FCC's rationale for its strict cable rules was that pay-TV should supplement conventional TV and not replace it.

The fact that theater-TV, a form of pay-TV, had sufficient resources to outbid commercial television for heavyweight boxing matches seemed to the commission to bear out the NAB's contention that when cable-TV achieved adequate penetration it would begin to outbid commercial television for many of its regular attractions—new movies, the Super Bowl and even its most popular series. Thus, the FCC rules for cable were designed to keep such a "siphoning" of programs from happening.

Aligning with the cable industry for a relaxation of the rules was the motion picture industry, which saw in pay-cable a whole new market to take the place of the neighborhood movie house that was eliminated by commercial television. To be able to deliver motion pictures into the home—with virtually no distribution or advertising costs and no extra production costs—and to reach the movie consumer who no longer went to theaters was highly appealing to the film studios, which saw also the possibility of producing films especially for pay-TV.

In 1975 the FCC did somewhat ease its pay-cable rules—not enough to suit the cable industry but too generously in the view of the NAB. Both went to court to appeal the rules. On March 25, 1977, the U.S. Court of Appeals for the District of Columbia held that the rules were unconstitutional and improper and ordered them vacated. The court said the FCC acted without knowing whether the siphoning threat was real or fanciful. The TV industry then mounted a legal challenge of the opinion.

Under the original rules, pay-cable was able to play movies only within two years of their theatrical release and otherwise not until ten years after their release. The new rules changed the formula to three and ten, on the premise that most movies purchased by commercial television are between three and ten years old.

In January 1976 the House Communications Subcommittee issued a staff report on cable-TV—the first Congressional assessment of the industry in 15 years—which favored cable's growth and which concluded that "constraints should not be imposed upon cable television simply to protect broadcasting from competition."

Pay-cable channels are operated in several ways, some by the owners of the cable systems who acquire programming from intermediaries; some by entrepreneurs who lease channels on systems for a straight rental fee; and some by "software" companies, such as Home Box Office (HBO), which enlist systems as affiliates and share the revenues with them.

In 1975 HBO, the leading pay-cable company, made a substantial advance and became in effect a network by distributing its offerings nationally via domestic satellite. By the following summer several of the major film companies had formed pay-cable services of their own in an effort to increase the revenues they might receive from the pay-cable market. Columbia Pictures established a pay-TV division for the licensing of movies, and 20th Century-Fox and

United Artists joined in the formation of Hollywood Home Theater to serve various pay operators.

Viacom's Showtime became the second largest pay-cable network in 1979, after Teleprompter's purchase of a 50% interest in the service.

PAY-TELEVISION—(known also as subscription television [STV] and toll-TV or tollvision) an alternate form of commercial TV which sells its programming directly to the consumer either over the air, through specially designated stations, or on cable-TV; both types involve a scrambled signal that may be decoded by a black-box device attached to the TV set. Although pay-TV promoters have sought the right to broadcast as early as 1950, the FCC did not authorize the service until Dec. 12, 1968. A further delay occurred when the National Assn. of Theater Owners and the Joint Committee Against Toll TV appealed to the courts to block the authorization, and that ended on Sept. 30, 1969, when the Supreme Court denied a petition for review of a lower court's decision upholding the FCC action.

Pay-cable began in 1972, but the over-the-air form was slower to get started. The FCC authorized three over-the-air systems—Zenith's Phonevision, Teleglobe's Pay TV Systems and Blonder-Tongue Laboratories' BTVision—as technically suitable for stations applying to broadcast pay-TV. Subsequently, UHF stations in Los Angeles, Chicago, Newark, Boston, Milwaukee, San Francisco and Washington, D.C., were granted approval to operate pay-television, but it was not until 1978 that STV caught fire. National Subscription Television's KSBC-TV, Corona–Los Angeles, started in April 1977 with 800 subscribers and a year later had more than 100,000 and was in the black. By mid-1979 the total swelled to 177,000 subscribers. Meanwhile, in the east, Wometco Broadcasting purchased Blonder-Tongue's ailing Newark STV station, WBTB (now WTVG) and similarly found the market highly receptive to its pay service, Wometco Television Theater. WTVG found its audience in the areas of metropolitan New York that were not yet wired for cable-TV. Its subscribers were charged $45 for installation, a $25 deposit for the decoding unit and $15 a month for the programming.

These twin successes, and those of a second Los Angeles station, sparked such interest in STV that Time Inc.— already heavily involved in pay-cable as owner of HBO— purchased 50% of Chicago UHF Station WSNS in 1979 with intentions of converting it to pay-TV. In July of 1979, NST's second STV station, WXON-TV Detroit, began its operations. Major commercial broadcasters, who once had fought pay-TV, now were eyeing it for their own expansion.

The vigorous opposition to pay broadcasting by theater owners and advertising-supported TV, which through their "Save Free TV" campaigns often won allies among the general public, posed a dilemma to policy-makers from the earliest years. Sporadically, the FCC authorized pay-TV tests but did not permit pay broadcasting to enter the marketplace until advertiser-supported television was firmly and securely entrenched.

The commission was obliged to permit over-the-air experiments because Section 303 of the Communications Act requires the agency to "study new uses of radio, provide for experimental uses of frequencies, and generally encourage the larger and more effective use of radio in the public interest.

In 1950 Skiatron tested its system briefly on WOR-TV New York; a year later, Telemeter experimented over KTLA Los Angeles and Zenith on its own experimental station in Chicago. Other tests were conducted on a Bartlesville, Okla., cable system (1957); on a broadcast station in Etobicoke, Canada, a suburb of Toronto; and in Hartford, Conn., on the RKO station, WHCT (Channel 18). The latter experiment, which involved the participation of Zenith with its Phonevision system, ran from 1962 to 1969.

The movie and TV industries marshaled their forces to frustrate an attempt in 1962 by an independent company, Subscription Television Inc., to wire up sections of Los Angeles and San Francisco for a pay-TV service. The well-organized campaign by the powerful forces resulted in a California referendum to prohibit pay-TV, which was passed by the state's voters in November 1964. By the time the state supreme court ruled the proposition unconstitutional, Subscription Television had lost $10 million and was in bankruptcy.

The rules for pay-TV finally adopted by the FCC late in 1968 were restrictive and designed to minimize the threat to the existing TV system. The commission's rationale was that it would not betray the millions of people who had purchased sets to receive free TV. Essentially, the rules specified that there be only one over-the-air pay-TV station licensed to any market; that only markets already receiving four commercial TV stations or more be authorized a pay-TV outlet; that the stations may not sell decoders but only lease them to subscribers; and that along with their pay services the stations must carry a schedule of free, unscrambled programs (which could carry advertising) a minimum of 28 hours per seven-day week.

Added to these were the "anti-siphoning" rules pertaining to programs. The initial rules barred the use of movies that were between 2 and 10 years out of theatrical release (later that was eased to 3 and 10) and required that at least 10% of the pay fare offered be something other than movies and sports. Also barred initially were continuing series, although that restriction was later lifted. Over-the-air pay-TV was prohibited from carrying sports events that had appeared locally on conventional TV the previous five years. Most of these rules were later abolished, and by the 80s several cities had two STV Stations and Dallas had three. See also HBO; Pay-Cable; Sagall, Solomon; Subscription Television.

PBS (PUBLIC BROADCASTING SERVICE)—public television's central program service, a network of sorts but differing from the BBC and the American commercial networks in lacking the power to produce programs or to exercise any authority over its affiliates. Created in November 1969 by the Corporation for Public Broadcasting, specifically to operate the national interconnection of stations, with a president and permanent staff based in Washington, PBS has assumed a number of configurations in the years since. Initially it had served as a network, in taking over from NET in New York the responsibility for providing national programs. In the fall of 1970, it began feeding out 12 hours of programs a week to the stations over network lines.

But after two years, PBS relinquished the central programming role and the designation of "network," as the entire system underwent a change to assert a local orientation. The change largely was prompted by forceful advice from the Nixon Administration to decentralize the system, underscored by the President's veto of a large federal appropriation for public TV. The Administration—at odds with the commercial networks and critical of what it perceived to be a liberal bias in PBS—called for a system dedicated to "grassroots localism."

PBS was reorganized in 1973 as the system's national membership organization, usurping the functions of the National Association of Educational Broadcasters (NAEB) as the congress of local stations. In 1974 the Station Program Cooperative was set up within PBS as the mechanism by which individual stations could participate in the selection of national programs.

In the meantime, when there was evidence that the Corporation was not providing the insulation from the Government it was supposed to provide, PBS mounted a challenge to CPB's supreme authority over the system. The confrontation resulted in the 1973 Partnership Agreement between CPB and PBS, whose most important provision was that specified amounts of the federal funds received by CPB be distributed directly to the stations (30% at the $45 million level, 40% at the $60 million level and 50% at an $80 million level). This not only served the purpose of decentralization, but severely limited the discretionary monies for CPB, which was responsible also for paying the costs of interconnection. CPB also agreed to consult with PBS on all program projects for which it intended to allocate funds.

Operating with a permanent staff and a number of advisory committees representing the 155 member licensees (who control 264 stations and satellite stations), PBS established program standards and created a national schedule from programs underwritten by private corporations and foundations, those selected from the Station Program Coop-

erative and the two series a year produced on grants from the CPB.

A 1979 reorganization was prompted in part by Carnegie II's call for more national programming but also by the system's switch-over to satellite distribution. Devised by PBS vice-chairman Hartford N. Gunn, the plan was designed to reconcile the differing needs and missions of public TV stations. PBS relinquished all its trade-association and lobbying functions to a new organization and concentrated on programming. .

However, a plan to divide PBS into three separate program services was abandoned in 1982 after the Reagan Administration's severe budget cutbacks in public television funding. The three were to have consisted of a network for cultural, public affairs, and children's programming; another for special interest and regional programming; and the third for educational fare. Lawrence K. Grossman, the embattled president of PBS, developed a plan for public television to earn money through pay television. Under this scheme, PBS and the major performing arts organizations would share in the operation of a pay-cable network whose proceeds would help support all the institutions involved, including PBS. He called the project the Grand Alliance.

See also British Broadcasting Corporation, Corporation for Public Broadcasting, Ford Foundation and Public TV, Public-Television.

PEABODY AWARDS—annual awards recognizing distinguished and meritorious public service that are regarded as among the most prestigious in broadcasting. Administered by the Henry W. Grady School of Journalism at the University of Georgia, with the aid of a national advisory board, the awards were established in 1940 to perpetuate the memory of George Foster Peabody, a native of Columbus, Ga., who became a successful New York banker and philanthropist. The awards are made to programs, stations, networks and individuals.

PEACOCK, IAN MICHAEL (MIKE)—British programmer who, after a spectacular career with the BBC, switched briefly to the U.K.'s commercial system and then came to the U.S. in 1974 as executive v.p. of network programs for Warner Bros. TV. He held that post for less than two years.

He rose at the BBC from a trainee in 1952 to program chief and head of the creative staff in the mid-60s. On his way up, he was a producer, then editor of BBC News and head of programs for BBC-2 when it was established. During his BBC career he was instrumental in developing such programs as *Panorama,* the prestigious weekly public affairs series, and the comedies, *Till Death Do Us Part* and *Steptoe and Son,* from which *All In the Family* and *Sanford and Son* were drawn in the U.S.

Peacock left the BBC to become managing director of London Weekend Television in 1967 but resigned within a year after a dispute with the board of directors. He became a consultant and independent producer for five years and then joined Warner Bros. Ltd., London, in 1972, transferring to the U.S. two years later.

PECARO, DANIEL T.—former president of WGN Continental Broadcasting Co., a group with headquarters in Chicago and other stations in Denver and Duluth. Pecaro rose through the programming ranks of WGN Radio and WGN-TV, became general manager of the latter in 1967 and succeeded Ward L. Quaal as head of the company in 1974. He left for early retirement in 1981.

PECKINPAH, SAM—famed movie director who dabbled in TV westerns during the early 60s, notably as producer-director of *The Westerner.*

"PENSIONS: THE BROKEN PROMISE"—NBC News investigative documentary on abuses in some private pension plans, narrated by Edwin Newman and broadcast Sept. 12, 1972. The program gave rise to a major Fairness Doctrine case, which was not resolved until February 1976, when the Supreme Court refused to hear the case, letting stand a lower court's decision in favor of NBC's position that the program was reasonably balanced for a journalistic expose.

The case began shortly after the broadcast when Accuracy In Media, a news-media watchdog group with a conservative orientation, filed a fairness complaint with the FCC arguing that the program maligned the private pension industry and was not properly balanced with material on the pension plans that had kept their promises to retired persons. Although Newman had mentioned several times in the program that most pension plans were honest, the FCC upheld the AIM petition and ruled that NBC was obliged to tell the positive side of the pensions story, under the Fairness Doctrine. Ironically, the day after the FCC staff deemed the program one-sided, *Pensions* won a Peabody award as "a shining example of constructive and superlative investigative reporting."

NBC appealed the FCC's ruling in the U.S. Court of Appeals in Washington and in September 1974 the court in a 2–1 opinion overturned the FCC's decision. Judge Harold Leventhal, who wrote the court's opinion, held that judgment in an investigative report intending to uncover abuses should not be disturbed if reasonable and in good faith.

AIM appealed to the court for a rehearing, arguing that the three-man panel had made factual and legal errors; and in December 1974 the court agreed to hear the case *en banc*, by the full bench of nine judges. Later, however, the court sent the case back to the original panel to decide whether the case had become moot because legislation had been enacted to regulate private pension systems. In July 1976 the panel directed the FCC to dismiss the complaint. AIM then carried the matter to the Supreme Court, appealing for a decision that would clarify the application of the fairness doctrine to documentaries. The court's denial closed the case, and the FCC vacated the complaint. See also AIM, Fairness Doctrine.

PENTHOUSE ENTERTAINMENT TELEVISION NETWORK—see Cable Networks.

"PEOPLE'S CHOICE, THE"—an early Jackie Cooper situation comedy vehicle in which he portrayed a small-town mayor with a talking basset hound named Cleo. On NBC (1955–58), it was via Norden Productions and featured Patricia Breslin.

PEPPIATT-AYLESWORTH—writing and producing team of Frank Peppiatt and John Aylesworth who have worked chiefly on variety shows and were creators and executive producers of *Hee-Haw.*

PERKINS, JACK—NBC News correspondent based in Los Angeles who became in 1979 one of the featured reporters on *Prime Time Sunday.* His storyteller's style earlier had earned him a regular spot with offbeat features for *Today.* He joined NBC News in 1961 as a correspondent for *The Huntley-Brinkley Report* and later was on general assignment for the *Nightly News.*

PERLMUTTER, ALVIN H.—documentary producer whose career swung between commercial and public television before he formed his own production company in 1977. As an independent, he produced the PBS series *Global Papers.* From 1975 to 1977, he was v.p. of NBC News in charge of documentaries. Before that, working in public TV for WNET, New York, he was associated with scores of documentaries and was executive producer of *The Great American Dream Machine, Black Journal, NET Journal* and *At Issue.* Earlier, he was director of public affairs and then program manager of WNBC-TV New York.

PERSKY-DENOFF—producing team of Bill Persky and Sam Denoff active mostly in situation comedy. After coming into prominence as writers on the first *Dick Van Dyke Show,* they became producers of *The Funny Side* and then executive producers of *That Girl, The Montefuscos, Big Eddie* and others.

PERSONAL ATTACK—that part of the Fairness Doctrine which lays down an affirmative obligation on the part of broadcasters to give persons attacked on the air free access to the airwaves to respond. A personal attack is defined as an attack upon the honesty, character, integrity or like personal qualities of an identified person or group during the presentation of a controversial issue of public importance.

The concept of "personal attack" was first put forth in 1949 in the Report on Editorializing, in which the commission noted that a personal attack situation might give rise to a more specific obligation on the part of the licensee other than the normal requirements of fairness.

A 1963 Public Notice mailed to licensees advised them that in instances of personal attack on an individual or organization, they had an obligation to transmit a text of the broadcast to the person or group, with a specific offer of time on the facilities.

In 1964, in its Fairness Primer, the commission reiterated this obligation with even more specificity. In 1967, during the middle of the Red Lion proceeding, it adopted rules placing an affirmative obligation on licensees when personal attack occurred, or when a licensee broadcast a political editorial.

The rules stated that when a personal attack was made, as defined above, the licensee was obliged, within a reasonable time, but no later than one week, to transmit to the person or group attacked notification of the date, time, and identification of the broadcast, a script or tape of the broadcast, and an offer of a reasonable opportunity to respond over the licensee's facilities.

When the commission promulgated the rules, it also included some exemptions for certain types of programming. Excluded from the positive obligations of notice and offer of time were attacks made on foreign groups and public figures; attacks made by legally qualified candidates and their spokesmen on other candidates and their spokesmen; attacks that occurred in bona fide newscasts, news interviews and on-the-spot coverage of news events, including commentary or analysis contained in these types of programs. This last provision is sometimes called the "Sevareid Ruling," since it seems to be designed to protect commentary such as that given by Eric Sevareid of CBS. Finally, the commission pointed out that while these exemptions may relieve the broadcasters of the specific procedural requirements of the Personal Attack Doctrine, they do not relieve him of his general Fairness Doctrine obligations.

The constitutionality of the Personal Attack Doctrine was upheld by the Supreme Court in 1967 (see Red Lion). While the doctrine continues in force, its administration remains difficult and imprecise with respect to definitional

problems, particularly with what is a "controversial issue of public importance." See also Straus Communications.

PERSONALITIES—performers with no exceptional musical, acting or comedy talent, and with no valued fund of knowledge, who fill a large, amorphous category in television for their ability to play host, converse with guests and draw affection from viewers. Most TV personalities, though not all, have the gift of glibness and a distinctive manner within the boundaries of conventional behavior; whatever else they may be, TV personalities are not "characters."

The Age of Personalities came in the 50s and early 60s, during the years before the networks gave over prime time entirely to film. Often a personality was strong enough to carry a show on the strength of his/her name, and the big ones were among the highest-paid persons in the medium. Arthur Godfrey and Garry Moore had sufficient appeal to spread over several shows. Other big names who "did television," rather than sing or dance, were Ed Sullivan, Dave Garroway, Faye Emerson, Arlene Francis, Art Linkletter, Ted Mack, Robert Q. Lewis, Allen Funt and Dick Clark.

A second group did have the ability to entertain, and occasionally contributed songs or monologues to the proceedings, but nevertheless achieved their prominence through an ability to project a likable personality. They included Jack Paar, Steve Allen, Tennessee Ernie Ford, David Frost, Johnny Carson, Merv Griffin, Mike Douglas and Dick Cavett.

Locally, stations thrived on personalities no less than the networks. Ruth Lyons, in Cincinnati, may have been the biggest local star anywhere in the 50s; national sponsors waited in line to get on her show. Bob Braun, who followed her, was also popular, as was the late Paul Dixon with his own show. Chicago had Irv Kupcinet, Studs Terkel, Paul Gibson, Lee Phillip, Jim Conway and a raft of popular "irritable" personalities—Tom Duggan, who went on to greater notoriety on the West Coast, Jack Eigen and Marty Faye. New York had Joe Franklin, Sonny Fox, Carmel Myers and dozens more.

Many of the personalities came to TV from the press or from radio—for many disc jockeys, the transition was natural. While they are far from a dying breed, personalities play a lesser role in TV today because the demand for them is confined to game shows and a handful of syndicated talk shows. Locally they have been usurped by newsmen and weathercasters.

"PERSUADERS, THE"—an attempt by ATV of London to create an action-adventure series that would be both American and British, with Tony Curtis and Roger Moore as the co-stars. The comedy-adventure series ran a single season on ABC (1971–72).

"PETROCELLI"—hour-long NBC dramatic series (1974–76) about a New York lawyer practicing in a southwest cattle town. The Paramount TV series featured Barry Newman, Susan Howard and Albert Salmi and was produced in association with Miller-Milkis Productions.

"PERRY MASON"—vastly successful hour-long series based on Erle Stanley Gardner's stories of a crime-solving criminal lawyer; it had a nine-year run on CBS (1957–66) and made a star of Raymond Burr. One of TV's most enduring series, it has never ceased running in syndication and is a staple in virtually every television market, although the 245 episodes have played numerous times and only a few are in color.

Featured were Barbara Hale and William Hopper, with William Talman portraying Hamilton Burger, the district attorney who was Mason's weekly opponent in the courtroom. The series was produced by Paisano Productions.

CBS attempted to revive the series in 1974 as *The New Adventures of Perry Mason,* with Monte Markham in the title role, but it failed and was canceled at midseason.

"PERSON TO PERSON"—a live weekly interview series on CBS (1953–1959) that was conducted informally at the homes of the famous and always involved a brief tour of those residences for what they might reveal of the subjects. Edward R. Murrow, the network's premier newsman, urbanely hosted from the studio, so that his vantage was the same as the viewer's.

Uniquely suited to TV, and highly popular, the show visited over the years such notable persons as former President Truman, Maria Callas, Marilyn Monroe, Jackie Robinson, Marlon Brando, Robert Kennedy, Elizabeth Taylor and Mike Todd.

Murrow left the program in 1959 and was succeeded for the duration of its run by Charles Collingwood. John Aaron was executive producer, and other production principals were Fred W. Friendly and Jesse Zoussmer.

"PETE AND GLADYS"—domestic situation comedy and one of TV's first spin-offs, the characters having been introduced on the successful *December Bride* series. Produced by CBS (1960–61), it featured Harry Morgan and Cara Williams.

"PETER GUNN"—stylish private eye series with outstanding jazz theme music by Henry Mancini. It helped launch its creator-producer, Blake Edwards, as a filmmaker. Featuring Craig Stevens, Lola Albright and Herschel Ber-

nardi, it was carried by NBC for two seasons (1958–59) and switched to ABC for one.

Mary Martin as Peter Pan

"PETER PAN"—one of NBC's perennial specials, with Mary Martin and Cyril Ritchard, which had its first telecast March 7, 1955, as a live production and was so popular it was repeated live the following January. After that, video tape made it possible to present a third mounting of the show annually for several years. In December 1976 NBC offered a new version of the James M. Barrie story, with Mia Farrow in the title role and Danny Kaye as Captain Hook; it was presented as a *Hallmark Hall of Fame* special, with a new score, under auspices of Britain's ATV/ITC.

The original production, which aired in the days when specials were called spectaculars, was billed as the first network presentation of a full Broadway production. Miss Martin had created a sensation as Peter Pan, the magical boy who wanted never to grow up. In addition to Ritchard as Captain Hook, Kathy Nolan was featured as Wendy. The producer was Richard Halliday, the musical score was by Moose Charlap and Carolyn Leigh and the special was staged, choreographed and adapted by Jerome Robbins.

The two-hour 1976 production was staged especially for TV by Michael Kidd from a new adaptation by Jack Burns and Andrew Birkin and with an original musical score by Anthony Newley and Leslie Bricusse. The featured cast included Paula Kelly, Virginia McKenna, Briony McRoberts and Tony Sympson. It was produced by Gary Smith and Dwight Hemion, with Hemion directing. Duane Bogie, whose Clarion Productions handles the Hallmark series, was executive in charge of production for ATV/ITC.

PETERSMEYER, C. WREDE—founder and chairman of Corinthian Broadcasting Corp.; he resigned Jan. 31, 1977, after 23 years at the helm. As a partner in John Hay Whitney's private venture capital firm, Petersmeyer in 1952 organized six CATV companies and then turned to broadcasting with the acquisition of KOTV Tulsa, in 1954. While serving as president and general manager of the station, he began to build the Corinthian group for the Whitney company, adding KHOU-TV Houston, WISH AM-TV Indianapolis and WANE AM-TV Fort Wayne in 1956. Three years later Corinthian acquired KXTV in Sacramento.

Under Petersmeyer's leadership, the company acquired Funk & Wagnalls Publishing in 1971 and the TVS Network in 1973. Petersmeyer was one of the organizers of AMST in 1956 and served as a director of that organization and of TIO and the TV board of NAB.

PETITION FOR RULEMAKING—a formal request to the FCC for changes in existing rules or for the adoption of new rules relating to aspects of broadcasting. The commission requires that the petition set forth arguments and data to support the requested action in order to differentiate a proposal from a casual suggestion.

PETITION TO DENY—a detailed complaint to the FCC asking to deny the renewal of a station's license for having failed to meet its public trust. Petitions to deny are accepted by the FCC at license-renewal time and serve as a means by which citizens groups may act against stations which they feel are undeserving of the privilege of broadcasting. A persuasive petition may lead to an investigation by the commission, holding up the license renewal and conceivably result in the ultimate removal of the license.

The petition to deny differs from the license challenge in that the petitioners seek only to cause the operators to lose their license, without bidding to receive the license themselves. Grounds for such competitions range from neglect of significant elements of the community to failure to fulfill the promises of the license application.

PETRIE, DANIEL—veteran director associated with David Susskind in his drama productions in the early 60s who later concentrated on movies-for-TV and drama specials.

"PETTICOAT JUNCTION"—part of the powerful phalanx of bucolic situation comedies on CBS during the 60s, concerning three well-endowed girls helping their mother run a small rural hotel. It premiered on CBS in 1963 and still had a sizable audience when CBS canceled it in 1970 in a general housecleaning to change the network's rural image.

Bea Benaderet played the mother the first five seasons and, after her death in 1968, June Lockhart joined the cast as a country doctor. Edgar Buchanan, Mike Minor, Frank Cady and Rufe Davis were featured as local color. The original three daughters were Linda Kaye, Jeannine Riley and Pat Woodell, the latter two later replaced by Meredith MacRae and Lori Saunders. The series was produced by Filmways.

PETTIT, TOM—executive v.p. of NBC News, moving suddenly and surprisingly into management after some 17 years as a correspondent. His appointment came in 1982 when Reuven French was named to succeed Bill Small as president of the division. Frank immediately named Pettit his lieutenant. Pettit had been Washington correspondent covering the Senate since 1975, after more than a decade of working from the Los Angeles bureau. He was noted for his investigative reports on the NBC newsmagazine, *First Tuesday,* such as one on U.S. chemical-biological warfare experiments and another on the country's nuclear establishment. He was also noted for his hard-boiled interviewing style in his stints on *Today.*

"PEYTON PLACE"—successful prime-time soap opera on ABC (1964–69) based on Grace Metalious's best-selling novel about sex in an American small town. The 30-minute episodes initially aired twice weekly; when they were expanded to three a week the series began to falter. *Peyton* was the career springboard for actress Mia Farrow, who left the series after two years. Other regulars were Ryan O'Neal, Ed Nelson, Barbara Parkins, Dorothy Malone, Tim O'Connor, Patricia Morrow, James Douglas and Chris Connelly. It was by 20th Century-Fox TV.

In 1973 NBC installed a daytime soap opera, *Return to Peyton Place,* with new characters in addition to those created by Miss Metalious and with three actors who had appeared in the ABC nighttime version—Miss Morrow, Evelyn Scott and Frank Ferguson. The venture was not successful, however.

PFEIFFER, JANE CAHILL—chairman of the National Broadcasting Company and a director of its parent, RCA Corp. (1978–80). In that post, the highest ever attained by a woman in the broadcast industry, she reported to Fred Silverman, president and chief executive officer of the company. Mrs. Pfeiffer had helped recruit Silverman for NBC while she was a consultant to RCA and then was hired by him and RCA president Edgar H. Griffiths to help reorganize and manage the company. While Silverman concentrated on the various broadcast entities, and particularly on rehabilitating the sagging program schedule of NBC-TV, Mrs. Pfeiffer looked after administration, employee relations, legal affairs and government relations. Her career at NBC ended explosively in 1981 when Silverman, apparently on orders from Griffiths, discharged her. The firing was poorly handled and resulted in Pfeiffer and Silverman exchanging insults and criticisms in the press.

Mrs. Pfeiffer had become acquainted with Silverman while she was v.p. of communications for IBM, a job that occasionally put her in the role of TV sponsor. Silverman was program chief of CBS-TV at the time. Mrs. Pfeiffer gave up her job at IBM, which she had held since 1972, after marrying a top IBM executive in 1975.

PHILBIN, JACK—executive producer or producer of virtually all the Jackie Gleason series and specials since the early 60s.

PHILIPS, LEE—actor-turned-director who did two seasons of *The Andy Griffith Show* and episodes for such other series as *The Dick Van Dyke Show, Bracken's World, Peyton Place* and *The Waltons.*

PHILLIPS, IRNA (d. 1974)—the leading creator and writer of daytime radio and TV serials, for which she earned the title, "Queen of the Soaps." Beginning her career as a radio actress in Chicago in 1930, after five years of teaching school in Dayton, she switched to writing with the serial, *Today's Children,* for WGN. This led to a succession of others: *Women in White, Road of Life, The Guiding Light, The Right to Happiness, Lonely Women, Young Dr. Malone* and *The Brighter Day.* When she began writing for television in 1949, *Guiding Light, Dr. Malone* and *Brighter Day* switched over with her.

She became show doctor for a number of other serials, but her most glorious achievement was to be *As the World Turns,* the first 30-minute soap opera (sharing the distinction with *The Edge of Night,* which started the same day in 1956) and the most successful TV serial ever. For more than a decade it was the No. 1 daytime show, and Miss Phillips continued to write it until a year before her death.

"PHIL SILVERS SHOW, THE"—See *You'll Never Get Rich.*

"PHYLLIS"—CBS situation comedy (1975–77) spinning off the character played by Cloris Leachman on *The Mary Tyler Moore Show;* it had a successful premiere season with buttressing from the network's traditionally strong Monday night lineup. Organic changes were made for the second year, with the essentially unlikable title character made more

appealing. The series concerned a newly widowed woman and her daughter re-locating in a new city (San Francisco) and entering the career world. It was by MTM Enterprises and featured Henry Jones and Lisa Geritsen.

Cloris Leachman as Phyllis with Dick Van Patten

PHONEVISION—Pay-TV system developed by Zenith Radio Corp. in 1947 that endured on an experimental basis until 1969, when the FCC finally authorized it for Los Angeles and Chicago over UHF stations. The system involves the transmission of a scrambled signal which becomes unscrambled by a device attached to the subscriber's set when it is activated by the insertion of a ticket. The major experiment with Phonevision took place in Hartford, Conn. (1962–69), on a UHF channel licensed to RKO General. The system offered an average of six new programs a week, with fees ranging from 50 cents for movies to $3 for certain sports and cultural events. With access to 500,000 households, the pay-programs never achieved a larger audience than 7,000. The typical subscriber spent only $1.20 per week, and the operation was unprofitable.

PIERCE, FREDERICK S.— the executive who presided over ABC's surge to prominence during the 70s. He became president of ABC Television in November 1974, advancing from senior v.p. to succeed Walter A. Schwartz in a period when ABC's competitive standing in prime time had deteriorated seriously. Almost immediately, after emergency program changes instituted by Pierce, the network began to climb in the ratings. A year later it gained parity with its rivals and in 1976 forged ahead into first place.

In 1979 he was named also executive v.p. of ABC Inc., the No. 3 post in the company, putting him prominently in line of succession. His corporate duties focused on expansion into the new business frontiers opened by the advancing technologies of cable-TV, home video recorders and the video disc, and Pierce's first moves in these directions were to create a new ABC theatrical film division and to discover and produce other forms of programming for the new video media. In 1981, he relinquished his post with the broadcast group to concentrate on corporate matters.

One of Pierce's greatest coups was to hire away from CBS the leading program expert in network television, Fred Silverman. The combination of Pierce and Silverman became the most formidable in television and resulted in such hits as *Happy Days, Laverne and Shirley, Charlie's Angels, The Six Million Dollar Man, Bionic Woman, Starsky and Hutch, Welcome Back, Kotter, Family, Rich Man, Poor Man* and *Barney Miller.*

Frederick S. Pierce

Although the ABC news division was not at the time under his supervision, Pierce played a role in the negotiations with Barbara Walters that brought her over from NBC in the spring of 1976 as co-anchor with Harry Reasoner of the *Evening News.* Whether or not she increased the program's ratings, the signing of Miss Walters, in an arrangement that would pay her $1 million a year, was deemed beneficial since it would add to the prestige of ABC News.

With ABC's ratings rising on all fronts except news, ABC early in 1977 broadened Pierce's responsibilities to include news and engineering—two divisions that previously had reported directly to Rule. The responsibility for news carried the implied mandate to make the changes that would improve ABC's competitive position.

Pierce joined the company in 1956 with a background in accounting. He moved up through the areas of audience research, sales development, sales management and network planning and was recognized as ABC's leading program strategist before his promotion to president of all television operations.

PIERPOINT, ROBERT—CBS White House correspondent since 1957, which makes him television's senior man on the beat; he has covered every President since Eisenhower and has traveled extensively with all of them. Pierpoint, who received his first professional broadcast news experience with the Swedish Broadcasting Corporation, joined CBS News in 1949 as Scandinavian correspondent. Next he was based in Korea and then in Tokyo, before receiving the White House assignment. His coverage of Presidents was interrupted twice, both times briefly: For a time in 1961 he covered Congress, and in 1964 he was assigned to Barry Goldwater's Presidential campaign.

PIERSON, FRANK R.—writer-producer primarily working in movies. His occasional work for TV has included *Have Gun, Will Travel* in the 60s and James Garner's *Nichols* series in the early 70s.

PIGGY-BACK—a commercial unit purchased by an advertiser who uses it to promote two products, one after the other. In the years when one-minute was the shortest commercial unit a network would sell, many companies that manufactured several products divided the minute for two messages, in effect *piggy-backing* the advertisements. The piggy-back proliferated in the late 60s when advertisers discovered that, through new editing techniques, they could present an effective announcement in half the time they previously consumed.

Variations were the *Split 30* and the *Matched 30.* In the first case, the advertiser agreed to purchase a single minute in a given program if the two 30-second portions could play separately, that is, in separate commercial breaks. The networks permitted the practice as a refinement of the *Matched 30* principle, under which two different advertisers agreed to mate halves of their one-minute commercials, so that each had 30 seconds at the start of the show and 30 seconds near the close.

In 1970 the networks and stations yielded to advertiser pressures and made the 30-second spot the standard unit of purchase. The effect was, of course, to increase the number of commercials viewers would be subjected to, without increasing commercial time.

PILKINGTON COMMISSION—See Britain's Broadcasting Commissions.

PILOT—a sample program, on film or tape, of a projected TV series, usually created to serve as a first episode but in any case establishing the continuing characters, the character relationships, the essential situation and the style of the proposed series. Pilots are tested by the networks for their appeal to viewers, and the testing scores are important factors in the decision-making process. In most seasons, only one of every four or five pilots produced becomes a series.

The cost of producing a pilot is usually twice that of a series episode because there is no amortization of the sets and props and because of the premiums paid for putting a hold on the services of actors and creative personnel. Rejected pilots recoup a small portion of their costs when they are played off in summertime anthologies, but otherwise they are useless to television.

In recent years, to reduce the financial waste in pilot-making, pilots have been produced as made-for-TV movies or as episodes in anthology series such as *Police Story.* Pilots for variety shows are usually created as one-hour specials. An attempt was made by ABC in the 60s to substitute 5- or 10-minute demonstration films for pilots, but the results were highly unsatisfactory. The current pilot preference is the shortflight series of four or six episodes. The theory is that production of actual episodes will show flaws or strengths better than a single pilot would.

PINKHAM, RICHARD A. R.—one-time program chief for NBC (1955–56) but better known for his subsequent career at Ted Bates & Co., a leading ad agency. As v.p. in charge of TV and radio for Bates, Pinkham exerted a strong influence on TV programming as a major sponsor of shows in the late 50s and early 60s. He rose to become vice-chairman of the agency.

Pinkham joined NBC in 1951 and became a program executive when the department was under the dynamic leadership of Sylvester Weaver. He became executive producer of the *Today, Home, Tonight* parlay (1952–55), then v.p. in charge of programs and later v.p. in charge of advertising (1956–57), until Bates hired him away.

"PINKY LEE SHOW, THE"—NBC children's show of the early 50s featuring impish comedian and storyteller, Pinky Lee.

"PLANET OF THE APES"—hour-long series on CBS (1974) based on the popular series of motion pictures, which scored powerful ratings when shown on TV. Produced by 20th Century-Fox TV, the video version featured Roddy McDowell, Booth Colman, Ron Harper and James Naughton. It fizzled and was pulled after 14 weeks.

"PLAY OF THE WEEK, THE"—a series of stage plays mounted for television from 1959 to 1961 by WNTA (Chan-

nel 13 in New York City, when it was a commercial station owned by National Telefilm Associates) and syndicated by NTA to a number of commercial stations around the country. Despite critical acclaim and topnotch actors working for scale, the series drew low ratings and had difficulty getting sponsors. When it was canceled, WNTA received thousands of letters from loyal viewers protesting the action. Among the 65 productions were *The World of Sholem Aleichem,* with Gertrude Berg, Sam Levene and Nancy Walker; *The Iceman Cometh,* with Jason Robards, Jr.; and *Medea,* with Judith Anderson.

In 1962, under the sponsorship of the Esso Oil Co. but without commercials, WNEW-TV in New York televised some of the reruns of the plays in prime time.

"PLAYING FOR TIME"—CBS drama special which raised a controversy in 1980 because of indiscretion in casting. The three-hour drama was based on a memoir by Fania Fenelon, a French Jew who survived Auschwitz, the Nazi concentration camp, by performing in the women's orchestra. Jewish groups assailed the show as an offense to their people, because the role of Miss Fenelon was being played by Vanessa Redgrave, who in private life was an active supporter of the Palestinian cause. Despite an intense campaign to prevent the show from airing, or perhaps because of it, the drama drew an exceptionally large audience for the Sept. 30, 1980 telecast.

The script was adapted for television by Arthur Miller, the novelist and playwright. Daniel Mann directed, and Linda Yellen produced the show for Syzygy Productions Ltd. Jane Alexander co-starred with Miss Redgrave, and the cast included Maud Adams, Marisa Berenson, Verna Bloom, Viveca Lindfors, Melanie Mayron, and Shirley Knight.

But what was to have been an artistic triumph that would enhance CBS's prestige had backfired because of the controversy. The CBS executive who approved the casting said it was done without knowledge of Miss Redgrave's political activism for causes inimical to the interests of Israel. The network argued that it could not accede to the demands of the Jewish groups that Miss Redgrave be removed from the cast without establishing a precedent that would allow all pressure groups to exert censorship on programming.

"PLEASE DON'T EAT THE DAISIES"—situation comedy on NBC (1965–66) based on the Jean Kerr best seller and movie about a happy family coping with a large, run-down house. Produced by MGM-TV, it featured Pat Crowley, Mark Miller, Kim Tyler, Brian Nash and Joe and Jeff Fithian.

PLIMPTON SPECIALS—series of ABC one-hour nonfiction specials (1970–73) featuring George Plimpton, a writer of derring-do, whose technique is to get on the inside of the activity. Thus, in *Plimpton! The Great Quarterback Sneak*, he posed as a professional football player for a month in the Baltimore Colts training camp. In *Plimpton! At the Wheel* he entered the world of auto racing. In others, he went on a safari, prepared to do a comedy act at Caesar's Palace, trained for the circus on the flying trapeze, rode in the steeplechase and became a guard at Buckingham Palace. William Kronick produced and directed the documentaries for the Wolper Organization.

PLUGOLA—the term for the commercial use (or abuse) of television by means other than advertising, through a prearrangement with producers. Once fairly common in television, the selling of product plugs in programs became illegal after the quiz show and payola scandals late in 1959, when Congress amended the Communications Act to require stations to indicate when money or other consideration was received for broadcast material.

Plugs continue to abound in giveaway shows, but always there is the disclaimer that the prizes are donated. In the years when plugola flourished, companies went openly into business to make contact with producers on behalf of manufacturers. A producer might be paid a set fee for agreeing to use a brand-name product in a scene or might be given the use of an automobile for making it the car driven by the hero in his show. At least one bowling manufacturer created a bowling series that was distributed free to stations, because it showed the company's name every time a ball was lifted or the automatic pin-spotter was lowered.

When cigarette advertising was banned from television, cigarette plugs became a problem. In some cases cigarette companies created sporting events irresistible to television that carried the brand name; in others they purchased billboards at ball parks where they would inevitably fall into camera range.

"PM EAST/PM WEST"—ambitious syndicated series produced by Westinghouse Broadcasting (1961) as competition to NBC's *Tonight*, with talk-variety segments produced on the East and West Coasts. Mike Wallace and Joyce Davidson (wife of David Susskind) were the co-hosts of *PM East*, which was produced in the studios of WNEW-TV New York, and Terrence O'Flaherty was host of *PM West*, produced at Westinghouse station KPIX San Francisco. The five Westinghouse-owned stations formed the nucleus of the lineup, but the program lasted around six months.

POCKETPIECE—the definitive bi-weekly network rating report, which earned its nickname from the fact that it was designed to fit in the inside coat pocket of the network salesmen. Officially the *Nielsen National TV Ratings Report*, its audience estimates are based on data from both the Nielsen SIA meters and diaries.

Information in the pocketpiece includes household audience estimates for all sponsored network programs, audience composition estimates by a variety of age groups, season-to-date averages for program performance, program-type averages, overall TV usage compared with the previous year and TV usage by time period. See also AA Ratings, Demographics, Rating.

"POLDARK"—swashbuckling drama produced by the BBC in two separate serials and presented on the PBS *Masterpiece Theater* series in 1977 and 1978. The romantic adventure story, based on four novels by Winston Graham, centers on the life of Ross Poldark, a British veteran of the American Revolutionary War. It was enormously popular in Britain and gained a stout following here among followers of public television.

The series was adapted for television by Jack Pulman, Paul Wheeler, Peter Draper and Jack Russell and was directed by Paul Annett. It was filmed in Cornwall. Robin Ellis portrayed Poldark, and the supporting cast included Jill Townsend, Clive Francis and Anghard Rees.

John Ericson & Christopher George in *Police Story*

"POLICE STORY"—NBC anthology series (1973–77) realistically portraying police work. Joseph Wambaugh, former policeman and author of several best-selling novels about police, created the concept and served also as production consultant. The Columbia TV series, which had no recurring stars, spun off *Police Woman* and *Joe Forrester*. David Gerber was executive producer and Stanley Kallis producer.

"POLICE SURGEON"—one of the first dramatic series produced expressly for the prime time–access rule in 1971 as a barter vehicle for Colgate-Palmolive. Filmed in Canada and featuring Sam Groom and Len Birman, under auspices of the Ted Bates ad agency, the series was intended to look like a network show in competition with lesser syndicated fare. It fooled no one and, although it lasted three seasons, was never a big success.

"POLICE TAPES, THE"—90-minute documentary by independent producers Alan and Susan Raymond portraying the urban battleground of the South Bronx, New York's highest crime area, as seen during a summer from a police squad car. The program had the distinction of being aired both on public and commercial television. It was originally broadcast on WNET New York, then was acquired by ABC News to air in the *ABC Close-Up* series on Aug. 17, 1978. The documentary won Peabody, DuPont and Emmy awards that year.

Police Woman Angie Dickinson with Earl Holliman

"POLICE WOMAN"—successful NBC hour-long series (1974–79) which grew out of an episode of the dramatic anthology, *Police Story*. Angie Dickinson, Earl Holliman, Ed Bernard and Charlie Dierkop comprised the regular cast. It was produced by Douglas Benton and was by David Gerber Productions and Columbia Pictures TV.

POLICY STATEMENT—the FCC's interpretation of a rule or law in a given area, or the articulation or the commis-

sion's prescription for broadcast performance that stops short of being framed as regulation. Policy statements have been issued on questions concerning UHF, children's programming and programming in general.

POLITICAL ADVERTISING RATES—by law, a station may never sell political time at higher rates than it charges commercial advertisers for comparable time. Moreover, during the 45-day period before a primary election and the 60-day period before a general election, political candidates cannot be charged more than a station's lowest rate in a given time period, provided that the candidate personally appears in the programs or spots. The lowest rate generally is the *end rate,* the heavily discounted price earned by commercial advertisers for frequency buys; political candidates, however, are not required to buy spots in volume to receive it. Broadcasters may, if they choose, offer special political discounts beyond the requirements of the law.

POLITICAL CONVENTIONS—long known as the "Olympics of television journalism" for the fact that the quadrennial events of the Democratic and Republican parties are the only predictable times when the network news divisions have the run of the airwaves in prime time and compete with each other head-on in covering the same running story. Journalistic careers often are advanced in these Olympics, and occasionally stars are born—Huntley & Brinkley, a 1956 phenomenon, the notable example. Moreover, the performances of the news divisions at these events may pay dividends in increasing the credibility, stature and popularity of a network's evening newscast when the conventions are over.

With prestige and company pride at stake, the networks invest huge amounts to cover the four-day events, despite the fact that only 30% of the television homes in the U.S. watch the conventions on a typical night. The cost of coverage is so great that the networks rarely recoup more than a third of their expenses from the advertising. Each constructs temporary but elaborate studios, booths and control rooms at the convention site and employs more than 500 news workers and clerical personnel in the coverage.

In 1976, CBS and NBC each spent around $10 million to cover the political parleys, ABC somewhat less. The two older networks have continued to adhere to the tradition of gavel-to-gavel coverage, but ABC, since 1968, has offered abbreviated coverage—a nightly digest usually running two or three hours.

Each network's principal cast consists of an anchorman or anchor team, one or more commentators and analysts and four floor reporters who, with minicam equipment, circulate among the delegates for interviews. Dozens of other reporters cover the hotels of the candidates and activities outside the convention hall.

Convention coverage by the networks began in 1952 when the manufacturers of TV sets hitched onto politics for their advertising thrusts. Philco Corp. that year spent $3.8 million for full sponsorship of the campaigns, the conventions and election night on NBC TV and radio. Admiral then bought similar coverage on ABC, TV and radio, and Westinghouse bought the coverage of CBS and DuMont.

POLITICAL EDITORIALIZING—broadcast activity for which FCC rules were adopted in 1967, concurrent with the rules relating to personal attacks. Before formulating the rules, the FCC had noted that between 1960 and 1964 the number of radio and TV stations that broadcast political editorials had increased from 53 and 2, respectively, to 103 and 13, and that there was some indication of failure to comply with the obligation to afford time for answer by disfavored candidates.

The 1967 rules state that when a licensee opposes legally qualified candidates in an editorial, the licensee has an obligation within 24 hours to transmit to the other candidates notification of the date and time of the editorial and script or tape of the broadcast and that it offer a reasonable opportunity to respond. The rule also notes that if an editorial is broadcast 72 hours before the election, the licensee shall comply with the provisions sufficiently far in advance of the broadcast to enable the other candidates to prepare and present a response.

It should be noted that this provision applies only to licensee endorsements and does not affect the licensee's obligation under the Equal Time provision of Section 315 (a) of the Communications Act. While not directly at issue in the Red Lion case, it was clearly implied that by upholding the constitutionality of the Personal Attack Doctrine, the court also implicitly passed on the constitutionality of the political advertising rules.

"POLITICAL OBITUARY OF RICHARD NIXON"—controversial telecast on ABC (Nov. 11, 1962) following Richard M. Nixon's gubernatorial defeat in California and his "final" press conference at which he said "You won't have Dick Nixon to kick around anymore." Conducted by Howard K. Smith, the telecast featured an interview with Alger Hiss, the former government official whose conviction for perjury in 1950 in a famous subversion probe by then Congressman Nixon, served to launch Nixon's national political career.

The reaction to the program was stormy. ABC stations as well as the network were besieged by protests, much of them apparently organized by right-wing groups, and Kemper Insurance retaliated by canceling its sponsorship of ABC's

Evening Report. James Hagerty, v.p. of news, went on the network Nov. 18 to defend ABC's right to present the news as it saw fit.

The Kemper action was challenged in the courts by ABC, and in 1968 the Supreme Court ruled that the company had no right to cancel sponsorship of its program because it disapproved of another program on the network. The court ordered Kemper to pay ABC $298,800 for sponsorship time it had not used.

POLITICAL SPENDING ON TELEVISION—a cost of campaigning that, for national offices collectively, has quadrupled in 20 years and now represents the major expense in running for office.

POMPADUR, I. MARTIN—former corporate executive and board member of ABC Inc. who appeared destined for the higher reaches of management when he resigned suddenly, late in 1976, apparently over differences with the president, Elton H. Rule. Pompadur then joined Ziff-Davis Publishing Co. as a corporate officer, and engineered its acquisition of Rust Craft Broadcasting. He eventually became president of Ziff-Davis.

Pompadur's legal and organizational skills contributed greatly to the revitalizing of the ABC network in the late 60s. Having been key aide to Rule while he was president of ABC-TV from 1968–72, Pompadur moved up in the company when Rule became president of the corporation, and in 1975 he was named corporate v.p. and assistant to the president. A lawyer, Pompadur spent two years in private practice in Connecticut before joining ABC in 1960.

"POP UP"—a series of one-minute educational spots for children designed to teach reading skills such as letter sounds, left-to-right and top-to-bottom eye movements. Beginning on NBC in 1971, the spots were interspersed with Saturday morning shows as part of the network's effort at public service. The short films were conceived by Dr. Caleb Gattegno, produced by Paul Klein and developed jointly with NBC.

PORTER, PAUL A. (d. 1975)—chairman of the FCC from 1944 to 1946 who then became a founding partner with Thurmond Arnold and Abe Fortas in what was to become one of the largest law firms in Washington—Fortas and Porter (now, Arnold and Porter). For a brief period in the late 30s, he served as Washington counsel for CBS, but his close ties to the Democratic party moved him into a number of high government posts with the Roosevelt and Truman Administrations. An imposing speaker famed for his anecdotes, he was a frequent toastmaster at Washington dinners. His second wife was Kathleen Winsor, author of the successful and spicy 1944 novel, *Forever Amber.*

POWELL, DICK (d. 1963)—screen actor who embraced TV in the early years when other movie stars eschewed it and who flourished in the medium as star, producer and executive. After playing roles in several drama anthologies, Powell joined with three other movie actors—Charles Boyer, Rosalind Russell and Joel McCrea—to form their own program in 1952, *Four Star Playhouse,* which was to involve a rotation. Miss Russell and McCrea withdrew after a time, and David Niven joined as a third partner (there was not to be a fourth again). The venture led to the creation of an independent production company, Four Star Studio, which began producing other shows. The company later passed into other hands but still is active as Four Star Entertainment.

Powell had no idle years in TV and went from one program series to another, with a scattering of guest shots, as well. He had a long run with *Dick Powell's Zane Grey Theatre,* a western anthology drawing from the Zane Grey stories, which Powell hosted and occasionally starred in. It ran on CBS from 1956 to 1961. When it ended, Powell began a new series, *The Dick Powell Show* (1961–63), an anthology of action-adventure films.

"PRACTICE, THE"—NBC situation comedy (1976–77) featuring Danny Thomas as an elderly New York City physician who refuses to join his son's (David Bedford) posh Park Ave. practice. The series was a January replacement that won a renewal for fall. Paul Junger Witt was executive producer, Tony Thomas supervising producer and Ron Rubin producer. The series was a Danny Thomas Production in association with MGM-TV.

PRECHT, ROBERT—director and then producer of *The Ed Sullivan Show* until it went off the air in 1970. He remained associated with Sullivan, his father-in-law, in a TV production company until Sullivan's death. Since then Precht has been executive producer of *The Entertainer of the Year Awards,* along with other specials, and of the 1973 situation comedy, *Calucci's Department.*

PREFREEZE STATIONS—pioneer TV stations, 108 in all, that were either already on the air or had been granted construction permits before Sept. 30, 1948, when the FCC halted all license awards. The prefreeze stations had the TV airways to themselves for four years while the commission developed an orderly table of assignments. See also Freeze on Licenses, Sixth Report & Order.

PREMIERE—short-lived pay-cable company formed in April 1980 by Getty Oil, Columbia Pictures, MCA Inc. (Universal Pictures), Paramount Pictures, and Twentieth Century-Fox Corp. Its creation shocked other cable entrepreneurs, and especially HBO, since as an alliance of major film studios Premiere would have had a competitive advantage over all other satellite program services for movie titles. A few months later, the Justice Department filed an antitrust suit against Premiere, arguing that the film companies partnered in the new company would control the production, distribution, and exhibition of motion pictures and that a pay-cable network formed of such companies would be anti-competitive.

As a result of the Justice Department action, the companies dissolved Premiere in June 1981. They more than recouped their start-up losses from selling the satellite transponder they had leased for Premiere at a huge profit. Premiere died before it could begin operation.

PRERELEASE—the practice by U.S. film studios and distributors of selling Canadian networks and stations the rights to televise programs in advance of their showing at home. This is done because, for most programs, Canada represents the most lucrative foreign market; yet its proximity to the U.S., coupled with the spread of cable-TV, gives most of the Canadian population daily access to the U.S. networks and independent stations. Thus, without the ability to offer the programs first, Canadian broadcasters would have no reason to buy American shows.

U.S. border stations and ABC, among the networks, have objected to the practice and have urged the FCC to protect them from the economic harm they claim it causes, but the distributors contend that the foreign business of the program production industry is outside the commission's authority.

PRESTEL—British Post Office videotext system which went online in 1979, thus becoming the first publicly available data base on home television sets (with, however, expensive adapters). Prestel did not enjoy the growth expected and soon found that its penetration of the home market was slim: the great majority of subscribers were business people. Besides the high cost of adapters, the system was limited by a rigid "tree" structure which made it difficult to use. Prestel was introduced quietly into the U.S. market in mid-1981.

Prestel is also the name of the British videotext standard. See Antiope, Telidon.

PRICE, FRANK—president of Columbia Pictures since 1978, after having been president of Universal TV since 1974. Earlier he was executive producer or producer of such series as *The Virginian, Ironside* and *It Takes a Thief.* He entered TV in the early 50s as a story editor for CBS and later for NBC while dramatic anthology series were in vogue. He then became a writer and eventually a producer of series programming.

"PRICE IS RIGHT, THE"—popular game show of the late 50s which was revived by Goodson-Todman in 1972, both as a Monday-Friday strip for CBS and in an evening version once a week for prime-access syndication. The original program began in daytime in 1956 and a year later won a nighttime berth. Bill Cullen had been emcee. For the revived daytime version, the emcee was Bob Barker and for the syndicated edition, Dennis James.

"PRIMAL MAN"—four-part series of specials on ABC (1973–74) dealing with the behavior patterns of prehistoric man to reveal what is atavistic in the human race. The series was produced by Jack Kaufman and directed by Dennis Azzarella for the Wolper Organization.

PRIMARY AFFILIATE—a station which has allied itself with one network for its main service, although it also carries programs of another. Distinctions between primary and secondary affiliations are chiefly made in the one- or two-station markets, where only one or two networks can have substantial exposure. The third network must rely on part-time, or secondary, affiliations to gain an outlet for its most popular shows.

PRIME TIME—the evening hours which, as the period of heaviest viewing, command the highest advertising rates. The FCC defines the prime hours as 7–11 P.M. (6–10 P.M. in the central and mountain time zones). On a typical evening during the main fall-winter season, there are likely to be around 85 million viewers at 8 P.M. close to 100 million at 9 P.M., and around 75 million at 10 P.M.

PRIME TIME-ACCESS RULE—an FCC rule which limited the networks' use of the peak viewing hours to three hours per night. Established in 1970 and put into force in the fall of 1971, it effectively shaved off 30 minutes of prime-time programming from the networks each night and returned it to the local stations in the top 50 markets for their own use. Although the rule permitted the network to use any 3-hour period between 7 and 11 P.M., the FCC infor-

mally requested that the networks all use 8 to 11 P.M. to minimize the confusion.

The rule was adopted for a number of reasons: to break the network monopoly over prime time, to open a new market for independent producers who complained of being at the mercy of three customers, to stimulate the creation of new program forms, and to give the stations the opportunity to do their most significant local programming in the choicest viewing hours.

Stations below the top 50 markets had no restrictions on their programming, but since the networks had ceased sending out programs at 7:30 P.M., most chose to use off-network reruns. Network affiliates in the major markets, however, had to use the time for first-run programming, either locally produced or from syndication. Virtually all of it came from syndication, as it proved, and the rule prompted a rebirth of the game/giveaway show in prime time, chiefly for economic reasons. Game shows could be taped five in a day, making them far cheaper than most filmed series, and they achieved acceptable ratings on the whole.

The rule also fostered a return of bartered programming, shows provided to the stations gratis in return for their carrying several commercials. It also increased the number of commercials, since stations are permitted to use more commercial minutes than networks under the television code. Moreover, around 30 per cent of the stations found it easiest to cope with the rule by *stripping* a game show, carrying it five nights a week instead of offering a different program each night.

Several others met the program gap by extending their local newscasts an additional 30 minutes each night.

On weekends, the most popular access rule shows were two of 60-minute length that had been canceled by the networks, *Lawrence Welk* and *Hee Haw.* Weekdays, the programs that succeeded were largely revivals of old game shows, such as *Truth or Consequences, To Tell the Truth* and *Beat the Clock,* or extra editions of such daytime programs as *Masquerade Party, The Hollywood Squares* and *Let's Make a Deal.*

Some of the prime-access programs were imported from abroad, and many of the new programs were made in other countries for economy reasons.

Two of the big access hits in the 80s were *The Muppet Show,* produced in England, and *PM Magazine,* a series by Group W Productions to which subscribing stations contributed locally-produced segments. Although, at first, most stations deplored the rule, fearing it would cut into profits, by the second year a majority wished for its retention because their profits were improved. The rule also benefited the networks, particularly third-ranked ABC, which was able to discard seven uncompetitive programs in returning time to the stations. This enabled ABC to move closer to its rivals in the rating race.

NBC and CBS gained from the reduction of the program overhead, since it costs as much to produce a program for 7:30 as for 9 o'clock, although the audience is much smaller in the earlier slot. Moreover, the reduction in commercials by 63 minutes a week among the three networks created, for them, a sellers' market which drove prices up for the rest of the schedule.

The rule was cumbersome and caused numerous problems. There were incessant requests for waivers, such as for sports runovers, Olympic coverage and news events, which added to the FCC's work load. Specials for children could not be presented by the networks until 8 P.M., and if they ran longer they interfered with bedtime schedules. Parents who did not understand the rule frequently complained to the networks.

In February 1974 the FCC revised the rule to allow the networks to present children's shows, documentaries and news at 7:30 and to prohibit the use of movies that had already been shown on the networks as prime-access fare. The revision was challenged in court by an ad hoc organization of syndicators, which argued that the agency had suddenly truncated their market after having encouraged them to invest in the development of shows for the fall of 1974. The District Court of Appeals in Washington upheld the syndicators and ruled that the FCC should have allowed more lead time for its revision. The court's ruling forced the networks to juggle the schedules that had already been planned under the FCC's amended rule; each had to drop two half-hour shows because their weekend schedules had shrunk again.

In November 1974 the FCC voted essentially to reinstate the rules that had been knocked down by the court, effective September 1975. The following January, the rules were put into official language. Thus, in the trade, the periods under the rule are classified as PTAR I, PTAR II (which never went into effect) and PTAR III.

To the dismay of much of the industry, the deregulation program of the 1982 FCC under Mark Fowler included the abolishing of PTAR. Stations and syndicators lobbied heavily to retain it.

The original rules grew out of a long-pending proposal by Donald McGannon, head of Westinghouse Broadcasting, which had an interest in syndication, to divide prime time between the networks and stations. This proposal was called the 50–50 rule. Limiting the networks to three hours was a compromise that had its detractors on the commission, notably the then chairman, Dean Burch.

PRIME TIME SUNDAY"—NBC's second venture into the weekly newsmagazine field, its first effort, *Weekend,* having attracted scant viewer interest during 1978–79, its first and only season as a prime-time entry. The failure of *Weekend* was a serious embarrassment to NBC News, since CBS was enjoying a huge success with *60 Minutes* and ABC News was having excellent results with its newsmagazine, *20/20.*

Prime Time Sunday, which debuted June 24, 1979, was different from *Weekend* in three conspicuous respects: It had a star TV personality, Tom Snyder, as host-anchorman; it concentrated on live television, incorporating elements of *Wide Wide World* and *Person to Person;* and it put the control room on the set, making the television process part of the show. The principal correspondents, Jack Perkins and Chris Wallace (son of Mike Wallace of *60 Minutes* and stepson of Bill Leonard, president of CBS News) each contributed a filmed news piece followed by a live segment in every telecast. Snyder himself conducted a live exploration of a current news story through interviews with various concerned persons, both in the studio and by satellite. The series was given a weekly berth in the NBC fall schedule for 1979–80 but was started during the summer to work out the kinks and, it was hoped, build a following.

Paul Friedman, who conceived the program and had worked with Snyder previously in live situations, such as the local New York newscast, *NewsCenter 4,* was executive producer. Wallace Westfeldt was senior producer and George Paul director.

Jessica Savitch joined the cast in October 1979, but the show continued to flounder in the ratings, and NBC began shunting it around in the schedule to minimize the ratings damage. The next season the series was scrapped in favor of a new version with David Brinkley entitled *NBC Magazine.*

See also *NBC Magazine.*

PRIMO, AL—news executive who is credited with originating the *Eyewitness News* concept, which spread from Philadelphia's KYW-TV to local TV stations across the country in the late 60s and early 70s.

Hired away from KYW-TV to develop the *Eyewitness* format for WABC-TV New York, Primo scored a success at the station that led to his being named v.p. of news for the ABC-owned stations in 1972. Three years later he shifted to ABC News as executive producer of *The Reasoner Report,* a half-hour weekly series. In 1976 he left ABC News to become a news consultant to local stations.

See also *Eyewitness News.*

"PRISONER, THE"—hour-long adventure series produced by Britain's ATV Ltd. which was carried as a summer replacement by CBS in 1968 and 1969. It featured Patrick McGoohan as prisoner of a mysterious group, struggling for his freedom while trying to learn why he was taken captive and by whom.

PRIVATE SCREENINGS—see Cable Networks.

PRIX ITALIA—one of the most prestigious of the international competitions for the recognition of excellence in radio and TV programs. Established in 1948 by RAI, the Italian broadcast system, "The Prix" is held annually in a major city of Italy, with more than 40 broadcast organizations of Europe, Asia and North America participating. Separate prizes equal to 13,000 Swiss francs are awarded for dramatic programs, documentaries, music programs and ballet, with additional special awards. The Prix is managed by a permanent staff based in Rome.

PRIX JEUNESSE (Youth Prize)—an international award given every two years by the Prix Jeunesse Foundation for excellence in children's television production. The selection is made at a biennial conference in Munich, held under auspices of the European Broadcasting Union and UNESCO, whose aims are to stimulate competition and improve communication and production standards in children's television. Participants are mainly from European and Canadian, Japanese and Australian television systems, with some attendance by representatives of less developed systems. The U.S. has been an occasional participant.

In 1964, the Free State of Bavaria, the city of Munich and the Bavarian Broadcasting Corporation established the Prix Jeunesse Foundation to promote good programming for the young.

"PRO BOWLERS TOUR"—ABC Saturday sports series since 1961, consisting of live coverage of professional bowling's big money tournaments for 16 weeks during the winter months. Each telecast represents the finals of a week's tournament, and each is conducted at one of the country's major bowling centers. The series concludes with the $125,000 Firestone Tournament of Champions, the richest event in the sport.

"PROFILES IN COURAGE"—series of documentary-dramas of men of courage in American history, many based on episodes in the Pulitzer Prize-winning book of that title by the late President John F. Kennedy. The series of 26 episodes produced by Robert Saudek Associates premiered on NBC in November 1964.

PSA—widely used abbreviation for public service announcement, a 20-, 30- or 60-second spot carried gratis by a station or network for its informational, moral or social assistance to the viewer, or to help promote a charity or a cause. Most PSAs are produced at the expense of the organizations presenting them—often at commercial production houses under the donated guidance of advertising agencies —and are distributed to stations for use at their discretion. Some are produced by the stations, however, as a service to groups unable to afford the expense of creating a spot.

PTAR—trade shorthand for the Prime Time–Access Rule. It is also represented as PTAR I, PTAR II and PTAR III, to denote the phases of the rule under the FCC's periodic modifications of it. See Prime Time–Access Rule.

PTL (People That Love)—see Cable Networks.

PUBLIC ACCESS—channels reserved for the exclusive use of the general public to produce or present uncensored program material on a first-come, first-served basis. The FCC required systems in the top 100 markets to provide such access channels under its 1972 rules, but the Supreme Court struck down the requirement in 1979. However, state authorities through legislation and municipalities through local cable franchises still have the power to require public-access channels and facilities.

PUBLIC BROADCASTING LABORATORY—a noncommercial production organization established by the Ford Foundation in 1967 to produce an innovative two-hour Sunday night news program, *PBL,* over the nation's ETV stations. Ford's grant of $10 million a year for the project provided for the interconnection of the stations (most of which had been inactive on Sunday nights at the time) as well as for the production. The program was conceived by Fred W. Friendly, former president of CBS News who had become broadcast consultant to the foundation, and he appointed his CBS protege, Av Westin, to head the project. Although it ran two years, *PBL* was not a success, but it did influence the commercial networks, which borrowed the format for *60 Minutes* and *First Tuesday* , which later became *Chronolog* and then *Weekend.* The Laboratory disbanded when the program ended its run.

PUBLIC HEARINGS—open hearings held by the FCC in the community of any station or stations. The general public is invited to attend, with interested parties acting as witnesses either for or against the licensee. The commission's authority to hold such hearings is stated in Section 403 of the Communications Act.

PUBLIC TELEVISION —noncommercial television, whose freedom from the constraints of the marketplace in theory permits it to strive to realize the full humanistic and social potential of the medium. It generally is looked to for a menu of cultural, informational, educational and experimental programming.

In the U.S., public TV is supported by federal and state funds, voluntary contributions from viewers and grants from foundations and corporations. In most other countries, the funds come from an annual tax on television receivers, sometimes supplemented by a limited sale of advertising. American public TV has been weaker than most of its foreign counterparts chiefly because it came into being after commercial TV had become firmly rooted as the primary system, the reverse of the pattern in most other countries. Also, PTV's penetration has been hampered by a predominance of UHF stations and its program vistas clouded during the first two decades by the uncertainty and inadequacy of federal funding. Finally, it has suffered from a history of discord within the system itself over the mission of public television.

The disharmony in the system traces to the fact that public television began as Educational Television in 1952, with each station independent and autonomous, serving, in one way or another, the purposes of education. Moreover, the public stations are not all of a kind but of four distinct types. Of the 264 stations that make up PBS, 99 are licensed to state authorities, commissions or boards of education; 73 are licensed to colleges and universities; 19 to municipal boards of education, school districts or agencies serving elementary and secondary education; and 73 to nonprofit civic corporations, most of them in the largest cities.

When the system was redefined, and redesignated as public television by the Public Television Act of 1967, the new label did not mean the same thing to all operators, notwithstanding the ideals for noncommercial TV put forth in the Carnegie Commission Report that year. Many stations continued to adhere to the original educational mandate, and others were opposed to surrendering any of their sovereignty to a national system.

Regional and political differences among the stations also contributed to the creation of factions. Many of the larger stations advocated a centralizing of the system to strengthen the national impact of PTV and make it a stronger force in American television, while others considered it heresy for PTV to aspire to reach larger audiences. There was also a particular aversion in middle America to public affairs programs emanating from the East Coast.

But despite the internal struggles that made public television a name without a concept, the noncommercial system has steadily been broadening its audience base and has exerted a positive influence on the commercial system. The networks have borrowed from PTV such program forms

as the newsmagazine , the miniseries and the serialized novel (*The Forsyte Saga*).

By 1973 public television had more than 1 million public subscribers voluntarily contributing $15 or more a year, testifying to its having developed a significant constituency.

Although its audiences have always been small by commercial television standards, PTV experienced a marked improvement in ratings during the early 70s with such well-produced native series as *The Adams Chronicles, Great Performances, Nova, Sesame Street, The Electric Company, Hollywood Television Theater* and *Live from Lincoln Center* and particularly with a raft of British imports such as *Upstairs, Downstairs, Civilisation, The Ascent of Man, Jennie* and *Monty Python's Flying Circus*. In 1975 a National Geographic special on the human body, *The Incredible Machine,* achieved higher ratings in some cities than the commercial network programs in competition. And in 1973 the PBS coverage of the Senate Watergate hearings proved to be one of the most potent programs ever for fund-raising.

Two developments in the mid-70s suggested that public television might at last become a unified system in the U.S. that would assert itself as a second force to commercial TV. One was the appointment of Lawrence K. Grossman, a former NBC executive, as president of PBS, who demonstrated immediately that he would provide aggressive leadership toward increasing the audience for PTV.

The other, in 1975, was the first authorization by Congress of five-year funding for public television, providing a total of $452 million from 1976 to 1980. The annual appropriations, graduating from $78.5 million the first year to $160 million the fifth, are made on a matching basis: one federal dollar for every $2.50 raised from non-federal sources. Thus, the government gave the system the incentive to compete for audience, since PTV will continually need greater public and corporate contributions to qualify for the full amount of the authorizations.

However, just as public TV was beginning to show strength as an alternative to commercial television, the Reagan administration—in its slashing of social programs—cut public television's budget severely for several years forward. Some stations were convinced that they could only survive through the sale of advertising, and the government okayed experiment with commercials at 10 PTV stations during 1982.

History. During the FCC's four-year freeze on station licenses (1948–52) a movement began among educators for channels that would be noncommercial and dedicated to education. Their cause had a champion on the commission in Frieda B. Hennock. When the freeze was lifted and the table of assignments issued in the FCC's historic Sixth Report and Order in 1952, there were 242 channels reserved for noncommercial TV—80 on VHF and 162 on UHF. Those initial allocations were later increased to 116 VHF and 516

UHF. In May 1953 KUHT Houston became the country's first educational TV licensee.

The Ford Foundation, which had been giving financial assistance to the organizations campaigning for educational television, made a profusion of grants to help the initial stations build their facilities and stay on the air. It also created the Educational Television and Radio Center in Ann Arbor, Mich., as a central agency to secure and distribute programs for the emerging system. The Center later moved to New York and became National Educational Television , supported chiefly by the Ford Foundation.

By 1962 there were 62 ETV licensees, although funds to build the stations and create the programs were a major problem. Then Congress enacted the Educational Broadcasting Facilities Act of 1962, which authorized HEW to disburse $32 million in matching grants over five years to assist in the construction of ETV facilities. In four years, the number of stations on the air more than doubled.

After facilities came the need for long-range funding for the programming, with some built-in mechanism to prevent the Government, as the prospective source of the money, from assuming any measure of control over the broadcast material. The Carnegie Corporation in 1965 established a 15-member commission, headed by Dr. James R. Killian of MIT, to study the ETV problem and make recommendations for its future development. The commission's work culminated in the publication of its report on Jan. 25, 1967, *Public Television: A Program for Action*. The document was to be regarded thereafter as Scripture for noncommercial TV. It was the Carnegie Commission that changed ETV to PTV, broadening its scope to include "all that is of human interest and public importance." The report emphasized local service and recommended that the system not strive to become the fourth network.

The Carnegie Commission's recommendations for the future support and development of public television formed the basis for the Public Broadcasting Act of 1967, which provided for the creation of the Corporation for Public Broadcasting to lead the system, distribute the federal funds for programming and serve as the insulation between the Government and the broadcasters. It was also charged with the responsibility for interconnecting the stations.

After negotiating an interconnection contract with AT&T, CPB then created the Public Broadcasting Service to manage the interconnection and guide the flow of programs over the national lines. PBS was chartered in November 1969 and began its transmissions as the central programming authority in October 1970.

A second Carnegie Commission to study the future of public broadcasting was created by the Carnegie Corporation in 1977 with an endorsement by President Carter (and recommended, among other things, greater Federal funding of the system and a reorganization of it in ways to provide more national programming and better insulation from political interference.

See also Carnegie Commission I and II, Corporation for Public Broadcasting, Ford Foundation in Public Television, Station Program Cooperative, Instructional Television, Public Broadcasting System.

Q

"QB VII"—3½ hour TV adaptation of the Leon Uris novel of that title, presented by ABC as a miniseries in April 1974. The success of both *QB VII* and the adaptation of Joseph Wambaugh's *The Blue Knight* the previous year spurred the networks into the development of short-term series based on novels. Produced by the Douglas S. Cramer Co. and Screen Gems, *QB VII* starred Ben Gazzara, Anthony Hopkins, Leslie Caron, Lee Remick and Jack Hawkins. Tom Gries directed, Edward Anhalt wrote the script and Cramer was producer.

Ben Gazzara & Lee Remick in *QB VII*

QUADRUPLEX (also 'QUAD')—type of videotape recorder most commonly used for 20 years at television stations and networks, so-called because it has four recording heads. The heads are arranged in a drum that spins at right angles to the tape's motion. The obsolescence of quad recorders was signaled in 1977 with the introduction of 1-inch VTRs capable of similar broadcast quality. See also Videotape Recording.

QUBE—trade name for a dazzlingly futuristic two-way cable system established experimentally in Columbus, Ohio, in December 1977 by Warner Communications Inc., the entertainment conglomerate. Under the design of the system, one cable line carries TV signals to the customer while the upstream line permits subscribers to send back responses to a central computer.

The system became reorganized as state-of-the-art technology during the franchising activities of the 80s, and it—or something comparable—was demanded by every large city. Qube helped Warner Amex cable win the franchises for Pittsburgh, Dallas, Houston, Cincinnati, and parts of New York.

Subscribers in Columbus receive a hand-held console resembling a large pocket calculator on which to make program selections from 30 channels. A row of ten buttons is dedicated to over-the-air television stations and the public-access channel. A second row of ten is for premium programming—movies, cultural events, games, entertainments, special sports events and self-help courses, each for a specific charge. This group includes an optional "adult" channel with soft-core pornography. Rates for the premium programs range from seventy-five cents to $9. The third row of ten channels, called the community channels, provides a varied menu of free programming (free, that is, after the $10.95 monthly charge for the basic service), including full-time channels for children's, religious, cultural and sports fare and a nostalgia channel offering old TV series. One channel in this group, provides the programs that utilize Qube's unique ability to ask questions of the home viewer and receive answers from the response pad which it can tabulate and flash on the screen in a matter of moments.

The system lends itself to public-opinion polls and voting on performances in amateur shows, but it also is adaptable to ordering merchandise and for multiple-choice examinations in video college courses. The systems built in Dallas, Pittsburgh and the other major cities are more elaborate than the Columbus installation, involving more than three times the number of channels and a more advanced tuning and response device.

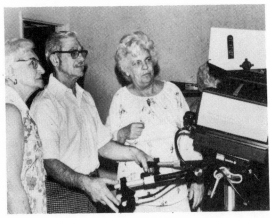

QUBE: Do-it-yourself tv

As an interactive system, Qube is also to be used for a variety of home security services, such as burglar- and fire-alarm protection and medical emergency alarms, each for additional monthly fees. Thus, even as the viewer watches television, the television set watches the home. Warner reportedly invested $20 million in the Columbus experiment. Attracted by Qube and its implications for credit cards, American Express bought a 50% interest in Warner Cable for $175 million in 1980. The company was then renamed Warner Amex Cable.

"QUEEN FOR A DAY"—an immensely popular daytime show which, between radio and TV, had a run of nearly 20 years (April 29, 1945 to Oct. 2, 1964), although widely criticized as a vulgar exploitation of human misery in an orgy of commercial plugs. It shifted from NBC radio to television on Jan. 1, 1955, and within months became the No. 1 daytime show, reaching a daily audience of around 13 million. When it was at the peak of its popularity, NBC increased the show's length from 30 to 45 minutes, in order to gain additional commercial time to sell at the then premium advertising rate of $4,000 per minute. In the fall of 1959 *Queen* moved to ABC where it spent its final five years.

Five contestants selected from the audience each day competed for the title of "queen" (and for the accompanying raft of prizes) by telling, usually through tears, why they wished for a particular item of merchandise. Always behind the need was a personal story of pathos. The studio audience then was asked to vote, by applause measured by an applause-meter, for the contestant most deserving to be "queen for a day."

Crowned, and decked in a sable-collared velvet robe, the weeping wretch, supported in her emotional state by emcee Jack Bailey, received her gifts from a procession of models as the announcer delivered the plugs for the donated merchandise. The show inspired two successful imitations, *Strike It Rich!* and *It Could Be You.*

Howard Blake, a sometime producer of the show, wrote in an article published in 1966: "Sure, *Queen* was vulgar and sleazy and filled with bathos and bad taste. That was why it was so successful: It was exactly what the general public wanted."

QUELLO, JAMES H.—FCC commissioner since 1974, appointed to a Democratic seat by President Nixon, despite objections by groups in the broadcast reform movement. A chief objection to Quello's appointment was that he had been a broadcaster with the Capital Cities group. He had risen from promotion manager of radio station WJR Detroit in 1947 to v.p. and general manager of that station in 1960. He retired from Capital Cities in 1972, after having simultaneously been a member of the House Housing and Urban Renewal Commission. He was reappointed by President Reagan in 1981.

"¿QUE PASA, USA?"—bilingual situation comedy on PBS (1977–) developed under a regional HEW grant for Cuban-American teenagers in South Florida. The crisply written series, centering on three generations of a Cuban family in Miami, broke new ground in programming for teenagers. Its production values were almost on a par with those of commercial network sitcoms, yet it was produced for approximately $25,000 an episode—about one-tenth the cost of a network show.

The program was produced by WPBT Miami, with Shep Morgan as executive producer and Jose Bahamonde as producer.

"QUEST, THE"—hour-long western series introduced by NBC in 1976 in an unsuccessful attempt to revive interest in the genre; it involved the adventures of two brothers in search of their sister, who was captured by Indians. Kurt Russell and Tim Matheson played the leads. The series was by David Gerber Productions in association with Columbia Pictures TV.

"QUINCY"—successful one-hour NBC drama series (1976—) featuring Jack Klugman as expert medical examiner for the Los Angeles Coroner's Office with a talent for solving cases through his ability to spot miniscule medical clues. With the series, Klugman overcame strong viewer identification with his previous TV role as the unkempt principal in *The Odd Couple*. The regular supporting cast includes Robert Ito, Val Bisoglio, Joseph Roman, Garry Walberg and John S. Ragin.

Via Glen A. Larson Productions, the series was created by Larson and Lou Shaw and produced by Peter Thompson.

QUIZ SHOW SCANDAL—a trauma in the television industry which occurred in the fall of 1959 with the revelation that producers of several highly popular giveaway shows had "fixed" them by providing certain contestants with answers in advance. The episode, which ruptured the public's and the Government's trust in the integrity of commercial broadcasting, was to have lasting implications on network policies, industry codes and the character of television programming.

Months of rumors, fanned by several magazine articles and fueled by accusations of malpractice by a former contestant on *Twenty One,* Herbert Stempel, led to an investigation early in 1959 by a New York grand jury. The probe of quiz show "rigging" was pursued further in the House of Representatives that fall by the Special Subcommittee on Legislative Oversight, headed by Rep. Oren Harris. On Nov. 2 came the confession from the key witness, Charles Van Doren, who had previously denied having any knowledge of cheating on *Twenty One,* the program on which he became a national celebrity after defeating Stempel. An English instructor at Columbia University, Van Doren became an NBC personality, regularly featured on *Today* with a salary of $1,000 a week, as a result of his popularity on the quiz show, on which over a period of months he had won $129,000.

In his confession, he told of being persuaded by the producer, Albert Freeman, to accept help in the form of a briefing on questions and answers before each program. He said the request was made because Stempel was an unpopular champion, whose weekly successes were hurting the show. He said he was told that quiz shows were only entertainment and that giving help to quiz contestants was a common practice.

The subcommittee was then to learn from other witnesses that two other popular giveaway series, *The $64,000 Question* and *The $64,000 Challenge,* received periodic instructions from the sponsor, Revlon, to eliminate uninteresting contestants or to let attractive ones continue.

Top network executives denied knowing that the programs were manipulated and none was implicated, but quiz shows disappeared from the airways almost at once, and several of the producers were never to work again in the medium. Jack Barry, whose Barry & Enright Productions owned *Twenty One, Tic Tac Dough* and other hit quizzes, was exiled from television for 10 years but made his way back in 1970 as a producer of daytime gameshows for the networks.

The scandal led to widespread self-examination in the industry and to the adoption of policies against even milder forms of deceit. The networks were prompted to assume greater control over their programming, trusting less to the sponsor.

Coupled with the concurrent payola revelations in radio (involving disk jockey payoffs, likened to commercial bribery), the episode led to amendments to the Communications Act in 1960. One amendment made illegal the presentation of programs purporting to be contests of skill or knowledge where the result is in any way prearranged, another made more explicit a station's obligation to make known on the air when money or other consideration is received for broadcast material.

The passing of the quiz programs contributed to the spread of filmed series produced in Hollywood and changed the nature of prime-time competition among the networks. Executives who were adept at live television and the showmanship involved gave way to those astute at selecting and scheduling continuing film series. Thus, Louis Cowan was deposed as president of CBS-TV and replaced by James Aubrey, who was hired away from ABC-TV.

R

RADIO TELEFIS EIREANN—the state-controlled radio-TV service in the Republic of Ireland, which operates a single TV channel on 625-line PAL, with about 650,000 sets reported in the country. The system is supported by license fees and advertising. The fees amount to $27.36 a year for black & white sets and $76 for color. The Dail, or Irish parliament, has been considering a proposal that would permit nationwide reception of the BBC's first TV service as a competitive medium.

RAI (RADIOTELEVISIONE ITALIANA)—Italy's only network, state-operated with a mandate to serve the country's educational, social and cultural progress. Funded partly by the licensing of TV sets, it is also partially commercial, with a few blocks of time designated exclusively for advertising. Five per cent of network time is for public access, made available to various groups and organizations.

Until 1975 RAI was under control of the executive branch of government; a new law that year placed it under the control of Parliament, specifically a Parliamentary commission of 20 members of each house.

During the 70s RAI became active in the exporting of program series—*Leonardo da Vinci* a notable one—and in major co-productions with British and U.S. companies, e.g., *Moses: The Lawgiver* and *The Life of Jesus*.

See also *Italy, Television in*.

"RAMAR OF THE JUNGLE"—syndicated adventure series (1952–54) featuring Jon Hall and Ray Montgomery, and produced by Arrow Productions and TPA.

RANKIN-BASS PRODUCTIONS (Arthur Rankin, Jr., and Jules Bass)—independent New York-based company specializing in animation and noted for a number of holiday specials which play perennially on the networks. These include *Rudolph, the Red-Nosed Reindeer, Frosty, the Snowman* and *'Twas the Night Before Christmas*, all on CBS; *The Little Drummer Boy, Santa Claus Is Comin' to Town*, and *The First Christmas* on NBC; and *A Year Without Santa Claus* and *Rudolph's Shiny New Year* on ABC. Some have been on the air 10 years or more. The firm was acquired by Tomorrow Entertainment in the early 70s, but Rankin and Bass resumed operations on their own when that company disbanded. They also turned out such Saturday morning children's series as *The Jackson Five, The Osmond Brothers* and *Kid Power*.

RANSOHOFF, MARTIN—founder and chairman of Filmways, highly successful independent production company which had started in 1952 as a producer of commercials and industrial films. On branching into TV series, the company had a string of hits, among them *Mr. Ed, The Beverly Hillbillies, Petticoat Junction*, and *The Addams Family*. It also did well in motion pictures (*The Americanization of Emily, Boys' Night Out*).

RAPF, MATTHEW—producer of numerous TV series, including *The Loretta Young Show* in the 50s, *Ben Casey* in the 60s and *Kojak* in the 70s. His credits include *The Great Gildersleeve, Frontier, The Web, Jefferson Drum, Two Faces West, The Man from Blackhawk, Slattery's People* and *The Young Lawyers*. In the 70s, he has been executive producer of *Switch, Doctor's Hospital* and *Kojak*.

"RAT PATROL"—ABC series (1966—67) of fictional adventures in the North African desert during World War II, featuring Christopher George, Gary Raymond and Lawrence Casey. It was produced by Mirisch-Rich TV Productions, with Jon Epstein as producer.

Dan Rather

RATHER, DAN—anchorman of the *CBS Evening News* since March 1981, the successor to Walter Cronkite as the network's premier newsman. Rather's contract negotiations in 1980 set off a chain of events that led to Cronkite relinquishing the anchor post somewhat earlier than planned and Roger Mudd leaving the network for NBC News after having been passed over. With Rather at the helm, the *Evening News* lost audience share points the first year but began recovering them the second.

Before being elevated to anchor, Rather was a CBS News correspondent who distinguished himself first in covering the assassination of President John F. Kennedy in 1963, while he was chief of the southwest bureau in Dallas, and later in his coverage of the White House and the Watergate developments during the Nixon Administration. To millions of viewers, Rather came to symbolize the adversary press in his numerous bold news confrontations with the President.

After Watergate, Rather was replaced at the White House and given the new assignment of anchorman-corre-

spondent for *CBS Reports* and anchorman of the *CBS Weekend News*. At the end of 1975, however, when *60 Minutes* was moved into prime time as a 52-week series, Rather was selected as one of the three editors. In 1977, he doubled as a reporter on the program's spin-off, *Who's Who*.

Rather joined CBS News in 1962 in the Dallas bureau and became White House correspondent in 1964. A year later he went to London as bureau chief, and returned to his Washington post in 1966, remaining until 1974. Among the CBS special reports on which he served as anchorman or reporter were *The White House Tapes: The President's Decision, The Watergate Indictments, The Mysterious Alert* and *Nixon: A Full, Free and Absolute Pardon*.

RATING—established unit of audience measurement in TV, carried over from radio, which represents the percentage of households tuned to a given program in a time period from the universe of households equipped to receive television. For a national or network program, the universe is the total number of TV households in the U.S., while for local ratings it is the number of TV households in the station's coverage area, which normally encompasses several counties. Thus, a national rating of 20 means that 20% of all possible TV homes in the country were tuned in, while a local rating of 20 means that 20% of the households with TV receivers were watching in the geographic area to which the station is licensed.

Following is an explanation of how to read the information found on the pages of the Nielsen report. Two pages from a pocketpiece have been included in the appendices, as has a table of the top 50 programs which attracted the largest viewing audiences from the beginning of television through Jan. 30, 1982.

How to Read the Nielsen NTI Report: A page from the Nielsen pocketpiece (actually two facing pages) carries national TV audience estimates for a night of the schedule over two separate weeks. The date at the top of the page notes the first week and that at the bottom the second.

At the top, the boxed strip marked TIME indicates the quarter-hour time periods for which the data are reported. The strip at the bottom shows, by quarter hours, the HUT (Households Using TV) levels for both weeks, that is, the percentage of all U.S. households using television.

Below that strip is a footnote for the number of U.S. households equipped to receive television: 71.2 million, at the time of this report.

The networks are sectioned off alphabetically for Weeks 1 and 2. To the right of them are the program titles with arrowed lines to delineate the time span of each show. Titles that are boxed are specials.

Above the program titles is the line for Total Audience. The top figure (add 000) indicates the number of households that watched 6 minutes or more of the program during its full length; the figure directly below tells what those households represent as a percentage of all U.S. households.

The first column below the program titles carries four pieces of information apropos of Average Audience: (1) the number of households (add 000) tuned to the program in the average minute; (2) the percentage of all U.S. TV households that the figure represents; (3) the share of audience during the average minute of the program, expressed as a percentage of all U.S. TV households using television at the time; (4) the average audience rating by the quarter hour expressed as a percentage of all possible U.S. TV homes.

The figure generally cited as the program's rating, and the one by which programs are ranked for popularity, is the one represented above as (2): the percentage of all U.S. households tuned to the program in the average minute.

The AA rating for the full program appears in the first column. The figure on the same line in the second column represents the AA rating for the first half hour; that in the fourth column for the second half hour. The number that appears in the first column is the average of the other two.

To illustrate: On Sunday, Feb. 13, 1977, CBS broadcast *60 Minutes* from 7 to 8 P.M. It scored a TA rating of 29.6, indicating that 29.6% of all U.S. TV households viewed at least 6 minutes of the program. That TA rating represents 21,080,000 TV households. During the average minute of *60 Minutes*, 23.9% of all U.S. TV households were tuned to the program. The AA rating of 23.9—which is the critical rating—represents 17,020,000 TV households.

That rating for the entire program is the average of 23.3 for the first half hour and 24.5 for the second.

Among programs that were on at the same time, *60 Minutes* received a 38% share of the audience. That is the average of a 38 share for the first half hour and a 37 share for the second (it is customary to round off the decimals and report the share in whole numbers).

From 7 to 7:15 P.M., *60 Minutes* had an AA rating of 22.3 and from 7:15 to 7:30 an AA rating of 24.2. The average for the half hour, 23.3, is shown at the top of the second column.

The ratings for the first three quarter hours reveal a steady increase in audience and then a leveling off in the fourth, a healthier situation than one in which the audience steadily falls away.

Nevertheless, while the rating improved, the share dropped a point in the second half hour. This indicates that the rise in viewership at 7:30 tended to benefit the competing shows somewhat more.

60 Minutes won its time period handily. It should be noted that the shares for the three network programs add up to only 95%. The remaining 5% of the national audience is presumed to have tuned to independent stations and PBS during that hour.

Not included are programs of less than 5 minutes' duration, programs that carried no advertising, and telecasts that were carried jointly by all three networks, such as the first walk on the moon, Presidential speeches and coverage of President Kennedy's assassination. See also AA Rating, Arbitron, Pocketpiece, Share, Appendices.

"RAWHIDE"—action-filled western on CBS which enjoyed a long run (1959–65). Produced by CBS, its leads were Erick Fleming, Clint Eastwood, Paul Brinegar and Steve Raines.

RAYBURN, GENE—comedian, announcer and television host who was a member of the resident cast of the Steve Allen *Tonight* show on NBC from 1954–59, and then host of several game shows. Those included *Dough-Re-Mi* on NBC (1958–60), *The Match Game* on NBC (1962–69) and two revivals of *Match Game* on CBS, in 1973 and 1978. He came into prominence with a daily radio program on New York's WNEW, *The Rayburn and Finch Show,* a four-hour stint of ad-lib comedy interspersed with records, which ran six years. On splitting with his partner, Dee Finch, in 1953, Rayburn began doing local television in New York until he was tapped for the original *Tonight* show.

"RAY MILLAND SHOW, THE" ("MEET MR. McNUTLEY")—moderately successful comedy series on CBS (1953–55) which served as a vehicle for the movie star. Milland portrayed a professor of dramatics at an all-girls college. On CBS, the series was entitled *Meet Mr. McNutley,* but it became better known by its syndication title, *The Ray Milland Show.* It was by Revue Studios.

Martha Raye

RAYE, MARTHA—raucous, slapstick comedienne prominent in TV during the 50s, rounding out a career that spanned vaudeville, radio, movies and nightclubs. She began on television in 1951 in *All Star Revue* and then launched *The Martha Raye Show* on a monthly basis (1954–56). She was also a frequent guest on *The Steve Allen Show, The Colgate Comedy Hour, The Milton Berle Show* and others. She returned to television in the 80s with a semi-regular role in the CBS sitcom, *Alice,* playing Vic Tayback's mother.

RCA SATCOM—RCA's 24-channel domestic satellite, launched in December 1975 and followed in March 1976 by Satcom 2. RCA Americom was manufacturer as well as operator of the craft.

Satcom 1 became the primary satellite for cable-TV services, while Satcom 2 was used for commercial television and other communications. In 1978, demand for transponder space on Satcom 1 became so intense that RCA moved up the scheduled launch date for Satcom 3 to the fall of 1979. When launched, however, the satellite failed to go into orbit and became lost in space, creating a crisis for several cable networks that had booked transponders on the new satellite. A new satellite was successfully launched as Satcom 3 early in 1982.

See also Cable-Television; HBO; Satellites, Communications.

REAGAN, RONALD—actor-turned-politician whose twoscore movies and TV reruns *(Death Valley Days* and *GE Theater)* were barred from the air under the equal time rule during the periods of his various candidacies. He became Governor of California (1966–74)—barely missed unseating Gerald Ford as Republican nominee for the presidency in 1976, and won the nomination and the Presidency in 1980.

A radio announcer and sportscaster in the Midwest during the 30s, using the name "Dutch" Reagan, he broke into films with Warner Bros. in 1937 and played leads in numerous second-line Hollywood movies for the next two decades. For three years he was host-narrator of the syndicated *Death Valley Days* and then host, actor and production supervisor for the CBS anthology, *GE Theater.* He was president of Screen Actors Guild from 1947 to 1952 and again in 1959, and it was under his leadership that the union achieved residual payment for actors from the studios as well as a pension and welfare fund.

"REAL McCOYS, THE"—popular countrified situation comedy starring Walter Brennan which originated on ABC (1957–62) and switched to CBS (1963). Featured were Richard Crenna and Kathy Nolan. It was via Brennan-Westgate Productions and VT&L Productions.

"REAL WEST, THE"—distinguished NBC documentary (March 1961) presenting through still photographs and paintings an unglamorized and historically accurate account of the western movement in the U.S.. It exploded many hero-myths and revealed the cruel effects the movement had on the American Indian. The film was narrated by Gary Cooper, the noted actor who had appeared in numerous Hollywood westerns and who was dying of cancer at the time; Cooper asked to do the narration "to set the record straight." The documentary was created by the NBC Project 20 unit: Donald Hyatt, producer; Philip Reisman, Jr., writer; Daniel Jones, researcher.

"REALIDADES"—magazine-format PTV show by and about the American Latino populations. Largely funded by CPB, it was produced in the mid-70s by WNET New York at an annual cost of $400,000.

Harry Reasoner

REASONER, HARRY— principal anchorman for ABC News (1970–78) and since then correspondent for CBS News, resuming his previous assignment there as a regular in *60 Minutes.* Before ABC hired him away in December 1970 to bolster its lagging newscast, Reasoner had acquired a measure of national popularity over 14 years as a CBS correspondent and newscaster, and was thought to be heir-apparent to Walter Cronkite.

At ABC, he was teamed for several years with Howard K. Smith until, in 1975, he became the sole anchor. His disagreements with ABC began a year later when, in another effort to bolster the news ratings, the network spirited away Barbara Walters from NBC and gave Reasoner a new and unwanted partner. The chemistry between them was poor both on and off the screen, and when the newscast underwent a total restyling by ABC News president Roone

Arledge in the spring of 1978, Reasoner did not figure in the plan. CBS welcomed him back.

From 1973 to 1975, in addition to anchoring the evening news, he contributed a regular 30 minute news-analysis series, *The Reasoner Report*. Earlier, he hosted a series of TV essays for ABC, *Who Do You Think You Are?*

In his first hitch with CBS, where he was regarded the likely successor to Walter Cronkite as anchorman, Reasoner hosted the audience-participation test broadcasts (*The National Driving Test*, etc.), hosted or co-hosted such news series as *Dear CBS, Portrait, One of a Kind* and *Calendar* and delivered the notable light-hearted essays written by Andrew Rooney, such as *Essay on Doors* and *The Strange Case of the English Language*. He was also for a time White House correspondent (1965–66) and anchorman of the CBS *Sunday News*. In addition, he had major assignments in election campaigns and was a reporter on numerous news specials. In 1968, when *60 Minutes* began, Reasoner and Mike Wallace were the anchormen and editors.

Reasoner began his news career with the *Minneapolis Times,* where for several years he was drama critic. In 1950 he became a newswriter for WCCO Minneapolis and, after a stint with USIA, returned to that city as news director of KEYD-TV. In 1956 he joined CBS News in New York. Reasoner switched to ABC when it was clear that Cronkite would go on indefinitely and that his own career was stalled at back-up anchorman.

"REBEL, THE"—post-Civil War adventure series on ABC (1959–60) featuring Nick Adams and produced by Goodson-Todman in association with Celestial Productions and Fen-Ker-Ada Productions. NBC picked up reruns as a summer replacement in 1962.

"RED CHANNELS"—a paperback book, published in June 1950, which served to destroy, interrupt or retard numerous careers in radio and TV. It was issued at a time when hysteria was mounting over whether communists and communist-sympathizers were working in media and lending themselves to propaganda uses. The book listed the names of performers, writers, composers and producers—with brief dossiers on each—who were alleged to be friendly to communist causes or dupes of the Red conspiracy.

The listings served immediately as a basis for blacklisting: Advertisers, networks and program packagers backed away from the names that would create controversy and bring pressure upon them. *Red Channels* also listed organizations cited as subversive or as communist fronts by the House Un-American Activities Committee.

Subtitled "The Report of Communist Influence in Radio and Television," the book carried no author credit beyond that it was published by *Counterattack,* a newsletter dedicated to exposing the communist influence in American corporations. Both publications were products of American Business Consultants, an organization of communist-hunters whose principals were former FBI agents, John G. Keenan, Kenneth Bierly and Theodore Kirkpatrick. The unsigned introduction was by Vincent W. Hartnett, a one-time production assistant on *Gangbusters* who as a solo crusader kept files on subversive suspects in show business, some of which material was undoubtedly used in *Red Channels*. Hartnett later formed his own organization, Aware Inc., with a Syracuse, N.Y. owner of supermarkets, Lawrence Johnson. It too was to become effective in blacklisting. See also Faulk, John Henry; *Goldbergs, The*.

RED LION DECISION [Red Lion Broadcasting Co. v. FCC/359 U.S. 367 (1969)]—the Supreme Court opinion which upheld the constitutionality of the Fairness Doctrine. The clear implication of the Court's decision was that broadcasters would be held to a different First Amendment standard than newspapers because of the scarcity of broadcast frequencies and because the Government has the right to license these frequencies. In addition, the Court articulated a fiduciary First Amendment obligation on the licensee to present views and voices representative of his community. Red Lion carried the implication of a First Amendment right of access by pointing out that the right of viewers and listeners was paramount, and not the right of broadcasters.

The case derived its name from WGCB in Red Lion, Pa., a fundamentalist radio station which broadcast mainly conservative, anti-communist opinion. In November 1964 WGCB carried a syndicated program, *The Christian Crusade,* which featured an attack by the Reverend Billy James Hargis on a book by journalist Fred J. Cook, entitled, *Goldwater: Extremist of the Right*. Hargis alleged that Cook was fired from the *New York World-Telegram* after making false charges, that he worked for a left-wing publication, *The Nation,* and that he had written articles absolving Alger Hiss and attacking J. Edgar Hoover, the FBI and the CIA.

Upon learning of the attack, Cook asked WGCB for an opportunity to reply to Hargis under the personal attack feature of the Fairness Doctrine, and he was denied free reply time.

The personal attack rules were adopted officially by FCC in 1967. They state, essentially, that when an attack is made upon the honesty, character, integrity or other personal qualities of an identified person or group during the presentation of news on a controversial issue of public importance, the person or group attacked must be notified within a week of the broadcast and offered a reasonable opportunity to respond over the licensee's facilities.

When WGCB refused time to Cook, he took the matter to the FCC, which ruled that the station was required to give him the air time. WGCB then appealed to the D.C. Court of

Appeals claiming that the rules violated the First Amendment rights of the broadcaster. This court held that the Fairness Doctrine and the personal attack rules were constitutional. The station then took the matter to the Supreme Court.

At about the same time, the FCC adopted new rules detailing not only the personal attack doctrine but also the provisions for political advertising. RTNDA (Radio-Television News Directors Assn.) appealed these rules to the Seventh Circuit Court of Appeals in Chicago, which eventually held that the rules violated the First Amendment. In early 1969 the Supreme Court consolidated the two cases.

To the shock of the broadcast industry, a unanimous Supreme Court (seven judges voting) affirmed the Appeals Court's Red Lion decision and reversed the RTNDA decision. The court said that the personal attack rules and the Fairness Doctrine were consistent with the First Amendment. See also Fairness Doctrine.

REDMONT, BERNARD—veteran European correspondent who, after 11 years with Group W (1965–76) as Paris bureau chief, joined CBS News as manager of its Moscow bureau. A native of New York, Redmont had lived in Paris since 1946 working for various English-language news services and news magazines before joining Group W.

REID, CHARLOTTE T.—FCC commissioner (1971–76) appointed to a Republican seat by President Nixon after she had completed four terms as a congresswoman from Illinois. She was the first female on the commission in more than two decades and was appointed when the women's movement was in full sail, but she was not an activist commissioner and was a champion of no particular cause. Moreover, she was cited by the press for a poor attendance record. She resigned from the FCC two years before the end of her full term, for marriage. A professional singer early in her career, she performed under the name of Annette King on *Don McNeill's Breakfast Club* on NBC (1936–39).

REINER, CARL—comedian, writer and producer who made his mark in TV as a regular performer on *Your Show of Shows,* with Sid Caesar and Imogene Coca, in the 50s, and then launched a new career as producer, writer and director of the original *Dick Van Dyke Show* on CBS in the early 60s. He returned in a similar capacity for the revival of the Van Dyke situation comedy in the early 70s but resigned when CBS applied a "family" standard to the show's content and would not permit sophisticated subject matter. Specifically at issue was a censored scene in which a child opened the door to his parents' bedroom while they were apparently making love.

Carl Reiner demonstrates his toupe

Reiner, who also made films and recordings *(The 2000 Year Old Man),* returned to TV in 1976 as a comedy actor in *Good Heavens,* a sitcom of brief duration in which he portrayed an angel; he was also executive producer of that ABC series. He also contributed to TV comedy a son, Rob Reiner, who portrayed the son-in-law in *All In the Family.*

REINER, MANNY (d. 1974)—TV film executive prominent in international sales. His last post was executive v.p. of international sales for Paramount TV. Earlier he had been president of Four Star Television and Filmways Intl., foreign manager of Samuel Goldwyn Productions and managing director of the Selznick Organization in Latin America and Australia.

REINSCH, J. LEONARD—long-time president of Cox Broadcasting Co., Atlanta-based station group, who for many years was also active in Democratic politics. He retired as Cox president at the close of 1973 (succeeded by Clifford M. Kirtland, Jr.) but continued as a member of the board and as chairman of Cox Cable Communications Inc., a subsidiary. In 1978, he served as a member of the Carnegie Commission on the Future of Public Broadcasting.

During the Truman Administration, he became radio advisor to the White House and later TV-radio consultant to the Democratic National Committee. He was executive director of the Democratic national conventions of 1960 and 1964 and arrangements director of the 1968 conventions. Reinsch also was TV-radio director for John F. Kennedy's presidential campaign in 1960. He began his broadcast career with WLS, Chicago, in 1924 and then went on to help build the Cox group.

REITH, JOHN C.W. (LORD) (d. 1968)—first director-general of the British Broadcasting Corp. and the individual who has probably made the greatest impact on broadcasting in the United Kingdom. His paternalistic influence is felt to this day.

Reith regarded broadcasting not simply as a medium of entertainment but as a force that should help to shape the nation's values and aspirations. He believed that radio, and later television, should primarily inform and uplift. A Scotsman and stern Calvinist, he stressed the need for high standards and a strong sense of responsibility at the BBC, and it was under his long reign that the British concept of public service broadcasting developed. He was made a peer of the realm after World War II.

RELIGIOUS TELEVISION—a program area that has taken two forms: paid access, in which evangelists and church-affiliated organizations purchase air time for their broadcasts, and public service, in which networks and stations donate time and production assistance to the major faiths. Both forms involve spot announcements as well as continuing program series. In the main, unpaid religious programs are consigned to marginal time periods, usually Sunday mornings, when viewing levels are low and, ironically, when much of the audience is attending church.

Religious television took a significant turn in the early 70s with the outcropping of Christian TV stations on the UHF band in various cities in the country. With the advent of satellite distribution in the late 70s the station trend led to the creation of three evangelical networks—each of them achieving a more or less national spread through cable-TV. The largest of these full-time services is the Christian Broadcasting Network (CBN), built upon Pat Robertson's evangelical talk show, *The 700 Club*. The others are the PTL Network (said to stand for Praise The Lord and also People That Love) and the Trinity Network.

By 1980, religious evangelists paying for air time on local stations drove out virtually all the long-established programs produced for the networks or syndication by the mainline churches. Many of these television preachers made fantastic amounts of money, not from the sale of advertising but from the solicitation of contributions from viewers. Some of them, on becoming national figures, campaigned for political causes. Jerry Falwell, for example, who was host of *The Oldtime Gospel Hour*, also founded the Moral Majority, an organization with a distinct political point of view that became a leading force in the rise of the New Right. To offset the trend, Norman Lear, a political liberal, created a national organization in 1981, People for the American Way, to fight the repressive influence of TV's religious fundamentalists.

Meanwhile, the television pulpit was producing a raft of new stars, among them Jim Bakker of the PTL Network, who previously had worked for Pat Robertson. One minister

of the mainline Protestant churches, Robert Schuller, grew popular by serving psychological as well as spiritual needs (and steering clear of politics) in his syndicated series, *Hour of Power*.

While ABC, CBS, and NBC maintain religious programming departments operating under the wing of the network news divisions, they produce a relatively small amount of the religious programming in television. The vast majority of programs available nationally or regionally are syndicated by religious groups, some with production budgets of more than $1 million a year. A number of programs derive the income for their continuance from on-air solicitations for viewer contributions.

Approximately one-third of the more than 100 religious syndicators place their programs by purchasing air time, while a number of others use both free and paid time. Largely, fundamentalist groups engage in paid broadcasts. Surveys have found that about two-thirds of the TV stations in the U.S.—most of them outside the major markets—will accept paid religious broadcasts. The leading carriers in the larger markets are independent stations, those not affiliated with a network. Policies of the networks and their owned stations do not permit the use of "paid religion."

Network policies for religious broadcasting were forged during the radio era and were carried over, with some modifications, into television. An early NBC policy was not to sell time for religious broadcasting because "such a course might result in a disproportionate representation of those individual groups who chance to command the largest purse." The basic policy guidelines of all three networks are that they assume the entire cost of production, that they work in program development with "recognized" or "central" religious agencies and that they discourage the treatment of controversial subjects.

The major groups with which the networks regularly work are the Broadcasting and Film Commission, National Council of Churches; Department of Film and Broadcasting, Office of Communications, U.S. Catholic Conference; Jewish Theological Seminary in America; New York Board of Rabbis; and the Southern Baptist Convention. Others involved in program development with less frequency are The Greek Orthodox Church in North and South America; the Christian Science Church; the Church of Jesus Christ of Latter Day Saints; Lutheran Church, Missouri Synod; and the American Council of Churches.

Religious broadcasters turned to television early in the medium's development. In 1948 the Protestant Radio Commission founded *Look Up and Live* and *Lamp Unto My Feet* on CBS-TV. *Frontiers of Faith* began as a church service remote on NBC in 1952 and then switched to studio-oriented formats. That same year, *The Eternal Light,* the widely praised half-hour dramatic radio series produced by the Jewish Theological Seminary, was adapted to television as part of NBC's three-faith series.

This Is the Life, a film series by the Lutheran Church, Missouri Synod, began its syndication rounds in 1952 and by 1955 had an annual production budget of $750,000. ABC in 1954 began a studio program with Episcopal Bishop James Pike. The veteran radio preacher Norman Vincent Peale was featured with his wife in an NBC program, *What's Your Trouble?* And on the DuMont Network, Bishop Fulton Sheen began what was to become the most successful of the religious shows, often scoring higher ratings than commercial entertainment programs appearing opposite it in prime time.

As the television audience expanded, the evangelical broadcasters began to use the medium increasingly, purchasing better time periods than other religious broadcasters received gratis. The Rev. Billy Graham, Oral Roberts, Rex Humbard and the Reverend Ike all made highly sophisticated use of the medium with opulently produced prime-time specials featuring name performers. By the 70s so many paying religious programs were available to broadcasters that the general manager of a TV station in the east claimed to have pulled his station out of the red by accepting "paid religion" almost indiscriminately.

Most stations produce some local religious programming —if only the closing prayer—or give over time to religious groups in their communities, but almost all rely heavily on syndicated programming to fulfill their obligations to religious service as licensees. Among the leading regularly syndicated programs, paid or unpaid, are *Cathedral of Tomorrow, This Is the Life, Hour of Power, Revival Fires, Oral Roberts, Herald of Truth, Day of Discovery, Insight, Faith For Today, Christopher Close-Up, Davey and Goliath, Sacred Heart, The Answer,* and *I Believe In Miracles.*

REPLY COMMENTS—opposing statements or arguments filed by individuals or organizations, on invitation by the FCC, in response to the petitions or filings of others. Reply comments are solicited by the FCC in the interest of examining the various sides of an issue.

REPORT ON EDITORIALIZING BY BROADCAST LICENSEE [13 FCC 1246 (1949)]—first formal articulation of the Fairness Doctrine by the FCC, which also upheld the right of licensees to editorialize on their own airwaves. That right had been in question since 1940, when the commission in the Mayflower decision declared that the broadcaster could not be an advocate. In the Report on Editorializing, which established that there was no prohibition on taking positions, the commission imposed on the licensees an obligation to present all sides of opinion in the discussion of public issues. That was the concept that evolved into the Fairness Doctrine.

Basic to the commission's position was its view that broadcasters operate their facilities as a public trust under the public interest standard of the Communications Act and that the public interest cannot be served by a licensee who did not provide a medium of free speech.

The commission thus articulated the two-step formula of the Fairness Doctrine—the requirement to devote a reasonable amount of time to the presentation of controversial issues of public importance and the obligation to provide time for the expression of contrasting attitudes and viewpoints on those issues. The FCC recognized, however, that the licensee should have general discretion in determining the issues to be covered, the shades of opinion to be presented, the appropriate spokesmen and the amount of time to be offered.

"REPORTER, THE"—hour-long series on CBS (1964) starring Harry Guardino as a columnist for a New York paper who becomes involved in the stories he covers. Created by novelist Jerome Weidman, the series was produced by Richelieu Productions in association with CBS and lasted 13 weeks. Others in the cast were Gary Merrill, George O'Hanlon and Remo Pisani.

REPS (Station Representatives)—firms soliciting national business from the key advertising centers for their roster of station clients, thus serving as an extended sales force. Reps receive commissions of up to 15%, but there is no standard rate policy and some accept lower commissions to represent a family of stations.

The rep's principal function is to sell the station's time, from data and brochures furnished him by the station's sales manager, to advertisers making spot purchases in television. But because many firms represent stations throughout the country and have a broad view of the industry, they also provide consultative services to their client stations, recommending syndicated programs, personnel, operating procedures and promotional campaigns.

Among the scores of television rep firms are Edward Petry & Co. (which originated radio station representation in 1932), Blair Television, The Katz Agency, Avery-Knodel Television, HR Television, and Peters, Griffin & Woodward Inc.. The television networks maintain their own reps—or spot sales organizations—for their owned stations, as do such other broadcast groups as Metromedia, Group W, Storer and RKO.

RERUNS—programs repeated some time after their original presentation; in the plural, the reference is usually to entire series.

Networks began using reruns because the producers of filmed shows could not physically operate on continuous 52-week schedules and needed, besides, the supplementary income from repeats. Meanwhile, the reruns served the networks as economical programming during the summer months when viewing levels declined, since the cost of the repeats was only 25% of the firstruns. Programs came to be purchased on a pattern of 39 firstruns and 13 repeats.

But in the mid-60s the networks dealt with rising production costs by extending the rerun period. When it was perceived that second runs could compete effectively and that most viewers preferred a repeat of a favorite show to other available programming, the networks began to contract for 26 firstruns and 26 reruns. By 1970 the firstrun order was down to 22 episodes, and the time periods were filled out with specials and short-term summer replacements.

Program series that amassed a sufficient library of episodes over four or five years, each having been shown at least twice on the network, were then sold to individual stations in syndication, where they were again replayed, numerous times, in strip form.

Taped shows did not initially lend themselves to the rerun practice because the scale for residuals was prohibitive. Tape was under the jurisdiction of AFTRA, while film contracts for performers were covered by SAG. In time, adjustments were made so that the residual payments required by each union were similar.

Animated cartoons scheduled on Saturday mornings are mostly reruns, since each episode is contracted for six exposures over two years.

Blaming reruns for exacerbating the Hollywood unemployment crisis in the 70s, representatives of several unions petitioned the FCC to restrict network indulgence in repeats. The FCC, after studying the matter, denied the petition.

RESIDUALS—fees paid to performers and other creative talent for subsequent exposures of their filmed or taped programs and commercials. Under the residuals formula devised by the unions, performers receive 75% of their original compensation for the first and second replays of their work, 50% for the third through fifth replay, 10% for the sixth and 5% for all additional exposures. In addition, there are residuals for foreign use of the materials.

Residual provisions began to appear in motion picture contracts after 1960. Thus performers are compensated for most films made after that year when they are sold to TV but not for movies made before 1960.

"RESTLESS GUN, THE"—western series on NBC (1957–58) starring John Payne and produced by Window Productions and Revue.

RETRANSMISSION CONSENT—controversial and widely debated proposal relating to cable-TV's carriage of broadcast signals. The proposal, advanced in 1979 by both NTIA and the House Communications Subcommittee's bill to rewrite the Communications Act, would require cable systems to gain the permission of the TV station or the copyright owners of individual programs when they bring in outside signals. The proposal was vigorously supported by the motion picture, TV and sports industries in the belief that cable should bargain for the use of their programming; it was vehemently opposed by cable interests, who contend that they meet their copyright obligations by paying blanket fees to the Copyright Tribunal under an agreement forged after the passage of the new Copyright Act.

REVOCATION—an FCC action terminating a licensee's broadcasting privilege. Section 312 of the Communications Act of 1934 grants the commission the power to revoke any station license or construction permit for the following reasons:

• Knowingly making false statements to the FCC, particularly in the license renewal application;

• Repeatedly and willfully failing to operate substantially as promised in the license application;

• Willfully violating, or repeatedly failing to observe, provisions of the Communications Act or rules adopted by the FCC;

• Violating critical sections of the United States Code;

• Failing to observe any cease and desist order issued by the commission.

REWRITE OF THE COMMUNICATIONS ACT—initiatives taken by the House and Senate in 1979 to revise the Communications Act of 1934 so that it might be more relevant to the emerging technologies of the late 20th century. The original Act was deemed inadequate to deal with such breakthroughs as cable, pay-TV, microwave and satellite transmissions and laser-based fiber optics.

The prime mover behind the rewrite in Congress was Rep. Lionel Van Deerlin (D-Calif.). As chairman of the House Communications Subcommittee, Van Deerlin was the principal architect of Bill H.R. 3333 in 1979, a modified version of the "floor to attic" revision which he unsuccessfully introduced in 1978. In keeping with the "deregulation" mood in Congress and the general public, H.R. 3333 proposed to overhaul the FCC and diminish its regulatory powers, phase out fairness and equal time restrictions on broadcasters and extend license terms indefinitely. In addition, it would lift almost all regulatory barriers on cable-TV and establish a spectrum fee for all broadcast and nonbroadcast users. More moderate bills were introduced by Sen. Ernest Hollings (D-S.C.), chairman of the Senate Communications Subcommittee, and by Barry Goldwater (R-Ariz.), the subcommittee's ranking member.

The Van Deerlin Bill and its counterparts, in varying degrees, attempted to create an open communications market where the existing and new electronic media could compete relatively free of government regulation. Proponents of the rewrite bills argued that deregulation would result in improved technology, lower operating costs and better service to the public. The chief justification for regulation in the communications field had been that the airwaves constituted a scarce public resource, but the major premise behind the rewrite bills was that the new technologies promised an abundance of communications channels. Specific provisions in the bills tried to negotiate a trade-off between the entrenched commercial broadcasters and the entrepreneurs in the new technologies. In exchange for more broadcasting freedom and security of licensed operations, stations would be required to pay regular fees for operating rights and to compete on a more extensive basis with the cable and pay-television industries.

Although praised in some quarters as a long-overdue attempt at streamlining communications legislation, the House rewrite bill met with stiff resistance from the industries concerned and from public interest groups. Not surprisingly, television industry leaders opposed the imposition of fees and the loosening of restrictions on the cable industry. The cable industry objected to allowing telephone companies to own cable systems. And the media reform groups strongly objected to the absence of any provision requiring broadcasters to serve "the public interest, convenience and necessity." This phrase represented the keystone of communications law since the Radio Act of 1929, and media reformers protested that its omission put the public "at the mercy of unregulated monopolies." Mixed with an apparent indifference toward the rewrite bills in Congress and the public at large, these objections caused the broadcast portion of Rep. Van Deerlin's bill to be scuttled in July 1979.

Van Deerlin lost his bid for reelection in 1980, and ironically a Republican dominated FCC in the Reagan Administration made policies along the lines of his rewrite bill.

"REX HUMBARD MINISTRY"—one of the longest-running "paid religion" shows, dating to 1952 and produced weekly, featuring TV evangelist Humbard and members of his family in an hour format of music and sermons. The series is dubbed in eight languages and carried in 28 countries. Humbard pays each station for carrying his program

and finances his broadcast operations on contributions from viewers.

Frank Reynolds

REYNOLDS, FRANK— twice anchorman of the ABC evening newscast, first from 1968–70 and again in 1978 with the restyling of ABC's evening newscast, *World News Tonight.* Although it began as a three-anchor format, Reynolds emerged as the principal newscaster.

AS ABC News special correspondent, his assignments have included all major political conventions and campaigns since 1965, coverage of the U.S. manned spaceflight program and commentary and analysis of presidential speeches and press conferences. From May 1968 until December 1970 he was co-anchorman with Howard K. Smith of the *ABC Evening News,* losing that post to Harry Reasoner when he was brought from CBS. He joined the network news division from ABC's Chicago station WBKB (now WLS-TV), where for two years he had anchored two newscasts daily. For 12 years prior to that he was a newsman with WBBM-TV, the CBS station in Chicago.

REYNOLDS, GENE—co-producer with Larry Gelbart of *M*A*S*H* and *Roll Out.* Reynolds became executive producer of *M*A*S*H* in 1976. Later he became executive producer of another CBS hit, *Lou Grant.* Reynolds was also exec producer of *Room 222* and *Anna and the King.* He began in show business as a child actor in movies.

REYNOLDS, JOHN T.—president of CBS-TV from Feb. 9, 1966, to Dec. 15 of that year, said to have resigned from a preference to reside in Southern California, where he had spent most of his career. He immediately became president of the TV division of Golden West Broadcasters and general manager of its Los Angeles station, KTLA, an independent.

During his brief presidency at the network, Reynolds authorized the return of *Playhouse 90* in the form of two or three specials a year. Having been for many years in charge of administration of the network's Hollywood operations, and therefore close to the programming function, he had been selected for the post upon John A. Schneider's promotion to head of the CBS Broadcast Group. For 12 years prior to joining CBS in 1959, Reynolds had been a salesman for various stations and an executive with the Don Lee broadcast group.

Nancy Walker & Valerie Harper

"RHODA"—hit CBS situation comedy (1974–78) built upon a character spun off of *The Mary Tyler Moore Show,* the luckless New York oddball Rhoda Morgenstern, played by Valerie Harper. Watched over by the executive producers and creators of *Mary Tyler Moore,* James L. Brooks and Allan Burns, the series was a success from the start. Helping its popularity was a 60-minute special episode several weeks after the premiere, entitled *Rhoda's Wedding,* which scored a healthy rating. The 1976 season brought the new wrinkle of Rhoda's divorce, along with the loss of a key supporting player, Nancy Walker (in the role of Rhoda's mother), who began a sitcom of her own on ABC. When that series failed, Miss Walker returned to *Rhoda* in the fall of 1977.

Others in support were Julie Kavner, David Groh, Harold Gould and Lorenzo Music (unseen, as the voice of Carlton, the doorman). Music was also co-producer with David Davis. The series was by MTM Enterprises.

RIBBON—horizontal crawl superimposed on the TV picture, at the bottom of the screen, for news bulletins and special announcements.

RICH, JOHN—director of *All In the Family* during its first four seasons, after which he went into independent production. His first effort was *On the Rocks* (1975–76), an ABC situation comedy drawn from a British show, for which Rich was producer-director. He has been director and one of four executive producers on *Benson* since 1980.

RICH, LEE—head of Lorimar Productions, an independent that hit it big with *The Waltons* and went on to produce *Doc Elliott, Apple's Way, The Blue Knight* and *Dallas* series and a raft of TV movies and dramatic specials, including *Helter Skelter* and *Sybil.* Rich was an alumnus of the program department of Benton & Bowles Advertising, which during the 60s was an incubator of program executives for the networks.

Before organizing Lorimar, he was a partner in Mirisch-Rich Productions where he produced *Rat Patrol, Hey Landlord* and two Saturday daytime cartoons, *Super Six* and *Super President.*

Partnered in Lorimar with Merv Adelson since 1969, Rich was responsible for such shows as *Eight Is Enough, Dallas, Kaz* and *Studs Lonigan. Dallas* yielded him a raft of spinoffs, including, *Knot's Landing, Falcon's Crest, King's Crossing,* and *Flamingo Road.*

"RICH MAN, POOR MAN"—12-hour serialized adaptation of a novel by Irwin Shaw televised by ABC in February and March of 1976; the miniseries was such a smash in the ratings that it was resumed in the fall as a continuing prime-time serial under the title, *Rich Man, Poor Man—Book II.* Since one of the brothers dies at the end of the Shaw novel, the continuation invented new stories around the surviving brother, Rudy Jordache (played in both versions by Peter Strauss). Both the miniseries and the series were by Universal TV.

Along with Strauss, the original featured Susan Blakely, Nick Nolte, Edward Asner, Fionnuala Flanagan and numerous familiar TV actors in cameo roles. David Greene was director and Alex North composed the score.

With the serialization leading the ratings most of the weeks it was on the air, all three networks were primed for a plunge into short-term series dramatizing popular novels.

"RICHARD BOONE SHOW, THE"—an attempt in 1963 to establish a dramatic repertory company for a weekly anthology series headed by Richard Boone. NBC carried 25 episodes but the project failed. It was produced by Classic Films and Goodson-Todman in association with NBC.

"RICHARD DIAMOND, PRIVATE DETECTIVE"—30-minute action series (1957—60) which played first on CBS and then on NBC, with David Janssen as a debonair detective. A gimmick character in the series was a telephone girl, named Sam, who was seen on camera only as a pair of legs. The role was played by Mary Tyler Moore. The series was by Four Star.

RICHARDS CASE [KMPC (Richards) 14 Fed Reg. 4831 (1949)]—early fairness case in which the FCC ruled a misuse of signal against George A. Richards, owner of radio stations in Detroit, Cleveland and Los Angeles. Richards was accused of slanting and distorting the news about President and Mrs. Roosevelt, implying that they were communists; of promoting the candidacy of Gen. Douglas MacArthur for President; and of operating his stations in a manner that totally precluded presentation of views other than his own.

The FCC began an investigation of Richards, and after Richards died in 1951 the commission renewed the three licenses upon a written promise that the stations' deceptive policies would cease. The stations were eventually sold to other parties.

RICHELIEU PRODUCTIONS—independent TV company headed by former actor Keefe Brasselle who sold three programs to CBS in 1964, the failure of which were believed partly responsible for the downfall of then CBS-TV president, James T. Aubrey. The company took its name from the restaurant in New York, Chez Richelieu, where Brasselle and Aubrey had frequently dined. Its ill-fated programs had been *The Cara Williams Show, The Baileys of Balboa* and *The Reporter.*

RICHMAN, STELLA—freelance British producer and former program executive who dealt frequently in the U.S. Her credits include *Jennie: Lady Randolph Churchill* (1974) for Thames TV and *Clayhanger* (1975) for ATV.

RIFKIN, MONROE M.—prominent figure in the cable industry as a founder of American Television and Communications Corp. (ATC) and its president since the company's formation in 1968. In 1974, he added the title of chairman. Rifkin built it into the second largest of the cable MSOs (behind Teleprompter) and remained its chief executive officer when ATC was acquired by Time Inc. in 1978, since then it has become the largest MSO. He was also named a v.p. of Time. Rifkin, a certified public accountant, had been president of Daniels Management, a brokerage and consulting firm devoted to the cable-TV industry, prior to the formation of ATC.

He resigned in 1982 to form a partnership with oilman Marvin Davis, who purchased Twentieth Century-Fox. The new company, Rifkin-Fox Communications, was created to explore the opportunities in the cable industry.

Rifkin was succeeded as chairman and chief executive of ATC by Trygne E. Myhren. At the same time, Joseph J. Collins assumed the post of president.

"RIFLEMAN, THE"—popular western on ABC (1957–62) which featured Chuck Connors, Johnny Crawford and Paul Fix and was produced by Four Star-Sussex Productions.

RINTELS, DAVID W.—screen and TV writer whose TV credits include *Fear on Trial, Clarence Darrow* and *Gideon's Trumpet.* He is also president of Writers Guild of America-West.

RIP-AND-READ NEWSCAST—news roundups that are built entirely on wire service copy, ripped from the teletype machines and read whole or in edited-down form. Such newscasts were prevalent in television until the late 50s, when viewer interest in local news became manifest and when it became clear that an enterprising news operation was vital to a station's competitive standing in its market. Only the smallest and least affluent TV stations practice rip-and-read news today, although it is still common among radio stations.

RIPPON, ANGELA—first female newscaster on BBC-TV, joining the British network's 9 P.M. newscast in the spring of 1976, replacing one of three males. She had first been a newspaper journalist and gained her TV experience on BBC regional programs.

RITCHIE, DANIEL L.—chairman and chief executive officer of Westinghouse Broadcasting since 1981 when Donald H. McGannon retired. Ritchie had been president and CEO since early 1979. With expertise in finance and acquisitions, as a former finance executive for MCA Inc. (1960–70) and protege of MCA chairman Jules Stein, Ritchie was a surprise choice to succeed McGannon since his broadcast experience was scant. But it was clear at once that he represented the new generation of management that was to expand Group W, the Westinghouse broadcast company, and chart its course in the changing communications environment of the 80s. Soon after he was in the post, Ritchie purchased Group W's first UHF station, WRET-TV, Charlotte, N.C., for a reported $20 million. He began looking also into the expansion prospects afforded by over-the-air pay-television, cable and FM.

After two years with Lehman Bros., as a securities analyst, ten years with MCA and four years in his own health foods business, Ritchie joined Westinghouse in 1974 as head of its Learning and Leisure Time division. This put him in charge of a miscellany of businesses, ranging from the direct-mail marketing of the Longines Symphonette to Econocar auto rentals. Later he moved up to the parent company, Westinghouse Electric Corp. in Pittsburgh, as executive v.p. In 1978 he returned to Westinghouse Broadcasting as president of the corporate staff and strategic planning. A few months later, he was president and chief operating officer of the company and not long after president and chief executive officer.

Geraldo Rivera

RIVERA, GERALDO—storefront lawyer-turned-journalist known for muckraking documentaries and ombudsman-like reporting for the poorer classes. He built his reputation with WABC-TV New York as a member of its *Eyewitness News* team, starting in 1970, and gradually gained network assignments. In addition to making regular contributions to ABC's *Good Morning, America* he became, in 1978, special correspondent for the network's weekly newsmagazine, *20/20*. For a time, he did late-night specials for the network, *Geraldo Rivera: Good Night, America.* He achieved wide recognition for his local documentary, *Willowbrook: The Last Disgrace,* exposing the horrendous conditions at a state institution for retarded children, in the early 70s and made other films on children born to drug-addicted mothers (*The Littlest Junkie*) and northern migrant workers (*Migrants: Dirt Cheap*).

"RIVERBOAT"—action series set on the Mississippi River in the 1840s which starred Darren McGavin and lasted for 44 episodes on NBC, premiering in 1959.

RKO GENERAL—station group owned by General Tire and Rubber Co., formed in 1950 by the purchase of all stations of the Don Lee Broadcasting System for $12.3 million. In 1952 the group took the name of General Teleradio and changed it in 1955 to General Teleradio Pictures after purchasing RKO Pictures Corp. for $25 million. In 1959, when General Tire bought out the 10% interest in the company held by R.H. Macy & Co., the group was renamed RKO General.

During the 70s license challenges were directed at several of the RKO stations. In 1976 General Tire proposed a spin off of the broadcasting subsidiaries because illegal actions by the parent company had put all of the licenses, some $400 million worth, in jeopardy. Shareholders of General Tire were to be apportioned shares in the new separated company, but the FCC rejected the proposal.

The group's TV stations were WOR-TV New York and KHJ-TV Los Angeles, both independents; WNAC-TV Boston, a CBS affiliate; and WHBQ-TV Memphis, an ABC affiliate.

In 1978, WNAC-TV was to be sold for $54 million to a local group, New England Television Corp., which was made up of the two companies that had filed competing applications for the station's license in the spring of 1969. Eight of the 51 shareholders were blacks, who together would own 13% of the company—a significant representation in a time when minority ownership of stations held a high priority at the FCC. The FCC disallowed the sale, however, because of license challenges at WNAC-TV and other RKO stations. The issue before the FCC was whether RKO General had the character qualifications to hold the Boston license. That suggested it might not be qualified to hold any of its other TV and radio licenses, estimated to be worth in total around half a billion dollars. In 1982, RKO lost the license to WNAC-TV and its court appeals were unavailing. The license was lost chiefly on misrepresentation to the FCC, which left the other RKO stations unaffected.

ROADBLOCKING—the technique of force-feeding the audience by arranging for a commercial or a program to play simultaneously on all three networks, and sometimes also on independent stations. The object is to gain maximum exposure for a message or an address.

Since it has been found that most viewers will watch television no matter what is on, the principle of roadblocking is to reduce or eliminate the other program choices on the dial. The concept was adopted for the presidential campaigns of 1972, when television strategists for both political parties bought the same half hour of time on every station in a market to ensure that the presentations reached the uncommitted voter. In large markets like New York, they frequently monopolized only five of the commercial stations, leaving the viewers (children, mainly) the alternative programming on one of the independents. On those occasions, the unbought independent scored extraordinary ratings.

When presidential speeches are televised live by all three networks and PBS they, in effect, roadblock the medium and capture audiences of 70 million or more. Reply time given to the opposition party rarely is in the form of a simultaneous broadcast; carried separately by each of the networks, the televised replies may attract less than half the audience for the President. Equal time thus proves not to be the same as equal audience.

"ROADS TO FREEDOM, THE"—TV adaptation of fiction by Jean-Paul Sartre, produced by the BBC on video tape. The 13 episodes were offered in syndication by Time-Life Films in 1973 and played here on several public TV stations.

ROBBIE, SEYMOUR—veteran director whose credits range from *Omnibus, Studio One* and *Play of the Week* to *Kojak, Barnaby Jones* and *Cannon*. He also directed episodes for *Wonderful World of Disney, That Girl* and *Love, American Style,* among other prime-time series.

ROBERT WOLD COMPANY, THE—independent company specializing in the temporary interconnection of stations and the distribution logistics for special networks. Wold, a former West Coast ad man, set up the independent networks for such telecasts as the 12-hour Bicentennial Fourth of July extravaganza, the four Nixon-Frost interviews in 1977, Christmas Eve Mass from St. Patrick's in 1978, the annual nationally distributed documentaries of Capital Cities Broadcasting and scores of national and regional sports events.

ROBINSON, GLEN O.—FCC commissioner (1974–76) appointed to a Democratic seat by President Nixon. He came to the commission from the faculty of the University of Minnesota Law School (1967–74), where he taught administrative law and regulated industries, and in his brief term was the agency's resident intellectual. Earlier he worked in communications law, serving with the Washington firm of Covington and Burling (1961—62).

Having been appointed to serve out the remaining two years of a seven-year term, Robinson decided early in 1976 not to wait around to learn whether President Ford would reappoint him to a full term and instead returned to academia.

He was named by President Carter to head the U.S. delegation to the 1979 World Administrative Radio Conference, with the title of Ambassador. He also became special advisor to the Aspen Institute and Professor of Law at the University of Virginia School of Law.

ROBINSON, HUBBELL (d. 1974)—head of the CBS program department for 13 years, noted for his ability to provide the network with commercially popular shows as well as prestige dramas and cultural programs. As one of the architects of the schedule that made CBS the leader in audience ratings through two decades, Robinson is credited with developing such series as *Gunsmoke, I Love Lucy, You'll Never Get Rich, Climax* and *Playhouse 90*. He earned a reputation as an innovator.

A newspaper reporter and drama critic during the late 20s, he later became a radio producer for advertising agencies and, for a brief period, with ABC. CBS hired him away from Foote, Cone & Belding Advertising in 1947, and he remained as its program chief until 1959, when he established his own production company. Robinson returned to CBS in 1962 as senior v.p. for programming but left the following year after a clash with James T. Aubrey, Jr., then network president. On his own, he became executive producer of the short-lived series, *Hawk,* and then was engaged by ABC to oversee its ambitious experimental series, *Stage 67.*

ROBINSON, MAX—former local Washington newscaster on WTOP-TV who joined ABC News in June 1978 as one of three anchors on the revamped national newscast, *World News Tonight.* Based in Chicago and with the title of correspondent, Robinson headed the domestic desk for the program. He began as a news reporter on WRC-TV in Washington in 1966, then switched to WTOP-TV in 1969 to anchor its popular *Eyewitness News.*

ROBINSON v. FCC [334 F2d (D.C. Cir. 1964), cert. den./ 379 U.S. 843 (1964) affirming Palmetto/ 33 FCC 250 (1962)]—case in which a licensee was denied renewal for lying to the FCC in renewal proceedings. WDKD Kingston, S.D., was advised by the FCC, when it applied for license renewal, of complaints from listeners about the broadcasts of Charles Walker, whose monologues allegedly were saturated with sexual references verging on the obscene. The owner of the station, Edward G. Robinson, Jr., replied by letter that he was unaware of Walker's vulgarities, that he had deplored them and that he had fired Walker.

The commission held hearings in which Robinson restated the substance of his letter and said additionally that he had never heard any complaints about the Walker material. Numerous witnesses, however, testified that they had complained to Robinson. The FCC denied Robinson renewal of his license, resting its judgment on Robinson's misrepresentations to the FCC, as well as on the vulgar material, which it said was a ground for denial even in the absence of knowledge by Robinson.

The D.C. Court of Appeals affirmed the FCC's decision but rested its opinion only upon Robinson's misrepresentation. Robinson again appealed, but the Supreme Court refused to hear the case.

"ROCKFORD FILES, THE"—hour-long private-eye series created as a vehicle for James Garner in 1974, enabling him to resurrect aspects of the character he introduced in *Maverick* 17 years earlier. The series was a Friday night hit on NBC. Noah Beery and Stuart Margolin were featured, and production was by Cherokee Productions in association with Universal TV.

Julie Covington, Charlotte Cornwell & Rula Lenska in Rock Follies

ROCK FOLLIES—Thames Television's 1976 six-part mini series which chronicled the rise of a female rock group called "The Little Ladies" played by Julie Covington, Charlotte Cornwell, and Rula Lenska. Written by Brooklynite Howard Schuman, directed by Brian Farnham and Jon Scoffield, and produced by Andrew Brown, the show had a catchy score by Andy Mackay—a survey of pop styles of the seventies, and earlier. *Rock Follies* dealt with some interesting issues that don't usually get treated on television, and it did it with some wit and good music. Aired perfunctorily in the U.S. by PBS in 1977. A sequel was less effective.

ROCKY MOUNTAIN SATELLITE PROJECT—an experiment during 1974–75 with the use of domestic communication satellites to deliver educational materials to remote areas of the Rocky Mountains region. The project, which beamed a series of programs on career education to 56 schools, along with other instructional material, was conducted with NASA's powerful ATS-6 satellite.

RODDENBERRY, GENE—TV producer-writer best known for creating and producing *Star Trek*. He was also producer of the series, *The Lieutenant;* head writer for *Have Gun, Will Travel* and writer of more than 80 TV scripts for various series.

RODMAN, HOWARD—writer specializing in creating series and writing pilot scripts. His credits include the pilots for *Harry O* and *Six Million Dollar Man*.

ROGERS, RALPH D.—millionaire Texas businessman who became chairman of the PBS board of governors in the early 70s and was responsible for settling the wars and skirmishes between factions of the public television industry. It was Rogers who engineered the Partnership Agreement between CPB and PBS on May 31, 1973, resolving, at least for the time, the struggle between those organizations for leadership of the system. Rogers also reorganized PBS, clarifying the roles of lay officials and practitioners.

In 1976 he was the force behind the selection of Lawrence K. Grossman, a New York advertising executive, as president of PBS, a move which gave aggressive national leadership to the public TV system and revived notions of investing PBS with some programming authority. Rogers yielded the PBS chairmanship to Newton Minow in 1978.

One of public TV's lay officials, Rogers was chairman of Texas Industries Inc. He retired in January 1975 as chief executive officer of that company but retained the title of chairman.

"ROGUES, THE"—sophisticated adventure-comedy concerning well-bred international con men. It drew excellent reviews and poor ratings when it premiered on NBC in 1964. It lasted a single season. Produced by Four Star Television, its cast included two of the company's principals, Charles Boyer and David Niven, and featured also Gig Young, Gladys Cooper and Robert Coote.

ROLFE, SAM H.—veteran producer and writer whose credits span *Have Gun, Will Travel* in 1959 to *The Delphi Bureau* in the early 70s.

"ROMPER ROOM"—a long-running daily children's series dealing in games, playthings and simple lessons for preschoolers, whose format was widely syndicated by Bert Claster Productions. The program was designed to be presented as a local live show in each market, using children from the community and a resident hostess or "teacher." It premiered on independent stations in 1953 as early morning fare, in studios designed to resemble a classroom. Before long, the format was syndicated to more than 100 markets as well as to numerous countries abroad, and it maintained wide distribution through two decades.

Consumerists called into question some of the practices of the program, such as its persistent featuring of brand-name toys used by the children on the show, its commercials and plugs for so-called "Romper Room Toys" (manufactured by Hasbro) and its use of the "teachers" as commercial announcers. Such criticism, along with a tightening of the NAB code on children's advertising, were effective in somewhat altering the commercialized character of the shows.

The Rookies: George Stanford Brown, Bruce Fairbairn, Sam Melville with Gerald S. O'Loughlin

"ROOKIES, THE"—hour-long ABC police series (1972–76) about a trio of youthful officers. It succeeded as a Monday night entry at 8 ,P.,M. but was ousted from the timeslot by family viewing time when it was adopted in 1975. In its new Tuesday timeslot at 9 ,P.,M., *The Rookies* faltered and was canceled. Georg Stanford Brown, Sam Melville and Bruce Fairbairn (preceded the first two seasons by Michael Ontkean) played the police rookies. Featured were Gerald S. O'Loughlin and Kate Jackson. It was an Aaron Spelling-Leonard Goldberg Production.

"ROOM 222"—half-hour comedy-drama series on ABC (1969–73) on a racially integrated high school, focusing on a

Michael Constantine & Lloyd Haynes in Room 222

black male teacher portrayed by Lloyd Haynes. It featured Denise Nicholas, Michael Constantine and Karen Valentine as faculty and Heshimu, David Jolliffe and Judy Strangis as students. The series was by 20th Century-Fox TV.

ROONEY, ANDREW A.—veteran TV journalist and director who gained distinction as a writer of essays for the medium, chiefly for CBS News. Many of his pieces were delivered by Harry Reasoner and other on-camera talents, but in the 70s Rooney began appearing on the screen himself, notably in *Mr. Rooney Goes to Washington* (1975) and *Mr. Rooney Goes to Dinner* (1975). Among his other credits are *An Essay on War* (1971), *An Essay on Churches* (1972), *A Birdseye View of California* (1972) and *In Praise of New York City* (1974). Later he was given a regular spot as essayist in *60 Minutes*.

"ROOTIE KAZOOTIE"—early children's puppet series which began on NBC in 1950. Todd Russel's puppet cast included Rootie Kazootie, El Squeako Mouse and Polka Dottie.

"ROOTS"—one of TV's milestone programs, an ABC miniseries based on Alex Haley's novel, *Roots*, which not only scored the highest ratings in TV history for an entertainment program but also was considered to have marked turning points in the medium, both for its form and subject matter. Purchased by ABC two years before the book's publication and aired shortly after the book's release, when it had already reached the top of the best-seller lists, *Roots* emptied theaters, filled bars, caused social events to be canceled and was the talk of the nation during the eight consecutive nights it played on ABC, from Jan. 23 to Jan. 30, 1977— Sunday through Sunday.

Georg Stanford Brown

Heading the large cast were LeVar Burton, Ben Vereen, John Amos, Leslie Uggams, Maya Angelou, Cicely Tyson, Edward Asner, Harry Rhodes and Robert Reed.

The series was produced by David L. Wolper Productions, with Wolper as executive producer and Stan Margulies as producer. The scripts were by William Blinn, Ernest Kinoy, James Lee and Max Cohen, and the original music was by Quincy Jones.

Indicating that ABC had not expected the bonanza that resulted was the fact that the eight episodes were scheduled just before the sweep weeks, a time when the networks tend to play what they believe to be their strongest shows. Meanwhile, the success of *Roots* created a receptive syndication market for a BBC series on a similar subject, *Fight Against Slavery*, distributed by Time-Life Films.

The program's popularity was remarkable for the fact that its cast was predominantly black and its villains mostly white, dealing as it did with the history of a black family traced to the capture of a young African warrior by American slave traders. Only 10 years before its airing, blacks had been scarce on the TV screens except as tap dancers, musicians and ball players; according to the industry's conventional wisdom of the 50s and 60s, blacks would not attract either a mass audience or advertisers and would cause Southern stations to defect from the network lineup. The sad experience of Nat (King) Cole on NBC was all the proof that had been needed.

But none of the stations, northern or southern, rejected *Roots,* and the 12-hour miniseries (which contained four 2-hour episodes and four 1-hour) was a shared experience for most of the country, a book read electronically by everyone at the same time and at the same pace. The mayors of more than 30 cities proclaimed the week "Roots Week," and more than 250 colleges and universities offered, or proposed to offer later, courses based on the film and the book. In the cities that were capitals of the civil rights turmoil of the 60s, no crosses were burned on station lawns, and disturbances were negligible.

The eight episodes of *Roots* averaged a 44.9 rating and a 66 share of audience, far surpassing any previous series of programs. The seven episodes following the opening show took the top seven spots in the ratings for that week, and the final two-hour telecast notched a 51.1 rating and 71 share—making it the leader in the all-time number of TV homes, with 36.38 million as compared to 33.65 million for *Gone With the Wind—Part 1,* aired by NBC earlier that season. According to Nielsen, an average of 80 million viewers watched the final episode of *Roots,* and around 130 million—or 85% of the TV homes—watched all or part of the 12 hours.

"ROOTS: THE NEXT GENERATIONS"—14-hour sequel to the phenomenally popular TV adaptation of Alex Haley's *Roots,* airing two years after the original, over a period of eight nights, Feb. 18—25, 1979. While *Roots II,* as it came to be called, did not rival *Roots I* as an audience draw, it nevertheless did exceedingly well, with a 30.1 rating and 45 share for the entire run, or about two-thirds as well as the original. The total audience, watching all or part of the serial, was estimated at around 110 million. Moreover, the program won every one of its time periods, despite much tougher competition than *Roots I* encountered.

The opening episode was challenged by two big movies, *Marathon Man* on CBS and *American Graffiti* on NBC. During the week, it was up against episodes of *Backstairs at the White House* and *From Here to Eternity.* On the final night, the lowest-rated of the series, although it featured Marlon Brando and James Earl Jones, *Roots II* had to contend with the popular family movie, *The Sound of Music. Roots II* was televised during a sweeps period; *Roots I* was not.

Based on unused material from Haley's book, some of the content of a second book, *My Search for Roots,* and additional material supplied by the author, *Roots II* continued the chronicle of the family, from Chicken George to Haley's own emergence, or from Reconstruction to modern times.

The cast included Georg Stanford Brown, Diahann Carroll, Ossie Davis, Ruby Dee, Olivia deHavilland, Ja'net DuBois, Henry Fonda, Al Freeman Jr., Andy Griffith, Rafer Johnson, Claudia McNeill, Carmen McRae, Harry Morgan, Greg Morris, Della Reese, Beah Richards and Richard Thomas.

David L. Wolper was executive producer, Stan Margulies producer, and John Erman, Charles S. Dubin, Georg Stanford Brown and Lloyd Richards directors. Ernest Kinoy supervised the scripts and wrote the first three episodes. Other writers were Sydney A. Glass, Thad Mumford and Daniel Wilcox, and John McGreevey.

ROSE, REGINALD—one of the outstanding playwrights to emerge from TV's drama era of the 50s, writing for *Studio One, Philco Playhouse* and the other anthology programs. His notable TV plays include *Twelve Angry Men, Thunder on Sycamore Street* and *The Sacco-Vanzetti Story.*

Rose also created and wrote *The Defenders* series (1961) on CBS. His first TV play, *Bus to Nowhere,* aired in 1951.

ROSEMONT, NORMAN—personal manager turned TV producer and executive in charge of a number of music-variety specials, principally those starring Robert Goulet, his prize client. As his production company grew, it turned out such dramatic specials as *The Man Without a Country, The Count of Monte Cristo, The Man In the Iron Mask, The Hunchback of Notre Dame* and *Ivanhoe.*

ROSEN, KENNETH M. (d. 1976)—writer, producer and director who for a time operated his own company, Profile Productions, in California. As a writer, he collaborated with Marshall Flaum on scripts for several Jacques Cousteau and Jane Goodall nature specials and earlier wrote episodes for such series as *Naked City, Perry Mason* and *Days of Our Lives.* He also wrote the TV adaptation of the book, *Future Shock.* As a producer, he was responsible for 23 documentaries, among them a National Geographic special for David Wolper Productions, *Journey to the Outer Limits.*

ROSENBERG, META—personal manager of James Garner who has been executive producer of his series, *Nichols, The Rockford Files,* and *Bret Maverick.*

ROSENFIELD, JAMES H.—president of CBS-TV since October 1977; the position's responsibilities had been confined to network sales and affiliate relations since the reorganization that occurred at the time of his appointment. However, in another reorganization in 1981, Rosenfield gained authority over the program functions as well. Previously, Rosenfield had been v.p. and national sales manager of CBS-TV. After working briefly for NBC in sales, he joined CBS network sales in 1965 and was promoted steadily every two or three years. He was named a v.p. in 1972.

ROSS, MICHAEL—writer-producer who collaborated with Bernie West and Don Nicholl to create and produce *The Jeffersons* and *The Dumplings.*

ROSTEN, IRWIN— producer director-writer of informational nonfiction programs such as those for MGM's *GE*

Monogram Series on which he teamed with Nicolas Noxon in the early 70s. Those programs included *The Wolf Men, The Man Hunters, Once Before I Die* and *Dear Mr. Gable.* Earlier, while he was with Wolper Productions, Rosten produced and wrote *Grizzly, The World of Jacques Cousteau* and six episodes of *Hollywood and the Stars.* He began as a TV documentary producer in the news division of KNXT Los Angeles.

"ROUTE 66"—popular series on the exploits of two young adventurers traveling across the country, starring Martin Milner and George Maharis (Glenn Corbett replaced Maharis in the fourth season). Produced by Screen Gems, it ran on CBS from 1960 to 1964.

Dan Rowan & Dick Martin with Brian Bessler

"ROWAN & MARTIN'S LAUGH-IN"—a madcap comedy hour, laced with silliness and satire, which gave a new free-wheeling form to TV variety shows and was a prodigious hit and a Monday night fixture for NBC (1967–73). With the nightclub comedy team of Dan Rowan and Dick Martin as hosts, the show featured a "new faces" resident company of skit players and buffoons. A chaotic montage of skits, blackouts, fast-cut inanities, dances, songs, stand-up comedy and recurring routines, it was unlike anything else on television and, with its irreverence and double-entendres, seemed to catch the liberated spirit of the times.

The series poured out one catch-phrase after another—"sock it to me," "verrry interrresting," "here come the judge," "look that up in your Funk & Wagnall's"—and turned up a wealth of new young talent in its revolving repertory company, notably Lily Tomlin, Goldie Hawn, Ruth Buzzi, Teresa Graves and Arte Johnson. Others in the cast, for all or part of the five-year run, were Judy Carne, Joanne Worley, Alan Sues, Dave Madden, Gary Owens, Jud Strunk, Richard Dawson and Sarah Kennedy.

Laugh-In was produced by Romart Inc., the company owned by Rowan and Martin. George Schlatter and Ed Friendly were the executive producers for the first three seasons, Paul Keyes for the next two. A revival of the series was essayed in 1977 in the form of NBC specials.

"ROY ROGERS SHOW, THE"—singing-cowboy series on NBC (1955–6), predating the era of the "adult western." It starred Roy Rogers and his wife, Dale Evans, and was produced by their own company, Roy Rogers Productions. Featured were The Sons of the Pioneers, a vocal group; Trigger, a horse; and Bullet, a dog.

ROYAL, JOHN F. (d. 1978)—first v.p. in charge of television at NBC (1940–53) who remained with the network after retirement as a consultant, until his death at 88. A former general manager in vaudeville, and later v.p. in charge of programs for the NBC radio network, Royal was close to performers, and the network benefited from his relationships with Mary Martin, Bob Hope, Ed Wynn and other stars. As NBC's resident showman, he helped to shape the medium's early programming and to secure the stars for dramas, variety shows and what were then called spectaculars.

RTNDA (RADIO-TELEVISION NEWS DIRECTORS ASSN.)—organization of professional broadcast newsmen founded in 1946 to work toward improving the standards of broadcast journalism and to defend the rights of newsmen. It is also concerned with journalism education to meet the specific needs of radio and TV. Members are mainly station news directors, network news executives and academicians.

With a membership of about 1,000, RTNDA conducts an annual conference and periodic regional seminars and workshops. It also confers annual awards for news reporting, editorials and outstanding contributions to broadcast journalism.

RUBEN, AARON—comedy writer whose career dates to the radio shows of Burns and Allen, Fred Allen and Henry Morgan and extends to *Sanford and Son* and *C.P.O. Sharkey*. In the 50s Ruben wrote for the TV shows of Sam Levenson, Danny Thomas, Milton Berle and Sid Caesar, then became director of Silvers's *You'll Never Get Rich* (Sgt. Bilko) and in 1960 produced *The Andy Griffith Show*. He was also creator and producer of *Gomer Pyle* and later producer of *Sanford and Son. Sharkey* was via his own production company.

RUBENS, WILLIAM S.—v.p. of research for NBC since 1972, closely involved in the network's program process as a scheduling strategist and as head of concept testing. His responsibilities include overseeing research for the owned stations and radio division.

He joined NBC in 1955 as director of marketing services in the stations division, then shifted to the research department. In 1970, he was named v.p. of audience research for the TV network. Prior to joining NBC, he was a research associate with an ad agency and, earlier, a statistician for ABC.

Hughes Rudd

RUDD, HUGHES—ABC News correspondent since August 1979, after a 20-year career with CBS News, during the latter part of which he anchored *The CBS Morning News* (1973–77). Originally teamed with Sally Quinn, he became sole anchor on her departure several months later. A Texan, Rudd retained an appealing countrified style although his journalistic assignments—first for newspapers and then for CBS—took him to many parts of the world. Rudd also was a writer of fiction and just before joining CBS wrote and directed industrial films.

"RUDOLPH, THE RED-NOSED REINDEER"—perennial children's Christmas special on CBS, playing every year since its premiere in 1964. Based on the song by Johnny Marks and a story by Robert L. May, the hour animated special was produced by Arthur Rankin, Jr., and Jules Bass for Videocraft International Production. Burl Ives is narrator.

RUKEYSER, BUD (M.S., JR.)—executive v.p. of public information for NBC since 1974, after 11 years as head of the company's publicity operations. He became an influential figure in the administration of Herbert Schlosser and remained a key aide to Fred Silverman. He was discharged by Silverman in 1980 but was rehired a year later by Silverman's successor, Grant Tinker. In the interim he was a public relations executive for *Newsweek*. He joined NBC in 1958 as a staff writer in the press department, after having worked for Young & Rubicam Advertising.

Ben Gazzara

"RUN FOR YOUR LIFE"—hour-long adventure series on NBC (1965–67) which starred Ben Gazzara as a man determined to squeeze 20 years of living into the three years that the doctors tell him remain. It was produced by Roncom Film, with Roy Huggins as executive producer.

RUNAWAY PRODUCTION—programs or films produced abroad or outside the jurisdiction of U.S. unions, usually for reasons of economy. The practice has, of course, been fought and condemned by the unions, which have applied the term to give the programs the stigma of having deprived Americans of work. "Runaways" have exacerbated the decline in film activity in Hollywood—to which the networks' greater use of reruns and growing reliance on video tape have contributed—but the union campaigns against overseas production were not notably successful.

Elton H. Rule

RULE, ELTON H.—president and chief operating officer of the American Broadcasting Companies Inc. since January 1972, rising to that post after markedly improving the competitive position of the TV network and the owned stations as head of the broadcasting group.

Rule's ascent to the corporate tier was rapid after his appointment, in January 1968, as president of the TV network. Sixteen months later he was placed in charge of all broadcast activities for the corporation and in 1970 was elected to the board and given the title of president of the American Broadcasting Co., subsidiary of the parent corporation. An executive who had developed within the ABC system, Rule had been v.p. and general manager of KABC-TV Los Angeles when he was tapped for the network post. Prior to that he had worked in station sales.

RULEMAKING—the FCC's process for formulating, amending or repealing a rule. The commission, as a government agency, is required to give interested parties an opportunity to participate in rulemaking through submission of written data and arguments or through an oral presentation in hearings.

RUSH, ALVIN—president of the MCA Television group since 1980, after having been executive v.p. of NBC-TV and before that senior v.p. of program and sports administration, essentially the executive in charge of talent and program negotiations. Rush came to the network in 1973 after a long career as a talent agent and packager, first with MCA and then with Creative Management Associates.

RUSH, HERMAN—one of the most influential program packagers of the 60s as head of the television division of a major talent agency, CMA, and earlier as an agent for GAC. In the 70s, he went out on his own as an independent producer and then joined Marble Arch Productions as president of its TV division. Later he became president of Columbia Pictures Television.

RUSHNELL, SQUIRE D.—v.p. of children's and early morning programs for ABC Entertainment; he assumed the first post in 1974 and the second when he was given responsibility also for *Good Morning, America* in April 1978. During his tenure, significant changes were made in children's programming at ABC, notably the addition of the high-quality *ABC Afterschool Specials,* the *ABC Weekend Specials, Kids Are People, Too* on Sunday mornings and public service messages on nutrition and health. Previously, Rushnell was with the ABC owned-stations division, where he was responsible for guiding the *Rainbow Sundae* children's series, and before that with WLS-TV Chicago, where he was assistant program manager.

RUSKIN, COBY—director of *Here's Lucy, Gomer Pyle* and *The Bill Cosby Show.* He also directed episodes for *The Doris Day Show, Julia* and *Love, American Style.*

RYAN, J. HAROLD (d. 1961)—a co-founder in 1928, with George B. Storer, of the Storer Broadcasting Co., where he spent most of his career as senior v.p. He also served for a time as president of NAB.

RYAN, NIGEL—v.p. of documentaries and special programs for NBC News (1977–79) after having been editor and chief executive of Britain's Independent Television News. In his post with ITN, which he assumed in 1971, he guided the news organization through a period of innovation, introducing documentary techniques to the U.K., commercial network's popular *News at Ten* and creating a 30-minute lunchtime newscast, *First Report.* He left NBC in the Fred Silverman administration and returned to England as director of programs for Thames Television.

RYAN, TIM—CBS sportscaster since 1977, chiefly covering football, basketball and boxing, after a five-year stint with NBC Sports (1972–77). A Canadian, he broke into broadcasting with CFTO Toronto as a sportscaster, then moved to California to do the play-by-play for the Oakland Seals of the National Hockey League. In 1970, he shifted his base to New York to cover the Rangers hockey games on WOR-TV, serving at the same time as regular sportscaster for WPIX-TV. Two years later, he joined NBC when it acquired the rights to televise the NHL and, while there, also covered a variety of other sports.

"RYAN'S HOPE"—ABC soap opera which began July 7, 1975. It differed from others of the genre in that it was set in New York City, rather than the traditional small towns of daytime serials, and concerned itself with young love. The cast included Bernard Barrow, Helen Gallagher, Ron Hale, Nancy Addison, John Gabriel, Michael Levin, Tom Arledge, Earl Hindman and Peter Haskell.

S

SACKHEIM, WILLIAM—TV and motion picture producer whose TV credits go back to *Goodyear Playhouse* in the late 50s and extend to the *Delvecchio* series and pilot for *The Law* in the mid-70s.

Morley Safer

SAFER, MORLEY—CBS News correspondent who became one of the principals of *60 Minutes* in 1970 after distinguishing himself in covering the Vietnam war. A Canadian who had been a correspondent and producer with the Canadian Broadcasting Corporation before joining CBS, Safer served as head of the CBS Saigon bureau (1965–67) and then as chief of the London bureau (1967–70). In the fall of 1967 he and cameraman John Peters went into mainland China to film *Morley Safer's Red China Diary*. In 1969 he contributed the special, *The Ordeal of Anatoly Kuznetsov*, an exclusive interview with a Soviet writer who had defected to the West. He joined *60 Minutes* as replacement for Harry Reasoner, who quit CBS News for ABC.

SAG—see Screen Actors Guild.

"SAGA OF THE WESTERN MAN, THE"—an occasional series of 13 specials, each documenting a decisive year in the story of modern man. Starting on ABC in 1963, the series was produced and written by John Secondari for the ABC News Special Projects Division.

SAGAL, BORIS—director of movies-for-TV and drama specials. His credits include four episodes of the *Rich Man, Poor Man* miniseries.

SAGALL, SOLOMON—pioneer promoter of pay-TV in its over-the-air form whose first venture was with large screen pay-television in two London cinemas in 1938. That company, which he founded, was known as Scophony Ltd. It developed a system of coded television which was proposed for use in military operations during World War II and still is the basis for all pay systems. In 1957 Sagall formed Teleglobe Pay Systems Inc. in the U.S., which had to endure years of FCC delays in permitting over-the-air pay-TV to operate. All legal and political barriers were finally removed in 1975.

Sagall envisioned a system that would make every home a private theatre and that would, in effect, extend the seating for drama, opera and orchestral performances and for sports and auditorium events. Under his concept, an entire family would be able to watch any program for the price of a single ticket to the event. He developed a decoding unit—a "black box"—to be attached to TV sets which would perform the dual function of unscrambling the picture and billing the viewer. But by the time the FCC allowed pay-TV to proceed, only UHF channels were available for the operations and pay-cable had gotten underway. See also Pay-Television.

"SAINT, THE"—mystery series based on the famed stories of Leslie Charteris, with Roger Moore in the role of Simon Templar. Produced by ATV/ITC in England, it premiered on NBC in 1967 and returned at intervals in the following two years. Of the 114 episodes, 43 were filmed in color.

Richard S. Salant

SALANT, RICHARD S.—former president of CBS News who in April 1979, immediately on reaching the mandatory retirement age of 65, switched camps and became vice-chairman of the National Broadcasting Co. His new position gave him general supervision over news and responsibility for assessing the role of the new technologies in the future direction of the company. His switch of networks, which startled the industry, raised the issue of company loyalty and pitted it against the issue of the morality of mandatory retirement policies. Energetic and eager to keep working, Salant maintained that he considered himself a broadcast journalist by profession rather than partisan of a network. His role at NBC, however, diminished through the years of

the Fred Silverman administration, and by 1982—with Silverman gone—he was little more than a consultant under contract.

He had been twice president of CBS News, the first time 1961 to 1964, the second since February 1966. In the period between, while Fred W. Friendly headed the news division, Salant served as CBS v.p. of corporate affairs and as special assistant to Dr. Frank Stanton, president of CBS. A lawyer, Salant joined the company as a v.p. in 1952. Because he lacked a journalism background, he was at first resented by many within the news division; but he proved knowledgeable and courageous, and he had a distinguished tenure which reached its high point with the bold CBS News coverage of the Watergate events.

Soupy Sales

SALES, SOUPY—slapstick performer who was popular with children in the 50s and later gained a following with a somewhat older audience. A comedian of the pie-in-the-face school, he dealt also in puns, corny jokes and communication with animals. His popularity on a local station in Detroit won him a network show in the mid-50s. Later he worked at a New York station, which served as the base for a syndicated show and in 1976 he began a Saturday morning children's show on ABC, *Jr. Almost Anything Goes.*

SALOMON, HENRY, JR. (d. 1957)—distinguished producer and writer of documentaries whose crowning achievement was probably *Victory at Sea,* an NBC series on the U.S. naval forces in World War II using archive film clips and featuring an original score by Richard Rodgers. Salomon also produced the noted early documentary on the Soviet Union, *Nightmare in Red,* and another on Adolf Hitler, *The Twisted Cross.* He shared in the credit for a documentary on atomic energy, *Three, Two, One—Zero.*

"SAM BENEDICT"—1962 NBC series starring Edmond O'Brien as an attorney handling unusual cases. Produced by MGM-TV, it yielded 28 episodes.

SAMISH, ADRIAN (d. 1976)—producer associated with Quinn Martin Productions and Spelling-Goldberg Productions, joining the latter in 1975. With QM, where he worked for nine years, he was involved in the creation of such shows as *Cannon, Streets of San Francisco, Barnaby Jones, Dan August* and *Caribe*. He also produced pilots for a number of series, including *Manhunter, Travis Logan, D.A.* and *Crisis Clinic*. Before joining QM in 1966, he was a program v.p. for ABC in Hollywood.

SAMPLING—the industry's perception of how viewers examine the shows. The start of a new season is considered to be a *sampling* period for three or four weeks, during which viewers try out the new entries before deciding whether to follow them regularly or switch to an established show on a competing channel. Programs which sample favorably—that is, achieve high ratings initially—usually have a good chance of succeeding; those sampling poorly have slim chances because they are destined to go largely undiscovered. Viewing patterns over the years indicate that audiences generally do not return to sampling activities once the season is underway.

Sampling, in another sense, is part of rating methodology. The audience surveys base their viewing estimates on what they hold to be a representative sample of the audience measured. The Nielsen household ratings, for example, are derived from a national sample of 1,200 homes equipped with Audimeters.

"SAN FRANCISCO BEAT"—see *Lineup, The*.

"SANDBURG'S LINCOLN"—four one-hour specials derived from Carl Sandburg's biography of Abraham Lincoln, carried by NBC in 1974–75. Hal Holbrook portrayed Lincoln and Sada Thompson, Mary Todd Lincoln. The series was by David L. Wolper Productions, with Wolper as executive producer and George Schaefer as producer-director.

SANDERS, MARLENE—producer and correspondent for CBS News, working chiefly on *CBS Reports* since joining the network from ABC News in January 1978. She had been v.p. and director of documentaries for ABC News since January 1976, first woman to hold a vice-presidency in a network news division. For three years previous, she had produced such documentaries for the *ABC Close-Up* series as *Women's*

Health: A Question of Survival, The Right to Die, Lawyers: Guilty as Charged? and *Prime Time TV: The Decision Makers*. She joined ABC as a correspondent in 1964 after working in local TV and radio in New York as a reporter, writer and producer. For four years she anchored a daytime newscast on ABC.

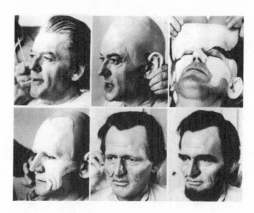

Hal Holbrook as Lincoln

SANDRICH, JAY—situation comedy director who figured importantly in the success of *The Mary Tyler Moore Show,* with which he was associated in its early seasons.

Redd Foxx as Fred Sanford

"SANFORD & SON"—hit NBC situation comedy (1972–78) about a querulous old junk dealer and his son, which was adapted with a black cast from the British hit, *Steptoe & Son*. Produced by Tandem Productions (Norman Lear and Bud Yorkin), it premiered as a midseason replacement and was instantly popular. Redd Foxx and Demond Wilson were the principals. Both *Sanford* and *Steptoe* were created by Ray Galton and Alan Simpson.

When Foxx left the series in 1977 to begin a weekly variety show on ABC, the program was continued under the title *Sanford Arms*. But it foundered and failed to last out the 1977–78 season. Foxx's ABC variety show also failed.

SAPERSTEIN, HENRY G.—president of UPA Pictures, which produced various Mister Magoo cartoon series for TV. Saperstein purchased UPA in the early 60s after producing a string of successful syndication shows from Chicago, among them *All Star Golf* (1958–62), *Championship Bowling* (1958–60) and *Ding Dong School* (1959–60).

"SARA"—one of several unsuccessful attempts to revive the western in the 70s; this one, introduced by CBS in January 1976, had the special angle of a female lead (Brenda Vaccaro). Featured were Louise Latham, Bert Kramer and Albert Stratton. It was by Universal TV.

"SARGE"—NBC series (1971) about a veteran detective who becomes a priest after his wife's mysterious death and continues to fight crime from his parish. A Universal TV production, which starred George Kennedy, it ran 13 weeks.

David Sarnoff

SARNOFF, DAVID (Gen.) (d. 1971)—one of the pioneers in the development of broadcasting who rose from a wireless operator to chairman of the Radio Corporation of America and founder of its subsidiary, the National Broadcasting Co. Advancing through the ranks of RCA in the 20s after it absorbed his employer, the Marconi Co., Sarnoff, on becom-

ing president, made the company a leader in the manufacture of radio sets (and later, TV sets as well) and of electronic parts and equipment.

He created the NBC network in 1926 to provide a program service that would stimulate the sale of radios and in the 40s organized the TV network to build a market for RCA television receivers. At various intervals, he doubled as president of the National Broadcasting Co. Not long after the FCC approved the RCA color system to serve as the basis for color TV in the U.S., Sarnoff had NBC-TV completely in color to help sell RCA's color TV line.

Sarnoff, a Russian immigrant, came into national prominence on the night of April 14, 1912 when, as a wireless operator in Manhattan, he allegedly received the first distress signal from the sinking S.S. Titanic and maintained contact with the ship. Subsequently, he worked for the Marconi Co. and held the title of commercial manager at the time it was taken over by RCA in 1919.

After nearly 20 years in the presidency, he became chairman in 1947. Sarnoff retired in 1970 but was elected honorary chairman, in which capacity he served until his death.

See also NBC.

SARNOFF, ROBERT W.—son of General David Sarnoff who, in January 1970, succeeded his father as chairman and chief executive officer of RCA, crowning a career that began in the NBC sales department in 1948 and included tenures as president and chairman of NBC. Sarnoff resigned from RCA in 1975, on pressure from members of the board, after a period of decline for the electronic manufacturing division.

Sarnoff became president of NBC Dec. 7, 1955, when Sylvester L. (Pat) Weaver moved up to board chairman. In 1958 Sarnoff became chairman and Robert E. Kintner president, beginning an era of great profitability for the TV network and stations and of the dramatic rise in stature of NBC News. In 1965 Sarnoff moved up to RCA as its president. Throughout his 10 years with the parent company, Sarnoff maintained direct supervision of NBC.

Under Sarnoff, RCA kept pace with the exploding electronic technology of the 70s and moved into the development and production of miniature integrated circuits, weather and communications satellites, control systems, portable moon cameras and the videodisc, among other products. But the company's share of the TV receiver market was seriously eroded by Japanese manufacturers, and RCA was forced to collapse its computer-manufacturing operations at a huge loss.

SARNOFF, THOMAS W.—executive v.p. in charge of NBC, West Coast (1962–77), and youngest son of the late *pater familias* of RCA and NBC, Brig. Gen. David Sarnoff. When he resigned in the spring of 1977 to form his own company, Sarnoff International Enterprises, engaged in the production of arena and stage shows, NBC was left for the first time without a Sarnoff in its management echelons.

Tom Sarnoff began his career with ABC as a floor manager in 1949 and two years later went to MGM Studios. In 1952 he joined NBC as assistant to the director of finance and operations and then began to move into executive posts in production and business affairs. He became a v.p. in 1957. As West Coast head of NBC, he was responsible for all functional operations but principally gave his time to administration, business affairs and labor relations.

SATCOM—See RCA Satcom.

An RCA Americom satellite

SATELLITE, COMMUNICATIONS—orbiting space vehicle used to relay electronically encoded communications services—including television, voice or teletype—over long distances. Modern synchronous communications satellites, used to relay television signals across oceans and domestically (as a substitute for microwave and coaxial relays), are precisely located at 22,300 miles above the equator, orbiting the earth, but traveling at the same rotation speed as the earth directly below them. Therefore, a synchronous satellite is considered to be in "stationary" orbit, since it always hovers at the same place in the sky.

All communications satellites launched since 1963, except the Soviet Union's Molniya series, are of the synchronous type. Communications satellites receive electromagnetic signals aimed at them from earth stations, amplify and retransmit them to other earth stations. Because electromagnetic waves of the bandwidth and frequencies used for television generally follow a "line-of-sight" pattern (that is, they do not bend with the curvature of the earth), the satellite must be "visible" from both the transmitting and receiving locations.

The first successful communications satellite, launched by NASA Aug. 12, 1960, was Echo I, a 100-foot balloon, placed in elliptical random orbit. It was a "passive" satellite—that is, it carried no electronic equipment and signals were "bounced" from its surface. The first relay of black-and-white and color television by satellite was accomplished by AT&T's Telstar in 1962. Telstar also followed a random orbit (although it was an "active" satellite) and therefore it could only be used for short and specific periods when both transmitting and receiving locations were in its "line of sight.

Syncom II, launched in 1963 was the first successful synchronous satellite. Commercial satellite communications became a reality with the orbiting of Early Bird (Intelsat I) in 1965, which was the first satellite owned by the International Telecommunications Satellite Consortium. Intelsat now has seven operational satellites and has 91 member countries. The U.S. participant is Comsat, or Communications Satellite Corporation.

The first domestic synchronous communications satellite was Canada's Anik 1, launched in November 1972. In North America, there currently are 10 communications satellites capable of relaying television. These include Western Union's two Westar satellites, launched in 1974 to become the first U.S. domestic satellites, RCA's Satcom 1 and 2, and two Comstar satellites owned by AT&T. The Anik and Westar satellites each are capable of relaying 12 television programs simultaneously, while the Satcoms and Comstars can carry 24.

The coming of domestic satellite relays has already started to revolutionize television networking by promising a lower-cost substitute for AT&T's land-line transcontinental television relay system. The public Broadcasting Service became fully interconnected by the Westar satellite in 1979.

Cable-TV systems currently make the most extensive use of domestic satellites for television transmission, and satellite interconnection was the key reason for the medium's surge of growth in the late 70s. The advent of satellites created the first national networking system for cable-TV, stimulating the development of a wide variety of new programs and services which made cable attractive in urban areas where retransmission of broadcast signals could not provide a sufficient subscriber base. See also Comsat.

SATELLITE STATION—a conventional TV station which receives and rebroadcasts programs from another station, essentially extending the signal of the parent station to communities beyond its reach. Unlike translators, satellites are full-fledged outlets and may operate at maximum powers permitted for TV stations. Each satellite counts against the number of stations a broadcaster may own.

Satellite stations maintain the merest facilities and operate with small staffs, chiefly technical. Most of them justify their licenses with local newscasts separate from those of the parent stations.

SATELLITE TELEVISION CORP. (STC)—subsidiary of Comsat proposing a pay-television service by direct-broadcast satellite to begin around 1985. The system would involve four operating satellites to cover regions of the country, each providing three channels of programming to be received by subscribers through rooftop dish antennas costing around $100 installed. The programming service would cost subscribers between $14 and $18 a month in 1981 dollars. STC maintains that it will provide a premium service comparable to that offered by cable-TV for the sparsely-populated areas that are unlikely to be wired for cable. In urban areas its direct-broadcast satellite service would provide consumers with an alternative to cable.

The Killer Bees on *Saturday Night:* George Coe, John Belushi, Chevy Chase & Gilda Radner

"SATURDAY NIGHT LIVE"—successful experiment by NBC (1975–) in creating a showcase for young comedians whose material might be too sophisticated or irreverent to be suitable for prime time. The live 90-minute program, originating from New York, was presented at 11:30 P.M. three Saturdays a month (yielding to the newsmagazine *Weekend* the fourth) and became immediately popular with youthful viewers, thus creating a new profit center for the network. The show premiered Oct. 11, 1975 with George Carlin as guest host (subsequent guests included Elliot Gould, Lily Tomlin and Louise Lasser) and with a resident cast that came to be headed by Chevy Chase. Others were Dan Aykroyd, John Belushi, Jane Curtin, Garrett Morris, Laraine Newman, Gilda Radner and Michael O'Donoghue. Chase developed a following for his satirical newscast and impressions

of a clumsy President Ford; he left the show in 1976 to pursue his career independently. Ackroy and Belushi left in the fall of 1979 to work as a team.

The show was developed for NBC by Dick Ebersol, a programming v.p., and was produced by Lorne Michaels and directed by Dave Wilson.

During the 1979–80 season, Belushi and Aykroyd left the show to work in movies separately and as a team, and Lorne Michaels gave notice of his own departure in 1980. This resulted in the rest of the cast and most of the production staff leaving at the end of the season. Jean Doumanian was chosen to produce the series for the 1980–81 season with an entirely new cast headed by Charles Rocket, Denny Dillon, and Joe Piscopo. This new version was a disaster from the start; the critics treated it harshly, and the ratings began declining with the initial show. In danger of losing what had become an extremely valuable attraction, NBC reacted by firing Doumanian and rehiring Dick Ebersol as producer in March 1981. He managed to get one program on the air before it was shut down for a complete overhaul.

By the start of the 1981–82 season, Ebersol had persuaded Michael O'Donoghue to return as head writer, and he also brought back a number of the original writers. He replaced everyone in the cast with the exception of Piscopo and Eddie Murphy and added Robin Duke, Mary Gross, Tim Kazurinsky, Christine Ebersole, and Tony Rosato. After a wobbly start, *SNL* seemed back on the right track, raised its ratings to a respectable level and won renewal for the 1982–83 season.

"SATURDAY NIGHT LIVE WITH HOWARD COSELL"—an attempt by ABC in the fall of 1975 to recapture the essence of the early *Ed Sullivan Show* with a live, topical variety series featuring the acts and celebrities currently in vogue, both in the U.S. and abroad. Scheduled at 8 P.M. on Saturdays—teen time—with the abrasive sportscaster Howard Cosell as host (supposedly fueling the debate between his fans and detractors), it originated from the Ed Sullivan Theater in Manhattan and was placed under the supervision of the network's resident expert in live television, Roone Arledge, president of ABC Sports. None of those touches, nor the hefty promotional campaign, availed. The program opened to poor ratings and died at midseason. It was by Jilary Enterprises, with Don Mischer as director, Walter Kempley as head writer and Arledge as executive producer.

"SATURDAY NIGHT REVUE, THE"—comedy-variety hour of the early 50s which showcased numerous comedians and was hosted first by Jack Carter and later by Eddie Albert and Hoagy Carmichael.

SAUDEK, ROBERT—producer identified with cultural programming, his credits including *Omnibus, Profiles in Courage* and telecasts of the New York Philharmonic. He had been a v.p. of ABC in 1951 when the Ford Foundation hired him away as director of its TV-Radio Workshop, out of which came *Omnibus*, a show that contained the vision of public television. In 1957 Saudek founded his own company, Robert Saudek Associates. He later became president of New York's Museum of Broadcasting, established in 1976 by CBS chairman William S. Paley. He yielded the post to Robert Batscha in 1982.

SAUTER, VAN GORDON—president of CBS News since March 1, 1982, succeeding Bill Leonard on his retirement. Sauter had varied experience as a CBS executive prior to the appointment, some of which was unrelated to news, and it was clear during the late 70s that he was on a fast track to a high post in the company. He was an anomaly in the executive echelons, because he did not dress in the conventional manner and cultivated a beard. In a broadcast career that began in Chicago in 1968, Sauter worked both in television and radio, variously as correspondent, producer, news director, anchorman, and executive. He moved from network broadcasting to local broadcasting and back again, and served as chief censor for the television network as readily as he headed the sports division. Few in television have had such diversified careers.

Sauter had worked as a reporter for several newspapers before joining WBBM Radio in Chicago as chief correspondent in 1968. He soon became news director of the station just as it was adopting the all-news format. In 1970 he moved to New York in the radio department of CBS News as executive producer of special events. Two years later he was back in Chicago as news director of WBBM-TV and then became anchor of the station's *5 O'Clock Report*. CBS News then tapped him as Paris bureau chief in 1975.

Sixteen months later he had a new job–vice president of program practices for the television network. Next, in September 1977, he was put in charge of KNXT, the CBS Los Angeles station that had developed serious ratings problems. Then, after revitalizing the station, he returned to New York in July 1980 as president of CBS Sports. In November the following year he was chosen to succeed Leonard in the news division. He served from that point until Leonard's actual departure as deputy president of CBS News.

SAVITCH, JESSICA— NBC News correspondent, who joined the network in 1977 from Group W. She covered the Senate for NBC (1977–79) and then became a general-assignment correspondent and principal writer-reporter for the Sunday edition of *Nightly News*. NBC recruited her from its Philadelphia affiliate, KYW-TV, where she had worked seven years on *Eyewitness News*, latterly as co-anchor.

"SAY GOODBYE"—David L. Wolper documentary on the disappearing animal species, telecast by NBC in 1970; it drew the wrath of the gun lobby for a scene in which a polar bear was shot from a helicopter and apparently killed. As it proved, the animal had only been shot with a drug so that it might be marked for research, and the pro-hunting forces charged that the scene was staged to discredit hunters. Giving credence to their argument was a sympathy-evoking scene of the mother bear's seemingly bereft cubs wandering about the frozen land.

The film and its outtakes were subpoenaed by the House Committee on Interstate and Foreign Commerce and were supplied by Wolper. But although the incident resulted in nothing more than an implicit reprimand, with NBC guiltless since it was not the producer, it nevertheless served to reinforce network policies against accepting nonfiction materials from outside producers because of the networks' inability to vouch for their factualness.

Wolper maintained that he was presenting an essay and not a news report and that therefore he was entitled to some dramatic license. Moreover, he noted, the whole one-hour program was unjustly discredited for the five minutes that were devoted to the vanishing of animals by means of hunting.

The program had been narrated by Rod McKuen and was sponsored by Quaker Oats.

SCALE—minimum wage permitted by the unions, AFTRA and SAG, for a TV performance or appearance. The rate varies with the type of performance and with whether the broadcast is local, regional or network. There is one scale for principal performances, another for actors speaking fewer than five lines and another for extras. The scale is different for voice-over and on-camera work, and there are special minimums for dancers, singing groups and specialty acts.

SCAN LINE—one line of the television picture. The NTSC television system in use in the U.S. has 525 lines to complete a picture; the European CCIR system has 625. The picture is scanned from left to right. See also Television (Technology of).

"SCARED STRAIGHT"—powerful hour-long documentary, syndicated in the spring of 1979, on a unique project at New Jersey's Rahway State Prison in which convicts sentenced to life confront juvenile offenders in "shock therapy" sessions to scare the crime out of them. Although it played on TV and not in movie houses, the film won an Oscar that year for short subjects; it also scored tremendous ratings, even on independent stations where it faced prime-time network competition. The Signal Companies, which sponsored the program without commercial interruptions, then scheduled a repeat a few months later. The documentary carried an advisory, because of the street language used and because of the brutal, searing nature of the confrontations.

Arnold Shapiro was producer and director, and Peter Falk the host and narrator.

"SCARLET LETTER, THE"—serial adaptation of Nathaniel Hawthorne's 19th-century classic, produced by WGBH Boston and presented on four consecutive nights in April 1979. Conceived and produced by Rick Hauser, it featured Meg Foster, Kevin Conway, and John Heard in the principal roles.

While it was not generally considered a critical success, the serial enjoyed the highest ratings of any PBS drama series in the 1978–79 season. Several aspects of the production were blamed for the artistic shortcomings of the project, among them the system of "step funding" by which the government underwrites public television programs. This system not only interfered with the creative processes but caused the project to take 4½ years to complete. The rationale for "step funding" is protection of public funds, but in this instance, as it has in several others, the process turned out to be an outrageously expensive form of insurance.

The four-step process follows this form: a planning grant to investigate the need for the project; then a Research & Development grant to cover script development costs; then a pilot grant to finance production of a pilot film; and finally a production grant to produce the work. A formal proposal is required at each of these steps, and often the proposals run longer, and are more time-consuming, than the scripts themselves. In the case of *The Scarlet Letter,* the R&D grant proposal alone ran 300 pages.

The National Endowment for the Humanities, a council of 26 Senate appointees, provided the seed money and reviewed the grant proposals for *The Scarlet Letter.* In the spirit of protecting public funds, the Endowment engaged 32 scholar-consultants to review the work from proposal to final script. They also reviewed the reviews of 16 scholar-consultants who were retained by the production team.

The Scarlet Letter went through two writers to qualify for the production grant. Allan Knee, who took the script through two years of proposals and scholar-dictated revisions, was sacrificed when the Endowment held up approval of the production request. Reviewers for the Endowment could not agree on the production's interpretation of the novel. Finally, Alvin Sapinsly was retained as writer, and his scripts were accepted and then produced.

SCATTER BUY—advertising time purchased in a manner that would disperse commercials for a product over several programs on different nights of the week, as opposed to straight sponsorship, wherein the ads are concentrated within a single show. See also Magazine Concept, Participating.

Erland Josephson & Liv Ullmann in *Scenes from a Marriage*

"SCENES FROM A MARRIAGE"—Ingmar Bergman's celebrated six-part series for Swedish TV on the disintegration of a once "ideal" marriage, broadcast in Sweden in the spring of 1973 and in the U.S. on public television in 1977. Before it played on PBS—in a dubbed version for the first outing and with subtitles for the repeats later in the week—it had been presented theatrically in America as an edited-down two-and-a-half-hour movie.

The leads were played by Liv Ullmann and Erland Josephson, with Bibi Andersson and Jan Malmsjo in the supporting cast. Bergman wrote and directed the series, and Lars-Owe Carlberg was executive producer.

SCHAEFER, GEORGE L.—producer-director of quality drama, both for TV and the Broadway stage, who had a long and distinguished association with *Hallmark Hall of Fame* (1955–68) and also produced for *CBS Playhouse.* His dozens of productions for TV, most of them under the aegis of his Compass Productions, included *Little Moon of Alban, Macbeth, Victoria Regina, Do Not Go Gentle Into that Good Night, In This House of Brede,* and the series, *Sandburg's Lincoln.* He was also executive producer of the short-lived *Love Story* series on NBC in the early 70s.

SCHAFFNER, FRANKLIN J.—producer and director of live TV drama in the "golden age" who went on to direct movies (*Patton, The Best Man, The Stripper*). He was one of the giants of video theater in the 50s and early-60s as director for *Studio One* (1948–56), *Ford Theater* (1951–52), *Person to Person* (1956–58) and *Playhouse 90* (1958–60), and producer-director of the *Kaiser Aluminum Hour* (1958) and *DuPont Show of the Week* (1963–65).

Chris Schenkel

SCHENKEL, CHRIS—ABC sports commentator whose wide range of announcing assignments included NCAA football, golf championships, the Professional Bowlers Tour and the summer Olympic games (since 1968). For nine years prior to his exclusive association with ABC, he regularly reported the Triple Crown events of horse racing on CBS and NBC, in addition to golf tournaments and heavyweight fights. He made his start in network television in 1952 when ABC hired him to replace the ailing Ted Husing for the Monday night boxing shows. That assignment lasted six years.

SCHERER, RAY—NBC correspondent in Washington and London (1947–75) who in 1976 became Washington v.p. for the parent company, RCA.

SCHERICK, EDGAR J.—movie and TV producer and former program chief of ABC (1963–66); he founded and became president of Palomar Pictures International in 1967, producing such films as *For Love of Ivy, Sleuth* and *They Shoot Horses, Don't They?*, along with numerous TV movies. As program v.p., Scherick was responsible for bringing to ABC such series as *The F.B.I., The Hollywood Palace, Peyton Place* and *Bewitched*.

He came to ABC in 1961 when the network purchased his independent sports packaging company, Sports Pro-

grams Inc., which was to serve as the foundation for the emerging ABC Sports. Scherick had started the company in 1957, after working two years in the CBS sports department. He became v.p. in charge of sales for the ABC network and then v.p. over programming.

In 1980, NBC's Fred Silverman talked him into taking over as the executive in charge of that network's Project Peacock series of prime-time children's specials. In the next year-and-a-half, he supervised the making of some 8 or 9 specials, but after Silverman's departure in 1981, Scherick left NBC and entered an agreement to return to independent TV production in association with Metromedia Producer's Corp.

SCHIEFFER, BOB—anchor of CBS's daily hour-long newscast, *Morning,* since January 1979 and previously a White House correspondent (1974–78) and the CBS Pentagon correspondent (1970–74). He doubled, while covering the White House, as anchorman of the Saturday edition of *The CBS Evening News* for two years and, prior to that, the Sunday edition.

Schieffer joined CBS News in 1969 from WBAP-TV Dallas-Fort Worth, Tex., and before that was with the Fort Worth *Star-Telegram*.

SCHILDHAUSE, SOL—first chief of the FCC's Cable Television Bureau who resigned early in 1974 after a 25-year career on the commission staff. In the 60s he headed the cable-TV task force that later became the bureau. His outspoken advocacy of cable was a source of concern to broadcasters and some members of the FCC.

SCHLOSSER, HERBERT S.—president of NBC (1974–78), after which he became executive v.p. of the parent company, RCA Corporation, responsible for developing the initial line of software for the forthcoming line of video disc playback units. In 1981, he was put in charge of a new RCA group consolidating all entertainment operations—RCA Records; SelectaVision; video discs and cassettes; the partnership with Rockefeller Center in RCTV, the new pay-cable network; and the partnership with Columbia Pictures in international disc and cassette distribution.

His rise to the top of NBC was steady, ever since his first job with the company, as an attorney for the syndication subsidiary in 1957. Highly competitive and outgoing, and with a background in the program area of television, Schlosser brought to the company an aggressive management style that had been lacking for nearly 10 years. He was named chief executive officer in February 1977.

On graduating from Yale Law School, Schlosser entered a Wall Street law practice and later joined Phillips, Nizer,

Benjamin, Krim and Ballon, a New York law firm with a large motion picture and television practice. He left in 1957 to become an attorney for California National Productions, then the syndication arm of NBC.

That led to a position as v.p. of programs on the West Coast for the TV network. In 1973 he was transferred to New York as executive v.p. of NBC-TV and in the summer of the year was named president of the network. Schlosser is credited with authorizing the successful NBC experimental program, *Tomorrow,* and *Saturday Night.*

SCHLATTER, GEORGE—variety show producer who made his mark in the late 60s as executive producer of *Rowan & Martin's Laugh-In.* Then, in the same capacity, he did *The New Bill Cosby Show* and the *Cher* series. He developed and became excutive producer of NBC's highly successful, *Real People.*

SCHMERTZ, HERBERT—possibly the leading TV impresario from the advertising side during the 1970s, as v.p. of public affairs for the Mobil Corporation. He was responsible for creating such PBS series as *Masterpiece Theater, The Nader Report, Mystery Mystery* and *The Way It Was* and for the annual ad hoc networks of commercial TV stations for such series as *Ten Who Dared, Between the Wars* and *Edward the King.* These were in addition to network specials and a number of national radio series.

Flamboyant and outspoken, unusual traits for the chief spokesman of a major corporation, Schmertz mounted his campaign of cultural television at a time in the national energy crisis when the soaring profits of oil companies made them highly suspect to the public at large. That negative image, in Mobil's case, was counterbalanced by the company's new projection as the good provider of such outstanding British series as *Cakes and Ale, Upstairs, Downstairs, Poldark, Lillie* and *I, Claudius.* Other oil companies such as Exxon, Gulf and Arco followed Mobil's lead into public television, as program underwriters, which inspired the PBS nickname of the Petroleum Broadcasting System.

See also Ad Hoc Networks, *Masterpiece Theater,* Public Television.

SCHMIDT, ROBERT —president of National Cable Television Assn. (1975–79). Previously, he had been a lawyer in Washington with clients in professional sports and earlier had been a public affairs executive for ITT. An accomplished college athlete, he had had a brief professional football career with the New York Titans, a team that preceded the Jets. From 1961 to 1964 he worked in various staff positions with the Democratic National Committee.

SCHNEIDER, ALAN—one of the leading directors of avant-garde plays who received his grounding in TV in the 50s. His work for television has included *Pullman Car Hiawatha* in 1951, *Oedipus Rex* in 1958, Samuel Beckett's *Eh, Joe?* in 1966 and Elie Wiesel's *Zalmen, or the Madness of God* in 1976.

SCHNEIDER, ALFRED R.—longtime ABC corporate executive who since 1972 has been v.p. with administrative responsibility for the Standards & Practices department and chairman of the Corporate Contributions Committee of ABC Inc. For the ten previous years he was v.p. and assistant to the executive vice president of ABC Inc. and before that v.p. in charge of administration for the TV network. A lawyer, he joined ABC in 1952 as a member of the legal department. He was with CBS from 1955 to 1960, first in business affairs and then as executive assistant to the president of the network. He rejoined ABC in a similar capacity.

John A. Schneider

SCHNEIDER, JOHN A.—president of the CBS Broadcast Group (1966–77) who was deposed in the management upheaval that made Gene F. Jankowski his successor in October 1977. Schneider, who had been one of the most powerful figures in broadcasting and who for a time was considered heir-apparent to the presidency if not the chairmanship of the corporation, left CBS in April 1978 to pursue other ventures. In 1979 he became president of Warner Amex Satellite Entertainment Corp (WASEC), the software subsidiary of Warner Amex Cable. There he has overall responsibility for the Movie Channel, Nickelodeon, Music Television, and other cable networks under development by the company.

Schneider rose from relative obscurity in 1965 when, as a station manager, he was named to succeed the legendary Jim

Aubrey as president of the CBS television network. A low-keyed executive of impeccable reputation, he was at the time the diametrical opposite of Aubrey in the company—businesslike, outgoing and ingratiating. He had been v.p. and general manager of WCBS-TV New York and before that of WCAU-TV Philadelphia. Until then his background had been entirely in broadcast sales, beginning in 1949 in his home town, Chicago, with WGN.

A successful year as network president earned him the promotion as head of the newly created broadcast group. In 1969 he appeared to be the leading candidate to replace Frank Stanton as corporate president when he was appointed executive vice-president of CBS Inc. But when the plans were changed, and the new president brought in from outside the company, Schneider returned to his previous post as president of the broadcast group.

His brush with controversy occurred during 1966 when Fred W. Friendly, then president of CBS News, charged Schneider with denying air time for the coverage of the Senate hearings on the Vietnam war in order to carry the commercially productive reruns of *I Love Lucy.* Friendly's resignation from CBS was tied to that episode. Schneider insisted that he was wronged, and that Friendly's departure was prompted by a change in his relationship with top management rather than by the denial of air time.

SCHOENBRUN, DAVID—foreign correspondent for CBS News (1945–62), based in France for most of those years, after which he became chief correspondent in the Washington bureau for a year. Since then, he has been a correspondent and commentator for broadcast groups such as Metromedia and a freelance journalist.

SCHONFELD, MAURICE W. (Reese)—president of Cable News Network (CNN) from its inception in 1980. Prior to that he was managing director of Independent Television News Association, which he founded in 1975; this was an organization that united the leading independent stations in a nightly news service. He began his news career with UPI and advanced to become v.p. of UPITN.

"SCHOOLHOUSE ROCK"—umbrella title for five-minute animated series on educational subjects designed for insertion in ABC's Saturday and Sunday morning children's schedules. The series began in 1972 with *Multiplication Rock* and added *Grammar Rock* and *America Rock,* on history, in subsequent seasons. General Foods and Nabisco were sponsors. It was produced by Scholastic Rock Inc., with Tom Yohe as executive producer and Phil Kimmelman and Associates directing the animation.

Daniel Schorr

SCHORR, DANIEL—veteran CBS newsman known for his investigative skills, whose frequent involvement in journalistic controversies climaxed with one that cost him his job at the network. In February 1976, after reporting on the contents of a secret report by the House Intelligence Committee on improper activities of the CIA and FBI, Schorr passed on the full document to the New York weekly, *The Village Voice,* for publication. Schorr and *The New York Times* had already covered the essentials of the report, but the House nevertheless voted to classify it as secret.

Schorr, who had been with CBS News for 23 years, had advised officials of the company that he meant to find publication somewhere for the copy of the report that had been leaked to him; yet, on the day it appeared in *The Voice,* he denied having been the source. A day later, he did admit to having passed it along. What occurred during the day of denial made it difficult for CBS to retain his services. He was suspended from reporting on Feb. 18, 1976, pending the resolution of a congressional inquiry into the leaking of the secret report.

CBS indicated at the time that it would defend Schorr's right to protect his sources and that it would pay for his defense in the event of a congressional subpoena (which came in August). But there was a strong hint in the terse official statement that he might not be reinstated at the network even if the probe were to end satisfactorily.

In the large sphere, the issue was whether a reporter had the right to override the judgment of Congress and to determine on his own that the people's right to know demanded publication. But at CBS, that issue came down to one of whether the reporter who received the material got it as an independent journalist or as an agent of CBS News; if as the latter, then it should have been up to CBS News and not

Schorr to decide whether to preempt the judgment of Congress and have the material published.

While that was a question that haunted the episode, it was not the critical issue in Schorr's relations with the network. On the day of publication, when Schorr denied having had a role in it, suspicion in the Washington bureau fell upon Lesley Stahl, a correspondent who was linked romantically with Aaron Latham, an editor of *The Voice* who had written the preface to the transcript of the House report. It was thought by some that Miss Stahl had taken the document from Schorr's desk and that she made photocopies of it for Latham. Not only had Schorr failed to put that rumor immediately to rest, he was said by some in the bureau to have initiated it.

Schorr later explained that his attorney had advised him to maintain silence that day until he could discuss his action with CBS officials in the presence of company lawyers. But CBS executives found it unforgivable that Schorr would allow a colleague to be unjustly accused of something as serious as stealing the materials.

Schorr's brilliant presentation before the congressional committee was hailed throughout the profession of journalism, and the committee thereupon dropped its investigation into the source of the leak. But just before he was to have entered into discussion with executives of CBS News on his future status, Schorr submitted to an interview on *60 Minutes* with Mike Wallace. Possibly to his surprise, the segment concentrated not on Schorr's great victory for journalism but rather on the tacky episode of his failure to disabuse CBS executives of their momentary suspicions about Miss Stahl. Thus the internal issue at CBS in the Schorr affair was externalized before a national audience.

Soon afterwards Schorr resigned. The arrangement kept his salary in force for the remainder of his three-year contract. He went on to teach, lecture and write a book. In 1979, he signed on as chief anchor of Ted Turner's Cable News Network.

SCHOUMACHER, DAVID—news correspondent for CBS and ABC who became anchorman for Washington station WMAL-TV in February 1976. A Chicagoan, he worked at TV stations there and in Oklahoma until he was hired to work out of the CBS Washington bureau in 1963. After a nine-year stint, which included a period in Vietnam, he switched to ABC News, again dividing his time between Vietnam and Washington, until WMAL lured him away. He had the reputation of a hard-working and scrupulous reporter and newsfilm producer who, unlike some, always did his own writing.

SCHULBERG, STUART (d. 1979)—executive producer for NBC News who for eight years (1968–76) was producer

of the *Today* show, the longest anyone has held that post. His assignment, on leaving *Today*, was to produce documentaries.

Schulberg, brother of novelist Budd Schulberg, joined NBC in 1961 as co-producer with Ted Yates of *David Brinkley's Journal*. In 1965 he became producer of *NBC Sports in Action*. His early documentaries include *The Angry Voices of Watts* (1966), *The Air of Disaster* (1966), *Losers Weepers* (1967), *The New Voices of Watts* (1968) and *Somehow It Works*.

SCHULLER, ROBERT—television preacher in the tradition of Norman Vincent Peale who preaches self-esteem and positive thinking on a weekly syndicated series, *Hour of Power*. The series, which began in 1970, airs on 149 stations and the Armed Forces Network. Like other TV evangelists, Schuller buys the air time on every station that carries the program; unlike the others, he belongs to one of the mainline Protestant churches, the Reformed Church in America, although his denomination does not pay for his telecasts. *Hour of Power* developed a strong following because of Schuller's charismatic personality and his ability to combine psychological with spiritual guidance.

Barbara Schultz

SCHULTZ, BARBARA—executive generally associated with dramas of substance, in both public and commercial TV. In 1974 she became head of the PTV *Visions* project, with the title of artistic director, after grants totaling $6.2 million had been made by the National Endowment of the Arts, the Ford Foundation and CPB for a three-year series of original dramas for TV by American playwrights. The series went on the air in the fall of 1976. Earlier she had been producer of video tape drama for Universal Television. For a number of years she was with CBS, variously as executive producer of *CBS Children's Hour*, executive producer of *CBS*

Playhouse and director of program development for the network. She had also been story editor on a number of CBS series, including *The Defenders, Trials of O'Brien* and *Armstrong Circle Theatre.* For a time, she served as consultant to CTW's *Electric Company.*

SCHULZ, CHARLES M.—creator and cartoonist of the *Peanuts* comic strip; he has written all the scripts for the *Charlie Brown* animated specials derived from that source.

SCHWARTZ, SHERWOOD—situation comedy producer who after a long association as writer for Red Skelton created and produced such series as *Gilligan's Island, It's About Time* and *Dusty's Trail.* He was also executive producer of *The Brady Bunch,* and *The Brady Girls Get Married.*

SCHWARTZ, WALTER A.—president of ABC Television from 1972 to 1974; he left ABC in December 1975 to become president of the station division of Blair Television, a rep firm.

First to hold the title of president of ABC Television, a new post that was created in a reorganization, Schwartz had been appointed to it after a successful five-year tenure as president of the four ABC Radio networks. The position put him in charge of all television divisions of the company—network, owned stations and production subsidiaries. Schwartz was relieved of it, and named president of ABC Scenic Attractions, a nonbroadcast unit, when ABC-TV made a poor showing in the ratings in the fall of 1974.

Schwartz came to ABC in 1963 from Group W, where he had worked in radio sales and management, to be general manager of its New York radio station, WABC.

SCHWIMMER, WALTER—Chicago packager and producer of syndicated programs, among them *Championship Bowling, Championship Bridge, All Star Golf* and *Let's Go To the Races.* He created a sports special event for the networks in *World Series of Golf* (1962) and attempted to make an annual TV series of the Nobel Prize awards, producing the network telecast in 1964 that proved to be the first and last. He sold his company, Walter Schwimmer Inc., to Cox Broadcasting in 1966.

"SCOTLAND YARD"—early British import which played on ABC (1957) for 26 weeks. Edgar Lustgarten hosted the episodes, which were based on cases from the Yard's files. It was by Anglo Amalgamated Film.

SCOTT, WALTER D.—chairman of NBC from April 1, 1966 to April 1, 1974, when he retired. He served briefly as president of the company and as president of the TV network. Scott maintained low visibility outside NBC and was not known for exceptional contributions to television except in the realm of sales. He rose in the company through the sales department, which he had joined in 1938, and became sales v.p. in 1955 and executive v.p. in 1959. He was president of the network for a year in 1965 before moving up to corporate.

SCOTTISH TELEVISION—commercial licensee for central Scotland with headquarters in Cowcaddens, Glasgow and other studios in Edinburgh.

SCOURBY, ALEXANDER—actor considered by many the ideal narrator, and he has performed that function hundreds of times on TV for documentaries and cultural and travel programs. He was used frequently by NBC for its *Project 20* specials and by David Wolper for the National Geographic documentaries. Essentially a stage actor, he had running roles in such daytime soap operas as *The Secret Storm* and *All My Children* and has had guest roles in a number of prime-time dramatic series.

SCREEN ACTORS GUILD (SAG)—union for performers in film, powerful in television because a majority of the programs are made on film. Many members of SAG also belong to AFTRA, the performer's union for live or video tape television, and the two have had numerous jurisdictional clashes since 1950. Talk of merging the unions, whose contracts have become fairly similar, has gone on since 1960 but has consistently been rejected by the SAG membership. SAG, which was formed in 1933 to represent performers against the abuses of motion picture studios and producers, has struck television four times. Its notable innovation has been the *residual*—the continuing compensation to actors and actresses as their TV films and commercials are reused in reruns. In 1960, after a six-week strike, SAG achieved a formula for payments to actors and actresses when their theatrical movies are sold to TV. A summertime strike in 1980 over the actor's share of home video rights disrupted the fall season and caused programs to premiere on staggered schedules.

The union's presidents have included Ronald Reagan, Charlton Heston, and Edward Asner. See also AFTRA.

SCREEN GEMS (now Columbia Pictures TV)—TV subsidiary of Columbia Pictures Corp. and one of the major program suppliers to the networks from the time the Hollywood film studios entered TV production in the late 50s. Founded in 1951, Screen Gems was also a pioneer in TV syndication.

The company name was changed in 1976 to Columbia Pictures TV (CPT). John Mitchell, who was one of the fabled "super salesmen" of the production field during the 60s, was the executive longest identified with *Screen Gems.* He resigned as president of CPT in 1977.

SCTV

"SCTV (SECOND CITY TV)"—improvisational comedy show produced out of Toronto, in syndication for several years before NBC acquired it for Friday late-night spot in fall of 1981, where it finally hit its stride. NBC money bought better production values (including improved laughtrack) and the network exposure seemed to help. *Second City* (distantly related to the Chicago improvisational troupe which provided a launching platform for Mike Nichols & Elaine May, Alan Arkin, and many others) combines some of the wittiest writing to be seen on TV in years with superb ensemble acting by John Candy, Joe Flaherty, Eugene Levy, Andrea Martin, Rick Moranis, Catherine O'Hara, and Dave Thomas. Moranis's and Thomas's extensive catalogue of parodies of comedians (Dick Cavett, Woody Allen, Rodney Dangerfield, Bob Hope, and many others) is of special interest. Moranis and Thomas gained some degree of cult fame in 1981–82 with their beer-swilling Canadian "Great White North" brothers, Bob and Doug Mackenzie.

SCULLY, VIN—veteran sportscaster who has broadcast the Brooklyn and then Los Angeles Dodgers games since 1950. In addition, as a member of CBS Sports since 1975, he does commentary for the National Football League telecasts and various golf tournaments. He also hosted the network's *Challenge of the Sexes* series and has co-hosted the Tournament

of Roses Parade telecasts. In the entertainment sphere, he was the unseen narrator of a 1966 NBC comedy series, *Occasional Wife,* the host of an NBC daytime panel show, *It Takes Two* (1969–70) and emcee of a 1973 variety program on CBS, *The Vin Scully Show,* which ran three months.

"SEA HUNT"—one of the most successful series made expressly for syndication (1957–61), notable also for its excellent underwater photography. The series concerned the adventures of an ex-Navy frogman whose skills are used for underwater searches. Lloyd Bridges starred and Ivan Tors was producer for Ziv-United Artists.

"SEARCH"—hour-long series about a securities firm of globetrotting agents who protect and insure banks, jewels and art collections.

Starring Hugh O'Brian, Doug McClure and Tony Franciosa on a rotating basis, with Angel Tomkins as romantic interest, the show premiered on NBC in September 1972 and remained 24 weeks. It was produced by Warner Bros. TV with Leslie Stevens Productions.

"SEARCH FOR TOMORROW"—oldest of the daytime serials, a carry-over from radio which had its TV premiere on CBS Sept. 3, 1951. The principal character has always been Joanne Tourneau, played by Mary Stuart, but after a quarter century the storylines have shifted to the younger generations in the fictive town of Henderson. Others in the cast have been Larry Haines, Rod Arrants, Nicolette Goulet, Sherry Mathis, Lisa Peluso, Millee Taggart, Marie Cheatham and Wayne Tippit.

On March 29, 1982, the series moved, without missing an episode, from CBS to NBC at the behest of its sponsor, Proctor & Gamble. P & G wanted the show to continue in its traditional timeslot, 12:30 P.M. which CBS could or would not supply.

Over the years the soap opera has featured such actors and actresses as Don Knotts, Sandy Duncan, Lee Grant, Tom Ewell, George Maharis, Roy Scheider, Barbara Baxley and Hal Linden.

SECA (SOUTHERN EDUCATIONAL COMMUNICATIONS ASSN.)—both a production center and a regional network in public TV, usually referred to in the industry as "the South Carolina Network" since it is based in Columbia, S.C. Among its productions have been William F. Buckley Jr.'s *Firing Line, Lowell Thomas Remembers, An Evening at the Dock Street Theatre* and the series of dramatized short stories, *Anthology.* SECA serves 87 public TV stations in 16 states, with instructional as well as PTV programming.

SECAM—Sequence à Memoire color system, developed by French commercial and government interests and first broadcast by France in 1967. It is incompatible with both the German-developed PAL system and the American NTSC. The SECAM system has also been adopted by most middle-eastern and African countries in the French sphere of influence and, in modified form, by the Soviet Union and most Eastern European nations.

SECONDARI, JOHN H. (d. 1975)—news correspondent for CBS and then ABC, who became an independent documentary producer with his wife, Helen Jean Rogers, in 1969. After heading the Rome bureau for CBS, he joined ABC in the late 50s as Washington bureau chief. In 1961 he organized the network's first documentary unit, serving as executive producer of *The Saga of Western Man* series and as the ABC News liaison with Drew Associates on the *Close-Up* documentary series. At the time of his death he was preparing a Bicentennial series through his own company.

"SECRET AGENT"—British series carried on CBS here (1965) about a professional spy on dangerous assignments around the world. By ATV, it starred Patrick McGoohan.

SECOND SEASON—the period between late December and late February, when the networks attempt to bolster their schedules by canceling their failures and installing new series which may be contenders for the following fall. The term "second season" was coined by Thomas W. Moore while he was president of ABC-TV to herald extensive program changes his network was forced to make during one of its lesser years in the 60s. By the early 70s, revising schedules at midseason had become common practice for all three networks, with each developing shows expressly for replacement service. Among the big hits that had been second season entries were *Batman, Rowan & Martin's Laugh-In* and *All In the Family.*

SECTION 315—See Equal Time Law.

"SEE IT NOW"—TV's first great documentary series and one that will probably never be exceeded for courage. While Sen. Joseph McCarthy was at the peak of his power, the series dared repeatedly to expose the injustices of McCarthyism and the viciousness of the senator in his zeal to discover communists and sympathizers. Indeed, the CBS series—a collaboration of Edward R. Murrow and Fred W. Friendly—contributed to the downfall of McCarthy, whose demagogu-

ery exposed itself when he went on the program to respond to Murrow's attacks.

See It Now, which made its debut on Nov. 18, 1951, grew out of a radio series, *Hear It Now,* which in turn was inspired by a successful Murrow-Friendly record album, *I Can Hear It Now,* a collection of recorded history. Sponsored by Alcoa, the half-hour prime-time TV program began with fairly standard documentary subjects but drifted gradually to the controversial. In October 1953 the series created a national stir with *The Case Against Milo Radulovich,* pointing up the injustice to an Air Force reservist who was deemed a security risk and asked to resign his commission because his father and sister were suspected by unnamed accusers of having subversive tendencies. This became the first of several programs concerning McCarthyism. On March 9, 1954, *See It Now* offered a report on the senator himself, consisting mostly of film clips of McCarthy which revealed his bullying tactics and reckless accusations of disloyalty and subversion. It was followed a week later by a program showing McCarthy badgering Annie Lee Moss, a suspected communist. On April 6 McCarthy went on the program to defend himself and denounce Murrow, and the newsman became a hero to some viewers and a traitor to others. The actual undoing of McCarthy occurred with the televised Army-McCarthy hearings, which began April 22, but the *See It Now* programs opened the way.

See It Now: Senator Margaret Chase Smith & Winston Churchill

The series lost its sponsor, gained another and lost that one in the fall of 1957. It continued to deal with other controversial subjects and was considered by most critics the most distinguished news program of its time. CBS canceled it in 1958, removed Murrow from the forefront and replaced *See It Now* soon afterward with *CBS Reports.* Along with Murrow, who was commentator and writer, and Friendly, executive producer, those connected with *See It Now* included Palmer Williams, producer; Don Hewitt, director; Joe Wershba, producer-reporter; Don Hollenbeck, reporter; and Charlie Mack, cameraman.

SEGAL, ALEX—drama director working between theater and TV, whose prominence in the latter dates to the early 50s with *Pulitzer Prize Playhouse, Celanese Theater, U.S. Steel Hour* and others. He directed *My Father's House* for *CBS Playhouse* in the 70s and, in the late 60s, *Death of a Salesman, The Crucible* and *Diary of Anne Frank,* among others.

SEGELSTEIN, IRWIN B.—vice chairman of NBC since 1981 after having served in a variety of executive capacities from the time he joined the company in 1976. Earlier he had been a high-ranking program executive with CBS (1965–73) and then president of CBS Records (1973–76). Segelstein proceeded up the ladder at both networks without apparent striving for higher office; always he was tapped for the next rung because he had the indicated skills, experience and integrity.

Two years after he joined NBC as executive v.p. of programs, his old friend and colleague Fred Silverman became president of the company; the two had worked together for many years and were neighbors in a Manhattan apartment building. Segelstein at once became Silverman's most trusted captain at NBC and took over large areas of corporate responsibility, including supervision of the radio and owned-stations divisions. It reflects Segelstein's abilities that when Silverman left the company, his successor, Grant Tinker, named Segelstein vice chairman, a higher post than he had held under Silverman.

Segelstein had been head of programming for Benton & Bowles Advertising when CBS recruited him in 1965 in a reorganization of the staff after the departure of James Aubrey as president. Segelstein signed on as v.p. of programs, New York, under Michael H. Dann, and in 1970 he became v.p. of program administration under Fred Silverman. Segelstein had been bypassed for Silverman to succeed Dann as head of programming for CBS.

When a scandal broke at CBS Records (the Columbia label) under the administration of Clive Davis, who subsequently was fired for alleged misuse of company funds, Segelstein was named to succeed him as president. And after NBC endured one of its most embarrassing seasons in the ratings (1975–76), it fired its program chief, Marvin Antonowsky, and hired away Segelstein to become the executive in charge, with Paul Klein, reporting to him as head of the department.

SELF, WILLIAM—president of the CBS Theatrical Films Group since March 1982, after having been a leading TV program figure in Hollywood through the 60s and 70s.

After working for CBS as an executive producer in the 50s, Self joined Twentieth Century-Fox in 1959 as head of its television division and continued in that post for 15 years. After leaving Fox, he teamed with Mike Frankovich in producing theatrical movies but rejoined CBS two years later as program v.p. for the West Coast.

Eventually he became v.p. in charge of made-for-TV movies and mini-series for the network, overseeing 40 to 50 titles a season. His adeptness in this field contributed to the revival of CBS in the prime-time ratings and its return to dominance in the 80s. It also established Self as the logical successor to Michael Levy as head of the theatrical films division.

During his years at Fox, Self was responsible for developing 44 television series that aired on the networks, including *M*A*S*H, Peyton Place, Room 222,* and *Batman.* In his early years with CBS, he was involved in developing *Twilight Zone.*

SELIGMAN, SELIG J. (d. 1969)—a top ABC executive until his death, as head of Selmur Productions, telefilm production arm of ABC which turned out such series as *Combat, Garrison's Gorillas, Day In Court* and *General Hospital.* Seligman was executive producer of those and other shows and a v.p. of ABC Inc.

With a law and theater administration background, he joined ABC in 1953 as a writer-producer and in 1955 became general manager of KABC-TV Los Angeles. Three years later he became a v.p. and soon afterward headed the network's production subsidiary.

"SELLING OF THE PENTAGON, THE"—CBS News documentary (Feb. 23, 1971) on the Pentagon's promotional activities. It had jarring repercussions and was assailed at high levels of government. An attempt by a House committee to investigate its editing procedures nearly brought a contempt of Congress citation upon the president of CBS Inc., Frank Stanton. In the midst of the furor, however, the film—produced by Peter Davis with Roger Mudd as reporter and narrator—won a special Peabody Award; Stanton later received RTNDA's Paul White Memorial Award for his courageous denial of Rep. Harley O. Stagger's subpoena for the outtakes, scripts and other materials used in the preparation of the documentary.

Although the documentary did not deal with the massive defense budget and its implications on the American economy, but concentrated on the use of federal funds to show off American weaponry at state fairs and other civic and social functions, the film was denounced by Vice-President Agnew as "a clever propaganda attempt to discredit the Defense establishment" and by Rep. F. Edward Hebert, chairman of the House Armed Services Committee, as "one of the most un-American things I've ever seen on a screen."

The program became vulnerable not for its central argument—that funds were being improperly used to glorify the tools of war at the grassroots and to promote American might in the Cold War—but for a number of liberties taken in the editing process. Those included the discarding of qualifying phrases by some of the persons interviewed and the joining of statements made in a number of contexts as the single answer to a question. This method of excerpting brought complaints from Col. John MacNeil and Daniel Z. Henkin, Assistant Secretary of Defense for public affairs, both of whom claimed to be personally embarrassed by it.

The editing issue prompted Rep. Staggers, chairman of the House Interstate and Foreign Commerce Committee and of its subcommittee on investigations, to subpoena the outtakes and background materials. Stanton insisted that they were equal to a reporter's notes and therefore privileged and said he would turn over only the completed film. Eventually, the House voted down Staggers's proposed citation of Stanton for contempt.

CBS aired *Pentagon* a second time in the *CBS News Hour* less than a month after the original telecast, ostensibly to let viewers see what the controversy was all about. The second broadcast reached a larger audience than the first had.

SENATE COMMUNICATIONS SUBCOMMITTEE (OF THE SENATE COMMERCE COMMITTEE)—the subcommittee most active in broadcast matters, with oversight of the FCC and the communications industries. It is perennially concerned with the possible need for legislation in such areas as cable-TV regulation, FCC regulatory reform, international satellite policy, broadcast license renewals, public broadcasting, UHF, copyright, and more. The subcommittee's hearings and investigations, which may lead to amendments to the Communications Act, have reflected its concern with sex and violence on TV, commercial practices in children's television, minority hiring practices in broadcasting and the lack of program diversity.

The subcommittee considers nominations to the FCC, the Corporation for Public Broadcasting and the board of directors of the Communications Satellite Corp. See also House Communications Subcommittee.

"SERGEANT PRESTON OF THE YUKON"—CBS series (1955–57) on the Royal Canadian Mounted Police, featuring Richard Simmons and his dog, King, and produced by Wrather Corp., the source of *Lassie*.

SERLING, ROD (d. 1975)—TV dramatist, producer and narrator whose credits range from the authorship of such outstanding plays as *Patterns* and *Requiem for a Heavyweight* to commercial spieling. After an apprenticeship as a radio and

TV writer for stations in Cincinnati, Serling quickly entered the front rank of playwrights in the medium's "golden age" when *Patterns* was produced on *Kraft Theater* in 1955.

For the various live drama series on the networks, Serling wrote, among other plays, *The Rack, A Town Turned to Dust* and *Requiem for a Heavyweight,* the latter for *Playhouse 90. Patterns* and *Requiem* were made into movies, and both, along with *Dust,* won Emmy Awards.

CBS signed Serling to contract in 1956. When live dramas began to fade, he created the series of occult tales, *Twilight Zone,* which he both wrote and hosted. The series ran from 1959 to 1964 on CBS and afterwards in syndication. A few years later, he created *Rod Serling's Night Gallery,*- another but less successful anthology of mystery tales, which ran on NBC.

Serling's image on camera was that of a well-educated tough guy, and his earnest delivery was so effective that he was used frequently as a commercial spokesman in the early 70s. He also, in his last years, narrated documentaries and nature programs, including the Jacques Cousteau oceanographic specials. Serling had been engaged as host of an ABC summer comedy series, *Keep On Truckin',* in 1975, but he died of complications following open-heart surgery before the series began.

Rod Serling

"SERPICO"—one-hour NBC police-adventure series (1976) loosely based on the film of that title concerning the real-life Frank Serpico, a New York City undercover detective who exposed corruption in the police department. The movie had been based on a book by Peter Maas. In the short-lived TV version, David Birney was featured in the title role. The series was by Emmett G. Lavery Jr. Productions, in association with Paramount TV, with Lavery as executive producer and Don Ingalls as producer.

SESAC (SOCIETY OF EUROPEAN STAGE AUTHORS AND COMPOSERS)—smallest of the three music-licensing organizations (the others being ASCAP and BMI) which originally represented European works but now is based in the U.S. and has compiled its own domestic repertory, heavily in the religious music and gospel field. It was established in 1931 and collects music licensing fees from networks and stations. See also ASCAP, BMI.

Big Bird & Friend

"SESAME STREET"—a revolutionary children's program which sought to teach numbers and letters (and later social concepts) to preschool children through TV entertainment, chiefly by harnessing the techniques used in commercials.

The brainchild of Joan Ganz Cooney, and the initial production of the Children's Television Workshop which she headed, it premiered as a daily hour-long show on PBS in November 1969, with an initial budget of $8 million, after two years of research and extensive piloting. The funds were raised from a variety of foundations and from CPB and HEW. The show received enthusiastic reviews, won numerous major awards and drew a regular U.S. audience of some 9 million children. Moreover, it gave PTV a show that frequently beat commercial TV competition in the ratings. And in setting a new standard for children's programming, it inspired some commercial imitation of its ability to teach through entertainment.

Employing a modular format, allowing for a mixture of fresh and repeated sequences in each program, *Sesame* adopted the pacing and razzle-dazzle of TV spot commercials, making virtuous use of their most arresting devices. Deriving its continuity from a street scene, the series intermingled a regular cast of adults and children with Muppets, animation and live-action film. The episodes cost approximately $50,000 each to produce, and 130 one-hour shows were produced the first year.

Sesame's target audience was disadvantaged inner-city children who lacked the language skills of preschoolers of the middle class. Evaluative tests by CTW and Educational Testing Service showed that youngsters who watched the program most learned most. The tests also seemed to validate the three important premises of the series: that children can learn their numbers and letters by film animation, that commercial techniques can be used constructively to gain the attention of 4-year-olds, and that repetition is effective.

Original cast members included a black husband and wife, played by Loretta Long and Matt Robinson, singer Bob McGrath as a white neighbor, Will Lee as the candy-store owner, along with Jim Henson's Muppets—Oscar the Grouch, Big Bird, the Cookie Monster and Bert & Ernie.

David Connell, formerly with *Captain Kangaroo,* produced the show, and Dr. Gerald Lesser, Harvard psychologist, served as chief advisor. Jeff Moss and Joe Raposo created the music, a vital element of the series.

Sesame Street has been adapted in Spanish as *Plaza Sesamo,* in Portuguese as *Villa Portuguese,* in German as *Sesamstrasse* and in French as *Bonjour Sesame.* Britain's BBC-TV rejected it as authoritarian in its aims and middle-class in its orientation, but the series was carried in the United Kingdom by ITV, the commercial system. BBC then accepted *Sesame's* sister program, *The Electric Company.* See also Children's Television Workshop; Cooney, Joan Ganz; *Electric Company, The.*

Eric Sevareid

SEVAREID, ERIC—CBS journalist through four decades as a member of the news team assembled by Edward R. Murrow in 1939; since the mid-60s he has been most familiar to viewers as news analyst on the evening newscast and as the network's premier political commentator. Sevareid's long experience as a war correspondent and political reporter combined with his gifts as an essayist made him one of the unique and most respected figures in all broadcast journalism.

After covering World War II in both the European and China-Burma-India theaters—scoring a major scoop in 1940 on the fall of France—he returned to the United States to report on the founding of the United Nations and then was assigned to the Washington bureau, where he began covering domestic political activities. In 1959, after becoming chief Washington correspondent, he went to London and served two years as roving European correspondent. He then became national correspondent for CBS News, maintaining contact with top government officials and diplomats.

In special broadcasts, he represented CBS News in two televised interviews with President Nixon, participated in the coverage of the resignations of Nixon and Vice-President Agnew, and was one of the correspondents interviewing President Ford in a special "conversation" program in 1975. He also gave his impressions of the Vietnam War in an "illustrated lecture" in 1966.

During his career he was moderator for such broadcasts as *Town Meeting of the World, Years of Crisis, The Great Challenge* and *Where We Stand.* During the summer of 1975 he conducted a series of televised interviews with noted world figures entitled *Conversations with Eric Sevareid.*

A native of North Dakota, he worked on the *Minneapolis Journal* and other newspapers before joining CBS. He reached the mandatory retirement age in November 1977, but remained active with numerous freelance assignments, including *Between the Wars,* a 1978 documentary series syndicated by Mobil. In 1982, he became the host of a weekly magazine series, *Eric Sevareid's Chronicle,* on Viacom's WVIT, Channel 30.

7–7–7 RULE—the FCC restriction of station ownership to no more than 7 AM, 7 FM and 7 TV stations by an individual or corporation. Of the 7 TVs, only five may be VHF stations.

"700 CLUB, THE"—90-minute daily talk show which is the keystone of the Christian Broadcasting Network and hosted by its president, Pat Robertson. The program is named for a fateful fund-raising telethon conducted by Robertson in 1963 to keep his frail, Christian-oriented UHF station, WYAH Virginia Beach, VA., on the air. Robertson's pitch was that the station could meet its monthly operating costs if 700 people would contribute $10 a month. The goal was met and the single station grew to a group of four television and five radio stations, a national satellite network carried mainly on cable systems, and an annual operating budget of $55 million.

The 700 Club began as a local WYAH show in 1966 with Jim Bakker as host. Bakker later went to Charlotte, N.C. as president of the PTL Network and host of *The PTL Club,* another evangelical talk show. Robertson then took over *The 700 Club,* with Ben Kinchlow, a Black, as his sidekick and substitute host.

See also Christian Broadcasting Network; Religious Television.

"77 SUNSET STRIP"—popular, fast-paced private-eye series which had a substantial run on ABC (1958–64) and helped establish an urban action-adventure vogue in the wake of the quiz scandals. By Warner Bros., its regular cast was headed by Efrem Zimbalist, Jr., Roger Smith, Edd ("Kookie") Byrnes and Jacqueline Beer.

SEVERINO, JOHN C.—president of ABC Television since May 1981, succeeding Frederick S. Pierce, who moved up to executive vice president of ABC Inc. Severino had spent the 16 previous years in the company's owned-stations division, where he had been v.p. and general manager of KABC-TV Los Angeles (1974-81) and of WLS-TV Chicago (1970-74).

Severino joined ABC in 1965 as a salesman at WABC-TV New York, having come from WBZ Radio in Boston. Eventually he became sales manager of WLS-TV and then WXYZ-TV Detroit, before returning to Chicago as general manager. As president of ABC Television, he oversees the ABC network, ABC Entertainment and the engineering and operations division.

SFM MEDIA SERVICES CORP.—a leading time-buying service involved also in program packaging and syndication. Thus, in 1976, while one division of the company was engaged in purchasing air time for President Ford's election campaign, another division was mounting a revival of *The Mickey Mouse Club* and re-packaging the old *Rin Tin Tin* programs for syndication. SFM also arranged the 50-station lineup for the so-called "Mobil Network" carrying the British series on explorers, *Ten Who Dared,* sponsored by Mobil Oil. On its own, the company mounted an ad hoc network for vintage family movies, which it called the SFM Holiday network. It also bought the air time for Ronald Reagan's successful Presidental campaign.

The company was founded in 1969 by Walter Staab, Robert Frank and Stanley Moger.

SGT. BILKO—classic character created by Phil Silvers in the 1955–58 series, *You'll Never Get Rich.*

"SHAFT"—unsuccessful TV series about a black, street-wise private detective, based on the hit movie of that title. Produced by MGM-TV as a 90-minute series designed to rotate with *Hawkins* on CBS (1973), it stopped at eight episodes.

SHAKESPEARE, FRANK J.—president of RKO General since 1975 and former director of the USIA during the Nixon Administration (1969–73). During the 60s, he was a fast-rising executive at CBS, moving from v.p.-g.m. of WCBS-TV in New York to executive v.p. of the TV network under James Aubrey. Later he became president of CBS Television Services, leaving in 1969 for the USIA post. In 1974 Shakespeare became vice-chairman of Westinghouse Broadcasting; the following year he was hired away by RKO General.

"SHAKESPEARE PLAYS, THE"—a massive coproduction by the BBC and Time-Life Films involving the full 37-play canon of William Shakespeare, to be presented in new productions over a six-year period. Offered during the first season, which began early in 1979, were *Richard II, Romeo and Juliet, Julius Caesar, As You Like It, Measure for Measure* and *Henry VIII.* Cedric Messina, who had been with the BBC since 1958 and was formerly producer of *Play of the Month,* was executive producer for the first two years, with Alan Shallcross as script editor and John Wilders as literary consultant. The directors, varying from play to play, include David Giles, Herbert Wise and Kevin Billington. Jonathan Miller was named producer for years three and four.

A controversy arose in the U.S., while the series was in the planning stages, when the Corporation for Public Broadcasting offered a grant of $1 million to the production to secure the plays for PBS. The CPB drew back when unions and American performers and producers protested the use of Federal funds to finance a foreign public television project. Eventually the necessary funds were raised in grants from Exxon Corp., Metropolitan Life Insurance and the Morgan Guaranty Trust Co. Later talent unions in the U.K. objected to the casting of an American, James Earl Jones, as Othello, but the BBC prevailed in the dispute.

SHALIT, GENE—resident wit and reviewer on NBC's *Today* show, starting in 1969 with occasional appearances as book and movie reviewer and graduating to a daily role a few years later. With black bushy hair and mustache as distinguishing features, and a staccato, slightly acerbic style, he broke into TV as a film critic for WNBC-TV New York, after having broadcast reviews on NBC Radio's *Monitor.*

"SHANE"—one of the so-called "adult westerns" of the early 60s, based on a successful movie and produced by Herbert Brodkin's Titus Productions. It had a brief run on ABC (1960) and featured David Carradine, Jill Ireland, Tom Tully and Chris Shea.

SHANKS, BOB—former ABC program executive (1972–78) and the initial executive producer of *20/20,* the ABC newsmagazine. When the pilot episode was poorly received and the format and production staff of *20/20* overhauled, Shanks was displaced, and he moved to the West Coast to resume an earlier career as an independent film and TV producer.

At ABC, he served in a variety of programming capacities and became, in 1976, v.p. in charge of all prime-time specials and the early-morning show, *Good Morning, America.* He also wrote a nonfiction book on television, *The Cool Fire.* From 1965 to 1970, he was producer of *The Merv Griffin Show* and earlier had worked for Bob Banner, *Candid Camera, Tonight* and *The Great American Dream Machine.*

SHAPIRO, MARVIN L.—executive v.p. of Westinghouse Broadcasting Co. and former president of Group W Broadcasting (1972–78). Shapiro's background had been in station representation; he had been president first of RAR (Radio Advertising Representatives) and then of TVAR (Television Advertising Representatives), both Group W subsidiaries. He had joined TVAR in 1961 as a salesman.

SHARE (Share-of-Audience)—a comparative evaluation of the ratings, representing in percentages how programs performed relative to the other programs in direct competition. In figurative terms, the share is the share of the pie. If all television viewing at a given time constitutes the pie, the share denotes the proportion of the pie each program has received.

More than a comparative evaluation, it is a competitive evaluation. In three-network (or three-station) competition, a program that receives approximately one-third of the available audience—or a 33 share—is considered to be competing adequately. Less than a 33 share usually spells trouble for a program because it means too much audience is being lost to the opposition. (Actually, in network competition today, a 30 share is considered the minimum index of success, since at least 10% of the prime-time tune-in goes to independent and public TV stations.

Among lay persons, there is often confusion about the difference between the rating and the share. A rating is the percentage of households tuned to a program from all the possible TV households in the service area. Since the rating is based on *all* households it can be directly translated into an absolute number of viewers for a program. The share, on the other hand, is based only on households using television at a specific time and therefore does not lend itself to estimating actual audience size.

Since viewing levels vary by the hour and by the day, the success of a program is told not in the absolute number of people it has reached but rather in its ability to command larger audiences than the programs opposite it. A program with a 20 share in the evening, when viewing levels are high, will normally play to a larger audience than one with a 40 share in the morning, when viewing levels are low. However, the evening program would in all likelihood be canceled for failing to compete adequately while the morning program would almost certainly be renewed.

Shari Lewis

"SHARI LEWIS SHOW, THE"—popular NBC children's program in the late 50s and early 60s which featured ventriloquist Shari Lewis and her hand puppets, principally Lamb Chop.

SHARNIK, JOHN—veteran executive producer of documentaries and specials for CBS News, who served briefly in the mid-70s as v.p. of public affairs broadcasts and earlier was senior producer of *CBS Reports.*

Sharnik, who joined CBS in 1954 as a writer for Eric Sevareid's *The American Week* broadcasts, later coproduced the *Eyewitness* series with Leslie Midgley and worked as a writer-producer for *60 Minutes.* He joined CBS News from the Sunday department of *The New York Times.*

SHEA, JACK—director of the Bob Hope specials and producer-director of such series as *The Glen Campbell Goodtime Hour* and *Hollywood Talent.* His directing credits also

include *The Waltons, We'll Get By, Sanford and Son, Calucci's Department* and *Death Valley Days.*

SHEAR, BARRY (d. 1979)—director of TV programs and movies. In the late 50s, when his specialty was comedy, he directed the Ernie Kovacs specials, the Edie Adams series, *Here's Edie,* occasional outings of *Tonight* and the series *The Lively Ones.* Later he turned to action-adventure and was regular director on *Police Woman* and *Starsky and Hutch,* with directing assignments in a raft of other series such as *Ironside* and *It Takes a Thief.* Before his death from cancer, he completed a four-hour miniseries for NBC, *Power,* for the 1979–80 season. The movies he directed included *Wild in the Streets.*

William Sheehan

SHEEHAN, WILLIAM—former president of ABC News (1974–77), demoted to executive v.p. in the reorganization that made Roone Arledge president of ABC News & Sports in June 1977. A year later, he left to become v.p. of public affairs for the Ford Motor Co. in Detroit.

Earlier, Sheehan had been London bureau chief for ABC News (1962–66) and co-anchorman of the *ABC Evening News* (1961–62). Sheehan came to ABC in 1961 from WJR Radio, Detroit, where he had been news director. He was hired by James Hagerty, then head of the network's news division. Hagerty, who earlier had been press secretary to President Eisenhower, had met Sheehan while he was covering the President's travels in Europe for WJR and was impressed with him. Before becoming president of ABC News, Sheehan had been senior v.p. under Elmer Lower.

SHEEN, FULTON J. (BISHOP) (d. 1979)—probably the most popular religious personality to have worked in TV and the only one to receive weekly prime-time exposure in a show of his own. Bishop Sheen's discourses on social and inspirational topics drew substantial ratings in the 50s against strong entertainment competition, enabling a five-year run. His original weekly program, *Life Is Worth Living,* premiered Feb. 12, 1952 on the DuMont Network in prime time. It moved to ABC on Oct. 13, 1955 and ran until April 8, 1957.

As national director of the Society for Propagation of the Faith (Catholic), Bishop Sheen later resumed the half-hour program in syndication, under auspices of that organization. Two separate series were produced on tape between 1961 and 1968.

"SHEENA, QUEEN OF THE JUNGLE"—syndicated series (1954–55) based on the comic strip and featuring Irish McCalla and Christian Drake. It was produced by Nassour Studios.

SHEPARD, HARVEY—CBS program executive skilled in scheduling strategies through his experience in audience research. He became v.p. of program administration, based in Hollywood, in 1978, after having been v.p. of programs in New York. When Bud Grant became president of ABC Entertainment in November of 1980, Shepard moved up to v.p. of the division. Shepard joined CBS from Lennen & Newell Advertising in 1967 as a research executive working in audience measurement.

SHERRY, ERNEST H.—director of *The Mike Douglas Show* for nine years, latterly doubling as producer. He also directed *Dinah's Place* and the Fabergé musical specials.

"SHINDIG"—half-hour rock 'n' roll series on ABC (1964–66) produced in Hollywood and hosted by Jimmy O'Neill, featuring youthful recording artists of the day and a resident chorus line. After one good season it spawned a second edition, and both faltered the next year. It was by Selmur Productions and Circle Seven Productions, with Leon I. Mirell as executive producer and Dean Whitmore as producer.

SHOOSHAN, HARRY M., III (CHIP)—chief counsel of the House Communications Subcommittee (1975—80) and a maverick in the eyes of most commercial broadcasters for his advocacy of cable-TV, his lack of sympathy for the industry's quest for a five-year license-renewal bill and his outspokenness for changes in the system. As head of the subcommittee's staff, he was chief writer of the controversial

cable report issued in 1976 while the late Rep. Torbert H. Macdonald was chairman of the subcommittee. Later, under Rep. Lionel Van Deerlin, he steered the panel's work on the proposed rewrite of the Communications Act of 1934. When Van Deerlin was defeated for reelection in 1980, Shooshan teamed with the committee's expert on technology, Chuck Jackson, and formed a consulting firm, Shooshan & Jackson.

Richard Chamberlain

"SHOGUN"—five-part adaptation of James Clavell's best selling novel of that title set in 17th Century Japan; it aired on NBC in the fall of 1980 and drew a large audience. With Clavell himself as exec producer, the program, which took three years to produce, grew into the most expensive mini-series ever; reportedly, the costs approached $21 million, including $1 million paid to Clavell for the rights. The five parts represented 12 hours of television.

Richard Chamberlain headed a cast that otherwise consisted of British and Japanese actors, including the great Japanese star Toshiro Mifune. The program was made by Paramount Television, with Eric Bercovici as script writer and producer.

SHORE, DINAH—one of TV's most attractive and enduring personalities, who began as a star of music-variety shows in the 50s and had a second career in the 70s as a talk-show hostess.

As a well-established recording artist and radio singer, she began working in TV on a fulltime basis in 1951 with *The Chevy Show Starring Dinah Shore,* a 15-minute variety series on NBC which aired twice weekly until July 1957. But it was the 15-minute form and not Miss Shore's popularity that had faded, and in October 1957 she began a weekly variety hour in color for NBC, *The Dinah Shore Chevy Show,* that ran five years. That series was produced by Bob Banner, written by Bob Wells and Johnny Bradford and choreographed by Tony Charmoli.

Although she dressed elegantly and stylishly, her warm folksy manner made her somewhat a symbol of grassroots America. Through the latter part of the 60s she appeared mostly in specials and briefly in a daytime variety show. But in 1972 she began on NBC a women's-oriented talk show, which later grew into a 90-minute daily syndicated series, *Dinah's Place,* carried by most subscribing stations in the afternoon or early evening.

Dinah Shore

SHORT-SPACING—the concept of lowering the FCC mileage standard between stations using the same channel position so that additional VHF or UHF stations can be added, or "dropped in," in certain markets. The FCC's allocations require stations with the same channel number to be separated by a minimum of 170 miles to avoid signal interference. See also Drop-Ins, Table of Assignments.

SHORT-TERM RENEWAL—a license renewal granted by the FCC for less than the normal three-year term (often for one year or less) in punishment to stations for transgressions of the rules. The FCC began imposing the short-term renewal in the late 50s, when it felt the need to impose sanctions for certain offenses that in its opinion did not justify the extreme punishment of license revocation.

SHOWTIME—national pay-cable service developed by Viacom and, since 1979, co-owned by Teleprompter, second largest of the cable MSOs. Teleprompter's purchase of a 50% interest and its adoption of the service for its systems helped break the monopoly position, though not the dominance in the field, of Home Box Office, owned by Time Inc. Show-

time's fare consists largely of movies and light entertainment specials, many of them based on cabaret or concert performances. In the fall of 1979, the pay-cable network began offering also videotaped productions of certain Broadway and off-Broadway plays. Mike Weinblatt, recruited from NBC, became president in 1980. See Cable Marketing.

SHPETNER, STAN—series producer whose credits include *The Sixth Sense* and *Kodiak.*

SHRINER, HERB (d. 1970)—popular TV personality and comedian of the 50s, with a low-key countrified manner and a harmonica as his trademark. He had his own show on ABC in 1952, *Herb Shriner Time,* then became emcee of the NBC quiz show, *Two For the Money.* Later he was signed by CBS, which gave him an evening show in 1957, *The Herb Shriner Show.*

SIA (STORAGE INSTANTANEOUS AUDIMETER)—an automatic and updated version (fall 1973) of the basic Audimeter used by the A.C. Nielsen Co. since 1942 to measure national tune-in for radio and TV programs. See also Audimeter.

SIDARIS, ANDREW W.—director specializing in sports who was with ABC for 12 years before starting an independent TV and film company in the early 70s. He directed numerous films for ABC's *Wide World of Sports* and was director in ABC's coverage of the Olympic Games from 1964 to 1972.

SIEGEL, SIMON B.—major figure in ABC Inc. from the time of the merger with United Paramount Theatres in 1953 until his retirement 20 years later. Having been comptroller for Paramount, he initially became treasurer of ABC and, as trusted aide to the president, Leonard Goldenson, was named executive v.p. of the corporation in 1961. Although he maintained low visibility outside the company, he was one of the most powerful figures in all broadcasting during the 60s as ABC's chief financial officer and overseer of the divisions.

SIEGENTHALER, ROBERT—veteran ABC News producer who in the fall of 1976 became executive producer of the *Evening News* when Barbara Walters began as co-anchor with Harry Reasoner. Since ABC had a $1 million-a-year investment in Miss Walters and was determined to improve its news ratings, Siegenthaler was chosen to make the news decisions and supervise the production because of his extensive experience in TV news. At the time of his appointment, he was executive producer of special events engaged in overseeing the political conventions of that year. With the return of Av Westin to ABC News in the summer of 1977, Siegenthaler yielded him the reins of the newscast and received a new assignment. After producing the series of latenight specials on the Iranian crisis in 1980, Siegenthaler was named executive producer of news specials.

"SIEMPRE EN DOMINGO" (ALWAYS ON SUNDAY)
—probably the longest format for a regular TV program anywhere, a seven-hour Sunday entertainment potpourri originating on Mexico's Televisa network and carried on 10 Spanish-language UHF stations in the U.S. The program, hosted by Raul Velasco, began in 1969 as a local entry in Mexico City but was soon picked up by Televisa. Its audience was estimated to be more than 20 million viewers a week.

SILLIPHANT, STIRLING—producer and writer who rose through the TV ranks to top screenwriting assignments (*In the Heat of the Night, Towering Inferno*). He wrote and produced *Naked City,* wrote the *Route 66* pilot in 1960 and created and was executive producer of *Longstreet*. A former New York publicist for 20th Century-Fox and Walt Disney, he broke into TV as a writer on the *Mickey Mouse Club* series (1955). In later years he wrote and produced the miniseries, *Pearl* for ABC and the pilot for *Fly Away Home*.

SILVERBACH, ALAN M.—veteran syndication executive who spent virtually his entire career with 20th Century-Fox, rising to executive v.p. in charge of worldwide syndication in 1975. He joined the company as a trainee in 1945 and shifted to the TV division in 1955. He left in 1976 for a private venture, and later became executive v.p. for syndication of Metromedia Producers Corp.

SILVERMAN, FRED—former program chief of CBS and ABC who became president of the National Broadcasting Company, in a blaze of publicity, in June 1978, a time when NBC was struggling desperately to climb out of third place. The price for Silverman's programming expertise was the position of chief executive officer, a three-year contract and a reported salary of $1 million a year. This put him, alone among network executives, in television's star class. Indeed, he became better known to the general public than most performers on the screen.

But Silverman's three years at NBC were a fiasco. His program strategies all failed, the ratings continued to decline in daytime as well as nighttime, a number of old-line

affiliates switched allegiances to competing networks, and dozens of key executives left the company, some voluntarily. When Silverman was ousted in 1981, the network was in a shambles. He went into independent production.

Silverman pledged in his public statements to take the high road and to move NBC forward with quality programs, without resort to the exploitation of sex, violence or the youth market and with higher priorities for news programming than had existed previously. But he found himself competing with the hit schedule he had left behind at ABC —one he now tacitly denounced—and also with a number of durable series he had launched at CBS, and the high road soon turned to the low road.

Fred Silverman

His first eight months at NBC were marked with frenzied revisions of the prime-time schedule, a blizzard of new program series and a raft of scheduling stunts, none of which availed. As NBC sank deeper into third place, Silverman altered his strategy, taking a more conservative approach. For the 1979 fall schedule, the first to be prepared entirely under his guidance, he preserved a number of marginally successful shows and mounted a schedule that concentrated the strongest series, such as they were, at 9 o'clock.

This was the "ridge pole" technique used by Frederick Pierce in 1975 to lift ABC out of the depths. Silverman also sought to rebuild NBC on the traditions of the network, and in this spirit he revived the NBC peacock symbol, and created the twice-yearly cultural series, *Live from Studio 8H*.

He also reorganized the company, separating the program division from the network, and brought in such former colleagues from the other networks as Lee Currlin, Jerry Golod, Irwin Moss, Perry Lafferty, Irv Wilson and Richard S. Salant. Later, he hired Lucille Ball to develop new comedy series for NBC.

Silverman's reputation as a master programmer grew out of his winning streak that carried over from CBS to ABC. A year after he left CBS in 1975 to become president of ABC Entertainment, his old network ended its 20-year dominance of prime time and his new one surged to the top. At ABC, Silverman shepherded a string of hits—some of which were on the network or in development before his arrival—that included *Happy Days, Laverne & Shirley, Donny and Marie, Family, Rich Man, Poor Man, Starsky and Hutch, Soap* and *Charlie's Angels.*

He was regarded by most program suppliers as the most knowledgeable and professional of the program chiefs, and his record of success with new programs was clearly the best for any programmer in many years.

A *wunderkind* in television, Silverman started with CBS as director of daytime programs in 1963 at the age of 25, after having been a minor program executive with WGN-TV Chicago and WPIX New York. He came to the notice of the networks while at Ohio State University, where he wrote a highly perceptive Master's thesis analyzing 10 years of programming by ABC.

When Mike Dann left CBS in 1970, Silverman who by then had become vice-president of program planning and development succeeded him as head of the program department. He was responsible for numerous hit shows including *Kojak, Cannon, Maude, Rhoda, The Jeffersons, Cher* and *The Waltons.*

After the program schedules were set for the fall of 1975, ABC hired away Silverman to replace Martin Starger, in hopes that he could nurse the network's schedule back to health. ABC's stock jumped almost two points the day the announcement was made.

Phil Silvers

SILVERS, PHIL—a nightclub and movie comedian who scored on television in a brilliantly funny situation comedy, *You'll Never Get Rich* (soon after retitled *The Phil Silvers Show*), written by Nat Hiken. It made its debut on CBS Sept. 20,

1955 and ran for four seasons, the reruns continuing many years after in syndication.

Silvers portrayed Master Sergeant Ernie Bilko, a lovable promoter and sometime con-man who was incurably occupied with money-making schemes, usually involving his subordinates at an Army camp in Kansas. The series(had a rich variety of comic Army characters, but it was Silvers's perfect realization of Bilko that made it one of TV's memorable shows.

Possibly from too strong an identification with Bilko, Silvers did not go on to other series. He starred in several specials and made guest appearances for a few years and then receded from the medium.

SIMMONS, CHESTER R. (CHET)—veteran network sports executive, who became head of NBC Sports in March 1977 and left the network suddenly, ending a 15-year association, in July 1979 to become president of ESPN, the 24-hour sports cable network. He was succeeded by Arthur Watson.

Simmons had previously been v.p. in charge of NBC Sports since March 1977, when he succeeded Carl Lindemann. He had previously been v.p. of NBC sports operations since 1973 and an executive of the network's sports division since leaving ABC in 1964. Simmons had been architect of NBC's weekend sports anthology, *Grandstand,* served as liaison to the various leagues with which NBC was associated and was overseer of the network's sports schedule and its production personnel. On leaving NBC, he became president of a new cable-TV network, backed by Getty Oil, called the Entertainment and Sports Programming Network.

SIMMONS *v.* FCCL [169 F2d 670 (D.C. Cir. 1948)/ cert. den. 335 U.S. 846 (1948)]—case in which the FCC denied license to a broadcast applicant who proposed merely to air the programing of a national network, with no locally produced programs.

The case arose when WADC in Akron, Ohio, and WGAR in Cleveland, both 5-kw radio stations, applied to the FCC for permission to increase power to 50 kw. Because of the proximity of the cities, the applications were mutually exclusive. But without comparing the proposals of the two stations, the FCC rejected the application of WADC, licensed to Simmons, because it proposed to air nothing but the programs of CBS and explicitly stated that there would be no locally produced shows. In denying the application, the FCC said the proposal was not in the public interest.

Simmons took the case to the D.C. Court of Appeals, which affirmed the FCC's decision, stating that although network programs were often of high calibre the Simmons proposal represented an abdication of the local licensee's responsibility to tailor broadcasting to the needs of the immediate community.

The court noted that even network-owned stations maintained local program staffs to serve local interests and needs. A broadcaster's responsibility to serve the public interest required nothing less. Finding the Simmons application defective, the court upheld the commission's grant of WGAR's application without a comparative hearing.

Simmons sought review by the Supreme Court, but the Court did not hear the case.

SIMON, NEIL—a former TV gag writer who became the leading comedy playwright on Broadway. Simon started his professional career writing routines and jokes for the television shows of Robert Q. Lewis, Red Buttons, Phil Silvers, Sid Caesar and Garry Moore. Two of his Broadway comedies, *Barefoot in the Park* and *The Odd Couple,* were turned into TV series. In 1975 he wrote one of four playlets for a comedy special, *Happy Endings.*

Neil Simon

SIMPSON, JIM—versatile NBC sportscaster who has covered NFL games, major league baseball, PGA tournaments, Wimbledon Open tennis and NCAA championships. Simpson holds the distinction of being the first television announcer to broadcast live via satellite from Japan while covering the 1972 Winter Olympics at Sapporo, Japan. He has covered nine Olympiads, beginning with the games in Helsinki, Finland in 1952.

SIN (SPANISH INTERNATIONAL NETWORK)—a consortium of 16 UHF stations in various parts of the U.S., all directing their programs at the Spanish-speaking population (Mexican-American, Puerto Rican and Cuban). They organized as a "network" in the early 60s to sell their combined audiences to national advertisers and for the purchase of programs from Mexico, Venezuela, Brazil and Colombia.

In the late 70s, most of the stations became interconnected by the Westar satellite and were receiving an average of 64 hours a week of network feeds. Tieing into the satellite, Mexico's Televisa was able to beam some of its programs live to the SIN affiliates in the U.S. on a regular basis. It was the first network to use a domestic satellite for affiliate interconnections. With cable systems, SIN now has one hundred sixty affiliates. In 1979, SIN began the first American newscast in Spanish, *Noiciero Nacional.*

The nucleus of the network is a group of stations owned by SIN: KWEX San Antonio, KMEX Los Angeles, WXTV New York, WLTV Miami, KDTV San Francisco, and KFTV Fresno, Calif. The other affiliates are concentrated in the southwest and at the Mexican border, although one is located in Chicago. Together, they claim a potential audience of 9 million. President of the network is Rene Anselmo, who has been with SIN since the first station, KWEX, went on the air in 1961. Anselmo owns 25% of SIN and Televisa, the Mexican communications conglomerate, the other 75%.

In 1979, the five stations owned by SIN were granted construction permits by the FCC for 1,000-watt translator stations to carry their signals to Philadelphia; Denver; Hartford, Conn; Austin, Tex. and Bakersfield, Calif. See Cable Networks.

SINATRA, FRANK—one of the great names in popular entertainment who failed to carve out a niche in television. Neither of two attempts at a regular series was successful, the first a half-hour variety series on CBS (October 1950 to April 1952), the second a variety hour alternating with dramatic plays on ABC (October 1957 to June 1958). Both were called *The Frank Sinatra Show.*

More successful were his specials, including one in 1974 that involved live coverage by ABC of his concert at Madison Square Garden. The production was assigned to the sports department because of its experience in covering auditorium events live.

"SING ALONG WITH MITCH"—musical hour of bouncy tunes on NBC (1961–65) hosted and conducted by Mitch Miller, noted producer for Columbia Records, and growing out of his popular series of *Sing Along* albums. The songs were performed mainly by The Sing Along Gang, a male chorus of 25, and by such resident female vocalists as Leslie Uggams, Diana Trask and Gloria Lambert. Miller wore a beard before beards were fashionable, Miss Uggams was featured before blacks became prominent in TV, and the series itself, for its form, had no precedent for the success it enjoyed. It was produced by All American Features.

SINGAPORE TELEVISION—started in 1963, and operating under the strict control of a government ministry, TS (Television Singapore) currently programs 110 hours a week on two channels in 625-line PAL. The service is supported both by advertising and license fees.

SIPES, DONALD—president of Universal Television since May 1978, rising to that position after a 20-year career as a business negotiator for programs and talent, working on both sides of the fence—buying and selling. He gave up private law practice in 1957 to join NBC in program negotiations, then went to CBS in a similar capacity and rose to v.p. of business affairs and planning. He was with CBS from 1963 to 1974, based in Hollywood for most of that period. He left to become senior v.p. of International Creative Management, the talent agency, and a year later joined Universal Television as v.p. of business affairs. Promotions came swiftly, and he was named president when Frank Price left to join Columbia.

SIPHONING—the drawing off of program fare by one medium from another, as occurred when television "siphoned off" radio's most popular programs. Commercial broadcasters have alerted the FCC to the likelihood of subscription TV and pay-Cable similarly siphoning its programs. Because this holds the prospect of people being asked to pay for what they now receive free, the FCC adopted anti-siphoning rules for the pay services. The rules were later struck down by the courts as unconstitutional and improper. See also Pay-Cable, Pay-Television.

"SIT IN"—a historic 1960 documentary on the sit-in movement for civil rights in the South, prepared by Irving Gitlin for NBC News. The 60-minute film concentrated on events in Nashville and covered the issue from both sides of the conflict. Gitlin was executive producer, Al Wasserman producer and Chet Huntley narrator.

SITUATION COMEDY—one of the program forms indigenous to television, the half-hour story comedy usually with an emphasis on sight gags. The form has proved both popular and durable, and the fact that CBS has favored it and developed it more successfully than the other networks largely accounted for the 20-year dominance of that network in the prime time ratings. Most programmers agree that hit sitcoms are harder to create and establish than hit adventure

series. The ratings supremacy of CBS, through the 50s, 60s and most of the 70s may be traced to a sitcom hit parade that has ranged from *I Love Lucy* and *The Honeymooners* to *The Mary Tyler Moore Show, M*A*S*H* and *All In the Family.*

In between have come twoscore hits in the situation comedy form, including *December Bride, Make Room For Daddy, Father Knows Best, Burns and Allen, Leave It to Beaver, Mister Ed, The Phil Silvers Show, Our Miss Brooks, Gilligan's Island, My Three Sons, The Andy Griffith Show, Family Affair, Green Acres, Hogan's Heroes, The Dick Van Dyke Show, Petticoat Junction, Maude, Rhoda, Phyllis, The Bob Newhart Show* and *The Jeffersons.* ABC and NBC have had sitcom hits of their own, but neither such an array.

The situation comedy derives its name from a dependency on plot to activate the comedy talent or the essential comedy premise. Vital to the sitcom is that the characters behave predictably; it is their consistency, their way of reverting to form, with each new situation—regardless how outlandish—that draws the laughs.

In the case of Lucille Ball, gifted at mugging and clowning, the bizarre predicaments she entered in each episode of *I Love Lucy* were not designed to be funny in themselves—although many were—but rather to turn the star loose at what she does best. With *Beverly Hillbillies,* the one-joke premise of the show—country bumpkins becoming millionaires—needed to be triggered anew by weekly circumstance in order to be told again.

Not to belittle the form, the closest equivalent to the sitcom in literature is the comic strip—it too with unchanging characters performing the predictable. Thus, NBC succeeded with *Hazel,* ABC with *The Addams Family* and CBS with *Dennis the Menace.*

The level of sophistication in situation comedy has changed markedly over the years but not the essential form or formula. Everything before *All In the Family* was wholesome; that program, in 1970, broke new ground for subject matter and made revolutionary changes in how sitcom protagonists addressed their fellow man. Bigotry, hysterectomies, homosexuality, menopause and impotency were aspects of life walled out of Lucy's experience; her raciest deed, in 1953, was to have a baby. Yet, although thematically the two shows seem eons apart, they both are at the core comedies about two more or less contrasting couples, and both have a central figure to make the funny faces. The fact that Jackie Gleason's *The Honeymooners* fits the identical pattern, and *The Dick Van Dyke Show* comes close, argues against coincidence.

The sitcom has never been precisely one thing but over the years has branched off into schools. Family comedy was one school, spanning *Mama, Father of the Bride, Trouble With Father, The Life of Riley, Ozzie and Harriet, My Little Margie, The Donna Reed Show, Good Times, Happy Days, The Brady Bunch, The Jeffersons* and even *Sanford and Son.* Related was "heart comedy," the widows and widowers of the late 60s and the divorcees of the 70s—*My Three Sons, Family Affair, Julia, The Doris Day Show, The Courtship of Eddie's Father, Fay* and *One Day At a Time.*

There was also the musical sitcom (*The Partridge Family, The Monkees*); the fantasy sitcom (*I Dream of Jeannie, My Mother the Car, Mr. Ed, Bewitched*); the service sitcom (*McHale's Navy, Hogan's Heroes, F Troop, M*A*S*H*); and the police sitcom (*Barney Miller, The Cop and the Kid*).

But all answer to the basic principle of external circumstances setting in motion a cast of appealing two-dimensional characters and releasing the built-in comedy component. The formula calls for complications in the story but always, always the matter is neatly resolved. And whether it involved the sophisticates of *M*A*S*H* or the witless inhabitants of *Gilligan's Island,* 30 minutes after the start the principals are happy.

SIXTH REPORT AND ORDER—momentous document issued by the FCC on April 14, 1952 lifting the four-year freeze on station licenses as of July 1 that year and establishing a permanent table of assignments for TV frequencies in 1,291 cities. The report set off an avalanche of competing applications for the choicest assignments.

The culmination of a long study looking toward an orderly and technically efficient plan for the allocation of TV channels around the country, the Sixth Report and Order provided for 2,053 stations, 617 of them on VHF and 1,463 on UHF. Of that number, 242 were designated noncommercial, 80 of those on VHF. The order also established three broad geographic zones, each with its own mileage separation standards and regulations for antenna height.

The FCC later revised the table, increasing the number of channels for both commercial and educational use. See also "Freeze" of 1948.

"SIXTH SENSE, THE"—ABC series (1972) in which Gary Collins portrayed a professor who was a leading investigator in cases involving ESP and psychic phenomena. Produced by Universal TV, it yielded 25 episodes.

"60 MINUTES"—CBS newsmagazine series which premiered in 1968 and overcame anemic ratings to develop into the most successful program of its kind. In the late 70s it became one of the most popular programs in television, consistently ranking among the top 3 in the Nielsen charts. For its first three seasons, with Mike Wallace and Harry Reasoner as co-editors, the one-hour series played in prime time: Tuesdays at 10 P.M., alternating with *CBS Reports.* The scheduling resulted in a ratings default for CBS, which probably helped the competing *Marcus Welby, M.D.* on ABC to become a major hit. In the fall of 1971 (with

Morley Safer having replaced Reasoner in the middle of the previous season) *60 Minutes* was moved out of prime time, to Sundays at 6 P.M., as a weekly series. But because NFL football spilled into that time period, the series went off the air in the fall until football season ended. Nevertheless, it drew larger audiences in Sunday fringe time than it had attracted in prime time, partly because it competed only with local or syndicated programs.

When a change in the Prime Time-Access Rule permitted the networks to begin programming at 7 P.M. Sundays in 1975, on condition that the shows were for children or in the news/public affairs sphere, CBS encountered clearance problems from its affiliates because of its hold on the preceding hour with *60 Minutes.* To resolve the problem of 7 o'clock clearances, CBS freed the 6 o'clock period by moving *60 Minutes* to the newly released prime-time hour. Having attracted a loyal following by then, *60 Minutes* began to draw prime-time audiences that were competitive with entertainment programs. By 1979 it was consistently a Top 10 program, ranking among ABC's phalanx of hit sitcoms. It grew so popular that ABC was moved to enter the newsmagazine field with *20/20,* while NBC redesigned *Weekend* for prime time and when it failed mounted a new entry, *Prime Time Sunday.*

With the shift to prime time, the series went on a 52-week schedule, and Dan Rather was added as the third editor-anchorman. When he rejoined CBS News from ABC in 1979, Harry Reasoner also rejoined *60 Minutes* as the fourth anchor. Dan Rather's departure for the *CBS Evening News* brought Ed Bradley as one of the four correspondents. Each program was made up of three documentary segments and a *Spectrum* insert or, occasionally, a discussion of letters from viewers. Don Hewitt guided the program for CBS News as executive producer.

Among others associated with the series as producers have been Joe Wershba, Paul Loewenwarter, Grace Diekhaus, Philip Scheffler, Barry Lando, Igor Oganesoff, David Buksbaum, Imre Horvath, Harry Moses, Marion Goldin, William McClure, Joe DeCola and Al Wasserman.

"$64,000 QUESTION, THE"—TV's first big-money quiz show, an adaptation in five figures of a radio success, *The $64 Question.* Packaged for CBS by Louis G. Cowan Inc., with Hal March as host and Steve Carlin as producer, it premiered June 7, 1955 and was a national sensation, making instant celebrities of many of its contestants, and a boon to its sponsor, Revlon. The program's trademarks were an isolation booth, tension music and plateaus which allowed the contestants a week to consider whether they would risk all for the next question. Although spawning a dozen competitors, including its own spin-off for Revlon, *The $64,000 Challenge,* it continued to attract a huge audience until it was forced off the air in 1958 in the quiz-show scandals of the time. The scandals primarily surrounded *Twenty-One* and *Dotto,* but the Revlon shows did not escape taint and were implicated in some of the testimony for questionable practices. *The $64,000 Question* was revived in the fall of 1976 for first-run syndication, with Carlin again the producer and Viacom the distributor. The prize money was doubled and the show retitled *The $128,000 Question,* but the syndicated version was short-lived.

"SIX MILLION DOLLAR MAN"—fantasy adventure series about a former astronaut with a super-human body—the result of its having been rebuilt by futuristic, cybernetic medical science at a cost of $6 million—who is given difficult special assignments by the Government. Lee Majors had the title role, with Richard Anderson and Alan Openheimer featured in the series produced by Universal TV. It premiered on ABC in the fall of 1973 as a monthly entry and the following January became a weekly hour series. It was cancelled in March 1978.

Based on Martin Caidin's sci-fi novel, *Cyborg,* the series became so popular that it spun off a program with a similar premise, *Bionic Woman.* Harve Bennett was executive producer of both series.

"SIX WIVES OF HENRY VIII"—BBC series of six 90-minute dramas, which found an appreciative audience when CBS offered them in the summer of 1971, two years after they aired in Britain. Anthony Quayle narrated and Keith Mitchell played King Henry. Separately featured in each episode were Annette Crosbie, Dorothy Tutin, Anne Stallybrass, Elvi Hale, Angela Pleasence and Rosalie Crutchley, as the wives.

Red Skelton

SKELTON, RED—a comedian and clown who was the CBS Tuesday night mainstay from 1953 to 1970, having successfully adapted his radio comedy-variety show to a weekly TV hour. His broad, slapstick style and numerous character creations appealed to the unsophisticated, but his artistry as a mime revealed itself in wordless vignettes, often laced with pathos, such as the classic of an old man watching a patriotic parade.

Skelton's most familiar characters were the rustic Clem Kaddidlehopper, the inebriated Willy Lump-Lump, the tophatted Freddie the Freeloader and the Mean Widdle Kid. In 1956 he appeared on *Playhouse 90* in a dramatic role in *The Big Slide.*

CBS canceled Skelton's Tuesday night series in 1970 while it was still high-rated, a move calculated to increase the network's share of the youthful, urban audience. NBC tried Skelton in a half-hour Monday night variety show the following season, but it was unsuccessful.

SKIATRON—an early pay-TV system demonstrated before World War II. A prototype system was tested in the 50s, but the company faded under the FCC's regulatory delays for pay-TV.

"SKIPPY, THE BUSH KANGAROO"—Australian series for children syndicated in the U.S. (1968–69) under Kellogg Co. sponsorship and produced by Norfolk International. Filmed in Australia's Waratah National Park, the series featured Garry Pankhurst, Ken James and Ed Deverraux.

"SKY KING"—a CBS children's series (1953–54) about a cowboy who flies a plane. Featuring Kirby Grant and Gloria Winters, it was produced by Jack Chertok Productions for National Biscuit Co.

SLADE, BERNARD—writer of situation comedies for Screen Gems (Columbia Pictures TV), including the pilot for *The Flying Nun.* He went on to write a smash hit for Broadway in 1975, *Same Time, Next Year.*

SLATER, GERALD—public television executive in a number of posts since 1967. In July 1975 he became executive v.p. of WETA Washington after having been v.p. of broadcasting for PBS. Earlier (1969–70) he was project specialist in communications for the Ford Foundation, a position preceded by one as director of operations for the Public Broadcasting Laboratory (1967–69). Between 1955 and 1967 he was with CBS, working chiefly in news administration.

"SLATTERY'S PEOPLE"—hour-long CBS series (1964) in which Richard Crenna portrayed a state legislator facing modern social and political challenges. Bing Crosby Productions produced 36 episodes.

SLOAN COMMISSION REPORT ON CABLE—major statement on cable-TV's potential to benefit society and a powerful endorsement of the medium, issued in 1971 by the Commission on Cable Communications, established by the Alfred P. Sloan Foundation in 1970. The commission's report, entitled *On the Cable,* was published Dec. 8, 1971 (McGraw-Hill).

The report portrayed cable-TV as "the television of abundance" supplanting the present "television of scarcity," and it noted that the nature of the medium would be transformed when it was no longer based on scarcity. In its conclusions, the commission held it in the public interest to encourage the growth of cable, and it predicted another communications revolution in the 1980s when, as it envisioned, 40 to 60% of U.S. homes would be "on the cable."

SMALL, WILLIAM J.—president of NBC News (1979–82). He left NBC News in March 1982 after management changes in the company made Robert Mulholland, an arch rival, his immediate superior as president of NBC. Small had brought a number of CBS traditions and personnel to NBC, and reportedly these irked Mulholland, a longtime executive of NBC News.

Before joining NBC, Small had had a 17-year career as a CBS News executive and was Washington v.p. for CBS Inc., serving as the corporation's chief liaison to the government and its agencies.

He had been director of hard news operations for CBS News since 1974, with the title of senior vice-president. During his preceding 12 years as Washington bureau chief for CBS, he established a reputation as a tough-minded and hard-driving new executive.

Before joining CBS in 1962, he had served as a news director of stations in his native Chicago (WLS-AM) and Louisville (WHAS-TV). He was president of RTNDA in 1960 and of Sigma Delta Chi, the professional journalism society, in 1974.

Small's books are *To Kill a Messenger: Television and the Real World* (1970) and *Political Power and the Press* (1972).

SMARTS—acronym for Selective Multiple Address Radio and Television Systems, which translates into a satellite distribution system for syndicated programming and commercials that RCA Americom founded in the spring of 1979. To create the system, whose essential purpose was to reduce the cost of tape distribution, RCA Americom offered to build a satellite earth-receive station free for the asking at commercial TV stations. Under the plan, when a sufficient number of earth stations were in place, customers could bring their tapes to RCA Americom indicating when they wished them broadcast to selected stations. The programs or commercials would then be scrambled at the uplink and decoded at the downlink. Viacom and the Post-Newsweek Stations initiated the SMARTS project in 1979 with experimental satellite distribution of a number of off-network syndicated shows.

SMITH, "BUFFALO" BOB—creator and host of *Howdy Doody,* early children's hit on NBC (1947–60). He began in broadcasting as a pianist and vocalist on radio stations WGR and WBEN, Buffalo, from the 30s through 1946. The following year he started a children's program on WNBC Radio in New York, which blossomed into TV's *Howdy Doody.* After a 16-year retirement from TV, Smith returned with a syndicated revival of the show in 1976, but it was short-lived.

SMITH, FRANK M., JR.—veteran sales executive for CBS-TV who rose to president of CBS Sports (1978–81). In all, he had spent 30 years with CBS, departing when Van Gordon Sauter was named to succeed him in Sports. For 11 years (1966–77), he was v.p. in charge of sales for the television network. Under the reorganization of the broadcast group in October 1977, he was promoted to group v.p., operational resources. He joined CBS in 1951 and held various posts in the TV film area until he joined the network's sales department in 1953.

SMITH, HOWARD K.—correspondent and commentator for ABC News (1961–79), after having spent 20 previous years with CBS News, more than half of them as chief European correspondent. Smith became co-anchorman of the *ABC Evening News* in May 1969 (teamed initially with Frank Reynolds and in 1970 with Harry Reasoner) and in September 1975 became the program's daily commentator. In a career of many distinctions, Smith had been moderator of the historic first *Great Debate* in 1960 between Kennedy and Nixon.

He resigned from ABC News in April 1979 when he perceived that his role in the newscast was being eliminated. He said that he did not want to be left without a real function.

Recognized by his peers as a political liberal through most of his career, Smith became an outspoken hawk during the Vietnam war and came to be one of the medium's more conservative voices during the Nixon Administration. Indeed, although it was Smith who presented the *Political Obituary of Richard M. Nixon* on ABC in 1962 after Nixon's defeat in the California gubernatorial race (in a program involving an interview with Alger Hiss), he was to become so trusted by President Nixon in the late 60s that he became the first broadcast newsman to interview the President in a one-to-one format. The live telecast (March 22, 1971) was entitled *White House Conversation: The President and Howard K. Smith*. Smith also publicly expressed some sympathy with Vice-President Agnew's charge that the broadcast press was politically biased.

A Rhodes Scholar, Smith began his journalism career with the United Press in Europe during World War II. He joined CBS as Berlin correspondent in 1941 and remained in Europe through most of the war. In 1946 he covered the Nuremberg war crimes trials. He became Washington correspondent for CBS News in 1957 and four years later was named chief correspondent and manager of the bureau. After a dispute with management, he switched allegiances to ABC News.

Howard K. Smith

SMITH, KATE—singer and major radio star who made the transition to TV with two shows in the early 50s—a daily program in the afternoons and a weekly series at night. They ended midway in the decade, and her work in TV since has largely been confined to guest shots and occasional specials.

SMITH, SIDNEY F. R.—producer-director of TV specials ranging from the Metropolitan Opera *Salute to Rudolf Bing* to the Miss U.S.A. and Miss Universe pageants. His credits

include *V.D. Blues* for PBS, *Elizabeth Taylor in London, Victor Borge at Lincoln Center, The Bell Telephone Hour* and the Emmy Awards telecast.

"SMITHSONIAN"—series of nonfiction specials on CBS (1974–) produced by Wolper Productions in cooperation with The Smithsonian Institution. The documentary programs, generally on natural history, follow the tradition of the National Geographic, Jacques Cousteau and Jane Goodall series. George Lefferts is executive producer.

SMOTHERS BROTHERS DISPUTE—an episode in the spring of 1969 in which a pair of star performers clashed with CBS over the censorial practices of the network's standards and practices department and lost both the battle and their Sunday night program. As grounds for their dismissal, CBS cited their tardiness in delivering tapes of the shows in time for closed-circuit previews for the affiliates, but it was already clear to the network that the comedians would not conform to the CBS acceptance standards and that the conflict would go on indefinitely. The Smothers Brothers had also committed the unforgivable: They spoke abusively of CBS in public and took their case to Washington. It was a classic case of creative expression stifled by the rigid rules of the corporation. But the rules had their origins in the broadcast laws that hold the individual stations, and not the artist, responsible for what goes out over the air.

Tom & Dick Smothers

The Smothers Brothers Comedy Hour on CBS was an unexpected smash when it began early in 1967 as a midseason replacement to compete with NBC's high-rated *Bonanza*. The youthful comedians took the measure of the established western hit with a style of comedy that inclined to social satire, topical jokes and political shafts. Their brand of humor and their mild identification with the youthful protest movements of the times were largely what brought on their success—*Rowan & Martin's Laugh-In* was a concurrent phenomenon on NBC—but it was also what brought down the censors.

Religious jokes, conversations about dissenters to the Vietnam War and references to the female anatomy (all mild by the liberated standards of the 70s) were continually cut from the scripts or ordered softened. The comedians took their complaints to the press, and through most of 1968 the CBS deletions were grist for a TV reporter's mill. The brothers also enlisted support from Nicholas Johnson, an FCC commissioner, and visited congressmen with their complaint against the network.

Ironically, less than two years later, CBS brought in another midseason replacement *(All In the Family)* that not only violated all the rules but forced all the networks to broaden their acceptance standards. Meanwhile, the Smothers Brothers were given a summertime show on ABC, which failed, then had a short-lived syndicated program.

SMPTE (SOCIETY OF MOTION PICTURE AND TELEVISION ENGINEERS)—professional organization of specialists in the closely allied visual arts which does major work in establishing technical standards for lighting, equipment and film.

"SNOOP SISTERS, THE"—one of the rotating series in the NBC *Wednesday Mystery Movie* lineup in 1973 which featured Helen Hayes and Mildred Natwick as quaint old ladies who write mystery stories and solve crimes. Produced by Talent Associates-Norton Simon Inc., it yielded six films.

Tom Snyder

SNYDER, TOM—NBC News personality and host of the *Tomorrow* program, (1973–81)who in June 1979 also became anchor of the network's new newsmagazine, *Prime Time Sunday,* which lasted one season. Snyder made the leap from local newscaster to the network in the fall of 1973 after his three-year stint as anchorman for KNBC, the NBC o&o in Los Angeles, gave the station news leadership in the market. Earlier, he had demonstrated his popular appeal and interviewing abilities at KYW-TV Philadelphia. Snyder began his broadcast news career at a Milwaukee radio station in 1957 and spent the next 16 years working at TV stations around the country until NBC selected him for *Tomorrow.*

For a time, Snyder doubled as anchorman of *News Center 4,* the WNBC-TV local newscast in New York, but in 1977 he returned to Los Angeles and made that the originating base for *Tomorrow.* In 1981, he balked at being teamed with Rona Barrett. Soon after, he lost his time period to a new program with David Letterman.

"SOAP"—unusual ABC prime-time series that interpreted daytime soap opera in situation comedy terms; it was launched successfully in September 1977 but only after provoking a storm of protest, while it was still in production, over its heavy concentration on sex and sexuality. The series ended its run in 1981.

Magazine reports on *Soap's* two-part pilot carried the litany of its themes: adultery, transvestitism, impotency, frigidity, voyeurism, premarital sex, and more. Angered that such a program was being prepared for the prime evening hours when young people would be watching, religious groups organized campaigns to ban the program. Some groups succeeded in driving *Soap's* charter advertisers out of the series by threatening to boycott their products. The furor subsided with the program's premiere, accompanied by ABC's public statements that subsequent episodes would be toned down and that the series would not, even by implication, condone aberrant behavior.

The show was built around the families of two sisters, one family upper middle-class, the other distinctly blue-collar. Each member of the families became distinctive through his or her hang-ups. Like soap opera, the various strands of plot continued from week to week, but like situation comedies the segments managed to stand as episodes in themselves. A new series, *Benson,* was spun off in the 1979–80 season from the character of the butler, played by Robert Guillame.

Katherine Helmond and Diana Canova headed the large cast, which included Robert Mandan, Richard Mulligan, Jimmy Baio, Billy Crystal, Cathryn Damon, Arthur Peterson, Robert Urich, Jennifer Salt and Ted Wass.

The series was created and written by Susan Harris, who also served as producer. Executive producers were Tony Thomas and Paul Junger Witt, and Jay Sandrich was director. It was via Witt/Thomas/Harris Productions.

SOCCER ON TV—event that attracts the greatest global audience, yet is one of the lowest-rated sports on U.S. television. The quadrennial World Cup Soccer championships, beamed around the world by satellite, reach a total TV audience estimated at 600 million viewers—a staggering number, made even more so by the omission of the U.S. from the viewing countries.

Soccer is indeed the world's most popular sport, but it has been almost unknown to U.S. spectators and had no tradition here beyond the weekend leagues that were organized by ethnic groups. Contributing to the American indifference to the sport was the fact that it has not been a gymnasium activity widely taught to youngsters.

In hopes that soccer might develop into the third major sport all the networks were seeking, CBS carried the games of the newly organized professional North American Soccer League in 1967 and 1968, but the ratings were disappointing in the extreme. CBS resumed coverage in 1976, after a concerted effort to promote the sport had been made by the league. That effort included the recruiting by the New York Cosmos of the highly publicized Brazilian superstar, Pele, and a six-year city-by-city campaign by the American Youth Soccer Organizations to establish more than 6,000 teams in various boys' leagues around the country.

Nevertheless, the 1976 ratings did not reflect a momentous change in U.S. attitudes toward soccer, and CBS again dropped its coverage of the sport. Subsequently, TVS Sports put together a schedule of professional games for syndication, and some public TV stations began carrying tapes of British soccer matches.

Professional soccer began drawing surprisingly large crowds in the summer of 1977, improving its prospects for another network TV contract.

SOCIETY OF FILM AND TELEVISION ARTS, THE—Great Britain's equivalent of the Motion Picture Academy of Arts and Sciences and the National Television Academy combined, and with much the same professional objectives. SFTA is a synthesis of the now-extinct British Film Academy and the Guild of Television Producers and Directors. Its most public activity has become the annual presentation of awards—known colloquially as "Stellas"—for excellence in the fields of television and theatrical motion pictures. The awards originally were presented separately by SFTA's predecessor organizations.

Other SFTA activities include sponsorship of meetings, symposia, screenings, maintenance of a film script library, and publication of a quarterly journal.

SOCOLOW, SANFORD—bureau chief of CBS News London bureau, since Jan. 1982, previously he had been executive producer of the *CBS Evening News with Walter*

Cronkite. His other positions have included news v.p. for Washington and executive editor in the Washington bureau, assuming the latter post in 1971. In the preceding eight years, he was coproducer of the Cronkite newscast.

Socolow joined CBS News in 1956 from INS, where he had served as Far Eastern correspondent.

"SOMERSET"—NBC soap opera (1970–77), as a spin-off of the successful serial, "Another World." The cast included Joel Crothers, Fawne Harriman, Gloria Hoye, Barry Jenner, Georgann Johnson, Audrey Landers and Michael Lipton.

"SONNY & CHER COMEDY HOUR, THE"—See Cher.

SOURCE, THE—online data base, the brainchild of William von Meister, and a precursor of videotext systems. In late 1980, von Meister sold out to Readers Digest. During 1981, The Source lost its early lead to its main competitors, CompuServe and the Dow Jones News Service.

SOUTHERN TELEVISION—British regional company licensed to serve the central southern and southeast areas of England, with facilities in Southhampton and Dover. Its shareholders are the Rank Organisation (37-1/2%), Associated Newspapers Group (37-1/2%) and D.C. Thomson Ltd..

SOVIET UNION, TELEVISION IN—governed by the Soviet Committee for Radio and Television, the Russian system is an arch-example of a state-controlled TV medium. Much of the fare is bluntly propagandistic, and all of the output, including simple entertainment shows, is stringently monitored for any possible deviation from Soviet political orthodoxy.

Russian TV uses the French 625-line SECAM color system. Many programs are still taped or filmed in monochrome, however. Latest estimates put the number of sets in all the U.S.S.R. at around 50 million, second only to the U.S. A sizable proportion are color receivers. Television reaches nearly every corner of the country, with the help of stationary satellites and ground relays, and officials claim there will be total saturation by the end of the 70s.

Program distribution ranges from a single channel in remote rural areas to at least two channels in the major cities and no less than four in Moscow, the capital. There is also a separate channel devoted exclusively to educational programming.

Besides outright propaganda (speeches, films of Soviet communist party functions, etc.), the mix of shows runs from drama, old movies, music and variety, to children's programs, travelogs and news. The prime news program is a 9 P.M. nightly wrapup called *Vremya* (Time), with two anchormen and sophisticated graphics.

There is regional programming in local dialects or tongues (Armenian, Lithuanian, etc.), but national programs are all in Russian.

SPACE FLIGHTS—priority special events for the networks' news divisions, televised since 1961 and reaching a spectacular climax on July 20, 1969, with the live coverage of men setting foot on the moon for the first time. The broadcast of the Apollo XI mission was called "the biggest show in history," and it attracted the largest total audience for any single telecast. An estimated 125 million American viewers watched all or part of the 2-hour, 21-minute moon walk by Neil A. Armstrong, commander of Apollo XI, and Col. Edwin E. Aldrin, Jr., of the Air Force. Worldwide, the audience was estimated at 723 million in 47 countries, or about one-fifth of the world's population.

The telecast from the moon was in black and white, and the historic walk took place after prime time, facts which make the viewing record all the more impressive. The lunar module had landed on the moon at 4:17 P.M., Eastern daylight time, but Armstrong did not emerge from it to touch the moon's surface until 10:56 P.M.

More than 40 million people in the U.S., in 65% of the households, witnessed the actual moment of Armstrong setting foot on the rock-strewn plain and saying, "That's one small step for man, one giant leap for mankind." But with the innumerable replays in the course of the coverage, the moon walk ultimately was seen by 93.9% of the U.S. households.

The networks among them devoted about 150 total hours to coverage of the full Apollo XI mission and spent $6.5 million to produce it, using approximately 1,000 personnel.

Not surprisingly, the second moon walk—in November 1969 by astronauts of Apollo XII—had less than half the audience of the first. This mission was to have provided color television from the moon, but it lost that novel aspect when the color camera failed to work on the moon's surface.

Apollo XIII, however, was a thriller, a real-life television melodrama that drew 75 million viewers. In the third day of its flight, as it was approaching the moon, the craft was forced to abort its mission by a mysterious blast that caused the command module to lose almost all its oxygen and power supplies and, for a time, its television contact. The return trip by the crippled craft—full of suspense, with three lives in danger—held the nation in thrall.

The TV networks began covering the manned space missions in earnest with the suborbital flights of astronauts John Glenn and Alan B. Shepard in 1960 and 1961. Initially, the National Aeronautics and Space Administration was

reluctant to permit coverage of the flights, but it eventually yielded to the reasoning that the public had a right to know how its tax money was being spent. The telecasts proved to be excellent public relations for NASA.

Each of the networks has developed experts in space coverage—NBC's Roy Neal, for example, has done reporting for every space flight—but those who covered the essential story of each mission were Walter Cronkite at CBS, Frank McGee at NBC (after his death, Jim Hartz had the assignment) and Jules Bergman at ABC.

"SPACE: 1999"—syndicated hour-long sci-fi series produced in England by ATV (1975–77) and distributed in the U.S. by its subsidiary, ITC. Produced on a large budget, part of which was justified by its showing on Britain's independent network, *Space* unabashedly sought to cash in on the *Star Trek* rage. Its cast was headed by American actors, Martin Landau and Barbara Bain, a husband-wife team that had gained notice in *Mission: Impossible*. Rejected by the American networks, the show found a receptive market among individual stations, many of which were network affiliates. *Space* made a strong threat in syndication early in the fall of 1975, but the ratings soon flagged. Organic changes were made in the second year to improve its chances.

The changes included the addition of Catherine Schell to the cast and the hiring of Fred Freiberger, who had been associated with *Star Trek,* as producer. The second skein involved more action and greater use of special effects.

Space: 1999

"SPACE PATROL"—futuristic adventure series (1955–56) in which an earth-based patrol group protects the United Planets against the perils of the galaxy. The program featured Ed Kemmer, Lyn Osborn, Virginia Hewitt, Rudolph Anders and Nina Bara.

SPAIN, TELEVISION IN—consists of two government-controlled channels operated by Television Espanola. Both accept a limited amount of spot advertising. Light entertainment (variety and comedy) and costume drama are among the more ambitious staples of the medium. There are an estimated 6 million sets in the country and the color standard is 625-line PAL.

SPARGER, REX—a former government investigator who was caught trying to rig the Nielsen ratings in 1966 to boost four programs. Sparger, who had worked for the House Investigations Subcommittee during its probe of the ratings services (1961–63), had learned the identity of Nielsen homes and contacted them with promotional matter for the shows. Foiled by the Nielsen security system, Sparger explained that he was just trying to demonstrate that it could be done.

SPECIALS—programs created singly rather than in series form and inserted into a network or station schedule as preemptions of regularly scheduled episodic programs. The increased use of such one-shot programs since the late 60s has served to expand the range of fare—making feasible the presentation of news, drama and cultural programs in prime time on a limited basis, as well as entertainment extravaganzas—and overall has improved the quality of television. Indeed, the special has generated larger audiences for TV by drawing the selective viewer to the set.

Additionally, the special enabled advertisers to continue the practice of sponsorship when the cost of weekly series became prohibitive for single advertisers, and it has permitted programmers to try out new ideas and formats. Such successful series as *The Waltons, Laugh-In, The Flip Wilson Show, The Untouchables* and *Family* all had their origins as specials.

The special evolved from the *spectacular*, an NBC innovation in the early 50s and a form to which that network remained devoted for many years before CBS and ABC became active on a comparable scale. An aversion to the special by those networks had grown out of the perception that heavy viewers of TV were creatures of habit who wanted the same shows to occur at the same times each week. Such leading programming figures as James Aubrey, when he was president of CBS, believed that any disruption of the fixed schedule might tend to break the viewer's habit and send him/her to another network.

However, when demographics came to matter more than a head-count of total viewers, in the mid-60s, the concentration on habit viewers gave way to the quest for an upscale, young adult audience. Research had found that the habit viewer tended to be an older person or a young child and that the audience most drawn to specials was the demo-

graphic group advertisers were most eager to reach. By the end of the 60s, each network was committed to presenting 70 or more specials a season.

Contributing to the rise of the special was the growing disinclination among veteran star performers, such as Bob Hope and Perry Como, to continue the punishing work of providing a weekly series. Such stars might have quit the medium but for the ability to cut back their contributions to four or five specials a season.

Specials gained in importance when the rising costs of producing weekly series brought a steady reduction in the number of firstrun episodes a network would buy each year. With series contracts shrinking from 39 original episodes to 22, the networks found that they could fill out the time periods with preemptions for specials before embarking on the rerun cycle.

See also Made-for-TV Movies, Miniseries, Spectaculars.

SPECTACULARS—the original television *specials,* introduced in 1954 by NBC as one-time-only extravaganzas, usually 90 minutes in length, which preempted regular series. There had been sporadic one-shots previous to 1954 —chiefly Christmas programs *Amahl and the Night Visitors* in 1951) and telethons—but the program that broke the ground for the spectaculars was the two-hour *Ford 50th Anniversary Show,* produced by Leland Hayward and televised simultaneously by CBS and NBC on June 15, 1953. Although it featured a number of stars and celebrities, the program's excitement came from a duet between Ethel Merman and Mary Martin, performing on stools against a bare stage. The pairing of the musical-comedy stars continues to be remembered as one of television's glamorous moments.

On March 29, 1954, NBC's new president, Sylvester L. (Pat) Weaver, noted for his innovations in programming, announced that NBC was developing a number of spectaculars, to be produced by Max Liebman, and that the network was committed to the form. From there, the once-only show achieved a permanent place in the television scheme. Liebman's first spectacular was a 90-minute version of the musical comedy, *Satins and Spurs,* starring Betty Hutton, which was telecast Sept. 12, 1954.

Because the term *spectacular* was a hyperbole that verged on the ridiculous as the form proliferated, it gave way to the more modest *special.*

SPELLING, AARON—consistently successful independent producer since 1960 when he launched *Zane Grey Theater.* He followed with *Burke's Law* in 1964, then entered into partnership with Danny Thomas on *Mod Squad* and other series. Later he teamed with Leonard Goldberg, former head of programming for ABC-TV, to produce *The Rookies, S.W.A.T., Family* and *Charlie's Angels,* along with numerous entries for ABC's *Movie of the Week* series. After Goldberg left to go his separate way, Spelling teamed up with Douglas S. Cramer in a production company called Aaron Spelling Productions, which has produced *Strike Force* and *Dynasty,* among others.

He had begun his TV career as an actor in 1953, then drifted into writing (including scripts for *Playhouse 90*) and then producing. His early credits include the *Lloyd Bridges Show* and *The June Allyson Show.* In the mid-60s, he was executive producer of *The Smothers Brothers Comedy Hour.*

SPIN-OFF—a new program series derived from an existing one, usually through the appropriation of characters. Thus, *Maude* grew out of a single episode of *All In the Family,* while *Good Times* was developed around a regular character in *Maude. The Mary Tyler Moore Show* spun off two characters, resulting in two more popular series for CBS, *Rhoda* and *Phyllis.*

Police Woman had its beginnings in the anthology series, *Police Story,* and *Dirty Sally,* which did not last long, was built from characters introduced in a single episode of *Gunsmoke.*

SPONSOR—the advertiser who buys an entire show, rather than participation spots in it, and therefore exercises some control over it in addition to dominating the commercial time.

During the 50s, when most television programs were sponsored, the advertiser paid the cost of production as well as the network's time charges. It was common for sponsors to own the shows, developing them through their advertising agencies and placing them with a network. The steadiest advertisers were able to reserve the choicest time periods year after year. Programs were scheduled without concern for audience flow or other elements of modern strategy.

Rising program costs led to shared sponsorship, usually by two advertisers, one form of which was alternating sponsorship. This applied mainly in half-hour series, which afforded three commercial minutes. In the alternating arrangement, the advertisers took turns as major and minor advertisers, the major using two of the minutes and the minor the remaining one.

Sponsorship went out of fashion in the mid-60s, giving way to the less risky and more efficient purchasing of minutes (and later half-minutes) dispersed over a network's schedule. Mainly, sponsorship survives with specials. Some advertisers prefer it to the "scatter plan" because of the benefits inherent in identification with a quality program and/or glamorous star and in the viewer's presumed gratitude. See also Participating, Scatter Buys.

SPONSOR IDENTIFICATION RULES—FCC requirement that stations and networks reveal on-air which announcements are paid for and by whom; also, that they disclose the sources of all gratis program material and details of the acquisition.

SPORTS—considered by many the perfect program form for television, at once topical and entertaining, performed live and suspensefully without a script, peopled with heroes and villains, full of action and human interest and laced with pageantry and ritual. The medium and the events have become so intertwined that playing rules often are altered for the exigencies of TV, contests are created expressly for TV, sports stars graduate from the playing field to the broadcast booth and new stadiums are built with giant screens on their scoreboards to provide paying customers with the instant replay enjoyed by viewers at home. Television is such an integral part of major sports, and so vital to their economics, that many of the professional leagues probably could no longer exist if for some reason they were cast out of the medium.

By the mid-70s, advertisers were spending $315 million a year for commercial spots in the 1,100 hours of sports coverage on the three networks, and millions more in local and syndicated sports.

Because sports programming tends to attract an upscale male audience, it has become a primary advertising vehicle for male-oriented products and services—beverages, cigars, cosmetics, automotive products and fuels, banks, insurance companies and airlines, with beer companies usually pacing the field.

The lively advertising market for sports only partly accounts for the spirited competition among the networks for television rights to the premium events—the Olympics, Super Bowl, World Series, Kentucky Derby, major college bowl games and the season-long pro football and baseball "games of the week." Along with attracting large audiences, these events carry prestige for a network and are sources of pride for its affiliates. Moreover, the network with the best reputation for sports coverage has a decided edge on its rivals in bidding for the rights to an event.

Football quickly surpassed baseball as the medium's premier sports series partly because it involves much more action but also because the NFL was willing early to adapt its rules to the convenience of television (creating time-outs, for example, for TV commercials).

By the mid-70s the rights for the various professional and college football packages—both network and local—

brought more than $80 million a year from TV and radio. Baseball's total was around $45 million in 1975.

Since the early 60s, the networks' sports divisions have sought to build up a third major sport with scant success. Basketball, hockey, golf and tennis all have intense but limited audiences; on the whole they are lucrative offerings for the networks but are not in the premium class. TV interest in boxing was revived in 1976 in hopes that it would provide the occasional blockbuster event. After having been ubiquitous on the dial in the 50s, along with wrestling, boxing had faded from prominence in TV for almost 15 years.

Scores of lesser sports received exposure on the popular weekend sports anthologies on the networks—ABC's *Wide World of Sports,* NBC's *Grandstand* and CBS's *Sports Spectacular* — and in syndicated programs.

Although sports programs attract advertisers readily, they do not always attract large audiences. Soccer fared poorly on CBS for two seasons in the 60s, but the network tried again in the mid-70s. The National Hockey League games failed to justify the $6 million NBC paid for the rights and were dropped. The World Hockey Assn. flopped on the ad hoc TVS network, and Canadian football did not develop a following in the U.S. in several syndication ventures.

During the 70s, the networks were each spending $80 million or more a year for sports rights and production, their investments rising as they stayed abreast of technological developments that would enhance coverage.

Basketball and hockey are relatively inexpensive to cover, requiring only four or five cameras. Professional football calls for the use of 10 cameras. Golf involves the biggest production effort of all, some tournaments requiring as many as 20 cameras, several of them operating from cranes or scaffolds for a vantage high above the course.

SPOT TELEVISION—the field of television advertising concerned with the placement of commercials for national or regional products on a station-by-station basis, rather than by network distribution. Although networks make it possible to achieve national circulation with a single transaction, spot TV thrives because the country is not homogeneous and most advertisers recognize that their products do not sell in a uniform manner across the country.

Some advertisers deal exclusively in spot, placing their business only in markets where there is apt to be greatest need for the product, but most buy a combination of network and spot, using the latter as a way to "heavy up" in cities where greater advertising support is indicated. In spot, it is also possible for commercials to address the rural viewer with a bucolic approach and the city-dweller with an urban approach.

Individual stations engage station representative firms to secure spot advertising for them, while large broadcast groups, like the network o&os, maintain their own spot sales forces in key advertising centers, such as New York, Chicago, Los Angeles, Detroit, Minneapolis, Atlanta, Dallas, St. Louis, Boston, Memphis and Philadelphia.

Spot is classified by stations as national business, as distinct from network and local revenues.

STANDARDS & PRACTICES DEPARTMENT—unit at each of the networks responsible for clearing all material to be aired, in accordance with industry codes and the company's own standards of acceptability and good taste; in effect, the network censors. On the station level, such a department may go by the name of Continuity Acceptance. The department's staff reads all scripts, monitors programs in production and screens the completed shows, as well as all commercials, for violations of broadcast policy. Neither programs nor commercials may be aired without the department's approval, which often requires that producers delete scenes or words or even, in the script stage, whole episodes. Although similar sets of standards have been formalized by the three networks, they vary enough so that a commercial rejected by one network is, on occasion, found acceptable by the other two. Similarly, one network may be a trace more permissive with sexual references or political satire than the others.

STANFILL, DENNIS C.—chairman and chief executive officer of 20th Century-Fox Film Corp. since 1971, gaining the post in a proxy battle which toppled the regime of Darryl F. Zanuck. Under Stanfill, who formerly had been v.p. of finance for the Times-Mirror Co. and prior to that a banking executive, the company made a dramatic financial recovery and enjoyed a resurgence in motion picture and TV series production. Fox had owned a single TV station, KMSP-TV Minneapolis-St. Paul.

STANTON, FRANK—famed president of CBS Inc. (1946–72) who not only helped shape that company and raise its prestige but also became the broadcast industry's leading statesman and its most effective witness before congressional committees. A perfectionist and tireless worker, he formed the perfect compliment to the showmanly, socially prominent chairman, William S. Paley, through the decades of the company's critical growth. Stanton received the title of vice-chairman the year before reaching the mandatory retirement age of 65, opening the presidency to his successor. But he remained a director of CBS and a consultant to the corporation, under a long-term contract, on his departure in 1973.

As president, Stanton rarely involved himself with light entertainment programming (an area of Paley's expertise) but concentrated on organizational and policy questions and on the political and cosmic issues growing out of the network's news function. In 1951 he reorganized the company along divisional lines—creating separate administrations for radio, TV and CBS Laboratories—and the plan served as a model for other broadcast companies.

A staunch defender of broadcasting's First Amendment rights, he led campaigns on behalf of the industry for broadcast access to Congress and the courts, equal to that accorded the printed press, and for the elimination of Section 315 of the Communications Act, the Equal Time rule. It was through Stanton's efforts that Congress suspended the rule in 1960 to permit the Kennedy-Nixon "Great Debates."

One of Stanton's greatest contests with the Government in defense of journalistic principles occurred late in his career, in 1971, during a controversy surrounding the CBS News documentary, *The Selling of the Pentagon*. Against the threat of being held in contempt of Congress, he steadfastly defied a subpoena from the House Commerce Committee for outtakes, work prints and written scripts for the documentary, arguing that such materials are, as a reporter's notebook, protected as privileged by the freedom of the press guarantees of the First Amendment. Stanton's position found a large body of support in and out of Congress, and the House voted on July 13, 1971, to refuse the Commerce Committee's request for a contempt citation.

Stanton came to CBS as a researcher in 1935, soon after earning a PhD in statistical psychology (the source of his honorific, "Dr."); certain officials at the network had been impressed with his doctoral dissertation, a critique of methods for studying radio listening behavior. In 1937, with Dr. Paul Lazarsfeld of Columbia University, he developed the Stanton-Lazarsfeld program analyzer (nicknamed "Little Annie"), a system for testing proposed programs that has remained in use at CBS ever since. Stanton rose to higher positions and was 38 when Paley appointed him president in 1946.

On leaving CBS, Stanton became chairman of the American Red Cross. He had also served as chairman of the U.S. Advisory Commission on Information and of the RAND Corporation and was a trustee of the Rockefeller Foundation and the Carnegie Institution in Washington. He was also a director of the Lincoln Center for the Performing Arts.

STANTON-LAZARSFELD PROGRAM ANALYZER
—system used by CBS to test the appeal of programs with a randomly selected audience, which has primary value to the network in determinations made with pilots. The system was devised for radio in 1937 by Dr. Paul Lazarsfeld of Columbia University and Dr. Frank Stanton, a PhD in statistical psychology who later became president of CBS

Inc., but it continues to be used for television. Viewers participating in the testing are invited in off the street in New York and Los Angeles to a screening room where they are given seats with knobs on the arms of their chairs. They are asked to press the knob on the left arm when something they dislike appears in the program, and the knob on the right arm when something strongly pleasing to them appears. The buttons are wired to a control room, where a graph records what is liked or disliked from each of around a dozen chairs. The participants are also asked to fill out questionnaires, specifying what they approved or disapproved of in the shows and which characters they especially liked or disliked. Testing results are drawn from several such screenings and have led either to the outright rejection of a pilot or to recommendations for cast changes. CBS maintains that the system, which is nicknamed "Little Annie," has an accuracy rate of 85%.

STAR—billing for a performer, used more loosely in television than in most other forms of show business. In theater tradition, the stars are those whose names appear above the title of the play in letters as large as those for the title, or even larger. All others are considered featured players, in their contracts, even if their names are preceded on the programs by the word "starring." Many plays have no actual stars at all. Star billing is more than an honor; it designates performers of great stature or those with such proven popular appeal that their names on the marquee prompt the sale of tickets.

In television, however, star billing is given freely and often indiscriminately, and it has become common to speak of the leads—even if they are relative unknowns—as the stars. Credits in television shows frequently list the principal cast as "starring," the supporting cast as "also starring," and well-known character actors as "guest stars."

The promiscuous use of the word "star" probably made necessary the coinage of "superstar" in later years to assume the original and largely lost meaning of stardom.

"STAR TREK"—TV science-fiction series with an extraordinary history: a failure, in mass audience terms, during three seasons on NBC despite a staunch cultist following, but a bonanza in syndication and merchandise-licensing long after the network run. *Star Trek* premiered on Sept. 8, 1966, was tried in a new timeslot the second season and was saved from cancellation only by a tremendous mail campaign from its fans. After an unimpressive third season, it ended its run in March 1969.

Ten years after its debut, however, its reruns were being carried on more than 140 stations and in 47 other countries, and a *Star Trek* animated cartoon series was running Saturday mornings on NBC (having begun in 1973) with the likenesses and voices of the original cast. Three national *Star Trek* societies drew thousands of "Trekkies"—devoted fans—to their annual conventions; a raft of *Star Trek* products were being marketed and a full-length movie, featuring the TV cast and produced by the series creator, Gene Roddenberry, had been ordered by Paramount.

The series concerned the space voyages of the star ship, *USS Enterprise*, and its crew, in reconnaissance missions to other worlds 200 years in the future. William Shatner, Leonard Nimoy and DeForest Kelly were principal characters. It was by Norway Productions, in association with Paramount TV and NBC.

Martin Starger

The crew of the USS Enterprise

STARGER, MARTIN—head of programming for ABC-TV (1969–74) who became an independent producer for the network when Fred Silverman was hired away from CBS to replace him. His term was marked not by successful ratings campaigns but rather by a steady improvement in ABC's credibility with quality specials, particularly in the area of drama. Among his contributions was the introduction of documentary-drama, a form in which real events of recent history were enacted, typified by *Pueblo* and *The Missiles of October*. Starger made several unsuccessful attempts to produce hit shows for the U.S. in Britain, mainly through Lord Lew Grade's company, ATV. As an independent, he produced the widely praised and high-rated drama, *Friendly Fire*, for ABC in 1979. He became also president of Marble Arch Productions, a motion picture company, in a partnership with Britain's Lew Grade.

He was hired by ABC in 1966, as east coast v.p. of programs, from BBDO, where he had been a v.p. in the television production department. In 1969 he became v.p. in charge of programs for the network and in 1972 became president of ABC Entertainment when the program department was established as a separate division.

"STARSKY AND HUTCH"—successful ABC hour-long detective series (1975–79) about a free-wheeling, wise-cracking two-man team operating in southern California. The leads were Paul Michael Glaser and David Soul and the series was by Spelling-Goldberg Productions, with Aaron Spelling and Leonard Goldberg as executive producers. It grew into one of the more violent series in prime time but was ordered toned down by the network for the 1977–78 season.

David Soul & Paul Michael Glaser

"STATE OF THE UNION/'67"—the first nationwide telecasts on ETV, made possible by a $10 million grant to NET by the Ford Foundation in 1966 for an experimental interconnection of the stations. That public affairs series was followed by the regular Sunday night interconnection of the stations in November 1967 for *PBL,* a two-hour news-magazine produced independently of NET by the Public Broadcasting Laboratory.

"STATE TROOPER"—syndicated fictional series (1957–59) on the Nevada State Police, produced by Revue Studios and starring Rod Cameron.

STATION—a broadcast entity licensed to a community by the FCC, specifically to serve "the public interest, convenience and necessity." In theory, a station secures and produces programming that best serves local needs, but in actuality most stations affiliate with networks because it is economically advantageous to do so. Network service covers approximately 60% of the broadcast day. While in the early years of TV the stations produced much of their own programming, now they produce little besides the local newscasts because syndicated programming is cheaper and easier to sell to advertisers. Station licenses are subject to renewal every three years.

Key officials at a station are the general manager, the operations or station manager, the sales manager, program director, news director, traffic manager and chief of engineering.

STATION COMPENSATION—monies paid by a network to each affiliated station for carrying its programs. Approximately 10% of a network's advertising revenues for any program is distributed to the stations as compensation. Since each affiliate has its own compensation agreement with the network, the formulas for payment vary. But as a rule of thumb, the amount a station receives from a network comes to around 25–30% of its ratecard price for the time. What is paid in station compensation bears no relation to the rates charged for the advertising by the networks.

STATION IDENTIFICATION—brief advisories, periodically broadcast, giving the call sign of a station and its city of license, either visually (with a slide) or aurally. The FCC requires such IDs at the beginning and ending of each broadcast day and once an hour in between, as close to the hour as feasible.

In practice, station IDs usually precede every program and occur at a greater frequency than is required, because stations consider it valuable for viewers to be reminded of the channel they are watching. While there is no requirement that the ID carry the channel number, most stations give more prominence to the channel position than to the call sign.

The networks usually provide 10– or 12–second station breaks at half-hour intervals. Since the identification can be accomplished in two seconds, many stations share the break with a commercial advertiser, while some tie in public service or self-promotion messages.

STATION PROGRAM COOPERATIVE—the public TV program market, administered by PBS, and also the mechanism by which all PTV stations participate in the selection of national programs for the system through their purchasing power. SPC sprang from an idea by Hartford N. Gunn, then president of PBS, after the Nixon Administration called upon the system to decentralize and to reorganize on the "bedrock of localism" while White House support for a long-range funding bill for public-TV was hanging in the balance. The SPC, which began operating in 1974, serves in place of a network and makes possible the local determination of approximately one-third of the national programming for the system.

The process begins, approximately 10 months in advance of a new season, with the submission of program proposals and appropriate budget data by individual stations and production entities. Several rounds of elimination balloting reduce the initial catalog of more than 200 proposals to a list of finalists. With funding assistance from the CPB and occasionally from foundations, the stations then vote for the programs to which they would lend their own financial support. Programs are considered purchased when the number of stations desiring the show is sufficient to cover the costs. Each station pays a prorated amount according to the size of its potential audience. While PBS distributes all the programs selected, each station may put on the air only the programs it has purchased. Thus, a program like *Sesame Street* would be carried by virtually every station, while other national shows would have only 20 or 30 outlets. See also Public Television.

STATION REPRESENTATIVES—See Reps.

STEINER, GARY A. (Dr.) (d. 1966)—author of a landmark book, *The People Look at Television,* the first major study of the public's attitudes toward television, conducted in 1960. The book was published by Knopf in 1963. Among its findings was that most viewers were satisfied with TV fare, although the average viewer wished TV were more informative; still, when given a choice between informative and escapist fare, most viewers selected the latter. The study, underwritten by CBS, was the first to note the difference between what people said they wanted of television and what they actually watched. Steiner, a University of Chicago psychology professor, died by suicide, apparently as a result of marital problems. Ten years after the publication of his book, CBS brought out another book through its publishing subsidiary, Holt, Rinehart and Winston, which updated Steiner's one-city study in Minneapolis-St. Paul. The sequel, *Television and the Public,* by Robert T. Bower, based on research conducted by the Bureau of Social Science Research in Washington, found that the superfans of TV–the devoted, uncritical viewers–had decreased in number from one-third to about one-quarter of the total audience.

STEP DEAL—the networks' standard arrangement with series producers for the development of new shows, which involves a progression of submissions so that a project may be discarded as impractical at any point, before too much money is invested in it. The process—from the presentation of the idea to its purchase as a series—may span a year.

Program development usually begins in the spring, following the formulation of schedules for the forthcoming season. As a first step, the producer or producing company submits the story idea in the form of a *treatment,* which is a detailed description of the premise and the characters. The networks pay up to $3,000 for an option on the idea, which then proceeds to the next stage, the preparation of a script. The script may undergo several revisions, as the various parties concerned make their recommendations. For his labors, depending on his professional stature, the writer receives between $10,000 and $30,000 for a situation comedy and twice that for an hour-long series.

The network may quit the project at any one of the steps or stages. Typically, two-thirds of the ideas are eliminated before the script stage and only one-third of those which advance to scripts are made into pilots.

The completed pilots are then studied by network programmers and tested, either on the air or in program testing theaters. In the final step, the network may purchase the series, reject it outright or (rarely) carry it over into further development. Generally, fewer than 25% of the pilots are accepted for TV series. Network programmers estimate that every 100 ideas for which step deals are made ultimately yield only two new TV series.

STEP FUNDING—procedure by which Federal agencies and endowments finance public TV projects. See *Scarlet Letter, The.*

"STEPTOE AND SON"—hit British situation comedy about a Cockney London junk dealer and his son, played with great style and charm by Wilfred Brambell and Harry H. Corbett. It led to the American adaptation by Yorkin-Lear for NBC-TV called *Sanford and Son,* with black principals instead of Cockney. It likewise became a ratings winner.

STERN, CARL—one of broadcasting's lawyer-journalists, assigned by NBC News since 1967 to cover the U.S. Supreme Court, the Federal judiciary and the quasi-judicial proceedings of the Federal agencies. His legal reporting on Watergate, the Berrigan case and the trials of Sam Sheppard, James Hoffa, Muhammad Ali, Arthur Bremer and Patricia Hearst were among his notable achievements. In 1973 he won a landmark Freedom of Information Act lawsuit requiring the FBI to disclose the details of its program to disrupt New Left political organizations, a program which Attorney General William Saxbe later described as "clearly illegal."

STERN, LEONARD B.—producer and creator of TV series. He had been executive producer of *The Governor and J.J., Diana, Faraday & Company, The Snoop Sisters, MacMillan and Wife* and *Holmes and Yo Yo.* He was also creator of *Get Smart.*

STEVENS, LESLIE—prolific writer for the TV drama anthologies that abounded in the 50s and later creator and producer-director of the *Stoney Burke* series (1962) and of the science-fiction anthology, *Outer Limits* (1963), both on ABC. His plays for *Playhouse 90* included *Invitation to a Gunfighter, Charley's Aunt, Rumors of Evening* and *The Second Man.* In the 70s he was executive producer of such series as *The Invisible Man* and *Gemini Man.*

STEWART, BILL (d. 1979)—ABC News correspondent who on June 20, 1979, was shot to death with his interpreter-driver by national guardsmen while covering the fighting in Managua, Nicaragua. Stewart's death was filmed from a distance by an ABC cameraman and was widely televised in news programs. Stewart had joined ABC News in 1976 after having worked as anchorman, commentator and investigative reporter at such major local TV stations as WCCO-TV Minneapolis, WCAU-TV Philadelphia and WNBC-TV and WNEW-TV in New York.

"STICKS AND BONES"—see Papp, Joseph.

"STOCKARD CHANNING IN JUST FRIENDS"—half-hour CBS situation comedy which debuted in midseason in 1979, did neither well nor poorly and went back to the drawing boards with a commitment from the network for a second chance as a 1979–80 midseason replacement. Miss Channing, who had won recognition for her performances in several motion picture comedies, was being groomed by CBS to inherit the mantle of Lucille Ball and Mary Tyler Moore. In the initial sitcom, she played a recent divorcee who moves from Boston to California to begin a new life working in a health spa.

STODDARD, BRANDON—ABC program executive who was responsible for developing such successful miniseries as *Roots, Roots: The Next Generations, Rich Man, Poor Man* and the three-hour drama, *Friendly Fire*. These triumphs earned him the promotion in June 1979 to president of ABC Motion Pictures, a new unit within ABC Entertainment that proposes to produce theatrical motion pictures, movies for television, dramas for *ABC Theater* and miniseries to air under the rubric, *ABC Novels for Television*.

Stoddard joined ABC from Grey Advertising in 1970 as director of daytime programs. He became a v.p. two years later and in 1973 added children's programs to his bailiwick. In 1974, he became v.p. of motion pictures for television, then added dramatic programs and novels for television. In 1978, he was elevated to senior v.p.

STONE, EZRA—former radio actor (the Henry Aldrich series) who became director of TV program development for CBS (1952–54) and then a prolific director of prime time series. He also worked in theater and film. Among the scores of series for which he directed episodes were *Julia, The Debbie Reynolds Show, The Flying Nun, The Sandy Duncan Show* and *Bridget Loves Bernie*.

STONE, SID—a regular on the early Milton Berle series. As the fast-talking pitchman, he contributed to popular speech his familiar line, "Tell ya what I'm gonna do.

"STONEY BURKE"—hour-long ABC series (1962) about a rodeo rider, with Jack Lord in the title role and featuring Robert Dowdell, Bruce Dern and Warren Oates. Created by Leslie Stevens, it was by Daystar Productions.

"STOREFRONT LAWYERS"—hour-long series for CBS (1970) in a season when the network was actively seeking "socially relevant" programs. Produced by Leonard Freeman, it concerned three young attorneys of the rock era recruited by a prestigious law firm to operate a storefront law office serving the needs of the underprivileged. It featured Robert Foxworth, Sheila Larken and David Arkin. After 13 episodes the format, locale and title were changed. As *Men at Law,* Foxworth played the young associate to Gerald S. O'Loughlin, senior partner in a large law firm. His former mates had lesser roles. The changes did not avail, and *Men at Law* ran 10 weeks.

STORER BROADCASTING CO.—major broadcast group with headquarters in Miami Beach. Publicly held, it was founded by George B. Storer, Sr., and J. Harold Ryan. TV stations are WSPD Toledo, an NBC affiliate; WSBK Boston, independent; WJBK Detroit, CBS; WAGA Atlanta, CBS; WJKW Cleveland, CBS; KCST San Diego, independent; and WITI Milwaukee, CBS. Officers are Peter Storer, chief executive, and Terry Lee, president and chief operating officer.

STORER, GEORGE B. (d. 1975)—a pioneer broadcaster and founder of the first independent station group, now known as Storer Broadcasting Co., which began in Toledo, O., in 1928 and grew to prominence in both radio and TV. One of the first radio broadcasters to enter television, Storer had three stations on the air by 1949: WSPD-TV, Toledo; WJBK-TV, Detroit; and WAGA-TV, Atlanta. Later, he added stations in Cleveland, Milwaukee, Boston and San Diego, the latter two UHFs. As head of a powerful group, Storer exerted an influence on network policies, until he went into semiretirement in the 60s.

The group was originally known as Fort Industry Broadcasting, taking its name from an oil company owned by the Storer family, of which the stations had been subsidiaries. But it was so widely identified with Storer personally that in 1952—some 20 years after the oil company was sold—the corporate name was changed. Three years later Storer Broadcasting went public, its stock listed on the New York exchange. So that it might be close to his estate, Storer moved the company headquarters to Miami Beach.

In the late 50s, hospitality that he had accorded FCC chairman John C. Doerfer led to charges against the official of fraternizing with the industry he regulated. When Doerfer resigned, Storer hired him as a vice-president of his company and employed him until his retirement.

STORKE, WILLIAM F.—former program executive for NBC who had been in charge of specials since 1968 and was ousted by the Fred Silverman administration in July 1979. He joined NBC in Hollywood in 1948 upon graduating from college, and his advancements in the sales ranks brought him to New York in 1955. Storke shifted to the program department in 1964 as v.p. of program administration. Currently, he is with Claridge Productions, a division of Trident Television.

"STORY THEATRE"—syndicated prime-access series (1970–72) based on the Broadway hit and employing the techniques and cast of the stage production, as well as the direction of Paul Sills. Produced by Winters-Rosen, it was shot in the countryside around Vancouver. It had limited acceptance from stations, however, during the first two years of the prime time–access rule.

STRAND MILES—the total number of construction miles in a cable-TV system.

STRAUS COMMUNICATIONS INC. *v. FCC—{30 F2d 1001 (1976)}*—personal attack case which was decided on the ground that the attack in question was not a violation because it did not occur during the discussion of a controversial issue of public importance. The complaint was brought against WMCA, a New York radio station owned by Straus Communications Inc., by a congressman who had been called a "coward" by a talk-show host because he had refused to go on the program. The congressman, at the time, was attempting to lead a national boycott of meat.

The congressman claimed in a complaint to the FCC that WMCA had violated the personal attack rule. The commission remanded a fine of $1,000 that had been imposed by the Broadcast Bureau but informed the station that it expected compliance with the personal attack rule. Straus appealed the case in the D.C. Court of Appeals.

The court pointed out a highly significant limiting principle to the personal attack rule, namely that not all personal attacks give rise to the reply time, and that the rule applies only when the attack occurs during the presentation of views on controversial issues of public importance. In addition, the court noted that the FCC will find a violation only where it determines that the licensee's actions and decisions have been unreasonable, or in bad faith, namely that it is the licensee in the first instance who decides what issue was involved and whether the issue is controversial and of public importance.

"STREETS OF SAN FRANCISCO, THE"—successful ABC hour-long law-enforcement series (1972–77) featuring Karl Malden as a veteran police detective and Michael Douglas as his young partner (Douglas was replaced in 1976 by Richard Hatch). Filmed on location in San Francisco, the series was by QM Productions, with Quinn Martin as executive producer and William R. Yates as producer.

STRIKE APPLICATION—See License Challenge.

STRIP—a program scheduled in the same time period five or more days a week. Weekly series that have concluded their network runs are frequently resold to individual stations for *stripping,* as a new means of presenting them. Stations that receive no network service prefer to strip programs, both for their own convenience and for the viewer's easy reference.

Karl Malden

STUART, MEL—producer-director associated with Wolper pictures whose TV work included *The Making of the President, 1960* (and the '64 and '68 sequels); *China: The Roots of Madness; Way Out Men;* and *Love From A to Z.*

"STUDS' PLACE"—an early TV comedy series from Chicago, which began locally and went on the network in 1950. The episodes took place in a small restaurant, whose proprietor was Studs Terkel. Other regulars were Beverly Younger, Win Stracke and Chet Roble.

STUNTING—trade term for the employment of inventive programming techniques to win an audience, techniques that do not reflect the true nature of the series or of the program schedule but are momentary "stunts."

In its simplest form, stunting involves the presentation of a special two-hour episode of a one-hour series to open the season, the front-loading of movie showcases, and early premieres of shows in weeks before the season officially opens. More complicated stunting involves the interchange of characters between an established series and a new entry, just for the opening weeks; the sharing of storylines, as when a two-part episode begins on *Six Million Dollar Man* and concludes on the next episode of *Bionic Woman;* and the introduction of shows in alien time periods, that is, on nights of the week other than those to which they are normally assigned.

Stunting was heavily practiced by the networks in the mid-70s and succeeded chiefly in confusing the issue of which network had the most potent schedule of series.

STV—See Pay-Television.

SUBSCRIPTION TELEVISION —company headed by Sylvester L. (Pat) Weaver, former president of NBC. It attempted to establish a major pay-cable operation in Los Angeles and San Francisco in 1962 but was beaten down two years later by a well-financed campaign by theater owners and broadcast interests. STV wired up homes expressly for its service, which largely consisted of baseball games and movies. The threatened interests proposed a state referendum prohibiting pay-television that California voters passed in November 1964, largely on the advice of the skillful "Save Free TV" campaign. By the time the state supreme court struck down the proposition as unconstitutional, STV had lost more than $10 million and gone bankrupt.

SUBLIMINAL ADVERTISING—messages transmitted with great rapidity in single frames with the intention of influencing viewers at levels below normal awareness. Experiments had been conducted with subliminal advertising in the 50s; broadcast codes now prohibit such attempts as devious persuasion.

"SUGARFOOT"—western about a serious-minded young cowboy which alternated for a time with *Cheyenne* on ABC. Starring Will Hutchins and produced by Warner Bros., it played from 1957 to 1961.

Ed Sullivan

SULLIVAN, ED (d. 1974)—Broadway columnist who hosted one of TV's longest-running and most successful variety series. *The Ed Sullivan Show* (originally *The Toast of the Town*) became a Sunday night habit in millions of households

and demolished dozens of programs that attempted to compete with it. Although Sullivan's popularity as a host was puzzling to many—since he was a wooden and unprepossessing performer given to nervous mannerisms and slurred speech—he brought to the medium a newsman's showmanship, the flair for presenting performing acts when they were most topical or when interest in them was high. Typically, Sullivan was first to present The Beatles in the U.S., and he gave Elvis Presley his first network exposure when he became a popular music sensation. His premiere show, on June 20, 1948, marked the TV debut of the comedy team of Dean Martin and Jerry Lewis. The Sullivan show spurred the advancement of scores of show business careers, and relatively obscure performers often parlayed an appearance on the CBS telecast into several months' worth of nightclub bookings.

Sullivan's pronunciation of the word "show" made a national joke of his familiar promise each Sunday of a "really big shew." The program was essentially a vaudeville hour, which used animal acts and circus acts, as well as popular singers and comedians, and the host made a practice of introducing celebrities in the studio audience from the spheres of politics, sports and entertainment. Through all his years on television, Sullivan continued writing his Broadway column for the New York *Daily News*.

Although the program placed in the Nielsen top ten for most of its years, its ratings went into a decline in the late 60s and the show was canceled in 1971. In part it was a casualty of the diverging entertainment cultures of youth and the older generations, made worse for shows like Sullivan's by the deteriorating patterns of all-family viewing as homes acquired their second or third TV sets. But some have also noted that the program's slip in popularity coincided with the decline of New York City as the nerve center of show business.

When the show ended after more than two decades, Sullivan and his producer (and son-in-law) Bob Precht formed Ed Sullivan Productions, which developed program series and specials with moderate success.

SUMMERALL, PAT—CBS sportscaster who began his association with the company in 1962 after a sparkling 10-year career as a star place-kicker for the Chicago Cardinals and New York Giants football teams. Summerall first was associated with the New York stations, WCBS Radio and WCBS-TV, and signed his network contract in 1971. In addition to NFL football, he has covered basketball, golf, tennis and other events.

"SUNDAY MORNING"—90-minute CBS news program which premiered on Jan. 28, 1979, with aspirations of becoming the television equivalent of the Sunday newspaper. To create a spot for the program, CBS canceled three highly respected long-running programs whose station clearances had shrunk to a paltry few: *Lamp Unto My Feet, Look Up and Live* and *Camera Three.*

Produced by Robert (Shad) Northshield, the program is anchored by Charles Kuralt, with key contributions from correspondents Richard Threlkeld and Lesley Stahl. It also has a staff of commentators: Blair Sabol on life-styles, Jeff Greenfield on television, Francis Cole on music and Heywood Hale Broun as critic-at-large.

See also *Morning;* Kuralt, Charles; Northshield, Robert.

"SUNDAY MYSTERY MOVIE"—umbrella title for four 90-minute rotating series (each expanded to two hours in 1975) carried by NBC (1971–77) and produced by Universal TV. Among the original series, which were continued in subsequent seasons, were *Columbo,* with Peter Falk; *MacMillan and Wife,* with Rock Hudson and Susan Saint James (retitled *MacMillan* in 1976, without Miss Saint James); and *McCloud,* with Dennis Weaver and J.D. Cannon. (*McCloud* actually had begun a year earlier as part of a rotating series called *Four in One.*)

Success eluded the fourth entry, however, and there was an annual turnover in the series. *Hec Ramsey,* with Richard Boone and Rick Lenz, was introduced in the fall of 1972 but was canceled. *Amy Prentiss,* for which its star, Jessica Walter, won an Emmy, replaced *Ramsey* but was also dropped. In the quest for a series that could maintain the rating levels of the other three, *McCoy,* with Tony Curtis, took its place in the rotation in 1975 and gave way the next year to *Quincy* with Jack Klugman. That went on to become a weekly series on its own the following year.

"SUNDAY NIGHT AT THE LONDON PALLADIUM" —the British equivalent of Ed Sullivan's CBS variety hour, and one of the pioneer rating hits for independent television in the U.K. Starting in September 1955 (the year of independent television's advent in Britain), the show ran live for 12 seasons and was a Sunday night national habit. It ran through a number of hosts, among them Tommy Trinder, Bruce Forsyth, Norman Vaughn and Jimmy Tarbuck.

Associated Television, the show's producer, reactivated the format on a pretaped basis in 1973 with actor Jim Dale as host, but by that time the excitement was gone and it only lasted one season.

"SUNRISE SEMESTER"—pioneer early morning adult education series fairly regularly carried on about 85 CBS stations. Produced in cooperation with New York University and airing Monday-Saturday at 6:30 P.M. since 1957, the series is said to have a loyal audience of around 2 million

viewers and has an annual production budget of $40,000, absorbed by CBS.

The courses, which may be taken for credit, have included Iranian Culture and Civilization with Dr. Peter Chelkowski (1970); Twentieth-Century American Art with Dr. Ruth Bowman (1972); Twentieth-Century Literature, Dr. Floyd Zulli (1973); The Meaning of Death, Dr. James P. Crase and others (1973); History of African Civilization, Dr. Richard W. Hull (1974); and Communication: The Invisible Environment, Prof. Neil Postman (1976). See also *Continental Classroom.*

"SUPER BOWL"—showdown game between the two conferences of the National Football League, a championship event literally made for television—indeed created *by* television—and one of the medium's top attractions since it was initiated in 1967. The game became possible when NBC's huge investment in the TV rights for the fledgling American Football League in 1964 ($42 million for five years) paid off eventually in teams worthy of competing with the long-established National Football League (whose games were carried by CBS). The Super Bowl became the catalyst for the merger of the two leagues, with separate conferences, when an AFL team—the N.Y. Jets—emerged the winner in 1969. But the game was created in the first place because it was bound to be lucrative for both leagues and promised to be a sure-fire spectacle for television.

Under the initial agreement, both CBS and NBC were to carry the first game and then would alternate coverage year by year. With two networks covering, Super Bowl I reached a combined total audience (viewing all or part of the telecast) of more than 77 million. Subsequent games have all had an audience of more than 51 million—larger than the audiences for most prime-time hits—although the Super Bowl was played in the usually low-rated hours of Sunday afternoon.

A rating history (NTI) of the Super Bowl for its first 16 years follows:

Year Game	Network	Total Ratings	Audience Share
1967 *Green Bay-Kansas City*	CBS	33.1	43
1968 *Green Bay-Oakland*	CBS	48.3	68
1969 *N.Y. Jets-Baltimore*	NBC	47.1	71
1970 *Kansas City-Minnesota*	CBS	52.7	69
1971 *Baltimore-Dallas*	NBC	50.7	75
1972 *Dallas-Miami*	CBS	55.0	74
1973 *Miami-Washington*	NBC	54.1	72
1974 *Miami-Minnesota*	CBS	51.9	73
1975 *Minnesota-Pittsburgh*	NBC	53.9	72
1976 *Dallas-Pittsburgh*	CBS	53.7	78
1977 *Oakland-Minnesota*	NBC	44.4	73
1978 *Dallas-Denver*	CBS	47.2	67
1979 *Dallas-Pittsburgh*	NBC	47.1	74
1980 *Los Angeles-Pittsburgh*	CBS	46.3	67
1981 *Oakland-Philadelphia*	NBC	44.4	63
1982 *Cincinatti-San Francisco*	CBS	49.1	73

"**SUPER CIRCUS**"—an early circus-variety on ABC (1949–56) produced at its Chicago station, WBKB (now WLS-TV). It featured Mary Hartline as band mistress, Claude Kirchner as ringmaster and clowns known as Scampy, Cliffy and Nicky.

"**SUPERMAN**"—syndicated series (1952–57) based on the comic book hero; it featured George Reeves and Noel Neill and was produced by National Periodical Publications Inc.

SUPERSTATION—the glamour child of satellite technology, an independent TV station that becomes a national station when distributed by satellite to cable systems around the country. Such stations are desired by cable operators because each fills a channel with programming heavily oriented to sports and old movies, and many broadcast around the clock.

The pioneer superstation was WTCG-TV (series changed to WTBS) Atlanta, owned by Turner Communications, which went on the RCA Satcom satellite in Dec. 1976. But the concept was actually invented by Bob Wormington, president of the Kansas City UHF station, KBMA-TV, in 1973. Cable systems were not equipped to receive satellite transmissions at that time, so Wormington fed out the KBMA signal in a large regional pattern—calling it Target Network Television (TNT)—by microwave relays. Although it had the acceptance of cable systems, TNT was shelved in less than a year for lack of advertiser support.

A few years later, after Home Box Office and other national program services made satellite earth stations standard equipment for modern cable systems, Atlanta's Ted Turner adopted the KBMA idea by arranging for the creation of Southern Satellite Systems as a common carrier to handle the satellite distribution of WTBS. This was necessary because under FCC rules broadcast stations may not lease satellite transponders to extend their signals. Common carriers, however, may lease satellite time and distribute a station nationally with or without its consent.

By mid-1979 there were two other commercial superstations on the RCA satellite—WGN-TV Chicago and KTVU San Francisco, both involuntarily—plus several evangelical religious stations, which were able thereby to call themselves networks.

Under the WTBS arrangement, Southern Satellite charges the cable systems taking the superstation a monthly fee according to their size, but not exceeding $2,000, while Turner's station makes its money through higher advertising rates reflecting the greater audience reach afforded by cable. By June 1979, WTBS was being carried on 750 systems in 45 states, plus another 120 that carried it part-time, at night. All told, the station claimed a potential cable audience of some 3 million households.

The superstation became a new category of broadcaster, falling between a local station and a network, and it posed a competitive threat to both those entities. It also created confusion in the syndication market. Because the superstation is able to transmit copyrighted programs into cities where the rights to those programs have not yet been sold to a local station, some syndicators have been reluctant to sell to superstations. Others have dealt with the problem by raising the prices of their programs. See Cable Networks.

"**SUPERTRAIN**"—high-budgeted suspense-cum-comedy series that was to have been the centerpiece of NBC's drastically revised midseason schedule in 1978–79 with which Fred Silverman, as the new president, hoped to reverse the network's prime-time ratings decline. The premiere was preceded by a huge publicity buildup, but the series flopped and became the symbol of Silverman's frustrating first year at the NBC helm.

Conceived as a sort of *Love Boat* on rails, with several stories going on at once in each episode, the series starred a huge, futuristic cross-country train whose luxurious appointments included a gymnasium, nightclub and disco. The key acting parts were played by guest stars.

Set for a midwinter debut to meet Silverman's timetable for the programing overhaul, *Supertrain* was rushed into production, with crews working around the clock to complete the sets and model train. Reportedly, the costs ran to $12 million for 13 episodes. After scoring a passable rating in its premiere, the series found itself in Nielsen quicksand and was yanked from the schedule for improvements. Dan Curtis was fired as producer during the doctoring period and was replaced by Robert Stambler. The program was brought back in a new time period, but the ratings remained abysmal.

"**SURFSIDE SIX**"—hour-long ABC series (1960–61) about a three-man team of detectives based on a houseboat. The Warner Bros. action-adventure series featured Troy Donahue, Van Williams, Lee Patterson, Diane McBain and Margarita Sierra.

SURGEON GENERAL'S REPORT ON VIOLENCE, THE—an $8 million 1972 study and report commissioned by Surgeon General William H. Stuart's office to determine whether a casual relationship existed between viewing TV violence and subsequent aggressive behavior. In five volumes, and formally titled *Television and Social Behavior—A mittee on Television and Social Behavior,* the report was broken into sections on Media and Content Control, Television and Social Learning, Television and Adolescent Aggressiveness, Television and Day-to-Day Life, and Further Explorations.

While the committee was not directly involved in the commissioning of new research, it made available $1 million for support of independent projects through the National Institute of Mental Health (NIMH). More than one hundred published papers representing 50 laboratory studies, correlational field studies, and naturalistic experiments involving 10,000 children and adolescents from every conceivable background all showed that violence-viewing produced increased aggressive behavior in the child and that remedial action in television programming was warranted.

Although confused newspaper headlines at first suggested that the commission found no relationship between TV violence and aggressive behavior, each member of the committee acknowledged finding such a correlation.

Further discounting of the report occurred when it became known that the committee had resulted from a procedure which systematically excluded some of the most distinguished researchers on the subject while including a number of network executives. Forty names had been proposed for the 12-person committee, and the television industry had been given the privilege of reviewing the list to make recommendations. The industry excluded Leo Bogart, of the Bureau of Advertising of the American Newspaper Publishers Assn. who had published a book on television; Albert Bandura, a psychology professor at Stanford and an expert on children's imitative learning; Lenard Berkowitz of the University of Wisconsin; and Leon Eisenberg, chairman of the Dept. of Psychiatry at Harvard University.

The television industry was represented, however, by Thomas Coffin, NBC; Joseph T. Klapper, CBS; and Gerhart D. Wiebe, formerly of CBS. See also Violence Hearings; Pastore, John (Sen.).

"SURVIVORS, THE"—an unsuccessful attempt by ABC to create the equivalent of a popular novel on TV, with a concentration on sex, glamour and wealth. Produced by Universal TV for the 1969–70 season, the series was created by novelist Harold Robbins and starred Lana Turner, George Hamilton, Ralph Bellamy, Kevin McCarthy and Rossano Brazzi. The episodes were shot in exotic foreign locales, and no expense was spared to establish the show as a Monday night mainstay, in the time period in which *Peyton Place* once flourished. The series was canceled after 15 weeks and gave way to *Paris 7000,* which was devised to honor Hamilton's 25-week contract. It too failed.

SUSSKIND, DAVID—prolific TV producer usually associated with quality drama, founder of the production company Talent Associates and host of his own syndicated interview program *The David Susskind Show* (originally entitled, *Open End*). His Talent Associates produced shows in a range from *Mr. Peepers* and *The Play of the Week* in the 50s to *Get Smart* and *Supermarket Sweeps* in the 60s. Susskind also produced the highly praised TV adaptations of *The Glass Menagerie, The Crucible* and *Death of a Salesman* and certain notable dramas for *Kraft Theater, Armstrong Circle Theater, Kaiser Aluminum Hour* and *DuPont Show of the Month.* He was also producer of the excellent CBS series with George C. Scott and Cicely Tyson, *East Side, West Side.*

In the early years, Susskind was an anomaly because he continued to sell programs to television while making newspaper copy around the country for his attacks on the medium and the low quality of its programming. His detractors in turn called him a fast-talking promoter who played it safe in the new medium by sticking to the classics and proven stage plays.

His own syndicated program, *Open End,* created a sensation when it began in 1958, both because it introduced a new form, the open-ended program with no set time limit, and because it featured such guests as Soviet Premier Nikita Khrushchev and then Vice-President Richard M. Nixon. *Open End* gave way in 1967 to *The David Susskind Show,* a discussion program with a finite running time. Susskind took on a partner in Talent Associates in 1967, when Daniel Melnick joined on leaving ABC as program v.p. The company was acquired in 1970 by Norton Simon Inc., with Susskind's services included, and became Talent Associates-Norton Simon. The company has been involved in motion picture as well as TV production.

SUSTAINING SHOWS—programs offered by networks or stations without advertising support or participation, generally in areas of news, public affairs and religion. Certain telecasts are designated sustaining for reasons of taste and propriety, such as presidential speeches, ceremonial events and religious observances; others simply because advertising is not available for them. The rating companies do not measure viewing for sustaining shows in their regular reports.

SWAFFORD, THOMAS J.—v.p. of program practices for CBS-TV (1972–76), a title producers consider a euphemism for chief censor. It was he who made the final determination of what was suitable for airing during a period when the networks were striving to be mature, sophisticated and *au courant* while much of the public and some in government were concerned about excesses of sex and violence on TV. The strain of making continuity-acceptance decisions in such a climate, exacerbated by the need to maintain a separate set of standards for family viewing hour, prompted Swafford to leave CBS in the summer of 1976. He was succeeded by a CBS newsman, Van Gordon Sauter. Swafford then became senior v.p. for public affairs for the NAB, but he held the position only briefly.

Swafford joined CBS as a radio salesman on the West Coast in 1951 and held a number of positions with stations in San Francisco, Los Angeles, Philadelphia and New York, including that of v.p.–g.m. of WCBS Radio, New York (1966–70). Before being named head of program practices, he had been v.p. of the International Services Division of CBS-TV.

SWANN, MICHAEL—distinguished biologist and educator who since Jan. 1, 1973, has been chairman of the British Broadcasting Corp. Before that he was principal and vice-chancellor of the University of Edinburgh. He was knighted in 1972.

"S.W.A.T."—hard-action series on ABC (1974–76) about a special police tactical unit, considered the most violent series in a time when TV violence levels were high. By Spelling-Goldberg Productions, it featured Steve Forrest, Robert Urich and Rod Perry.

John Cameron Swayze

SWAYZE, JOHN CAMERON—one of television's first network newscasters, who began the popular 15-minute *Camel News Caravan* on NBC in 1948 and continued with it in the early evening until 1956. He was noted for his crisp style and distinctive articulation, developed in radio. When *The Huntley-Brinkley Report* replaced Swayze's broadcast, he remained in the medium as an announcer, panel show host, and commercial spokesman. See also *Camel News Caravan.*

SWEENEY, BOB—producer of action-adventure series, although he began in TV as a comedy actor, playing Fibber McGee in the TV version of *Fibber McGee and Molly.* He and

Bill Finnegan were co-producers of *Hawaii Five-0* until 1975 when they left to go into independent production. After producing two pilots that weren't sold, Finnegan left to go into motion pictures, and Sweeney remained in TV producing *The Andros Targets* pilot for CBS in New York.

SWEEPS—rating surveys conducted three times a year by Nielsen and Arbitron in some 200 individual TV markets. The local viewing data derived from the sweeps serve as the seasonal criteria for national spot advertising.

Each of the sweeps covers a four-week period, one spanning October-November, another February-March and the third April-May. Since these survey periods provide the only ratings most stations receive for the entire year, and since they directly affect revenues, the sweeps have tempted stations to substitute surefire programming, such as movies, for low-appeal normal programming during the sweep weeks. Such deliberate manipulation of the ratings, known as "hypoing," is held illegal by the FCC although the regulations forbidding it have not been aggressively enforced.

Moreover, affiliated stations constantly beseech their networks to schedule their most potent movies, specials and series episodes during the sweep periods, and the networks usually cooperate. This in part explains why the highly appealing programs appear in bunches on television and then give way to arid stretches.

To eliminate the abuses, some network research executives have proposed that the sweep surveys cover eight weeks rather than four, but the idea was rejected by the stations because it would substantially increase the cost of the ratings.

In 1979, with network competition at a high pitch, the sweeps produced the most expensive week and the most expensive single night of programming in television history. The week of Feb. 18 began with the opening episode of ABC's powerhouse miniseries, *Roots II.* In attempting to blunt its impact, NBC countered with the hit movie *American Graffiti* and CBS with another big film, *Marathon Man.* During the week, NBC threw against *Roots* episodes of the high-budgeted miniseries *Backstairs at the White House* and *From Here to Eternity,* and on the final night scheduled the blockbuster, *The Sound of Music.*

Feb. 11 was the most expensive single night, with CBS airing *Gone with the Wind,* NBC *One Flew over the Cuckoo's Nest* and ABC a three-hour television movie on the life of Elvis Presley. The total cost of those ventures was estimated at close to $10 million, or about three times the cost of regular programming for a single night.

SWERLING, JO, JR.—generally closely associated with Roy Huggins in his independent production firm, Public Arts. He was producer of numerous series, executive producer of *Baretta* in its first season and exec producer of various pilots by Huggins. His father was the screen writer, Jo Swerling.

SWIFT, LELA—director, active in the 50s and early 60s with such programs as *Studio One, Suspense, Justice, Hotel Cosmopolitan, House on High Street* and the *Purex Specials for Women.*

"SWITCH"—hour-long CBS private-eye series with comedy touches patterned after the popular movie, *The Sting,* and introduced in the fall of 1975 with Robert Wagner and Eddie Albert in the leads. The series, by Glen Larson Productions and Universal TV, was moderately successful. Larson was executive producer.

SYNDICATION—the marketing of programs on a station-by-station basis, with distribution achieved by locomotion (mail or air express) instead of electronically. Syndicated, or "canned," programs essentially take the place of local programing in the time periods not used by the networks: 6–7 A.M., 9–10 A.M., 4:30–8 P.M. and 11–11:30 P.M. daily (Eastern time), with slightly more time open on weekends.

The fare includes movies, off-network reruns, talk shows, revivals of game shows, programs imported from Britain or Canada, exercise shows, religious shows, children's shows, country music shows, some sports programs, specials placed by advertisers and a number of fairly ambitious series vying for spots in the prime-access half hour (7:30 P.M.). Although relatively few time periods are available to syndication on most stations, scores of distributors regularly compete for them, and new companies continually enter the market.

Syndicated shows are generally preferred to local programs by stations for two main reasons: (1) they cost less than homemade shows by as much as 80% and (2) are better known to national advertisers, therefore more likely to attract national spot business.

Syndication might have been a counterforce to networking but for the problems and expense of distribution. It is a high-risk business in which the companies try to hold the risk to a minimum, with obvious effects on the product. Producers of syndicated shows are subject to most of the union requirements that obtain for network shows, and they are faced besides with treacherous sales and distribution costs. While a network may instantly "clear" a show over most of its 200-odd stations simply by announcing it, a syndicator has to sell each market individually, often with the expense of sending out salesmen from city to city on personal calls.

In most cases, the syndicated film and tapes are *bicycled*— that is, shipped by one station to another by mail or other means, along a designated route. Although, in this manner, several stations are able to share a single print, a syndicator nevertheless must create numerous prints for satisfactory national distribution, while a network needs but a single print (and a back-up) to cover the entire country. The cost of making dozens of prints compounds the risk of making a show that may not sell very well.

While the networks may expect close to 200 stations to carry any of their shows, a syndicator usually does well to get 70 or 80, and his programs most often will be scheduled in the lower-viewing time periods. Obviously, the difference in economies of scale makes for a difference in production budgets. The producer of a syndicated game or panel show will schedule the production of a full week's worth of programs—those for Monday through Friday—on a single day. Many of the film shows are produced in Canada or abroad to hold down costs.

Most viewers are unaware of the difference between network and syndicated shows, and many no doubt assume that a daily celebrity talk show like that of Mike Douglas or Merv Griffin emanates from a network. Douglas, with a lineup of more than 150 stations for a program that was established in the early 60s, earns $2 million a year; this makes him probably the second highest-paid performer in TV after Johnny Carson and proves that network exposure is not essential to a huge success in the medium. Dinah Shore, Lawrence Welk, Andy Williams, *Hee-Haw* and *Wild Kingdom* all found a market in syndication after being dropped by the networks.

Syndication has been a flourishing field since the early 50s, when Frederick Ziv of Cincinnati extended his radio syndication business to TV with popular adventure films like *Highway Patrol* and *Sea Hunt*, but its entrepreneurs have risen and fallen with the trends. Most who speculated with costly, unproven shows soon went out of business.

Mary Hartman, Mary Hartman, the comedy soap opera introduced by Norman Lear in 1975, probably would not have survived or even been attempted if Lear were not enjoying the riches of a string of network hits. Although the program was the talk of television and enjoying high ratings, it was losing $50,000 a week its first season because of the brutal costs involved in duplicating, distributing and promoting a syndicated show. Unlike *Monty Python's Flying Circus,* which had already justified its costs in England and was picking up extra money in the U.S., and *Space: 1999,* which had amortized much of its production expense in England and the Commonwealth countries, *Mary Hartman* was all risk.

Syndication takes several forms, the most difficult of which (in terms of making sales) is "straight syndication"— the sale of programs to stations for cash. Mike Douglas, *Mary Hartman* and most of the off-network reruns were marketed in that manner. Much easier for a producer is the sale of a program to an advertiser, who then purchases time on stations in the markets he desires as the show's sole sponsor.

The dominant form of syndication in the 70s, however, especially for firstrun shows, has been "barter." This involves the purchase of the program by an advertiser or ad agency, which then gives it to the station free with two or three built-in commercials. The station makes its money by selling the remaining spots, incurring no expense in receiving what usually is a quite fancy show.

A variation on barter is "time bank syndication." Here the station receives the program in exchange for time credit to the advertiser, who may place his accumulated commercial minutes on other shows in the schedule. *Newsweek* magazine's weekly library of newscast features have been distributed to stations on such a plan.

Satellite communications promise in time to resolve the distribution problem in syndication. When earth stations proliferate in markets around the country, it will become possible for syndicators to send out their programs electronically, in the manner of the networks.

T

TA RATING—see AA Rating.

TAISHOFF, SOL J.—editor of *Broadcasting Magazine,* weekly trade journal for the radio and TV industries, who has been such an effective agent for broadcasters in Washington that he was given the NAB's Distinguished Service Award in 1966. Not only has he been a staunch defender of commercial broadcast interests in the capital, and an influential voice in the Government's communications policies, but he also arranged for individual broadcasters to make personal contact with legislators on their visits to Washington. Although his publication is strongly biased in favor of the industry, and frequently attacks its critics, *Broadcasting* is respected for the thoroughness of its coverage of industry news and for its sophistication in examining industry issues. Its editorials are distinctly conservative and generally oppose government regulation.

Taishoff and Martin Codel founded *Broadcasting* in 1931. About 14 years later, after a disagreement between them, Taishoff bought out his partner and became sole owner.

TAFT BROADCASTING CO.—prominent station group, publicly owned and based in Cincinnati. Founded by Hulbert Taft in 1939 it was initially incorporated as WKRC/ Cincinnati Times Star Co., then in 1948 as Radio Cincinnati Inc. and as Taft Broadcasting in 1959. TV stations are WKRC Cincinnati, an ABC affiliate; WTVN Columbus, Ohio, ABC; WBRC Birmingham, Ala., ABC; WGR Buffalo, NBC; WDAF Kansas City, NBC; and WTAF Phila-

delphia and WDCA Washington, both independents. John L. McClay is executive v.p. of broadcasting and Donald L. Chapin, v.p. of sales.

Taft acquired Hanna-Barbera productions in 1966 and QM Productions in 1979. In the Fall of 1979, it acquired Worldvision, a major syndication company, for about $13 million. In 1982, Taft bought into Robert Johnson's cable network, Block Entertainment Television, vastly expanding the hours of programming.

"TALES OF THE TEXAS RANGERS"—western series on CBS (1957–59) based on a radio series. It featured Willard Parker and Harry Lauter and was by Screen Gems. General Mills sponsored.

"TALES OF WELLS FARGO"—half-hour western series on NBC (1957–60) expanded to an hour in its final season (1961–62). It concerned a troubleshooter for the stage lines, portrayed by Dale Robertson. Overland Productions and Universal produced the half-hour version, Juggernaut Inc. the hour program.

"TALL MAN, THE"—western on NBC (1960–62) featuring Barry Sullivan and Clu Gulager and produced by MCA-TV.

"TAMMY GRIMES SHOW, THE"—a situation comedy in the daffy mode that became legendary as one of the fastest flops in TV history. Premiering on Sept. 18, 1966, it received its cancellation notice from ABC after three weeks. The series was meant to be a vehicle for Miss Grimes on the heels of her success on the Broadway stage, but the concept was weak and the program widely ignored by viewers. Dick Sargent and Hiram Sherman were featured; William Dozier was executive producer; and the series was made by Greenway Productions and 20th Century-Fox.

"TARGET! THE CORRUPTORS"—series on ABC (1961–62) about newspapermen devoted to exposing criminal connections in high places. By Four Star, it featured Steve McNally and Robert Harland.

TARTIKOFF, BRANDON—president of NBC Entertainment, based in Burbank. The youngest division president in NBC history, he was named to the post in 1980 at the age of 33.

As a protege of Fred Silverman, Tartikoff held high positions in programming at ABC and NBC while only in his twenties. He caught Silverman's attention through several imaginative on-air promotions he mounted while director of advertising and promotion at Chicago ABC station, WLS-TV (1973–76). During that time he was also producing and writing comedy specials. Silverman, then president of ABC Entertainment, recruited Tartikoff for his program staff. A year later, Tartikoff was hired away by NBC as director of comedy programs and in July 1978 became vice president of programs for the West Coast. When Silverman became president of NBC, he made Tartikoff his program chief.

"TARZAN"—hour-long adventure series based on Edgar Rice Burroughs's tales of the jungle man. It premiered on NBC in September 1966 and ran through 1968, with Ron Ely as Tarzan and Manuel Padella as an orphan and was produced by Banner Productions.

"TAXI"—ABC sitcom in the blue-collar genre which was almost instantly successful when introduced in September 1978, partly because it was inserted in the network's powerhouse Tuesday night lineup. The series centers on a group of colorful New York City cabbies, some of whom are part-time actors and boxers, and was created by the *Mary Tyler Moore Show* team of James L. Brooks, Stan Daniels, David Davis and Ed Weinberger, who also serve as writers and producers. James Burrows directs.

The cast is headed by Judd Hirsch and includes Tony Danza, Jeff Conaway, Randall Carver, Marilu Henner, Andy Kaufman and Danny DeVito.

Arthur R. Taylor

TAYLOR, ARTHUR R.—president of CBS Inc. (1972–76); he was forced to resign in October 1976 by chairman William S. Paley, who reportedly decided that, despite his outstanding financial record, Taylor was not the man he wanted as his successor. The action surprised Wall Street and stunned executives of CBS, with whom Taylor had been popular. The quasi-official explanation was that there was a poor "personal chemistry" between Paley and Taylor. John D. Backe, head of the CBS publishing division, was immediately named to succeed Taylor.

In 1980, he began organizing a new pay-cable network in which Rockefeller Center, and later RCA, were principal investors. Taylor, as president of RCTV, secured the American rights to the British Broadcasting Corporation's programming for the new network. This suggested to many that the network would be cultural in orientation, a pay-cable version of PBS. To counter that impression, Taylor called the new program service The Entertainment Network. It made its debut in the Spring of 1982.

Taylor, who was 37 when he joined CBS in 1972, had been hired away from the International Paper Co., where he had been executive vice-president, chief financial officer, a director and a member of the executive committee. Discovered for CBS by executive headhunters, he was brought in to succeed Dr. Frank Stanton, although the late Charles Ireland had served in the post briefly in between. The selection made Taylor heir-apparent to Paley as chairman.

Having been chosen for his skills in finance and acquisitions, Taylor brought order to a company that had been diversifying haphazardly, maximized the main businesses of CBS and improved the profit performance of the company as a whole. Meanwhile, he acquired a background in broadcasting sufficient to bid for Stanton's mantle as the industry leader.

He led the broadcast industry's campaign against pay-cable, became the champion of the "family viewing time" concept in 1975, and the executive most closely identified with it, and made the most concerted effort in all broadcasting—albeit unsuccessfully—for a suspension of Section 315 of the Communications Act (the equal time law) that would have permitted the networks to mount a set of presidential debates in 1976.

Taylor's business career began in 1961 with the First Boston Corp., where he rose rapidly in nine years to become a v.p. and director. He joined International Paper in 1970 as v.p. of finance.

TBC—see Time Base Corrector.

TBN (Trinity Broadcasting Network)—see Cable Networks.

TEBET, DAVID W.—NBC executive in charge of talent relations (1960–79), who with the title of executive v.p. in the program department was responsible for maintaining liaison with the network's stars and for acquiring new talent. He resigned early in the Fred Silverman administration and promptly became a consultant to the Lew Grade-Martin Starger company, Marble Arch Productions.

Before joining NBC, Tebet had been a theatrical press agent for a number of Broadway shows and theatrical production companies. Later he formed his own publicity firm, handling, among other clients, Max Liebman Productions which produced *Your Show of Shows* and numerous TV specials. He joined NBC as a general program executive in 1956 and received the talent post after the death of Manie Sacks.

TEDESCO, LOU—director, principally of game shows and specials. His credits include *Eye Guess, Personality, The Face Is Familiar, Miss Teenage America, Miss U.S.A., Miss America Pageant, Johnny Carson Discovers Cypress Gardens* and *A Salute to Oscar Hammerstein.*

TELCOS—telephone companies, regulated by the FCC as common carriers and also, to a degree, in their activities with cable-TV. Because they own the poles and ducts on or through which the wires would be strung for cable-TV, telcos play a critical role in the cable scheme, a role complicated by the fact that the phone companies would find it desirable to control the second wire on the pole since it both

affords opportunity for growth and, in certain respects, poses competition to telephone services.

Telcos have in some situations impeded the construction of cable systems by charging excessively high pole or conduit fees, creating delays or even flatly refusing to let their poles be used for cable. AT&T and General Telephone both agreed in 1969 to permit the rental of their poles and ducts by cable entrepreneurs, but disputes have arisen in various localities over attempts to raise the rental rates.

When the FCC adopted a rule barring telcos from owning, directly or through affiliates, cable systems in their operating territories, it was challenged in the Court of Appeals for the Fifth Circuit by General Telephone Co. of the Southwest [449 F2d 846 (Fifth Circuit 1971)].

The court recognized that if phone companies had a proprietary interest in cable systems, competing CATV systems could be excluded from the area by the phone company's refusal to carry their wires or lease its lines. The court specifically rejected General Telephone's claim that it had been unconstitutionally deprived of its right to own a CATV system by pointing out that a phone company was free to do so outside its local service area.

In an earlier case, the D.C. Court of Appeals upheld an FCC rule that required telephone companies, as common carriers in interstate commerce, to apply for special permission to purchase or build CATV systems. General Telephone of California had challenged this ruling [413 F2d 390 (D.C. Cir. 1969)/cert. den. 396 U.S. 888 (1969)] on the ground that cable-TV was confined to individual states and was therefore intrastate in nature. It said that certificates of public convenience and necessity, required under Sec. 214 of the Communications Act, concerned interstate matters only and were not applicable here.

The Court of Appeals noted that CATV systems were engaged in interstate communications even when the intercepted signals emanated from stations located in the same state as the subscriber. Additionally, it said, television programming was to a large extent designed for national audiences. The court spoke of an "indivisible stream" of interstate broadcast communication, of which the telephone companies were an "integral part." Since telcos were found to be engaged in interstate communications, they were clearly covered by Sec. 214 of the Act, the court said.

TELECINE CHAIN—a film projector and camera combination used to originate filmed material in television.

TELECOMMUNICATIONS—the transmission and reception of signals, sounds, images or written matter by electromagnetic means, whether over radio frequencies or wire. It has become an all-embracing term for television, cable, telephone, computers, and satellites.

TELEMATION PROGRAM SERVICES (TPS)—a pay-Cable supplier which acquires programming from distributors for individual cable systems operating their own pay channels. Unlike HBO, which distributes a single package of pay-TV fare to all its affiliated systems in the manner of a network, Telemation provides cable operators with a choice of fare, allowing them to customize their pay service.

TELEMETER (INTERNATIONAL TELEMETER CORP.)—pay-TV subsidiary of Paramount Pictures active during the 50s with pay experiments in Palm Springs, Calif., Etobicoke (a suburb of Toronto) and London, England. It became dormant when the five-year Etobicoke test ended in 1965, pending a full go-ahead for pay-TV by the FCC.

TELEPROMPTER CORP. *v.* CBS [415 U.S. 394 (1974)] —CATV copyright case in which the Supreme Court ruled that a network's copyrights were not violated when its signals were imported by a cable-TV system.

CBS had brought suit against Teleprompter Corp. charging infringement of copyright on the CATV systems that brought in distant signals, as differentiated from those systems that merely amplified the signals of local stations which could be received without cable service. The district court dismissed the suit, but this was, in part, reversed by the Court of Appeals.

The Court of Appeals held that where a CATV system simply transmitted local signals, there was no "performance" within the meaning of copyright laws and thus no violation (Fortnightly Corp. *v.* United Artists Television). The court found, however, that transmission of distant signals amounted to a "performance" because in that case the CATV system was the functional equivalent of a broadcaster.

But the Supreme Court refused to make a distinction between the importation of distant signals and the carriage of local signals, at least for the ambit of copyright laws. The Court specifically rejected the proposition that choosing which distant signal to import was analogous to a broadcaster choosing which program to broadcast in the first instance. Instead, the Court said, the selection of signals by CATV was analogous to a TV viewer deciding which station or program to watch.

Finally, the Court explained that the nature of the broadcaster is to supply the original signal, which can be received as images and sounds. The nature of the viewer is to receive the broadcast signal. Since the CATV service did not alter this relationship, its transmission of the signals did not constitute a "performance" and therefore did not violate the CBS copyright in network TV shows. See also Fortnightly Decision.

TELESAT, CANADA—corporation which owns and operates the Canadian domestic satellite system, used by the CBC for interconnection.

TELETEXT—supplementary home television broadcast service which stores and displays on-screen written and graphic material. Now in experimental use in the United Kingdom, Teletext is a compromise between two systems— Ceefax, developed by British Broadcasting Corporation , and Oracle, by Independent Broadcasting Authority (IBA). Teletext material is transmitted by television stations during the vertical, or blanking, interval between TV pictures, in a number of constantly changing "pages." Viewdata is a similar system which uses telephone lines instead of the vertical interval. An optional attachment to the home television receiver permits the storage of these pages and dialing of any individual page, for up-to-date information. Teletext service also includes captions which may be displayed while the picture is on the screen as a service to deaf or foreign-language-speaking viewers. Similar services are in use in France, Germany and Sweden. See also Viewdata.

TELEVISA—large, privately owned network that operates four national channels in Mexico. Since its inception in 1973, Televisa's four channels have cornered 90% of the Mexican market; its exports in 1975 totaled 18,000 hours of programming with chief outlets being Ecuador, Puerto Rico, Central America and Spanish International Network in the U.S. SIN was the biggest buyer with 9,000 hours. Installation of a satellite station on the edge of Mexico City is now underway.

Although Televisa is commercially operated, it devotes 31 hours a week to educational programming. Its four channels are designed to reach different elements of the audience. Thus, while Channel 2, which covers most of Mexico (4 million homes) with programs for the middle class (news, soaps, sports), Channel 4 broadcasts films, documentaries and history series to the urban public. Another channel is geared to youth and the fourth to the intellectual elite.

Televisa's facilities on two sites include three theaters, nine fully equipped studios, eight microwave channels, nine mobile units and four emergency transmitters. The permanent staff numbers around 2,000. See Mexico, Television in.

TELEVISION CODE—a set of standards for proper program and advertising practices, created as a mechanism for industry self-regulation by the National Assn. of Broadcasters for its member stations. Transgressions by subscribers are punishable only by the loss of the right to display the NAB seal of good practices. The Code is administered by the

Code Authority, a full-time office based in New York, and the industry-appointed Television Code Review Board.

Code standards for commercials extend to time limits, specific content and techniques. The number of commercial minutes networks and stations may carry in each day part is ordained by the Code, and liquor advertising and attempts with subliminal perception are among the items barred by it.

In programming, the Code forbids "profanity, obscenity, smut and vulgarity" and specifies that illicit sex relations are not to be treated as commendable. Sex crimes and sex abnormalities are deemed generally unacceptable as TV material. In news presentations, murder may not be represented as justifiable nor suicide as acceptable. Horror for its own sake and the indulgence in morbid detail are to be avoided.

Concerning their special responsibilities toward children, broadcasters are admonished by the Code to treat sex and violence without emphasis and to take "exceptional care" with subjects such as kidnapping or crime episodes involving children.

The industry's adoption of "family viewing time" in 1975, in response to the concern of government officials with excessive violence and mature subject matter to which children were being exposed in prime time, was engineered through the Code by means of an amendment requiring that programs aired between 7 and 9 P.M. (6 to 8 P.M. Central Time) be suitable for all-family viewing.

Amendments to the Code, reflecting either changing mores or pressure from consumerist and special-interest groups, are recommended by the Television Code Review Board to the NAB's Television Board of Directors. The Review Board also considers appeals by the Code Authority and is responsible for preferring formal charges where there are serious violations of the Code.

The NAB Code Authority consists of a manager, an assistant, several editors and secretaries, and it concerns itself mainly with Code interpretation and the reviewing of commercials for both radio and TV. However, when there is unusual pressure or heightened sensitivity to an issue, other surveillances may be initiated. For example, in 1971, when Action for Children's Television exposed unconscionable practices by toy advertisers, all toy commercials were reviewed by the Authority before broadcast. In 1968, after the assassinations of Martin Luther King and Robert F. Kennedy, the Review Board ordered the Code Authority to increase scrutiny of violence in TV programs.

Within the compass of the Code, the networks and station groups maintain their own policies concerning standards. Approximately 70% of TV stations in the U.S. subscribe to the Code. *Justice Department Suit.* In June 1979, the Justice Department announced that it was seeking a court order to overturn the NAB's Television Code on the ground that the portions of it relating to the limits on commercials were in violation of Federal antitrust laws. The suit alleged that the code guidelines for the number of commercials permitted per hour, and for the limits on the length of the advertisements, were anticompetitive and served to inflate the cost of commercials. The suit startled broadcasters and puzzled regulatory agencies and media reform groups, since the concept of industry self-regulation had been endorsed by Congressional leaders and the courts. Moreover, Justice was apparently calling for more commercials to bring prices down at a time when advertisers and members of the public were complaining that TV commercials were excessive. An official of the Justice Department explained in the press that reductions in the cost of TV spots might be passed on to the consumer in reductions in the price of advertised products. Most advertising agencies considered this a naive view and doubted it would happen.

TELEVISION INFORMATION OFFICE (TIO)—an organization formed in 1959 as an arm of the NAB to counteract the criticism of television in the press by asserting, through advertisements and forums, the positive contribution of the commercial medium, socially and culturally. Funded primarily by the three major networks, TIO also maintains a reference library in New York City and produces spots for local stations to carry which contribute to a favorable image for television.

TELEVISION, TECHNOLOGY OF—transmission and reception of pictures or images electronically, generally by electromagnetic radiation (radio). All television systems currently in use (and all of their successfully demonstrated predecessors) utilize the phenomenon of persistence of vision, which makes it possible to break a scene into tiny segments at the origination point and reassemble it in sequence at the receiver.

At the origination point or studio, the pickup tube in the television camera converts an optical image into a corresponding "image" composed of varying electrical charges. This image is scanned by an electron gun, with 525 scanning lines (in the NTSC system) making up a single image. Thirty individual 525-line pictures are scanned each second. The scanning process roughly corresponds to reading a page in a book—each line is read, one at a time, from left to right, from top to bottom. In television scanning using the principle of interlace, alternate lines are scanned in sequence— lines 1, 3, 5, and so forth, until 525 are scanned in 1/60 of a second. Then the electron beam moves to line 2 and scans the even-numbered lines, in effect filling in the spaces between the odd-numbered ones. This "interlace" system helps reduce flicker and is used in all broadcast television systems. A complete 525-line picture (1/30 of a second) is called a "frame." A half-picture—either odd or even lines (1/60 of a second)—is a "field.

The electrical signal scanned within the camera is amplified and, along with a synchronizing signal and a sound signal, is superimposed on a carrier frequency—the frequency of the channel on which the signal is to be transmitted. The entire carrier and signal is fed to the transmitter, which amplifies it to many times its original power. The broadcast antenna emits the carrier and signal as powerful electromagnetic radiations.

The receiving antenna intercepts the radiations, which are fed to the tuner in the television set. The tuner selects only the desired frequencies (channel). The signal is removed from the carrier, amplified and fed into the picture tube, where the image is reconstructed by reversing the pickup operation. A black-and-white picture tube has a fluorescent face, which glows at the spot where it is struck by an electron beam. An electron gun in the neck of the tube is the source of the beam, which is deflected for scanning by magnetic or electrostatic devices mounted around the tube's neck, while brightness is controlled by the amount of voltage on the gun. The beam then "paints" a recreated 525-line picture, again from left to right, top to bottom, line by line.

In color television, the camera may have one or more scanning tubes, but in any case the scanned image is, in effect, broken into three individual pictures by means of filters — one containing only red elements of the picture, one only green and the third only blue. These are television's three primary colors, and when the three are combined at the same strength they result in white. The color signals are scanned and combined mathematically into a chrominance signal, which, in effect contains the key to the coloring of the elements of the picture when it is reconstructed at the receiver. Also derived from the camera is the luminance signal, which controls brightness, and is a complete black-and-white picture. These signals, together with a burst signal to keep the colors in phase, are impressed on the carrier frequency.

At the color television receiver, the information in the signal is decoded, and the signals representing each color as the luminance (brightness) signals are amplified separately. The color picture tube contains three guns (or three "barrels" in a single gun) in its neck — one for the electrical impulses representing each of the three primary colors. The phosphor screen contains an array of tiny dots, lines or rectangles (depending on the type of tube) in sets of three, each with a different chemical composition to cause it to glow red, green or blue when excited by the electron beam. Just back of the faceplate, toward the electron guns, is mounted a thin perforated metal mask, known as a shadow mask or aperture grille. The holes, slits or slots in the mask are positioned so that each of the three electron beams can strike only the proper colored phosphor on the screen.

If a black-and-white picture is being transmitted, all three electron guns are activated to the same extent, thereby exciting all three colored phosphors, resulting in a black-and-white picture. When a black-and-white television set is receiving a color picture, it ignores the chrominance signal, using only the luminance, or black-and-white (brightness) information.

This description applies to the American (NTSC) black-and-white and color systems. Other systems, such as the European (CCIR) black-and-white, PAL and SECAM color systems, operate in a similar, but not identical manner. See also Color Television.

TELIDON—videotext system developed by the Canadian Department of Communications which can be transmitted over broadcast, cable, or telephone. More sophisticated than most other viewdata systems, Telidon has exceptional graphics capability and higher resolution than the kindred European technologies; moreover, it is capable of displaying information from other viewdata systems. It is thus somewhat slower and more expensive than other videotext standards. The system uses a computerized data base to store information and a key-pad device for retrieval. It lends itself also to electronic mail and tele-shopping. Telidon tests have been conducted in the U.S. by Times-Mirror and Time Inc.

TELSTAR I—experimental communications satellite operated by the National Aeronautics and Space Administration which relayed the first live transatlantic transmission on July 10, 1962—a fluttering American flag outside the sending station at Andover, Me. Two weeks later Telstar I began carrying more panoramic telecasts exchanged between the U.S. and European countries. Because it had a random elliptical orbit, transmission times were sharply limited, and the first telecast across the ocean lasted only 18 minutes. A second Telstar was launched March 7, 1963. See also Satellites.

TENNIS ON TELEVISION—a sport long ignored by TV as appealing to the elite, until the mid-70s, when a boom in the sale of rackets and apparel echoed in the profuse coverage of tournaments and in the creation of events especially for TV. One such event, billed as *The Battle of the Sexes* and pitting Billie Jean King against Bobby Riggs (Sept. 20, 1973) at the Houston Astrodome, drew a TV audience of 65 million. This fanned hopes among TV's sports programmers that tennis might become their long-sought third national sport, but by 1976—a year when the networks and public TV covered more than 40 tournaments among them—it was clear that the enthusiasm for playing tennis did not carry over to viewing it. Ratings on the whole were low, except for the major events, such as the Wimbledon and the U.S. Open.

Among the problems with tennis as a TV spectator sport was that it was not a league sport in which teams could be tracked for their progress over a season. Since tournaments do not spring from a central organization, viewers were often confused to find the same tennis star playing different opponents on different channels simultaneously, one a live telecast the other on tape.

Nevertheless, tennis earned a permanent place on TV in the 70s, and like other sports the game adapted itself to the medium. A notable concession to TV was the switch to "optic yellow" balls and colorful shirts from the traditional tennis whites. In turn, TV's interest in tennis served to boost the prize money in tournaments to levels unprecedented for the sport. See also "Winner Take All" scandal.

"TESTIMONY OF TWO MEN"—See Operation Prime Time.

"TEXAN, THE"—western on CBS (1958–59) starring Rory Calhoun, via Rorvic and Desilu Productions. Reruns were carried by ABC in 1960.

"TEXAS"—NBC soap opera which began Aug. 4, 1980 as an attempt to introduce to daytime the themes and style of CBS's prime-time hit, *Dallas.* The locale is primarily Houston, but the show is shot in Brooklyn. During its first 18 months on the air, the serial did little to bolster NBC's sagging daytime ratings. The cast is headed by Elizabeth Allen, Carla Borelli, Phillip Clark, Daniel Davis, Randy Hamilton, Jerry Lanning, and Pam Long.

THAMES TELEVISION—commercial licensee for London on weekdays (Mondays to Fridays at 7 P.M.), yielding the franchise on weekends to London Weekend Television. With production centers at Euston in Central London and Teddington on the River Thames, the company is a leading supplier for the ITV network in the U.K. and has exported such critically successful shows as *Jennie—Lady Randolph Churchill, The World at War* and *Rock Follies.*

"THAT GIRL"—ABC situation comedy (1966–70) about an attractive young woman in the business world. Via Daisy Productions, it starred Marlo Thomas and featured Ted Bessell, Rosemary DeCamp and Lew Parker.

"THAT'S MY MAMA"—ABC situation comedy (1974–75) about a black family that runs a barber shop in Washington, D.C. Via Columbia Pictures TV, it featured Theresa Merritt and Clifton Davis.

TW3: David Frost, Elliot Reid & Nancy Ames

"THAT WAS THE WEEK THAT WAS"—or "TW3," as it came to be abbreviated, an innovative British sketch series that employed satire and ridicule on current events and personalities. Created and produced by Ned Sherrin (a prolific English producer, air personality and stage dramatist), and with a regular company that numbered among others David Frost, the show was one of a number of "breakthroughs" for the British medium in the 1960s. It enjoyed wide publicity and tall ratings. A year after its debut in 1964, the title, format (and Frost) transferred to the U.S. airlanes as an ill-fated NBC-TV series produced by Leland Hayward, with actor Elliot Reid as anchorman.

Blythe Danner in Williams's *Eccentricities of a Nightingale*

"THEATER IN AMERICA"—public TV series, started in 1974, devoted to presenting the work of regional theaters around the country. Among those represented were the Long Wharf Theater in New Haven; the Guthrie Theater, Minneapolis; the Phoenix Theater, New York; and the Arena Stage, Washington, D.C. Most of the productions had dual directors—the original stage director and a co-director for television to restage the work for the cameras. Jac Venza was executive producer for the originating station, WNET New York. The series was funded for PBS chiefly by Exxon.

"T.H.E. CAT"—series about an acrobatic detective on NBC (1966–67) featuring Robert Loggia, R. G. Armstrong and Norma Benzell. It was produced by Boris Sagal for NBC Productions.

"THEN CAME BRONSON"—hour-long adventure series on NBC about a free spirit traveling across America on a motorcycle; it premiered in 1969 and lasted one season. It featured Michael Parks and was produced by MGM-TV.

"THIN MAN, THE"—urbane husband-wife detective series with Peter Lawford and Phyllis Kirk, based on the classic movie series which had starred William Powell and Myrna Loy. Both versions featured a dog named Asta. Produced by MGM-TV, it played on NBC (1957–59), with the reruns stripped in daytime the following two seasons.

"30 MINUTES"—junior version of *60 Minutes* for the teenage audience, airing Saturday afternoons on CBS at 1:30, since September 1978. The youth newsmagazine produced by CBS News deals in topical issues, such as why teenagers run away from home, teenagers with babies and the high accident rate for 16- and 17-year-old drivers. On camera principals are Christopher Glenn and Betsy Aaron. Joel Heller is executive producer, Allen Ducovny executive editor and Vern Diamond director.

"THIS IS THE LIFE"—a pioneer (September 1952) entry in religious TV; this Lutheran Church Missouri Synod series consists of 30-minute dramas dealing with problems of contemporary society—alcoholism, adultery, death, desertion, drugs, gambling and the generation gap—and illuminates the church's position toward them.

"THIS IS TOM JONES"—music-variety series produced by ATV in London for ABC (1969–71) with a rock singer popular on both sides of the Atlantic. But despite his youth-ful following in the U.S., and his reputation as a male sex symbol, Tom Jones was not a hit on American TV.

"THIS IS YOUR LIFE"—one of the big prime-time hits of the 50s, a sort of TV surprise party in which the honored guest, caught unawares, was subjected repeatedly to voices from his/her past and in the grand finale encircled by old friends, colleagues, relatives and teachers. Ralph Edwards, host and creator of the series, did his sentimental utmost to induce tears in the subject and presumably also in the viewer, who was made an eavesdropper on the celebrated person's fondest memories. The program premiered in 1952 and later inspired a British version, hosted there by Eamonn Andrews.

The series was revived in the U.S. in 1971 as a syndicated entry for the prime-access time periods, again with Edwards as host and producer. Lever Bros. placed it in 50 markets, but the ratings were disappointing and the series dropped after two seasons.

THOMAS, DANNY—active in TV as comedian, actor and producer since 1953, when he began his first situation comedy *Make Room For Daddy* (later *The Danny Thomas Show*), which ran on CBS until 1965. By then he had his own production company, which also turned out *The Bill Dana Show* and *The Tycoon*, featuring Walter Brennan, and also had a producing partnership with Sheldon Leonard, which had a hand in *The Andy Griffith Show* and its spin-offs.

Thomas soon returned on screen in the 1967 season with *The Danny Thomas Hour* on NBC, a potpourri of one-hour musicals, comedies and dramas; he appeared in all of them, either as a performer or storyteller. Meanwhile, he had made a new producing partnership with Aaron Spelling; as Thomas-Spelling Productions they were responsible for such series as *Mod Squad, The Guns of Will Sonnett, Rango, The New People* and *Chopper One*.

After a performing layoff of two years, Thomas was back on air in *Make Room For Grandaddy*, an attempt by ABC in 1970–71 to revive the original sitcom in an updated version. The following season he did a pair of specials for ABC, and in a new sitcom on NBC, *The Practice*, in which he played an aging doctor. His daughter, Marlo Thomas, was launched on a TV career of her own through *That Girl*.

THOMAS, LOWELL (d. 1981) —legendary journalist, author and broadcaster whose radio career, mostly with CBS, spanned almost half a century and spilled over into TV. During the experimental period with television, his radio newscasts were simulcast in the medium. He also was a key participant in the CBS-TV coverage of the political conventions of 1952, 1956 and 1960. In 1957 Thomas began a

TV travel series, *High Adventure,* and even in the 70s, as an octogenarian, he conducted a syndicated series, *Lowell Thomas Remembers,* and narrated a retrospective television series on the early newsreels of Movietone News.

He was the author of 54 books, a noted world traveler, narrator for innumerable films, a founder of Cinerama and one of the founders of, and principal stockholder in, Capital Cities Broadcasting, a major group with six TV and 13 radio stations.

Lowell Thomas

THOMOPOULOS, ANTHONY D.—president of ABC Entertainment, the programming arm of ABC-TV, since February 1978 when he succeeded Fred Silverman, who resigned to become president of the National Broadcasting Company. At the time, Thomopoulos had been v.p. of ABC Television and assistant to the president, Frederick S. Pierce. Previously he had served in a number of executive capacities in programming and administration.

Thomopoulos joined ABC in 1973 after two years with Tomorrow Entertainment, where he was involved with the production and marketing of *The Autobiography of Miss Jane Pittman.* Earlier he was with the RCA SelectaVision division as director of programming and before that with Four Star Entertainment Corp. for five years. He began his career in broadcasting with NBC's International Division, working in sales.

THORPE, JERRY—producer-director (son of the late motion picture director Richard Thorpe) who was executive producer of *The Untouchables, Kung Fu, The Chicago Teddy Bears* and *Harry O* and with Garry Marshall, co-exec producer of *The Little People.*

THREE ROCKS, THE—the three TV networks, specifically their headquarters buildings along the Avenue of the Americas in Manhattan. NBC long had the nickname "30 Rock," short for its formal address, 30 Rockefeller Plaza. CBS acquired the somewhat derisive nickname of "Black Rock" when it moved into its granite skyscraper in 1965. To complete the scheme, wags in the trade then dubbed ABC "Hard Rock," for its trials as the third-ranked network.

"THREE'S COMPANY"—situation comedy based on a popular British series, *Man about the House,* which turned into a huge hit for ABC after it was introduced in March 1977. During the 1978–79 season it was consistently one of the top three programs in the weekly Nielsen reports and frequently was No. 1.

The series explores the comic possibilities of two women sharing an apartment with a single man, upstairs from the suspicious landlord and his wife. Episodes are freighted with sexual allusions and double entendres. The buxom Suzanne Somers, who became an overnight star with the series, left it after a contract dispute at the end of the 1980–81 season. Somers was replaced by Priscilla Barnes who was featured along with the original cast members, Joyce DeWitt and John Ritter. Norman Fell and Audra Lindley, who play the people downstairs, were spun off into their own series, *The Ropers,* in the middle of the 1978–79 season. Don Nicholl, Michael Ross and Bernie West developed the series and produce it for the NRW Company, in association with T.T.C. Productions.

"THRILLER"—NBC anthology series (1960–62) of suspense dramas about people caught in unexpected situations, each of which was hosted by Boris Karloff. It was by Hubbell Robinson Productions.

TICCIT (TIME-SHARED, INTERACTIVE, COMPUTER-CONTROLLED, INFORMATIONAL TELEVISION)—an experiment in the educational use of two-way cable developed by the Mitre Corp. and begun in the fall of 1975 on cable systems in Amherst, Mass. The system uses the screen of a color TV set for the presentation of printed instructional materials from a computer, and permits the viewer to respond by means of a keyboard.

"TILL DEATH US DO PART"—high-rated BBC situation comedy of the late 60s, written by Johnny Speight, which inspired an American adaptation, *All In the Family.* A series about a working-class family whose breadwinner, Alf Garnett, was intended as a satire of mindless bigotry, it was regarded a breakthrough program in that for the first time in

British TV the storyline comedy form was used to tackle sacred cows and controversy. As *Till Death* rocked Britain, *All In the Family* rocked the U.S. and altered the mores of prime-time television in the 70s.

Till Death originated as a single program in a BBC summertime anthology and began as a short-term series in June 1966. The series was revived for short runs virtually every year since, with Speight as its only writer. Warren Mitchell portrayed Alf Garnett, Dandy Nichols his wife, Anthony Booth the son-in-law and Una Stubbs the daughter. Besides the U.S., TV systems in several countries, including Germany and Holland, created series based on *Till Death*.

TIME BANK—system used by many stations to exchange commercial time for programs or other bartered goods. Through time-buying companies, barter houses or other intermediaries, stations are able to purchase office equipment, restaurant charge accounts and even automobiles for air time instead of cash. The time, which is given a price value, then is "banked" on behalf of an advertiser to be used at his discretion. Some syndicated programs are also "sold" to stations in that manner, the advertiser choosing to bank his commercials rather than have them appear in the program he provides the station free. See also Barter.

TIME BASE CORRECTOR (TBC)—a device which converts the output of a simple videotape recorder/player or other videoplayer to a signal suitable for broadcast or for playing through a standard television receiver or monitor. Using sophisticated techniques, the TBC compensates for the lack of precise timing control in small and low-cost videoplayers. The development of time-base correctors in the early 70s by several firms made possible electronic newsgathering with portable videotape recorders. See also Alternate TV, ENG, TVTV.

TIME-BUYING SERVICES—companies that sprouted during the early 70's specializing in purchasing local station time for a variety of advertisers, or advertising agencies, whose leverage enabled them to achieve highly favorable rates. Some of them also engage in the placement of programs for advertisers, as Vitt Media International did in 1974 with ITT's children's series, *The Big Blue Marble*.

TIME INC.—major communications corporation that has probed more aggressively than most others, and also more successfully, the business frontiers presented by the new communications technologies. By mid-1979, Time owned the largest of the pay-television networks, Home Box Office;

the largest cable-TV company in the U.S., ATC, plus franchises in Manhattan and Queens in New York City; 50% of WSNS, an STV station in Chicago; David Susskind's Talent Associates; Time-Life Video, a company active in marketing prerecorded video cassettes; and station WOTV in Grand Rapids, Mich. Of no less significance are the long-term leases it holds on four RCA Satcom transponders.

These ventures, which put Time Inc. in the forefront of the advancing technologies, represent a rebounding from embarrassing decisions made in the early 70s. Time sold off its TV station group to McGraw-Hill just before the new boom in television that sent broadcast revenues and station prices soaring. It applied the money from that sale to the acquisition of cable-TV systems just as the first glittering promise of cable faded; and after a few years those systems were sold. Next the company made a large financial commitment to hotelvision, the sale of movies over closed-circuit to hotel guests, and suffered losses with it. Yet it was the delivery of current movies in another form, over pay-cable channels via satellite, pioneered by Time Inc. with Home Box Office, that gave a new life to cable-TV and reason for Time to reenter the field of cable ownership with the acquisition of ATC.

TIME-LIFE FILMS—a wholly owned subsidiary of Time Inc. until it was disbanded in 1980 as an unprofitable venture. The company made a name for itself as the U.S. distributor and coproducer of prestige British programs through its deals with the British Broadcasting Corporation. It brought over many of the shows in *Masterpiece Theater*, in addition to Civilisation, The Ascent of Man, Monty Python's Flying Circus, Elizabeth R and *The Six Wives of Henry VIII*. The company has also co-produced series to the extent of investing in the production.

Several executives of Time-Life Films formed Lionheart Productions and continued to distribute British productions.

The company purchased David Susskind's Talent Associates, Inc., a major independent production company for about three million dollars in 1977.

"TIME TUNNEL"—hour-long science fiction series on ABC (1966–67) about three scientists who find themselves transported into the past and future. Produced by Irwin Allen, it featured James Darren, Robert Colbert and Lee Meriwether.

TIMES-MIRROR CO.—publicly owned station group based in Los Angeles. TV stations are KDFW Dallas and KTBC Austin, Tex., both CBS affiliates; WVTM Birmingham, Alabama, WETM Elmira, and WSTM Syracuse,

N.Y., all NBC affiliates; and KTVI St. Louis, Missouri, and WHTM Harrisburg, Pennsylvania, both ABC affiliates. The company, which once also owned KTTV Los Angeles (now licensed to Metromedia), also publishes the *Los Angeles Times,* the *Dallas Times* and *Newsday,* Long Island, N.Y., among other papers. In 1979, Times-Mirror acquired Communications Properties Inc., now Times-Mirror Cablevision, the nation's seventh largest cable-TV company, adding its 46 systems to 15 already owned by the company. Otis Chandler is chairman, Robert F. Erburu president and chief executive officer, Phillip L. Williams senior v.p., newspapers and television, and Dow W. Carpenter senior v.p., cable communications.

TIMES *v.* **SULLIVAN**—See First Amendment and Defamation.

TINKER, GRANT—Hollywood producer who became chairman of the National Broadcasting Company in 1981, following the Fred Silverman debacle. Previously he had been president of MTM Productions, a leading independent TV studio in the 70s which grew out of the success of *The Mary Tyler Moore Show* on CBS. Tinker—who had been a high-ranking executive with Universal and then 20th Century-Fox—became involved with the situation comedy after its first season on the air and proceeded to build the company that was formed to produce it. In a relatively brief time, he placed on the networks *The Bob Newhart Show, Rhoda* and *Phyllis,* among others less successful, and MTM (Miss Tyler's initials) began to rival the Norman Lear organization for primacy among Hollywood's independents.

Considered a topnotch program executive throughout his career, Tinker was an NBC programmer from 1961 to 1967, after having spent the 50s in the TV program departments of such ad agencies as McCann-Erickson and Benton & Bowles. Although he had a promising career with NBC, Tinker chose to join Universal because his wife's work made it difficult for him to live in New York. They were later divorced.

"TODAY"—the prototypical TV talk show, introduced by NBC on Jan. 14, 1952, as a sort of electronic morning newspaper served with music, chitchat and comedy. The brainchild of Sylvester (Pat) Weaver, while he was NBC president, the early morning entry inspired such subsequent NBC variations as *Home* and *Tonight.* Now a TV institution and a major profit center for NBC, *Today* is second only to *Meet the Press* as the network's longest-running commercial show; but, as a five-a-week strip carried for two hours a day, it is TV's uncontested champion in total broadcast hours.

Today drew its style and pacing—and indeed its first host —from *Garroway at Large,* a Chicago origination (1949–51) which featured a bright, low-pressure personality, Dave Garroway. Assisted in the early years by Betsy Palmer, Jack Lescoulie, Frank Blair (as news reader) and a chimpanzee, J. Fred Muggs, Garroway conducted interviews, delivered commentary, reported feature stories, took part in comedy skits and read commercial copy in addition to knitting the segments of the show together as emcee. He remained with the program until June 16, 1961.

By then, the NBC news division had taken over production responsibility from the program department, and the comedy skits and variety show gambits gave way to a greater news orientation, with cutaways to Washington and remote broadcasts from other cities and foreign countries. John Chancellor, a journalist, was installed as Garroway's successor for 15 months. Then on Sept. 10, 1962, the former announcer on *Tonight,* Hugh Downs, became host and held the spot for nine years. Downs was succeeded by Frank McGee, another newsman, whose term ended with his death in April 1974.

Barbara Walters, one of the featured reporters, was then named a co-host, and after working with a series of guest partners, she was joined full-time in July by newsman Jim Hartz. Both were later succeeded by Tom Brokaw. The turnover in supporting cast brought to the show, at various times, such personalities as Charles Van Doren, Helen O'Connell, Joe Garagiola, Edwin Newman, Bill Monroe, Aline Saarinen, Paul Cunningham, Louise O'Brien, Maureen O'Sullivan, Gene Shalit, Ron Hendren, Lew Woods, Phil Donahue, Jane Pauley and Floyd Kalber. In August 1979, Kalber was succeeded as regular newscaster by Tony Guida. Later Brokaw did the newscast as well as the hosting.

In February 1982, after Brokaw had been tapped as co-anchor of the *Nightly News,* Bryant Gumbel became co-host with Jane Pauley and Chris Wallace. Based in Washington, Wallace delivered the newscasts.

Today began to be beamed coast-to-coast in 1954 despite the three-hour time difference. The following year, fast kinescopes were fed to the West Coast, enabling the program to maintain a *clock time* schedule (7 to 9 A.M.) across the country. Video tape eased the problem in the late 50s. The show was designed so that its two hours were interchangeable. In the Central time zone, where there is a one-hour time difference, stations carried the second hour of the broadcast live at 7 A.M. and then followed with the first hour, which is fed out again on tape. In 1979, however, satellite distribution made it possible for stations in all time zones to carry *Today* in the order as broadcast.

"TOMA"—hour-long series on ABC based on the exploits of a real-life police detective, David Toma, whose reliance on disguises helped him establish an outstanding record of arrests and convictions. The series, produced by Universal TV in association with Public Arts Inc., drew marginally acceptable ratings in its first season (1973) and would have been renewed by ABC for a second year had its star, Tony Musante, not refused to continue. Susan Strasberg and Simon Oakland were featured. The series was revived a year later under a new title, *Baretta,* with Robert Blake as the lead, and became a hit.

"TOMBSTONE TERRITORY"—western on ABC (1957–59) produced by United Artists TV and featuring Pat Conway and Richard Eastham.

"TOMORROW"—post-midnight hour-long strip initiated by NBC on Oct. 15, 1973, as a companion to the *Today* and *Tonight* parlay, with Tom Snyder as host. With a 1 A.M. starting time on the east and west coasts, the program explored an uncharted frontier for network television: the middle of the night. The programs consisted mainly of interviews with controversial guests, newsmakers and experts on topical subjects. Originating initially at NBC's Burbank studios, the series moved to New York in 1975 and then back to the Coast in 1977. It was produced and directed by Joel Tator.

"TONIGHT"—the premier desk & sofa show, which started on NBC Sept. 27, 1954, as a 90-minute latenight vehicle for comedian Steve Allen, growing out of his local New York show, although the ground was broken for it on the network by the earlier *Broadway Open House.* For more than two decades it has dominated latenight viewing (11:30 P.M. to 1 A.M.) despite periodic changes of host and format and despite imitative competing shows by CBS and ABC. Since the Steve Allen original, the nightly entry has taken such titles over the years as *Tonight: America After Dark, The Jack Paar Tonight Show* and *The Tonight Show Starring Johnny Carson.*

Allen held forth until January 1957. His show was notable for developing a resident company of vocalists and comedians, many of whom went on to become stars themselves. The singers were Andy Williams, Eydie Gorme, Steve Lawrence and Pat Kirby; the comics Don Knotts, Bill Dana, Louis Nye, Pat Harrington, Jr., Tom Poston and Gabe Dell. Skitch Henderson was bandleader and Gene Rayburn the announcer.

When Allen left the format was redesigned for an ambitious nightly sweep of the country, with cutaways from anchorman Jack Lescoulie in New York to show-business newspaper columnists in various cities—Hy Gardner, Bob Considine, Earl Wilson, Irv Kupcinet, Vernon Scott and Paul Coates, among others. This version fared poorly and lasted only a few months. Jack\Paar took over on July 29, 1957, returning *Tonight* to a studio show with informal talk, comedy and musical entertainment.

Paar's forte was the amusing interview, which called for a stream of guests rather than a resident company; but some guests appeared with such frequency as to comprise a regular supporting cast—Genevieve, Dody Goodman, Alexander King, Zsa Zsa Gabor and Cliff Arquette. Hugh Downs was the announcer and Jose Melis the bandleader.

Paar retired from the show in March 1962, and on Oct. 1 (with guest hosts and Paar reruns serving in the interim) Johnny Carson took over, using a similar format but a different school of celebrity guests. Carson's style was cooler and less emotional than Paar's, and his fast quips and schoolboy antics brought *Tonight* to its peak of popularity. In 1971 he moved the show from New York to Hollywood, ostensibly for access to more glamorous guests. To keep the host from tiring of the grind NBC provided him with frequent vacations, filling in with guest hosts. Carson's resident team consisted of announcer Ed McMahon and bandleader Skitch Henderson, later replaced by Doc Severinsen. See also Carson, Johnny; Parr, Jack.

TONY AWARDS TELECAST—annual springtime television event since 1967, the presentation of the Antoinette Perry Awards for excellence in the Broadway theater. Produced by Alexander H. Cohen, the noted theatrical producer, the two-hour telecasts have been true spectaculars, studded with excerpts and big musical numbers from the current shows. But because the Broadway theater is essentially a local industry, albeit a great tourist attraction, the Tony telecasts are never the ratings blockbusters that the Oscar and Emmy telecasts are. The awards were founded in 1947 and administered at first by the American Theater Wing. In 1966, they were taken over by the League of New York Theaters, and Cohen made them a national event the following year with a one-hour telecast on ABC. The event has switched among the networks, but it is always presented on a Sunday night.

"TONY BROWN'S JOURNAL"—weekly series on black issues that originated on public television and, after a long run, switched in 1978 to commercial syndication under sponsorship of the Pepsi-Cola Company. The program is produced by Tony Brown Productions and distributed by Show Biz Inc.

"TONY ORLANDO AND DAWN RAINBOW HOUR, THE"—CBS hour-long music-variety series (1974–76) built around a pop recording trio—Tony Orlando and Dawn

(Telma Hopkins and Joyce Vincent Wilson)—whose big hit song had been "Tie a Yellow Ribbon 'Round the Old Oak Tree." Originally a summer series, it proved popular enough to be brought back as a midseason replacement in December 1974. The series was doctored for the 1976 season with a greater emphasis on comedy and the inclusion of a regular segment with George Carlin, but it was canceled at midseason. It was produced by Ilson-Chambers Productions and Yellow Ribbons Productions, with Saul Ilson and Ernest Chambers as producers and Bill Foster director.

Joyce Vincent Wilson, Telma Hopkins & Tony Orlando

"TONY RANDALL SHOW, THE"—ABC situation comedy (1976) created as a vehicle for the veteran comedy actor, previously featured in *Mr. Peepers* and *The Odd Couple*. In the series by Mary Tyler Moore's MTM Enterprises, Randall portrayed a widowed municipal judge with two children, a pushy housekeeper and a wise-cracking secretary. The program was created by Tom Patchet and Jay Tarses, who also served as producers. It was switched to CBS the following season, but had a brief life there. In 1981 Randall starred in another sitcom, *Love, Sidney*, on NBC, which initially was controversial because he was identified in the pilot as a homosexual. This detail was muted in the series as a result of the controversy.

TOOBIN, JEROME—public-affairs producer and executive working in public television since the mid-60s, having previously been manager of the *NBC Symphony of the Air* for 10 years. After serving as a producer of *The Great American Dream Machine* and later as executive producer of *Bill Moyers' Journal*, Toobin in 1974 was named director of public affairs for WNET New York, a principal supplier of programs for PBS.

Toobin had joined NET in 1964 as exec producer of *The World of Music*. Later he worked for Group W as a producer of public affairs and cultural specials, and then returned to the public TV outlet in New York. He is married to Marlene Sanders, v.p. of documentaries for ABC News.

"TOPLESS RADIO" CASE—[Sonderling Broadcasting Corp./27 P&F Radio Reg. 2d 285 (1973)]—instance in which a station was fined for violating the obscenity and indecency standard laid out in the United States Code. The focus of the case was WGLD-FM Oak Park, Ill., which employed a format utilized by a number of stations around the country called "topless radio." In this format, an announcer took calls from listeners to discuss sexual topics. In 1973, on a program called *Femme Forum* on WGLD, explicit exchanges took place by female callers on their experiences with oral sex.

The FCC held that while sex per se is not a forbidden subject for broadcasting, the licensee had violated the obscenity and indecency standard of the United States Code. The commission also found additional support in a Supreme Court case, Ginzburg *v.* U.S./383 US 463, in which the court maintained that it is relevant to look at commercial exploitation or pandering in determining liability under an obscenity prosecution. The FCC imposed a forfeiture of $2,000.

"TOPPER"—early situation comedy about ghosts based on the Thorne Smith stories, which played one season on each of the three major networks (1953–55). Produced by Loverton-Schubert Productions, it featured Leo G. Carroll, Lee Patrick, Anne Jeffreys, Robert Sterling and a dog named Neil.

"TO ROME WITH LOVE"—situation comedy about a handsome American widower working in Rome and raising his children there. It ran on ABC (1969–71) and was produced by Don Fedderson Productions. Featured were John Forsythe, Joyce Menges, Susan Neher, Melanie Fullerton and Kay Medford. Walter Brennan joined the cast in the second season. The series was always only marginally successful.

TORS, IVAN—TV and film producer who specialized in fictional animal adventure series, notably *Flipper* and *Daktari* in the mid-60s. Earlier, his Ivan Tors Productions had been responsible for such shows as *Sea Hunt, Aquanauts, Man and the Challenge* and *Ripcord*. In the 70s he had a syndicated series about animal life, *Last of the Wild*. Tors began as a playwright in Europe (he is a native of Hungary) and wrote screenplays in the U.S. before forming his company.

"TO TELL THE TRUTH"—Goodson-Todman panel show which made its debut in 1956 and had a 10-year run on CBS, popularizing the line, "Will the real . . . please stand up." The format involved the panel's confrontation with a person of some achievement and two impostors; after rounds of questions the panelists were to attempt to identify the real guest. Various celebrities performed as panelists over the years, among them Kitty Carlisle, Orson Bean and Tom Poston, with Bud Collyer as moderator.

TOTSU COMPANY—probably the major independent production facility in Tokyo for television producers. More recently it entered into a collaboration with London-based Video Communications for the production of TV shows, feature films and stage projects.

"TOUR OF THE WHITE HOUSE WITH MRS. JOHN F. KENNEDY, A"—hour-long telecast carried by all three networks early on a Sunday evening, Feb. 18, 1962. In the program the President's wife guided the tour through the public rooms of the White House and discussed the efforts she and a committee of citizens were making to restore furnishings that had been purchased by earlier presidents.

TRADE-OUT—a station's acceptance of goods and services in lieu of money for its commercial time. This may involve contest merchandise, furniture and equipment for the station or accumulated credit at hotels and restaurants, the latter useful for entertaining sales clients. Related to the trade-out are reciprocal advertising agreements with other media.

TRADES—the industry's principal trade papers and magazines, which include *Advertising Age, Broadcasting Magazine, Cablevision, Daily Variety, Hollywood Reporter, Multichannel News, Television Digest, Television International, TV/Radio Age, Variety,* and *View.* Others peripherally connected are *Backstage, Editor and Publisher, Gallagher Report, Media Decisions, M/R Media Report, Show Business,* and *Video Publisher.*

TRAFFICKING—misuse of a broadcast license or construction permit by acquiring it with intentions of making a quick sale to another party at a profit. To curb such "trafficking" in TV and radio, the FCC adopted rules requiring licensees to operate a station a minimum of three years before they may transfer (i.e., sell) the property—except under conditions of clear economic distress, in which case the seller may not realize a profit. Construction permits may not be

sold at a profit, nor may companies secure and then sell cable franchises without having built the system.

TRAMMEL, NILES (d. 1973)—colorful president of NBC (1940–49) and then its chairman (until 1952) whose dynamism in the trial-and-error years of TV resulted in million-dollar advertising deals and program schedules of broad appeal. Trammel was the consummate TV executive who could also devise strategies to outfox the competition, pacify nervous sponsors and soothe temperamental stars.

Trammel spent most of his career with NBC, joining in 1928 as a salesman in Chicago and moving rapidly into executive positions there. In 1939 he was transferred to New York as a v.p. At the radio network he was instrumental in developing such hits as *Amos 'n' Andy* and *Fibber McGee and Molly.*

He resigned as chairman in December, 1952, at the age of 58, to become president of Biscayne Television Corp., which won the license to build Channel 7 in Miami.

TRANSLATORS—low-powered relay facilities used by stations to carry their signals beyond the normal coverage area into remote areas. Usually situated in high terrain, the translator receives the over-the-air signal of a station and retransmits it on another unused channel in a prescribed direction. Unlike a satellite station, a translator maintains no studios, originates no programming of its own and is not required to have an engineer present while it is operating.

Some translator systems are owned by television stations to extend their systems but most are built and supported by rural communities which are otherwise beyond the reach of normal television signals.

TRANSMISSION LINE—a coaxial cable, waveguide or other system used to carry a television or other signal.

TRANSMITTER—the physical facility used by a television station to send the signal out on the air. The transmitter superimposes the signal on the carrier, amplifies it and feeds it into the transmitting antenna.

TRANSMITTER SITE—physical location of the transmitter, tower and sending-antenna of a station. TV station transmitters are often clustered in antenna "farms," so that all stations in the area may be received by orienting reception antennas in one direction.

In TV, the tower acts as a supporting structure for the antenna, which actually transmits the signal (in AM radio, the entire tower radiates). TV towers atop mountains are usually 100–200 feet in height, including the 50-150 feet of the antenna structure itself. Self-supporting towers are rarely more than 700 feet in height, with "guyed" towers usually about 1,000 feet tall but occasionally twice that height. The candelabra tower is one supporting more than one TV antenna at equal heights.

TRAVIESAS, HERMINIO—longtime head of broadcast standards for NBC until he retired in 1979. He was succeeded by Ralph Daniel who once was president of the CBS owned TV stations.

Traviesas joined the network in 1937 as traffic manager in its international shortwave radio department, went to CBS 10 years later, then to BBDO and returned to NBC in the mid-60s as west coast director of standards & practices. He headed the department that decides what program material and commercials may be broadcast, under NBC policies for acceptance and the NAB Code.

"TREASURE HUNT"—game show created and hosted by Jan Murray in the late 50s. It was revived for prime-access syndication as *The New Treasure Hunt* by Chuck Barris Productions in 1974, with Geoff Edwards as host.

TRETHOWAN, IAN—director-general of the British Broadcasting Corp. since September 1977, successor in that post to Sir Charles Curran. The position is the highest operating post in the BBC hierarchy.

Trethowan was a print journalist before joining the BBC in 1963 as a news and public affairs anchorman. He entered the management ranks as managing director of radio and later became managing director of television only a year before assuming overall command of the company.

TREYZ, OLIVER E.—dynamic president of ABC-TV (1957–62) who took bold and sometimes erratic steps to keep ABC competitive with the larger, better established networks in the 2.5-network economy that prevailed at the time. Although his background had been in research and advertising, he functioned as a super-salesman, programmer and promoter, the most energetic of the wheeler-dealers in a colorful era. He was fired not long after the network was censured by the Senate Juvenile Delinquency Subcommittee for carrying a particularly violent episode of *Bus Stop.*

As network president, Treyz frequently studied the rating reports for local markets in search of programming leads. Thus, when he found that old Bugs Bunny cartoons were succeeding in prime time on a Chicago independent station, he ordered a new Bugs Bunny series created for ABC. Its popularity prompted Treyz to build a stable of animated

programs in the early evening: *The Flintstones, The Jetsons, Top Cat* and *Johnny Quest,* among others. Treyz also made heavy use of the Nielsen MNA ratings to demonstrate that in the markets where all three networks had full-time outlets, ABC's programs were equal in popularity to those of its rivals.

After leaving ABC Treyz tried to promote a fourth network and became president of the abortive Overmyer Network, which had hoped to link UHF and VHF independents. He later opened a consultancy serving TV advertisers.

"TRIALS OF O'BRIEN"—series about a flamboyant lawyer with domestic problems. It was the first regular vehicle for Peter Falk but failed in the ratings. It was produced by Filmways for CBS in 1965.

TRIDENT-ANGLIA—production company formed in 1977 by two of Britain's regional independents, Trident Television (parent of Yorkshire Television) and Anglia Television, principally with an eye to penetrating the U.S. market. Initially, the new company put into syndication in America the 13-week series *Raffles,* produced by Yorkshire; *Dickens of London,* a 13-week series by Anglia; and *West End Anthology,* a series of TV dramas produced by both companies. After selling *The Lunatic Express* and *Four Feathers* to NBC, Trident-Anglia set up New York offices headed by John F. Ball, who also heads Anglia's production company specializing in wildlife documentaries, Survival Anglia Ltd. See also Anglia TV.

"TROUBLE WITH FATHER"—one of the early situation comedies about bumbling fathers. It was introduced in 1953 in syndication, with Stu Erwin playing a high school principal and June Collyer his wife. Later it was retitled, *The Stu Erwin Show.* Produced by Hal Roach, Jr., and Roland Reed Productions, it ceased after 126 episodes in 1955.

TROUT, ROBERT—broadcast newsman noted for his calm delivery and ability to extemporize during such major events as national political conventions. Throughout most of his career, which began in the early 30s, he was associated with CBS, but he was with NBC for a period (1958–62) as host of the quiz show, *Who Said That?* Trout reported from Washington until World War II, when he was assigned to London. In 1964 CBS teamed Trout with Roger Mudd to cover the Democratic National Convention in an effort to stem the growing popularity of NBC's Huntley-Brinkley tandem, but without great success. Trout then became anchorman for WCBS-TV New York, and during the 70s went into partial retirement as special roving correspondent for CBS.

"TRUMAN TOUR OF THE WHITE HOUSE"—first of the First Family tours for television, carried by all three networks in May of 1952. President Truman conducted an informal tour of the renovated White House and played several piano selections. He was accompanied by a news representative of each network: Walter Cronkite, Bryson Rash and Frank Bourgholzer.

TRUNK CABLE—the major distribution coaxial cable used in a cable-television system. Feeder cables branch out from the trunks to the *drop line* cable, which is connected to the subscriber's residence for service.

"TRUTH OR CONSEQUENCES"—low-budget audience participation show that has spanned three decades in television, after having originated in radio during the 1940s. Created, produced and initially hosted by Ralph Edwards, the show is essentially a party game in which contestants are either asked to perform bizarre stunts or are subjected to elaborate practical jokes. The series was an NBC staple for 15 years (1950–65), was revived in syndication by Edwards in the late 60s and then had a new life in the 70s when the prime time–access rule went into effect. During the NBC run, Edwards was succeeded as host by Jack Bailey, but Bob Barker took over in 1956 and remained with the show for 18 years.

The Tunnel

"TUNNEL, THE"—bold 90-minute NBC News documentary (Dec. 10, 1962) on a tunnel dug beneath the Berlin Wall by a group of European students to help 59 friends trapped in East Berlin to escape. NBC acknowledged that a few parts of the suspenseful program were re-created, but most of the filming was done as the tunnel was being dug. The program was produced by Reuven Frank, who was also co-writer with Pers Anderton. Ray Lockhart was director, and the filming was by Peter and Klaus Dehmel.

"TURN ON"—shortest TV series in the medium's history, canceled after its first telecast in February 1969 because of complaints from affiliates and the public. The half-hour show was to be ABC's version of the popular *Laugh-In* on NBC, but its playfulness and irreverence overstepped the bounds of good taste.

TURNER, R.E.—flamboyant Atlanta sportsman, promoter and broadcaster who upset the broadcast industry's sense of order in December 1976 by using the RCA Satcom satellite to distribute the signal of his obscure Atlanta UHF station—WTCG-TV (now WTBS), Channel 17—to cable systems around the country. In a period of two years, the station gained access to 3 million additional households with 24-hour programming that leans heavily on sports and old movies, making it the biggest independent station in the country. Turner thus pioneered the "superstation" and became the center of a raging controversy in government and the sports, motion picture and broadcast industries over licensing fees.

In June 1980, Turner created the Cable News Network, a 24-hour-a-day all-news channel for cable systems, with Daniel Schorr as principal anchorman and Reese Schonfeld of ITNA as news chief. It is based in Atlanta. To finance the project, Turner sold his Charlotte, N.C., UHF station, WRET-TV, to Westinghouse Broadcasting for a reported $20 million.

A champion yachtsman, he won the America's Cup in 1977 with his craft, *Courageous*. He also owns, or has ownership interest in, a number of professional sports teams, including the Atlanta Braves, Falcons, Hawks and Flames, many of whose games he televises on his superstation.

TURTELTAUB-ORENSTEIN—comedywriting and producing team of Saul Turteltaub and Bernie Orenstein. Their credits include *The New Dick Van Dyke Show, Sanford and Son* (since the 1974–75 season), *Grady, What's Happening, Carter's Country*, and *One of the Boys*.among many others.

TVB (TELEVISION BUREAU OF ADVERTISING)—organization supported by commercial broadcasters devoted to promoting the advantages of local television advertising over that of other media. It also provides extensive promotional and research tools for its member stations, and holds periodic clinics for sales executives. TVB is based in New York and is headed by Roger Rice.

TV GUIDE—phenomenally successful weekly magazine devoted to TV listings and feature articles on programs, stars and developments in the industry; its growth has paralleled the growth of the medium itself. Despite the fact that most newspapers provide extensive daily and weekly program listings, *TV Guide* achieved a circulation in 1976 of 20 million copies a week. Highlight listings in the magazine, in featurettes known as "Close Up," generally are a boon to the ratings for the programs selected for such special attention.

TV Guide has ruled the program information field since its founding in 1953 by Walter Annenberg. The magazine's large editorial staff is dispersed over 27 regional offices, which among them publish 88 separate editions in the United States. Eight Canadian editions were sold to a Canadian publisher in 1977.

The magazine has its headquarters in Radnor, Pa., home base of Annenberg's Triangle Publications. Over the years, numerous national and regional magazines attempted to compete with *TV Guide*, but all failed. It remains the largest-selling magazine in the U.S.

TVN (TELEVISION NEWS INC.)—syndication company formed in 1973 to distribute electronically a daily package of national and international news to subscribing stations. With only 80 stations (many of them Canadian) participating at TVN's peak, the company folded in November 1975 after running up losses of $2 to $3 million a year. A group of independent stations then formed a cooperative, ITNA, using the Westar satellite, to provide a similar news service for member stations.

TVN's principal backing came from Joseph Coors, ultraconservative head of the Adolph Coors Brewing Co., who felt a need (and a desire among stations) for a national news service in TV that would offset the networks. While it was a foregone conclusion that major independent stations would subscribe, TVN was organized in the belief that many affiliated stations would drop their network newscasts if they could receive national and foreign news on the same day from an independent, "unbiased" source. That proved not to be the case. TVN succumbed not to the cost of maintaining news bureaus throughout the U.S. and abroad but to the expense of daily transmission over AT&T lines. A key factor in the resistance of stations to buying the service was the additional cost it would have entailed for them to rent *local loops*, their own connections to the cross-country lines, like-

ned to cloverleaf exits off super-highways. See also Coors, Joseph; ITNA.

TvQ—periodic studies conducted by Marketing Evaluations Inc. which attempt to define the actual appeal of network and syndicated programs, over and above their rating performance. Sometimes spoken of as qualitative ratings, the TvQ reports are utilized by networks and advertisers as an advance indicator of shifts in program popularity or as evidence that a potentially successful program has been assigned the wrong timeslot.

Questionnaires are mailed to a representative panel, asked to indicate from a list of familiar programs how they would rate them on a scale from "one of my favorites" to "poor." The TvQ score is the percentage of respondents who rated the program a "favorite." The scores are also broken down demographically, to reveal the level of appeal by age or sex. Programs with low ratings but high TvQ scores are occasionally renewed by the networks on the theory that the shows would be favored by viewers who had not yet examined them. See also PiQ.

TVS TELEVISION NETWORK—an occasional, or ad hoc, network for sports events operated as a division of Corinthian Broadcasting and headed by Eddie M. Einhorn. The network chiefly carries games of major college basketball conferences. In 1973 it carried the World University Games from Moscow and U.S.-China basketball from Peking.

TVTV (TOP VALUE TELEVISION)—organization of alternate video journalists who produced the first documentaries on portable video equipment ever to be televised. With foundation funding, and using relatively inexpensive cameras, the group produced unorthodox topical documentaries in a "scrapbook" style, as a kind of counterculture answer to conventional network documentaries.

TVTV attracted notice in 1972 for its video tape pieces on the two political conventions of that year. They were purchased by Teleprompter for cable and by Group W for its TV stations. In 1974, with the broadcasting of half-inch tape vastly improved by the development of the time base corrector, TVTV placed *Lord of the Universe* and *Gerald Ford's America* on public TV, the latter under auspices of WNET. In 1975 public TV also carried the group's documentary on Cajuns, *The Good Times Are Killing Me.* The group was later disbanded, and at least one of its principals, Michael Shamberg, became a producer of commercial movies.

"12 O'CLOCK HIGH"—World War II series on ABC (1964–67) based on the 20th Century-Fox motion picture of that title about an Air Force bombardment group based in England. Robert Lansing starred in 31 episodes and Paul Burke in the other 47. It was produced by Quinn Martin in association with 20th Century-Fox TV.

"20/20"—ABC's entry in the newsmagazine derby (1978–   ), mounted when CBS's long-running *60 Minutes* began to crack the Nielsen Top 10. The program's format includes an in-depth story on current affairs, a personality profile, an investigative report and occasional features on science, medicine, the arts and other specialized fields.

The first episode, airing in June 1978, was almost universally derided for its tastelessness and its showmanly approach to news. ABC responded rapidly to the criticism, appointing a new producer and a new host and adopting a more sober style. The program was scheduled irregularly on a monthly basis through most of its first season but was given a regular Thursday night slot in the 1979–80 schedule.

Bob Shanks, whose previous experience had been with talk shows and entertainment programming, was the original producer. Harold Hayes, former editor of *Esquire,* and Robert Hughes, art critic of *Time,* were the hosts, before they were replaced by the TV veteran, Hugh Downs. The program's regular correspondents are Sylvia Chase, Tom Jarriel and John Stossel, with Geraldo Rivera and Barbara Walters assigned to special reports. The executive producer is Av Westin.

"TWILIGHT ZONE"—dramatic anthology series on CBS (1959–64) concerned with tales of the supernatural; it enjoyed great popularity and turned its creator, Rod Serling, previously a behind-the-scenes playwright, into a TV star. Serling served as host and narrator, providing what programers call "the glue" for the series, and gave it the serious, mysterious tone that made it a winner. He also wrote several of the playlets. The series originated in half-hour form, expanded to an hour in January 1963 and returned to the half-hour format that fall.

TWO-WAY CABLE—an interactive or bi-directional cable system with the ability to carry signals upstream and downstream, so that communication is possible from the subscriber's set to the head-end, usually in a digital mode. This feature enables the subscriber to respond to incoming messages for purposes of polling, at-home shopping and audience-participation entertainment.

Other forms of two-way cable allow for full video and audio communications between two points. Such a form was used between an office at Mount Sinai Hospital in New York and a pediatrics clinic in East Harlem, enabling doctors to study patients and provide consultation to nurses and paramedics by cable.

In 1978, Warner Communications' Qube system began operation in Columbus, Ohio, as the first fully two-way commercial cable system. All previous interactive installations were financed by grants.

TYNE TEES TELEVISION—British commercial licensee for the region in northeast England, with its production centers in Newcastle and Leeds. The company is a subsidiary of Trident Television Ltd., which acts as a station rep in selling air time.

U

"UFO"—hour-long syndicated science-fiction series (1970) from England, produced by ATV, and featuring Ed Bishop, George Sewell, Peter Gordon and Gabrielle Drake. Consisting of 26 episodes, it was found suitable by many U.S. stations for programming under the new prime time-access rule but was not impressive in the ratings.

UHF (ULTRA HIGH FREQUENCY)—the television band in the electronic spectrum from 470 to 890 mHz encompassing channels 14 to 83 (in the U.S., Canada and some other Western countries). In theory, the UHF band could make possible up to 3,000 stations in the U.S. beyond the 650 that can be accommodated by the VHF band. A total of 1,400 UHF channels were allocated by the FCC in its April 1952 Sixth Report and Order, but there has been no rush to apply for many of the frequencies; the history of UHF has been a struggle for attention against the competition of the easier-to-receive VHF stations.

In retrospect, a glaring omission in the Sixth Report and Order was a requirement that all allocated channels be tuned with equal ease. Most of the early TV sets were not equipped to receive UHF; viewers who wanted to watch those channels had to purchase and attach to their sets special tuners with loop antennas. These proved neither efficient nor reliable. Reception was often marginal.

There was also a manifest need for a much higher power output at the UHF transmitters than the early transmitters were able to attain. UHF requires 5 million watts to cover about the same area that a low-band VHF station (channels 2–6) can cover with only 100,000 watts. The high-band channels, 7–13, are allowed to have 316,000 watts. Even as the technology developed to reach maximum power, cost was a major obstacle to the growth of UHF.

Still, many of the UHF frequencies were claimed when the freeze on allocations ended because those applications were likely to be processed immediately, while bids for VHF channels carried the prospect of long and costly comparative hearings with other applicants. The early UHF operators hoped to become established before the VHF stations could get on the air or, if they did sustain losses, to receive preferential treatment from the FCC for an available VHF channel in return for their pioneering efforts.

The first commercially licensed UHF station was KPTV (Channel 27), Portland, Ore., which began broadcasting Sept. 20, 1952. It went dark on April 17, 1957, after four VHF stations had been activated.

CBS and NBC both were early UHF operators, bringing their total owned-stations complement up to the maximum seven in 1953–54 with two UHF properties each. Both had stations in Hartford, Conn. NBC's other "U" was in Buffalo and CBS's in Milwaukee. All four of those stations were either abandoned or sold by 1959.

By 1954 approximately 120 UHFs were on the air. Six years later approximately half those stations had ceased operating. After reaching its nadir in 1960, UHF began to grow again in 1961. It took until 1964 for the number of stations to reach 120 again.

The resurgence was, in part, attributable to the growth of educational (now public) television, which had to settle for UHF channels in most markets, including such critical ones as Los Angeles and Washington. Also contributing to growth was the fact that many markets had been allocated only one or two VHF stations, leaving UHF to accommodate the other networks. But the most important impetus was the passage by Congress in 1962 of the All-Channel Law, requiring all sets sold in interstate commerce to have the ability to receive both VHF and UHF by 1964.

The national boom in color TV in 1965 and the demand for portable sets accelerated the sale of new sets and hastened the penetration of UHF. But although close to 90% of the television homes were capable of receiving UHF channels in the 70s, a "U" in general was considerably inferior to a "V" in ability to deliver audience.

Gradual technological improvements in UHF have eased the so-called "UHF disadvantage," among them more efficient transmitters and more sophisticated dial tuners. Also, the FCC required set manufacturers to provide "comparable tuning" for UHF and VHF by the 1975 model year, meaning that all stations could be brought in on a "click" (or detent) dial, relieving UHF of the handicap of having to be tuned differently from VHF, in the manner of radio. New tuners are now being manufactured featuring digital electronic controls similar to hand calculators. These fully equalize UHF and VHF tuning.

Spurs to UHF growth in the 70s were the emergence of the independent station as an alternative to the three network-affiliated stations in any market and the fact that by 1975 UHF stations as a whole were profitable. Network-affiliated UHF stations had been profitable prior to that year, but 1975 marked a turning of the corner for most of the independent UHFs. The success, in 1978, of several over-the-air pay-television stations put a focus on UHF as a facet of the advancing technologies.

Approximately 350 UHF stations were operating by mid-1976. Of these, 195 were commercial and 155 public TV outlets. Of the commercial UHFs, 120 were network affiliates and 65 independents. Of the independents, 34 were conventional—that is, dedicated to serving predominantly English-speaking general audiences, with program schedules consisting of movies, off-network reruns, live sports, talk shows and regularly scheduled primetime newscasts. The 31 specialty stations programmed primarily in foreign languages to serve Spanish, Japanese, Chinese and other minorities of a community. A few specialized in religious programming or in stock market reports.

Many of the UHF stations were built by major corporations and established broadcasters with the financial ability to keep the stations going through the difficult years. Among those companies were AVC Corp., Kaiser Industries, Storer, Trans America Corp. (through its subsidiary, United Artists) and Taft Broadcasting. Metromedia had had a UHF station in San Francisco but donated it to the public TV licensee after several years of unprofitable operations. In 1979, Group W began its program of expansion by purchasing a UHF station in Charlotte, N.C., for $20 million. It

was taken as a sign that UHF was finally coming into its own.

During the 50s there had been serious talk at the FCC of moving all television to the UHF band. While the idea was abandoned, most of Western Europe and other countries in the world have concentrated their television on UHF, many of them combining the move to the new band with the inauguration of color. In the U. S., meanwhile, some of the UHF channels have either been reallocated for nonbroadcast use or authorized to be "shared" with nonbroadcast services. Channels 70–83 are now allocated to land-mobile, which shares this portion of the band in some parts of the country with TV translators—low-powered repeaters carrying the signal of the nearby TV station to areas where TV service is hard to receive.

Additionally, channels 14-20 are shared with land-mobile in some of the largest urban areas of the country.

Future growth and investment in UHF seems likely, particularly in view of the favorable trends in recent years. But impediments to such growth could be the "dropping in" of other possible VHF channels in many communities (as a result of proposals to the FCC by OTP and the United Church of Christ) and the reallocation of additional UHF channels to nonbroadcast service. On the other hand, a fourth commercial network, built largely on UHF affiliates, would prove a boon to the further development of the band.

ULSTER TELEVISION—commercial ITV licensee for Northern Ireland with principal studios in Belfast.

UNDA (INTERNATIONAL CATHOLIC ASSN. FOR RADIO AND TELEVISION) — agency based in Middlesex, England, representing Catholic radio and TV at the international level and through which its members make contact and share research. Its members consist of autonomous Catholic bodies in more than 100 countries. Founded in 1928 as the International Catholic Bureau for Broadcasting, it took its present name in 1947.

"UNDERSEA WORLD OF JACQUES COUSTEAU, THE"—series of oceanographic documentaries played at the rate of four a year on ABC since 1968. They center around the contemporary scientific expeditions of Capt. Jacques Cousteau and the crew of his specially equipped vessel, the *Calypso*. Episodes have examined such phenomena as the Great Barrier Reef and the lobster migration along the Bahama Bank. The series was by Marshall Flaum Productions in association with The Cousteau Society, Metromedia Producers Corp. and ABC News. ABC dropped the series in 1976 but it was continued on PBS on underwriting by Atlantic Richfield Corp.

Jacques Cousteau

UNDERWRITING—in public television, grants from private corporations or foundations to cover all or part of the cost of presenting specific programs or series. The underwriter is identified as donor by a slide and voice announcement at the beginning and end of each program. While in theory it is not commercial advertising, in many instances it serves as institutional advertising, and many corporate underwriters buy magazine and newspaper space to promote their public TV shows and boast their contributions to culture and public affairs. During the energy crisis in the early 70s, when public opinion had turned harsh toward the major oil companies, Exxon, Mobil, Gulf and Atlantic Richfield became the principal underwriters of cultural programs on PBS.

Underwriters make their grants to one of the local stations in the PTV system (called "points of entry") and usually are able to designate the day of the week and the time period for their program to be carried by the entire system.

Although corporate underwriting verges on sponsorship and lends itself to the kind of advertiser control the commercial networks were subject to in the early years, the funding of programs by commercial companies in the 70s gave the financially starved PTV industry a terrific boost. Some of the system's most distinguished and best-watched shows were provided by underwriters.

UNGER, ALVIN E. (d. 1975)—syndication executive who for five years, until his death, was v.p. of domestic sales for Warner Bros. TV. He began in the syndication field in 1939 with the Frederic Ziv organization, becoming v.p. of sales and then v.p. of Chicago for Ziv TV. He moved on to other companies and, before joining Warners, was syndication v.p. for Independent Television Corp.

UNIT MANAGERS SCANDAL—a demoralizing episode at NBC in which several members of the unit managers department were discovered, in the spring of 1979, to have been systematically conducting acts of embezzlement over a period of years. The widespread improprieties, uncovered by the new management of NBC, led to an extensive investigation by Federal prosecutors to determine the extent of the white-collar crimes and whether higher officials of the company had collaborated in the schemes. In a matter of weeks, a total of 15 unit managers were fired, including Stephen Weston, who had been v.p. and supervisor of the department.

NBC's unit managers, in a system borrowed from the motion picture industry, travel with news, sports and program production units to handle the living arrangements and financial matters. Because it is sometimes necessary to bribe people—to shoot film on the streets, smooth the way through customs or otherwise expedite the assignment—unit managers frequently carried satchels full of cash with the tacit understanding of their superiors that it might have to be dispensed in ways for which there could be no receipts.

The system invited abuses that extended to falsified expense accounts, fictitious vouchers, kickback schemes and the purchase of airline tickets charged to the network and then returned for cash. The abuses were thought to be flagrant during the coverage of major sports events and political conventions, which involved time pressures and hundreds of personnel.

When the improprieties were uncovered, NBC moved immediately to tighten fiscal controls over the department, which employs about 50 unit managers in New York and Washington. Operating with somewhat different systems, CBS and ABC maintained that they had no comparable problems. Previous top officials of NBC claimed to have been totally unaware of the illegal activities in the company.

UNITED CHURCH OF CHRIST, OFFICE OF COMMUNICATION—a leading organization in the broadcast reform movement, concentrating chiefly on minority rights in the licensed media. While most other religious organizations worked at securing air time to promote their views and their churches, the United Church of Christ chose to work behind the scenes on behalf of the public's rights in radio and TV. The Office of Communication began its work in 1954 under the leadership of the Rev. Dr. Everett C. Parker and has been funded by the Ford and Markle Foundations for some of its programs.

It was notably successful in bringing about the license revocation of WLBT-TV Jackson, Miss., for reasons of race discrimination in programming and employment, and in obtaining rules from the FCC forbidding discriminatory practices. The Office has also helped black groups to purchase stations and assisted racial minorities and women in their negotiations for representation at local stations. See also Parker, Everett C.; Broadcast Reform Movement; WLBT Case.

UNITED STATES v. RADIO CORP. OF AMERICA

[358 U.S. 334 (1959)]—case in which the Supreme Court held that action which is authorized by the FCC is not exempt from prosecution under the antitrust laws. At issue was an exchange of stations between NBC (a subsidiary of RCA) and Westinghouse Broadcasting Co.

Each of the companies owned five VHF television stations, the maximum number allowed by the FCC. NBC, with a station in Cleveland, was desirous of trading up for one in Philadelphia, a much larger market. Westinghouse, which had a Philadelphia station, was affiliated with NBC in Boston and Pittsburgh. NBC allegedly threatened to terminate those affiliations if Westinghouse did not agree to swap the Philadelphia station for NBC's Cleveland outlet and $3 million. The FCC approved the transaction without holding a hearing.

Subsequently, the Justice Department filed suit against RCA and NBC for violating the antitrust laws. The defense of RCA and NBC was that approval of transfer by the commission cleared them of any antitrust violation, and the court agreed, dismissing the action. The case was then appealed to the U.S. Supreme Court, which reversed the decision of the trial court.

The Supreme Court held that the statutory grant of power to the FCC in the Communications Act did not give it authority to pass on formal antitrust questions, which were within the province of the courts alone. The FCC only had power to determine if the transfer had been in the "public interest, convenience and necessity." And although the FCC had the authority to find that certain combinations of media control were against the public interest, the Court insisted that this was not the same as a finding that the antitrust laws had been violated. Therefore, the Court remanded the case to the District Court to determine whether the antitrust laws had been violated.

But before the District Court heard the case, Westinghouse and RCA entered into a consent agreement to annul the exchange. The Philadelphia license was returned to Westinghouse, and NBC moved back to Cleveland and was repaid its $3 million.

UNITED STATES v. SOUTHWESTERN CABLE CO.

[392 U.S. 157 (1968)]—case in which the Supreme Court upheld the FCC's jurisdiction over cable-TV's impact on broadcasting.

When Southwestern Cable, operating in San Diego, began importing signals from Los Angeles to attract additional subscribers, a local TV station (KFMB-TV) operated by Midwest Television Inc. complained to the FCC that the importations were adversely affecting San Diego broadcasters and that they were contrary to the public interest.

After considering both sides of the argument, the FCC ordered Southwestern, pending a decision on the merits, to restrict its services to areas it had operated in prior to Feb. 15, 1966. Southwestern appealed, and the Court of Appeals for the Ninth Circuit reversed. The court held that the FCC had no jurisdiction over CATV systems and that it therefore could not require Southwestern to restrict its services. The case was then appealed to the Supreme Court which reinstated the commission's initial order.

The Court found substance to the commission's argument that regulation of cable-TV was necessary to regulate its broadcast licensees effectively. The Court noted that the Communications Act of 1934 was explicitly applicable to "all interstate and foreign communication by wire or radio" and that since CATV systems were interstate communications by wire, the FCC had the power to regulate cable systems generally.

Further, it reasoned, since the commission has the power to issue "such orders, not inconsistent with (the Act), as may be necessary in the execution of its functions," the order limiting Southwestern's services pending a full hearing on the merits was proper.

UNITED STATES v. STORER BROADCASTING CO.

[351 U.S. 192 (1956)]—case that affirmed the right of the FCC to limit the number of licenses any one person or firm could have. It also gave the FCC the power to summarily dismiss any application which was, on its face, in violation of the commission rules.

In 1953 the commission promulgated rules which placed maximum limits on the number of television, AM and FM licenses any one person or company could control. The rules also stated that no future licenses would be granted if any applicant already held the maximum number of licenses. Storer Broadcasting Co., which held the maximum complement of licenses in each category, challenged the rules because it felt there might be situations in which the public interest dictated an extra station.

The Court of Appeals vacated the rules insofar as they placed absolute limits on the number of stations a licensee could control. The FCC then appealed the case to the Supreme Court, which reversed the judgment of the Court of Appeals.

The Supreme Court noted that an applicant always had a right to apply for a waiver of the rules if it could demonstrate that the public interest would be served. Thus, broadcasters still had the ability to seek and obtain a full hearing for an additional license, despite the rules, even if it already had the maximum number of licenses.

UNIVERSAL TV—leading supplier of TV series to the networks since the mid-60s, usually having three times the number of weekly hours on the air as the runner-up Hollywood studio. A subsidiary of Universal Pictures, UTV has been responsible for such series as *Ironside, Kojak, Columbo, Baretta, Emergency, The Rockford Files, Rich Man, Poor Man* and the *NBC World Premiere* movies, among scores of other shows. Its sister company, MCA-TV, engaged in domestic and foreign syndication, habitually turns huge profits from the network hits through the sale of reruns.

Universal Pictures was a foundering company when the talent agency MCA (originally, Music Corp. of America) bought control of the studio in 1962. MCA then divested itself of the talent agency and concentrated on making movies, television series and recordings (it had also purchased Decca Records). Earlier, MCA had bought the old Republic film studios and created Revue Productions, which turned out low-budget TV series. With Universal, MCA Inc. had the largest studio in Hollywood and the best-equipped backlot, and the show-business savvy of its principal officers, Jules Stein and Lew Wasserman, quickly turned the facilities into a giant TV production source. Adding to the networks' trust in the studio is the fact that it maintains a large roster of contract producers and writers.

Sid Sheinberg, who had been president of UTV, became president of MCA Inc. in the early 70s and was succeeded as head of the TV company by Frank Price. When Price left for Columbia Pictures in 1978, he was succeeded by Donald Sipes. See also MCA.

UNIVERSITY OF MID-AMERICA—a four-state project developed by the University of Nebraska and the Nebraska Educational Telecommunications Center which represents the first attempt in the U.S. to establish an open-learning system offering a four-year college degree program via public television. Formed in the early 70s, the system expanded when it received major federal funding in 1974. In addition to TV, the system involves the use of radio, telephone and books and provides for personal consultation.

"UNTAMED WORLD"—syndicated (1969) documentary series on the struggle for survival of animals in the wild, via Metromedia Producers Corp. and sponsored by Kellogg Co.

"UNTOUCHABLES, THE"—landmark police series on ABC (1959–63) which became notorious for its escalation of violence in prime time and its association of Italian names with the gangland crime of the Prohibition Era. On the positive side were its quasi-documentary style, with Walter Winchell narrating in the staccato idiom of headlines, and the careful attention to authentic detail in depicting the period. Robert Stack played Eliot Ness, head of the "untouchable" government agents cracking down on the mobs.

Although it was highly popular, the series was attacked by Italian-American groups for ethnic defamation and by others for the sensationalism of machine-gun murders and acts of brutality. It was produced by Quinn Martin for Desilu, in association with Langford Productions.

Robert Stack as Eliot Ness

UPITN—a leading international newsfilm agency jointly owned by United Press International (UPI), Britain's Independent Television News and Panax Corp., a midwest newspaper chain which acquired Paramount Picture's 50% interest in 1975. The company, with principal offices in London and New York, claims to supply 17 million feet of newsfilm a year to 100 clients in 70 countries. In 1976 UPITN expanded by taking over the foreign newsfilm service from ABC News, which withdrew from film syndication outside of North America. Under the arrangement ABC News supplies its U.S. newsfilm in exchange for UPITN's foreign materials.

For about two years in the early 70s, UPITN attempted to provide a daily electronic feed to subscribing stations in the U.S. but ran head-on against another company, TVN Inc. (backed by Joseph Coors), which was attempting to do the same thing. Neither company was able to enlist enough clients to justify its costs, and in 1974 UPITN bowed out because the venture had driven the entire company into the red.

UPITN came back into the picture, but in a different capacity, when TVN disbanded in November 1975. Major independent stations, needing a daily national and international news feed, then created a cooperative, Independent Television News Assn. (ITNA), which purchased its foreign news from UPITN.

Downstairs at the Bellamy household with Jean Marsh & Gordon Jackson

"UPSTAIRS, DOWNSTAIRS"—taped British series, via London Weekend Television, about the intrigues of the wealthy and the servant classes living under the same elegant roof in Edwardian England at the turn of the century. The series was a great success in the U.K. and also developed a staunch following in the U.S. on PBS, where it played as a subseries of *Masterpiece Theatre.* It featured Jean Marsh (who became well known in the U.S. through the series and was winner of a 1974 Emmy), David Langton, Rachel Gurney,

Gordon Jackson and Maggie Wells, among others. In 1975, CBS attempted an American version, *Beacon Hill,* which had a brief run. The Robert Stigwood Organisation produced both the original and the imitation.

Upstairs, Downstairs spent its firstrun episodes on PBS in April 1977.

UPSTREAM/DOWNSTREAM—designating the directions traveled by TV signals on two-way cable systems. Upstream signals are those transmitted by the subscriber to the head-end, usually in digital form; downstream signals are those traveling from the head-end to the receiver, generally in picture form.

USA Network—see Cable Networks.

UTLEY, GARRICK—NBC News correspondent, working chiefly on foreign assignments. In 1980 he became a regular weekly correspondant on *NBC MAGAZINE.* *He joined the network in Brussels in 1963, later was sent to Saigon to cover the Vietnam War and then to Berlin and Paris, and in 1973 became NBC's London bureau chief. His father, Clifton Utley, was a noted NBC correspondent and local anchorman in Chicago during the 50s, and his mother, Frayn, an NBC reporter.*

UTV (Ultimedia Television)—see Cable Networks.

VALENTI, JACK—former special assistant to President Lyndon Johnson who became, in 1966, president of the Motion Picture Assn. of America and its related agencies, the Assn. of Motion Picture and Television Producers and the Motion Picture Export Assn. As the film industry's representative in Washington, he became one of the leaders of the campaign to persuade the FCC to loosen or discard the rules restricting pay-Cable. Before joining the White House, he headed a Houston ad agency, Weekly & Valenti.

VAN DEERLIN, LIONEL (Rep.)—chairman of the House Communications Subcommittee (1976–80), succeeding the late Rep. Torbert Macdonald. A nine-term Democratic congressman from San Diego, Van Deerlin, a cable-TV advocate and former broadcast newsman, proposed to rewrite the Communications Act of 1934, making it more applicable to the television era and the emerging technologies.

Van Deerlin's proposals, first issued in 1978 and revised in 1979, created a storm of controversy in the communications industry and among public interest groups. He advocated a free-marketplace approach to Federal communications policy, eliminating most government regulation of radio, television and cable-TV. He proposed an easing of broadcast licensing requirements and a virtual elimination of the public interest standard by which the FCC regulates stations, but in exchange for that he proposed that broadcasters pay a fee for their use of the public airwaves. For lack of support, either on his subcommittee or in the concerned industries, Van Deerlin was forced in July 1979 to scrap the rewrite provisions for broadcasting, although he proceeded with a bill for changes in the communications laws affecting common carriers. He was defeated in his bid for reelection in 1980, but many of his proposals were adopted in the Reagan administration.

In the 50s, before he was elected to Congress, Van Deerlin had been a newscaster in San Diego and news director for stations KOGO and XETV.

VANDERBILT TELEVISION NEWS ARCHIVE—a videotape collection of the evening newscasts of ABC, CBS and NBC begun in August 1968 and augmented daily with tapings of the broadcasts on network affiliates in Nashville, Tenn. The collection, which includes 1-inch and ¾-inch tape recordings of presidential addresses and coverage of political conventions, is administered through the Joint University Libraries of Vanderbilt University, Peabody College and Scarritt College in Nashville.

The archive was begun because complete video tapes of the news programs were not being saved elsewhere either for study purposes or as a record for the future. It was created on the suggestion of Paul C. Simpson, a Nashville insurance executive, with grants from individuals and foundations. CBS subsequently copyrighted its evening newscast and then sued the Archive for appropriating copyrighted material.

However, the new copyright law enacted in 1976 protected Vanderbilt's activities. After the law was passed, the U.S. District Court dismissed the suit, with the agreement of the parties. The new law specified that libraries and archives which met certain standards were permitted to record audiovisual news programs and to make a limited number of copies available on loan.

The Vanderbilt Archive does not sell but rents either complete programs or compilations of news reports on a given rebroadcast. The fees, based on a half-hour minimum charge, are nominal ($30 per hour of compiled material, $15 per hour of duplicated broadcast and $5 for audio only). The copies are presumed to be used by authors, educators and students.

The Archive employs a small permanent staff for recording, indexing and administration, and in 1972 it began publication of a monthly index to television news, detailing the coverage of stories each night on each network, to facilitate access to the collection. See also Copyright Revision.

VAN DOREN, CHARLES—See Quiz Show Scandals.

VAN VOLKENBURG, J.L. ("JACK")—first president of the CBS television network, appointed July 16, 1951, and credited with lining up many of the most desirable early stations as CBS affiliates and many of the early advertisers. A tough and shrewd executive, he lasted four years in the post, to cap a CBS career of 23 years.

Van Volkenburg joined the company as sales manager of KMOX St. Louis in 1932 and a year later became president and general manager of the radio station. In 1945, he moved to New York as general sales manager of CBS Radio Sales, later becoming director of sales administration for all the owned stations. In 1948 he was made v.p. of CBS television operations and two years later v.p. in charge of CBS-TV sales. Before taking his first job in broadcasting, he had worked in a Chicago advertising agency and rose to head of its radio department in 1928.

VANE, EDWIN T.—president of Group W Productions Inc. since August 1979 and previously a senior program executive with ABC for 15 years. With the network he was instrumental in introducing such programs as *Good Morning, America, Family Feud, The Dating Game, The Newlywed Game, Dark Shadows, Concentration, Jeopardy* and many of the notable dramas in the *ABC Theater* series.

Starting as director of daytime programs, he rose to a number of higher posts through several network administrations and in 1976 was named v.p. and national program director for ABC Entertainment, the network's programming arm. Vane began his broadcasting career with NBC as a page while still at Fordham University. He later advanced through several departments and became manager of daytime programming in 1961. In heading Group W Pro-

ductions, Vane changed his base from New York to Los Angeles.

VANOCUR, SANDER—NBC newsman during the 60s, later senior correspondent for NPACT and then TV critic for the *Washington Post*. In June 1977, he was recruited by Roone Arledge to head a special unit for investigative and political reporting for ABC News. But little came of that project, and Vanocur's visibility was scant during his first years at ABC. At NBC Vanocur came to prominence as one of the "four horsemen" (floor reporters) in its political convention coverage and as one of the reporters closest to President Kennedy and his family. His shift to public TV in the late 60s, where for most events he was teamed with Robert MacNeil, brought charges from the Nixon Administration of a liberal bias in PTV. He left NPACT when the organization was criticized by Government officials, including Democrats, for paying the star newsmen salaries comparable to those paid by commercial networks. Vanocur was reported receiving close to $85,000 a year from PTV. For two years, before joining the *Post* as TV editor and critic, he taught communications at Duke.

VANOFF, NICK—producer-director of variety shows, whose credits include the second *Sonny and Cher Show, The Julie Andrews Show* and the *Perry Como Kraft Music Hall.* He was also a partner in Yongestreet Productions, packagers of *Hee Haw,* and other TV specials and series.

Beginning his career as a musical performer with the New York City Center Opera Company and in *Kiss Me, Kate* on Broadway, he entered television as co-producer of the Steve Allen *Tonight* Show and added to his credits *The King Family* series and *Hollywood Palace.*

VAST WASTELAND—phrase that rocked the broadcast industry when it was used by a new FCC chairman, Newton N. Minow, in 1961 to describe the state of TV programming. Minow coined the phrase in his first speech to the industry, at the NAB convention in May, five months after becoming FCC chairman in the Kennedy Administration. To broadcasters, a regulator passing judgment on the quality of programming suggested Government interference with their right of free speech. But the press gave prominent coverage to Minow's assessment of television, and the "vast wasteland" speech reverberated throughout his productive two-year tenure as chairman and inspired broadcasters to increase and improve their public affairs programming.

The famous phrase was used in the following context:

"I invite you to sit down in front of your television set when your station goes on the air and stay there without a book, magazine, profit and loss sheet or rating book to distract you—and keep your eyes glued to the set until the station signs off. I assure you that you will observe a vast wasteland.

"VD BLUES"—a PBS special (Oct., 1972) which attempted with a fair degree of success to convey information on venereal disease to the young through the light entertainment devices of songs and sketches. Produced by WNET New York, the program was hosted by Dick Cavett, produced by Don Fouser, directed by Sidney Smith and written by Fouser and Gary Belkin. Special material and skits were by Jules Feiffer, Israel Horovitz and Clayton Riley. The underwriter was 3M. Some public TV stations refused to carry the telecast because of strong language used and because of the open discussion about sex as the medium for transmitting VD, but WNET alone received 15,000 calls from viewers asking for the guidance promised by the program.

VECCHIONE, AL—TV news executive who, after 20 years with NBC, joined public television's NPACT in Washington in the early 70s. He served variously as exec producer, director of production operations and general manager. In 1975 he was named executive director of public affairs programming for NPACT and WETA Washington but left the following year to serve as broadcast advisor to the Democratic National Committee. He returned to public television after the elections as executive producer of *The MacNeil-Lehrer Report.*

VENZA, JAC—a leading producer of performing arts programs for public TV since 1966, having previously been with CBS. Venza was executive producer of *The Adams Chronicles* and the *Great Performances* series and held the title of director of performance programs for WNET New York. Earlier, he helped develop such PTV series as *Playhouse New York, NET Playhouse, Song of Summer, To Be Young, Gifted and Black, Hogan's Goat* and *American Ballet Theatre: A Close-Up in Time.*

Venza began as a scenic designer at CBS in 1949 and eventually moved into the production of public affairs programs there. He became associate producer of the *Adventure* series in 1953 and later worked on the *Roots of Freedom.* In addition to his TV work, he has been a scenic designer for theater, opera and ballet.

VERTICAL INTERVAL—a series of 20 or 21 horizontal lines transmitted between pictures. These lines normally carry no video information but have recently been used to convey test and monitoring data for automatic monitoring of start and end of commercials, or for color reference signals (vertical interval reference or VIR). In the United Kingdom, the vertical interval is being used experimentally to transmit written data (Teletext) which may be displayed on the screens of specially equipped receivers. See also Ceefax, Teletext.

"VEGA$"—hour-long ABC detective series (1978–1981) whose action takes place against the background of Las Vegas hotels and casinos. Robert Urich stars in the Spelling-Goldberg Production. Featured are Tony Curtis, Bart Braverman, Judy Landers and Naomi Stevens.

VERTUE, BERYL—British producer working in American TV as co-deputy chairman of the Robert Stigwood Group and head of its television operations. Her U.S. credits include *Beacon Hill,* the miniseries *Charleston* and the special, *The Entertainer,* which starred Jack Lemmon. In association with Bob Banner, she produced the series *Almost Anything Goes,* and she was executive producer of the movie *Tommy.* She made her entry in American TV handling the rights arrangements for the British shows on which *All in the Family* and *Sanford and Son* were based.

VHF (VERY HIGH FREQUENCY)—the FCC's designation for the radio spectrum band 30 to 300 megaherz. The original 12 television channels in the United States extend from 54 to 72 MHz (Channels 2ND4), to 76 to 88 MHz (Channels 5 and 6) and 174 to 216 MHz (Channels 7–13).

Ralph Baruch, President of Viacom

VIACOM INTERNATIONAL—company formed as a spin-off of CBS in June 1971 as a result of an FCC order requiring the networks to divest themselves of their program syndication and cable-TV divisions. The spin-off included executives and staff as well as properties, and the company's first president was Clark B. George, who formerly had been president of CBS Radio. Later he was replaced by Ralph Baruch. The syndication operation, Viacom Enterprises, no longer sustained by the off-network reruns of CBS, expanded into program production, both for the networks and the syndication market. In the cable field, Viacom was the 6th largest multiple system operator.(In addition, Viacom in 1976 entered the pay-TV business with Showtime, now the second largest pay-TV firm, behind Home Box Office, with over 600,000 subscribers. In 1979, Showtime became a joint venture of Viacom and Teleprompter Corp., the largest cable-TV company. In the late 70s, Viacom began to expand into station ownership, first acquiring WVIT-TV Hartford, Conn., and then the Sonderling Broadcasting group which includes, in addition to a number of radio stations, WAST-TV Albany, N.Y. The first TV series produced by Viacom for prime time was *The Lazarus Syndrome,* with Lou Gossett and Ron Hunter, which made ABC's 1979 fall schedule.

VICTOR, DAVID—veteran executive producer associated with Universal TV whose credits include *Marcus Welby, M.D., Owen Marshall, The Man and the City, Griff, Lucas Tanner* and *Kingston.* During the 60s he was producer of several series, among them *Dr. Kildare.*

Victory at Sea

"VICTORY AT SEA"—a classic documentary series of 26 half-hour episodes on the exploits of the U.S. Navy during World War II, utilizing combat footage and a stirring musical score by Richard Rodgers. It had its first presentation in 1952, was aired numerous times afterwards and remains a model for the form of historical documentary. Produced by Henry Salomon and narrated by Leonard Graves, it won dozens of awards.

VICTORY, JIM—syndication executive heading his own firm. He had been president of NBC Films (1971–73) and for 10 years previously had been a v.p. of CBS Enterprises.

VIDEO—generic for the field of television outside the formal broadcast system that has grown up around portapak equipment since its introduction into the consumer market by Japanese manufacturers in 1970. The field embraces student groups, individual techno-artists, community video centers and a range of collectives and groups dedicated to innovative television. Some of them function as the video equivalents of the underground press.

Nam June Paik, a Korean-born artist who had been experimenting with abstract images on television since 1963, is generally credited with having fathered the Alternate Television movement when he exhibited, at the Cafe Au Go Go in Greenwich Village in 1970, scenes shot from a taxi with his newly purchased portapak gear.

The portapak units, initially costing around $2,000 for a complete system, are built around lightweight hand-held cameras that record on 12-inch tape in much the way that movie cameras expose film. An attractive feature of the videotaping unit is its playback capability, which allows the operator to review immediately what he has shot. Portapak technology thus permits one person to achieve a video result which in studio television involves a clutch of professionals, from director to lighting engineer.

Usually without expectation of remuneration for their efforts and with few outlets for their work beyond public access channels on cable television or varied forms of video theaters, several schools of Alternate Television began to grow, some devoted to *service video* (serving the communications needs of communities too small for broadcast TV to serve, such as ethnic ghettos, schools, banks and industries); some to explorations of optical effects; some to producing nonfiction programs, such as documentaries; and some to perfecting "street video" as an art form.

A leader in the documentary field has been TVTV (Top Value Television), a San Francisco-based group whose candid two-part report on the way the formal media covered the 1972 political conventions was carried by the Group W television stations and several cable systems. This was followed by the sale to public television of two more documentaries, *Adland,* on the advertising industry, and *Lord of the World,* on the 16-year-old guru Mahara-ji in his 1973 appearance at the Houston Astrodome. Both were well-received by the critics.

The development of the Time-Base Corrector by Consolidated Video Systems in 1973, chiefly for the expensive minicameras that had begun to come into use in commercial television for newsgathering, made it possible for Alternate Television productions to infiltrate broadcast television. The corrector regenerates the signal of ½-inch, ¾-inch and 1-inch videotape, bringing it up approximately to the standard of 2-inch quadruplex used in studios.

Alternate Television groups took such names as Raindance, the Videofreex, the Ant Farm, Alternate Media, Peoples Video Theatre and Global Village. The hundreds of groups operating in various parts of the U.S., with hundreds more throughout the world, gave rise to networks-by-mail, video festivals and exhibitions and a periodical, *Radical Software*. See also Nam June Paik, TVTV.

VIDEO FEEDBACK—the optical effect produced by aiming a television camera at a television monitor, so that it simultaneously sees and displays its own electronic image. Since the movement of the camera's position in relation to the monitor alters the imagery, kaleidoscopic effects are possible. Video feedback has been basic to the development of electronic television art and yields infinite other variations when the process is augmented by additional cameras, reflectors and time-delays.

VIDEOCASSETTE—a cartridge containing video tape, generally on two reels, designed for simple operation in video recording and playback units. The most widely used videocassette system in broadcast and closed-circuit TV is Sony's U-Matic (with ¾-inch tape). Videocassette systems have been widely adapted at TV stations for electronic newsgathering operations.

Home videocassette recorders, such as Sony's Betamax and Matsushita's VX-2000 and its Video Home System—all using 12-inch tape—were introduced in Japan and the U.S. during 1975 and 1976.

See also Videotape Recording.

VIDEODISC, VIDEO DISC—a disc with the general appearance of a phonograph record, on which is stored video and audio information for playback through a home television receiver in the same way a phonograph plays back an audio signal.

The first videodisc was demonstrated in 1927 in Great Britain by John Logie Baird, using a waxed phonograph record, but the signals were of such low resolution that the project was abandoned.

The first videodisc system actually placed on the market was the TeD system developed by Telefunken of Germany, Decca of the United Kingdom and Teldec, a joint subsidiary of the two companies. TeD is known as a mechanical system and, like other systems, plays pre-recorded material in color

when attached to the antenna terminals of a color television receiver. TeD uses a "pressure transducer" stylus which makes direct contact with a flexible record on which the electronic impulses in the television picture and sound are represented by hills and valleys. Changes in the pressure on the transducer are translated into picture and sound elements. TeD videodisc players and records were introduced on the German-speaking European market in 1975 by Telefunken without notable success. Each eight-inch disc provides up to 10 minutes' playing time.

Several optical videodisc systems have been demonstrated. The only one definitely scheduled for production is a compromise between two similar systems—the Philips VLP and MCA Disco-Vision. This system employs a 12-inch reflective disc upon which electrical impulses are represented by a series of pits of various lengths and spacings. The beam from a low-powered laser is aimed at the disc, reflected into a transducer and reconstructed into television picture and sound.

A variation on this system was developed by Thomson-CSF of France, using a transparent, rather than a reflective, disc. Another optical system has been built in prototype by Hitachi of Japan, using the principle of holography. Other optical systems which have been proposed are based on film techniques. A common characteristic of optical systems is lack of physical contact between the pickup and the disc.

A capacitanco system, known as SelectVision VideoDisc, has been developed by RCA. This also uses a disc containing pits representing picture and sound elements. The disc is metallized, and when placed on the turntable becomes, in effect, one plate of a capacitor. The stylus, riding in a pre-cut groove in the disc, is the other plate. The differing small voltages on the stylus are reconstructed into color television pictures and sound.

The discs of the TeD, Philips/MCA and Thomson-CSF systems revolve at 1,800 rpm for the American NTSC television system and 1,500 for the CCIR-PAL and SECAM systems. The RCA capacitanco disc spins at 450 rpm for the American system. Hitachi's holographic system uses a speed of six rpm. With the exception of TeD, all major systems announced to date are capable of providing playing time of at least 30 to 40 minutes per side of a 12-inch disc.

A magnetic videodisc system, designed to record as well as play back, has been developed by Erich Rabe in West Germany.

VIDEOTAPE RECORDING—the recording of television signals on magnetic tape for later replay. The videotape recorder (VTR) is the principal device used for storing television programs; in fact, recorded shows have virtually replaced "live" television except for sports events.

Three principal types of VTRs are currently in use or under development: longitudinal-scan, helical-scan and quadruplex.

The first VTRs were of the longitudinal-scan type; on these, the tape passes at high speed before fixed recording and playback heads, in a manner similar to an audio recorder. The first VTR to be publicly demonstrated was built by Bing Crosby Laboratories in 1951. The tape ran at 100 inches per second, with a total of 16 minutes recording time per reel. The picture had poor resolution, flickered and displayed a diagonal pattern. RCA displayed longitudinal-scan monochrome and color VTRs in 1953; they had similar deficiencies.

The type of VTR that revolutionized television, and is in use at virtually all television stations today, was demonstrated for the first time by Ampex Corporation in 1956, where it created a sensation at the annual convention of the National Association of Broadcasters in Chicago. The quadruplex, or quad, VTR, rather than recording information longitudinally, records it transversely, with four heads rotating at high speed at right angles to a two-inch-wide tape moving at 15 inches per second, depositing a diagonal track. The quad VTR can accommodate about 90 minutes of programming on a 14.5-inch reel. Slowed down to 7.5 inches per second, it can record for three hours on a reel.

Color versions of the quad VTR were developed in 1958. Ampex and RCA are the only American producers of this standard VTR. It is estimated that 5,400 quad recorders are in use in the U.S., 9,200 worldwide.

The helical-scan, or slant-track, VTR was introduced in the early 60s, providing significant cost reductions as compared to quad units, although for many years they did not compare in picture quality.

In the helical VTR, the tape is wrapped in a spiral (or helix) around a drum containing one or more rotating heads. The heads produce a diagonal track on the tape. Helical-scan VTRs were popularized by Sony Corporation of Japan for closed-circuit use, but in 1973 and later International Video Corp., Ampex, Bosch-Fernseh, Sony and others were to introduce helical recorders of broadcast quality. By 1977 this simpler and more economical format was threatening to replace quad as the standard broadcast VTR.

All of the videocassette recorders introduced to date are of the helical-scan variety. These recorders, while not producing a signal of broadcast quality, are frequently used in electronic newsgathering (ENG) with signal-correction provided by time-base correctors. The so-called home video recorders, introduced by Cartridge Television Inc. (now dissolved in bankruptcy proceedings) in 1972 and by Sony in 1975 (Betamax), use the helical-scan principle, as do those of a number of Japanese manufacturers that entered the market in 1976.

There has recently been a revival of interest in longitudinal-scan recording, led by Germany's BASF, which has developed a longitudinal home unit, not yet on the market.

VIDEOTEXT—generic for the various electronic text systems that allow printed information to be called up on a television screen or a separate computer terminal.

Teletext is the form that operates on broadcast frequencies, specifically the vertical blanking interval of television transmissions. It is a one-way service, flashing pages of text in a prearranged order and in a cycle that is repeated continuously. The information travels at the rate of four pages a second, which means that the consumer may have to wait half a minute or more for the information requested. Teletext may be called up on a blank screen or can be superimposed on a program in progress. Closed captioning is a form of teletext. The cable counterpart, *Viewdata*, operating with broadband technology, can accomodate more than ten times the number of pages a teletext system provides. Because of this, consumers may retrieve sports scores, airline schedules, and television listings almost at once. Moreover, with two-way cable, the viewer may interact with the data bases and perform, for example, banking and shopping transactions.

Videotext is the term sometimes used for electronic text systems that work either through telephone or cable. The British systems are known as Prestel, Ceefax, and Oracle. The French have developed Antiope, and the Canadians Telidon. Also known as "Videotex" without the "t," reportedly because someone made a printing error at a European conference–an error, which until now, has been perpetuated.

VIEWDATA—keyboard device which transforms an ordinary television set into a computer terminal, permitting subscribers to call up a central data bank over telephone lines for a wide range of alphanumeric information. Systems already in limited use in Britain and Canada can store up to a million pages of information, ranging from financial news to kitchen recipes, for retrieval on demand. Eventually the systems are expected to be used also for electronic mail, making department store purchases and booking airline tickets. Britain's viewdata system is known as Prestel, Canada's is called Telidon. General Telephone & Electronics has obtained the rights to market Prestel in the U.S., and Knight-Ridder Newspapers has been preparing to test another system.

"VILLA ALEGRE"—bilingual (English and Spanish) PBS series for children produced on grants from HEW, Exxon Corp. and the U.S.A. and Ford Foundations. It premiered in 1974 as a daily half-hour series, with Mario Guzman as producer for BC/TV, Oakland, Calif. The program's prime purpose was to ease the transition from home to school for millions of Spanish-speaking children.

VIOLENCE HEARINGS—formal inquiries by congressional bodies and commissions into the question of whether the viewing of violence on television contributes to actual violent behavior and to the increase of crime in the U.S.

In chronological order, the principal hearings were the following:

• In 1954, the Kefauver hearings, prompted by the view of Sen. Estes Kefauver that television was a contributor to the growing crime rate. He was chairman of the Senate Subcommittee on Juvenile Delinquency at the time. In 1961 and 1964, the Dodd hearings in which Sen. Thomas Dodd of Connecticut reacted to what he felt was the industry's rampant and opportunistic use of violence.

• In 1968, hearings by the National Commission on the Causes and Prevention of Violence, led by Milton Eisenhower, on the role of the mass media. Network executives were questioned about research which they had promised, which was not forthcoming.

• In 1969, the Surgeon General's Inquiry on Violence with the National Institute of Mental Health, was ordered to establish what harmful effects, if any, televised crime, violence and antisocial behavior have on children. The study resulted in a 5-volume report, *Television and Social Behavior—A Technical Report to the Surgeon General's Scientific Advisory Committee on Television and Social Behavior.*

• In 1972, shortly after publication of the Surgeon General's report, the Pastore Hearings sorted out some confused statements and compromise language that had made the summary of the report indecisive concerning the relationship of TV violence to subsequent violent behavior. In these sessions Sen. John Pastore questioned all 12 members of the Surgeon General's Committee, and all of them — with varying views and backgrounds — agreed that the scientific evidence indicated that viewing of television violence by young people causes them to behave more aggressively.

After these hearings, the networks made some attempt to reduce the violence quotient, but the problem persisted. Pastore's and other committees continued to pressure the networks, and they expanded the issue to sex and violence. The networks responded to the concern by removing the more violent cartoons from the Saturday morning children's block and later by adopting, with the NAB, the concept of "family viewing time," designating the period from 7 to 9 P.M. for programming suitable for all age groups. See also Family Viewing Time; Pastore, John O. (Sen.); Surgeon General's Report.

VIP (VIEWERS IN PROFILE)—title of the periodic Nielsen reports containing, for each local market, audience data (size and composition) to assist the buyers and sellers of national spot and local TV advertising. The NSI equivalent of the NTI "pocketpiece," the VIP reports are issued from a minimum of three times a year to a maximum of eight times, depending on the size of the market.

Nielsen measures the audiences for individual stations in most markets through the use of diaries. These are placed in sample homes by telephone contact and remain in each home for one week. What is reported in the VIP is audience data for the average week, measured over four weeks.

The first of three sections in the VIP report is the Day Part Audience Summary, which breaks down audience estimates by day parts (e.g., 7:30–11 P.M., Sunday through Saturday; 9 A.M.–Noon, Monday through Friday, etc.) and includes age and sex demographics and each station's average weekly cumulative audience.

Another section, the Program Audience Averages, reports audience data for individual programs. The third section, Time Period Audiences, reports audience data by half-hour or quarter-hour segments. See also Demographics, NSI.

"VIRGINIA GRAHAM SHOW, THE"—syndicated hour-long daily talk show (1970–72) devoted chiefly to topics of interest to women. It was produced by RKO General and Gold Key Entertainment.

"VIRGINIAN, THE"—popular Western based on Owen Wister's classic, which was one of the first TV series in the 90-minute form. Produced by Universal for NBC, it ran for eight seasons (1962–70) and featured James Drury, Doug McClure, Lee J. Cobb and Pippa Scott. Cobb was replaced by Charles Bickford and then by John McIntire. In its final season, the series took the title of *The Men from Shiloh* and added Stewart Granger and Lee Majors to the principal cast.

"VISION ON"—pantomime series produced by BBC-TV in the early 70s, which was developed with the hearing-impaired child as its prime target, although it proved popular with children of normal hearing, as well. Three sets of the half-hour series were produced, totaling 74 episodes, which were syndicated to commercial stations in the U.S. by Time-Life Films. *Vision On* won a Prix Jeunesse as "best children's program in the world." The series, which combined live performances with animation sequences by Tony Hart, was carried by the ABC-owned stations and other groups, such as Post-Newsweek, Capital Cities, Cox and Storer.

"VISIONS"—series of original TV dramas by new American playwrights introduced by PBS in the fall of 1976 on grants totaling $7 million from the Ford Foundation, the National Endowment for the Arts and the Corporation for Public Broadcasting. The two-year series was created in hopes of revitalizing TV playwrighting, with the additional purpose of bringing new ideas and new creative talent into the medium. With Barbara Schultz (former executive producer of *CBS Playhouse*) as artistic director, the 14 plays for the first season—each running about 90 minutes—were produced on budgets averaging $210,000 per show at the studios of KCET Los Angeles.

VISNEWS—London-based international TV newsfilm agency owned by a consortium of the BBC, Reuters, the Canadian Broadcasting Corp., Australian Broadcasting Commission and New Zealand TV. Stuart Revell of Australia as chairman. See also UPITN.

"VOICE OF FIRESTONE"—long-running classical music series in prime time (1949–63) sponsored by Firestone Tire and Rubber Co. after 21 years on radio, with Howard Barlow as conductor of the orchestra. Regular vocalists over the years included Thomas L. Thomas and Vivienne della Chiesa, with John Daly as host. The weekly hour began on NBC and in 1954 switched to ABC. Although the size of the TV audience satisfied the sponsor, the program was dropped from the schedule because its comparatively low ratings were felt to have diminished the audience for adjacent shows.

"VOYAGE TO THE BOTTOM OF THE SEA"—hour-long science fiction series based on the hit motion picture. It had a moderately successful run on ABC (1964–68). With Richard Basehart and David Hedison in the leads, it was produced by 20th Century-Fox in association with Irwin Allen Productions.

VTR (VIDEOTAPE RECORDER)—See Videotape Recording.

WACC—see WASEC.

"WACKIEST SHIP IN THE ARMY, THE" — comedy series in the World War II motif which was based on a popular 1960 motion picture of that title. Premiering on NBC in the fall of 1965, it lasted 29 weeks. Heading the cast of the Screen Gems series were Jack Warden, Gary Collins and Mike Kellin.

WADSWORTH, JAMES T.—FCC commissioner (1965–69), a New York Republican who considered himself moderate-to-liberal and who voted against the Boston *Herald-Traveler* in the WHDH license case but for the renewal of WLBT in Jackson, Miss., a vote later overturned by the D.C. Court of Appeals. Prior to joining the commission he was a member of the New York state legislature and worked in the Federal Government for agencies concerned with international relations.

WAGNER, ALAN—longtime CBS program executive, who left in 1982 to become president of The Disney Channel, a new pay-cable network scheduled to begin in 1983. Wagner had been with CBS 21 years, starting in 1961 after four years in the network program department at Benton & Bowles Advertising. He had a brief career earlier as a standup comedian, the highlight of which was a performance on the *Ed Sullivan Show* in 1956. At CBS, he held a variety of posts in the program department, the last of them v.p. for enter-

tainment programs from New York and Europe. The Disney Channel, a partnership of Walt Disney Telecommunications and Group W Satellite Communications, is to be a 16-hour a day subscription service of family entertainment.

WAGNER, JANE—writer-producer who won one Emmy in 1974 as writer of a Lily Tomlin special and two more the following year as writer and producer of another Tomlin special. She was also executive producer of *People,* an NBC latenight entry introduced in 1976, having created and developed the program for Time-Life Films.

"WAGON TRAIN"—successful dramatic western series about families moving to settle the West. It began on NBC in one-hour form (1957–61) and switched to ABC in a 90-minute format (1962). The following year it returned to the hour format, and its reruns were stripped during the daytime by ABC. Produced by Universal TV, it originally had starred Ward Bond and Robert Horton as the wagon train scouts. They were replaced in the later version by John McIntire and Robert Fuller.

"WAIT TILL YOUR FATHER GETS HOME"—animated comedy series produced by Hanna-Barbera for prime-access syndication (1972–74) with characters suggestive of *All In the Family* and dealing primarily with the generation gap. It drew moderate ratings. The voices were by Tom Bosley, Joan Gerber, Tina Holland and David Hayward.

WALD, RICHARD C.—former president of NBC News (1973–77) who left after a clash with NBC president Herbert Scholosser, briefly worked for the Times-Mirror Co. in Los Angeles and then became, in 1978, executive v.p. of hard news operations for ABC News, supervising the regularly scheduled news programming and the worldwide news bureaus.

Wald came to NBC in 1967 after a career in print journalism, most of it with *The New York Herald Tribune,* where he had been variously political reporter, foreign correspondent and associate editor and managing editor (during the paper's final year, until its demise in 1966). He subsequently became Sunday editor of the short-lived *World Journal Tribune* in New York, then assistant managing editor of the *Washington Post* and, prior to joining NBC, v.p. of Whitney Communications Corp.

WALKER, NANCY—during 1976 and 1977, a popular TV comedienne in search of a vehicle. Under contract to ABC, she opened the fall season in 1976 in *The Nancy Walker Show,* a situation comedy produced by Norman Lear's T.A.T. Productions, in which she portrayed a Hollywood talent agent. When that series showed scant promise, ABC scuttled it and immediately found a new vehicle for her, *Blansky's Beauties,* via Garry Marshall's production organization at Paramount TV. In this, she played house mother to a group of Las Vegas chorines and was teamed with Roz Kelly, the young actress who had scored a hit in a guest role on *Happy Days* as Pinky Tuscadero, a momentary sweetheart of the Fonz.

Blansky's entered the ABC schedule as a January replacement. Thus, Miss Walker, who had never before starred in a TV series, became one of few performers to venture two series in a single season.

Although she had extensive stage and screen credits, the diminutive actress came into prominence on TV late in her career, first in commercials for Bounty Towels and then as the mother in the *Rhoda* series and concurrently in a regular featured role in *MacMillan and Wife*. She returned to *Rhoda* when both her ABC series failed.

WALKER, PAUL A. (d. 1965)—one of the most distinguished members of the FCC who served 19 years (1934–53), the last 14 months of that long term as chairman. A lawyer who had a brief fling with politics in his native Oklahoma and then became a member of the state utilities commission, he was appointed to the FCC by President Roosevelt just after the agency was formed. Walker worked toward regulating monopolistic practices in broadcasting and network control over station programming, and he led the hearings on the FCC proposals for a new table of allocations during the license freeze. He also supported the adop-

tion of the 1946 Blue Book, which attempted to set minimum public interest standards for broadcast licensees.

In February 1952, at the age of 71, he was finally appointed chairman by President Truman and presided over the commission during the frenetic period when the freeze was lifted and applications had to be processed. Walker lost the chairmanship to Rosel Hyde after the election of President Eisenhower, and he retired in 1953 to Oklahoma.

"WALL $TREET WEEK"—weekly 30-minute series on PBS concerned with issues and trends in business and finance and featuring interviews with stock market experts. The series began in January 1972 and has been hosted by Louis Rukeyser, former financial correspondent for ABC News. It originates at WMPB Baltimore, with Anne Truex Darlington as producer.

Mike Wallace

WALLACE, MIKE—CBS News correspondent noted for his penetrating interviews and journalistic enterprise. It was he, in fact, who caused the style of TV interviewing to change from blandly cordial to tough and searching, during the mid-50s, with a local New York program, *Nightbeat,* which later became a weekly ABC prime-time series, *Mike Wallace Interviews*. His technique of hard-boiled persistence became widely imitated. Wallace joined CBS News in 1963 and in 1968 became co-editor (with Harry Reasoner) of *60 Minutes*.

He began his broadcast career as a radio newsman in Chicago during the 40s, but became prominent in television initially as a quiz-show host *(The $100,000 Big Surprise)* and game-show panelist *(Who Pays?)*. His interview series led him back into journalism. Wallace anchored the *CBS Morning News* and the *CBS Mid-Day News* from 1963–1966. He also became one of the key CBS reporters covering political conventions and elections. Among his documentary credits for *CBS Reports* are *Marijuana* and *The Homosexual and the Law*.

WALLIS, W. ALLEN (Dr.)—chancellor of the University of Rochester who became chairman of the Corporation for Public Broadcasting in March 1977 and quit abruptly, without explanation, in July 1978. He had been appointed to the CPB board, as a political independent, by President Ford in 1975. Wallis served on numerous boards and government commissions and for a time had been a special assistant to President Eisenhower.

WALSH, BILL (d. 1975)—producer for Walt Disney. In TV he was co-producer of Disney's first Christmas special in 1950 and producer of *The Mickey Mouse Club* and the *Davy Crockett* series in the Disney anthology.

Barbara Walters

WALTERS, BARBARA—first female anchor on a network newscast and highest-paid news performer in television, achieving both distinctions in 1976 when ABC wooed her away from NBC with a five-year contract guaranteeing $1 million a year. In 1981, Roone Arledge, president of ABC News, announced that Ms. Walters contract had been renewed for another 5 years, and that she would be appearing regularly on ABC's 20/20.

Before joining ABC, Ms. Walters had been a personality on NBC's *Today* show, noted for her incisive interviews with world figures; previously, for three years, she had been on the *Today* staff of writers.

In 1964, while *Today* was in search of a regular female cast member, she was given the assignment on a trial basis and won the spot permanently with an aggressive, intelligent

performance. She was elevated to co-host, with Jim Hartz, in April 1974. Meanwhile, on the death of Aline Saarinen in 1971, Miss Walters took over the syndicated program, *For Women Only* (later retitled, *Not For Women Only*), and continued as daily host until the fall of 1975 when, to reduce her workload, she alternated weeks on the show with Hugh Downs. She also did reporting assignments for NBC News and hosted several prime-time specials.

Miss Walters became a star in the medium despite a speech fault which turned "r" into "w" and despite the relatively low visibility of early-morning broadcasts. She did, however, gain more prominent exposure in covering certain major events, such as President Nixon's first visit to the People's Republic of China, and in substituting occasionally for Johnny Carson on *Tonight*.

ABC's offer of $1 million a year, and NBC's counter-offer at a similar fee, was frontpage news around the country in the spring of 1976. Miss Walters accepted the ABC offer because it involved the co-anchoring with Harry Reasoner of the evening newscast; NBC was not prepared to give her such an assignment. Although she accepted the offer in April, her contract with NBC did not permit her to join ABC until September. The contract also called for a number of prime-time interview specials for ABC. On the first of them, Miss Walters interviewed President and Mrs. Carter.

Miss Walters is the daughter of the late Lou Walters, a show-business impresario who for many years operated the Latin Quarter in New York.

The Waltons

"WALTONS, THE"—highly popular CBS dramatic series concerning the struggles of a large, close-knit family in Appalachia during the Depression. Drawn from an autobiographical TV special, *The Homecoming,* by Earl Hamner, Jr., the series premiered in 1972 in a time period considered hopeless — that opposite *The Flip Wilson Show* on NBC, then one of the medium's big hits. After a slow start, *The Waltons* eventually conquered Wilson and ruled the time period for several years.

The series featured Richard Thomas as John-Boy (he left the cast in 1977), Michael Learned as the mother and Ralph Waite as the father. Will Geer played the grandfather until his death in 1978, and Ellen Corby played the grandmother. The children were played by Jon Walmsley, Judy Norton, Mary Elizabeth McDonough, Davis S. Harper, Eric Scott and Kami Cotler. It was via Lee Rich's Lorimar Productions.

WALWORTH, THEODORE H., JR.—president of the NBC Television Stations division since 1971 and member of the NBC board. He rose through the sales ranks after joining NBC in 1953, with earlier experience at Edward Petry & Co. and ABC. He held sales posts at the Cleveland and Philadelphia stations, then became v.p. and general manager of the flagship, WNBC-TV in 1961.

WAMEX—see WACC.

WANAMAKER, SAM—noted stage and film actor-director who settled in England; his TV directing efforts in the U.S. were for the pilots of the *Hawk, Custer* and *Lancer* series during the 60s.

"WANTED: DEAD OR ALIVE"—western series that propelled Steve McQueen to stardom. It played on CBS (1958-61) via Four Star-Malcolm Productions.

WAREHOUSING—the networks' tactic of purchasing movies from studios with exclusivity provisions in the contracts that effectively barred the films from pay-cable, or that diminished the period of time they were available to pay-cable under the FCC's original rules. (Those rules had permitted pay-cable to use a movie only within the first two years of its theatrical release; later that was increased to three years.) The rules and the practice of warehousing were knocked out by a D. C. Court of Appeals ruling in 1977 that held the restrictions on pay-cable to be unconstitutional.

WARNER AMEX CABLE—See QUBE.

WASEC—see WAMEX.

WASHBURN, ABBOTT M.—FCC commissioner appointed by President Nixon in 1974 for a two-year term in a Republican seat. After establishing a special concern about television for children, Washburn was reappointed to a full 7-year term in 1976. Prior to joining the commission, Washburn had headed the department of public services for General Mills. Earlier he was deputy director of USIA (1953–61) and had served with Radio Free Europe (1950–51).

Cliff Robertson, Jason Robards & Harold Gould

"WASHINGTON: BEHIND CLOSED DOORS"—12-hour serialized drama, based in part on John Ehrlichman's novel, *The Company,* presented on ABC-TV on six consecutive nights, Sept. 6–11, to open the 1977–78 season. The program, about corruption spawned in the pursuit of political power, was fiction with thinly disguised characters from the Nixon Administration. The fictive President, Richard Monckton, was played by Jason Robards. Other principals in the cast were John Houseman, Cliff Robertson, Harold Gould, Robert Vaughn, Andy Griffith and Stefanie Powers.

David W. Rintels was the creator, and also cowriter with Eric Bercovici, of the opening and closing episodes. Bercovici wrote the episodes between. Stanley Kallis was executive producer for Paramount Television, Norman Powell was producer and Gary Nelson director.

"WASHINGTON WEEK IN REVIEW"—long-running PBS series (since February 1967) on which a panel of journalists examine and comment upon recent developments in the capital. Produced by WETA Washington, the half-hour series has regularly featured such print media correspondents as Peter Lisagor (d. 1977), Elizabeth Drew, Hedrick Smith,

Charles Corddry and Neil McNeil, with Paul Duke as moderator.

WASILEWSKI, VINCENT T.—president of the National Assn. of Broadcasters since 1965, succeeding Gov. Leroy Collins, and in that capacity the leading spokesman for the interests of the industry. Unlike Collins, whose attempts to inspire broadcasters to greater achievement were controversial with station operators, Wasilewski, as a veteran of the NAB staff, understood the broadcast industry and how it wished to be represented in the capital.

He joined NAB in 1949 after receiving his law degree from the University of Illinois, became chief attorney in 1953, manager of government relations in 1955 and v.p. for government affairs in 1960. Wasilewski was serving as executive vice-president of the association when he was selected to be president.

WASSERMAN, ALBERT—independent documentary producer-director who had worked for both NBC and CBS News. For the NBC *White Paper* series he wrote, directed and produced such documentaries as *The U-2 Affair, British Socialized Medicine* and *Oswald and the Law;* for CBS, *Out of Darkness, Biography of a Cancer* and *Hoffa and the Teamsters.*

WASSERMAN, DALE—dramatist active in TV during the 50s. His TV play, *I, Don Quixote* for the *DuPont Show of the Month* in 1959, led to the script for his best-known work, the stage musical, *Man of La Mancha*. Among his other TV credits were *The Fog, The Power and the Glory* (starring Sir Laurence Olivier), *The Lincoln Murder Case, The Eichmann Story* and *The Stranger.*

WATERGATE—a squalid episode in American history whose chapters unfolded for the entire nation on network television and whose developments dominated network news from early 1973 until President Nixon's resignation in August 1974. Although the heroic work in uncovering the story and bringing the scandal to light was done by the *Washington Post*'s Bob Woodward and Carl Bernstein, it was television—entering somewhat late, to be sure—that disseminated it nationally and illuminated the issues and the new disclosures for the citizenry.

Television itself became part of the fabric of the story as the medium for the President's speeches and news conferences, and as carrier of two extraordinary Congressional proceedings: the Senate Select Committee's inquiry into the role of the Nixon Administration in the Watergate break-in and subsequent cover-up (begun in the spring of 1973 and concluded in the fall) and the House Judiciary Committee's climactic debate on the impeachment of President Nixon (televised July 24–30, 1974).

If TV had a hero in the Watergate affair to compare with the *Washington Post* it was easily CBS News, which offered the most intrepid and intensive coverage, often to the despair of its affiliates who were either sympathetic to the President or still responsive to the Administration's admonitions that local broadcasters would be held responsible for what they carried of biased or nonobjective reporting by the networks. Indeed, it was CBS News that gave the *Post* the validating support it needed, at a critical time, when it devoted 23 minutes to a two-part report on *The Evening News with Walter Cronkite* in October 1972 laying out, for the first time to a national audience, the Woodward and Bernstein disclosures. After this, CBS made its own contributions to the story through distinguished reporting by Dan Rather, Daniel Schorr, Fred Graham, Roger Mudd and others.

As to the fears of local stations, they were heightened by the license challenges against two Florida TV stations—WJXT Jacksonville and WPLG Miami—in 1973. The challenging groups were identified as Nixon loyalists, in the main, and the stations in jeopardy were owned by Post-Newsweek, parent company of the *Washington Post.*

By far, the biggest television event of the period was the coverage by all three networks and PBS of the Senate Watergate Hearings, which introduced Americans to the full cast of characters in the running story and produced the bombshell, from Alexander Butterfield, that Nixon had made audio tape recordings of all conversations in the Oval Office. The tapes, when they were eventually surrendered by Nixon, paved the way for the impeachment proceedings.

For a total of 11 weeks during the spring, summer and fall of 1973 the Senate Watergate Hearings were a huge attraction on daytime TV and also reached large audiences at night in taped replays on PBS. Because of the length of hearings and the expense involved in presenting them, the coverage was rotated daily among the commercial networks.

The hearings began on May 17 and ran for 37 weekdays. They recessed for seven weeks and were resumed in the fall for 16 more days, ending on November 15. Another round was set for the following January, but it was canceled.

The almost 300 hours of rotated coverage by the three networks was estimated to have cost them a combined total of $10 million in lost advertising revenues and air time. Audience surveys found that 85% of the nation's households watched all or part of at least one of the sessions. CBS estimated that viewers spent 1.6 billion total home-hours watching the daytime coverage on the three networks and an additional 400 million home-hours watching at night on PBS. Public television stations, meanwhile, enjoyed a financial windfall from their twice-a-day coverage. Viewers responding to the PTV appeals for new subscribers made contributions of more than $1.5 million, which was taken as a show of appreciation for the nighttime repeats.

While normally the preemption of daytime soap operas and game shows produces a flood of angry mail from viewers, in the case of Watergate the mail to the networks was overwhelmingly in favor of the preemptions.

On May 9, 1974, the House Judiciary Committee began its debate on the possible impeachment of President Nixon. Only 18 minutes of the opening session was allowed to be televised. For the next 11 weeks the meetings were closed to all outsiders, but after numerous disturbing leaks they were opened to broadcast on July 24. The networks again covered on a rotating basis, and PBS again carried nighttime repeats as well as the live(daytime coverage. The committee's deliberations were on the air for 6 days—a total of 46 hours of coverage.

When the hearings ended on July 30, three articles of impeachment had been voted against the President. They were to be presented before the full House on August 19. The House voted to permit TV coverage of the floor debate, and the Senate was expected to do the same, but it was all academic. On August 8, Nixon went on the air to announce his resignation. That speech, running 16 minutes, was watched by 110 million people, more than had watched any Presidential speech in history and was exceeded only by the TV audience for the first walk on the moon.

Watergate served to raise the prestige of television and, more importantly, restored the credibility of American journalism, which had been under severe attack by the Nixon Administration even before the scandal developed. The ratings for network news grew steadily as the Watergate story unfolded, the evening newscasts gaining more than 2 million additional viewers a night during the first year. Watergate also stimulated a new interest in investigative reporting at broadcast stations and was the spur for the reentry of ABC News in the field of investigative documentaries with the monthly series, *ABC Close-Up*.

WATSON, ARTHUR—broadly experienced NBC executive who has held high posts in sports, radio, sales, the television network and the owned-stations division. He succeeded Chet Simmons as president of NBC Sports in July 1979, after having been executive v.p. of NBC-TV for six months. Prior to that, he was executive v.p. of the owned-stations division for three years. From 1969 to 1971 he was president of NBC Radio. After that, he served five years as executive v.p. and general manager of WNBC-TV New York. He caught the notice of NBC management as a bright young executive while at the helm of the company's Cleveland station, WKYC-TV (1965–69).

"WAY IT WAS, THE"—sports retrospective series on PBS (1974–76) involving film of memorable sports events followed by a studio discussion with athletes who took part in those events. Curt Gowdy was host. The series was underwritten by Mobil Oil and produced by Gerry Gross Productions at KCET Los Angeles, in association with Syndicast Services.

WAYNE & SHUSTER—Canadian comedy team that made numerous guest appearances in the 50s and 60s after exposure on the *Ed Sullivan Show* but who failed with their own U.S. series.

WBAI INDECENCY CASE—case in which the D.C. Court of Appeals reversed the FCC on its ruling that indecent language ("dirty words") be barred from the air during the hours when children were likely to be in the audience. The Supreme Court then reversed the Appeals Court decision and upheld the FCC's right to suppress foul language. The Court's ruling was widely considered a severe blow to the First Amendment rights of broadcasters.

The FCC's action had come in response to a single citizen's complaint about a broadcast on WBAI-FM, a listener-supported station in New York, which had featured a recording of a comedy monologue by George Carlin who discussed "seven dirty words you can never say on television." The program aired on the afternoon of Oct. 30, 1973, and the complainant said his young son had heard it.

Citing its authority under the obscenity-indecency statute, the FCC held that the seven words used by Carlin were indecent and said that language describing "sexual or excretory activities and organs," used in a way that is offensive under community standards, could not be broadcast. The commission said the words might be aired late at night but only on condition that the context in which they were used have serious artistic, scientific or political value. WBAI then appealed on constitutional grounds.

In March 1977, the Court of Appeals ruled 2 to 1 that the FCC violated Section 326 of the Communications Act, the provision prohibiting the commission from censoring by interfering with the licensee's discretion. Judge Edward A. Tamm, who wrote the opinion, said also the FCC's position on the WBAI indecency question was "overbroad and vague" and that its attempt to channel the allegedly offensive material into the late evening still constituted censorship.

In a 5–4 decision in July 1978, the Supreme Court supported the FCC's position in ruling that the First Amendment does not bar the government from prohibiting broadcasts of words that are "patently offensive," although they may fall short of the Constitutional definition of obscenity. FCC chairman Charles D. Ferris assured the broadcast industry that he would not use the power given him by the Court to bar the use of bad language if it were used legitimately, as in news documentaries.

WBBM "POT PARTY" INCIDENT—a 1967 controversy focusing on the issues of "news staging" and deception of the public, growing out of a two-part documentary by WBBM-TV Chicago entitled *Pot Party at a University* (Nov. 1 and 2, 1967). Presented as an investigative report on the pervasiveness of marijuana use on college campuses, the film covered an actual marijuana party of Northwestern University students at a campus rooming house. Northwestern filed a complaint with the FCC, charging news staging, on the ground that a reporter for the station had arranged for the party to be held so that he could film it.

The FCC ruled that it was an authentic party, not one staged by actors but involving regular marijuana smokers gathered in an apartment where similar pot parties had been given. However, the FCC reprimanded WBBM-TV on two counts: first for representing itself to the public as having been invited to the party when in fact it had induced the party, therefore misleading its audience; and second for having induced the commission of a crime, the smoking of marijuana. "The licensee has to be law-abiding," the commission said.

WEATHER CHANNEL, THE—see Cable Networks.

WEAVER, SYLVESTER L., JR. —one of television's most creative executives, whose relatively brief term as president of the National Broadcasting Co. (Dec. 4, 1953–Dec. 7, 1955) was marked by innovation and a profound understanding of the new medium's natural program forms. He is the acknowledged father of the TV talk show—the desk & sofa format represented by the programs he created, *Today, Home* and *Tonight*—and of the "spectacular," now more modestly called the special. Weaver was also responsible for the *Wide Wide World* concept. Dave Garroway, Steve Allen and Arlene Francis were among those caught by Weaver's keen eye for talent.

Weaver was elevated to chairman of NBC in 1955, but he resigned a year later in a dispute with management of the parent company, RCA. Meanwhile, he developed program executives such as Richard A. R. Pinkham, Mort Werner and Michael H. Dann who were later to become the program chiefs of NBC and CBS. On leaving NBC Weaver formed his own broadcast company and then became an advertising executive with McCann-Erickson.

From 1963–1966, he headed Subscription Television Inc., the company that attempted to wire Los Angeles and San Francisco for pay-TV but was demolished by a well-organized campaign of theater owners and commercial broadcasters. Although STV finally established in the courts its right to promote pay-TV, its funds were dissipated by the time the court handed down its decision. After a brief fling with drawing program proposals for a proposed fourth network that never came to pass, Weaver returned to the advertising field as consultant to the Wells, Rich, Greene agency.

He had joined NBC in 1949 as head of its new television operations after two decades in broadcasting and advertising. He had been a writer-producer for CBS and the Don Lee radio network in the 30s, and in the 40s he was advertising manager for American Tobacco and v.p. of radio-TV for Young & Rubicam Advertising.

Sgt. Joe Friday a.k.a. Jack Webb

WEBB, JACK—enormously successful TV producer, after a career as a popular actor (the 1952 *Dragnet,* which he also directed). He formed an independent production company, Mark VII Ltd., and had a string of hits, including *Dragnet* (a revival), *Adam-12* and *Emergency.* Other series by Webb were *The D.A., The Rangers, O'Hara, U.S. Treasury, Hec Ramsey* and *Mobile One.*

Webb's patented style, carried over from *Dragnet* (which began on radio), was to play down the melodramatic glamor of the perilous assignments of his protagonists and to give their execution an all-in-a-day's work flavor. The shows seemed more realistic and authentic than most others of the genre, but the heroes were no less heroic.

WEBSTER, EDWARD MOUNT (Commodore) (d. 1976)—FCC commissioner (1947–56). He had joined in 1934 as assistant chief engineer and remained on staff until his appointment as commissioner by President Truman. He received his military title with the U.S. Coast Guard, serving as chief communications officer.

WEBSTER, NICHOLAS—director active in documentaries and primetime entertainment series. With the Wolper organization he directed *Showdown at O.K. Corral* and *The*

Last Days of John Dillinger in the docu-drama form for the CBS *Appointment With Destiny* series. His documentaries included *The Long Childhood of Timmy, Walk in My Shoes* and *The Violent World of Sam Huff.* He also directed episodes of *The Waltons, The FBI, Big Valley, Dan August, Get Smart* and *East Side, West Side.*

"WEEKEND"—monthly 90-minute latenight news magazine introduced by NBC in 1974 to alternate with rock concerts and the then experimental comedy series, *Saturday Night,* at 11:30 P.M. on Saturdays. As an attempt by NBC to stay even with CBS News with its popular *60 Minutes* news magazine, *Weekend* adapted to its inferior time period by putting its accent on nonfiction of particular interest to youth. *Weekend,* whose progenitors were *First Tuesday* and *Chronolog,* was developed by Reuven Frank, senior producer for NBC News, shortly after he stepped down as president of the division. Producers of the program's segments included Bill Brown, William B. Hill, Peter Jeffries, Clare Crawford, James Gannon, Karen Lerner, Sy Pearlman, Anthony Potter and Craig Leake. Gerald Polikoff was the director.

"WEEKEND SPECIALS"—Saturday morning dramatic series for children on ABC that began in 1977 under the title of *ABC Short Story Specials.* The weekly half-hour series consists of multipart serializations of book and short-story adaptations, as well as some original dramas.

WEIL, SUZANNE—head of programming for PBS since Chloe Aaron's departure in 1981, before which she served under Ms. Aaron as director of arts and humanities programming. Earlier she had been director of the dance program for the National Endowment for the Arts and coordinator of performing arts for the Walker Arts Center in Minneapolis.

WEINBERGER-DANIELS—writing-producing team of Ed Weinberger and Stan Daniels, whose credits include producing *The Mary Tyler Moore Show* and *Phyllis.*

WEINBLATT, MIKE (MYRON)—president of Showtime, the pay-cable network, since 1980. Before that he was a key NBC executive with wide-ranging experience at the network who became president of ABC Entertainment when that division was created in September 1978. The division encompasses programming, advertising and promotion, and program and talent negotiations. Weinblatt had previously been executive v.p. and general manager of the television network for a year and prior to that was senior v.p. of sales.

Earlier titles, in a career with the network that began in 1957, included v.p. of talent and program administration; v.p. of eastern sales; director of pricing and financial services and a number of others in business affairs. He was especially skilled in talent negotiations and the acquisition of programs.

Mike Weinblatt

WEINRIB, LEONARD—writer and director who wrote all the scripts for NBC's children's shows, *H.R. Pufnstuf* and *Dr. Doolittle.* He also wrote episodes of *All In the Family, Love, American Style* and other prime-time series.

WEITMAN, ROBERT M.—a leading studio executive, head of production for MGM from 1962 to 1967, then for Columbia Pictures. He had come up through television, having been v.p. in charge of programs for ABC (1953–56) and v.p. of program development for CBS (1956–60). He joined MGM as a TV executive and two years later became production chief for the motion picture studio, as well.

"WELCOME BACK, KOTTER"—successful ABC situation comedy (1975–79) which sprang John Travolta to stardom, although it was created as a vehicle for comedian Gabriel Kaplan. Kaplan played a young teacher assigned to the Brooklyn high school from which he had graduated, but the series derived much of its appeal from the group of obstreperous misfits in the class, The Sweathogs, of which Travolta was one. Others were played by Ron Palillo, Lawrence-Hilton Jacobs and Robert Hegyes. Also featured were John Sylvester White and Marcia Strassman. Created by Kaplan and Alan Sacks, the series was produced by The Komack Co. and Wolper Productions.

WELLS, ROBERT—FCC commissioner (1969–71) whose appointment had been controversial because he was, at the time, an active broadcaster in Kansas. Wells was at the center of an even stormier controversy in 1975 when President Ford had planned to nominate him as director of the Office of Telecommunications Policy. That prospect raised a furore among citizens groups, the cable industry and certain influential members of Congress because, on leaving the FCC, Wells had returned to his broadcast post and reacquired his minority interest in four radio stations owned by Harris Enterprises. Those who opposed his appointment argued that a commercial broadcaster could not be objective in assessing the country's communications priorities. Their efforts ended his candidacy for OTP. He became chairman of Broadcast Music Inc. in 1979.

WELPOTT, RAYMOND W. (d. 1973)—president of the NBC-owned TV stations from 1965 until his retirement in 1971. He had been general manager of the NBC Philadelphia stations, WRCV-AM-TV, until they were moved to Cleveland in 1962. He then became v.p. of the stations division and spot sales, and three years later was named president.

WENDKOS, PAUL—director of such TV movies as *A Death of Innocence, The Woman I Love, Haunts of the Very Rich, Brotherhood of the Bell, The Family Rico* and *The Mephisto Waltz.*

Mort Werner

WERNER, MORT—program chief for NBC from 1961 to 1972. Earlier he had helped to develop *Today, Home* and *Tonight* for the network and became executive producer of all three. Werner joined NBC in 1951 after a varied show business career in which he had been a radio actor, band-

leader, singer and summer stock manager. In 1957 he left the network to become an executive of Kaiser Industries and later went with Young & Rubicam Advertising as director of radio and TV. NBC rehired him in 1961.

WERSHBA, JOSEPH—long-time CBS News producer and reporter whose career spanned the noted *See It Now* documentaries of the early 50s and the popular news-magazine of the mid-70s, *60 Minutes.* He worked with Edward R. Murrow and Fred W. Friendly on such famed *See It Now* broadcasts as the expos(e on Sen. Joseph McCarthy, *Desegregation in North Carolina* and *The Milo Radulovich Story.* During the 60s he became a producer of *CBS Reports.* Wershba joined CBS News in 1944 as a writer and later became news director of WCBS-TV and then a Washington correspondent. He was co-editor with Don Hollenbeck of the original *CBS Views the Press* and was also a reporter-director on the *Hear It Now* radio series that later became *See It Now.*

WESH-TV LICENSE RENEWAL CASE—a precedent-setting case concerning the license renewal of a Daytona Beach, Fla., television station in a comparative proceeding with a challenging applicant. In the early 1970s, the FCC ruled, in the face of the challenge, that WESH-TV had provided sufficient service to the community, and it renewed the operating license of the station's owner, Cowles Communications Inc. But in September 1978, a three-judge panel of the Court of Appeals in Washington sent the case back to the FCC for further consideration, declaring that the incumbent operator would have to demonstrate a record of performance indicating its clear superiority to the competing applicant, Central Florida Enterprises Inc.

The full Court of Appeals subsequently denied the FCC's request for a review of the panel's decision but did eliminate the explicit requirement that the record of the station must be *superior* to assure renewal. Although the court clearly indicated that it objected to the "renewal expectancy" of most licensees, station operators were somewhat relieved that the language referring to superior performance was dropped. Broadcasters were motivated to lobby harder for the various proposals in Congress that would lead to legislation relaxing performance standards and extend TV and radio license terms beyond the present three-year period.

The court had previously overruled the FCC in a landmark case when it denied the license renewal of WLBT Jackson, Miss., after various citizen groups appealed the case. The WESH-TV case, however, marked the first time that the court rejected the FCC's judgment in a license renewal proceeding involving a competing applicant.

WEST, BERNIE—See Ross, Michael.

WESTFELDT, WALLACE—producer for ABC News since summer of 1977. Previously he was executive producer of public affairs for WETA, the Washington public TV station, and for 15 years before that a producer and reporter for NBC News. With WETA, Westfeldt was executive producer for numerous special events carried by PBS and of the 1976 series, *USA: People and Politics*.

At NBC News, which he had joined in 1961 as a writer, he served variously as reporter, producer, exec producer of the *Nightly News* (1969–73) and exec producer for documentaries (1973–76). His documentaries included *How Watergate Changed Government, The Meaning of Watergate, The Nuclear Threat to You, The Man Who Changed the Navy* and *1975: The World Turned Upside Down*. Before joining NBC, he was a reporter for *The Nashville Tennessean* and a correspondent for *Time* and *Life*.

WEST GERMANY, TELEVISION IN—regarded by some as one of the world's best services, in programming respects. It is a complicated three-channel structure mixing regional production with prime time national network shows. Though ultimately under control of the regional state republics, the companies operating the channels have a degree of independence within guidelines imposed on them; these cover program patterns and the number of hours devoted to various kinds of shows—light entertainment, drama, news, etc.

Two of the channels, the first and second, are run by the older ARD (Arbeitgemeinschaft der Oeffentlich-Rechtlichen Rundfunkanstalten der Bundes-Republik Deutschland). This is the supervising authority for a cooperative system encompassing regional stations, each of which programs individually up to 8 o'clock at night, when a single network pattern takes over.

The third channel, known paradoxically as Zweites Deutsches Fernsehen (ZDF), or Second German Television, furnishes a single networked schedule for the whole of the country. It was created in 1961 under a treaty adopted by the various regional states which imposes on ZDF and ARD an obligation to furnish viewers with a complementary service in all time periods: Action drama may not compete with action drama, or comedy with comedy.

By common consent, the strengths of German television are in drama and public affairs. The latter in particular enjoys a kind of sacrosanct status in the scheme of things. West German television generally, and ZDF in particular, is one of the more active co-producing services, frequently as a financial participant in British and Anglo-American production.

The three channels are funded mainly by license fees, with spot advertising permitted on a limited basis. The color standard is, of course, Germany's own 625-line PAL, and there are close to 4 million color sets out of more than 18 million total receivers.

WESTAR—first domestic satellite in the U.S., launched by Western Union in April 1974, with a second craft, Westar 2, sent aloft in October of that year. Both were manufactured by Hughes Aircraft. Westar provides the interconnection for the Public Broadcasting Service and Spanish International Network and has tended to carry much of the commercial television and radio traffic, in contrast to RCA Satcom, which has been saturated with cable-TV services.

WESTERN—a program type that for a decade had such a grip on the medium that no American would have predicted its virtual disappearance from the airwaves by 1975. The western in fact had been a TV staple from the earliest days of the medium. In the first wave were the horse operas of the movie matinee genre, low-budget formula fare aimed largely at the juvenile audience: Hopalong Cassidy, Gene Autry, Roy Rogers, The Lone Ranger and others. 1955 brought the second wave, the so-termed "adult western," typified by *Cheyenne, Gunsmoke, The Life and Times of Wyatt Earp* and *Tales of the Texas Rangers*. These not only dealt with hard-riding cowboys but also with social and philosophical themes, relatively complex characters and offbeat heroes. Frequently in the adult western, the demarcation between good and evil was blurred.

Nevertheless, they were essentially actionfull shows, and they served to introduce violence to television on a grand scale. Because there were so many westerns on TV in a concentrated period—a total of 32 in the 1959 season, for example—each strived to succeed against the field by doubling the gunplay and driving up the body count. Teasers featured the most violent scenes to lure the viewer, for example, an outlaw using his foot to push a man's head into the campfire. The heightened violence quotient was inherited by the next action-adventure trend, the private-eye series, and subsequent ones, the spy and then the police series.

Of the adult westerns that rode into prime time in the 1955–56 season, *Cheyenne* was probably the most significant since it played in the *Warner Bros. Presents* showcase, which represented a breakthrough in the production of TV series by the major Hollywood film studios. Its profitability to WB prompted 20th, Paramount, MGM and Universal to get in on the television action, and they too broke in with westerns. The form was appealing because it enabled the studios to use stock footage for many scenes.

Some believe the Eisenhower Presidency contributed to the rash of oaters, the former war hero fanning a hero-consciousness in the land. Eisenhower was also a native of Abilene, Texas—cowboy country—and made it known that he fancied the western story and legend. At any rate, the time was apparently right, and season after season the westerns came thick and fast to television.

Every producer looked for a new wrinkle — the cowardly hero, *Maverick;* the fancy-dan with a cane, *Bat Masterson;* the ingenue, *Sugarfoot;* the knife-slinger, *Adventures of Jim Bowie;* the frontier newspaperman, *Jefferson Drum.* Some worked, some didn't. *The Virginian's* game was to run 90 minutes, and with it NBC stole half an evening on its rivals who tried but couldn't match it for success. *Frontier* attempted to outclass the field by pressing into the service of westerns Worthington Miner, one of the big names of the "golden age" of TV drama. Serious actors like Lee J. Cobb became involved, although Cobb dropped out of *The Virginian* after a few seasons.

A lot of nobodies were pitched into starring roles—Will Hutchins, Scott Forbes, Wayde Preston, Jeff Richards, Tony Young and Don Durant, among scores of others—but quite a few of the nobodies made it to actual stardom. Steve McQueen sprang from *Wanted: Dead or Alive;* Clint Walker from *Cheyenne;* Lorne Greene, *Bonanza;* James Garner, *Maverick;* Gene Barry, *Bat Masterson;* Chuck Connors, *Rifleman;* Jim Arness and Dennis Weaver, *Gunsmoke;* Richard Boone, *Have Gun, Will Travel;* Hugh O'Brian, *Wyatt Earp;* Clint Eastwood, *Rawhide.*

Others who rose from obscurity were Ty Hardin, Nick Adams, Dale Robertson, the late Dan Blocker, Michael Landon, Robert Horton, Robert Culp, Neville Brand, Robert Conrad, Ryan O'Neal and James Drury. They mingled among established stars who went western, such as Henry Fonda, Joel McCrea, William Bendix, Dick Powell, Ward Bond, George Montgomery, Preston Foster and Ronald Reagan.

The titles smacked of machismo and action: *Rawhide, Broken Arrow, Death Valley Days, Lawman, Colt .45, Wagon Train, Johnny Ringo, Shotgun Slade, Cimarron City, Restless Gun, Wells Fargo, Wild Wild West.*

Wild Wild West was cut down in its prime by congressional hearings on violence, and in the early 70s all that remained were *Gunsmoke* and *Bonanza.* They too disappeared by 1975 and, ironically, the networks tried sporadically to revive an interest in the western in order to find a way to cut back on the violence of police shows. *Barbary Coast, Dirty Sally, Sara* and *The Quest* all were tried, and all failed but the networks were not deterred. A handful of new westerns were in the schedule for the fall of 1977.

WESTIN, AV—news executive prominent at ABC during the 70s and previously in public TV and at CBS News. He left ABC in a dispute with management in 1976 and became an independent producer and news consultant. But he returned to the network in summer of 1977 as a key figure in the rebuilding of the division by Roone Arledge, president of ABC News and Sports. Westin was put in charge of the evening newscast with the title of v.p.

During his previous hitch with ABC, he redesigned the evening newscast and organized the *ABC Close-Up* documentary operation; in February 1975 both vital areas of the news division were put in his charge.

Westin began in broadcast journalism with CBS News in the 50s and rose rapidly from newswriter to editor, director and producer of news programs as a protege of Fred W. Friendly, then president of the division. In 1967 Westin left to join the newly formed Public Broadcasting Laboratory as producer of *PBL,* the experimental Sunday night newsmagazine. Two years later he signed on with ABC News.

Av Westin

WESTINGHOUSE BROADCASTING CO.—See Group W.

WESTWARD TELEVISION—regional commercial station licensed to serve southwest England, an area surrounded on three sides by the sea, giving a nautical character to many of its programs. Principal studios are located in Plymouth.

WGBH—Boston public TV station and one of the leading production centers for PBS, the source of such series as *The French Chef, The Advocates, Nova, Zoom World* and *Evening at Pops,* among others. Its prestige was heightened in the 70s when it acquired and assembled for PBS the *Masterpiece Theatre* series of dramatic productions from Britain.

Blessed with a Channel 2 position, WGBH was one of the first PTV stations to establish a substantial following in its community, and this may partly account for its vitality as a producer of national programs from public TV's earliest times. Known for being innovative and resourceful, WGBH has been responsible for such PBS fringe-time programs as *Crockett's Victory Garden,* a how-to-garden series; *Erica,* offering instruction in needlework; *Joyce Chen Cooks,* Chinese-cooking instruction; and *Maggie and the Beautiful Machine,* featuring an exercise therapist.

Other notable productions have been a 1975 dramatization of the Watergate coverup; *Arabs and Israelis,* a seven-part series which attempted to explain the Middle East situation; and *Grand Prix Tennis.* WGBH was the first station to use captioning for the deaf.

The station is licensed to the WGBH Educational Foundation and has a radio counterpart.

WGN CONTINENTAL BROADCASTING CO.—station group of the Tribune Co. of Chicago, publisher of *The Chicago Tribune* and New York *Daily News,* all its stock privately held. The group has been built around WGN Chicago, one of the leading and most prestigious independents in the country. It also owns KWGN Denver, an independent. The WGN group has an interlocking relationship with WPIX New York, another property of the Tribune Company, but operating autonomously on the premises of the New York newspaper. Daniel T. Pecaro is president and chief executive officer of WGN Continental.

"WHAT'S HAPPENING!"—ABC sitcom (1976–79) about three black high school students who live in a middle-class neighborhood. Originally presented as a four-week summer replacement, the series returned on a regular basis in November of the same year and scored excellent ratings. Ernest Thomas, Haywood Nelson and Fred Berry played the leads.

"WHAT'S IT ALL ABOUT?"—CBS public affairs specials for children examining events in the news, thus such individual titles as *What's Apollo-Soyuz All About?, What's the Middle East All About?* and *What's The CIA All About?* The series began in 1972. Joel Heller was executive producer and Walter Lister producer-writer.

"WHAT'S MY LINE?"—popular CBS panel show which ran in prime time for nearly two decades, starting in 1950. Mechanically, it involved nothing more than a panel of celebrities trying to guess the occupation of a guest, but its appeal derived from the wit and personality projection of the

four panelists. After some early experimentation with panelists, the regular group came to consist of Dorothy Kilgallen (until her death in 1965), Arlene Francis and Bennett Cerf, with John Daly as moderator. The fourth seat on the panel was temporary. The series helped to launch Goodson-Todman Productions.

WHDH CASE—first in which a competing applicant was awarded a broadcast license over an incumbent with no substantial negative record. The decision by the FCC in January 1969 to award Channel 5 in Boston to a group known as Boston Broadcasters Inc., after the Boston Herald-Traveler Corp. had operated the station for more than a decade, sent shockwaves through the industry and spurred a drive for legislation that would protect broadcasters from license challenges.

The WHDH case was a unique one since the Herald-Traveler Corp. had been operating the station only on a temporary authorization from the FCC, except for a brief period when it was awarded a license for four months. Herald-Traveler, which had been one of four mutually exclusive applicants for Channel 5 in 1954, had been granted a construction permit for the station in April 1957. The award was challenged by two of the other applicants in federal courts.

A new issue of improper *ex parte* contacts entered the case when Robert Choate, president of the corporation, was charged with entertaining FCC chairman George C. McConnaughey at several social luncheons. The meetings were assumed to be attempts by Choate to influence the commission outside the normal adjudicatory processes. In July 1958 the U.S. Court of Appeals remanded the case to the FCC to investigate the *ex parte* question.

The commission found that although the president of WHDH had demonstrated an attempted pattern of influence, this did not disqualify WHDH but warranted setting aside the construction permit and holding new comparative hearings with the four original applicants. WHDH, meanwhile, was permitted to continue broadcasting under a special temporary authorization. In September 1962, the FCC determined that WHDH was the most desirable applicant and awarded it a short-term license. On its expiration, three applicants (only one of them one of the original applicants in 1954) challenged the license with competing bids.

During the long comparative hearings that followed, the FCC applied its own 1965 policy statement on Comparative Broadcast Hearings, which ruled out the broadcaster's past performance from consideration unless it had been exceptional. With this concept eliminated, the commission in January 1969 voted 3 to 1 to award the license to Boston Broadcasters Inc.

A chief reason for denying the station to Herald-Traveler was its ownership of a Boston newspaper (albeit a financially ailing one) and an FM station in the market. The FCC, concerned about media concentration in general and having adopted a policy supporting diversity of ownership, favored BBI for its lack of other media properties and for the greater integration between ownership and management that it promised to provide. BBI renamed the station WCVB-TV.

In its appeals, Herald-Traveler noted that the newspaper was being kept alive only by the profits from the TV station and that if it were to lose the station it would be forced to cease publication of the *Boston Herald-Traveler.* When the appeals failed, the newspaper did collapse.

To calm an industry fearful of the precedents that might be set by the WHDH case, the FCC issued a policy statement early in 1970 saying that, in license challenges, it would give preference to the incumbent licensee if he could demonstrate that the programming had substantially served the community needs and interests. But the D.C. Court of Appeals struck down the policy statement as overly protective of the incumbent broadcaster.

WHEELER, THOMAS—president of the National Cable Television Assn. since August 1979, named to that post after the resignation of Robert L. Schmidt. For three years previous, he had been executive v.p. of NCTA concentrating chiefly on government relations. He joined the trade association from Grocery Manufacturers of America, where he was a v.p. and lobbyist in Washington.

WHELDON, HUW —Welsh careerist with the British Broadcasting Corp. who started as an air personality and producer, and who did a great deal to popularize programming that covered the arts. He later became a program executive and then managing director of BBC-TV, retiring on Dec. 31, 1975, at the mandatory age of 60. Shortly thereafter he was knighted by Queen Elizabeth II.

His retirement as an executive allowed him to pick up, from time to time, his earlier career as a performer. He served as host and narrator for a number of BBC documentaries, including those on the U.S. Library of Congress and Britain's royal family.

"WHEN THINGS WERE ROTTEN"—situation comedy created and produced by Mel Brooks which failed to survive its first season on ABC (1975). A spoof of the Robin Hood fable, it drew encouraging ratings initially and then faded. Dick Gautier, Dick Van Patten, David Sabin and Misty Rowe were featured. It was by Paramount TV.

"WHIRLYBIRDS"—syndicated adventure series concerning helicopter heroics, produced by Desilu (1957–59) and featuring Kenneth Tobey, Craig Hill and Nancy Hale.

"WHISPERING SMITH"—western on NBC (1961) featuring Audie Murphy and Guy Mitchell. It was by Whispering Smith Co., in association with NBC and MCA.

WHITAKER, JACK—senior sports commentator for ABC Sports since 1982. He joined that network after a 20-year career with CBS Sports, coming from the Philadelphia o&o, WCAU-TV, in 1961. Initially hosting the *CBS Sports-Spectacular,* he later did play-by-play for NFL football and then became host of the pregame, postgame and half-time shows. He was also a principal commentator for CBS on golf events.

WHITE, BETTY—actress who starred in minor situation comedies of the 50s, *Life with Elizabeth* and *A Date with the Angels,* then in panel shows *(Make the Connection).* She was also regularly a commentator on the Rose Bowl Parade on New Years Day. Her acting career caught fire again in the mid-70s when she became a featured player in *The Mary Tyler Moore Show.* When that show folded she became star of a new series, *The Betty White Show,* on CBS in 1977.

WHITE, FRANK—president of NBC from January to August 1953, succeeding Joseph H. McConnell who left to become president of Colgate-Palmolive-Peet Co. He resigned after seven months because of ill health, and in December Sylvester L. (Pat) Weaver was named president. White entered broadcasting in 1937 as treasurer of CBS, later became president of Columbia Records and then of the Mutual Broadcasting System. He joined NBC in 1952 and soon became v.p. and general manager of the network.

WHITE, JOHN F.—first president of National Educational Television (1958–69) through the period when it served as the network for educational television. He left to become president of Cooper Union in New York, thus resuming an academic career that had begun in 1941. Before joining NET, he had been general manager of WQED, the educational station in Pittsburgh.

WHITE, LAWRENCE R.—program chief for NBC (1972–75) who later became executive v.p. of production for Columbia Pictures Television. In the brief period between he had formed Larry White Productions, which developed the ABC series, *Feather and Father.*

White began his broadcast career as a producer-director for the DuMont Network in 1948. Three years later he became director of programming for Benton and Bowles Advertising, from which post he served as executive producer of the soap opera, *The Edge of Night.* He left the agency to join CBS in 1959 as v.p. of daytime programming, then moved to Goodson-Todman Productions and then back to CBS as director of program development. He joined NBC in 1965 as v.p. of daytime programs and after several promotions became v.p. in charge of programs.

Lawrence White

WHITE, MARGARETA (MARGITA)—FCC commissioner from 1976 to 1978, appointed by President Ford for two years to complete the term of Charlotte Reid. Mrs. White, a Republican, had previously been assistant press secretary and director of the Office of Communication at the White House.

Initially, the President had nominated her for a full seven-year term and Joseph Fogarty, a Democrat, for the remainder of Commissioner Reid's term. That package ran into opposition in the Senate not only for political reasons but also for a conflict-of-interest issue raised by the fact that Mrs. White's husband, Stuart White, a lawyer, was with a firm that represented clients before the FCC. President Ford then switched the terms, proposing Mrs. White for the two-year and Fogarty for the seven. The Senate, adopting a two-year waiver of the conflict-of-interest question, then was able to confirm the appointments.

The conflict-of-interest question was resolved in 1977 when Mrs. White's husband switched firms, joining one not engaged in communications law.

WHITE, PAUL (d. 1955)—newsman who made his mark principally in radio, building the CBS News bureau into the most prestigious of the network news organizations. He founded the bureau in 1933 when he was named v.p. and general manager of Columbia Broadcasting News. During his 13-year administration, White directed a staff that included Edward R. Murrow, William Shirer, Robert Trout, Elmer Davis, Cecil Brown, H. V. Kaltenborn, Quincy Howe and other prominent newscasters.

In 1945 White won a Peabody Award for outstanding news coverage, then left CBS to write a book, *News on the Air.* He became associate editor of the *San Diego Journal* (1948–50) and then joined KFMB and KFMB-TV in San Diego as news director.

"WHITE SHADOW, THE"—CBS dramatic series, which began Nov. 1978, concerning the white coach of a basketball team at a racially mixed school. Although marginally successful, it showed steady improvement in the ratings and was renewed for the 1979–80 season. Ken Howard stars as the coach, and the featured cast includes Jason Bernard, Joan Pringle, Kevin Hooks, Eric Kilpatrick, Nathan Cook, Robin Rose and Jerry Fogel. The series is via MTM Productions, with Bruce Paltrow as executive producer and Mark C. Tinker as producer.

WHITEHEAD, CLAY T.—first director of the White House Office of Telecommunications Policy (1970–74) who helped both to establish the agency and almost to destroy it by using it politically. The OTP, under Whitehead, became a base for attacks on network news, one of President Nixon's favorite targets.

The post, created in September 1970, made Whitehead the ranking White House advisor in the field of broadcasting, responsible for developing policy on present and emerging communications technologies and for regulating the federal government's own communications system. As a brilliant young management specialist and electrical engineer, with three degrees from M.I.T., he was eminently qualified for the office. But he began from the first to rail against network journalism, calling it "ideological plugola" and "elitist gossip" for its alleged liberal slant. His cautionary speeches to commercial and public broadcasters—made while Administration-sponsored legislation hung in the balance—were widely interpreted as bold moves by the White House to bring broadcasters ideologically in line and to win sympathetic treatment on the air for the President's policies.

Indeed, Whitehead's speeches carried the strong suggestion that legislation desired by the broadcasters could be traded off for their compliance with the Administration's view of good journalism. The tactic backfired, however, as Congress and the press read sinister motives into the bills drafted by the OTP, principally the one in 1972 that would have extended the broadcast license period from three years to five.

Only when Watergate destroyed the credibility of the White House media campaign did Whitehead and the OTP function as they were supposed to. Whitehead was responsible for an admirable cabinet committee report on cable-TV in January 1974, which received a great deal of praise, although Dean Burch of the FCC noted that it came two years too late to be useful. Whitehead also produced a long-range funding bill for public TV. His last act in government service, before leaving to return to academia (as a fellow at Harvard's Institute of Politics), was to join a small team in preparing the transition from the Nixon to the Ford Administration when it became apparent that Nixon would have to resign. He later became president of Hughes Satellite Corporation which owns the Galaxy Satellite system.

Although valuable studies and reports began to come from OTP, Whitehead will undoubtedly be better remembered for his attempts to shape broadcast news policy than for forging the more cosmic telecommunications policy.

In his "ideological plugola" speech, delivered at a Sigma Delta Chi luncheon in Indianapolis on Dec. 18, 1972, Whitehead said that stations which tolerate bias by the networks can only be considered willing participants in the bias. Not only did the speech seem designed to drive a wedge between the networks and their affiliated stations, it also suggested a carrot-and-stick ploy, since it was coupled with his announcement that OTP had drafted a bill to safeguard and extend broadcast licenses. The speech created such a furore over attempted government censorship that the bill was never introduced.

At an NAEB convention in Miami in the fall of 1971, Whitehead criticized the growth of PBS in the direction of a "fourth network," expressed the Administration's displeasure with PTV's liberal leanings, and called for the system to deemphasize national programming and to embrace "grassroots localism."

As if to punctuate that speech, President Nixon in 1972 vetoed a two-year funding bill for PTV that called for $165 million on the ground that the industry's power was too centralized. Hungry for federal funding, public broadcasters promptly dissolved PBS as a network and adopted a plan of overproducing national shows, to allow stations to select what they wished, following Whitehead's suggestion. "Localism" became the PTV byword.

Whitehead's use, or abuse, of the office gave it a low priority in the executive branch, and when the Ford Administration tried to reduce the executive budget early in 1975 the OTP was one of the agencies ticketed for extinction. The office was saved when John Eger, Whitehead's successor as acting director, alerted Congress to the plan. However, the office was dismantled in 1977 by the Carter Administration. See also Geller, Henry; National Telecommunications and Information Administration; Office of Telecommunications Policy; Public Television.

"WHO'S WHO"—nonfiction magazine series devoted to featurettes on newsmakers, celebrities and interesting obscure persons essayed by CBS as a prime-time entry in 1977. The series was a spin-off of *60 Minutes* in a year when that program achieved Top 20 popularity, and it was signaled by the trend to "people" magazines on the newsstands and the resurgence of gossip columns in the press.

Don Hewitt, executive producer of *60 Minutes,* and Dan Rather, a key correspondent on that show, both doubled on *Who's Who.* Along with Rather, the regular *Who's Who* correspondents were Barbara Howar and Charles Kuralt.

The ratings for the series were generally unimpressive its first months on the air, but that was largely a function of its having to compete with *Happy Days* and *Laverne and Shirley,* the top-rated series in TV at the time. *Who's Who* was given another test in a new period during June 1977 but failed to win a renewal for the fall.

"WIDE WIDE WORLD"—novel and much-praised Sunday afternoon series on NBC (1955–58) which traveled about the country electronically (and also to Canada and Mexico) to develop a subject—aging, holiday preparations, etc. Dave Garroway was host of the 90-minute program, which also occasionally used film from abroad. The title had such an attractive ring that ABC later borrowed it for *Wide World of Sports* and *Wide World of Entertainment.*

"WIDE WORLD OF ENTERTAINMENT"—umbrella title for the potpourri of latenight shows instituted by ABC-TV in the fall of 1972 as a new attempt to compete with NBC's *Tonight* and CBS's movies, following three years of third-place standing with *The Dick Cavett Show.*

Initially, *Wide World* rotated a week of Cavett, a week of original TV movies, a week of Jack Paar in his television return and a week of music-variety shows from 11:30 P.M. to 1 A.M. (EST). Within a year Paar was dropped and Cavett cut back to one program a month, as the format loosened to accommodate a diversity of comedy, variety, dramatic, news and event programs—in effect, a special every night, providing an opportunity to test new formats and talent. When Cavett left ABC in December 1974 for opportunities that CBS might afford, *Wide World*'s only regular commitment was to a bimonthly rock concert on Friday nights, entitled *In Concert.*

Among the special programs offered on *Wide World of Entertainment* were *Alan King at Las Vegas* (Jan., 1973), *David Frost Presents the Guinness Book of World Records* (Nov., 1973), *Dick Clark Presents the Rock of the 60s* (Jan., 1974), *Geraldo*

Rivera: Good-Night America (five programs, 1973–74), *Marilyn Remembered* (Dec., 1974), and a variety of testimonials, salutes and celebrity tours. The concept was abandoned in 1976 as ABC gave over the late-night hours increasingly to reruns of prime-time action-adventure shows. These proved more effective in the ratings than the *Wide World* originals.

Nadia Comaneci

"WIDE WORLD OF SPORTS"—ABC sports anthology on Saturday afternoons, begun in 1961, which became the keystone of the network's sports programming. With a kind of magazine format, *WWS* covered a wide range of fringe competitive sports that would not otherwise have received network exposure, events involving figure skating, barrel jumping, Ping Pong, motorcycle feats, auto racing, gymnastics and a variety of track and field competitions. During ABC's leanest years, *WWS* was one of the network's most consistently popular programs and always heavily sponsored.

The series was started on ABC by Edgar Scherick, then head of sports for the network, who had to build a series quickly when Gillette quit NBC for dropping Friday night boxing and was prepared to give ABC $9 million in billings for a suitable sports showcase. *Wide World* was threatened with cancellation after 13 weeks but won a reprieve from Tom Moore, then program chief, and established itself soon afterwards. Scherick hired away a young producer from NBC to take charge of the series, and it became the career springboard for Roone Arledge.

Arledge, who remains executive producer of *Wide World of Sports,* rose swiftly to become president of ABC Sports and the most dynamic and successful sports executive in television. The series was also springboard for Jim McKay, its host, and has been a prime showcase for all of ABC's sports commentators. In 1975 ABC created an additional Sunday edition, which did approximately as well as the Saturday edition.

"WILD BILL HICKOK"—early syndicated western series (1952) about the adventures of a U.S. Marshal and his sidekick in the old west. The program featured Guy Madison and Andy Devine. It was revived by ABC (1957–58) when TV's western cycle was in flower.

"WILD KINGDOM"—natural history series actually billed as *Mutual of Omaha's Wild Kingdom,* which began on NBC in January 1963 and ran through the 1970 season when the network dropped it in program cutbacks under the prime time–access rule. Since then, with the same host, Marlin Perkins; the same producer, Don Meier Productions; and the same sponsor, Mutual of Omaha, the series has continued in syndication as a prime-access entry.

"WILD, WILD WEST, THE"—hour-long series about special government agents in the Old West, one of whom specialized in disguises. Produced by Bruce Lansbury Productions for CBS (1965–70), the series featured Robert Conrad and the late Ross Martin. It was canceled not for declining ratings but because government officials were voicing concern about television violence. Having pledged to Congress that CBS would reduce violence, Dr. Frank Stanton, then president of CBS Inc., ordered the program off the schedule.

"WILD, WILD WORLD OF ANIMALS" — half-hour nature series on patterns of animal survival, produced by the BBC in association with Windrose-Dumont-Time, and distributed here for prime time–access syndication in 1973. William Conrad did the narration.

WILEY, RICHARD E.—chairman of the FCC (1974–77), noted for his efficiency, intelligence and pragmatic approach to regulation. Although the commission acted on dozens of controversial and politically sensitive issues during his administration, none of those matters was ever resolved in a way that created drastic changes in conventional broadcasting. A possible exception was "family viewing time" as an answer to the sex and violence question, but technically this was adopted voluntarily by the industry and was not an official FCC action. On leaving the FCC, he joined the Washington law firm of Kirkland and Ellis.

A conservative Republican from Illinois who came to the FCC in 1970 as general counsel, Wiley was named commissioner in 1972 and chairman in March 1974. He maintained an unshakable aversion to government intervention in business and believed that most broadcast problems were best dealt with through industry self-regulation. In his first speech to the industry as chairman, he called for a "new ethic," urging broadcasters to dedicate themselves voluntarily to excellence and public service and to resist the temptation of fraudulent practices. And he immediately took steps to "re-regulate" the industry, doing away with regulations found to be obsolete.

His sympathy toward business and his philosophy of nonintervention permeated his commission, which was made up entirely of Nixon appointees (Robert E. Lee, the sole holdover, having been reappointed to a fourth term by Nixon). Thus, a set of FCC rules sought by ACT to govern children's programming resulted only in a policy statement offering guidelines; the Justice Department push for an FCC policy on media cross-ownership yielded a mild rule that essentially prohibited the creation of *new* newspaper-broadcast combinations and did not break up most existing ones; and a plea by the cable industry for more liberal pay-Cable rules brought only a slight relaxation of the rules. The courts rebuffed Wiley's commission both on the cross-ownership and cable rules.

Wiley's commission declined to impose limits on network reruns, to bar the use of strip programming in prime time–access slots or to juggle station assignments to satisfy New Jersey's plea for a VHF station of its own. Perhaps its most radical action was a reinterpretation of the equal time rule, permitting TV to cover debates between major party candidates if they are bonafide news events—that is, organized by nonbroadcast organizations and presented outside the studios.

But if he was not a reformer as a regulator, he brought reforms on the commission itself, altering the bureaucratic process to permit greater efficiency and to handle more work. A prodigious worker himself, he set a timetable on all matters before the commission and pushed the staff and other commissioners to meet the deadlines. He is believed to have called more *en banc* sessions of the commission than any chairman before him.

Wiley was also adept at bargaining and jawboning, using those skills frequently to achieve consensus votes from the commissioners. He believed near-unanimous votes to carry greater force than the decisions reached by the narrowest 4–3 margin. His jawboning talent, however, involved him in a lawsuit by Hollywood producers in 1976 who charged him with forcing "family viewing time" upon the industry while he was under pressure from three congressional committees to do something to control sex and violence on TV. Wiley maintained that he and his staff merely visited with the heads of the networks, made suggestions on how they might deal with the problem and left the rest to industry self-regulation. See also Cross-Ownership, Family Viewing

Time, Federal Communications Commission, Pay-Cable Rules.

WILLIAM MORRIS AGENCY—largest of the talent agencies, engaged in motion picture and television program packaging and in the representation of performers, writers, directors and producers. Founded in 1898 by William Morris, the agency originally represented talent working in vaudeville and theater. Progressively it included other forms of show business—motion pictures, radio, records, concerts, fairs, night clubs and television. The agency at present operates with 30 agents and receives 10% of the fees for the individuals it represents or 10% of the package (producer, writer, director, stars) it may place with a network.

The agency assists the film studios in selling to the networks shows in which it has an interest.

WILLIAMS, PALMER—legendary veteran of CBS News, who began in 1951 as a key member of the Edward R. Murrow–Fred W. Friendly *See It Now* team and remained to become a producer of the popular *60 Minutes,* a post he has held since the premiere in September 1968. He joined Murrow and Friendly as a newsreel photographer and was credited by them with guiding their transition from radio to the visual medium. Later he became director of operations for *CBS Reports* and then executive producer. He worked on numerous CBS documentaries, including Jay McMullen's famed *Biography of a Bookie Joint.*

WILSON, IRV—v.p. of specials for NBC Entertainment (1979-81) former head of production for Viacom Enterprises. He was executive producer of *The Missiles of October* and *The Harlem Globetrotters Popcorn Machine,* among others.

WINANT, ETHEL—casting director for CBS through most of two decades, regarded the most powerful in her field, until she left in July 1975 to join the Children's Television Workshop as executive producer of a new history series designed for adult viewers, *The Best of Families.* Later she joined NBC as head of casting, resigning in 1981 to join Metromedia Producers Corp. as an independent producer.

After 9 years as director of casting for CBS Hollywood, she became in 1973 vice-president of talent and casting for the network, responsible for casting all network pilots, series and specials and for maintaining the personal relationship between the network and its stars, producers and directors.

With a background in various production jobs in Broadway theater, she entered television in the mid-50s to work on *Studio One;* then became head of casting for Talent Associates, which produced *Armstrong Circle Theater* and *Philco Playhouse;* then casting director of *Playhouse 90* for CBS. She served as associate producer of various TV shows and of four movies produced by John Houseman. In the early 60s, she produced *The Great Adventure,* an American history series, for CBS.

WINCHELL, WALTER (d. 1972)—newspaper gossip columnist and radio personality who probably realized his greatest TV success, after a number of tries, as off-camera narrator of the hit crime series, *The Untouchables.* His first effort in television, on ABC in 1952, was a personalized news program in his inimitable clipped style. It lasted three years. In 1956 he essayed a variety series and the following year hosted a dramatic series, *The Walter Winchell File.* But the reportorial antics that suggested deadline pressure in print and on the radio were betrayed as affectation on camera.

"WINKY DINK AND YOU"—a CBS children's series (1953–57) that encouraged viewer participation. It featured Jack Barry and youngster Harlan Barnard. Winky Dink, an elfin character, was a drawing.

"WINNER TAKE ALL" TENNIS SCANDAL—situation that developed after CBS Sports promoted a series of four tennis specials in the late 70s as "winner take all" competitions when in fact the participants were all guaranteed substantial fees, whether they won or not. The FCC determined in 1978, after a ten-month inquiry, that CBS deliberately attempted to deceive its viewers for the sake of ratings. It also found the network in violation of its sponsor-identification rules for giving plugs to the luxury hotels hosting the matches without noting on the air that these were in exchange for a variety of special considerations by the hotels.

The matches were presented on television under the umbrella title of *The World Heavyweight Championship of Tennis,* and all four featured Jimmy Connors.

The scandal apparently precipitated the departures of Robert Wussler and Barry Frank, who were president and v.p. of CBS Sports, respectively. Meanwhile, CBS was censured by the FCC and punished with a short-term (one year) license renewal for its Los Angeles o&o, KNXT.

WINTHER, JORN H.—producer-director of comedy-variety series and specials, including *The Phyllis Diller Show, Candid Camera, The Jonathan Winters Show, The Story Theatre* syndicated series, *Hollywood Palace, Shindig* and *California Jam.*

WISEMAN, FREDERICK—filmmaker specializing in *cinéma vérité* documentaries on American institutions; virtually all of them were made for public TV. Indeed, during the early 70s, the presentation of the latest Wiseman work conventionally marked the opening of a new PBS season.

A lawyer turned filmmaker in 1967, Wiseman produces, directs and edits all his documentaries, working with a cameraman. His works, which often run two hours or longer, depict the daily routines of the institutions being examined and employ no narration to explain or interpret the events. His social analysis is therefore implicit.

Titicut Follies, a controversial film shot at the Massachusetts State Hospital for the Criminally Insane, was his first effort. It was followed by *High School* (1968), *Law and Order* (1969), *Hospital* (1970), *Basic Training* (1971), and then in successive years *Essene, Juvenile Court, Primate, Welfare* and *Meat,* the last five commissioned by WNET New York.

WITT, PAUL JUNGER—producer whose credits include *The Partridge Family, The Rookies* and the TV movie, *Brian's Song.* As head of Production for Danny Thomas Productions, he was also executive producer of the situation comedies, *Fay* and *The Practice.* As partner in a new independent company formed in 1977, he was coproducer of the controversial ABC comedy, *Soap,* and the spinoff, *Benson.*

WITTING, CHRIS J.—a leading figure in television in the early years as managing director of the DuMont Network (1950–53) when it owned three stations and had 65 affiliates. Witting is credited with having brought Sid Caesar, Imogene Coca, Jackie Gleason, Dennis James and Bishop Sheen to network TV. He joined DuMont in 1947 and left to become president of Westinghouse Broadcasting Co. In 1955, he became v.p. and general manager of the Westinghouse consumer products division, later went to ITT and in 1965 became president and chief executive officer of the Crouse-Hinds Co., an electrical equipment manufacturer.

"WIZARD OF OZ, THE"—in certain respects, the most successful single program in TV history, a 1939 theatrical movie that was presented 18 times between 1956 and 1976 and never received a rating lower than 20.2. The MGM film, based on L. Frank Baum's book and starring Judy Garland, has passed between CBS and NBC, earning millions with each transaction for long-term licensing permitting one play a year. There being no reason to assume after two decades of exposure that *Oz* will be any less an attraction in the future, the film bids fair to become the first on TV to amass an accumulated U.S. audience of one billion.

The film made its TV debut Nov. 3, 1956, on CBS's *Ford Star Jubilee,* airing 9–11 P.M., too late for very young children. Yet it scored a 33.9 rating and 53 share. CBS offered it again three years later, after which it has had an unbroken string of annual showings. The second outing, at 6 P.M. on a Sunday, garnered a 36.5 rating and 58 share. For the initial nine telecasts on CBS—from 1956 to 1967—*The Wizard of Oz* averaged an incredible 53 share.

NBC then acquired the film, and for the next eight years it averaged a 40 share. CBS regained *Oz* in 1976, scoring a 42 share with a showing in March. On all but three occasions, the movie was scheduled on a Sunday evening, either at 6 or at 7 P.M.

In the initial sale of the film to television, MGM was paid $250,000 a showing. The value of the rights has increased steadily, and CBS's five-year purchase in 1976 was at the rate of $800,000 a showing.

Jack Haley, Judy Garland & Ray Bolger in The Wizard of Oz

"WKRP IN CINCINNATI"—CBS sitcom introduced twice during the 1978–79 season, the second time successfully. The program, about the people working in a small Top 40 radio station, opened the season as a companion to another new series, *People*. After a few weeks, CBS canceled *People* and put *WKRP* on the shelf, having decided that the program needed a stronger lead-in to succeed. It was returned at midseason in the hammock slot between *M*A*S*H* and *Lou Grant* and began to flourish. During the summer of 1979, the *WKRP* reruns went to the top of the Nielsen ratings.

Produced by MTM Enterprises, and resembling in style and concept MTM's *The Mary Tyler Moore Show*, it features Howard Hesseman, Gary Sandy, Gordon Jump, Sylvia Sydney, Richard Sanders and Loni Anderson.

WLBT LOSS OF LICENSE [United Church of Christ *v.* FCC/ 359 F2d 994 (D.C. Cir. 1966)/ 425 F2d 543 (D.C.

Cir. 1969)]—case in which Lamar Life Insurance Co. was denied license renewal for WLBT Jackson, Miss., in 1969 when citizens groups proved that the station used its air time to promote a segregationist philosophy and in general denied the use of its airwaves to blacks and proponents of the civil rights cause.

Apart from the Fairness Doctrine considerations, the station was in violation of its license pledge to serve the needs, interests and convenience of its community, since around 40% of the Jackson population was black. This was the first time a TV license was lost on substantive grounds, but it was brought about by an action of the court rather than by the FCC.

The FCC had repeatedly warned the station during the 60s that it was required under the Fairness Doctrine to inform black leaders when they or their causes were attacked. Eventually, the Office of Communication of the United Church of Christ decided to bring a test case that would force access to broadcasting for blacks in the South.

After monitoring the station to accumulate evidence, the UCC, along with some local residents, filed a petition in 1964 to deny renewal of the WLBT license. The petition claimed that WLBT discriminated against blacks in its commercials and entertainment programming and was unfair in its presentation of controversial issues. The commission held, in 1965, that the UCC and other petitioners had no legal standing to intervene. While recognizing the seriousness of the allegations, the FCC, with two members dissenting, concluded that WLBT was entitled to a one-year renewal on the condition that the station comply strictly with the established requirements of the Fairness Doctrine.

The UCC then appealed to the Court of Appeals for the District of Columbia. In 1966 a three-judge panel unanimously ruled that the UCC and citizens groups did have standing because they were consumers with a genuine and legitimate interest. This proved a landmark decision, making it possible for private groups around the country to file petitions to deny against local stations.

The Court of Appeals also held that the grant of a one-year renewal to WLBT was in error, and it directed the FCC to conduct hearings on the station's renewal application. During the next two years, while the FCC was investigating WLBT, the station tried to improve its record. In June 1968, the FCC granted WLBT a full three-year renewal (again, two members dissented).

Again the UCC appealed to the court, and again the court reversed the FCC, vacating the grant of the license and directing the commission to invite applications for the license. The court held that the commission's conclusion had not been supported by substantial evidence. Lamar Life then asked the Supreme Court to grant certiorari, and the request was denied.

WLBT did not go off the air when the license was lost in 1969 but continued to operate under supervision of a biracial, nonprofit caretaker group, Communications Improvements Inc., while the FCC considered five new applicants for the license, one of which was the original owner, Lamar Life. Lamar meanwhile, was paid a monthly fee of $35,000 by the operating group for the use of its facilities and equipment.

The case had the effect of propelling the broadcast reform movement by establishing the right of citizens groups to have their views made part of the official license-renewal proceedings. It also proved the effectiveness of the petition to deny as an instrument for change.

See also Office of Communication of the United Church of Christ; Broadcast Reform Movement; Dilday, William Jr.

WNTA—the commercial station which preceded WNET, the public TV station, on Channel 13 in New York. Assigned in 1952 to Newark, N.J., it was originally operated by a TV syndication concern, National Telefilm Associates, and was noted for the *Play of the Week* series which it produced and its parent company sold nationally. Having run up losses of $3 million, it was sold in 1961 to Educational Broadcasting Corp. to become the New York metropolitan area's first educational TV station, WNDT. In 1970, when WNDT merged with National Educational Television, the station became WNET.

WOLD CO., THE ROBERT—Los Angeles-based company formed in 1970 which specializes in setting up part-time regional and national networks for clients through the wholesale leasing of satellite time and AT&T land lines. Wold is involved in the transmission of 6,000 individual broadcast programs annually, including the regional distribution of sports telecasts and the live distribution of programs to Hawaii. Its clients include the major networks. In 1979, Wold produced a live teleconference between Brazil, Japan, Great Britain, and the U.S. for the World Soybean Conference in Atlanta through an interconnection of international satellites. A subsidiary, Satellink of America, arranges for temporary satellite use by providing clients with portable earth stations and other technical facilities.

WOLFE, JAMES D.—head writer for *That Was the Week That Was* and *The Bill Cosby Show* and comedy consultant for *Laugh-In.* He was also producer-director and writer of *Opryland, USA* and served in a creative capacity on numerous specials.

WOLFF, PERRY (SKEE)—veteran writer and producer of documentaries for CBS News, who, in a career that began in 1947, produced more than 200 hours of documentaries for the network. He was named executive producer of CBS News specials in 1976.

Wolff's credits as producer include *The Selling of the Pentagon, Voices from the Russian Underground, You and the Commercial, The Italians, The Japanese* and *The Israelis.* He also wrote and produced a 1961 news special on the building of the Berlin Wall and in 1962 *A Tour of the White House with Mrs. John F. Kennedy.*

He came to CBS from its Chicago o&o, WBBM Radio, in 1947, where he worked in the dual capacity of music director and investigative reporter. He transferred to the network in the early 50s and created the public affairs series *Adventure* with the American Museum of Natural History. In 1956 he took over the documentary series, *Air Power.*

WOLFF, SANFORD I. (BUD)—chief executive of the American Federation of Television and Radio Artists since February 1968, with the title of executive secretary. He is responsible for administering AFTRA's national office and negotiating the union contracts for the 31,000 AFTRA members working in TV, radio, phonograph recordings and slide films. He is also a board member of the Council of AFL-CIO Unions for Professional Employees and executive secretary of the Associated Actors and Artistes of America. Before joining the union as its top executive, Wolff was a lawyer in Chicago representing the Chicago local of AFTRA, the Screen Actors Guild and the Directors Guild, along with a number of performers.

WOLPER, DAVID L.—independent producer whose company, Wolper Productions, specialized in documentaries and other nonfiction programs and launched the careers of numerous young film directors and producers. Wolper succeeded better than any other independent at getting around network policies requiring documentaries to be produced by their own news divisions.

But in the 1970s, frustrated by the resistance of the networks to outside documentaries, he switched fields and began producing entertainment programs. With James Komack, he produced the situation comedies *Welcome Back, Kotter* and *Chico and the Man,* both successful. He hit the jackpot, however, when he purchased the TV rights to Alex Haley's *Roots,* reportedly for $250,000, and produced the 12-hour miniseries that became, in January 1977, the highest-rated program in TV history. He also produced the sequel *Roots: The Next Generations,* a big rating hit in 1979.

In the documentary field, Wolper broke the network barrier initially with *Race For Space* (1958) and followed with numerous others: *The Rafer Johnson Story* (1960), *Biography* (1962), *The Story of Hollywood and the Stars* (1963), *Escape to Freedom* (1963), *The Rise and Fall of American Communism* (1963) and *The Legend of Marilyn Monroe* (1964).

One of Wolper's tactics was to secure sponsorship before taking a program to the networks. With two sponsors in tow, he placed the National Geographic specials—four a year—with CBS in 1964. He also produced the quadrennial series, *The Making of a President,* based on the Theodore H. White books; *The Undersea World of Jacques Cousteau;* the *Smithsonian* series; and a number of theatrical movies. See also *Roots.*

"WONDER WOMAN"—a "floating" series while on ABC (1974–77) but given a regular berth by CBS, which acquired it for the fall of 1977. The series was piloted for ABC in 1974 and then was re-cast and piloted again in 90-minute form. Although it drew encouraging ratings wherever ABC chose to play it, the series continued to hold bench-warmer status with the network, which ordered only a limited number of episodes each season. CBS won it with an offer of a weekly slot.

Based on a super-heroine of comic books, the series featured Lynda Carter, glamorous and shapely, in the title role. Lyle Waggoner, Richard Eastham and Beatrice Colen were featured. *Wonder Woman* was produced by Warner Bros. TV and the Douglas S. Cramer Co., with Cramer as executive producer and W. L. Baumes as producer.

Walt Disney & friend

"WONDERFUL WORLD OF DISNEY, THE"—sturdy Sunday evening family series on NBC (since 1961), unlike any other series on TV in that it offers a varied menu of nature films, animated cartoons, dramas, comedies, serials and movies, making it almost a network unto itself. Virtually all the Disney forms have succeeded, and the series—which has gone by other titles, such as *Walt Disney's Wonderful World of Color* and, while it was on ABC (1954–61), *Disneyland*—has never seriously been threatened by compet-

ing programs on the other networks. Meanwhile, it spawned such successful subseries as *Davy Crockett, Elfego Baca* and *Zorro.* It was also an expandable series, lending itself to two-hour presentations when that served NBC's purposes.

The historical significance of the series is that, under its original title, *Disneyland,* it provided the opening wedge in 1954 for the entry of the major Hollywood studios into TV production.

After the 1980-81 season, NBC dropped the long running series, which had been having trouble as competition to CBS's *60 Minutes* on Sunday nights at 7, but CBS immediately signed the Disney Studio to supply it with specials during 1981-82. This deal soon turned into the Walt Disney series on Saturdays at 8 P.M., which did quite well in the ratings even though the programming mix was not that different from the mix used on *Disney's Wonderful World,* which had been its title during the last two years on NBC.

Robert D. Wood

WOOD, ROBERT D.—president of CBS-TV (1969–76), the executive chiefly responsible for the bolder direction taken by television in the decade of the 70s. He resigned early in 1976, saying he could no longer endure the pressures of the job, and moved to Hollywood as an independent producer in a new company financed by CBS. In 1979, he was named president of Metromedia Producers Corp.

Wood was an effective network president, noted for his candor and his willingness to take chances. His first significant action, a year after he took office, was to rebuild the prime-time schedule—even though CBS was first in the ratings—with entertainment programs keyed to the times and capable of dealing with mature themes. He startled the industry by discarding a number of programs that were still popular—those of Jackie Gleason, Red Skelton, Ed Sullivan and Andy Griffith, along with *Beverly Hillbillies, Petticoat Junction, Green Acres* and *Hee-Haw*— and replaced those programs, all of distinct rural appeal, with new urban-oriented shows.

Although most of the new 1970 series that purported to what Wood termed "contemporary relevance" were failures, CBS managed to hold onto its ratings leadership. And, as Wood had intended, the network improved its audience profile with respect to age levels, attracting more viewers in the 18–49 year age range than it had previously. Those were the young adults advertisers were most eager to reach.

Wood's most courageous—and, as it proved, most significant—decision was made at midseason in 1970–71, when he purchased for the schedule *All In the Family,* after ABC had twice rejected its pilots. The series, whose central character was a bigot, opened the way to the new era of outspokenness and permissiveness in prime-time programming. It was, of course, a prodigious hit and led to other programs that quickly put many of the long-standing taboos of broadcasting to rest.

Until he became president of the network, where he displayed a knack for programming, Wood had spent his entire career at CBS in station management and sales. He joined KNX, the CBS radio station in Los Angeles, following graduation from USC in 1949, and shortly afterwards moved to the TV station there in a sales capacity. In 1955 he became general sales manager of KNXT (TV) and five years later v.p. and general manager of the station. He moved to New York in 1966 as executive v.p. of the CBS Television Stations and the following year became president of that division. Two years later he became head of the network. Wood held the office longer than any previous president.

WOODS, MARK—early president of ABC after it was formed from NBC's Blue Network. He was elected chairman in 1950, when Robert E. Kintner became president, and became vice-chairman in 1953 in the merger with United Paramount Theatres. He resigned soon afterwards.

"WORLD AT WAR, THE"—series of 26 documentary films produced by Thames TV of Britain and narrated by Sir Laurence Olivier, covering critical episodes of World War II. It was syndicated in the U.S. in 1973.

"WORLD NEWS TONIGHT"—the ABC evening newscast developed in the Roone Arledge administration, under Av Westin's supervision, which made its debut in July 1978. As the successor to the floundering *ABC Evening News with Harry Reasoner and Barbara Walters,* a program whose problems included a widely publicized feud between the two anchors, the new entry had as its distinguishing feature a three-anchor format that involved switching among three cities—Washington, Chicago and London. The anchors

were Frank Reynolds in the capital, Max Robinson in the heartlands and Peter Jennings in Europe.

With Reasoner gone back to CBS, Miss Walters represented the New York contingent as a contributor of special reports and interviews. Howard K. Smith was also part of the regular cast as commentator, although he resigned in the spring of 1979 on learning that his segment was being eliminated.

The "whiparound" system, with its electronic gimmickry and profusion of introductions that consumed precious news time, was jeered at by critics and other broadcast journalists. But after less than a year, the newscast settled into a more orthodox mold with Reynolds emerging as the dominant figure and principal anchorman. By spring of 1979, the ratings began to inch upwards to where they eventually challenged NBC's and, on occasion, even pulled ahead. In part, the gains were a reflection of ABC's growing popularity in entertainment programming and its improved affiliate lineup, resulting from the spiriting away of major affiliates from the other networks. But it was clear also that ABC News had gained credibility with viewers and that the changes in the newscast were effective.

"WORLD OF SURVIVAL, THE"—syndicated animal documentary series produced in England by Survival-Anglia Ltd. in association with the World Wildlife Fund. With John Forsythe hosting the American version, the series was bartered for prime-access time periods by the J. Walter Thompson agency. It began in 1971.

"WORLD PREMIERE" MOVIES—NBC's umbrella title for two-hour features made expressly for television by Universal TV during the late 60s when theatrical motion pictures suitable for television were in short supply. The films, most of which were produced for $800,000, were essentially standard TV entertainment but were introduced—successfully at first—as movie premieres. ABC later was to beat NBC at this game by scheduling 90-minute TV movies, called *The Movie of the Week,* produced by a variety of companies for half the price of the NBC films. In 1974 NBC made *The World Premiere Movie* a weekly 90-minute series, like ABC's, but returned the following year to the occasional two-hour type, shuffled in with theatrical movies.

"WORLD PRESS"—PBS discussion series on international affairs originating at KQED San Francisco since 1969. Moderated by John Boas, the half-hour series featured such journalists as Paul Zinner, Leslie Lipson, Chalmers Johnson, Michael Nabti, Gerald Feldman and Maurice Jonas. Zev Putterman is executive producer.

WORLDVISION—syndication company formed in 1972 from what had been ABC Films, when ABC was forced to divest its syndication division by FCC rules. ABC sold the division, reportedly for around $10 million, to a group formed by its own syndication executives, Kevin O'Sullivan, Jerry Smith, Neil Delman, Colin Campbell and Howard Lloyd. O'Sullivan, who had been president of ABC Films, became president of Worldvision, an international distribution company which, on its own, financed the production of a number of syndication properties. The division was sold off under FCC rules requiring the networks to refrain from engaging in program syndication by June 1, 1973. In 1979, Worldvision was acquired by Taft Broadcasting, under which it continues to operate as a subsidiary.

WPIX LICENSE CHALLENGE—stormy ten-year contest for the license to operate Channel 11 in New York; it was resolved finally in May 1979 in an out-of-court settlement that secured the license for the incumbent, WPIX Inc., owned by the *New York Daily News*. The settlement involved the purchase by WPIX of all the shares in Forum Inc., the company formed expressly to file for the license, for approximately $10 million.

The proceeding, which took a number of twists and turns in its course, was begun in May 1969 after reports appeared in the press that WPIX-TV had engaged in a number of news deceptions, such as superimposing the words "via satellite" over stock military training footage and mislabeling the origins of other news reports to suggest that they were eyewitness accounts. These abuses seemed serious enough to make the license vulnerable to a challenge, and Forum was created with a widely representative group from the locale, including singer Harry Belafonte. The president of Forum was Lawrence K. Grossman, a former NBC v.p. who headed his own advertising agency, but he withdrew from Forum when he was named president of PBS.

At various points, each side scored victories that were subsequently overturned. In 1975, the competing parties worked out a settlement, but the FCC disallowed it on the ground that the public benefits to be derived from it were vague. In June 1978, the FCC voted 4-3 to renew the WPIX license, but the three who voted nay—FCC chairman Charles W. Ferris and commissioners Joseph Fogarty and Tyrone Brown, all lawyers—filed a stinging 83-page dissent that formed the basis for a court appeal by Forum. The dissenters argued that offenses that occur during a three-year license period cannot be forgiven by the positive record of a station before or after that period. Legally, they said, everything was extraneous but the station's performance during the license period in question.

Eleven months later Forum's appeal was withdrawn and the case closed by the settlement. The FCC could not oppose the payment of cash to be rid of a challenger because the matter was no longer in its jurisdiction.

The entire proceeding involved hundreds of hours of hearings and millions of dollars in legal fees, and early on it cost Fred Thrower his job as president of the station.

WRATHER-ALVAREZ BROADCASTING *v.* FCC [248 F2d 646 (D.C. Cir. 1957)]—case in which the Court of Appeals held that a U. S. network may affiliate with a foreign television station but that the FCC must first consider whether the station's locally originated programming merits such affiliation.

ABC had petitioned the FCC for permission to supply its programs to XETV in Tijuana, Mexico, a station that transmits a city-grade signal to San Diego, Calif. Since San Diego had only two local stations, and they carried CBS and NBC, the commission deemed it in the public interest for residents of San Diego to receive the third network. Over the objections of the two San Diego stations, the commission granted ABC's petition.

Wrather-Alvarez Broadcasting, licensee of one of the San Diego stations, appealed the case. The D. C. Court of Appeals held that there was nothing inherently wrong in permitting a foreign station transmitting into the U. S. to affiliate with an American network. The fact that this foreign station might have an advantage over nearby American stations because of its freedom from American taxes and regulatory schemes was of no moment.

However, the court said, since network affiliation would immediately make XETV's programming more prominent, the FCC should have determined whether the added exposure of the local programming was in the public interest. The court remanded the case to the commission for a determination of the local program issue but held that until decided otherwise ABC could feed its programs to the Mexican station.

WRATHER, JACK—president of Wrather TV Productions (*Lassie, The Lone Ranger, Sgt. Preston of the Yukon*) and of Wrather-Alvarez Broadcasting Co., a station group (KFMB-TV San Diego, and others). In the late 60s he was appointed to the board of the Corporation for Public Broadcasting by President Nixon. Wrather was also one of the founders in the 50s of two prominent syndication companies, Television Programs of America and Independent Television Corp. (ITC), the latter now owned by Britain's ATV, as its American arm.

An oil producer before he entered movie and TV production, Wrather's many interests include several hotels in the west, including the Disneyland Hotel, and he is chairman of Muzak Inc. His wife, the actress Bonita Granville, served as producer of the *Lassie* series through its long run on CBS and later in syndication.

WRITERS' GUILD OF AMERICA (EAST, WEST)—
union formed in 1954 to represent writers in TV, radio and movies as a synthesis of the old Screen Writers', Radio Writers' and Dramatists guilds, which had operated as branches of the Authors' League. With a total membership of more than 4,000, the guild is made up of two units, Writers' Guild of America West (WGAW) and Writers' Guild of America East (WGAE), determined geographically by the Mississippi River. Each has its own council and its own branches for TV-radio and screen, but the two cooperate closely.

Writers' Guild broadcast contracts cover staff writers, news writers and freelancers. They do not prohibit nonmembers from selling a first script, but those new writers are required to apply for membership within 30 days of script acceptance. In addition to establishing minimum fees and residual payments for writing, the guild specifies how writers' credits must appear on the screen and provides for his/her artistic rights.

The Writers Guild staged a fairly successful 13-week strike against the networks in the spring of 1981 to gain residual concessions for disc and cassette uses of television material.

WRYE, DONALD—director whose credits include *Born Innocent, The Man Who Could Talk to Kids* and *Death Be Not Proud*. He also wrote and produced the latter.

Robert Wussler

WUSSLER, ROBERT—executive vice president of Turner Broadcasting System since 1980. Earlier he was president of

CBS-TV (1976–77) and twice head of CBS Sports who resigned in March 1978 to become an independent producer. Wussler's rapid rise in the executive echelons of the network, after 15 years with CBS News, came to an abrupt end with the massive reorganization of CBS in November 1977 prompted by the network's decline in the ratings. In the reorganization, Wussler was downgraded to president of CBS Sports, and he left several months later in the wake of the "winner take all" tennis scandal, for which CBS was reprimanded by the FCC.

On his own, Wussler created ad hoc networks for several major events, including the live telecast by satellite of the Cannes Film Festival and the 1979 pre-Olympics summer series from Moscow, *Spartacade*.

In a 20-year period at CBS, Wussler rose from mailroom clerk to president of the network. He was groomed for the presidency by John A. Schneider, then president of the CBS Broadcast Group, who was impressed with the executive abilities and production savvy Wussler displayed at CBS News as director of special events (space flights, political conventions, state funerals, the coverage of assassinations, etc.). Schneider lifted him from the news division and assigned him to WBBM-TV Chicago as v.p. and general manager, at a time when the station was at a low point with the audience. After two years, the station was back on the upswing, and Wussler was rewarded with a promotion as v.p. in charge of CBS Sports, although he had little background in that field. Less than two years later, he was named president of CBS-TV when Robert D. Wood resigned.

WXUR CASE [Brandywine-Main Line Radio Inc. v. FCC/473 F2d 16 (D.C. Cir. 1972)]—case notable for the fact that the FCC revoked a license for the first time because of Fairness Doctrine implications. The station involved was WXUR in Media, Pa. (a small town near Philadelphia), owned by Dr. Carl McIntire, head of the Faith Theological Seminary, a fundamentalist religious organization.

McIntire purchased the radio station in 1964, promising to operate in accordance with the NAB code and to comply with the Fairness Doctrine. Eighteen organizations opposed the purchase because of alleged attacks by McIntire on other religious organizations, governmental agencies and political figures. In 1965 the FCC, with one dissent, permitted the transfer of license on the ground that Faith Theological Seminary guaranteed it would provide reasonable opportunity for presentation of contrasting viewpoints on controversial issues.

A year later McIntire filed for a normal three-year license. The groups that had opposed the transfer filed against the renewal, charging that WXUR had been one-sided, unbalanced and weighted on the side of extreme right-wing radicalism. Various other religious and civic organizations attacked McIntire, and the Pennsylvania House of Representatives passed a resolution calling for an FCC investigation into its extremist views.

The FCC ordered that the renewal application be denied because the licensee had not afforded reasonable time for presentation of contrasting views on controversial issues of public importance, that it ignored the personal attack rule and that the licensee misrepresented the manner in which the station would be operated.

A petition for reconsideration was filed which claimed that the FCC had applied the Fairness Doctrine in a manner which violated the constitutional protections of the First Amendment. When this petition was denied by the FCC, McIntire appealed to the D.C. Court of Appeals.

In the fall of 1972 the Court of Appeals affirmed the FCC's right to terminate WXUR's permit on the general ground that WXUR knowingly misrepresented itself in its transfer application and deliberately deceived the commission and the people of the Philadelphia area by not adhering to the promise of fairness.

An important aspect of the case was the dissent by Chief Judge David L. Bazelon, who wrote that this application of the Fairness Doctrine did infringe on the First Amendment rights of broadcasters. Bazelon argued that the decision imposed a double standard on broadcasters and print journalism, and he stated that the "democratic reliance on a truly informed American public is threatened if the overall effect of the Fairness Doctrine is the very censorship of controversy which it was promulgated to overcome."

After a refusal by the Supreme Court to hear McIntire's appeal, WXUR ceased operation in July 1973. When it went off the air, McIntire transferred broadcast operations to a "pirate ship" off the New Jersey coast. This operation was stopped by an order from a federal court. See also Fairness Doctrine.

WYLIE, MAX (d. 1975)—writer who created the situation comedy *The Flying Nun,* and earlier wrote and produced shows for *Omnibus, The March of Time* and *Wide, Wide World,* among others. In radio days (1933–40) he worked for CBS as script director, producer and writer, then worked for Lennen & Newell Advertising in TV programming.

WYMAN, THOMAS H.—president and chief executive officer of CBS Inc. since May 1980, and thus heir-apparent to the chairmanship long held by William S. Paley. Recruited by Paley from the Pillsbury Co., where he was vice chairman, Wyman succeeded John D. Backe, who was suddenly dismissed after it had appeared that Paley had placed the future of the company in his hands. The firing of Backe took the industry and the financial community by surprise, but equally surprising was the selection of Wyman for the post, since his background was packaged foods (Nestle's, Green Giant, Pillsbury) and not in media. He seemed an odd choice to lead CBS into the promising but treacherous new age of telecommunications. Wyman, however, has the qualities Paley admired: steeped in business and finance but also cultured, he is worldly, but also a midwesterner; he is tall, handsome, intelligent, dignified, personable, and a good executive. He would learn about broadcasting and media. Wyman was 50 at the time of his appointment.

WYNN, ED (d. 1966)—venerable vaudevillian who was prominent on TV in the 50s, first as star of a variety show and later as a lead in a family situation comedy. He also played serious dramatic roles, notably on *Playhouse 90* in *Requiem for a Heavyweight,* and made guest appearances on scores of shows.

Billed in vaudeville as "The Perfect Fool," he was a visual comedian who served the new medium well in his initial series on CBS, *The Ed Wynn Show* (1949–50). When that concluded, he appeared on a rotating basis in NBC's *Four Star Revue* (1950–51). He also starred in *The Pet Milk All Star Revue* and appeared on *Shower of Stars* and *Hallmark Hall of Fame.*

The Ed Wynn Show, produced by Screen Gems in 1958, was a family situation comedy in which Wynn portrayed the grandfather. It had a brief run. In September 1957 *Texaco Command Performance* on NBC saluted Wynn for 55 years in show business.

WYNN, TRACY KEENAN—TV writer (son of actor Keenan Wynn) whose credits include *The Autobiography of Miss Jane Pittman, Tribes* and *The Glass House*—all powerful and widely praised TV dramas on film. He also wrote the pilot for the 1976 western series, *The Quest,* and made his directing debut with a 1975 TV movie, *Hit Lady.*

Y

YALE BROADCASTING CO. *v.* **FCC [478 F2d 594 (D.C. Cir. 1973)/cert. den. 414 U.S. 914 (1973)]**—case in which the Court of Appeals upheld the right of the FCC to require management-level personnel of radio stations to be aware of the meanings of drug-related recorded music.

Having become aware in the early 70s that numerous popular songs glorified or promoted the use of drugs, couching their messages in the esoteric language of the drug culture, the FCC issued a Notice advising management-level personnel to make "reasonable efforts" to determine the meaning of the lyrics of popular recordings. The commission subsequently clarified the Notice by issuing a Memorandum stating that it was not prohibiting the playing of any records and that no reprisals would be taken against stations that played "drug-oriented" music but that it was still necessary for a station to know the content of the records played and make a judgment regarding the wisdom of playing such records.

The Yale Broadcasting Co. brought suit to challenge the Notice and Memorandum as an unconstitutional infringement upon free speech. The D.C. Court of Appeals rejected the argument, noting that to require a licensee to have knowledge of its program content was merely a form of reminder to act within the public interest standard. There was no burden upon free speech, the court said, because licensees were not required to pre-screen all programming material, and even if they were, such a burden would not be overwhelming.

Chief Judge Bazelon, who had not heard the case, proposed that the entire Circuit hear the case en banc; when the other judges rejected his proposal, Judge Bazelon wrote a dissent. He argued that the record clearly indicated that broadcasters had every reason to interpret the Notice as a prohibition of certain types of songs, and this, he said, amounted to an unconstitutional censorship. He cited also congressional testimony given by the chairman of the FCC in which he explicitly stated that he would vote to deny the renewal application of any licensee who played songs containing lyrics that promoted drug use.

"YANKI NO!"—a notable hour news documentary of the *Bell & Howell Close-Up* series on ABC which aired in 1960. Produced for ABC News by an outside company, Robert Drew Associates, it examined anti-American sentiment in Cuba and other Latin American countries and revealed the widening gulf in relations between the U.S. and its neighbors. Joseph Julian narrated and William Worthy and Quinera King were reporters. Drew was executive producer and the filmmakers were Richard Leacock, Albert Maysles and Donn Pennebaker.

YATES, TED (d. 1967)—producer and correspondent for NBC News who died of gunshot wounds while covering the Middle East War.

"YOGI BEAR"—popular cartoon character created for TV by Hanna-Barbera in 1958. Initially, the series of 30-minute cartoons were sold to Kelloggs', which placed them in markets of its choosing in early evening time. The series went to

ABC in 1961 for several seasons and has been in syndication ever since. In 1973 ABC began a new series on Saturday mornings, *Yogi's Gang.*

YORKIN, BUD—partner with Norman Lear in Tandem Productions, the company which broke new ground in TV in the early 70s with such explosive hits as *All In the Family, Sanford and Son, Maude* and their several spin-offs. Yorkin was the executive producer of *Sanford* from its inception and also of the short-lived spin-off, *Grady.* He was also executive creative consultant to the *What's Happening!* series.

Before teaming with Lear to concentrate on making movies in the mid-60s, Yorkin had made his reputation in TV as a producer-director of variety specials, such as *Another Evening with Fred Astaire,* and of the Tony Martin and Tennessee Ernie Ford weekly series. He and Lear returned to TV after securing the rights to produce American versions of the controversial but popular British series, *Till Death Us Do Part* and *Steptoe and Son;* they yielded *All In the Family* and *Sanford and Son,* respectively.

YORKSHIRE TELEVISION—British commercial station for the Yorkshire region, with its chief production center in Leeds. Like Tyne Tees Television, the company is a subsidiary of Trident Television Ltd.

You Are There: Geraldine Brooks as Amelia Earhart

"YOU ARE THERE"—CBS public affairs series (1954–57) and revived (1971–72) which presented dramatic re-creations of history as news events, using contemporary TV reporting techniques upon such episodes as the final hours of Joan of Arc, the death of John Dillinger and the Boston Tea Party. Walter Cronkite served as anchorman, cutting away to various correspondents on the scene, as it were. Burton

Benjamin was executive producer and Vera Diamond producer of the most recent series of half-hours.

"YOU ASKED FOR IT"—half-hour novelty series (1955–59) which involved unusual people performing unusual stunts, with Smiling Jack Smith and Art Baker as emcees. It was revived in syndication (1971–73) by Markedon Production, with Smith hosting 56 new episodes.

It was revived again in the 1981-82 season as *The New Your Asked for It,* hosted by Rich Little and syndicated by Sandy Frank.

Groucho Marx & George Fenneman

"YOU BET YOUR LIFE"—quiz show offering modest prizes which really functioned as a vehicle for the specialized comedy of Groucho Marx. It provided him opportunities for his patented insults, wisecracks and witticisms. The show had a long run on NBC (1951–61) and in later years changed its title to *The Groucho Show.* Selected filmed episodes of the series were later syndicated as *The Best of Groucho* and aired in certain larger markets as recently as 1976. The series was by John Guedel Productions.

"YOU'LL NEVER GET RICH"—see Silvers, Phil.

"YOUNG AND THE RESTLESS, THE"—CBS daytime serial dealing often with contemporary issues and the problems of modern youth, as well as with the standard romantic snarls common to the genre. It premiered March 26, 1973, and its success was a factor in the network's promotion of Donald (Bud) Grant from daytime program v.p. to program chief of CBS-TV.

Featured in the cast are Robert Colbert, Jaime Lyn Bauer, Eric Braeden, Peter Brown, Dennis Cole, Jeanne Cooper, Michael Damian, Steven Ford, Brett Hadley, David Hasselhoff, Tom Ligon, and Howard McGillin.

YOUNG, COLLIER—a creative force in the early years of TV as producer of the *Alcoa Presents* drama series. Later, in the 60s, he produced the series *The Rogues,* whose ratings did not reflect its critical acclaim.

"YOUNG LAWYERS, THE"—hour-long series on ABC (1970) about a prominent Boston attorney working with young law students at a legal aid society. It starred Lee J. Cobb and featured Judy Pace and Zalman King; Paramount TV produced. The series ran 26 weeks.

"YOUNG PEOPLE'S CONCERTS"—CBS series of music education programs with the New York Philharmonic, which began in 1958 while Leonard Bernstein was conductor of the orchestra. Bernstein's exceptional ability to explain the intricacies of music and the purposes of the composers gave *Concerts* wide appeal and helped make it a long-running series. Michael Tilson Thomas succeeded Bernstein in the early 70s. Roger Englander was producer-director.

YOUNG, ROBERT—former movie star who entered TV in 1954 with a continuation of his radio situation comedy *Father Knows Best,* and began a second career in the medium in 1969 as star of *Marcus Welby, M.D.,* an hour-long medical melodrama. *Father* ran six years on TV and *Welby* seven. For one season, in between, he starred in *Window on Main Street* and then did guest shots in other dramatic series.

"YOUR HIT PARADE"—weekly series carried over from network radio spotlighting the seven most popular songs of the week, according to trade surveys, with a buildup to the song that was No. 1. Sponsored by Lucky Strike cigarettes, it ran 9 seasons on TV, starting in 1950. Along with the top tunes, the shows featured "extras," which usually were production numbers staged by Tony Charmoli initially, and then by Ernie Flatt and Peter Gennaro.

For the first two seasons, the cast consisted of Eileen Wilson, Dorothy Collins and Snooky Lanson. Two years later, Lanson and Miss Collins were joined by June Valli and Russell Arms, with Raymond Scott as music director. Gisele MacKenzie joined in 1953, replacing Miss Valli. In 1957 there was a complete change of cast: Jill Corey, Virginia Gibson, Tommy Leonetti and Alan Copeland. The show was canceled in 1958 but was mounted again the following year with Dorothy Collins and Johnny Desmond, running only a few months. It was considered to be a victim of the new trend in pop music, rock.

Howard Morris, Sid Caesar, Imogene Coca & Carl Reiner

Your Hit Parade

"YOUR SHOW OF SHOWS"—historic and remarkable 90-minute live Saturday night comedy series on NBC (1949–54) which starred Sid Caesar and Imogene Coca, featured numerous others who went on to stardom and hatched from its writing stable such talents as Neil Simon, Mel Brooks, Woody Allen, Mel Tolkin and Larry Gelbart.

Originally entitled *Admiral Broadway Revue,* the series took its new title soon afterwards and developed a large following for the brilliant pantomimes, satirical sketches and burlesques that became hallmarks of the Caesar-Coca artistry. Regularly featured were Carl Reiner, Howard Morris, Marguerite Piazza, the Hamilton Trio, the Billy Williams Quartet and dancers Bambi Linn and Rod Alexander. Max Liebman was the producer-director.

In 1975, kinescopes of ten of the skits were fashioned into a theatrical movie, *Ten From Your Show of Shows,* and in 1976 Liebman compiled eight 90-minute specials from other kinescopes that were syndicated by the J. Walter Thompson agency.

Z

ZAMORA TRIAL—a widely publicized Florida murder case in 1977 in which television was named "an accomplice" by the defense. Ronald Zamora, the 15-year-old defendant, pleaded that heavy viewing of television violence had incited him to kill an 82-year-old woman neighbor. During the week-and-a-half-long trial in Miami during October 1977, Zamora's lawyers argued persistently that television had "intoxicated" him to the point of causing him to lose his ability to distinguish right from wrong. The judge, however, refused to hear the opinions solicited by the defense from social scientists concerning the general behavioral effects of television violence. Zamora was found guilty of first-degree murder and sentenced to 25 years. After his conviction, Zamora filed a multimillion-dollar damage suit against the three networks.

Another notable aspect of the trial was the ironic presence of a television camera in the courtroom. Courts in some other states had allowed television coverage of trials, but the precedent in the Zamora case was the most far-reaching because the State Supreme Court ruled that cameras could be placed in the courtroom regardless of the desires of the prosecution or the defense.

"ZANE GREY THEATRE"—successful "adult western" anthology on CBS (1956–60), hosted by and occasionally starring Dick Powell. The episodes used more than 154 guest stars. It was by Four Star, Zane Grey and Pamric Productions.

ZAPPLE RULE—the FCC's 1970 extension of the Fairness Doctrine as it applies to political campaigns, which resulted from a series of hypothetical questions posed by Nicholas Zapple, then staff counsel to the Senate Commerce Committee. In a letter to Zapple, the commission established that broadcasting's equal time obligations would apply to *supporters* of legally qualified candidates as well as to the candidates themselves. Thus, a broadcaster who sells air time to any person or organization supporting a candidate in an election period cannot deny the sale of comparable time to supporters of the opponents.

"Z" CHANNEL—successful pay-cable channel on the Theta Cable system in Los Angeles. It began operating in January 1974 and within four months was the largest pay-cable channel in the country. At the end of the first year it had 20,000 subscribers for a penetration of 42% of Theta Cable households.

ZDF (Zweites Deutsches Fernsehen)—See West Germany, Television in.

"ZIM BOMBA"—a 13-episode syndicated TV series created by editing down to hourlong episodes old *Bomba, the Jungle Boy* (Allied Artists) theatrical movies, which had starred Johnny Sheffield. Syndication in that form began in the early 60s, but the series' unique success was limited by the availability of only 13 films.

ZINBERG, MICHAEL—producer-director for MTM Productions who in June 1979 joined NBC-TV as v.p. of comedy development on the West Coast. In his eight years with MTM, Zinberg was executive producer of *The Bob Newhart Show* and at various times a director and writer of episodes for *The Mary Tyler Moore Show, Rhoda, The Tony Randall Show* and *Phyllis*. Before MTM, he worked for a time with Talent Associates and with James Garner's company, Cherokee Productions.

ZIV, FREDERICK W.—founder of the largest and most potent syndication company in the history of television, Ziv Television Programs, the leading programming force outside the networks during the 50s. Ziv-TV produced and distributed telefilms for local station use—chiefly for 7:30 and 10:30 P.M., before those half-hours were claimed by the networks—releasing them at the rate of one every two or three months. Among its big hits were *Highway Patrol, Sea Hunt* and *Whirlybirds,* shows that eventually brought the networks into the action-adventure field. Employing scores of salesmen around the country, Ziv-TV turned out to be the Harvard Business School of television syndication, since most of its staff went on to form companies of their own or to take executive positions with a host of syndication companies that sprang up in the late 50s and 60s.

Ziv built his colossus from what had been a transcription company, World Broadcasting System, which he acquired from Decca Records in August 1948 for $1.5 million. Before long, he split the company into two divisions, one selling firstrun programs, the other off-network reruns. In 1959 Ziv sold 80% of the company to Wall Street investment firms for $14 million. A year later the entire company went to United Artists for approximately $20 million and took the new name of Ziv-UA Television. John L. Sinn, who had been president of Ziv-TV, became president of United Artists TV. But as the conditions of the market changed, the company in 1962 gave up first run production and concentrated on selling movies and reruns. With the sale to U.A., Ziv retired to reside in his native Cincinnati.

ZONES—three geographical areas designated by the FCC in its 1952 frequency allocation plan, each with a separate set of requirements for mileage separation and antenna height.

Zone I embraces Massachusetts, Rhode Island, Connecticut, New Jersey, Maryland, Pennsylvania, Delaware, District of Columbia, Ohio, Indiana, Illinois and parts of New Hampshire, Vermont, New York, Virginia, W. Virginia, Michigan, Wisconsin and Maine. In this zone, the minimum co-channel separation is 170 miles for VHF and 135 miles for UHF stations.

Zone II has a minimum co-channel separtion of 190 miles for VHF and 175 for UHF. It covers Kentucky, Ten-

nessee, N. Carolina, S. Carolina, Missouri, Iowa, Minnesota, Arkansas, Kansas, Nebraska, Oklahoma, N. Dakota, S. Dakota, Utah, Idaho, Arizona, New Mexico, Montana, Wyoming, Nevada, Colorado, Oregon, Washington, California, Alaska, Hawaii and parts of Maine, New Hampshire, Vermont, New York, Virginia, W. Virginia, Georgia, Alabama, Mississippi, Louisiana, Michigan, Wisconsin and Texas.

In Zone III, the separation is 220 miles for stations on the same VHF channel and 205 for those on UHF. This zone includes Florida and parts of Georgia, Alabama, Louisiana, Mississippi and Texas. See also "Freeze" of 1948.

"ZOO GANG, THE"—miniseries by ATV of London, based on Paul Gallico's best-selling book about members of a Resistance group who worked underground against the Nazis in France and meet thirty years later. Stars were Brian Keith, John Mills, Lilli Palmer, Barry Morse, Michael Petrovitch and Seretta Wilson, with music by Paul McCartney. NBC carried the 6-part series during the summer of 1975.

"ZOOM"—public TV series of weekly half-hour programs designed to capture the whimsy and humor of children in the 6–12 age group and to encourage the active participation of young viewers. A magazine of skits, songs, dances, jokes and puzzles was presented by a resident cast of seven nonprofessional children from material submitted by viewers. The cast was changed every 26 weeks.

Originating on WGBH Boston, and produced initially by Christopher Sarson, the show premiered on PBS in January 1972, went off for lack of funding after three years and was revived in 1976.

"ZOO PARADE"—early children's series on NBC (1950–57) involving visits with animals behind the scenes at Chicago's Lincoln Park Zoo. The long-running show featured Marlin Perkins, the zoo's director, and newsman Jim Hurlbut, who asked the layman's questions. The shows included travel film to Amazon jungles and snake farms. The series began as a local Chicago program in 1949 and went on the network the following year.

"ZORRO"—briefly popular ABC adventure series (1957–59) featuring a mysterious masked rider who defends the weak and oppressed and signs his work with the initial "Z," usually carved out by sword. Set in California in 1820, the series was a Walt Disney Production. It starred Guy Williams and featured Gene Sheldon, Britt Lomond and Henry Calvin.

ZOUSMER, JESSE (d. 1966)—news executive with all three networks but usually associated with the Edward R. Murrow unit at CBS as co-producer, with John A. Aaron, of *Person to Person* and editor and writer for *Edward R. Murrow and the News*. In his 19 years with CBS, he had also been associated with Murrow's *See It Now* and its radio precursor, *Hear It Now*. In 1959 he and Aaron resigned from CBS in a dispute over the adoption of a network policy, after the quiz-show scandals, requiring an advisory that the shows were rehearsed. He then joined NBC as a producer and in 1963 went to ABC as director of television news. He and his wife died in the crash of a commercial plane near Mt. Fuji in Japan.

ZWORYKIN, VLADIMIR K. (Dr.)—U.S. electronics engineer and inventor considered the father of modern television for inventing the iconoscope (electronic camera) and the kinescope (picture tube). Although the iconoscope was eventually replaced by the orthicon and image orthicon tubes, Zworykin's invention was the basis for further developments in television cameras. The modern TV picture tube, however, is essentially Zworykin's kinescope.

A radio expert in the Russian Army Signal Corps during World War I, Zworykin worked on X-rays and electrical and gaseous discharges and was associated with the Russian Society of Wireless Telephone and Telegraph. He emigrated to the U.S. in 1919 and a year later joined the Westinghouse Electric Corporation. It was there, while working with the cathode-tube principle he had developed in 1907, that Zworykin invented the first practical television transmission tube, his iconoscope. The device used photoelectric effects (i.e., the ejection of electrons from metals by the action of light) as the basis for scanning and converting images into electric currents.

In 1924, a year after filing a patent application for the iconoscope, he filed another for the kinescope, his TV receiver. While the older systems were electromechanical, usually involving a rapidly rotating perforated disc, Zworykin's inventions formed an all-electronic television system. Westinghouse officials were unconvinced by Zworykin's early demonstrations of his system, but in 1929 they impressed an RCA official, and the inventor accepted a position as director of electronic research at RCA. Zworykin's system became the basis for the electronic system developed by RCA.

Zworykin was named honorary vice president of RCA when he retired in 1954 and continued to remain active as a consultant. From then until 1962, he served also as director of the medical electronics center of the Rockefeller Institute for Medical Research (now Rockefeller U.) in New York City. In 1967, the National Academy of Sciences awarded him the National Medal of Science for his contributions to science, engineering and television.

Bibliography

Adler, Richard and Baer, Walter S., *The Electronic Box Office: Humanities and Arts on the Cable,* Praeger Publishers, New York, 1974.

Barnouw, Eric, *The Image Empire* (Vol. 3, *History of Broadcast in America*), Oxford University Press, 1971.

Barrett, Marvin, *The Politics of Broadcasting,* Thomas Y. Crowell Co., New York, 1973.

Bluem, A. William, *Religious Television Programs,* Television Information Office and the National Assn. of Broadcasters, New York and Washington, 1969.

Brown, Les, *Television: The Business Behind the Box,* Harcourt Brace Jovanovich, Inc., 1971.

Campbell, Robert, *The Golden Years of Broadcasting: A Celebration of the First 50 Years of Radio and TV on NBC,* Charles Scribner's Sons, New York, 1976.

Cater, Douglass and Strickland, Stephen, *TV Violence and the Child: Evolution and Fate of the Surgeon General's Report,* Russell Sage Foundation, New York, 1975.

Diamond, Edwin, *The Tin Kazoo,* The MIT Press, Cambridge, Mass. 1975.

Emery, Walter B., *Broadcasting and Government,* Michigan State University Press, 1971.

Friendly, Fred W., *Due to Circumstances Beyond Our Control,* New York, Random House, 1967.

Green, Timothy, *The Universal Eye,* Stein and Day, New York, 1972.

Jennings, Ralph and Richards, Pamela, *How to Protect Your Rights in Radio and Television,* Office of Communication, United Church of Christ, New York, 1974.

Kanfer, Stefan, *A Journal of the Plague Years,* Atheneum, New York, 1973.

Lesser, Gerald S., *Children and Television: Lessons from Sesame Street,* Random House, New York, 1974.

Lichty, Lawrence W. and Topping, Malachi C., *A Source Book on the History of Radio and Television,* Hastings House Publishers, 1975.

Mayer, Martin, *About Television,* Harper & Row, 1972.

Metz, Robert, *CBS: Reflections in a Bloodshot Eye,* Playboy Press, New York, 1975.

Mitchell, Curtis, *Cavalcade of Broadcasting,* Follett Publishing, Chicago, 1969.

National Association of Broadcasters, *Political Broadcast Catechism,* NAB, 1976.

Quaal, Ward L. and Martin, Leo A., *Broadcast Management,* Hastings House Publishers, New York, 1968.

Ray, Verne M. ed., *Interpreting Broadcast Rules and Regulations,* TAB Books, 1966 and 1972.

Rivers, William L. and Slater, William T., *Aspen Handbook on the Media (Aspen Institute Program on Communications and Society), 1975.*

Rivers, L. and Nyhan, Michael J., eds., Aspen Notebook on Government and the Media, Praeger Publishers, New York, 1973.

Shanks, Bob, *The Cool Fire,* W.W. Norton & Co., New York, 1976.

Shulman, Arthur and Youman, Roger, *How Sweet It Was,* Shorecrest, Inc., New York, 1966.

Sloan Commission on Cable Communications, *On the Cable: The Television of Abundance,* McGraw-Hill Book Company, New York, 1971.

Smith, Anthony, *The Shadow in the Cave: The Broadcaster, His Audience, and the State,* University of Illinois Press, Urbana, Ill., 1973.

Stanley, Robert H. and Steinberg, Charles S., *The Media Environment,* Hastings House Publishers, New York, 1976.

Terrace, Vincent, *The Complete Encyclopedia of Television Programs 1947–1976* (Vols. 1 and 2), A.S. Barnes and Co., Cranbury, N.J. 1976.

Who's Who In America (Marquis Who's Who Inc.).

Wilk, Max, *The Golden Age of Television,* Delacorte Press, New York, 1976.

Advertising Age/BIB Source Book, *Series, Serials and Packagers*

The Broadcasting Yearbook
Broadcasting Magazine
TV/Radio Age
Television Digest.
Television Digest Fact Book
TV Guide
Television Internct Book
TV Guide
Television International
Variety

Appendices

The Neilsen "Pocket Piece"

A-14 NATIONAL *Nielsen* TV AUDIENCE ESTIMATES — EVE. SUN. FEB. 13, 1977

TIME	7:00	7:15	7:30	7:45	8:00	8:15	8:30	8:45	9:00	9:15	9:30	9:45	10:00	10:15	10:30	10:45	11:00
ABC TV		Hardy Boys / Nancy Drew Mysteries (1)			Six Million Dollar Man				Oscar's Best Movies (9:00-11:24PM)								
TOTAL AUDIENCE Households (000)	15,810				23,570				25,560								
TOTAL AUDIENCE %	22.2				33.1				35.9								
AVERAGE AUDIENCE (000)	12,250				19,650				14,880								
AVERAGE AUDIENCE %	17.2		16.1*	18.3*	27.6	25.8*		29.4*	20.9		24.3*		22.4*		20.9*	19.3*	
SHARE OF AUDIENCE %	27		26*	28*	40	38*		43*	35		36*		33*		34*	35*	
AVG. AUD. BY 1/4 HR. %	15.5	16.7	17.4	19.1	24.8	26.9	29.1	29.7	24.4	24.2	21.9	22.9	21.3	20.5	19.6	19.0	
CBS TV	60 Minutes				Rhoda		Phyllis		Switch				Delvecchio				
TOTAL AUDIENCE Households (000)	21,080				15,660		14,170		20,580				17,230				
TOTAL AUDIENCE %	29.6				22.3		19.9		28.9				24.2				
AVERAGE AUDIENCE (000)	17,020				13,880		13,170		16,590				14,450				
AVERAGE AUDIENCE %	23.9		23.3*	24.5*	19.5		18.5		23.3		22.1*	24.4*	20.3		20.7*	19.9*	
SHARE OF AUDIENCE %	38		38*	37*	29		27		35		33*	36*	35		34*	36*	
AVG. AUD. BY 1/4 HR. %	22.3	24.2	24.6	24.5	19.8	19.3	18.3	18.6	21.2	23.0	24.0	24.9	24.0	21.1	20.3	19.9	
NBC TV	Wonderful World of Disney — "THIS IS YOUR LIFE, DONALD DUCK" (R)				Big Event Part I — "2001: A SPACE ODYSSEY" (8:00-11:00PM) (OP) "2001: A SPACE ODYSSEY" (8:00-11:00PM)												
TOTAL AUDIENCE Households (000)	18,010				24,140												
TOTAL AUDIENCE %	25.3				33.9												
AVERAGE AUDIENCE (000)	13,530				12,530												
AVERAGE AUDIENCE %	19.0		18.0*	20.0*	17.6	20.9*		18.9*	18.2*				17.2*		16.5*	14.0*	
SHARE OF AUDIENCE %	30		29*	30*	27	31*		27*	27*				26*		27*	25*	
AVG. AUD. BY 1/4 HR. %	17.3	18.6	19.7	20.3	21.2	20.5		18.5	19.3	17.7		18.7	17.0	17.5	16.1	15.4	12.6

EVE. SUN. FEB. 20, 1977

U.S. TV Households: 71,200,000

ABC TV

	Hardy Boys / Nancy Drew Mysteries(1)	Six Million Dollar Man	ABC Sunday Night Movie "SECRETS" (9:00-11:00PM)
TOTAL AUDIENCE (Households (000) & %)	18,010 / 25.3	21,790 / 30.6	26,560 / 37.3
AVERAGE AUDIENCE (Households (000) & %)	14,030 / 19.7	17,660 / 24.8 ; 17,800 / 25.0	18,940 / 26.6

AVERAGE AUDIENCE % / SHARE % (half-hour ratings marked *) and AVG. AUD. BY ¼ HR. %:

- Hardy Boys: 18.0* / 30* ; 21.5* / 33* — ¼ hr: 16.8, 19.1, 21.3, 21.8
- Six Million Dollar Man: 24.5* / 37* ; 25.0* / 38* ; 25.1* / 38* — ¼ hr: 24.4, 24.4, 25.4, 25.5
- Movie: 21.7* / 32* ; 25.0* / 38* ; 25.9* / 40* ; 26.5* / 43* ; 25.9* / 45* — ¼ hr: 21.6, 24.8, 25.5, 26.4, 26.3, 26.6, 26.7, 25.3

CBS TV

	60 Minutes	Rhoda / Phyllis	Switch	Delvecchio
TOTAL AUDIENCE (Households (000) & %)	19,580 / 27.5	14,520 / 20.4	14,950 / 21.0	18,940 / 26.6 ; 15,380 / 21.6
AVERAGE AUDIENCE (Households (000) & %)	15,660 / 22.0	13,810 / 19.4 ; 14,520 / 20.4	14,170 / 19.9 ; 14,950 / 21.0	13,030 / 18.3

AVERAGE AUDIENCE % / SHARE % (half-hour ratings marked *) and AVG. AUD. BY ¼ HR. %:

- 60 Minutes: 22.0* / 37* — ¼ hr: 21.5, 21.2, 22.7, 22.6
- Rhoda / Phyllis: 22.0* / 34* — ¼ hr: 19.5, 19.2, 20.3, 19.3
- Switch: 19.8* / 30* ; 21.0 / 32 — ¼ hr: 19.5, 20.3, 22.1, 19.3
- Delvecchio: 22.2* / 34* ; 18.3 ; 18.4* / 30* ; 18.2* / 32* — ¼ hr: 17.9, 18.7, 18.8, 17.7

NBC TV

	Wonderful World of Disney "GO WEST, YOUNG DOG"	Big Event Part I "THE SPELL" (8:00-9:30PM)	Big Event Part II "LIVE FROM THE MARDI GRAS, IT'S 'SATURDAY NIGHT' ON SUNDAY" (9:30-11:00PM)
TOTAL AUDIENCE (Households (000) & %)	14,950 / 21.0	21,150 / 29.7	16,520 / 23.2
AVERAGE AUDIENCE (Households (000) & %)	10,960 / 15.4	15,880 / 22.3	9,260 / 13.0

AVERAGE AUDIENCE % / SHARE % (half-hour ratings marked *) and AVG. AUD. BY ¼ HR. %:

- Disney: 14.1* / 24* ; 16.8* / 26* — ¼ hr: 13.5, 14.7, 15.9
- Big Event Part I: 21.3* / 33* ; 22.7* / 34* ; 23.0* / 34* — ¼ hr: 21.0, 22.8, 23.6, 22.6
- Big Event Part II: 15.6* / 24* ; 12.9* / 21* ; 10.7* / 19* — ¼ hr: 16.9, 13.7, 12.0, 10.9, 10.4

HOUSEHOLDS USING TV (See Def. 1)

WK 1	59.6	63.4	65.1	66.5	67.2	68.4	69.1	68.8	67.9	67.7	67.5	66.6	62.0	59.6	57.2	53.9
WK 2	58.0	60.9	63.6	64.7	65.1	66.0	66.7	67.1	66.8	66.7	66.6	65.3	61.5	60.5	58.8	56.2

---

*Half-hour ratings (for immediately preceding and subject quarter-hour).   (R) Repeat, see page B.   (OP) See Other Programs Section: Page A-36

(1) "ABC MINUTE MAGAZINE", ABC, (7:58-7:59PM)(SUS.).

A-15

HIGHEST RATED PROGRAMS OF ALL TIME

RANK	PROGRAM NAME	TELECAST DATE	NETWORK	DURATION MINUTES	AVERAGE AUDIENCE (%)
1	Dallas	Nov. 21, 1980	CBS	60	53.3
2	Roots	Jan. 23, 1977	ABC	115	51.1
3	Super Bowl XVI Game	Jan. 24, 1982	CBS	213	49.1
4	Gone With The Wind-Part 1	Nov. 7, 1976	NBC	179	47.7
5	Gone With The Wind-Part 2	Nov. 8, 1976	NBC	119	47.4
6	Super Bowl XII Game	Jan. 15, 1978	CBS	218	47.2
7	Super Bowl XIII Game	Jan. 21, 1979	NBC	230	47.1
8	Bob Hope Christmas Show	Jan. 15, 1970	NBC	90	46.6
9	Super Bowl XIV Game	Jan. 20, 1980	CBS	178	46.3
10	Roots	Jan.28, 1977	ABC	120	45.9
10	The Fugitive	Aug. 29, 1967	ABC	60	45.9
12	Roots	Jan. 27, 1977	ABC	60	45.7
13	Ed Sullivan	Feb. 9, 1964	CBS	60	
14	Bob Hope Christmas Show	Jan. 14, 1971	NBC	90	45.0
15	Roots	Jan. 25, 1977	ABC	60	44.8
16	Super Bowl XI	Jan. 9., 1977	NBC	204	44.4
16	Super Bowl XV	Jan. 25, 1981	NBC	220	44.4
18	Super Bowl VI	Jan. 16, 1972	CBS	170	44.2
19	Roots	Jan. 24, 1977	ABC	120	44.1
20	Beverly Hillbillies	Jan. 8, 1964	CBS	30	44.0
21	Roots	Jan. 26, 1977	ABC	60	43.8
21	Ed Sullivan	Feb. 16, 1964	CBS	60	43.8
23	Academy Awards	Apr. 1970	ABC	145	43.4
24	CBS NFC Championship Game	Jan. 10, 1982	CBS	195	42.9
25	Beverly Hillbillies	Jan. 15, 1964	CBS	30	42.8
26	Super Bowl VII	Jan. 14, 1973	NBC	185	42.7
27	Super Bowl IX	Jan. 12, 1975	NBC	190	42.4
27	Beverly Hillbillies	Feb. 26, 1964	CBS	30	42.4
29	Super Bowl X	Jan. 18. 1976	CBS	200	42.3
29	Airport	Nov. 11, 1973	ABC	170	42.3
29	Love Story	Oct. 1, 1972	ABC	120	42.3
29	Cinderella	Feb. 22, 1965	CBS	90	42.3
29	Roots	Jan. 29, 1977	ABC	60	42.3
34	Beverly Hillbillies	Mar. 25, 1964	CBS	30	42.2
35	Beverly Hillbillies	Feb. 5, 1964	CBS	30	42.0
36	Beverly Hillbillies	Jan. 29,1964	CBS	30	41.9
37	Miss America Pagent	Sept. 9, 1961	CBS	150	41.8
37	Beverly Hillbillies	Jan. 1, 1964	CBS	30	41.8
39	Super Bowl VIII	Jan. 13, 1974	CBS	160	41.6
39	Bonanza	Mar. 8, 1964	NBC	60	41.6
41	Beverly Hillbillies	Jan. 22, 1964	CBS	30	41.5
42	Bonanza	Feb. 16, 1964	NBC	60	41.4
43	Academy Awards	Apr. 10, 1967	ABC	150	41.2
44	Bonanza	Feb. 9, 1964	NBC	60	41.0
45	Gunsmoke	Jan. 28, 1961	CBS	30	40.9
46	Bonanza	Mar. 28, 1965	NBC	60	40.8
47	Bonanza	Mar. 7, 1965	NBC	60	40.7
47	All in the Family	Jan. 8,1972	CBS	30	40.7
49	Roots	Jan. 23, 1977	ABC	120	40.5
49	Bonanza	Feb. 2, 1964	NBC	60	40.5
49	Beverly Hillbillies	May 1, 1963	CBS	30	40.5
49	Gunsmoke	Feb. 25, 1961	CBS	30	40.5

TELEVISION STATIONS ON AIR
1946 - 1981
As of January 1 of each year

Year	VHF Coml.	VHF ETV	Total VHF	UHF Coml.	UHF ETV	Total UHF	Total Coml.	Total ETV	Grand Total
1946			*6			-			*6
1947			12			-			12
1948			16			-			16
1949			51			-			51
1950			98			-			98
1951			107			-			107
1952			108			-			108
1953			120			6			126
1954	233	1	234	121	1	122	354	2	356
1955	297	8	305	114	3	117	411	11	422
1956	344	13	357	97	5	102	441	18	459
1957	381	17	398	90	6	96	471	23	494
1958	411	22	433	84	6	90	495	28	523
1959	433	28	461	77	7	84	510	35	545
1960	440	34	474	75	10	85	515	44	559
1961	451	37	488	76	15	91	527	52	579
1962	458	43	501	83	19	102	541	62	603
1963	466	46	512	91	22	113	557	68	625
1964	476	53	529	88	32	120	564	85	649
1965	481	58	539	88	41	129	569	99	668
1966	486	65	551	99	49	148	585	114	699
1967	492	71	563	118	56	174	610	127	737
1968	499	75	574	136	75	211	635	150	785
1969	499	78	577	163	97	260	662	175	837
1970	501	80	581	176	105	281	677	185	862
1971	503	86	589	179	113	292	682	199	881
1972	508	90	598	185	123	308	693	213	906
1973	510	93	603	187	137	324	697	230	927
1974	513	92	605	184	149	333	697	241	938
1975	514	95	609	192	152	344	706	247	953
1976	511	97	608	190	162	352	701	259	960
1977	515	101	616	196	160	356	711	261	972
1978	515	102	617	201	164	356	716	266	982
1979	515	107	622	209	167	367	724	274	988
1980	516	109	625	218	168	386	734	277	1011
1981	519	111	630	237	171	408	756	282	1038

*Does not include 1 CP operating intermittently.

TV HOUSEHOLDS, SETS & % SATURATION
Monochrome and Color
1946 - 1981

Year	Total U.S. Homes (000)	TV Homes % of Total	% of Homes with Color TV	TV Homes (000)	Secondary Sets in TV Homes (000)	Total Non-Home TV Sets (000)	TV Sets (000)
1946	37,825	0.02		8	-	2	10
1947	38,575	0.04		14	-	2	16
1948	39,950	0.4		172	1	17	190
1949	41,475	2.3		940	10	50	1,000
1950	43,000	9.0		3,875	50	75	4,000
1951	43,888	23.5		10,320	165	115	10,600
1952	44,760	34.2		15,300	315	185	15,800
1953	45,640	44.7		20,400	505	295	21,200
1954	46,660	55.7		26,000	800	500	27,300
1955							
Mono only		64.5		30,695	998	792	32,485
Color		#	.02	5	2	8	15
Total	47,621	64.5		30,700	1,000	800	32,500
1965							
Mono only		87.7		49,890	11,470	2,775	64,135
Color		4.9	5.3	2,810	40	225	3,075
Total	56,900	92.6		52,700	11,510	3,000	67,210
1970							
Mono only		57.9		36,300	23,100	2,500	61,900
Color		37.3	39.2	23,400	1,400	1,600	26,400
Total	62,700	95.2		59,700	24,500	4,100	88,300
1974							
Mono only		31.4		21,850	40,500	2,100	64,450
Color		64.7	67.3	44,950	4,700	2,900	52,550
Total	69,500	96.1		66,800	45,200	5,000	117,000
1975	71,100	97.1	74	69,600	29,900		121,100
1976	72,800	98.0	77	71,200	32,000		131,500
1977	74,100	98.0	78	72,900	33,500		138,200
1978	76,000	98.0	81	74,500	35,800		144,500
1979	77,000	98.0	82.5	75,900	36,500		150,400
1980***	77,300	98.0	83	76,300	38,000		154,200
1981***	79,000	98.0	85	77,800	39,700		158,300

(In 1981, cable subscribers comprised 25% of total, or 19,727,290 homes; non-cable: 75%, 59,000,000 homes.)

-    The number of secondary sets in homes is insignificant during these years.

\#   Less than 0.1%.

†   Preliminary.

*   ARF-Census Report — National Survey of Television Sets in U.S. Households, January, 1969.

**   Revised. Note: Households having color sets may also have monochrome sets. Source: NBC Corporate Planning; A.C. Nielsen Co.

***   The apparent reduced growth between 1979 and 1980 results from a change in time of estimates. Before 1980, figures were based on data as of January 1. From 1980, figures are based on September 1 data.

The Public's Expenditures for TV Sets
(In Billions)

	1959	1965	1970	1976	1979	1980
Purchases in year	$ 1.0	$ 2.9	$ 3.0	$ 4.5	$ 5.3	$ 5.4
Cumulative expenditure	$17.0	$27.7	$46.9	$74.1	$90.2	$95.6

Merchandising

Percent of TV Homes Able to Receive
Multiple Stations

No. receivable	1968	1970	1972	1974	1976	1979	1981
15 or more stations	N.A.	N.A.	N.A.	N.A.	N.A.	N.A.	8%
13 or more stations	N.A.	N.A.	N.A.	N.A.	N.A.	N.A.	17%
12 or more stations	N.A.	N.A.	8%	9.0%	10%*	11%*	25%
11 or more stations	N.A.	N.A.	14%	17.6%	20%*	22%*	33%
10 or more stations	N.A.	17%	20%	25.5%	27%	38%	43%
9 or more stations	24%	26%	31%	34.7%	38%	47%	54%
8 or more stations	34%	37%	42%	44.0%	47%	56%	61%
7 or more stations	53%	57%	60%	63.1%	65%	66%	71%
6 or more stations	65%	66%	70%	76.3%	77%	79%	81%
5 or more stations	79%	79%	82%	86.6%	88%	89%	91%
4 or more stations	90%	90%	93%	94.9%	96%	96%	97%
3 or more stations	97%	97%	98%	98.5%	98.8%*	99%	98%*

A.C. Nielsen   (N.A. = Not available)   *TIO estimate

Performance of Community Institutions

"In every community, the schools, the newspapers, the television stations, the local government, each has a different job to do. Would you say that the local schools (the ones you are familiar with) are doing an excellent, good, fair, or poor job? How about the local (newspapers, etc.)?"

Rated Excellent or Good	1959 %	1963 %	1968 %	1972 %	1976 %	1978 %	1980 %
Television	59	60	57	60	70	68	68
Newspapers	64	55	51	51	59	59	60
Churches	—	—	—	—	66	67	65
Police	—	—	—	—	65	66	62
Schools	64	61	58	50	47	48	47
Local Government	44	43	41	36	41	37	36

The Roper Organization Inc.

Increase in Media Voices

	1959	1965	1970	1975	1977	1981
Commercial TV st'ns.	520	672	690	709	725	864
Educational TV st'ns.	43	115	182	242	258	292
AM radio st'ns.	3,377	4,058	4,269	4,436	4,502	4,729
FM radio st'ns.	776	1,301	2,471	3,405	3,840	3,466
Daily newspapers	1,755	1,763	1,761	1,768	1,762	1,745

FCC Broadcasting Magazine. Editor & Publisher Yearbook

Source of Most News

"Where do you usually get most of your news about what's going on in the world today?"

	1959 %	1961 %	1963 %	1968 %	1972 %	1976 %	1980 %
Television	51	52	55	59	64	64	64
Newspapers	57	57	53	49	50	49	44
Radio	34	34	29	25	21	19	18
Magazines	8	9	6	7	6	7	5
People	4	5	4	5	4	5	4
All Mentions	154	157	147	145	145	144	135
DK/NA*	1	3	3	3	1	—	—

*Don't Know/No Answer

The Roper Organization Inc.

Preferred Sources of Control Over Television

"Which of the people or groups on this list should have the most to say about what people see and hear on television?"

	News	Enter-tainm'nt	Com'ls	What's on TV	Pro-fanity	Sex	Vio-lence
Individual Viewer (by program selection)	64	69	49	71	61	67	67
TV Networks & stations (by programming)	36	28	25	26	22	19	20
Advertisers (by sponsorship)	9	14	35	12	10	9	9
Federal Govt. (by regulation)	9	4	6	9	15	12	15
Social & religious groups (by recommendation)	6	8	4	9	15	15	13
None	1	1	—	1	1	1	1
DK/NA	3	3	3	2	3	2	2

The Roper Organization, Inc., 12/78

Believability

"If you got conflicting or different reports of the same news story, which would you be most inclined to believe?..."

	1959 %	1961 %	1963 %	1967 %	1968 %	1972 %	1974 %	1976 %	1980 %
Television	29	39	36	41	44	48	51	51	51
Newspapers	32	24	24	24	21	21	20	22	22
Radio	12	12	12	7	8	8	8	7	8
Magazines	10	10	10	8	11	10	8	9	9
DK/NA	17	17	18	20	16	13	13	11	10

The Roper Organization Inc.

Source of News about National Candidates

"From what source did you become best acquainted with the candidates running in the national election — for President, Vice President, Senate, House of Representatives?"

	1972 %	1976 %
Television	66	75
Newspapers	26	20
Radio	6	4
People	5	3
Magazines	5	5
Other	2	1
Total Mentions	110	108

The Roper Organization Inc.

Attitude Toward Principle of Having Commercials

"...Everything considered, do you agree or disagree that having commercials on TV is a fair price to pay for being able to watch?"

	1963 %	1967 %	1971 %	1976 %	1980 %
Agree	77	80	80	74	72
Disagree	14	9	10	20	24
DK/NA	9	11	10	6	4

The Roper Organization Inc.

Source of News about Local Candidates

"From what source did you become best acquainted with the candidates running in local elections — like mayor, members of the state legislature, etc., — from the newspapers or radio or television or magazines or talking to people or where?"

Local elections	1971 %	1972 %	1974 %	1976 %	1978 %	1980 %
Newspapers	41	41	41	44	45	36
Television	27	31	30	34	39	44
Radio	6	7	8	7	10	6
People	19	23	14	12	15	11
Magazines	1	1	1	2	1	2
Other	5	5	5	6	7	5
Total Mentions	99	108	99	105	117	104

The Roper Organization Inc.

Government Control Over TV Programs

"...Do you think the government should exercise more, about the same or less control over the programs that are on TV?"

	1963 %	1964 %	1967 %	1971 %	1972 %	1974 %	1976 %	1978 %
The government should exercise more control over TV programs	16	19	18	12	17	15	24	24
There is about the right amount of government control over TV now	43	41	40	48	38	36	34	38
The government should have less control over TV programs	27	26	28	31	39	41	36	30
DK/NA	14	14	14	9	6	8	6	7

The Roper Organization Inc.

Children's Comprehension of Purpose of Commercials

"Does your child know the purpose of commercials on TV, that it is, to get him or her or you to buy something?"

Whether child understands the purpose of commercials	Total children who know the difference %	3 %	4 %	5 %	6 %	7 %	8 %	9 %	10 %
Yes, know purpose	79	43	69	69	85	80	85	93	96
No, do not	17	55	28	26	9	12	13	7	3
DK/NA	4	2	3	5	6	8	2	—	3

The Roper Organization Inc., 12/78

The publisher would like to thank the Television Information Office for providing the TV Facts Tables